Non-Marine Permian Biostratigraphy and Biochronology

Geological Society books refereeing procedures

The Society makes every effort to ensure that the scientific and production quality of its books matches that of its journals. Since 1997, all book proposals have been refereed by specialist reviewers as well as by the Society's Books Editorial Committee. If the referees identify weaknesses in the proposal, these must be addressed before the proposal is accepted.

Once the book is accepted, the Society Book Editors ensure that the volume editors follow strict guidelines on refereeing and quality control. We insist that individual papers can only be accepted after satisfactory review by two independent referees. The questions on the review forms are similar to those for *Journal of the Geological Society*. The referees' forms and comments must be available to the Society's Book Editors on request.

Although many of the books result from meetings, the editors are expected to commission papers that were not presented at the meeting to ensure that the book provides a balanced coverage of the subject. Being accepted for presentation at the meeting does not guarantee inclusion in the book.

More information about submitting a proposal and producing a book for the Society can be found on its web site: www.geolsoc.org.uk.

It is recommended that reference to all or part of this book should be made in one of the following ways:

Lucas, S. G., Cassinis, G. & Schneider, J. W. (eds) 2006. *Non-Marine Permian Biostratigraphy and Biochronology*. Geological Society, London, Special Publications, **265**.

Virgili, C., Cassinis, G. & Broutin, J. 2006. Permian to Triassic sequences from selected continental areas of southwestern Europe. *In*: Lucas, S. G., Cassinis, G. & Schneider, J. W. (eds) *Non-Marine Permian Biostratigraphy and Biochronology*. Geological Society, London, Special Publications, **265**, 231–259.

GEOLOGICAL SOCIETY SPECIAL PUBLICATION NO. 265

Non-Marine Permian Biostratigraphy and Biochronology

EDITED BY

SPENCER G. LUCAS
New Mexico Museum of Natural History, USA

GIUSEPPE CASSINIS
Università di Pavia, Italy

and

JOERG W. SCHNEIDER
TU Bergakademie Freiberg, Germany

2006
Published by
The Geological Society
London

THE GEOLOGICAL SOCIETY

The Geological Society of London (GSL) was founded in 1807. It is the oldest national geological society in the world and the largest in Europe. It was incorporated under Royal Charter in 1825 and is Registered Charity 210161.

The Society is the UK national learned and professional society for geology with a worldwide Fellowship (FGS) of over 9000. The Society has the power to confer Chartered status on suitably qualified Fellows, and about 2000 of the Fellowship carry the title (CGeol). Chartered Geologists may also obtain the equivalent European title, European Geologist (EurGeol). One fifth of the Society's fellowship resides outside the UK. To find out more about the Society, log on to www.geolsoc.org.uk.

The Geological Society Publishing House (Bath, UK) produces the Society's international journals and books, and acts as European distributor for selected publications of the American Association of Petroleum Geologists (AAPG), the Indonesian Petroleum Association (IPA), the Geological Society of America (GSA), the Society for Sedimentary Geology (SEPM) and the Geologists' Association (GA). Joint marketing agreements ensure that GSL Fellows may purchase these societies' publications at a discount. The Society's online bookshop (accessible from www.geolsoc.org.uk) offers secure book purchasing with your credit or debit card.

To find out about joining the Society and benefiting from substantial discounts on publications of GSL and other societies worldwide, consult www.geolsoc.org.uk, or contact the Fellowship Department at: The Geological Society, Burlington House, Piccadilly, London W1J 0BG: Tel. +44 (0)20 7434 9944; Fax +44 (0)20 7439 8975; E-mail: enquiries@geolsoc.org.uk.

For information about the Society's meetings, consult *Events* on www.geolsoc.org.uk. To find out more about the Society's Corporate Affiliates Scheme, write to enquiries@geolsoc.org.uk

Published by The Geological Society from:
The Geological Society Publishing House, Unit 7, Brassmill Enterprise Centre, Brassmill Lane, Bath BA1 3JN, UK

Orders: Tel. +44 (0)1225 445046
Fax +44 (0)1225 442836
Online bookshop: www.geolsoc.org.uk/bookshop

British Library Cataloguing in Publication Data

A catalogue record for this book is available from the British Library.

ISBN-10: 1-86239-206-4

ISBN-13: 978-1-86239-206-9

Typeset by Charlesworth Ltd, Wakefield, UK

Printed by Cromwell Press, Trowbridge, UK

Distributors

North America
For trade and institutional orders:
The Geological Society, c/o AIDC, 82 Winter Sport Lane, Williston, VT 05495, USA
Orders: Tel +1 800-972-9892
Fax +1 802-864-7626
Email gsl.orders@aidcvt.com
For individual and corporate orders:
AAPG Bookstore, PO Box 979, Tulsa, OK 74101-0979, USA
Orders: Tel +1 918-584-2555
Fax +1 918-560-2652
Email bookstore@aapg.org
Website http://bookstore.aapg.org

India
Affiliated East-West Press Private Ltd, Marketing Division, G-1/16 Ansari Road, Darya Ganj, New Delhi 110 002, India
Orders: Tel. +91 11 2327-9113/2326-4180
Fax +91 11 2326-0538
E-mail affiliat@vsnl.com

CONTENTS

Non-marine Permian biostratigraphy and biochronology: an introduction

SPENCER G. LUCAS[1], JOERG W. SCHNEIDER[2] & GIUSSEPE CASSINIS[3]

[1]*New Mexico Museum of Natural History and Science, 1801 Mountain Road N.W., Albuquerque, New Mexico 87104-13YJ, USA (e-mail: spencer.lucas@state.nm.us)*

[2]*TU Bergakademie Freiberg, B.v. Cotta-Strasse 2, D-09596 Freiberg, Germany*

[3]*Pavia University, Earth Science Department, via Ferrata 1, 27100 Pavia, Italy*

Abstract: The Permian time scale based on marine rocks and fossils is well defined and of global utility, but non-marine Permian biostratigraphy and chronology is in an early phase of development. Non-marine Permian strata are best known from western Europe and the western United States, but significant records are also known from Russia, South Africa, China and Brazil. Global time terms based on non-marine Permian strata, such as Rotliegend, Zechstein, Autunian, Saxonian and Thuringian, are either inadequately defined or poorly characterized and should only be used as lithostratigraphic terms. Macro- and microfloras have long been important in non-marine Permian correlations, but are subject to limitations based on palaeoprovinciality and facies/climatic controls. Charophytes, conchostracans, ostracodes and freshwater bivalves have a potential use in non-marine Permian biostratigraphy but are limited by their over-split taxonomy and lack of well-established stratigraphic distributions of low-level taxa. Tetrapod footprints provide poor biostratigraphic resolution during the Permian, but tetrapod body fossils and insects provide more detailed biostratigraphic zonations, especially in the Lower Permian. Numerous radioisotopic ages are available from non-marine Permian sections and need to be more precisely correlated to the global time scale. The Middle Permian Illawarra reversal and subsequent magnetic polarity shifts are also of value to correlation. There needs to be a concerted effort to develop non-marine Permian biostratigraphy, to correlate it to radio-isotopic and magnetostratigraphic data, and to cross-correlate it to the marine time scale.

In 1840, British geologist Roderick Murchison (1792–1871) visited Russia as a guest of the Czar. East of Moscow, he examined strata in the Perm region of the western Urals and applied the name 'Permian System' to a 'vast series of beds of marls, schists, limestones, sandstones and conglomerates' (Murchison 1841, p. 419) that overlie the Carboniferous strata in a great arc that extends from the Volga River in the west to the Ural Mountains in the east, and from the White Sea in the north to Orenburg in the south. Thus was born the Permian System, and, in western Europe, it soon came to be equated to the British New Red Sandstone and Magnesian Limestone, and to the German Rotliegend and Zechstein.

Much of Murchison's type Permian section in Russia, and a significant portion of its equivalents further west in Europe, encompass strata of non-marine origin. However, by the twentieth-century, stratigraphers agreed that a global time scale (here referred to as the standard global chronostratigraphic scale, or SGCS) needs to be based on marine fossils in marine strata, not on non-marine rocks and fossils. For this and other reasons outlined here, the development of non-marine Permian biostratigraphy and correlation has lagged behind developments in the marine realm.

The Permian period, as currently conceived, extends from about 251 to 299 Ma and encompasses nine ages (stages) arranged into three epochs (series) (Wardlaw *et al.* 2004) (Fig. 1). Most of this time scale (the Permian SGCS) has been defined by ratified global stratotype sections and points (GSSPs) for the stage boundaries (Henderson 2005). As the work of formally defining the Permian SGCS draws to a close, it is logical to push forward into developing non-marine Permian biostratigraphy and correlation. This volume is part of that push forward, as its articles present new data, analyses and understanding of the chronology of the non-marine Permian. In this introduction, we assess the state of the-art of non-marine Permian biostratigraphy and correlation, and place the articles in this book into that context.

From: LUCAS, S. G., CASSINIS, G. & SCHNEIDER, J. W. (eds) 2006. *Non-Marine Permian Biostratigraphy and Biochronology*. Geological Society, London, Special Publications, **265**, 1–14.
0305-8719/06/$15.00

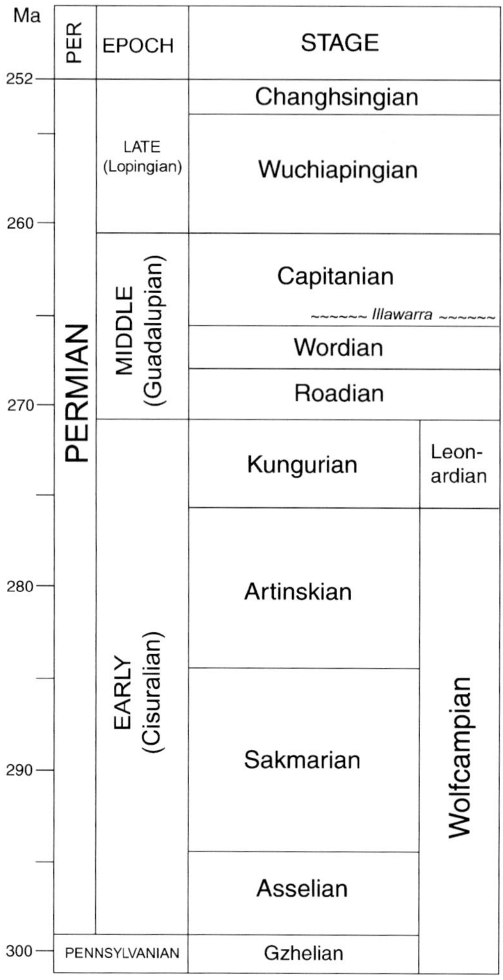

Fig. 1. The standard global chronostratigraphic scale (SGCS) for the Permian (after Henderson 2005).

The non-marine Permian world

The Pangaean supercontinent came together (accreted) at the end of the Carboniferous, when Laurasia and Gondwana were sutured along what has been termed the Hercynian megasuture (Fig. 2). Very old mountain ranges mark the collision boundaries of the Pangaean blocks: the Appalachian Mountains of eastern North America, the Ural Mountains of European Russia and the Variscan (Hercynian) mountain ranges of southern Europe and Mauretanid ranges of North Africa. The Late Carboniferous was a time of vast coal swamps in the tropical latitudes and a steep temperature gradient from icy poles to hot tropics, more similar to today's world than perhaps at any other time in the Earth's history. In Gondwana, ice ages occurred that pushed glacial ice streams to within 30°S of the palaeo-equator.

Once assembled, at the beginning of the Permian, Pangaea stretched from pole to pole in a single hemisphere (Fig. 2). The ocean of Panthalassa covered the other hemisphere (two-thirds of the earth's surface). Permian Pangaea was a relatively diverse place in terms of climate and topography. Glacial deposits found in South America, Africa, India, Australia and Antarctica are evidence of the continuation of glaciations in southern Gondwana during the Early Permian (Veevers 2004). Along the sutures of Pangaea, huge mountain ranges towered over vast tropical lowlands. During the Mid- and Late Permian, interior areas included dry deserts where dune sands accumulated. Evaporites (particularly gypsum and halite) deposited in the southwestern USA and northern Europe document the evaporation of hot shallow seas and formed the most extensive salt deposits in the geological record. Perhaps the best testimony to the diversity of Permian Pangaea can be seen in its fossil plants, which identify several floral provinces across the vast supercontinent.

The Late Carboniferous (Pennsylvanian) and the Permian are distinguished by a degree of continentality only matched by the last 5 million years of Earth's history. This extensive continentality underlies numerous problems of Permian stratigraphy that range from local geological mapping to global correlations. These problems are well reflected in the recently published SGCS: most of the Pennsylvanian and Permian stages have been adopted relatively recently, compared to the much longer accepted stages of the Devonian, Mississippian and the Triassic (Ogg 2004). This is the result of conditions that are unique to the Phanerozoic.

One of these was the formation and break-up of the Pangaean supercontinent, which was surrounded by a single ocean (Panthalassa) with only one intervening seaway, the Tethys. Of the two largest components of the supercontinent, Gondwana covered an area of about 73 million km^2 but was only about 15% covered by epi-continental seas, while Laurasia covered an area of about 65 million km^2 but was only about 25% covered by epi-continental seas. This exceptionally low sea level was due to the accumulation of water in polar and inland icecaps during the late Palaeozoic glaciations, little to no spreading activity of the mid-oceanic ridges and, possibly, to the elevation of the geoid because of thermal shielding by the huge landmass of Pangaea.

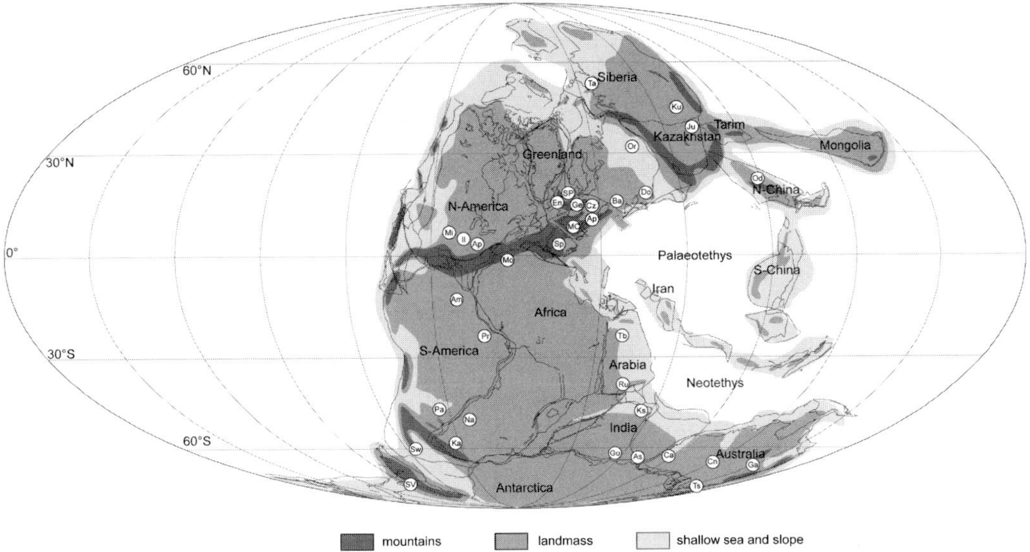

Fig. 2. Map of Pangaea at 270 Ma (modified from Golonka 2000) with the most important continental Permo-Carboniferous basins indicated. **Al**, Alpine basins, e.g. Carnic Alps, Collio Basin, Salvan-Dorénaz Basin; **Am**, Amazon Basin; **Ap**, Appalachian Basin; **As**, Assam-Arakan Basin; **Ba**, Balkan Basins, e.g. Moesian Basin, Resita Basin, Sirina Basin; **Ca**, Carnarvon Basin; **Cn**, Canning Basin; **Cz**, Czech Basins, e.g. Intra-Sudetic Basin, Boskovice Graben, Bohemian basins; **Do**, Donetsk Basin; **En**, localities of Hopeman and Elgin; **Ga**, Galilee Basin; **Ge**, German Basins, e.g. Saar-Nahe Basin, Thuringian Forest Basin, Saale Basin; **Go**, Godavari Valley Basin, Mahanadi Valley Basin; **Il**, Illinois Basin; **Ju**, Junggur Basin; **Ka**, Karoo Basin; **Ks**, Kashmir Basin; **Ku**, Kuznetsk Basin; **MC**, basins of Massif Central and surroundings, e.g. Lodève Basin, Autun Basin, Bourbon l'Archambault Basin, Commentry Basin; **Mi**, Midcontinent Basin; **Mo**, Moroccan Basins, e.g. Chougrane Basin, Khenifra Basin, Tiddas Basin, Souss Basin; **Na**, Namibia region; **Od**, Ordos Basin; **Or**, Orenburg region (Cis-Urals); **Pa**, Paraná Basin; **PB**, Northern and Southern Permian Basins; **Pr**, Parnaíba Basin; **Ru**, Rub Al Khali Basin; **Sp**, Spain Basins, e.g. Puertollano Basin, Cantabrian Mountains; **SV**, South Victoria Land, Trans-Antarctic Mountains; **Sw**, SW South America Basin, e.g. San Rafael Basin, Paganzo Basin, Golondrina Basin; **Ta**, Taimir Basin; **Tb**, Tabuk Basin; **Ts**, Tasmanian Basin.

Diverse tectonic processes, such as the collision of plates, closing of ocean basins, build-up and collapse of orogenic belts, compressional/extensional tectonics and the Late Permian onset of rifting that initiated the break-up of Pangaea, led to a wide variety of Permian basin types, ranging from marine to paralic foreland basins to intramontane and perimontane basins (Fig. 2). Linked to this, diverse tectonomagmatic processes – from synorogenic to post-orogenic magma intrusions, underplating and volcanism to rift-bounded upper mantle basalt extrusions – also influenced basin formation, the rejuvenation of topographic relief and basin reorganization. For example, in Europe, the Mid- to Late Permian cooling of the crust produced the huge southern Permian basin with an areal extent of about 2000×700 km, which was the embryonic stage of the Mesozoic and Cenozoic central European basin.

In this volume, **Roscher & Schneider** discuss the interference of these processes with the transition from the Palaeozoic icehouse to the Mesozoic greenhouse – a multi-stage process of wet and dry phases that created a very broad spectrum of facies types and facies architectures, and an array of evolving environments that triggered biotic evolution. Interrupted by the Permo-Triassic crisis, Mesozoic types of continental biota began to develop during the Permian (e.g. Kerp 1996; DiMichele *et al.* 2001; Kerp *et al.* 2006).

As a consequence, the continental biofacies patterns are as differentiated and complicated as are the lithofacies patterns. Each of the nearly 100 Permian continental basins in Euramerica therefore has its own lithostratigraphic subdivision, which can often only be correlated over several hundreds of square kilometres. Correlations are hampered by the distinctive development of single basins that lack interbasinal lithological marker horizons or marine intercalations and the sparse and scattered fossil content of many of the non-marine Permian strata.

Non-marine Permian rock record

The non-marine rock record of the Permian System is very diverse and widespread. Long known and best studied are the western European Rotliegend and correlative strata. This is well reflected by several papers in this volume (**Arche & López-Gómez; Durand; Roscher & Schneider; Virgili *et al.***) that focus on problems of tectonism, volcanism, sedimentation and ultimately correlation in the non-marine strata of the classical Permian basins in Germany, France, the Czech Republic, Italy and Spain. Also, in this volume, **Hmich *et al.*** present new stratigraphic and biostratigraphic data on the North African Permian strata.

Since the end of the eighteenth century, in the classical Permian basins of Europe, palaeontological research and sampling by scientists and private collectors focused on the Upper Pennsylvanian and lowermost Permian coal-bearing grey facies. This was driven by economic interest in Rotliegend coals (of poor quality) that were explored and exploited well into the twentieth century. In the French Autun Basin, extensive Upper Permian bituminous black shales have been exploited since the middle of the nineteenth century. Until 2002, open-cast coal-mining in the Bourbon–l'Archambault Basin of the French Massif Central yielded one of the most diverse late Lower Rotliegend or Upper Autunian faunas of lacustrine bituminous black-shale type. In the nineteenth century, the well-known fossiliferous geodes of the Rotliegend Lebach in the Saar–Nahe Basin were exploited for iron ore, and the fish coprolites of the Goldlauter lake horizon in the Thuringian Forest were a source of sulphide ores.

From the approximately 150 years of research history comes a detailed knowledge of the fossil biota of Permian swamp and lake environments. Nevertheless, modern palaeobiological research in combination with advanced sedimentology have recently added much new information on ecosystems, as exemplified by Boy (1998) and Boy & Schindler (2000).

The seemingly barren Permian red beds have not been regarded as of interest for palaeontology, other than tetrapod track sites, such as the Tambach locality of the Thuringian Forest, known since the end of the nineteenth century. In the 1960s, gas exploration in the Rotliegend dune sandstones began. From this, arose a demand for palaeozoological biostratigraphy of grey to red clastics, poor or barren in plants. New biostratigraphic methods, based on insects, conchostracans, small amphibians and isolated fish remains, were developed (see below). Sediments of restricted lakes, grey and red alluvial plains and playa and sabkha deposits were intensively investigated for the first time. They yielded a thus far unknown fossil content, changing the picture of those environments, which had long been regarded as unfossiliferous in Germany, or 'azoique' in France (e.g. Schneider & Gebhardt 1993). The famous Tambach vertebrate site, with an upland tetrapod fauna of North American affinity, was also discovered at this time (Eberth *et al.* 2000), and arthropod tracks and freshwater jellyfishes of the genus *Medusina* were discovered to be common in floodplain pool and playa fine clastics (e.g. Walter 1983; Gand *et al.* 1996, 2000). For the first time in Europe, Early Permian aistopod-like amphibian remains were discovered in greyish micritic pond-limestones of red alluvial plain deposits with coarse channel fills and intensively rooted calcisols (Schneider & Rössler 1996).

During the last decade, one of the most remarkable discoveries in the western European Permian red beds is the fossil content of the Lodève playa at the southern border of the French Massif Central. Besides the well-known tetrapod tracksite of La Lieude (e.g. Gand *et al.* 2000), a very rich arthropod fauna with mass occurrences of conchostracans and triopsids, as well as a diverse insect fauna, was discovered in decimetre-wide, clayey-silty channel fills (Gand *et al.* 1997*a*, *b*; Bethoux *et al.* 2002). Higher in that section, sheetflood deposits containing carbonate nodules of reworked calcisoils have yielded skeletal remains of an approximately 3-m-long caseid pelycosaur, similar to *Cotylorhynchus*, as well as the minute vertebral column of an embolomerous amphibian, both clearly of North American affinity.

In this volume, **Hmich *et al.*** report on Permian tetrapod bones, the first in Morocco, discovered in reworked calcisols as well as in channel conglomerates, and new Lower to Upper Permian tetrapod tracksites. In contrast to North America and the eastern European platform, grey to red deposits of Permian alluvial environments were not investigated in detail in western Europe and North Africa during the last three decades, because research traditionally focused on coal-bearing sequences. The sediments of Permian alluvial plain environments are therefore a very promising prospect for future research.

According to **Virgili *et al.*** (this volume), in the continental domains of southwestern Europe, from the Carboniferous to Triassic, palaeontological investigations of the macroflora, microflora and tetrapod footprints, as well as radio-isotopic data, essentially indicate the presence of three main 'tectono-stratigraphic sequences,' separated by marked unconformities and gaps of as yet uncertain duration. The most significant geological episode began at about the Early Middle Permian boundary and persisted throughout the Middle Permian (Guadalupian) time. This episode was characterized by specific tectonic, magmatic, thermal and basinal features, which marked the presumed change, suggested by some authors (e.g. Muttoni *et al.* 2003), from a Pangaea 'A' to a Pangaea 'B' configuration.

The ancestral Rocky Mountain foreland of western North America has extensive non-marine Permian deposits, and their biostratigraphic records of tetrapod footprints and body fossils are discussed in this volume by **Hunt & Lucas, Lucas & Hunt**, and **Lucas**. The Ural foreland basin in Russia (Tverdokhlebov *et al.* 2005) and the Karoo foreland basin of southern Africa are also of importance to tetrapod biostratigraphy. In Brazil, the intracratonic Paraná Basin, and the Ordos and Junggur basins of northern China are some of the better-known locations that contain significant non-marine Permian rock and fossil records.

Non-marine Permian time scales

The first attempt to establish time units and correlations based primarily on non-marine Permian rocks was by Jules Marcou (1859), who introduced the term 'Dyas' for the '*nouveau grès rouge en Europe dans l'Amerique du Nord et dans l'Inde.*' At about the same time, the Germans Veltheim (1821–24) and Geinitz (1861) attempted to delineate the red beds of the Rotliegend and to correlate them with the classical Permian of the former Russian department of Perm in the Urals. Geinitz (1869) and Suess (1869) also correlated the Rotliegend of central Europe with some alluvial to lacustrine, plant-bearing dark shales of the upper Val Trompia, in the Southern Alps. The term 'Rotliegend' was then applied to strata in the Southern Alps by German geologists (e.g. Lepsius 1878; Gümbel 1880; Vacek & Hammer 1911; Heritsch 1939).

Several formally proposed subdivisions of Permian time have also been based on non-marine strata. These include:

(1) Rotliegendes, used by Harland *et al.* (1990) as a series to encompass the Asselian, Sakmarian, Artinskian and Kungurian Stages (but now abandoned and replaced by Cisuralian);
(2) Autunian, Saxonian and Thuringian, which form a threefold division of Permian time based largely on French and German non-marine strata;
(3) Tatarian, based on non-marine Middle–Upper Permian strata in the Russian Urals is a regional stage for the latest Middle Permian and the Upper Permian of the East European platform.

However, no global time scale based on non-marine Permian rocks and fossils is in use.

Thus, the 'parastratigraphic' subdivision into Autunian, Saxonian and Thuringian is neither a suitable chronostratigraphical division of the Permian System nor useful for correlation with non-European domains or with the international scale. This stratigraphic trinity can be compared with the Cisuralian, Guadalupian and Lopingian Series, but, based on current data, their temporal ranges and lack of proper stratotypical definition make the terms 'Autunian', 'Saxonian' and 'Thuringian' unsuitable for use in a global time scale. Despite this, Broutin *et al.* (1999) indicate that significant 'Autunian' floras (such as the mass presence of *Autunia*, *Rhachiphyllum*, *Lodevia*, *Arnhardtia* and *Gracilopteris*) are widespread and well known in many continental basins of Europe. Therefore, even though the temporal boundaries of this fossiliferous succession are not defined in the typical Autun Basin, the floral abundance and its biostratigraphical value cannot be underestimated. Indeed, Broutin *et al.* (1999) suggested that the fossil content of 'Autunian' should be regarded as characteristic of the basal and immediately older time interval (latest Ghzelian to the early Sakmarian) of the continental Permian.

The use of 'Saxonian', due to its chronostratigraphical inconsistency, should be rejected (e.g. Kozur 1986, 1993; Cassinis *et al.* 1992, 1995). In the type area, it seems to include more than six stages, from Sakmarian to basal Dzhulzfian. Thus, 'Saxonian' is too extensive temporally and lacks palaeontological characterization.

'Thuringian' has up to now been used in different countries of Europe to refer to both marine and continental strata, which often bear a number of palaeontological elements that generally pertain to Late Permian time. The type area of this unit, which represents an equivalent of the German Zechstein, corresponds to the Dzhulfian and early Changshingian (Kozur 1988, 1993). However, the term has been subject to different stratigraphic interpretations, also equating it to the Kazanian and Ufimian.

Since Geinitz (1861), the German Rotliegend, and its literal translations – 'Cervená Jalovina' (Czech) or 'Èzerwonego Spagowca' (Polish) – have been regarded as of Permian age, as has the New Red Sandstone in England. But it is primarily understood as a lithostratigraphic unit. Indeed, 'Rotes Todes Liegendes' (or 'red dead underlayer': Freiesleben 1795) was a miner's term for the ore-free continental red beds below the copper ore-bearing marine black shale, the approximately 40-cm thick Kupferschiefer (or 'Copper slate').

The Kupferschiefer and its equivalents, marine sandy to conglomeratic clastics as well as limestones and marls (e.g. the Marl Slate of the United Kingdom), mark the base of the second Permian unit: the Zechstein. Again, 'Zechstein' is an old miner's term for 'firm ground' – the rock (*Stein*) on which the mine (*Zeche*) for the exploitation of the Kupferschiefer was built. The type area of both terms is the southern foreland of the Harz Mountains (Hercynian Mountains) in Germany at the northern border of the Thuringian basin.

The French term 'Autunien', later used as 'Autunian' in a chronostratigraphical sense, is comparable to the Rotliegend and is nothing other than a lithostratigraphical term. It was introduced by Mayer-Eymar (1881) as 'Autunin' for a sequence of sandstones with characteristic bituminous black shales in the Autun Basin of the French Massif Central. As often discussed (e.g. Kozur 1978; Broutin *et al.* 1999; Schneider 2001), the Autunian has no chronostratigraphical content, and the base and the top have not been biostratigraphically defined (Broutin *et al.* 1999).

The terms 'Saxonian' and 'Thuringian' were introduced by the French workers Munier-Chalmas & De Lapparent (1893). 'Saxonian' should refer to the red sandstones below the marine Zechstein in the Mansfeld area of the southern foreland of the Harz Mountains (former Saxony, or Sachsen, now Sachsen-Anhalt). Those red sandstones have been understood as Upper Rotliegend (Oberrotliegendes). But, by the end of the nineteenth century, it became apparent that they are only partily Rotliegend and that some of these red beds belong to the Mansfeld Subgroup, which is of Late Carboniferous age (see **Roscher & Schneider**, this volume).

Because of translation problems into French (Oberrotliegendes = Saxonian), the German term 'Zechstein' was replaced by 'Thuringian' because the Late Permian Zechstein crops out along the borders of the Thuringian basin. Clearly, both terms primarily indicate lithostratigraphical units. Later attempts to give them a chronostratigraphical meaning were pointless because of the sparse fossil content of the continental red beds of the 'Saxonian' and the lack of exact correlation of the restricted-marine 'Zechstein' or 'Thuringian' to global marine scales (Legler *et al.* 2005; Roscher & Schneider 2005). Both terms are no longer in use in Germany, either as lithostratigraphical or as chronostratigraphical terms.

The term 'Thuringian' has often been used in palynostratigraphic publications to indicate microfloras of Late Permian Zechstein affinity. This is very misleading because it has been overlooked that, in the type area of the Zechstein of Lopingian age (the forelands of the Harz and Thuringian Forest Mountains), the youngest palynological records come from Lower Rotliegend ('Autunian') grey sediments of Early Permian (Asselian/Sakmarian) age (Schneider 2001; Roscher & Schneider 2005). The Early to Late Permian Upper Rotliegend ('Saxonian') red beds in this area are barren of microfloras because of oxidation. Thus, microfloras of 'Thuringian' affinity occur long before the Zechstein transgression took place. It follows that the terms 'Saxonian' and 'Thuringian' should be abandoned as chronostratigraphical units and that 'Autunian' should be used in its primary lithostratigraphical sense as 'Autunin' or 'Autunien', equivalent to the Lower Rotliegend.

Non-marine Permian biostratigraphy

Microfloras and macrofloras

The biostratigraphy of Permian continental deposits has long been based on microfloras (palynomorphs) and macrofloras (e.g. Kozur 1993; Utting & Piasecki 1995). Macrofloras are most valuable as environment indicators and for palaeogeography, as shown in many publications, for example, by Ziegler (1990) in his classical paper '*Phytogeographic patterns and continental configurations during the Permian period*', and by Wang (1996), LePage *et al.* (2003), Berthelin *et al.* (2003) or Cleal & Thomas (2005).

An explosive diversification in the microfloral record is observed during the Late Carboniferous and the Permian, which enables concurrent range zonations based on the first appearances, acmes and last occurrences of different associated forms. Although the environmental influence on macrofloras is reflected in the microflora from local edaphic conditions up to floral provinces as well, regional palynostratigraphical correlations within the same floral provinces or biomes are possible in the Permian, but correlations between different floral provinces remain speculative. Palynostratigraphy is used in many southwestern European continental deposits, as is shown in this volume by **Virgili *et al.*** and **Arche & López-Gómez**.

Roscher & Schneider, in this volume, discuss the decreasing marine transgressions during the Pennsylvanian and the aridization during the Late Pennsylvanian and the Permian that generated a change from inter-regionally balanced wet macro- and mesoclimates (with some degree of maritime imprint) to increasingly drier continental climates with stronger seasonality and stronger accentuation of meso- and microclimatic effects. This resulted in a strong edaphic differentiation of the floral associations. The persistence of conservative Carboniferous hydro- to hygrophilous floral elements into Permian (local) wet biotopes and the local appearance of modern typical Permian meso- to xerophilous floral elements in the Carboniferous underlies the well-known problems of Permian floral biostratigraphy (e.g. Broutin *et al.* 1990, 1999; DiMichele *et al.* 1996; Kerp 1996).

In this volume, **Rössler** provides an overview of two remarkable Permian petrified forests, those of Chemnitz, Germany, and northern Tocantins, Brazil. These essentially contemporaneous forests represent seasonally influenced, tree-fern-dominated plant communities in the Northern and Southern Hemispheres of Permian Pangaea. The outstanding three-dimensional preservation of particularly large fossil remains, made possible by siliceous permineralization, provides the opportunity to study the gross morphology, anatomy and internal organization of plant tissues in a way not allowed by other preservational states and will force a re-evaluation of the taxonomy and reconstructions of Early Permian floras.

Although the paper by **Rössler**'s is the only one in this volume devoted strictly to palaeobotany, many of the other papers (especially those by **Virgili *et al.***; **Arche & López-Gómez**, **Durand**, and **Hmich *et al.***) rely heavily on fossil plant-based age determinations. The fact is that, despite the caveats and pitfalls, many age determinations in non-marine Permian strata continue to be plant based.

Charophytes

The oogonia of freshwater characeous algae (gyrogonites) fossilize and have some utility in the correlation of non-marine strata, particularly in the Cretaceous and Cenozoic. However, the Permian record of charophytes is very poorly known (Feist *et al.* 2005). The only substantial record is from China (e.g. Wang & Wang 1986), and the biozonation based on it (four Permian assemblages: Feist *et al.* 2005, table 4) needs to be tested with data from other regions.

Ostracodes

Non-marine ostracodes are very common in various lacustrine deposits of the Permian, ranging from black shales and limestones of perennial lakes to claystones and micritic limestones of temporary ponds and pools. They can even be common in strata deposited by brackish waters or environments of higher salinity. Mass occurrences in shales and limestones, sometimes rock-forming, may be linked to ecological factors that prevent the co-occurrence of other inhabitants of the same or similar guild, as well as the occurrence of ostracode-feeding predators.

The use of non-marine ostracodes in biostratigraphy is hampered by two factors. Firstly, freshwater ostracodes are very simple in terms of morphological features of the shell, and the state of preservation (lack of preserved muscle scars, deformation up to complete flattening during sediment compaction) very often prevents any precise identification. The second factor is their nearly hopelessly over-split alpha taxonomy, which may have been resolved by the taxonomic revisions carried out by Molostovskaya (2005). As observed in modern semiarid and arid environments in Africa and Arabia, the minute eggs of freshwater ostracodes are drought resistant. They were easily distributed over hundreds of kilometres by air currents. Therefore, Permian non-marine ostracodes could have promise for biostratigraphy.

Conchostracans

Conchostracans are bivalved crustaceans whose fossils have been employed in some non-marine Permian correlations (e.g. Martens 1982; Schneider *et al.* 2005). They have a very high distribution potential because of their minute, drought-resistant and wind-transportable eggs, and they often form mass accumulations in lacustrine lithofacies. Hence, conchostracans are some of the most common animal fossils of the continental Permian.

Nevertheless, the time ranges of many Permian conchostracan species have not been well established, and much alpha taxonomy needs to be resolved (e.g. Martens 1983; Schneider *et al.* 2005). If these obstacles can be overcome, then conchostracans may contribute to regional and, perhaps, global non-marine Permian correlations.

Insects

As early as 1879, Scudder attempted to use insect wings for Permian biostratigraphy. Thus, he recognized the common occurrence of genera and species of blattid insects (cockroaches) in North America and Europe and their potential for 'delicate discriminations of the age of rock deposits' (Scudder 1885). Later, Durden (1969, 1984) proposed blattid zonations for the Pennsylvanian and Permian, but his correlations were doubtful because of inadequate taxonomy. Schneider (1983) published a revised classification of Pennsylvanian and Permian blattids, from which came the first proposal of spiloblattinid zones (Schneider 1982; Schneider & Werneburg 1993) and later of archimylacrid/spiloblattinid/conchostracan zones, for the Early Pennsylvanian (Westphalian A) through the late Early Permian (Artinskian).

In this volume, **Schneider & Werneburg** present an updated insect zonation for the Late Pennsylvanian to Early Permian with a time resolution of 1.5–2 Ma. The zonation is based on the morphogenetic evolution of lineages of time-successive species of three genera of spiloblattinids. All three genera are widely distributed in the palaeo-equatorial zone from Europe to North America. As **Schneider & Werneburg** note, new reports of spiloblattinid zone species in non-marine strata, intercalated with conodont-bearing marine strata in the North American Appalachian, Midcontinent and West Texas basins could be one key to the direct biostratigraphical correlation of continental Permian strata to the SGCS.

Bivalves

During the Carboniferous–Permian, non-marine bivalves, including anthracosiids, palaeomutelids, and some myalinids (brackish water), had a worldwide distribution. Records include (but are not limited to): the Middle–Upper Permian Karoo Supergroup of southern Africa and Madagascar; the Permian Mount Glossopteris Formation of the Ohio Mountains, Antarctica; the Upper Carboniferous of Nova Scotia and the Upper Carboniferous to Lower Permian of the eastern United States; the Upper Permian (Tartarian) strata of the Oka–Volga River Basin in Russia; and the Upper Carboniferous Coal Measures of northern France, England and Ireland. Additional assemblages are known from southern Asia and South America. Some biostratigraphic correlations have been based on these bivalves (e.g. Eagar, 1984), but their alpha taxonomy (taxonomic names of genera and species) seems extremely over-split, as most variation is ecomorphophenotypic, not interspecific, in origin.

Furthermore, it seems unlikely that the stratigraphical ranges of all non-marine Permian bivalves are well established. Thus, for example, Lucas & Rinehart (2005) recently documented *Palaeanodonta* in the Lower Permian of North America, whereas the genus is otherwise known from the Middle or Upper Permian of Antarctica, South Africa, Kenya, Russia, Myanmar and Siberia, among other places. This substantial range extension suggests to us that the true distributions in time and space of all late Palaeozoic freshwater bivalves are not well known. This and the taxonomic problems should make us very cautious in using non-marine bivalves for Permian biostratigraphy.

Fishes

Fishes have never provided a robust biostratigraphy in non-marine strata. This is because of the limitations of these fishes and their fossils to specific lithofacies and locations, so that their record is dominated by facies-control and endemism. Permian xenacanth shark teeth have been applied to regional correlations between some neighbouring European basins, but their wider use is limited because the migration of fishes is restricted to joint river systems connecting the basins (e.g. Schneider 1996; Schneider *et al.* 2000). But, these fishes do deliver valuable information about basin interconnections and drainage systems (Schneider & Zajic 1994), as well as about ecological changes and events (Boy & Schindler 2000). Thus, the fish zonation of Zajic (2000) is actually a local ecostratigraphy of some Bohemian basins, not a robust biostratigraphy.

In this volume, **Stamberg** reviews the actinopterygian fishes from the continental Westphalian to Lower Permian basins of the Czech Republic and compares them to those of equivalent age in some other European basins. Nine genera of actinopterygians belonging to eight families are known from the Westphalian–Stephanian sediments, and five genera belonging to four families are known from the Lower Permian sediments of the Bohemian Massif. Very close relationships of the actinopterygian fauna between the Bohemian Massif and the basins of the French Massif Central are evident, but the resulting correlations are neither precise nor really extensive.

Tetrapod footprints

Permian tetrapod footprints are known from localities in North America, South America, Europe and Africa, and attempts to use footprints to correlate non-marine Permian strata have a long tradition, especially in Europe (Voigt 2005). This is reflected in the papers by **Gand & Durand, Hunt & Lucas** and **Lucas & Hunt** in this volume, which document the distribution of Permian tetrapod footprints and their use in biostratigraphy.

In this volume, **Hunt & Lucas** assign Permian tetrapod footprints to four ichnofacies, the *Chelichnus* ichnofacies from aeolianites and the *Batrachichnus, Brontopodus* and *Characichichnos* ichnofacies from water-laid (mostly red bed) strata (Hunt & Lucas 2006). In this volume, **Lucas & Hunt** conclude that Permian track assemblages of the *Chelichnus* ichnofacies are of uniform ichnogeneric composition and low diversity, range in age from Early to Late Permian, and are thus of no biostratigraphical significance.

In contrast, footprints of the *Batrachichnus* and *Brontopodus* ichnofacies represent two biostratigraphically distinct assemblages:

(1) Early Permian assemblages characterized by *Amphisauropus, Batrachichnus, Dimetropus, Dromopus, Hyloidichnus, Limnopus* and *Varanopus*;
(2) Mid- to Late Permian assemblages characterized by *Brontopus, Dicynodontipus, Lunaepes, Pachypes, Planipes*, and/or *Rhynchosauroides*.

Few Permian footprint assemblages are demonstrably of Middle Permian (Guadalupian) age, and there is a global gap in the footprint record equivalent to at least Roadian time. Permian tetrapod footprints thus only represent two biostratigraphically distinct assemblages: an Early Permian pelycosaur assemblage and a Mid- to Late Permian therapsid assemblage. Therefore, footprints provide a global Permian biochronology of only two time intervals, much less than the 10 time intervals that can be distinguished with tetrapod body fossils.

Tetrapod body fossils

Permian tetrapod (amphibian and reptile) body fossils have long provided a basis for non-marine biostratigraphy and biochronology (see reviews by Lucas 1998, 2002, 2004). The most extensive Permian tetrapod (amphibian and reptile) fossil records come from the western United States (New Mexico–Texas) and South Africa. In this volume, **Lucas** uses these records to define 10 land-vertebrate faunachrons ('ages') that encompass Permian time. These faunachrons provide a biochronological framework with which to determine and discuss the age relationships of Permian tetrapod faunas. Their correlation to the SGCS and its numerical calibration is relatively straightforward in the Early Permian, as the Texas Lower Permian red-bed section has marine intercalations that yield fusulinids, conodonts and/or ammonoids that allow for marine ages to be assigned. Correlation of the Mid- to Late Permian tetrapod record with the SGCS is much less certain.

In Europe, biostratigraphical zonations of aquatic or semi-aquatic amphibians were presented by Werneburg (e.g. 1996, 2001, 2003). In this volume, **Werneburg & Schneider** present a revised amphibian zonation (nine amphibian zones) for the European Pennsylvanian and Lower Permian. The index fossils belong to species chronoclines of two or three closely related species. The time resolution of these amphibian zones is about 1.5 to 3.0 Ma, and biostratigraphic correlations based on them are applicable to 16 basins in the Czech Republic, Poland, France, Italy and Germany.

Isotopic ages

Relatively few reliable isotopic ages can be used to calibrate the Permian SGCS. Thus, the numerical calibration is imprecise, being based largely on interpolating between a cluster of radio-isotopic ages near the Carboniferous–Permian boundary, an Artinskian U–Pb age from Russia of 280.3 ± 2.5, an U–Pb age of the Capitanian base of 265.3 ± 0.2 Ma, and a cluster of ages near the Permo-Triassic boundary (Wardlaw *et al.* 2004) (Fig. 1).

Many more radio-isotopic ages are available in non-marine Permian stratigraphic successions, especially in the German Rotliegend and related strata in France and Italy (e.g. Cadel 1986; Zheng *et al.* 1991–1992; Del Moro *et al.* 1996; Schaltegger & Brack 1999; Cassinis & Ronchi 2001; Deroin *et al.* 2001; Cassinis *et al.* 2002; Pittau *et al.* 2002; Bargossi *et al.* 2004; Roscher & Schneider 2005; Lützner *et al.* 2006). Some of these ages are old K–Ar ages of questionable precision, but recent work is providing more reliable Ar/Ar and U–Pb ages for some igneous rocks intercalated with non-marine Lower Permian strata. These numbers thus provide direct calibration of the non-marine fossil biostratigraphies of the Lower Permian rocks. Fewer Middle and Upper Permian isotopic ages are available in the non-marine succession, but some are emerging from work on the Karoo Basin of South Africa (Bangert *et al.* 1999; Wanke *et al.* 2000).

These ages hold great promise for yielding a more precise numerical calibration of non-marine Permian biostratigraphy than can be directly achieved for the Permian SGCS. The challenges thus lie in cross-correlating non-marine Permian biostratigraphy to the SGCS so that all the ages can be combined to produce a more precise numerical time scale for the Permian.

Magnetostratigraphy

Most of Permian time has long been considered an interval when there was little or no reversal in the activity of the Earth's magnetic field. Thus, all of Early Permian and some Middle Permian time comprise the latter part of the Carboniferous–Permian reversed polarity superchron (also called the Kiaman superchron). The field began to reverse frequently during the Middle Permian, and this begins the Permo-Triassic mixed superchron. The initiation of the superchron is usually referred to as the Illawarra reversal.

The Illawarra reversal thus has been taken to provide an important datum for intercontinental correlation. Menning (2001) concluded that the Illawarra reversal is Early Capitanian: ~265 Ma on the SGCS. Thus, for example, its presence in the Russian Tatarian has been used to directly correlate the Russian non-marine section to the SGCS.

In this volume, **Steiner** presents a synthesis of Mid- to Late Permian and Early Triassic magnetostratigraphy, much of it from non-marine sections. From this, she recognizes a polarity pattern for the Mid- and Late Permian of two normal polarity intervals and, just below the Permo-Triassic boundary, a distinctive short-duration reversed-normal-reversed polarity pattern. According to **Steiner**, the oldest normal polarity in the Middle Permian occurred during the Wordian Stage, established by results from three global sequences. Therefore, the resumption of geomagnetic field reversals after the ~50 Ma-long Carboniferous–Permian reversed-polarity superchron was during the Mid- to Late Wordian, or ~267 Ma. If this is correct, then many accepted correlations based on the Illawarra reversal will have to be revised.

Conclusions

Despite nearly two centuries of study, non-marine Permian biostratigraphy and correlation remain poorly understood. However, a diversity of biostratigraphical methods are available that need further development. In particular, sound alpha taxonomy and well-established stratigraphical ranges are needed for many fossil groups. A plethora of radio-isotopic ages in non-marine Permian rocks can be directly related to much non-marine Permian biostratigraphy. Furthermore, in Middle–Upper Permian strata, magnetostratigraphy provides another correlation tool. All three datasets for the correlation of non-marine Permian strata – biostratigraphy, radio-isotopic ages and magnetostratigraphy – need to be integrated and cross- correlated to the marine time scale. Only then can a better understanding of Permian earth history on land and sea be achieved.

We thank all the authors for their contributions to this volume. M. Roscher provided Figure 2.

References

BANGERT, B., STOLLHOFEN, H., LORENZ, V. & ARMSTRONG, R. 1999. The geochronology and significance of ash-fall tuffs in the glaciogenic Carboniferous–Permian Dwyka Group of Namibia and South Africa. *Journal of African Earth Sciences*, **29**, 33–49.

BARGOSSI, G. M., KLÖTZLI, U. S., MAIR, V., MAROCCHI, M. & MORELLI, C. 2004. The Lower Permian Athesian volcanic group (AVG) in the Adige Valley between Merano and Bolzano: a stratigraphic, petrographic and geochronological outline. *32nd International Geological Congress 'Florence' 2004 'Scientific Sessions': Abstracts*, in, 187.

BERTHELIN, M., BROUTIN, J., KERP, H., CRASQUIN-SOLEAU, S., PLATEL, J.-P. & ROGER, J. 2003. The Oman Gharif mixed paleoflora: a useful tool for testing Permian Pangaea reconstructions. *Palaeogeography, Palaeoclimatology, Palaeoecology*, **196**, 85–98.

BETHOUX, O., NEL, A., GAND, G., LAPEYRIE, J. & GALTIER, J. 2002. Discovery of the genus *Iasvia* Zalessky, 1934 in the Upper Permian of France (Lodéve Basin) (Orthoptera, Ensifera, Oedischiidae). *Géobios*, **35**, 293–302.

BOY, J. A. 1998. Möglichkeiten und Grenzen einer Ökosystem-Rekonstruktion am Beispiel des spätpaläozoischen lakustrinen Paläo-Ökosystems. 1. Theoretische und methodische Grundlagen. *Paläontologische Zeitschrift*, **72**, 207–240.

BOY, J. A. & SCHINDLER, T. 2000. Ökostratigraphische Bioevents im Grenzbereich Stephanium/Autunium (höchstes Karbon) des Saar-Nahe-Beckens (SW-Deutschland) und benachbarter Gebiete. *Neues Jahrbuch für Geologie und Paläontologie, Abhandlungen*, **216**, 89–152.

BROUTIN, J., CHATEAUNEUF, J. J., GALTIER, J. & RONCHI, A. 1999. L'Autunian d'Autun reste-t-il une référence pour les dépôts continentaux du Permien inférieur d'Europe? Apport des données paléobotaniques. *Géologie de la France*, **2**, 17–31.

BROUTIN, J., DOUBINGER, J. *et al.* 1990. Le renouvellement des flores au passage carbonifère-permien: approches stratigraphique, biologique, sédimentologique. *Comptes Rendus – Academie des Sciences, Paris, Série II*, **311**, 1563–1569.

CADEL, G. 1986. Geology and uranium mineralization of the Collio basin (Central-Southern Alps, Italy). *Uranium*, **2**, 215–240.

CASSINIS, G. & RONCHI, A. 2001. Permian chronostratigraphy of the Southern Alps (Italy): an update. *In*: WEISS, R. H. (ed.), *Contribution to Geology and Palaeontology of Gondwana in Honour of Helmut Wopfner*. Geological Institute, University of Cologne, 73–88.

CASSINIS, G., TOUTIN-MORIN, N. & VIRGILI C. 1992. Permian and Triassic events in the continental domains of Mediterranean Europe. *In*: SWEET, W. C., YANG, Z. Y., DICKINS, J. M. & YIN, H. F. (eds) *Permo-Triassic Events in the Eastern Tethys*. Cambridge University Press, Cambridge, 60–77.

CASSINIS, G., TOUTIN-MORIN, N. & VIRGILI, C. 1995. A general outline of the Permian continental basins in southwestern Europe. *In*: SCHOLLE, P. A., PERYT, T. M. & ULMER-SCHOLLE, D. (eds) *The Permian of Northern Pangaea*. 2; *Sedimentary Basins and Economic Resources*. Springer-Verlag, Berlin, 37–157.

CASSINIS, G., NICOSIA, U., PITTAU, P. & RONCHI, A. 2002. Palaeontological and radiometric data from the Permian continental deposits of the central-eastern Southern Alps (Italy), and their stratigraphic implications. *Mémoire de l'Association des Géologues du Permien*, **2**, 53–74.

CLEAL, C. J. & THOMAS, B. A. 2005. Palaeozoic tropical rainforests and their effect on global climates: is the past the key to the present? *Geobiology*, **3**, 13–31.

DEL MORO, A., DI PISA, A. & OGGIANO, G. 1996. Relationship between an Autunian volcano-sedimentary succession and the Tempio massif granites (Northern Sardinia): geochronological and field constraints. *Plinius*, **16**, 94–95.

DEROIN, J.-P., BONIN, B. *et al.* 2001. The Permian of southern France: an overview. *In*: CASSINIS, G. (ed.) *Permian Continental Deposits of Europe and Other Areas. Regional Reports and Correlations*. Natura Bresciana, Monografie, **25**, 189–202.

DIMICHELE, W. A., MAMAY, S. H., CHANEY, D. S., HOOK, R. W. & NELSON, W. J. 2001. An Early Permian flora with Late Permian and Mesozoic affinities from north-central Texas. *Journal of Paleontology*, **75**, 449–460.

DIMICHELE, W. A., PFEFFERKORN, H. W. & PHILLIPS, T. L. 1996. Persistence of Late Carboniferous tropical vegetation during glacially driven climatic and sea-level fluctuations. *Palaeogeography, Palaeoclimatology, Palaeoecology*, **125**, 105–128.

DURDEN, C. J. 1969. Pennsylvanian correlation using blattoid insects. *Canadian Journal of Earth Sciences*, **6**, 1159–1177.

DURDEN, C. J. 1984. North American provincial insect ages for the continental last half of the Carboniferous and first half of the Permian. *Neuvième Congrès International de Stratigraphie et de Géologie du Carbonifère, Nanking 1979, Comptes Rendus*, **2**, 606–612.

EAGAR, R. M. C. 1984. Late Carboniferous–Early Permian non-marine bivalve faunas of northern Europe and eastern North America. *Neuvième Congrès International de Stratigraphie et de Géologie du Carbonifère, Washington and Champaign-Urbana, 1979, Comptes Rendus*, **2**, 559–576.

EBERTH, D. A., BERMAN, D., SUMIDA, S. S. & HOPF, H. 2000. Lower Permian terrestrial paleoenvironments and vertebrate paleoecology of the Tambach Basin (Thuringia, central Germany): the upland holy grail. *Palaios*, **15**, 293–313.

FEIST, M., GRAMBAST-FESSARD, N. *et al.* 2005. Charophyta: *Treatise on Invertebrate Paleontology. Part B: Protoctista 1. Volume 1.* The Geological Society of America Boulder and the University of Kansas, Lawrence.

FREIESLEBEN, J. C. F. 1795. *Bergmännische Bemerkungen über den merkwürdigsten Teil des Harzes. Zweiter Teil.* Schäferische Buchhandlung, Leipzig.

GAND, G., GARRIC, J., SCHNEIDER, J., SCIAU, J. & WALTER, H. 1996. Biocoenoses à Méduses du Permien francais (Bassin de Saint-Affrique, Massif Central). *Géobios*, **29**, 379–400.

GAND, G., GARRIC, J. & LAPEYRIE, J. 1997*a*. Biocénoses à Triopsidés (Crustacea), Branchiopoda) du Permien du bassin de Lodève (France). *Géobios*, **30**, 673–700.

GAND, G., LAPEYRIE, J., GARRIC, J., NEL, A., SCHNEIDER, J. & WALTER, H. 1997*b*. Découverte d'Arthropodes et de bivalves inédits dans le Permien continental (Lodèvois, France). *Comptes Rendus de l'Academie des Sciences, Paris, Serie II*, **325**, 891–898.

GAND, G., GARRIC, J., DEMATHIEU, G. & ELLENBERGER, P. 2000. La palichnofaune de vertebres tetrapodes du Permien superieur du bassin de Lodève (Languedoc, France). *Palaeovertebrata*, **29**, 1–82.

GEINITZ, H. B. 1861. *Dyas oder die Zechsteinformation und das Rothliegende. Heft 1. Die animalischen Ueberreste der Dyas*. Verlag Wilhelm Engelmann, Leipzig.

GEINITZ, H. B. 1869. Über fossile Pflanzenreste aus der Dyas von Val Trompia. *Neues Jahrbuch für Mineralogie, Geologie and Paläontologie*, **1869**, 456–461.

GOLONKA, J. 2000. *Cambrian-neogen: plate tectonic maps*. Habilitation thesis, Uniwersytet Jagiellonski.

GÜMBEL, C. W. 1880. Geognostische Mitteilungen aus den Alpen. VI: Ein geognosticher Streifzug durch die Bergamasker Alpen. *Sitzbungberichte der König Akademie der Wissenschaftliche, Mathematische-Natalische Klasse, München*, **10**, 164–240.

HARLAND, W. B., ARMSTRONG, R. L., COX, A. V., CRAIG, L. E., SMITH, A. G. & SMITH, D. G. 1990. *A Geologic Time Scale 1989*. Cambridge University Press, Cambridge.

HENDERSON, C. M. 2005. International correlation of the marine Permian time scale. *In*: LUCAS, S. G. & ZIEGLER, K. E. (eds) *The Nonmarine Permian*. New Mexico Museum of Natural History and Science Bulletin, **30**, 104–105.

HERITSCH, F. 1939. Karbon und Perm in den Südalpen und in Südosteuropa. *Geologische Rundschau*, **30**, 529–588.

HUNT, A. P. & LUCAS, S. G. 2006. Tetrapod ichnofacies: A new paradigm. *Ichnos*.

KERP, H. 1996. Post-Variscan late Palaeozoic Northern Hemisphere gymnosperms: the onset to the Mesozoic. *Review of Palaeobotany and Palynology*, **90**, 263–285.

KERP, H., HAMAD, A. A. & BANDEL, K. 2006. Typical Triassic Gondwanan floral elements in the Upper Permian of the paleotropics. *Geology*, **34**, 265–268.

KOZUR, H. 1978. Beiträge zur Stratigraphie des Perm. Teil III (1): Zur Korrelation der überwiegend kontinentalen Ablagerungen des obersten Karbons und Perms von Mittel- und Westeuropa. *Freiberger Forschungshefte, Hefte C*, **342**, 117–142.

KOZUR, H. 1986. Boundaries, subdivision and correlation of the marine and continental Permian. *In*: Italian IGCP–203 group (ed.) *Field Conference on 'Permian and Permian-Triassic Boundary in the South-Alpine Segment of the Western Tethys, and Additional Regional Reports', 4–12 July 1986, Brescia, Italy, Abstracts*, 29–31.

KOZUR, H. 1988. The age of the central European Rotliegendes. *Zeitschrift für Geologische Wissenschaften*, **16**, 907–915.

KOZUR, H. 1993. Application of the 'parastratigraphic' and international scale for the continental Permian of Europe. *In*: HEIDTKE, U. (compiler) *New Research on Permo-Carboniferous Faunas*. Pollichia-Buch, Bad Durkheim, **29**, 9–22.

LEGLER, B., GEBHARDT, U. & SCHNEIDER, J. W. 2005 Late Permian non-marine–marine transitional profiles in the central Southern Permian Basin, northern Germany. *International Journal of Earth Sciences*, **94**, 851–862.

LEPAGE, B. A., BEAUCHAMP, B., PFEFFERKORN, H. W. & UTTING, J. 2003. Late Early Permian plant fossils from the Canadian High Arctic: a rare palaeoenvironmental/climatic window in northwest Pangaea. *Palaeogeography, Palaeoclimatology, Palaeoecology*, **191**, 345–372.

LEPSIUS, R. 1878. *Das Westliche Sud-Tirol, geologisch dargestellt*. Verlag W. Hertz, Berlin.

LUCAS, S. G. 1998. Toward a tetrapod biochronology of the Permian. *New Mexico Museum of Natural History and Science Bulletin*, **12**, 71–91.

LUCAS, S. G. 2002. Tetrapods and the subdivision of Permian time; *In*: HILLS, L. V. & BAMBER, E. W. (eds) *Carboniferous and Permian of the World.* Canadian Society of Petroleum Geologists, Memoir, **19**, 479–491.

LUCAS, S. G. 2004. A global hiatus in the Middle Permian tetrapod fossil record. *Stratigraphy*, **1**, 47–64.

LUCAS, S. G. & RINEHART, L. F. 2005. Nonmarine bivalves from the Lower Permian (Wolfcampian) of the Chama Basin, New Mexico. *In*: LUCAS, S. G., ZEIGLER, K. E., LUETH, V. W. & OWEN, D. E. (eds) *Geology of the Chama Basin.* New Mexico Geological Society, Field Conference Guidebook, **56**, 283–287.

LÜTZNER, H., LITTMANN, S., MÄDLER, J., ROMER, R. L. & SCHNEIDER, J. W. 2006. Radiometric and biostratigraphic data of the Permocarboniferous reference section Thüringer Wald. *XVth International Congress on Carboniferous and Permian Stratigraphy, Utrecht, Proceedings*.

MARCOU, J. 1859. *Dyas et Trias ou le Nouveau Grès Rouge en Europe dans l'Amerique du Nord et dans l'Inde*. Archiv des Sciences de la Bibliothèque Universelle, Ramboz et Schuchardt, Genève.

MARTENS, T. 1982. Zur Taxonomie und Biostratigraphie neuer Conchostrakenfunde (Phyllopoda) aus dem Permokarbon und der Trias von Mitteleuropa. *Freiberger Forschungshefte, Hefte C*, **375**, 49–82.

MARTENS, T. 1983. Zur Taxonomie und Biostratigraphie der Conchostraca (Phyllopoda, Crustacea) des Jungpaläozoikums der DDR, Teil I. *Freiberger Forschungshefte, Hefte C*, **382**, 7–105.

MAYER-AYMAR, C, 1881. Classification internationale naturelle, uniforme, homophone et pratique des terrains de sediments. *Société Géologique de la France, Archives*, 21402, 1–15.

MENNING, M. 2001. A Permian time scale 2000 and correlation of marine and continental sequences using the Illawarra Reversal (265 Ma). *In*: CASSINIS, G. (ed.) *Permian Continental Deposits of Europe and Other Areas. Regional Reports and Correlations*. Natura Bresciana, Monografie, **25**, 355–362.

MOLOSTOVSKAYA, IJA I. 2005. Prospects of using non-marine ostracodes for distant correlations of Permian continental formations. *32nd International Geological Congress, Florence, 2004, Scientific Sessions, Abstracts*, **1**, 747.

MUNIER-CHALMAS, E. & DE LAPPARENT, A. 1893. Note sur la nomenclature des terrains sédimentaires. *Société Géologique de la France, Bulletin*, **3**, 438–493.

MURCHISON, R. I. 1841. First sketch of the principal results of a second geological survey of Russia. *Philosophical Magazine, Series 3*, **19**, 417–422.

MUTTONI, G., KENT, D. V., GARZANTI, E., BRACH, P., ABRAHAMSEN, N. & GAETANI, M. 2003. Early Permian Pangaea 'B' to Late Permian Pangaea 'A'. *Earth and Planetary Science Letters*, **215**, 379–394.

OGG, J. G. 2004. Status of divisions of the International Geological Time Scale. *Lethaia*, **37**, 183–199.

PITTAU, P., BARCA, S., COCHERIE, A., DEL RIO, M., FANNING, M. & ROSSI, P. 2002. Le bassin permien de Guardia Pisano (Sud-Ouest de la Sardaigne, Italie): palynostratigraphie, paléophytogéographie, corrélations et age radiométrique des produits volcaniques associés. *Géobios*, **35**, 561–580.

ROSCHER, M. & SCHNEIDER, J. W. 2005. An annotated correlation chart for continental Late Pennsylvanian and Permian basins and the marine scale. *In*: LUCAS, S. G. & ZIEGLER, K. E. (eds) *The Nonmarine Permian.* New Mexico Museum of Natural History and Science Bulletin, **30**, 282–291.

SCHALTEGGER, U. & BRACK, P. 1999. Excursion 1: The continental Permian from eastern Lombardy to the southwestermost part of Trentino. 4.1. Radiometric age constraints on the formation of the Collio Basin

(Brescian Prealps). *In*: CASSINIS, G., CORTESOGNO, L., GAGGERO, L., MASSARI, F., NERI, C., NICOSIA, U. & PITTAU, P. (eds) *Stratigraphy and Facies of the Permian Deposits between Eastern Lombardy and the Western Dolomites. Field Trip Guidebook. 23–25 September 1999*. International Field Conference on. The Continental Permian of the Southern Alps and Sardinia (Italy) regional reports and general correlations, Brescia, 15–25 September 1999, Earth Science Department, Pavia University, 71.

SCHNEIDER, J. 1982. Entwurf einer Zonengliederung für das euramerische Permokarbon mittels der Spiloblattinidae (Blattodea, Insecta). *Freiberger Forschungshefte, Hefte C*, **375**, 27–47.

SCHNEIDER, J. 1983. Die Blattodea (Insecta) des Paläozoikum, Teil 1 : Systematik, Ökologie und Biostratigraphie. *Freiberger Forschungshefte, Hefte C*, **382**, 106–145.

SCHNEIDER, J. W. 1996. Xenacanth teeth: a key for taxonomy and biostratigraphy. *Modern Geology*, **20**, 321–340.

SCHNEIDER, J. W. 2001. Rotliegendstratigraphie: Prinzipien und Probleme. *Beitrage zur Geologie von Thüringen*, **8**, 7–42.

SCHNEIDER, J. & GEBHARDT, U. 1993. Litho- und Biofaziesmuster in intra- und extramontanen Senken des Rotliegend (Perm, Nord und Ostdeutschland). *Geologisches Jahrbuch, A*, **131**, 57–98.

SCHNEIDER, J. & RÖSSLER, R. 1996. A Permian calcic paleosol containing rhizoliths and microvertebrate remains from the Erzgebirge Basin, Germany: environment and taphonomy. *Neues Jahrbuch für Geologie und Paläontologie, Abhandlungen*, **202**, 243–258.

SCHNEIDER, J. & WERNEBURG, R. 1993. Neue Spiloblattinidae (Insecta, Blattodea) aus dem Oberkarbon und Unterperm von Mitteleuropa sowie die Biostratigraphie des Rotliegend. *Naturhistorisches Museum Schoss Bertholdsburg, Schleusingen, Veröffentlichungen*, **7/8**, 31–52.

SCHNEIDER, J. W. & ZAJIC, J. 1994. Xenacanthodier (Pisces, Chondrichthyes) des mitteleuropäischen Oberkarbon und Perm: Revision der Originale zu Goldfuss 1847, Beyrich 1848, Kner 1867 und Fritsch 1879–1890. *Freiberger Forschungshefte, Hefte C*, **452**, 101–151.

SCHNEIDER, J. W., HAMPE, O. & SOLER-GIJÓN, R., 2000. The Late Carboniferous and Permian aquatic vertebrate zonation in southern Spain and German basins. *In*: BLIEK, A. & TURNER, S. (eds) *Palaeozoic Vertebrate Biochronology and Global Marine/Non-Marine correlation.* Courier Forschungsinstitut Senckenberg, **223**, 543–561.

SCHNEIDER, J. W., GORETZKI, J. & RÖSSLER, R. 2005. Biostratigraphisch relevante nicht-marine Tiergruppen im Karbon der variscischen Vorsenke und der Innensenken. *In*: Deutsche Stratigraphische Kommission & WREDE, V. (eds) *Stratigraphie von Deutschland. V. Das Oberkarbon (Pennsylvanian) in Deutschland.* Courier Forschungsinstitut Senkenberg, **254**, 103–118.

SCUDDER, S. H. 1879. Paleozoic cockroaches: a complete revision of the species of both world, with an essay toward their classification. *Memoirs of the Boston Society of Natural History*, **3**, 23–134.

SCUDDER, S. H. 1885. New genera and species of fossil cockroaches, from the older American rocks. *Proceedings of the Academy of Natural Sciences, Philadelphia*, **1885**, 34–39.

SUESS, E. 1869. Über das Rothliegende im Val Trompia. *Sitzungberichte der König Akademie der Wissenschaftliche, Mathematische-Naturalische Klasse, Wien*, **59**, 107–119.

TVERDOKHLEBOV, V. P., TVERDOKHLEBOVA, G. I., MINIKH, A. V., SURKOV, M. V. & BENTON, M. J. 2005. Upper Permian vertebrates and their sedimentological context in the South Urals, Russia. *Earth-Science Reviews*, **69**, 27–77.

UTTING, J. & PIASECKI, S. 1995. Palynology of the Permian of northern continents: a review. *In*: SCHOLLE, P., PERYT, T. M. & ULMER-SCHOLLE, D. (eds) *Permian of northern Pangaea.* Geological Society of America, Boulder, 236–261.

VACEK, M. & HAMMER, W. 1911. *Erläuterungen zur Geologischen Karte, SW-Gruppe Nr. 79 Cles.* Verlag K.K. der Geologischen Reichsanstalt, Wien.

VEEVERS, J. J. 2004. Gondwanaland from 650–500 Ma assembly through 320 Ma merger in Pangaea to 185–100 Ma breakup: Supercontinental tectonics via stratigraphy and radiometric dating. *Earth-Science Reviews*, **68**, 1–132.

VELTHEIM, W. V. 1821–1824 [1940]. Geognostische Betrachtung der alten Sandsteinformation am Harz und in den nördlich davon gelegenen Landstrichen (unpublished manuscripts 1821–1824, 1826; edited by H. Freydank 1940). *Jahrbuch des Halleschen Verbandes für die Erforschung Mitteldeutscher Bodenschätze, Neues Folgos*, **18**, 15–292.

VOIGT, S. 2005. *Die Tetrapodenichnofauna des kontinentalen Oberkarbon und Perm im Thüringer Wald: Ichnotaxonomie, Paläoökologie und Biostratigraphy.* Cuvillier Verlag, Göttingen.

WALTER, H. 1983. Zur Taxonomie, Ökologie und Biostratigraphie der Ichnia limnisch-terrestrischer Arthropoden des mitteleuropäischen Jungpaläozoikums. *Freiberger Forschungshefte, Hefte C*, **382**, 146–193.

WANG, Z. 1996. Past global floristic changes: the Permian great Eurasian floral interchange. *Palaeontology*, **39**, 189–217.

WANG, Z. & WANG, R. 1986. Permian charophytes from the southeastern area of the North China platform. *Acta Micropaleontologica Sinica*, **3**, 273–278.

WANKE, A., STOLLHOFEN, H., STANISTREET, I. G. & LORENZ, V. 2000. Karoo unconformities in NW Namibia and their tectonic implications. *Geological Survey of South West Africa/Namibia, Communications*, **12**, 259–268.

WARDLAW, B. R., DAVYDOV, V. & GRADSTEIN, F. M. 2004. The Permian period. *In*: GRADSTEIN, F. M., OGG, J. G. & SMITH, A. G. (eds) *A Geologic Time Scale* 2004. Cambridge University Press, Cambridge, 249–270.

WERNEBURG, R. 1996. Temnospondyle Amphibien aus dem Karbon Mitteldeutschlands. *Naturhistorisches Museum Schloss Bertholdsburg, Schleusingen, Veröffentlichungen*, **11**, 23–64.

WERNEBURG, R. 2001. Die Amphibien- und Reptilien-Faunen im Permokarbon des Thüringer Waldes. *Beiträge zur Geologie von Thüringen, Neue Folge*, **8**, 125–152.

WERNEBURG, R. 2003. The branchiosaurid amphibians from the Lower Permian of Buxières-les-Mines, Bourbon l'Archambault Basin (Allier, France) and biostratigraphic significance. *Société Géologique de France, Bulletin*, **174**, 343–349.

ZAJIC, J. 2000. Vertebrate zonation of the non-marine Upper Carboniferous–Lower Permian basins of the Czech Republic. *In*: BLIECK, A. & TUMER, S. (eds) *Palaeozoic Vertebrate Biochronology and Global Marine/Non-Marine Correlation.* Courier Forschungsinstitut Senckenberg, **223**, 563–575.

ZHENG, J. S., MERMET, J. F., TOUTIN-MORIN, N., HANES, J., GONDOLO, A. MORIN, R. & FERAUD, G. 1991–92. Datation $^{40}Ar–^{39}Ar$ du magmatisme et de filons minéralisés permiens en Provence orientale (France). *Geodinamica Acta*, **5**, 203–215.

ZIEGLER, A. M. 1990. Phytogeographic patterns and continental configurations during the Permian Period. *In*: MCKERROW, W. S. & SCOTESE, C. R. (eds) *Palaeozoic Palaeogeography and Biogeography*. Geological Society of London, Memoir, **12**, 363–379.

The magnetic polarity time scale across the Permian–Triassic boundary

MAUREEN B. STEINER

Department of Geology/Geophysics, University of Wyoming, Laramie, Wyoming 82071, USA (e-mail: magnetic@uwyo.edu)

Abstract: Early Triassic and Late to Middle Permian magnetostratigraphic investigations are numerous and span the globe. More than 20 magnetostratigraphic sequences have documented all or part of the Early Triassic geomagnetic field polarity, and > 27 have examined the Late and Middle Permian; 13 span the Permian–Triassic boundary. In order to assess the exact polarity sequence in the time period surrounding the Permian–Triassic boundary, the sequences have been compared diagrammatically. Four distinctive intervals of geomagnetic polarity characterize the Early Triassic, and have been named for discussion purposes: Gries N, Diener R-N, Smith N, and Spath N. A polarity pattern for the Mid- and Late Permian is also recognizable. The Mid- and Late Permian are characterized by two normal polarity intervals (Chang N and Capitan N) of greater apparent duration than those of the Early Triassic. Below the Permo-Triassic Gries N, a distinctive short duration reversed-normal-reversed polarity pattern characterizes the uppermost Changhsingian. The oldest normal polarity in the Middle Permian occurred during the Wordian Stage, established by results from three global sequences. Therefore, the geomagnetic field resumed reversing behaviour after the ~50 Ma-long constant polarity of the Kiaman Reversed Polarity Superchron ('Illawarra reversals') during the Mid- to Late Wordian, or ~267 Ma.

Very significantly, the magnetostratigraphic summary from this work indicates that the Siberian Traps were active in the Late Permian and spanned the Permian–Triassic boundary. This new geomagnetic polarity dating of the massive Siberian flood basalt activity suggests long-term eruption and environmental degradation, therefore making this igneous activity the most likely cause of the end-Permian mass extinctions. Magnetostratigraphy suggests that eruptions probably commenced in the Late Guadalupian; therefore, the eruptions of two large igneous provinces, Emishan and Siberian, were probably partly simultaneous during part of the Mid- to Late Permian. Environmental havoc throughout the late Mid- and Late Permian is easy to imagine, stressing the environment prior to probably more voluminous eruptions at the end of the Guadalupian and Permian. Siberian eruptions continued through the early Early Triassic, and probably contributed to the slow biotic recovery.

The greatest demise of life forms on this planet occurred at the close of the Permian Period. The quest to understand the cause of this largest of mass extinctions would be aided by the knowledge of the exact sequence of geomagnetic field polarity reversals around that time, because, when preserved, the magnetic polarity sequence is a constant, independent of facies, climatic zonation, endemism, and other variables that have hampered global Permian–Triassic biostratigraphical correlation.

The most accurate assessment of the geomagnetic field polarity changes during the Late Permian, the Early Triassic, and their mutual boundary will be one determined from examination of all data sources, that is, marine and terrestrial sedimentary strata and igneous rocks (Siberian Traps, Emishan Basalts). Despite the fact that marine sequences allow better age control because of their faunal content, terrestrial and igneous sequences commonly record the geomagnetic field more faithfully because of their greater magnetic mineral content and less chemically reactive depositional setting.

This investigation, therefore, has compiled nearly all of the global magnetostratigraphic results of this time interval. The compilation was restricted to data, excluded publications that consisted solely of summaries, but has included two sets of unpublished data because of their importance in assessing the magnetostratigraphy of the Middle Permian. Some published data have not been included because of the difficulty of including the many results that presently exist into a single clearly readable diagram. Because of this, older studies in which stratal age control was poorly known beyond the period assignment (e.g. Valencio *et al.* 1977) are not included in the diagrams. Furthermore, studies published in regional journals not readily accessible in North

From: LUCAS, S. G., CASSINIS, G. & SCHNEIDER, J. W. (eds) 2006. *Non-Marine Permian Biostratigraphy and Biochronology*. Geological Society, London, Special Publications, **265**, 15–38.
0305-8719/06/$15.00

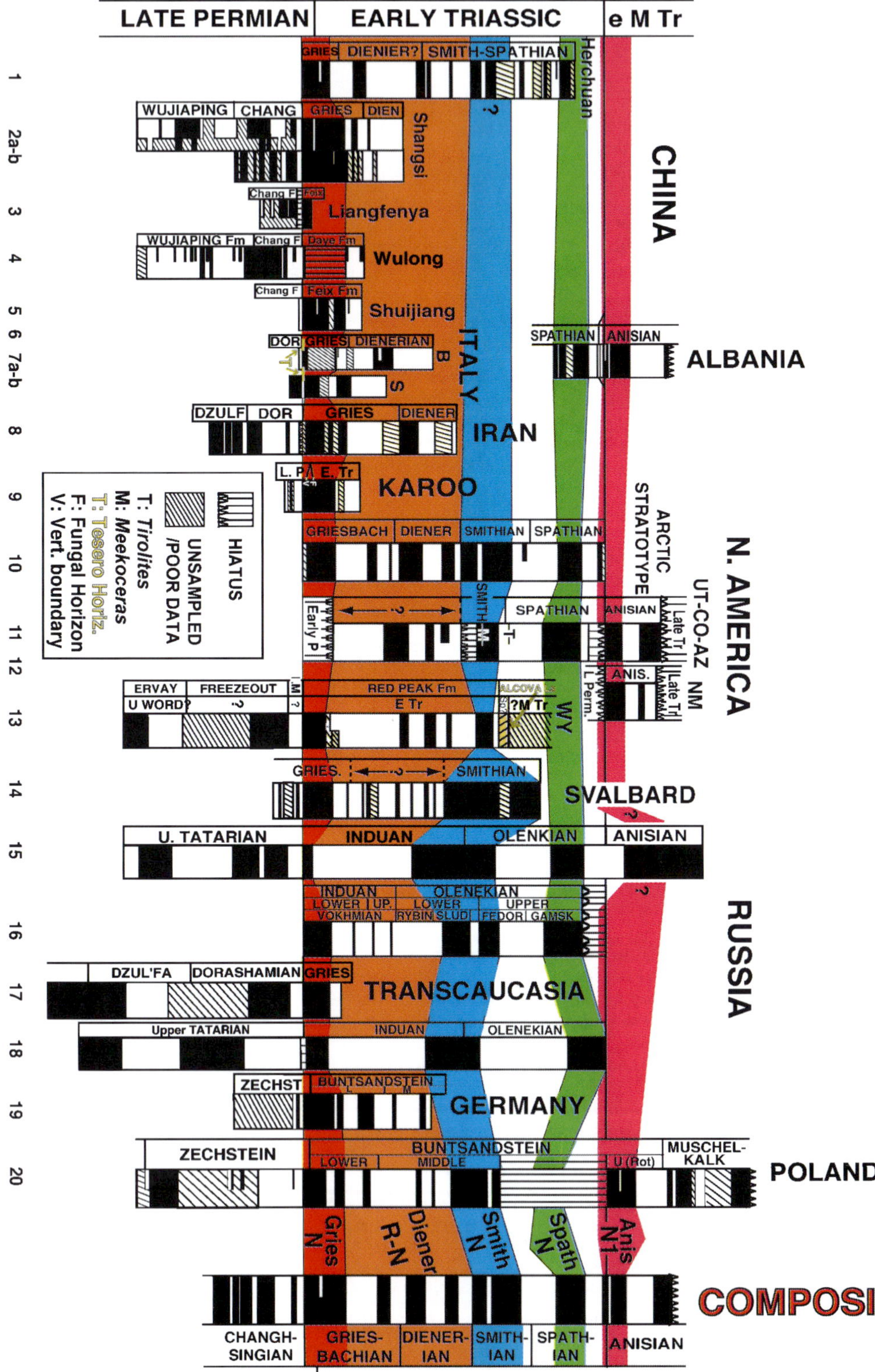
LATE PERMIAN
EARLY TRIASSIC
e M Tr
CHINA
Herchuan
Shangsi
Liangfenya
Wulong
Shuijiang
ALBANIA
ITALY
IRAN
KAROO
N. AMERICA
ARCTIC STRATOTYPE
UT-CO-AZ
NM
WY
SVALBARD
RUSSIA
TRANSCAUCASIA
GERMANY
POLAND
COMPOSITE
ANISIAN
SPATHIAN
SMITHIAN
DIENERIAN
GRIESBACHIAN
CHANGHSINGIAN
OLENEKIAN
INDUAN
U. TATARIAN
Upper TATARIAN
BUNTSANDSTEIN
ZECHSTEIN
MUSCHEL-KALK
DZULFA
DORASHAMIAN
WUJIAPING
Anis N1
Spath N
Smith N
Diener R-N
Gries N
HIATUS
UNSAMPLED /POOR DATA
T: Tirolites
M: Meekoceras
T: Tesero Horiz.
F: Fungal Horizon
V: Vert. boundary
1
2a-b
3
4
5
6
7a-b
8
9
10
11
12
13
14
15
16
17
18
19
20

American libraries (and possibly not referenced in North American databases) may have been unintentionally excluded from this compilation.

The compilation is displayed as two diagrams. Results from the Early Triassic are displayed in Figure 1, and those from the Mid- and Late Permian in Figure 2. The Early Triassic is examined first in this assessment of the Permian–Triassic geomagnetic polarity sequence, because a common polarity pattern is apparent in most Early Triassic sequences and easily correlated among them, in sharp contrast to that of the preceding Mid- and Late Permian time interval.

The Early Triassic

The Early Triassic polarity pattern is generally readily recognizable among global sequences. All or part of the Early Triassic magnetic polarity sequence has been investigated in at least 19 locations globally. Of these magnetostratigraphic sequences, six span approximately the entire Early Triassic; and 13 span the boundary between the Permian and Triassic.

Numerical ages are not used in Figure 1, because no radiometric ages have been determined for the Early Triassic except that of the Permian–Triassic boundary. Lacking Early Triassic numerical ages, various scaling methods have been applied (e.g. Sweet & Bergstroem 1986; Ogg, in Gradstein *et al.* 2004, chapter 17). Ogg assigned Early Triassic radiometric ages based on extrapolation between the Permian–Triassic boundary age and U–Pb-dated tuffs at the Mid- Triassic Anisian–Ladinian boundary, and employed assumptions about the length of the Early Triassic and the length of ammonite zones. However, the Anisian–Ladinian U–Pb dates are not entirely in agreement, varying by ± 1.0 Ma, and have error limits of ~0.5 to 1 Ma. Because these numerical ages assigned to the Early Triassic by Ogg have not been verified by geochronological studies, that numerical time scale is employed in Figure 1.

Individual magnetostratigraphic sequences were plotted in the figures by stretching or compressing each entire polarity sequence to approximately fit the master biostratigraphical scale of the figure. No allowances were made for changes in sedimentation rate within a measured section; these changes become evident when the magnetostratigraphies are compared. Sections without biostratigraphical control were simply scaled to the best match of their displayed polarity patterns to those with biostratigraphical constraints.

Early Triassic magnetostratigraphic sequences agree quite well (Fig. 1) and display no polarity-biostratigraphy conflicts. This agreement, in such marked contrast to the Mid- and Late Permian or the Late Triassic, is probably due to relatively accurate recording of the geomagnetic field polarity by most Early Triassic sequences. The Early Triassic geomagnetic field polarity behaviour appears to be no different in reversal frequency than that of, for example, the Late Triassic. The most likely reason for the great similarity among Early Triassic records of magnetic stratigraphy is the fact that most of the Early Triassic sequences have significant terrestrial sedimentary contributions: Early Triassic strata are either terrestrial deposits, marginal marine deposits, or represent marine depositional environments that received significant terrestrial detritus. Detrital terrestrial sediment provides far more magnetic mineral grain carriers (magnetite, hematite, etc) than are commonly available in wholly marine settings.

A second factor is the fact that Fe^{2+} ions are readily mobile in solutions that have reducing pH values. Magnetite (Fe_3O_4) is dominantly composed of Fe^{2+} ions; hematite (Fe_2O_3) also contains a portion of Fe^{2+} ions, as do iron sulphides. The biogenic activity in shallow marine environments creates a reducing pH. Therefore, under common marine sedimentary rates of deposition, magnetic grains deposited in the shallow marine environment are readily dissolved; hence the magnetic recorders and their information are removed from the (magneto) stratigraphical record. However, the added magnetic mineral input from the influence of greater terrestrial detritus could compensate for the process of dissolution of magnetic mineral grains; that is, the Earth's field

Fig. 1. Global magnetostratigraphic correlation of the Early Triassic. Normal (reversed) polarity is black (white); diagonal lines represent unsampled section or poor data. Magnetostratigraphic sequences shown were taken from: **1.** Steiner *et al.* 1989; **2a.** Steiner *et al.* 1989; **2b.** Heller *et al.* 1988; **3.** Steiner *et al.* 1989; **4. & 5.** Heller *et al.* 1995; **6.** (Albania) Muttoni *et al.* 1996, revised Spathian-Anisian boundary from Bachman & Kozur 2004; **7a-b.** Bella and Siusi, Scholger *et al.* 2000; **8.** Besse *et al.* 1998 and Gallet *et al.* 2000; **9.** Ward *et al.* 2005, Steiner *et al.* 2003; **10.** Ogg & Steiner 1991; **11.** Steiner *et al.* 1993; **12.** Molina-Garza *et al.* 1991; Steiner & Lucas 1992; **13.** Steiner *et al.* 1993; unpub. data; **14.** Hounslow *et al.* 1996; **15.** Khramov 1987; **16.** Lozovsky & Molostovsky 1993; **17.** Kotylar *et al.* 1984; **18.** Gurevich & Stautsitay 1985; **19.** Szurlies *et al.* 2003; **20.** Nawrocki 1997.

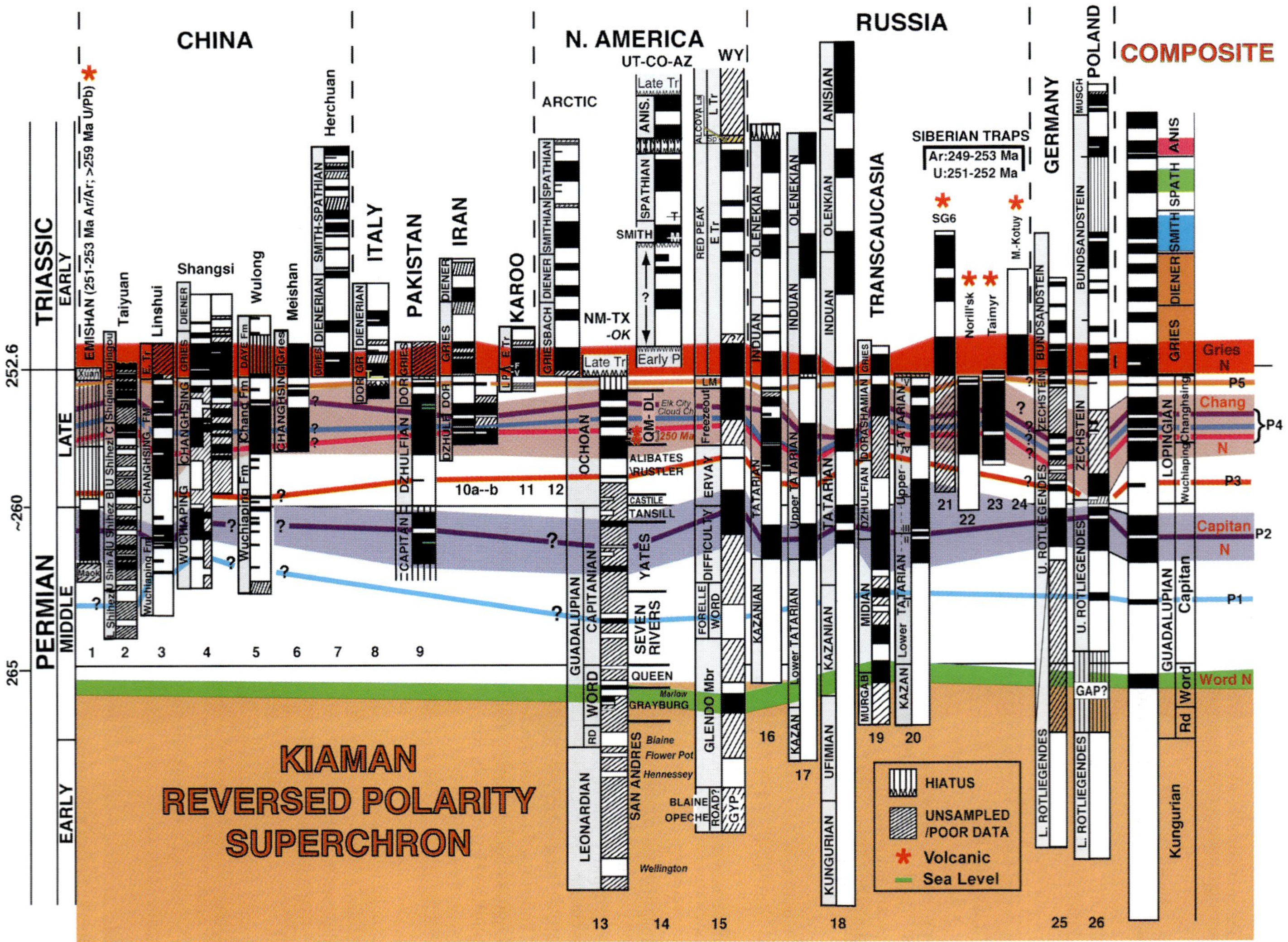
CHINA
N. AMERICA
RUSSIA
COMPOSITE
ARCTIC
UT-CO-AZ
WY
NM-TX
-OK
TRANSCAUCASIA
GERMANY
POLAND
ITALY
PAKISTAN
IRAN
KAROO
SIBERIAN TRAPS
Ar:249-253 Ma
U:251-252 Ma
SG6
Norill'sk
Taimyr
M.-Kotuy
EMISHAN (251-253 Ma Ar/Ar; >259 Ma U/Pb)
Taiyuan
Linshui
Shangsi
Wulong
Meishan
Herchuan
TRIASSIC
PERMIAN
EARLY
LATE
MIDDLE
252.6
~260
265
Gries N
P5
Chang N
P4
P3
Capitan N
P2
P1
Word N
GRIES
DIENER
SMITH
SPATH
ANIS
LOPINGIAN
GUADALUPIAN
Capitan
Word
Rd
Kungurian
HIATUS
UNSAMPLED /POOR DATA
Volcanic
Sea Level
KIAMAN
REVERSED POLARITY
SUPERCHRON

record of the time survives because the greater input of magnetic grains ensures that not all carrier grains are dissolved.

This greater amount of terrestrial sediment input in the Early Triassic may be an effect of the preceding mass extinction; the extinction of life forms, particularly floral, may have allowed greater erosion of the land areas. Because the terrestrial records are the most reliable magnetic data, the terrestrial sequences are described first in the following discussions and followed by the marine records.

North America

Two sequences of largely red-bed sedimentary strata, but with marine strata interfingering, have been studied; both the Moenkopi Formation and the Chugwater Group span most or all of the Early Triassic. In addition two deep-water, near-shore sedimentary sequences of Arctic North America of the Triassic have been investigated, including the global Early Triassic stratotype.

Despite the fact that deposition in the Moenkopi Basin was largely terrestrial red beds, marine incursions into the western part of the basin provide marine faunal control to the red-bed magnetostratigraphy (see Fig. 1, column 11). Ammonite faunas of Smithian (*Meekoceras*) and Spathian (*Tirolites*) ages are present in the western Moenkopi (McKee 1954; Poborski 1954); these are indicated by the letters 'M' and 'T' in Figure 1. Strata overlying the Spathian Moqui Member contain a vertebrate fauna that demonstrates that the Moenkopi Formation (Fig. 1, column 11) continues into the early Anisian Stage of the Middle Triassic (Lucas Morales 1985; Morales 1987). Magnetostratigraphy has been measured in numerous (16) sections across the entire Moenkopi depositional basin (some published – see Steiner *et al.* 1993– and some collected into an archive and obtainable from Steiner). The combined litho- and magnetostratigraphic information has permitted basin-wide correlation of facies and polarity zones.

Column 11 of Figure 1 is a composite of all the Moenkopi studies. Basin-filling began in the eastern part, and the section adjacent to the Uncompahgre uplift, the eastern border of the basin (Helsley & Steiner 1974) spans the entire time of basin-filling, whereas the western part of the basin was filled only in the later half of the Early Triassic and the early Mid-Triassic. Two hiatuses within the Moenkopi deposition were revealed by the magnetostratigraphy. Erosion removed most of the Smithian deposits in the western part of the basin, leaving a conglomeratic facies containing *Meekoceras* ammonites in the clasts; these deposits form the lowest Moenkopi deposits in the western basin, lying directly on Early Permian (Kaibab) limestones. Comparison with the Arctic stratotype sequence (Ogg & Steiner 1991; Fig. 1, column 10) suggests that part of Smithian time is missing in the Moenkopi Formation, even in the westernmost marine deposits (e.g. Shoemaker's Virgin River section; see Steiner *et al.* 1993). The magnetostratigraphy of the eastern basin sections reveals that Moenkopi deposition did not begin at the beginning of the Griesbachian, but slightly later in the Early Triassic, so that the lower part of the Griesbachian Substage (normal polarity) is not represented in the Moenkopi Formation. The second hiatus revealed by magnetostratigraphy is late in Moenkopi deposition, in the latest Spathian, up to and/or including the earliest Anisian. Strata of this time interval were removed across the entire basin. Magnetostratigraphic correlations at this horizon in the western part of the basin display a progressive lateral thinning and, in places, a complete removal by erosion of a relatively short-duration reversed-polarity interval (Grey Mountain magnetozone of Purucker *et al.* 1980) beneath the Anisian vertebrate-bearing strata (Purucker *et al.* 1980; Steiner *et al.* 1993). Again, comparison with the Arctic stratotype sequence suggests that only a short amount of time is missing.

Further north in North America, another red-bed depositional basin also recorded the geomagnetic field polarity during the Early Triassic. The Chugwater Group of western Wyoming contains the marine Griesbachian Dinwoody

Fig. 2. Global magnetostratigraphic correlation of the Late and Middle Permian. Normal (reversed) polarity is black (white); diagonal lines represent unsampled section or poor data. Magnetostratigraphic sequences are from: **1.** Huang & Opdyke 1998 (Xuan, Xuanwei Fm; Maok, Makou Fm); **2.** Embleton *et al.* 1996; **3.** Heller *et al.* 1995; **4.** Steiner *et al.* 1989, Heller *et al.* 1988; **5.** Heller *et al.* 1995; **6.** Zhu & Liu 1999, Ziu *et al.* 1999; **7.** Steiner *et al.* 1989; **8.** modified from Scholger *et al.* 2000; **9.** Haag & Heller 1991; **10a.** Besse *et al.* 1998; **10b.** Gallet *et al.* 2000; **11.** Ward *et al.* 2005, Steiner *et al.* 2003; **12.** Ogg & Steiner 1991; **13.** Steiner 2001*a* & *b*; unpub. data; **14.** unpub. data; Steiner *et al.* 1993; **15.** Steiner *et al.* 1993; **16.** Molostovsky 1983, fig. 20 and Lozovsky & Molostovsky 1993; **17.** Gurevich & Slautsitay 1985; **18.** Khramov 1987; **19.** Kotylar *et al.* 1984; **20.** Gialanella *et al.* 1997; **21.** estphal *et al.* 1998; **22.** Fedorenko & Czamanske 1997; **23.** Lind *et al.* 1994; **24.** Gurevich *et al.* 1995; **25**. Menning 1988 and Szurlies *et al.* 2003; **26**. Nawrocki 1997.

Formation shales and limestones, overlain by the red beds of the Red Peak Formation, in turn overlain by the probably marine Alcova Limestone. The marine Dinwoody Formation contains Griesbachian conodonts (Paull & Paull 1986); the Alcova Limestone is thought to be Spathian (Picard *et al.* 1969; Carr & Paul 1983; Storrs 1991), although no diagnostic biostratigraphic proof exists. The Alcova Limestone is overlain by the probably Middle Triassic Crow Mountain or Jelm formations (High & Picard 1967; Lucas 1994). The Red Peak Formation has been studied in a number of sections in western Wyoming; a summary of the magnetostratigraphic results (Steiner *et al.* 1993) is shown as column 13 in Figure 1. The western Wyoming magnetostratigraphy agrees well with the Moenkopi Formation results from further south, as well as with other sections globally. The common polarity pattern recorded in the Red Peak Formation and the lower Moenkopi Formation indicates a common depositional time frame and an essentially accurate recording of the Early Triassic geomagnetic field.

The grey shales and limestones of the Griesbachian Dinwoody Formation were sampled below the Red Peak Formation in the southern Wind River Mountains of western Wyoming, but gave only a Cretaceous normal polarity magnetization (Steiner, unpub. data); this magnetization probably reflects remagnetization associated with the Cretaceous uplift of the Wind River Range during the Laramide orogeny. However, in south-central Wyoming, the lowest ~6 m of the red beds of the Red Peak Formation overlying the Permian Goose Egg Formation were sampled and these recorded normal polarity. The match between the magnetostratigraphies of the western Wyoming Red Peak Formation and the Moenkopi Formation (Steiner *et al.* 1993), combined with the presence of a normal polarity interval in the basal Red Peak beds of southeastern Wyoming, strongly suggests that the lowest Red Peak red beds in southeastern Wyoming are probably lateral equivalents of the Dinwoody Formation. Their normal polarity, therefore, is a representation of the normal polarity interval observed globally in the lower Griesbachian.

The Arctic portion of North America contains the Early Triassic stratotype sequence (Tozer 1967). The strata are marine, and their sequence stratigraphy demonstrates a number of highstands and lowstands of sea level (Ogg & Steiner 1991). Therefore, the relative thicknesses of the polarity intervals are biased by sea-level changes and, to compensate, Ogg (pers. comm. 2004) revised the stratotype polarity sequence in order to minimize the thickness changes related to sea-level changes (Ogg in Gradstein *et al.* 2004, chapter 17). The revised polarity sequence is that displayed in column 10 of Figure 1.

Although Svalbard is no longer part of North America, it was laterally adjacent in Early Triassic time to the Early Triassic stratotype-bearing Canadian Arctic portion of North America. The Svalbard Early Triassic strata were studied (Hounslow *et al.* 1996), and a polarity interpretation was published, but without any supporting palaeomagnetic data on which to base the polarity interpretation. More recently, this same polarity interpretation was used in a sedimentological study (Mork *et al.* 1999), but again without any supporting palaeomagnetic behavioural data. The Svalbard polarity sequence (Fig. 1, column 14) is generally similar to the magnetostratigraphics of the Arctic stratotype and the red-bed sequences of North America; however, numerous additional tiny normal polarity intervals were interpreted in the lower part of the Svalbard record that are not seen in any other sequence, and the lack of published magnetic behavioural data makes it impossible to assess the reality of this large number of short normal polarity intervals.

Europe – non-marine

The terrestrial magnetostratigraphy studied in Europe is that of the Buntsandstein Formation of the Central European Basin (CEB), studied in both Poland and Germany. The Buntsandstein Formation consists of a sequence of red beds that has been divided into three parts ('lower', 'middle', and 'upper') that lie between the Mid-Triassic Muschelkalk and the Permian Zechstein beds. The entire thickness of the Buntsandstein beds has been studied in Poland (Nawrocki 1997), using a combination of outcrop and core samples.

Magnetostratigraphic investigation was subsequently continued through the overlying Middle Triassic Muschelkalk Formation (Nawrocki & Szulc 2000). The polarity sequence Nawrocki (1997) obtained from the Buntsandstein strata is somewhat similar to that of other Early Triassic magnetostratigraphic sequences (Fig. 1, column 20), but appears to lack some polarity intervals. The upper division of the Buntsandstein, the Röt Formation, constitutes a lithological change from terrestrial red beds to hypersaline and marine deposition. Its age is Early Anisian, the faunal basis of which was reviewed in detail by Bachman & Kozur (2004). Bachman & Kozur (2004) also pointed out that

Nawrocki & Szulc's (2000) magnetostratigraphic correlation of the Upper Buntsandstein Röt beds cannot be correct in view of their biostratigraphy. The lithological change represented by the Röt strata probably constitute a sequence-stratigraphic boundary; therefore, the Mid- to Upper Buntsandstein boundary is suggested to represent a hiatus, so that Spathian age strata are absent in the Polish Buntsandstein sequence. The Buntsandstein polarity sequence matches well to that of other global Early and early Mid-Triassic sections if Spathian strata are truly absent (Fig. 2).

In Germany, the 'lower' and the lower portion of the 'middle' Buntsandstein strata have been studied magnetostratigraphically (Szurlies *et al.* 2003; Szurlies 2004; Fig. 1, column 19). The German Buntsandstein polarity results are similar to those of the Polish lower and lower middle Buntsandstein and to the global results.

Russia and Transcaucasia

A very large number of terrestrial red-bed strata have been studied in Russia. Khramov (1987) summarized the results of his abundant investigations, and this summary is shown as column 15 of Figure 1. Molostovsky (1983) also extensively studied the Russian Early Triassic terrestrial strata; the summary presented by Lozovsky & Molostovsky (1993) is shown in column 16 of Figure 1. A clastic sequence transitioning from marine to non-marine deposition in Transcaucasia straddles the Permian–Triassic boundary; the sequence was studied by Kotylar *et al.* (1984) and is shown as column 17 in Figure 1. (The exact same magnetostratigraphy, with the minor exception of moving the basal Griesbachian normal interval down into the Dorashamian, was also published by Zakharov & Sokarev in 1991, but without any supporting palaeomagnetic data from which the magnetostratigraphy was derived.) Gurevich & Slautsitays (1985) studied the marine to non-marine transitional sequence of Upper Permian and Lower Triassic strata on Nova Zemlya (See Fig. 1, column 18). These sections all show the same general pattern, but it had been the practice of Russian authors to eliminate fine-scale details in their magnetostratigraphic summaries. Thus, summary Russian and Transcausian magnetostratigraphies have commonly shown less polarity structure than other sections globally. However, if individual section results are examined (e.g. Molostovsky 1983), the presence of the same short-polarity intervals observed elsewhere is commonly observed.

Eurasia – marine

The Tethys Early Triassic Griesbachian and Dienerian strata have been investigated in Iran (Besse *et al.* 1998; Gallet *et al.* 2000) and in Italy (Scholger *et al.* 2000). In addition, a continuous sequence of the upper part of the Spathian and basal Anisian was studied in Albania (Muttoni *et al.* 1996).

Early Triassic strata were studied in Iran at the Abadeh section (Gallet *et al.* 2000). The Griesbachian and Dienerian Substages were identified by the conodont fauna, which was accompanied by a sparse ammonite fauna. The Griesbachian is characterized by normal polarity succeeded by reversed polarity (Fig. 1, column 8), as is observed globally. The Dienerian begins with normal polarity and is succeeded by a reversed interval. When the gaps in sampling are taken into account (highlighted as diagonal lines in column 8 of Fig. 1), the Griesbachian and Dienerian Substages of the Iranian section agree well with the Arctic stratotype section.

In Italy, two sections were sampled 8 km apart (Scholger *et al.* 2000). The Bulla section is a roadcut, and the Siusi section is a river-cut exposure. Each section contained sampling gaps, and Scholger *et al.* (2000) combined the magnetostratigraphic results from the two into a single polarity column based on the intervals of common polarity and lithostratigraphy. The overall polarity pattern of the Griesbachian and Dienerian Substages is relatively similar to other global sections, although, relative to the well-dated Arctic and Iranian sequences, the Griesbachian–Dienerian boundary appears to be lower, not unlike the situation in the Chinese Herchuan section.

However, the Tesero Horizon (basal Early Triassic Werfen Formation), which is the designated base of the Griesbachian in Italy, displays different polarities at the two sections sampled by Scholger *et al.* (2000). At Bulla, the Tesero Horizon is normal in polarity, whereas at Siusi, the Tesero Horizon displays reversed polarity (Scholger *et al.* 2000, figs 6 & 8). Sedimentation is argued to be continuous between the Permian Bellerophon and the Griesbachian Werfen Formations (see discussion by Scholger *et al.* 2000). However, the polarity difference of the Tesero Horizon between two sections only 8 km apart appears to suggest otherwise. Scholger *et al.* (2000) did not discuss this polarity difference in the Tesero Horizon results, but they claimed that their data supported the concept of continuous deposition between the Bellerophon and the Werfen formations. In fact, they refer to the Tesero Horizon as a synchronous 'boundary event horizon'. But, the difference in polarities indicates that either the Tesero beds are not a synchronous horizon between these two neighbouring sections, or that the remanent magnetization of at least one of the sections does not represent the geomagnetic field at the time of Tesero deposition.

In an attempt to investigate this issue further, the magnetic study of the Gartnerkofel core (Zeissl & Mauritsch 1991) was re-examined. Examining the inclination data, the polarity was interpreted very stringently, that is, only the normal or reversed polarity results clearly fitting the palaeogeographic location of the locality at the time of deposition were considered. This stringent polarity interpretation suggests that the upper part of the Tesero Horizon above the black clay layer at the Gartnerkofel core site has reversed polarity, while the lower part has normal polarity, but a direction that resembles that of the recent geomagnetic field and thus may be a recent secondary magnetization. However, of far greater significance, the data from the entire core displayed far more reversed polarity than is known for the Early Triassic – Late Permian time interval, suggesting the possibility that a reversed-polarity secondary magnetization may have overprinted the strata cored. These complications prevented the issue of the polarity of the Tesero Horizon from being determined from the core data. Therefore, the problem remains that the differences between the Bulla and Siusi sections suggest that the Tesero between Bulla and Siusi is not a synchronous horizon or has been remagnetized in at least one section.

The uppermost Permian strata also are different between the two sections. At Bulla, only the uppermost beds of the Permian Bellerophon Formation beneath the Tesero Horizon, ~1.7 m thick, were sampled (probably because that was the extent of the available exposure in the roadcut); these beds display reversed polarity, consistent with most Late Permian sections globally. In the natural outcrop at Siusi, the Tesero Horizon and underlying uppermost 0.5 m of the Bellerophon Formation showed noisy, reversed polarity. Below, another ~15.5 m of Bellerophon strata were sampled and displayed only normal polarity; however, these normal polarity samples exhibit a slightly different palaeomagnetic direction than that of the normal polarity samples in the overlying Werfen Formation. If the magnetizations at both sections are original magnetizations dating from deposition of the strata, the normal polarity of the Bellerophon strata at Siusi (except for the uppermost 0.5 m) and reversed polarity of the uppermost 1.7 m at Bulla may indicate erosion of the uppermost Bellerophon strata beneath the Tesero Horizon. The issue of erosion has been much debated in the literature, with most authors presently considering that no depositional break exists. But, for these magnetic results to represent the original magnetization at the time of deposition, some strata must be missing at Siusi, or alternatively, the Tesero Horizon at Siusi has been remagnetized. Even so, the distinctive reversed polarity of the Late Changhsingian (discussed in the following Permian section) is missing at Siusi.

A section in Albania investigated the uppermost Early Triassic and the early Mid- Triassic (Muttoni *et al.* 1996). The Spathian–Anisian boundary was studied in a fossiliferous marine sequence at Kçira (Fig. 1, column 6). The Spathian of the Kçira section is similar to that of the Arctic stratotype sequence. Recent revision of the conodont biostratigraphy of the Kçira section (Bachman & Kozur 2004) suggests that the Spathian–Anisian boundary lies within a short reversed interval (Fig. 1, column 6). The Kçira section represents the only complete sequence across the Spathian–Anisian boundary; in both the Arctic and the Chinese Herchuan sections, sampling stopped short of the boundary. North American terrestrial sections in Arizona and New Mexico (Fig. 1, columns 11 & 12) both sampled Anisian strata above an erosional unconformity. Only vertebrate fossils provide any age control, and vertebrate faunal age resolution presently is inadequate to determine whether the strata of these sections straddle the Anisian–Spathian boundary, begin at the boundary, or begin within the Anisian. However, comparison of these two North American sections with the Kçira section indicates that North American deposition probably began a bit above the Spathian–Anisian boundary.

China

Only marine strata have been studied in China thus far; five marine sections in the Sichuan Basin on the South China Block sampled some portion of the Early Triassic. Early Triassic deposition at the Herchuan section (Fig. 1, column 1; Steiner *et al.* 1989) was in a marginal marine setting, whereas Late Permian strata were deposited in a wholly marine environment. The lower portion of the Early Triassic, the Feixianguan Formation, consists of alternating red and grey shales and limestones. The Feixianguan strata gave unambiguous palaeomagnetic polarity results. At the time of the study, the age of only the lower 100 m of the 400 m of Feixianguan strata were known; these contained Griesbachian conodonts. The upper 300 m contain no age-diagnostic fauna, but conformably overlie the lower Feixianguan beds and are conformably overlain by the Jialingjiang Formation. Conodonts found throughout the 600 m of the Jialingjiang Formation were identified as long-ranging species of the Smithian–Spathian Substages.

The Jialingjiang strata consist of grey, thin-bedded limestones and brown dolomite. The dolomite has experienced considerable dissolution of interbedded evaporites. Solution breccia and appreciable vuginess in the dolomites resulted in a poorly preserved record of the geomagnetic field polarity, particularly in the upper part of the formation; the many hachured intervals in column 1 of Figure 1 indicate the poor quality of the record. Although the biostratigraphical age information at the time of the study was limited, comparison of the magnetic polarity results with other global sections suggests that the Griesbachian may extend considerably higher than the highest Griesbachian conodonts, and that the lower part of the Jialingjiang strata may actually be Dienerian. In other words, the lower Jialingjiang strata up to the two larger normal polarity intervals in the middle of the formation, which were thought to be Smithian–Spathian, may be Dienerian. Those two thicker normal polarity intervals and the overlying vuggy interval with poor or no data may represent the complete Smithian Substage. Despite the problems of the upper part of the Herchuan sequence, the fit of the preserved magnetostratigraphy with the pattern from other global sections is reasonably good.

The Shangsi section (Fig. 1, column 2) of the northern Sichuan Basin has been sampled twice, first by Heller *et al.* (1988), who sampled the Upper Permian Dalong Formation and a portion of the overlying Lower Triassic Feixianguan Formation. Shortly thereafter, Steiner *et al.* (1989) also studied this sequence because Chinese colleagues considered this to be a very important section and insisted upon additional sampling. Steiner *et al.* (1989) sampled the entire Upper Permian section overlying the Middle Permian Maokou Formation at this site, that is, the Wuchiaping and Dalong formations, as well as the exposed portion of the overlying Feixianguan strata. Both sampling teams found that the original magnetization was preserved in only a portion of the samples in the Permian part of the section. When the portions of the stratigraphy that lacked stable magnetization (Fig. 1, column 2: diagonal lined areas in each of the sections) are taken into account, the resulting magnetostratigraphies are notably similar. Both indicate a basal normal polarity interval in the Feixianguan Formation overlain by reversed polarity, and the uppermost Changhsingian beds have reversed polarity in both studies.

Heller *et al.* (1995) sampled two sections in central Sichuan located near one another, Wulong and Shuijiang (Fig. 1, columns 4 & 5). The Upper Permian formations were sampled at Wulong, but because of better exposure and less weathering near the boundary, and the Lower Triassic Feixianguan Formation sampled at Shuijiang; the results were then combined into a single polarity sequence (Heller *et al.* 1995). However, Steiner *et al.*'s (1989) Liangfenya section is located very close to the Wulong section of Heller *et al.*; Steiner *et al.* noted that the Permian–Triassic boundary was a tectonic boundary in this area. Boudinage structures in the clay layer at the boundary indicate that slip has occurred along this horizon during a folding episode in that area. Two factors, the proximity of the Wulong section to that the Liangfenya section and reversed polarity characterizing the Lower Triassic lower Daye Formation at the Wulong section, suggest the tectonic disruption of the stratigraphic sequence was even greater at the Wulong outcrop, because the lowest Triassic strata are always normal in polarity elsewhere. For these reasons, a large hiatus is shown in the Wulong section in Figure 1 (column 4). Although the Shuijiang section also is near the Liangfenya and Wulong sections, the Shuijiang section shows the standard polarity pattern of the Early Triassic sequences elsewhere, suggesting a lack of disruption of the stratigraphic section there.

Africa

The location of the Permian–Triassic boundary is not precisely known within the terrestrial Karoo Group strata, because there are two possible indicators of the extinction event. Vertebrate remains indicate a faunal change from the Permian *Dicynodon* to the dominantly Triassic *Lystrosaurus*, but their ranges overlap (Smith 1995). An abundance of fungal/algal remains is present in the Karoo Supergroup (Steiner *et al.* 2003), and this horizon presents another biostratigraphic indicator of the Permian–Triassic boundary. Furthermore, although the magnetostratigraphy of the Karoo Supergroup has been studied in several locations (Schwindt *et al.* 2003; de Kock & Kirschvink 2004; Ward *et al.* 2005), different conclusions have been reached. Schwindt *et al.* (2003) studied the magnetostratigraphy throughout the Upper Permian and lowest Triassic strata at the locality in which the fungal horizon was found; only magnetizations strongly overprinted by Early Jurassic Karoo igneous activity were observed. Much of the magnetization was either normal Jurassic polarity or multicomponent, unstable magnetization (Schwindt *et al.* 2003). The unstable multicomponent magnetization was commonly observed in green mudstone strata, which constitute most of the section; isotopic data (Tabor & Schwindt pers. comm.) suggest that the area may have been a swamp environment during deposition. The reducing geochemical environment within a swamp would explain the paucity of stable remanent magnetization in much of the strata, because of the dissolution of magnetic carriers under these conditions. Carbonate nodules also were collected in an attempt to locate the Permian–Triassic boundary by its delta-C^{13} anomaly, but the nodules gave only a recent climatic signature.

Subsequently, Ward *et al.* (2005) published a magnetostratigraphic interpretation for part of Schwindt *et al.*'s (2003) section, but without any supporting palaeomagnetic data. Column 9 of Figure 1 is their interpretation of the magnetostratigraphy; they indicated the relative positions of the change in vertebrate faunal change ('V' in column 9) and their interpretation of the location of Steiner *et al.*'s (2003) fungal horizon ('F' in column 9). These potential Permian–Triassic boundary markers do not coincide, therefore the Permian–Triassic boundary position in the Karoo strata still is not precisely known, although both potential Permian–Triassic boundary markers apparently lie within the lower portion of the Griesbachian normal polarity interval recognized globally. Based on the fact that the marine Permian–Triassic boundary indicator, the FAD of the conodont *Hindeodus parvus*, lies about one third of the distance above the base of the Griesbachian normal polarity interval, the fungal spike may most accurately represent the Permian–Triassic boundary in the Karoo Group.

Early Triassic polarity: summary

Figure 1 suggests that Early Triassic geomagnetic polarity was slightly dominated by reversed polarity; normal polarity intervals occur within the more extensive reversed polarity and form a distinctive pattern, identifiable in most of the sections and providing good correlation among them. Four distinctive portions of the Early Triassic polarity pattern repeat in the global magnetostratigraphic sequences; these are highlighted by different colours in Figure 1 and labelled according to the characteristic polarity of the interval and the substage that each occupies: 'Gries N', 'Diener R-N', 'Smith N', and 'Spath N'. Very latest Late Spathian time probably began in the 'Anis N1' interval.

Originally, the base of the Triassic was considered to be the base of the Griesbachian Substage (Tozer 1967), but now, based on more refined conodont zonations, the lower portion (one third) of the Griesbachian Substage is thought to be Late Permian. The Griesbachian Substage of the Arctic stratotype contains normal polarity in its lower part (Ogg & Steiner 1991), corresponding to the Gries N interval of Figure 1. Following this normal polarity interval, the Upper Griesbachian and Dienerian display a distinctive interval of reversed polarity punctuated by a number of shorter normal polarity intervals; this is the interval Diener R-N. Many of the sections display three to four relatively short normal polarity intervals, although other sequences suggest the possibility of more. Overall, the Griesbachian and Dienerian Substages of Tozer (1967) are comprised of a normal interval overlain by a dominantly reversed interval in which three or more relatively short normal polarity periods are interspersed. In the stratotype sequence, the Smithian Substage is dominantly of normal polarity (Fig. 1: Smith N); Smith N begins near the base of the Smithian and persists for most of the duration of the Smithian. The Lower Smithian boundary of the stratotype section begins near the polarity boundary between Diener R-N and Smith N. A reversed polarity interval of relatively long duration existed during the Late Smithian through the early half of the Spathian. In the Mid-Spathian, the polarity changed to normal; the upper half of the Spathian is characterized by slightly shorter normal and reversed intervals. In the stratotype section, normal polarity resumes before the end of the Spathian.

The Late and Middle Permian

An even larger number of magnetostratigraphic sequences have been studied in Upper and Middle Permian strata (Fig. 2) than in Early Triassic beds. Chinese strata contain the Late Permian (Lopingian) global stratotype, whereas the Middle Permian (Guadalupian) stratotype is in the United States. Figure 2 includes, in addition to the marine and terrestrial sedimentary sequences, the igneous sequences of the Siberian Traps and Emishan Basalt.

In general, Permian magnetostratigraphic sequences do not agree well. However, most of the sedimentary sections show that a relatively short duration of reversed polarity characterized the very latest Changhsingian Stage, preceded by a longer duration interval of normal polarity.

A hiatus near the Permian–Triassic boundary is described or discussed in the section descriptions from the western Tethyan magnetostratigraphic investigations in Pakistan and Italy, but no allowance for a hiatus was made in the published polarity interpretation columns. In Figure 2, inferences of missing time have been added from the lithological descriptions and inferences from the magnetostratigraphy.

China

Five marine sections (Fig. 2, columns 3–6), one marginal marine (column 7), one terrestrial (column 2), and one basalt section (column 1) have been studied in China. One of the marine sections, the Liangfenya section, is not shown in Figure 2, because it is depicted in Figure 1 and has only minor data content and because of the space constraints of Figure 2. Most of the Chinese sections show that normal polarity prevailed in the lower Griesbachian Substage. Generally, Changhsingian-age beds indicate that a reversed polarity interval of relatively short duration preceded the Early Griesbachian normal polarity. Below this relatively short reversed interval, a considerably longer period of normal polarity commonly is observed. The oldest polarity signature in the marine Chinese sections is an even longer duration of reversed polarity, occupying the Wuchiapingian Stage (*not* formation). But, within this lengthy Wuchiapingian reversed interval, the Linshui (Heller *et al.* 1995) and the Shangsi (Steiner *et al.* 1989) sections indicate the presence of another (shorter) normal polarity interval. In the GSSP Meishan section, a very brief reversed-polarity interval was observed (Liu *et al.* 1999; Zhu & Liu 1999); this reversed polarity is not observed in any other Permian–Triassic sequence. Bachman & Kozur (2004) state that this interval of the Meishan section was restudied, and the reversed polarity was concluded to be an overprint (see Bachman & Kozur 2004, pers. comm. by Yin Hongfu to Kozur).

The non-marine Taiyuan sequence (Embleton *et al.* 1996) has no firm age control, but is reported to lie conformably between Upper Carboniferous to Lower Permian strata and Lower Triassic strata. Because the sequence is reported to contain no breaks in sedimentation, Embleton *et al.* (1996) approximated ages for it by assuming a uniform sedimentation rate. Their interpretation indicates that the earliest normal polarity stratigraphic horizons (U Shihezi Formation Member A) are in the earliest Middle Permian, essentially in the Roadian and approximately middle Wordian. However, the dominance of normal polarity in the lower and upper parts of this undated sequence most resembles the upper and lower normal polarity intervals observed globally in the Middle and Upper Permian. Therefore, in Figure 2, the Taiyuan section has been uniformly compressed in an attempt to make it fit the global pattern; the fit is not outstanding, suggesting that sedimentation rates were not constant. But, this representation of the approximate age of the Taiyuan section makes its polarity sequence agree moderately well with the global pattern.

The magnetostratigraphy of the Emishan basalts is shown in Figure 2, column 1; the basalts exhibited a lengthy normal polarity interval succeeded by reversed polarity (Huang & Opdyke 1998; Fig. 2, column 1). The basalts lie disconformably on Maokou limestones (Capitanian–Kazanian age: Huang & Opdyke 1998; Lo *et al.* 2002) and are overlain by the Changhsingian age Xuanwei Formation. Recent Ar^{40}/Ar^{39} dating of the basalts indicated that they are 251–253 Ma (Lo *et al.* 2002). Lo *et al.* (2002) also argued that the proximity of the Emishan basalts to the Late Permian to Early Triassic marine strata of southern China was consistent with the Emishan extrusions being the source of the numerous tuffaceous beds in the southern China strata. However, U–Pb dating of a sill intruding the Emishan basalts gave a 259 Ma age for the intrusion, implying that the basalts are older than 259 Ma. The conflicting radiometric data make it difficult to correlate the Emishan basalt magnetostratigraphy with other Permian sequences. The basalts exhibit dominantly normal polarity, which could represent either the upper lengthy normal polarity interval of the Late Permian or the older normal polarity interval of the upper Middle Permian. The second correlation is shown in Figure 2 because of the possibly greater reliability of U–Pb radiochronometry.

Western Tethys

Two sections were studied in the Late Permian of Iran (Fig. 2, column 10a & b; Besse *et al.* 1998; Gallet *et al.* 2000). A Middle and Upper Permian section was studied in Pakistan (Fig. 2, column 9; Haag & Heller 1991). The Pakistan section subsequently has been dated in detail with conodont biostratigraphy (Wardlaw & Pogue 1995). In Italy, three localities in Permian strata have been investigated (Zeissl & Mauritsch 1991; Scholger *et al.* 2000), but only the Dolomites sequences (Scholger *et al.* 2000) gave reliable data (Fig. 2, column 8).

Both the Iranian and Pakistani sequences display reversed polarity in the uppermost Changhsingian, interrupted by a very short normal event (Fig. 2, columns 9, 10a & b). Both show that this triplet of R-N-R succeeded a lengthier normal polarity interval. In China, this normal polarity interval is within the Changhsingian Stage, but in Iran Krystyn dated it with conodonts (Gallet *et al.* 2000) as Dorashamian. Wardlaw & Pogue (1995) also identified sequence stratigraphical horizons indicating transgressions and regressions in this section; these sequence boundaries are shown by green lines in Figure 2, column 9.

The two sections in the Italian Dolomites (Scholger *et al.* 2000) have already been discussed; the results were differing magnetic polarity for the Tesero Horizon and the upper Bellerophon Formation from these sections located 8 km apart. The minimal exposure of the Permian Bellerophon Formation at the possibly more reliable Bulla section yielded 1.7 m of reversed polarity below the normal polarity of the Tesero Horizon (Fig. 2, column 8: T in yellow) and Mazzin Member of the lower Werfen Formation. This magnetostratigraphy is consistent with sections globally. The *c.* 16 m of upper Bellerophon strata sampled at the Siusi outcrop exhibit normal polarity, but with a direction slightly different from the overlying Werfen Formation normal polarity at this section. Column 8 of Figure 2 displays the Siusi Bellerophon results below those of Bulla (after scaling for an apparent difference in sedimentary accumulation between the two sections). The amount of reversed section sampled at Bulla is too little to represent the whole short reversed interval of the Late Permian that is indicated by most global magnetostratigraphic results, therefore, a hiatus was placed between the Bulla and Siusi Bellerophon results in column 8 of Figure 2. Erosion of the uppermost Bellerophon Formation at Siusi by the 'current event' (Dolomites-wide erosion event) discussed by Scholger *et al.* (2000) would explain the differences in the Permian Bellerophon strata between the two localities and make the Dolomites results compatible with global Permian sections.

North America

Two sequences of Late and Middle Permian strata in North America have been studied magnetostratigraphically, those in Texas–New Mexico (Fig. 2, column 13) and in Wyoming (Fig. 2, column 15). Both consist of terrestrial red beds interbedded with marine carbonates.

In Texas, the uppermost Permian formation, the Quartermaster Formation (also widely known as the Dewey Lake Formation, a name proposed somewhat later), has been investigated. The formation is truncated everywhere by the Late Triassic Santa Rosa Formation, hence Quartermaster sections vary appreciably in thickness from one location to another. The thickest outcrop section known is that in Caprock Canyon State Park, studied in its lower part by Molina-Garza *et al.* (1989) and in its entirety by Steiner & Renne (1996). Molina-Garza *et al.* (1989) observed four polarity intervals in the lower 32 m. Steiner & Renne (1996) sampled the same section more densely every 0.1 to 0.5 m throughout the entire 93 m; the results were largely normal polarity, but punctuated by three distinct, short reversed polarity intervals (Fig. 2, upper column 13). Two volcanic ash beds are present in the lower part of the Dewey Lake beds, separated by 20 m of fine-grained red sediment. Preliminary $^{40}Ar/^{39}Ar$ dating gave ages of ~250 Ma for each ash bed; however, the zircon populations show evidence of detrital contamination (Renne *et al.* 1996; Steiner & Renne 1996, 1998; Steiner 2000; Steiner 2001*a*, *b*). A short section (22 m) in the middle part of the same sequence, but with only one ash bed, was sampled 120 km to the south, and the same magnetostratigraphy was observed (Steiner 2001*a*, *b*). Beneath the Dewey Lake beds (4 m below the lower ash bed), lie frequently alternating beds of anhydrite and red siltstone, locally called the Alibates Beds, forming the lower part of the Quartermaster Formation. The Alibates Beds exhibited entirely reversed polarity with the exception of a short normal polarity interval within the lower part of the reversed polarity (Steiner 2001*a*, *b*).

In the subsurface of the sampling area, the Dewey Lake Formation overlies the Rustler Formation. Much farther south (southeastern New Mexico), the Dewey Lake Formation overlies the Rustler Formation in outcrop. Molina-Garza *et al.* (2000) sampled the Dewey Lake beds and observed dominantly normal polarity with one short reversed polarity interval in approximately the middle of the normal polarity. No ash beds were observed by Molina-Garza *et al.* (2000).

The underlying Rustler Formation consists of five members of alternating red beds and dolomite. The fourth member from the base, a 12 m thick dolomite, was studied in five sections (Steiner 2001*a*, *b*). The dolomite exhibited dominantly reversed polarity with a short normal interval in the upper part (Steiner 2001*a*, *b*). Palaeogeography of the Permian Basin during the Late Permian consisted of the sea to the south and terrestrial environments to the north, west and east. Because both bedded anhydrite and dolomite require standing water from which to precipitate, it is probable that the interbedded anhydrite and red beds of the Alibates Beds below the Dewey Lake beds in northern Texas are the lateral equivalent of the interbedded dolomite and red beds of the Rustler Formation lying below the Dewey Lake in southwestern Texas–New Mexico. Both probably represent frequent marine incursions of the sea into the terrestrial depositional environment; that is, they probably represent deposition during approximately the same time period. Moreover, the same magnetostratigraphic signature is displayed by both sets of strata underlying the Dewey Lake Formation. The Alibates Beds and the upper Rustler Formation exhibit very similar magnetostratigraphies: dominantly reversed polarity encompassing a short normal polarity interval. On these bases, Steiner (2001*a*, *b*) concluded that the Alibates Beds and the Rustler Formation are likely to be lateral equivalents.

Below the Rustler Formation lies the thick, laminated (varved?) anhydrite sequence of the Castile Formation. Strata below the Castile Formation are the back-reef facies time equivalent of the reef and fore-reef facies that make up the global stratotype section for the Middle Permian. Reconnaissance sampling was conducted in the Castile evaporites and in the underlying back-reef facies formations: the Tansill, Yates, Seven Rivers, Queen and Grayburg formations, and the Cherry Canyon Member of the San Andres Formation. Short intervals in each of the formations were studied to assess suitability for palaeomagnetic investigation; much of the back-reef strata gave reliable palaeomagnetic results (Steiner, unpub. data).

Much earlier, Peterson & Nairn (1971) had studied sites in the Middle and Late Permian of Oklahoma and Texas–New Mexico; they investigated sites in stratigraphical equivalents of the Quartermaster Formation, the Elk City and Cloud Chief formations, a site in the New Mexico Yates Formation, three sites in the Seven Rivers Formation, and sites in the Oklahoma terrestrial equivalents of the San Andres Formation (Blaine, Flowerpot, Hennessey, and Wellington formations). Peterson & Nairn's (1971) study was only concerned with obtaining palaeopole positions; they did not sample for magnetostratigraphy. Thus, they sampled only short stratigraphical intervals of one to several metres per site, and did not specify the stratigraphical locations of their sampling sites within the formations, nor the stratal thickness sampled. Nevertheless, their polarity results have been widely quoted in the search for the beginning of geomagnetic field reversals after the lengthy duration constant polarity of the Carboniferous–Permian. Both their results and those of Steiner (unpub. data) are displayed in column 13 of Figure 2, although the stratigraphic positions of Peterson & Nairn's (1971) polarity results are relatively arbitrary; column 13 displays Texas–New Mexico formation names on the left and Oklahoma names in italics on the right.

Below the Rustler Formation, a single hand sample from the underlying Castile was investigated; surprisingly however, definite Permian reversed polarity was observed (Steiner, unpub. data). The Tansill Formation below was sampled almost in its entirety in a roadcut, but it was only weakly magnetized. Some suggestions of reversed polarity were observed, but at this locality, the formation appears to retain little of an original magnetic signature. The very top of the underlying Yates Formation, at the Yates/Tansill contact, exhibited definite normal polarity; a roadcut site in the middle of the formation was generally poorly magnetized, but a few samples displayed reversed polarity. Peterson & Nairn's (1971) single site in the Yates yielded normal polarity; the minimal outcrop description and the absence of coring holes at either of Steiner's sites suggests that Peterson and Nairn's locality was not the same location as the present author's Yates/Tansill contact site. Therefore, Peterson & Nairn's (1971) result is arbitrarily placed lower within the formation in Figure 2 (the lower normal polarity shown in the Yates Formation in column 13), although it could be almost anywhere in the formation.

Peterson & Nairn (1971) and Steiner (unpub. data) both sampled the Seven Rivers Formation. Steiner (unpub. data) sampled two duplicate short stratigraphical sections (4.5 m) about 6 m above the base of the formation in a roadcut; these exhibited a magnetostratigraphy of 1.6 m of normal polarity, overlain by 3 m of reversed polarity. Peterson & Nairn (1971) sampled three sites at one locality in the Seven Rivers Formation and obtained only reversed polarity; their result is represented by the lower Seven Rivers reversed polarity in Figure 2, which again, could be stratigraphically anywhere within the formation.

The uppermost Queen Formation, at its contact with the Seven Rivers Formation, displays reversed polarity magnetization over 3 m (Steiner, unpub. data). A continuous exposure, consisting in part of roadcuts, exposes the lowest Queen Formation through the upper half of the

Grayburg formations. The lowermost Queen Formation, at its contact with the underlying Grayburg Formation, contains a reversed to normal polarity sequence. The uppermost part of the Grayburg Formation is a light grey limestone/dolomite and is surprisingly well magnetized (Steiner, unpub. data); the upper 18 m display reversed polarity. The lower part of the Grayburg Formation, overlying the Cherry Canyon Sandstone Member of the San Andres Formation, is exposed on a hill slope. The Grayburg beds have a yellowish hue on the natural outcrop; the lowest 21 m generally exhibited relatively poor magnetization. However, a number of samples indicate the presence of a low-inclination normal polarity magnetization; because of the poorer quality of the magnetization, it cannot be certain whether this normal polarity dates from deposition without further investigation. Therefore, the normal polarity observation is shown as half bars of polarity among the diagonal-ruled pattern indicating poor data in column 13 (Fig. 2). The directly underlying Cherry Canyon Member of the San Andres Formation yielded indecipherable data throughout ~19 m (Steiner, unpub. data). However, Peterson & Nairn (1971) sampled San Andres-equivalent terrestrial strata in Oklahoma; their sites in the Blaine, Flowerpot, Hennessey, and Wellington formations yielded only reversed polarity (Fig. 2, column 13). Furthermore, magnetostratigraphic sequences of many tens of metres at both the top and bottom of the several hundred metres thick Blaine Formation of western Texas also recorded only reversed polarity (Steiner, unpub. data).

The Wyoming Permian–Triassic sequence has been studied in magnetostratigraphic reconnaissance from the lowermost Triassic down through 75 m of Permian strata; the investigation was conducted 48 km west of Laramie, Wyoming, as part of several palaeomagnetic class projects. In western Wyoming, the Permian section consists of limestones, shales, and cherts deposited in the Phosphoria seaway that covered western Wyoming, Idaho, Montana, and Utah in the Middle and Late Permian, the result of which was the deposition of the Phosphoria Group. Much of Wyoming east of the sea was a broad flat region on which red-bed deposition took place, but which was occasionally was invaded by marine waters, creating a stratal succession of intertonguing red beds and marine carbonates and gypsum: the Goose Egg Formation. Conodonts and gastropods in the marine Phosphoria Formation indicate an age of Middle Permian, Roadian through earliest Capitanian (Wardlaw & Collinson 1986). The highest Phosphoria beds are overlain by marine shale of the Griesbachian Dinwoody Formation. Hence, a gap must exist in western Wyoming between the lower Capitanian uppermost Phosphoria (Wardlaw & Collinson 1986) and the Griesbachian Dinwoody beds (Paull & Paull 1986).

Eastward in the dominantly terrestrial strata, no indication of a physical break can be found between the Goose Egg and overlying Triassic Red Peak formations, and the Permian–Triassic boundary cannot be identified in this area, in either surface exposures or subsurface data. In fact, a thesis was aimed specifically at locating the Permian–Triassic boundary in these strata, by measuring gamma ray stratigraphy on outcrop and correlating the results with the gamma ray logs of subsurface wells (Renner 1988; Renner & Boyd 1988). Renner was unable to find any evidence of a depositional break, and he concluded that no significant break in sedimentation existed, but that possibly extensive deposition of loess, perhaps containing a number of small breaks in sedimentation, might explain the apparent continuity.

The largely Permian Goose Egg Formation and lowermost part of the Triassic Red Peak Formation were sampled magnetostratigraphically at Renner's (1988) Red Mountain section in southeastern Wyoming. Approximately 80 m of stratigraphic section have been sampled in part, from the top of the locally named 'Blaine Gypsum' into the lower beds of the Red Peak Formation. Red-bed strata and three carbonate members were sampled, all of which gave clearly defined magnetic data and unambiguous polarity interpretations; the summary of these data is shown in column 15 of Figure 2. Although the magnetic stability was excellent, the ages of these strata are embarrassingly poorly known. The age of the Goose Egg Formation had been assigned by lateral tracing of the carbonate tongues into the main body of the Phosphoria Formation. (e.g. Thomas 1934). However, outcrop exposures between that wholly marine province and that of dominantly terrestrial deposition are not continuous; therefore, the exact connections are not well established and, in some cases, have been much debated (Boyd & Maughan 1973). It was hoped that the magnetostratigraphic results would help to remove some of the age uncertainties, by allowing magnetic correlation of the Goose Egg strata to better-dated Permian strata.

At the sampling locality, the *c*. 6 m of lowermost Red Peak Formation red beds displayed normal polarity, and they rest directly on gypsum at the top of the Little Medicine Member of the Goose Egg Formation; *c*. 2 m of gypsum overlying *c*. 2 m of dolomite constitute the Little Medicine Member. The dolomites exhibit reversed polarity. The age of the Little Medicine Member has been argued to be Late Griesbachian from lateral tracing into the top of the Dinwoody Formation of western Wyoming (Thomas 1934). However, subsequently it has been argued (Paull & Paull 1990) that the Little Medicine Member is not physically continuous with the Dinwoody strata. The Little Medicine Member has no fauna, so its age really is not known, only hypothesized. Beneath its dolomite lie fine-grained red beds of the Freezeout Member.

The Freezeout red beds also have no age control. They exhibited well-defined normal polarity in the portion sampled. Interbedded dolomite, gypsum and red beds of the Ervay Member underlie the Freezeout Member; the Ervay dolomites can be traced laterally into the Ervay Member of the marine Park City Formation. The Ervay Member forms the top of the Park City Formation, and therefore the top of the Permian in western Wyoming. The Ervay Member is directly overlain by the Griesbachian Dinwoody Formation. Wardlaw & Collinson (1986) found that the highest Ervay beds of the Park City Formation contain conodonts of late Wordian to early Capitanian age; however, Henderson & Mei (2000) reviewed provinciality in conodont identifications and questioned the identification of certain conodonts in the upper Ervay strata, suggesting that these beds may be as young as Lopingian. In a later publication, Wardlaw (2003) stated that those conodonts are late Wordian.

The Ervay Member exhibits largely reversed-polarity magnetization, but changes to normal polarity in its lowest portion. The normal polarity continues in the underlying fine-grained Difficulty Member red beds. Below the Difficulty red beds, another marine tongue, the Forelle Member, gave reversed polarity for the portion sampled; the Forelle Member is traced into Wordian strata of the Park City Formation. The underlying Glendo Member red beds were sampled in their middle portion, and these displayed normal polarity overlain by reversed polarity. The base of the Glendo Member was also sampled, down to the top of a 20-m-thick gypsum deposit, the local ‘Blaine Gypsum’. Whether this gypsum correlates to the type Blaine gypsum of Oklahoma and Texas has not been established, but it does occur at approximately the same stratigraphical position in the Wyoming sequence as the Blaine Formation in the Texas and Oklahoma sequences. Moreover, comparison of the magnetostratigraphic results from the entire Wyoming section between the Blaine gypsum and the Red Peak red beds with the global results appears to suggest that the Wyoming Blaine gypsum is correlative with the Texas–Oklahoma Blaine Formation.

The magnetostratigraphy of the succession exposed in southeastern Wyoming is combined with the composite of entire Early Triassic Red Peak Formation from its contact with the Dinwoody Formation of western Wyoming to the Alcova Limestone in column 15 of Figure 2; the juncture between the western and southeastern Wyoming sections is marked with an unsampled interval by a ‘?’, because the exact correlation has not been studied. The incomplete magnetostratigraphic results from the southeastern Wyoming Permian section display three prominent normal polarity intervals between the base of the Red Peak Formation there and the top of the gypsum. Despite the poor age control, these normal polarity intervals are located in positions similar to, and with similar thicknesses (possibly indicating comparable durations), to the normal polarity intervals observed in the global collection of Middle and Late Permian sequences of Figure 2.

The positions of these normal polarity intervals below the Red Peak Formation suggest that the Little Medicine and Freezeout members of the Goose Egg Formation of southeastern Wyoming are Late Permian in age, and not Early Triassic as has been supposed. The correlation suggests that the Little Medicine Member represents the uppermost Changhsingian in Wyoming, and the Freezeout Member represents the lower Changhsingian. Other correlations between the Wyoming Permian and Triassic composite and the global results, based on the locations of these normal polarity intervals, have been tested, but that shown in Figure 2 provides the best fit of the magnetostratigraphy and the limited biostratigraphy. However, the biostratigraphy of the upper portion is significantly at odds with the global data, and may suggest that Henderson & Mei’s (2000) conodont identifications are correct. The field of conodont identification has undergone much revision since 1986. If the magnetostratigraphic correlation is correct, the polarity sequence certainly permits the possibility of relative continuity of deposition across the Permian–Triassic boundary in southeastern Wyoming that Renner (1988) concluded.

Russia and Transcaucasia

An extremely large number of Upper Permian magnetostratigraphic sections have been studied in Russia, largely in terrestrial red-bed sequences; five representative magnetostratigraphies columns are shown in Figure 2 (columns 16–20). Khramov has studied the Permian and Triassic magnetostratigraphy of Russia since 1960; his most recent summary (Khramov 1987) is shown in column 18 of Figure 2. Molostovsky (1983) has studied the eastern Russian Platform magnetostratigraphy extensively; his figure 20 is reproduced in column 16 of Figure 2 (beneath the summary of the Early Triassic by Lozovsky & Molostovsky 1993). Molostovsky's (1983) figure 20 is illustrated instead of his summary figure representing all of his studies of the Upper Permian, in order to show the fine-scale polarity details he observed in that sequence. These details are eliminated from his summary, but display the same fine-scale polarity details observed globally.

On Novaya Zemlya, Gurevich & Slautsitays (1985) investigated a sequence of strata that consists of grey marine clastics grading upward into variegated clastic rocks and into terrestrial red beds, without an observable break in deposition. Tuffs occur in the base of the variegated deposits and comparison of the magnetostratigraphy (Fig. 2, column 17) to global sequences suggests that the Permian–Triassic boundary lies at the base of the variegated deposits with their tuffaceous content.

The type Tatarian is a terrestrial section originally investigated by Khramov (1963). Recently, a portion of this sequence was restudied by Gialanella *et al.* (1997). The result of all investigations is displayed in column 20 of Figure 2. The lithologic subdivisions of the Tatarian are indicated by Roman numerals on the right side of the age column.

Kotylar *et al.* (1984) studied a marine sequence in Transcaucasia, the stratotype region for the Dorashamian and Dzhulfian, and at the location of the type Midian; the results are shown in column 19 of Figure 2. An identical magnetostratigraphy was published by Zakharov & Sokarev (1991), with the exception that the normal interval designated as basal Triassic in Kotylar *et al.*'s (1984) magnetostratigraphy is shown as uppermost Dorashamian. The Transcaucasian sequence displays normal polarity low in the Middle Permian, like the North American sequences.

Central European Basin

The Rotliegend and Zechstein formations have been studied in Germany (Menning 1986, 1988) and in Poland (Nawrocki 1997). Fairly similar results were obtained by both studies (Fig. 2, columns 26 & 27). The frequent reversals observed by Menning (1980, 1988) in the uppermost Rotliegend, although incomplete, are similar to the short reversed intervals observed in Iran and western. Texas within the lengthy Changhsingian normal polarity interval (Fig. 2, columns 10 & 13). This correlation is suggested in Figure 2, but the uncertainty of its validity is indicated by numerous question marks. The limited magnetostratigraphic data do not allow a definite correlation of the Zechstein strata to the rest of the world. However, in the underlying upper Rotliegend beds, the Capitanian normal polarity interval appears to be well represented in both sequences. Neither sequence displayed any normal polarity below the Capitanian normal interval, normal polarity that would be correlative with that observed in a few Wordian-equivalent strata of other sequences. However, its absence might be due to a hiatus between the upper and lower Rotliegend strata, as hypothesized by Nawrocki (1997).

Permian polarity: summary

Just prior to the Permian–Triassic massive extinctions, the latest Permian exhibits a distinctive polarity pattern: a short duration R-N-R. The uppermost Changhsingian strata of a number of magnetostratigraphic sequences exhibit this observed pattern, including the reliable and biostratigraphically well-dated Iran and Pakistan sections, as well as the Russian Platform sequences, including the type Tatarian section, and the German sequence (Fig. 2, columns 9, 10a & b, 16, 20 & 25). Other sections (Shangsi, Wulong, Meishan and Poland) also appear to display this polarity structure, although possibly compromised by either poor magnetic recording or inadequately-detailed sampling.

Most of the Middle and Upper Permian sequences exhibit three prominent normal polarity intervals: the Permo-Triassic 'Gries N', the Late Permian 'Chang N' and the Middle Permian 'Capitan N' (Fig. 2). The recognition of this common polarity sequence provides important correlation markers for the Middle and Late Permian. In addition, five shorter duration polarity intervals appear recurrently among the various sequences and, when recognized, provide potential for fine-scale correlations. These are highlighted by coloured lines in Figure 2 and labelled in decreasing age as 'P1' – 'P5'. P5 is the short normal interval already discussed in the latest Changhsingian R-N-R sequence. The relatively long duration normal polarity interval, Chang N, immediately precedes the distinctive Changhsingian R-N-R interval; Chang N is observed in every Late Permian sequence. Within Chang N, the triplet of brief reversed-polarity intervals recorded in the Iranian and Texas sequences constitutes P4. Linshui may also preserve this triplet (Fig. 2, column 3), but many sequences display only one of two of these short reversed-polarity intervals. P3, a short normal interval in the Lopingian reversed polarity preceding Chang N, is observed in only a few sequences: Pakistan, Texas, Wyoming, and Germany. P2 is a brief reversed-polarity interval within Capitan N. P1 is a short normal polarity interval documented in 4–6 sections (Fig. 2).

The oldest normal polarity observed is in the Wordian. 'Word N' occurs in the Queen Formation (possibly also in the Grayburg Formation) of the Texas–New Mexico section, in the Russian Transcausian section, and in the North American Wyoming Goose Egg Formation; all three sections clearly exhibit a relatively short duration interval of normal polarity. Therefore, the oldest normal polarity lies very low in the Middle Permian, approximately in the middle to upper part of the Wordian Stage. This interval constitutes the earliest normal polarity of the geomagnetic field and terminates the ~ 50 million years of constant polarity of the Carboniferous and Early Permian Kiaman Reversed Polarity Superchron. Therefore, it presently appears that the 'Illawarra reversals' began in the middle–upper Wordian Stage.

The oldest normal polarity, the initiation of the Illawarra reversals, previously had been argued to occur in the early Capitanian (Menning & Jin 1998). However, no magnetostratigraphic data have ever been published from the Guadalupian global stratotype. The stratotype beds (the United States Permian Basin strata) have been heavily oil saturated, which consequently dissolved the magnetic carriers (Steiner, unpub. data); therefore, the Guadalupian stratotype is unlikely to ever yield a detailed magnetostratigraphic sequence. However, the strata deposited behind the reef, the backreef facies, have yielded good palaeomagnetic data (Peterson & Nairn 1971).

Menning & Jin (1998) based their conclusion that the Illawarra reversals began in the Capitanian on Peterson & Nairn's (1971) observation of normal polarity in the backreef Yates Formation and no normal polarity in any older strata. Correlation of the backreef strata with the biostratigraphically dated reef and forereef strata (Glenister *et al.* 1992) suggests that the age of the Yates Formation is probably Middle Capitanian. As discussed earlier, Peterson & Nairn (1971) did no magnetostratigraphy, only limited site sampling (commonly one site per formation) and their sites spanned a limited stratigraphic thicknesses (≤1 to 4 m). Peterson & Nairn (1971) sampled the Yates Formation at only one site. Consequently, the normal polarity they observed probably represents several metres or less from an unspecified stratigraphical location within a 130-m-thick formation. Therefore, the age of Peterson & Nairn's (1971) Yates normal polarity is constrained only as some level within the middle Capitanian Stage. Moreover, despite the fact that Peterson & Nairn's (1971) three sites in the underlying Seven Rivers Formation yielded only reversed polarity, a single site from very low in the formation yielded 1.6 m of normal polarity succeeded by 3 m of reversed polarity (Steiner, unpub. data).

Age of the Siberian flood basalts

Many investigators have concluded or speculated that the Siberian igneous activity caused the end-Permian mass extinction (recently summarized by Kamo *et al.* 2003). Magnetostratigraphy of the Siberian basalts has been published from four widely spaced areas (Lind *et al.* 1994; Gurevich *et al.* 1995; Fedorenko & Czamanske 1997; Westphal *et al.* 1998), but the results do not agree well at all (see Westphal *et al.* 1998). The polarity data were correlated with the uppermost Permian and early Early Triassic (Westphal *et al.* 1998). However, the present compilation and summary of global Late Permian and Early Triassic magnetostratigraphic data (Figs 1 & 2) suggests a better correlation of Siberian flood basalt magnetostratigraphic results, and consequently significantly different ages for the igneous activity.

Early $^{40}Ar/^{39}Ar$ results suggested an earliest Triassic age for the Siberian basalts (see Kamo *et al.* 1996 for tabulation of all Siberian radiometric data). In an early magnetostratigraphic assessment, Lind *et al.* (1994) correlated the basal reversed interval in their study of the Siberian igneous rocks with the youngest reversed interval in the Permian (Fig. 2, column 3). The most recent magnetostratigraphic study (Westphal *et al.* 1998) followed this scheme in correlating the results of all four studies among themselves and to the Early Triassic polarity sequence (Fig. 3, lower right). Westphal *et al.*'s (1998) correlation also preserved relative section thicknesses, whereas this constraint was relaxed somewhat in creating the new magnetostratigraphic correlation, because it is not reasonable to expect that the same rates of extrusion will occur in all parts of a volcanic field, and especially in one this large.

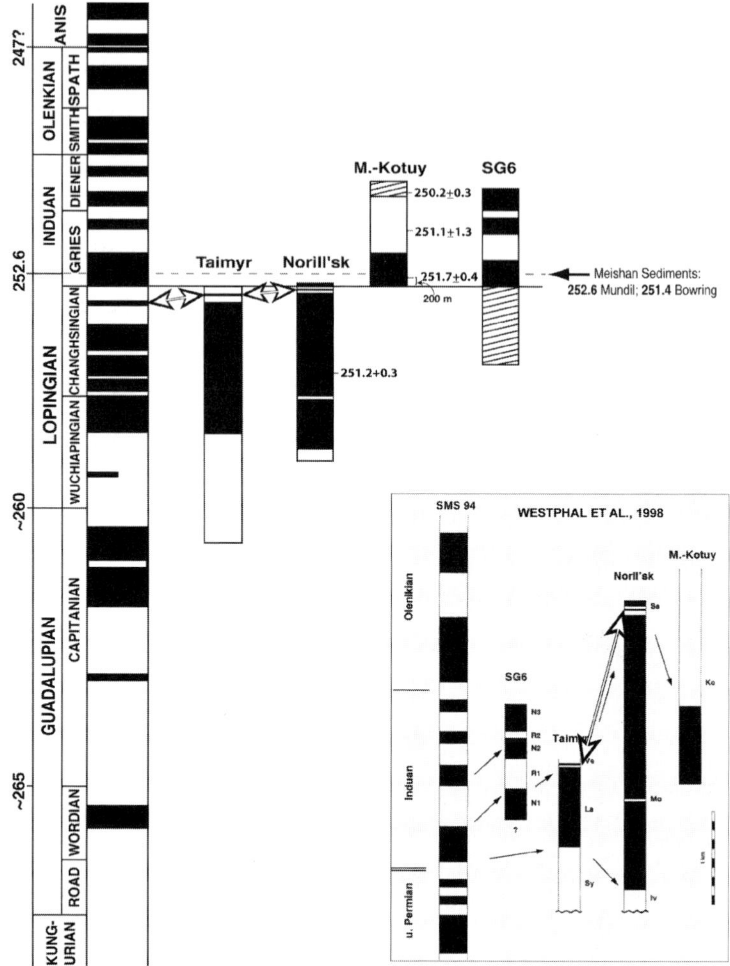

Fig. 3. The revised correlation of the magnetostratigraphy of the Siberian igneous episode and the Late and Middle Permian and Early Triassic magnetic polarity time scale derived by this study. The earlier magnetostratigraphic correlation of Westphal *et al.* (1998) is shown in the lower right. U–Pb radiometric ages (Kamo *et al.* 2003) for the igneous rocks are approximately in the stratigraphic positions indicated by Kamo *et al.* U–Pb dates determined for the Permian–Triassic boundary are on the right (Bowring *et al.* 1998; Mundil *et al.* 2004). Siberian magnetostratigraphy credits are listed in Figure 2. Normal (reversed) polarity is black (white); unsampled section indicated by diagonal lines.

The new magnetostratigraphic correlation is based on several characteristics of the Siberian igneous rocks that can be recognized as similar to features in the global Permian–Triassic magnetostratigraphic summary (Fig. 3, left side). Although the four studies were displayed in Figure 2 (columns 21–24), their characteristics are more clearly visible at the larger scale of Figure 3. In Westphal *et al.*'s (1998) correlation, the relationship of the SG6 sequence to the other sequences was considered problematic. As observed by Westphal *et al.* (1998), the SG6 sequence strongly resembles the Induan polarity sequence; however, the Taimyr and Noril'sk sections certainly do not.

One significant characteristic of the Siberian magnetostratigraphy is the record, in both the Taimyr and Noril'sk sections, of a very short normal polarity interval embedded within slightly longer duration, but also short, reversed polarity (Fig. 3, bold arrows). This signature is strikingly similar to that preserved in uppermost Changhsingian magnetostratigraphic sections, the interval 'P5' (Fig. 2). A second Siberian basalt characteristic is the lengthy normal polarity interval below this short R-N-R sequence in the Taimyr and Noril'sk sections. No lengthy normal polarity periods like that occur in the Induan (Fig. 1), but a relatively long duration normal polarity interval is widely observed in the Late Permian (Chang N, Fig. 2).

Once the concept is considered that the four Siberian Trap magnetostratigraphic sections may not represent the exact same time interval (and indeed, their magnetostratigraphy suggests that they do not), the correlation shown in Figure 3 becomes the more probable. The two characteristics in the Taimyr and Noril'sk sections, of a short R-N-R succeeding a lengthy normal polarity interval, suggest that these sections are Late Changhsingian in age. The SG6 sequence is probably Induan, but could also represent extrusion spanning the Induan–Olenkian boundary. The Maimecha–Kotuy (M.-Kotuy) polarity sequence resembles the lower Induan (Griesbachian), particularly if either a hiatus or increased extrusion volume is responsible for the absence of the short normal polarity interval of the upper Griesbachian. However, the non-distinctive M.-Kotuy polarity sequence also resembles other parts of the time scale; the enormous thickness of reversed polarity displayed at M.-Kotuy (Gurevich *et al.* 1995) could suggest that the age of this sequence is early Lopingian. The detailed U–Pb dating of Kamo *et al.* (1996) at M.-Kotuy suggests that the lower Induan correlation is most probable, despite the current uncertainty in U–Pb dates recently demonstrated by Mundil *et al.* (2004).

This revised magnetostratigraphic correlation of the Siberian igneous rocks has enormous significance, in that it indicates that the Siberian flood basalt volcanism did not occur simply at the time of the mass extinction, but was ongoing in the Late Permian, that it spanned the Permian–Triassic boundary, and that it continued into the Early Triassic. The magnetostratigraphy of the Siberian flood basalts in Figure 3 suggests that the Siberian volcanism began in the late Guadalupian. Palaeontologists have long emphasized dual late Permian extinctions, one at the end of the Guadalupian and a second at the end of the Permian (see review by Erwin *et al.* 2002); it is possible that the Siberian volcanism also caused the end-Guadalupian extinctions, although the Emishan flood basalt volcanism may also have been active at similar times. Perhaps two separate pulses of Siberian volcanism caused the two separate mass extinctions, but with an appreciably larger eruption at the end of the Permian. The complication of Emishan volcanism erupting at similar times, and the presently sparse radiometric data, make it difficult to determine which was responsible or more responsible for the extinction events. The fact that both erupted in a similar time frame certainly affected the biosphere more than the Siberian activity alone; if the Emishan eruptions preceded those of Siberia, the Siberian flood basalt activity added environmental stress to an already stressed biosphere. Moreover, because the SG6 magnetostratigraphy indicates that volcanic extrusion continued through the early Early Triassic, the Siberian volcanism may also have been responsible for the slow Early Triassic faunal and floral recovery.

The magnetostratigraphy of the Siberian flood basalts indicates that the volcanism was not short-lived. Approximately eight geomagnetic field polarity intervals (ignoring the two short events) are observed. By the fast reversal rates of the late Tertiary, the magnetostratigraphically sampled part of the Siberian igneous rocks might represent ~1 Ma, but early Tertiary rates would imply 8 Ma. Comparison with the sparsely dated Permian–Triassic time scale indicates that the Siberian magnetostratigraphy represents ~9 Ma. Furthermore, geomagnetic field behaviour after constant polarity periods displays lengthy polarity intervals initially: the Late Cretaceous was characterized by polarity intervals of 4–6 Ma for the first *c.* 11 Ma after reversals resumed, succeeded by shorter polarity intervals of 1–2 Ma during the latest Cretaceous and early Tertiary and shorter still in the late Tertiary (*c.* 250 000 years). The Middle and Late Permian polarity intervals are 4 to 2 Ma, white the Early Triassic intervals are appreciably shorter.

The entire time spanned by the Siberian flood basalts may not yet have been sampled magnetostratigraphically; thus the estimated 8–9 Ma duration may be a minimum. Nevertheless, this magnetostratigraphic correlation of the Siberian rocks demonstrates that the eruptions of this greatest of flood basalt volcanism on Earth entirely overlaps in time with the greatest of mass extinctions. Therefore, Siberian basalt magnetostratigraphy strongly indicates that this massive flood basalt volcanism was causally related to the Permian–Triassic boundary mass extinctions. The prolonged volcanism probably was responsible for the biotic decline in the late Middle and Late Permian, probably by continual injection of excessive fluxes of carbon dioxide and hydrogen sulphide into the atmosphere.

The exact mechanism(s) by which the immense flood basalt volcanism affected the biosphere and produced mass death is beyond the scope of this investigation, but has recently been considered by others (e.g. Knoll *et al.* 1996;

Kump *et al.* 2005). In addition to periods of blocked sunlight due to dust and aerosols injected into the atmosphere and resulting temperature declines (volcanic winters), the excessive carbon dioxide delivered into the atmosphere could subsequently cause greenhouse conditions and overheating. Heating may have reached the point of warming the ocean waters to the extent that clathrates melted and released additional large quantities of carbon dioxide into the atmosphere. Finally, massive volcanic eruptions would inject lethal fluxes of hydrogen sulphide into the atmosphere, destined to return to the Earth's surface as acid rain. The fact that two large igneous provinces (Siberian and Emishan) operated partially at the same time and partially sequentially doubles the volcanic environmental devastation surrounding the time of the greatest of mass extinctions.

Conclusions

The Early Triassic is characterized by four distinctive polarity intervals: Gries N, Diener R-N, Smith N and Spath N. The Upper and Middle Permian are characterized by two relatively longer duration normal intervals, Chang N and Capitan N, which are separated by two lengthy reversed-polarity intervals. Distinctive features of the Permian geomagnetic polarity time scale include a distinctive series of short reversed-normal-reversed polarity intervals in the uppermost Changhsingian and the earliest normal polarity in the Permian occurring in the Wordian Stage. The end of the 50 Ma of constant polarity in the Carboniferous–Permian is documented by three sequences as ending in the middle to late Wordian.

Correlations among the magnetostratigraphy of the large Siberian igneous province and with the composite Permian–Triassic magnetic polarity time scale developed in this study indicate that the Siberian volcanic activity not only spanned the Permian–Triassic boundary, but also began long before the end-Permian mass extinction and continued well into the Early Triassic. A total of at least eight polarity intervals are observed, suggesting that the duration of the volcanic activity may have been ± 9 Ma. The Siberian magnetostratigraphic record suggests that volcanism began in the late Guadalupian and continued through the early half of the Induan. The age of the volcanism inferred from the magnetic polarity strongly suggests that the effects of the voluminous volcanic activity (carbon dioxide, hydrogen sulphide, etc.) disturbed the biosphere sufficiently to degrade the environment and to cause the mass extinctions. Magnetostratigraphy indicates that Siberian volcanism may have overlapped the latter stages of Emishan flood volcanism. Their essentially sequential activity must have delivered a one-two (or double) punch to the environment, and hence to the biosphere. Their close relationship in time leaves no doubt that the greatest extinction of advanced life on Earth was caused by abnormal volcanic activity.

This study also demonstrates that a magnetostratigraphy can be developed for the Guadalupian global stratotype sequence, despite the oil-saturated nature of the stratotype rocks. The extensive passage of oil through the reef and forereef facies, and the consequent dissolution of magnetic carriers, prohibits development of a magnetostratigraphy directly from the Guadalupian stratotype strata, thus excluding the stratotype beds from ever yielding a magnetostratigraphy to complement its excellent biostratigraphy. However, reconnaissance magnetostratigraphy of the backreef strata marginal to the reef–forereef stratotype rocks indicates good preservation of magnetization in many formations. Through correlations established between the backreef and reef–forereef strata, a magnetostratigraphic sequence could be determined for the Guadalupian stratotype, thus facilitating a more precise correlation of its fauna to other faunas globally. Therefore, the backreef strata should be investigated with detailed magnetostratigraphy.

The magnetostratigraphic summaries of this study also indicate that the southwestern North American Ochoan time period is not as short a time as has been hypothesized. Ochoan magnetostratigraphy duplicates much of the global Lopingian polarity sequence, therefore its duration is equivalent to nearly all of the Lopingian.

The intertonguing terrestrial–marine Permo–Triassic sequence in southeastern Wyoming may represent largely continuous sedimentation across the Permian–Triassic boundary. The uppermost Permian of this area probably is represented by the upper two members of the Goose Egg Formation (Little Medicine and the Freezeout Shale members). The magnetostratigraphic record obtained from the entire, poorly dated Wyoming Middle and Late Permian to Early Triassic strata (Goose Egg and Red Peak formations) matches the global time scale if the upper two members of the Goose Egg Formation are Upper Permian (Changhsingian), not earliest Triassic, as has been speculated for almost a century. Also, the 'Blaine Gypsum' of southeastern Wyoming occurs at the same magnetostratigraphic and stratigraphical position as the Blaine Formation of Texas–Oklahoma, supporting long-time local speculation that this relatively thick gypsum deposit is approximately age equivalent to the Blaine Formation.

This work owes its existence to S. Lucas, who requested it and continually encouraged me when the task became overwhelming, as it frequently did. The very conscientious, constructive and detailed reviews by K. Ziegler and H. Közur greatly improved the quality of the manuscript and I sincerely and earnestly thank them. The study was supported by the Geology and Geophysics Department of the University of Wyoming.

References

BACHMAN, G. H. & KOZUR, H. W. 2004. The Germanic Triassic: correlations with the international chronostratigraphic scale, numerical ages and Milankovitch cyclicity. *Hallesches Jahrbuch für Geowissenschaften, Reihe B*, **26**, 17–62.

BESSE, J., TORCQ, F., GALLET, Y., RICOU, L. E., KRYSTYN, L. & SAIDI, A. 1998, Late Permian to Late Triassic palaeomagnetic data from Iran: constraints on the migration of the Iranian Block through the Tethyan Ocean and initial destruction of Pangaea A. *Geophysical Journal International*, **135**, 77–92.

BOWRING, S. A., ERWIN, D. H., JIN, Y. G., MARTIN, M. W., DAVIDEK, K. & WANG, W. 1998. U/Pb zircon geochronology and tempo of the end-Permian mass extinction. *Science*, **280**(5366), 1039–1045.

BOYD, D. W. & MAUGHAN, E. K. 1973. Permian–Triassic boundary in the Middle Rocky Mountains. *In*: LOGAN, A. & HILLS, L. V. (eds) *The Permian and Triassic Systems and their Mutual Boundary*. Canadian Society of Petroleum Geologists; Memoir, **2**, 294–317.

CARR, T. R. & PAULL, R. K. 1983, Early Triassic stratigraphy and paleogeography of the Cordilleran Miogeocline. *In*: REYNOLDS, M. W. & DOLLY, E. D. (eds) *Rocky Mountain Paleogeography Symposium.* Society of Economic Paleontologists and Mineralogists Denver, **2**, 39–55.

DE KOCK, M. O. & KIRSCHVINK, J. L. 2004, Paleomagnetic constraints on the Permian–Triassic boundary in the terrestrial strata of the Karoo Supergroup, South Africa: implications for causes of the end-Permian extinction event. *Gondwana Research*, **7**, 175–183.

EMBLETON, B. J. J., MCELHINNY, M. W., MA, X., ZHANG, Z. & LI, Z. X. 1996. Permo-Triassic magnetostratigraphy in China: the type section near Taiyuan, Shanxi Province, North China. *Geophysical Journal*, **126**, 382–388.

ERWIN, D. H., BOWRING, S. A. & YUGAN, J. 2002. End-Permian mass extinctions: a review. *In*: KOEBERL, C. & MACLEOD, K. G. (eds) *Catastrophic Events and Mass Extinctions*: Impacts and Beyond. *Geological Society of America Special Paper*, **356**, 363–383.

FEDORENKO, V. & CZAMANSKE, G. K. 1997. Results of new field and geochemical studies of the volcanic and intrusive rocks of the Maymecha-Kotuy area, Siberian flood-basalt province, Russia. *International Geology Review*, **39**, 479–531.

GALLET, Y., KRYSTYN, L., BESSE, J., SAIDI, A. & RICOU, L.-E. 2000. New constraints on the Upper Permian Lower Triassic geomagnetic polarity time scale from the Abadeh section (central Iran). *Journal of Geophysical Research*, **105**, 2805–2815.

GIALANELLA, P. R., HELLER, F. *et al.* 1997. Late Permian magnetostratigraphy on the eastern Russian platform. *Geologie en Mijnbouw*, **76**, 145–154.

GLENISTER, B. F., BOYD, D. W. *et al.* 1992. The Guadalupian, proposed international standard for a Middle Permian series. *International Geology Review*, **34**, 857–888.

GRADSTEIN, F. M., OGG, J. G. & SMITH, A. G. 2004. *A Geological Time Scale 2004*. Cambridge University Press, Cambridge.

GUREVICH, YeL & SLAUTSITAYS, I. P. 1985, A paleomagnetic section in the Upper Permian and Triassic deposits of Novaya Zemlya. *International Geology Review*, **27**, 168–177.

GUREVICH, YEL, WESTPHAL, M., DARAGAN-SUCHOV, J., FEINBERG, H., POZZI, J. P. & KHRAMOV, A. N. 1995. Paleomagnetism and magnetostratigraphy of the traps from western Taimyr (northern Siberia) and the Permo-Triassic crisis. *Earth and Planetary Science Letters*, **136**, 461–473.

HAAG, M. & HELLER, F. 1991. Late Permian to Early Triassic magnetostratigraphy. *Earth and Planetary Science Letters*, **107**, 42–54.

HELLER, F., LOWRIE, W., LI, H. & WANG, J. 1988. Magnetostratigraphy of the Permo-Triassic boundary section at Shangsi (Guangyuan, Sichuan Province, China). *Earth and Planetary Science Letters*, **88**, 348–356.

HELLER, F., CHEN, H., DOBSON, J. & HAAG, M. 1995. Permian–Triassic magnetostratigraphy – new results from South China. *Earth and Planetary Science Letters*, **89**, 281–295.

HELSLEY, C. E. & STEINER, M. B. 1974, Paleomagnetism of the Lower Triassic Moenkopi Formation. *Geological Society of America, Bulletin*, **85**, 457–464.

HENDERSON, C. M. & MEI, S. 2000. Preliminary cool water Permian conodont zonation in north Pangaea: A review. *Permophiles*, **36**, 16–23.

HIGH, L. R. JR. & PICARD, M. D. 1967, Rock units and revised nomenclature, Chugwater Group (Triassic), western Wyoming. *Mountain Geologist*, **4**, 73–81.

HOUNSLOW, M. W., MORK, A., PETERS, C. & WEITSCHAT, W. 1996. Boreal Lower Triassic magnetostratigraphy from Deltadalen, central Svalbard. *Albertiana*, **17**, 3–10.

HUANG, K. & OPDYKE, N. D. 1998. Magnetostratigraphic investigations on an Emeishan Basalt section in western Guizhou Province, China. *Earth and Planetary Science Letters*, **163**, 1–14.

KAMO, S. L., CZAMANSKE, G. K. & KROGH, T. E. 1996 A minimum U-Pb age for Siberian flood-basalt volcanism. *Geochimica et Cosmochimica Acta*, **60**, 3505–3511.

KAMO, S. L., CZAMANSKE, G. K., AMELIN, Y., FEDORENKO, V. A., DAVIS, D. W. & TROFIMOV,

V. R. 2003. Rapid eruption of Siberian flood-volcanic rocks and evidence for coincidence with the Permian-Triassic boundary and mass extinction at 251 Ma. *Earth and Planetary Science Letters*, **214**, 75–91.

KHRAMOV, A. N. 1963. Paleomagnetic investigations of Upper Permian and lower Triassic sections on the northern and eastern Russian platforn. *Trudy VNIGRI, Nedra, Leningrad*, **204**, 145–174. [In Russian]

KHRAMOV, A. N. 1987. *Paleomagnetology*. Springer-Verlag, Berlin.

KNOLL, A. H., BAMBACH, D. E., CANFIELD, D. E. & GROTZINGER, J. P. 1996. Comparative Earth history and the Late Permian mass extinction. *Science*, **273**, 452–457.

KUMP, L. R, PAVLOV, A. & ARTHUR, M. A. 2005. Massive release of hydrogen sulfide to the surface ocean and atmosphere during intervals of ocean anoxia. *Geology*, **33**, 397–400.

KOTYLAR, G. V., KOMISSAROVA, R. A., KHRAMOV, A. N. & CHEDIYA, I. O. 1984. Paleomagneitaya kharakteristika verkhnepermakikh otlozheniy Zakavkaz'ya. [Paleomagnetism of the Upper Permian Rocks of Transcaucasia.] *Doklady Akademii Nauk SSSR*, **276**(3), 669–674. [Translated by Scripta Technica.]

LIND, E. N., KROPOTOV, S. V., CZAMANSKE, G. K., GROMME, C. S. & FEDORENKO, V. A. 1994. Paleomagnetism of the Siberian flood basalts of the Noril'sk area: a constraint on eruption duration. *International Geology Review*, **36**, 1139–1150.

LIU, Y-Y., ZHU, Y-M. & TIAN, W-H. 1999. New magnetic results from Meishan section, Changxing County, Zhejiang Province. *Earth Science, Journal of China University of Geoscience*, **24**(2), 151–154.

LO, C-H., CHUNG, S-L., LEE, T-Y. & WU, G. 2002, Age of the Emishan flood magnetism and relations to the Permian–Triassic boundary. *Earth and Planetary Science Letters*, **198**, 449–458.

LOZOVSKY, V. R. & MOLOSTOVSKY, E. A. 1993, Constructing the Early Triassic magnetic polarity time scale. *In*: LUCAS, S. G. & MORALES, M. (eds) *The Nonmarine Triassic*. New Mexico Museum of Natural History and Science Bulletin, **3**, 297–300.

LUCAS, S. G. 1994. The beginning of the age of dinosaurs in Wyoming. *Wyoming Geological Association, 44th Annual Field Conference, Guidebook*, 105–113.

LUCAS, S. G. & MORALES, M. 1985. Middle Triassic amphibians. *In*: LUCAS, S. G. & ZIDEK, J. (eds) *Santa Rosa-Tucumcari Region*. New Mexico Geological Society Guidebook, **36**, 56–58.

MCKEE, E. D. 1954. *Stratigraphy and history of the Moenkopi Formation near Grey Mountain*. Geological Society of America Memoir, **61**.

MENNING, M. 1986. Zur Dauer des Zechstein aus magnetostratigraphischer Sicht. *Zeitschrifte für Geologische Wissenschaften*, **14**, 395–404.

MENNING, M. 1988. Magnetostratigraphic investigation of the Rotiegendes (200–252 Ma) of Central Europe. *Zeitschrifte für Geologische Wissenschaften*, **16**, 1045–1063.

MENNING, M. 1993. A revised Permian polarity time scale. *EUG VIII Abstracts*, Strasbourg.

MENNING, M. 2000. Magnetostratigraphic results from the Middle Permian type section, Guadalupe Mountains, West Texas. *International Geological Congress, Abstracts*, **31**, pp. On CD-ROM.

MENNING M. & JIN, Y. G. 1998. Comment on EMBLETON, B. J. J., MCELHINNY, M. W., MA, X., ZHANG, Z. & LI, Z. X. Permo-Triassic magnetostratigraphy in China: the type section near Taiyuan, Shanxi Province, North China. *Geophysical Journal International*, **133**, 213–216.

MOLINA-GARZA, R. S., GEISSMAN, J. W. & VAN DER VOO, R. 1989. Paleomagnetism of the Dewey Lake Formation (Late Permian), Northwest Texas, end of the Kiaman Superchron in North America. *Journal of Geophysical Research*, **94**, 17,881–17,888.

MOLINA-GARZA, R. S., GEISSMAN, J. W., VAN DER VOO, R., LUCAS, S. G. & HAYDEN, S. N. 1991. Paleomagnetism of the Moenkopi and Chinle formations in central New Mexico: implications for the North American apparent polar wander path and Triassic magnetostratigraphy. *Journal of Geophyscial Research*, **96**, 14,239–14,262.

MOLINA-GARZA, R. S., GEISSMAN, J. W. & LUCAS, S. G. 2000. Paleomagnetism and magnetostratigraphy of uppermost Permian strata, southeast New Mexico, USA: correlation of the Permian–Triassic boundary in non-marine environments. *Geophysical Journal International*, **141**, 778–786.

MOLOSTOVSKY, E. A. 1983. *Paleomagnetic Stratigraphy of the Eastern European Part of the USSR, Saratov, Russia*. University of Saratov, Saratov. [In Russian]

MORALES, M. 1987. Terrestrial fauna and flora of the Triassic Moenkopi Formation of the southwestern United States. *Journal of the Arizona-Nevada Academy of Science*, **22**, 1–19.

MORK, A., ELVEBAKK, G., FORSBERG, A. W., HOUNSLOW, A. W., NAKREM, H. A., VIGRAN, J. O. & WEITSCHAT, W. 1999. The type section of the Vikinghogda Formation, a new Lower Triassic unit in central and eastern Svalbard. *Polar Research*, **18**, 51–82.

MUNDIL, R., LUDWIG, K. R., METCALFE, I. & RENNE, P. R. 2004, Age and timing of the Permian mass extinctions: U/Pb dating of closed-system zircons. *Science*, v. **305**(5691), 1760–1763.

MUTTONI, G., KENT, D. *et al.* 1996. Magnetobiostratigraphy of the Spathian to Anisian (Lower to Middle Triassic) Kcira section, Albania. *Geophysical Journal International*, **127**, 503–514.

NAWROCKI, J. 1997. Permian to Triassic magnetostratigraphy from the Central European Basin in Poland: implications on regional and worldwide correlations. *Earth and Planetary Science Letters*, **152**, 37–58.

NAWROCKI, J. & SZULC, J. 2000. The Middle Triassic magnetostratigraphy from the Peri-Tethys basin in Poland. *Earth and Planetary Science Letters*, **182**, 77–92.

OGG, J. G. & STEINER, M. B. 1991. Early Triassic magnetic polarity time scale, integration of magnetostratigraphy, ammonite zonation and sequence stratigraphy from stratotype sections (Canadian

Arctic Archipelago). *Earth and Planetary Science Letters*, **107**, 69–89.

ORCHARD, M. J. & KRYSTYN, L. 1998, Conodonts of the Lowermost Triassic of Spiti, and new zonation based on Neogondolella successions. *Rivista Italiana di Paleontologia e Stratigrafia*, **104**(3), 341–368.

PAULL, R. K. & PAULL, R. A. 1986, Epilogue for the Permian in the western Cordillera: A retrospective view from the Triassic. *Rocky Mountain Geology*, **24**(2), 243–252.

PAULL, R. K. & PAULL, R. A. 1990, Persistent myth about the Lower Triassic Little Medicine Member of the Goose Egg Formation and the Lower Triassic Dinwoody Formation, central Wyoming. *Wyoming Geological Association, 41st Field Conference, Guidebook*, 57–68.

PETERSON, D. N. & NAIRN, A. E. M. 1971. Paleomagnetism of Permian red beds from the southwestern United States. *Geophysical Journal of the Royal Astronomical Society*, **23**, 191–206.

PICARD, M. D., AADLAND, R. & HIGH, L. R. JR. 1969. Correlation and stratigraphy of Triassic Red Peak and Thaynes formations, western Wyoming and adjacent Idaho. *American Association of Petroleum Geologists Bulletin*, **53**, 2274–2289.

POBORSKI, S. J. 1954. Virgin Formation (Triassic) of the St. George, Utah, area. *Geological Society of America, Bulletin*, **65**, 971–1000.

PURUCKER, M. E., ELSTON, D. P. & SHOEMAKER, E. M. 1980. Early acquisition of characteristic magnetization in red beds of the Moenkopi Formation (Triassic), Grey Mountain, Arizona. *Journal of Geophysical Research*, **85**, 997–1012.

RENNE, P. R., STEINER, M. B., SHARP, W. D., LUDWIG, K. R. & FANNING, C. M. 1996. ^{40}Ar/^{39}Ar and U/Pb SHRIMP dating of latest Permian tephras in the Midland Basin, Texas. *Eos, Transactions of the American Geophysical Union*, **77**(46, Supplement), F794.

RENNER, J. M. 1988, *Placement and paleogeographic significance of Permian–Triassic boundary, central and Southeast Wyoming*. MSc thesis, University of Wyoming.

RENNER, J. M. & BOYD, D. W. 1988, Placement and paleogeographic significance of Permian–Triassic boundary, Southeast Wyoming. SEPM Midyear Meeting, *Abstracts*, **5**, 45.

SCHOLGER, R., MAURITSCH, H. J. & BRANDNER, R. 2000. Permian–Triassic magnetostratigraphy from the Southern Alps (Italy). *Earth and Planetary Science Letters*, **176**, 495–508.

SCHWINDT, D. M., RAMPINO, M. R., STEINER, M. & ESHET, Y. 2003. Paleomagnetic results and preliminary palynology across the Permian–Triassic (P–Tr) Boundary at Carlton Heights. *In*: KOEBERL, C. & MARTINEZ-RUIZ, F. (eds) *Impact Markers in the Stratigraphic Record*. Springer, Heidelberg, 280–302.

SMITH, R. M. H. 1995. Changing fluvial environments across the Permian–Triassic boundary in the Karoo Basin, South Africa and possible causes of tetrapod extinctions. *Palaeogeography, Palaeoclimatology, Palaeoecology*, **117**, 81–104.

STEINER, M. B. 2000, Geomagnetic polarity between the P/Tr boundary and the Kiaman termination. Eos, *Transactions of the American Geophysical Union*, **81**, F361–362.

STEINER, M. B. 2001*a*, Magnetostratigraphic correlation and dating of West Texas and New Mexico Late Permian strata. *American Association of Petroleum Geologists Bulletin*, **85**, 1695.

STEINER, M. B. 2001*b*, Magnetostratigraphic correlation and dating of west Texas and New Mexico Late Permian strata. *In*: LUCAS, S. G. & ULMER-SCHOLLE, D. (eds) *Geology of Llano Estacado*. New Mexico Geological Society, 52nd Field Conference, Guide, **52**, 59–68.

STEINER, M. B. 2004, Global magnetostratigraphic correlation of the Middle and Late Permian and the Early Triassic. *Proceedings of the 32nd International Geological Congress*, Florence, 2004, Volume 1, 746.

STEINER, M. B. & LUCAS, S. G. 1992. A Middle Triassic paleomagnetic pole for North America. *Geological Society of America, Bulletin*, **104**, 993–998.

STEINER, M. B. & RENNE, P. R. 1996. Magnetic polarity at the end of the Permian: New insights from west Texas. EOS, *Transactions of the American Geophysical Union*, **77**(46, Supplement), F164.

STEINER, M. B. & RENNE, P. 1998. Late Permian and and Permo-Triassic boundary magnetostratigraphy; progress? and problems, in Gondwana-10. *African Journal of Earth Science*, **27**(1A), 190.

STEINER, M. B., OGG, J. G., ZHANG, Z. & SUN, S. 1989. The Late Permian/Early Triassic magnetic polarity time scale and plate motions of South China. *Journal of Geophysical Research*, **94**, 7343–7363.

STEINER, M. B., MORALES, M. & SHOEMAKER, E. M. 1993. Magnetostratigraphic, biostratigraphic, and lithologic correlations in Triassic strata of the western United States. *In*: AISSAOUI, D. M., MCNEILL, D. F. & HURLEY, N. F. (eds) *Applications of Paleomagnetism to Sedimentary Geology*. Society of Economic Paleontologists and Mineralogists Special Publication, **49**, 41–57.

STEINER, M. B., ESHET, Y., RAMPINO, M. R. & SCHWINDT, D. M. 2003. Fungal abundance spike and the Permian–Triassic boundary in the Karoo Supergroup (South Africa). *Palaeogeography, Palaeoclimatology, Palaeoecology*, **194**, 405–414.

STORRS, G. W. 1991. *Anatomy and relationships of Corosaurus alcovensis (Diapsida: Sauropterygia) and the Triassic Alcova Limestone of Wyoming*. Bulletin of the Peabody Museum of Natural History, Yale University, **44**.

SZURLIES, M. 2004. Magnetostratigraphy: the key to a global correlation of the classic Germanic Trias—case study Volpriehausen Formation (Middle Buntsandstein), central Germany. *Earth and Planetary Science Letters*, **227**, 395–410.

SZURLIES, M. B., GERHARD, H., MENNING, M., NOWACZYK, N. R. & KAEDING, K. C. 2003. Magnetostratigraphy and high-resolution lithostratigraphy of the Permian–Triassic boundary interval in central Germany. *Earth and Planetary Science Letters*, **212**, 263–278.

SWEET, W. C. & BERGSTROEM, S. M. 1986. Conodonts and biostratigraphic correlation. *Annual Reviews of Earth and Planetary Sciences*, **14**, 85–112.

THOMAS, H. D. 1934, Phosphoria and Dinwoody tongues in lower Chugwater of central and southern Wyoming. *American Association of Petroleum Geologists Bulletin*, **18**, 1655–1697.

TOZER, E. T. 1967. A Standard for Triassic Time. *Geological Survey of Canada, Bulletin*, **156**.

VALENCIO, D. A., VILAS, J. F. A. & MENDIA, J. E. 1977. Paleomagnetism of a sequence of red beds of the middle and upper sections of Pagnazo Group (Argentina) and the correlation of Upper Paleozoic-Lower Mesozoic rocks. *Geophysical Journal of the Royal Astronomical Society*, **51**, 59–74.

WARD, P. D., BOTHA, J. *et al.* 2005. Abrupt and gradual extinction among Late Permian land vertebrates in the Karoo Basin, South Africa. *Science*, **307**, 709–714.

WARDLAW, B. R. 2003. Global Guadalupian (Middle Permian) conodont correlation and distribution. *Permophiles*, **42**, 20–21.

WARDLAW, B. R. & COLLINSON, J. W. 1986. Paleontology and deposition of the Phosphoria Formation. *Rocky Mountain Geology*, **24**, 107–142.

WARDLAW, B. R. & POGUE, K. R. 1995. The Permian of Pakistan. *In*: SCHOLLE, P. A., PERYT, T. M. & ULMER-SCHOLLE, D. S. (eds) *The Permian of northern Pangaea, Vol. 2*. Springer-Verlag, Berlin, 215–224.

WESTPHAL, M., GUREVICH, E. L., SAMSONOV, B. V., FEINBERG, H. & POZZI, J-P. 1998. Magnetostratigraphy of the Lower Triassic volcanics from deep drill SG6 in western Siberia, evidence for long-lasting Permo-Triassic volcanic activity. *Geophysical Journal International*, **134**, 254–266.

ZAKHAROV, Y. D. & SOKAREV, A. N. 1991. Permian–Triassic paleomagnetism of Eurasia, *In*: KOTAKA, T., DICKENS, J. M., MCKENZIE, K. G., MORI, K., OGASAWARA, K. & STANLEY, G. D. JR. (eds) *Proceedings of the International Symposium on Shallow Tethys*, **3**, 313–323.

ZEISSL, W. & MAURITSCH, H. 1991. The Permian-Triassic of the Gartnerkofel-1 core (Carnic Alps, Austria), Magnetostratigraphy. *Abhandlungen der Geologischen Bundesanstalt*, **45**, 193–207.

ZHU, Y-M. & LIU, Y-Y. 1999. Magnetostratigraphy at the Permian–Triassic boundary at Meishan, Changxing, Zhejiang Province. *In*: YIN, H. & TONG, J-N. (eds) *Proceedings of the International Conference on Pangaea and the Paleozoic-Mesozoic Transition China University of Geosciences, Wuhan, September 1999*. Geoscience Press, 79–84.

ZIU, Y., ZHU, Y. & TIAN, W. 1999. New magnetostratigraphic results from Meishan section, Changxing County, Zhejiang Province. *Earth Science, Journal of China University of Geosciences*, **24**(2), 151–154.

Two remarkable Permian petrified forests: correlation, comparison and significance

RONNY RÖSSLER

Museum für Naturkunde, Moritzstrasse 20, D - 09111 Chemnitz, Germany
(e-mail: roessler@naturkunde-chemnitz.de)

Abstract: Two outstanding Permian petrified forests, those of Chemnitz, in Germany, and northern Tocantins, in Brazil, contribute to the understanding of the composition, peculiarities and dynamics of Early Permian wetland ecosystems. These assemblages represent seasonally influenced, essentially contemporaneous but quite comparable, tree-fern-dominated plant communities in the Northern and Southern Hemispheres. The Chemnitz fossils are embedded in coarse-grained pyroclastics of the Zeisigwald Tuff Horizon (Leukersdorf Formation, Erzgebirge Basin), whereas those of Tocantins occur in different lithofacies of a cyclic alluvial succession (Pedra de Fogo/Motuca formations, Parnaíba Basin). The outstanding three-dimensional preservation of particularly large fossil remains, made possible by siliceous permineralization, provides the opportunity to study the gross morphology, anatomy and internal organization of plant tissues, as well as taphonomical and ecological aspects of late Palaeozoic plants in a way not allowed by other preservational states. Recent studies of newly collected material permit a re-evaluation of the popular reconstructions of Early Permian floras. Various plant–plant and plant–animal interactions add to our understanding of two diverse lowland ecosystems that, irrespective of their different palaeogeographic position and taphonomic modes, show striking similarities.

There is no doubt that land plants have played an important role in shaping both terrestrial ecosystems and landscapes through time. They have not only played a striking role in the global carbon cycle (Berner & Kothavala 2001), they also affect weathering and soil-forming processes, constitute the basis of terrestrial food chains, frame habitats for all terrestrial animals, and, shape and develop the terrestrial biota in general. During the late Palaeozoic, plant life rapidly adapted to diverse terrestrial landscapes and conditions, thus creating a set of advanced ecosystems (e.g. Falcon-Lang & Bashforth 2004). Some differed more or less from the well-known tropical wetlands, which are generally favoured with respect to preservation.

In Permian times, one recognizes a considerable decrease in the fossil plant record compared with the Carboniferous. Most of the coeval ecosystems, including those of the hinterland areas, remain highly enigmatic, and the depositional sequence is usually less complete (Schneider 1989; Kerp 1996, 2000). Due to the tectonic and palaeogeographic evolution, which considerably influenced the depositional environments, accommodation space and climatic regimes on a variable scale, most of the terrestrial Permian is represented by hiatuses or largely unfossiliferous strata. Localities that provide a rich fossil record are rare and especially noteworthy. A few of them became known as so-called 'petrified forests', viz. localities where numerous petrified trees have been fossilized essentially in a growth position, making them particularly instructive.

Petrified forests are known from both volcanic deposits and clastic sediments from the Devonian onwards and are more widespread than commonly thought. However, only a few examples are known from the Palaeozoic, such as the regionally restricted fossil forest of Chemnitz, in Germany (Sterzel 1875, 1918; Barthel 1976; Rössler 2001) and the rather spatially extended fossil forest of Tocantins in northern Brazil (Dias-Brito & Castro 2005). Some currently well-known fossil generic names, such as *Psaronius, Arthropitys* and *Medullosa,* are based on finds from Chemnitz, the *locus typicus* and one of the richest localities of Permian permineralized plants. Scientific investigation of the Chemnitz site dates back to the early history of palaeobotany around the early nineteenth century (Sprengel 1828; Cotta 1832).

Tocantins, in contrast, is a long-known, but poorly studied site that has recently yielded a rich array of Lower Permian plant fossils (Herbst 1999; Rössler & Noll 2002). New finds from the Southern Hemisphere, considerably exceeding the long-known *Psaronius brasiliensis* material, have lead to notable progress in research on late Palaeozoic plants, such as different ferns,

From: LUCAS, S. G., CASSINIS, G. & SCHNEIDER, J. W. (eds) 2006. *Non-Marine Permian Biostratigraphy and Biochronology.* Geological Society, London, Special Publications, **265**, 39–63.
0305-8719/06/$15.00

sphenopsids and gymnosperms. Revisions of classical Permian permineralized organ taxa were necessarily initiated as a prerequisite to accommodate the newly discovered fossils (Rössler & Galtier 2002*a*; Rössler & Noll 2006).

Because of their noteworthy preservation and *in situ* to parautochthonous nature, both the Chemnitz and Tocantins occurrences enable the palaeobiological study of fossil plants and animals, and the interactions between them. In addition, the taphonomy of these deposits can be related directly to the peculiarities responsible for growth and preservation. In addition to the study of material in collections, special emphasis was placed on gaining new specimens from the field and studying the geological context.

Since comparable fossil ecosystems seem to be rare or remain poorly documented, the primary objective of this contribution is to discuss and compare the Permian age petrified forests of Chemnitz and Tocantins. The main purpose is to characterize their geological origin, taphonomy and original condition of growth, and also to pose a number of questions that remain to be solved by future investigations. Recent preparations stimulated new observations and gave unique insight into floral and faunal elements that lived in more or less restricted Early Permian wetland communities.

Localities and stratigraphy of the source strata

The Chemnitz petrified forest

Only a small number of petrified forests are as famous as the one found in the present-day city of Chemnitz, where nearly an entire forest of fossilized plants from the Permian was discovered (Sterzel 1875, 1918; Barthel 1976; Rössler 2001). The origin of the Chemnitz petrified forest is closely related to rhyolithic explosive volcanism that occurred almost 280 Ma ago (Roscher & Schneider 2005). One of the eruptions to the northeast of Chemnitz resulted in the formation of a pyroclastic sequence now referred to as the Zeisigwald Tuff Horizon (Fischer 1990), part of the Leukersdorf Formation, which consists of approximately 800 m of sedimentary and volcanic deposits.

The stratigraphic position of the Leukersdorf Formation corresponds to the Upper Asselian/ Lower Sakmarian, as indicated by a rich microflora obtained from the palustrine Rottluff Coal situated within the lower part of the Leukersdorf Formation (Fig. 1). Palynostratigraphic investigation carried out by Döring *et al.* (1999) further suggested close similarities of the Lower Leukersdorf Formation and the late Asselian Slavjanskaja Svita of the Donetsk Basin reference section. Correlations based on eryopid and diadectid vertebrate remains are similar (Werneburg 1993; Schneider *et al.* 1995). The Zeisigwald Tuff Horizon of the Upper Leukersdorf Formation corresponds to an absolute age of about 278 ± 5 Ma, based on SHRIMP U-Pb measurements on accessory zircons (Nasdala *et al.* 1998). Important fossil-bearing outcrops of this formation are situated throughout the city of Chemnitz (Erzgebirge Basin). Therefore, this set of localities has long been known as the Chemnitz petrified forest (for summary and multiple references see Rössler 2001).

The Tocantins petrified forest

Although Permian petrified plants from Brazil have been mentioned in the literature for some 140 years (Brongniart 1872), only the frequently found psaroniaceous tree ferns became known in any detail (Pelourde 1912; Derby 1913, 1915; Solms-Laubach 1913; Herbst 1992) or were distributed to collections around the world. The newly discovered set of localities in the northern part of the State of Tocantins in Brazil (Fig. 2) is sometimes designated the petrified forest of Araguaína (Dernbach 1996; Herbst 1999). As a consequence of the high potential of this area for both scientific investigation and popular tourism, this unique Fossillagerstätte was preserved and designated the Tocantins Fossil Trees Monument (Dias-Brito, pers. comm. 2005).

The fossil-bearing strata classically have been considered part of the Pedra de Fogo Formation (Barbosa & Gomes 1957; Lima & Leite 1978), which was introduced by Plummer (1948) and represents a 25 m to almost 200 m thick sedimentary sequence of dominantly terrestrial clastics of a well-drained depositional environment distributed within the Parnaíba sedimentary basin. This basin stretches across more than 600 000 km^2. Intercalated marine and continental sediments from the Silurian to the Cretaceous were deposited here (Petri & Fulfaro 1983).

The Permian Pedra de Fogo Formation consists of deposits of several fluvial cycles containing thick sets of fluvial sandstones and overbank fine clastics with palaeosol horizons up to lacustrine strata at the cycle tops. The whole environment may be characterized as a seasonally influenced alluvial plain setting with a variable base level. Pinto & Sad (1986) pointed out that the plant fossils are included in the base of the Permian Motuca Formation, which lies above the Pedra de Fogo Formation. Other authors are in agreement with this conclusion (e.g. Araujo 2001; Dias-Brito & Castro 2005).

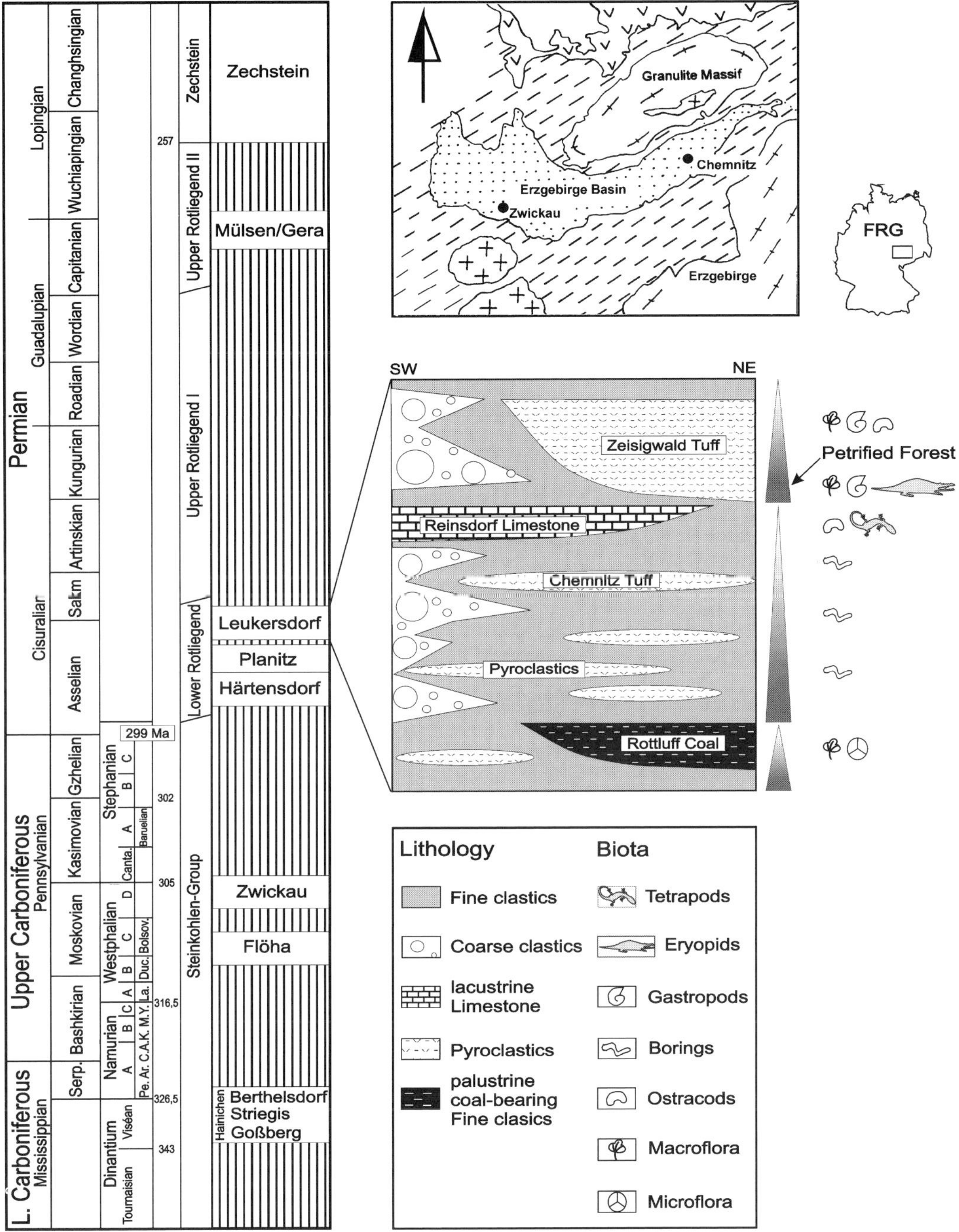

Fig. 1. Stratigraphical and geological framework of the petrified forest of Chemnitz.

However, further stratigraphic studies are needed to clarify this question.

The age of the fossil-bearing strata also remains a matter of discussion. Based on current palaeobotanical information, the Pedra de Fogo and Motuca formations are usually assigned to the Early Permian (Petri & Fulfaro 1983), although there are also palynological studies that suggest a Late Permian age (Dino *et al.* 2002). Based on tetrapods, such as the archegosaurid

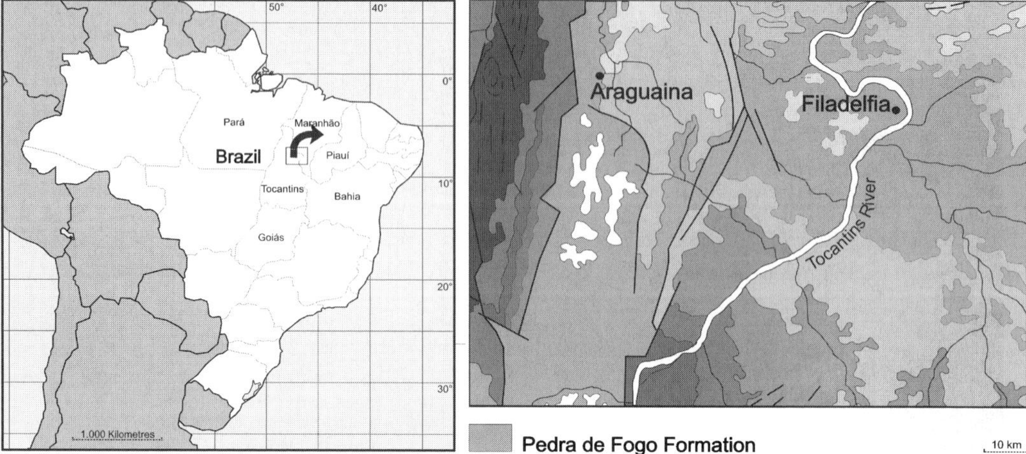

Fig. 2. Simplified location map showing the outcrops of the Permian Pedra de Fogo Formation in NE Brazil.

amphibian *Prionosuchus plummeri*, Price (1948) argued for an Early Permian age. Finds of other archegosaurids are recorded from the Early Permian of the Saar–Nahe Basin, Germany (type species *A. decheni*) and from the early Late Permian of Kashmir, India (Werneburg & Schneider 1996), as well as from the Middle Permian (Kazanian) of the southern Urals, Russia. Cox & Hutchinson (1991) argued for a Late Permian age of the Pedra de Fogo Formation because, in their interpretation, *Prionosuchus* represents a continuation of the archegosaur trend towards an elongated, narrow snout. Cox & Hutchinson (1991) provided further data, such as the new palaeoniscid *Brazilichthys macrognathus*, fragments of ctenacanth and xenacanth sharks, edestid holocephalians and dipnoans, but these remains do not significantly contribute to the age determination. Because xenacanth remains have been successfully used for dating purposes (Schneider 1996), the age of the Pedra de Fogo and Motuca formations will hopefully be resolved in the near future.

In addition to the classical vertebrate localities lying in the valley of the river Pedra de Fogo between Pastos Bons and Nova Iorque, important fossil-bearing outcrops of this widespread depositional complex are situated to the southeast of Araguaína, between Bielandia and Filadelfia (around 7° 28′ S, 47° 42′ W) or east of Colinas do Tocantins (around 8° 02′ S, 48° 11′ W).

Material, preservation, methods and storage

This study focuses on permineralized specimens collected from the Permian petrified forests of Chemnitz, Germany (Fig. 1) and northern Tocantins, Brazil (Figs 2 & 3). The Chemnitz fossils were found in coarse-grained, purple-red or pale-green coloured volcaniclastics of the surge sequence of the Zeisigwald Tuff Horizon. Many fossils are preserved *in situ* (Fig. 4a, b). However, depending on the completeness of silicification, the material sometimes shows imperfect anatomical details. Because organic remains usually have not been preserved in this kind of permineralized material, the sometimes low contrast can make it difficult to recognize cell walls in particular and characters below the cell level in general. Silica-rich fluids penetrated the remaining trunks and plant fragments in some instances, which subsequently led to several generations of dense quartz polymorphs (Witke *et al.* 2004). This may have happened when silicic acid leached out of the overlying sedimentary and volcanic deposits and filled the cells, which solidified and preserved their anatomical structure. In places, the surrounding tuffaceous matrix is also partially silicified. Aside from normal siliceous permineralizations and weakly stratified plant-bearing cherts, a considerable percentage of the Chemnitz petrifactions is preserved by purple-coloured calcium fluoride. This unique feature is attributed to the migration of volcanic fluids, but the process by which this happened remains unresolved.

The Tocantins fossil plant material is three-dimensionally preserved as siliceous cellular permineralizations. Depending on both the amount of red to purple-coloured ferric constituents and the completeness of silicification, the fossils generally show clear details of the plant

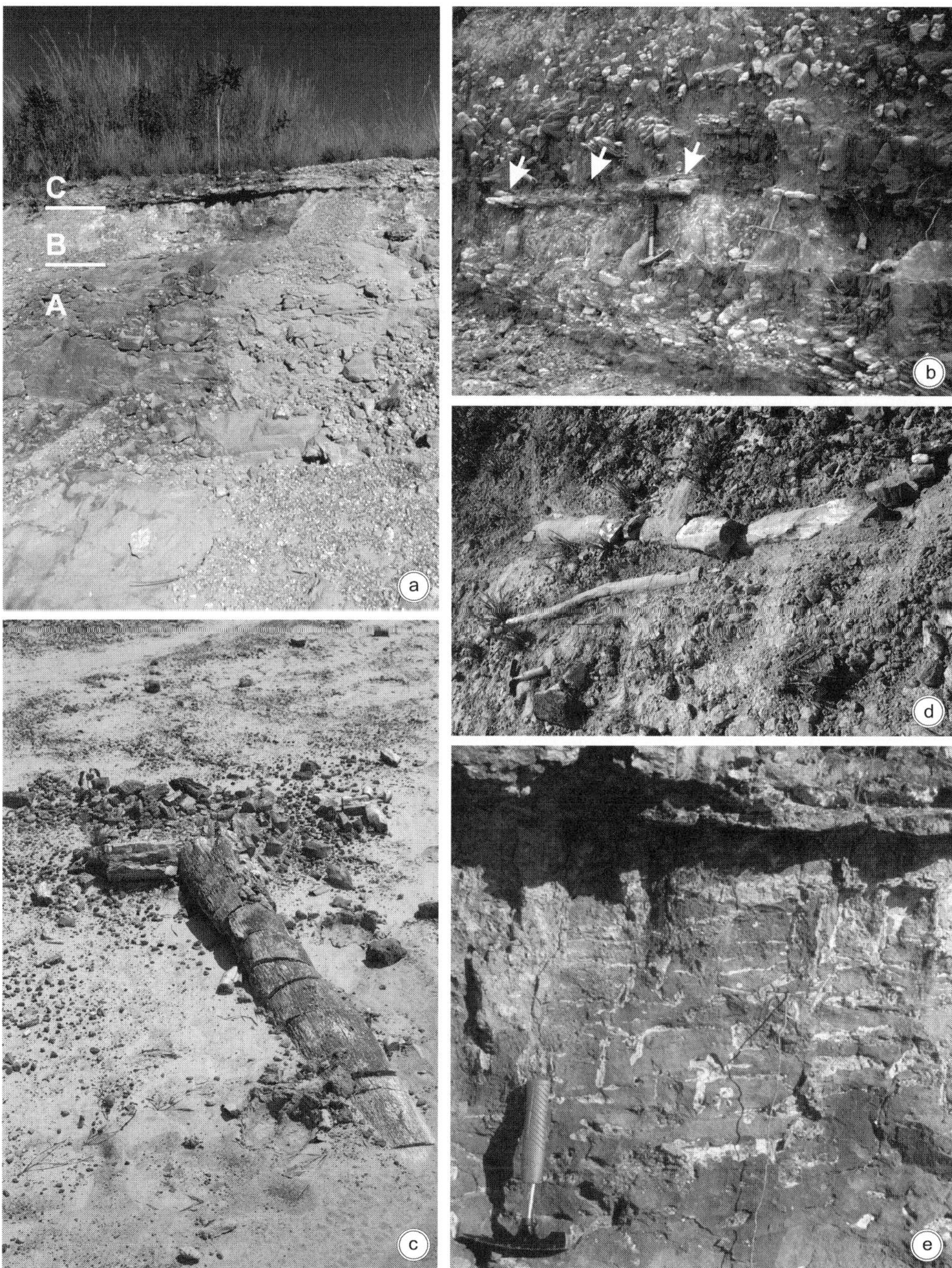

Fig. 3. (**a**) Cyclic fluvial succession of the Pedra de Fogo Formation on the road Araguaina-Filadelfia, 2003. Lithofacies (**A**) cross-bedded channel sandstones; Lithofacies (**B**) fine-clastics of the inter-channel areas; Lithofacies (**C**) fossil-rich lacustrine and sheet flood deposits of the cycle tops. (**b**) Gymnospermous stem (arrows) lying above pedogenicallyaltered overbank fine-clastics, at the base of the next fluvial cycle. Hammer is 33 cm long. (**c**) Almost complete psaroniaceus trunk preserved in fine-grained overbank deposits, Filadelfia region, 2005. (**d**) *Arthropitys*-type calamite trunk with large attached branch preserved parautochthonously within fine-grained overbank deposits, Filadelfia region, 2005. Hammer is 28 cm long. (**e**) Fine-grained deposits of the inter-channel areas showing post-depositional changes such as initial palaeosol formation. Note the light-coloured root traces. Hammer is 28 cm long.

tissues. However, the origin of the siliceous permineralization agent remains unclear, although it could be linked to the formation of pedogenic silcrete nodules, which are considerably enriched in discrete horizons of the sedimentary sequence. The outstanding size of many fossil trunks has allowed additional study of the branching characteristics of the plants (Fig. 3d). In addition to the trunks, these deposits also contain small remains of three-dimensionally preserved material preserved in irregularly shaped, densely silicified sandstone bodies (Fig. 5f), which compositionally represent quartz arenites.

The material described and illustrated in this contribution is mainly kept in the palaeontological collection (petrified wood collection) of the Museum für Naturkunde, Chemnitz, Germany (MfNC). Additional material was provided by Robert Noll, Tiefenthal, Germany (SNB) and Ulrich Dernbach, Bösdorf, Germany. The Dernbach collection is now curated at the Bavarian State Collection of Palaeontology, Munich, Germany.

The specimens were prepared with sandblasting, and subsequently cut with a trimming saw to reveal different diagnostic sections of the permineralized trunks. Exposed cut surfaces subsequently were ground, polished and examined using reflected light microscopy. Additional details were obtained from a few thin section preparations.

Sections were photographed under reflected or transmitted light using Nikon Eclipse ME 600 and Nikon SMZ 1500 microscopes attached to a Nikon DS-5M-L1 digital camera. Overview photographs were made by using a Nikon F 70 or Fujifilm FinePix S1 Pro camera combined with a Sigma Apo Macro 180 F 3.5 EX lens or an Epson Perfection 4870 scanner.

Geological setting and taphonomic implications

The plant-bearing pyroclastic horizon of the Chemnitz locality

The major plant-bearing layer is situated at the base of the so-called 'base surge deposit' of the Zeisigwald Tuff Horizon, a coarse-grained accretionary lapilli-bearing ash tuff (Fig. 4d), which originated from the deposition of several hyperconcentrated pyroclastic flows. The eruption preserved a species-rich association (Table 1) of tree ferns, arboreal sphenophytes, different pteridosperms, conifers, cordaitaleans and climbing or epiphytic plants. Only here does such a large diversity of Permian permineralized plants occur *in situ*. Wherever the base of the Zeisigwald tuff crops out at the surface, petrified trunks are still being found.

The spectrum of preservational types is diverse, ranging from so-called moulds and casts to almost completely preserved specimens, which reveal microscopic details of internal anatomy. A more detailed overview of the geological setting is presented in Rössler (2001). A list of the entire fossil record of the Chemnitz petrified forest is given in Table 1.

The genesis of the Chemnitz petrified forest can be reconstructed based on comparison with the effects produced by the eruption of Mount St Helens in the Cascade Range of Washington State, USA, in May 1980 (Hoblitt *et al.* 1981). Thorough study of the volcanic deposits (Wiatt 1981) and the appearance of the vegetation after this eruption greatly contributed to our understanding of the events that occurred more than 280 Ma earlier at Chemnitz.

The deposition of the Zeisigwald Tuff Horizon is the result of a multi-storey caldera eruption, that is, a depositional sequence of several devastating 'glowing-cloud eruptions', each accompanied by pyroclastic flow deposits (Eulenberger *et al.* 1995). Such hot, dense and unsorted particle systems are often connected with so-called surges – tremendous lateral blasts, whose force and direction are sometimes strikingly demonstrated by the parallel alignment of toppled large trees broken off at their bases (Hoblitt *et al.* 1981). Such a giant blast is interpreted as the key process for the origin of the Chemnitz petrified forest. The tops of the Chemnitz fossilized trees point towards the west, in the direction the annihilating blast was moving as it spread outwards from the volcano during the first few seconds of the eruption. Large tree trunks were snapped off and stripped of their branches and bark, but they fell and remained lying for subsequent petrification near where they had stood previously. In some cases (compare Fig. 4b), the tree trunks even remained standing, as shown, for example, by the recent discovery of the largest calamite trunk known to date (Rössler & Noll 2006). The rapidity and intensity of the processes related to volcanigenic fossilization caused preservation of some of the most complete and perfectly preserved fossil plant assemblages known (Rössler & Barthel 1998). Among them the Chemnitz locality provides favourable conditions for the task of investigating both the well-preserved Permian floral elements and the permineralization process itself (Witke *et al.* 2004).

Fig. 4. (**a**) Petrified gymnospermous trunk lying *in situ* at the base of the Zeisigwald Tuff Horizon, Chemnitz 1996. (**b**) Trunk base of *Arthropitys ezonata* (Goeppert) Rössler & Noll 2006, still standing upright, Chemnitz 2002, MfNCK5200. (**c**) Gymnospermous branch enclosed in coarse-grained ash-tuff of the Zeisigwald Tuff Horizon. MfNC F 12065; scale bar equals 50 mm. (**d**) Accretionary lapilli containing ash tuff of the Zeisigwald Tuff Horizon – the plant-bearing layer of the Chemnitz petrified forest. Scale bar equals 30 mm. (**e**) Chemnitz petrified forest in the Museum of Natural History, Tietz-Building, Chemnitz. Scale bar equals 1 m.

The Permian fossil-bearing clastic sequence of the northern Tocantins region

Although investigation of the petrified forest localities is still in progress (Dias-Brito & Castro 2005), some general points of the sedimentological and taphonomic context can be given here. These observations were obtained during three short field sessions in 2000, 2003 and 2005, and are of a preliminary character. It is beyond the scope of this note to clarify the existing problems with both the stratigraphical assignment and physical delimitation of the formations.

The majority of the fossil trunk remains of the northern Tocantins region come from alluvial deposits (Fig. 4a, b) that may have formed in a huge lowland network of ancient rivers draining the Parnaíba sedimentary basin (sometimes designated the Maranhão-Piauí Basin). The overall drainage direction generally is approximately SW–NE, as indicated by sedimentary structures such as bedding patterns, sole marks and the main direction of large trunks embedded in the sediment. Present-day erosion exposes the petrified trunks and, because of their toughness, they remain and become concentrated at the soil surface (Fig. 3c). For the open alluvial plain environments, there is a clear separation of lithofacies associations, as indicated by the composition of the sedimentary succession (Fig. 3a). The most likely model is of high-sinuosity sandbed streams associated with a generally variegated overbank suite, including sheetfloods and different ephemeral to perennial aquatic environments. Alluvial cycles are up to approximately 10 m thick and appear to be almost complete, since the cycle tops lying below the next erosional surface are characterized by very resistant, silicified strata. A fluctuating, but generally low base level may have resulted in repeated depositional phases, enabling accumulation and preservation of stacked alluvial lithofacies.

The channel facies association (lithofacies A) composes one half to three-quarters of each cycle, and is dominated by cross-bedded sandstones showing a slight upward-fining trend, but normally without distinct coarse-grained, pebbly lags (Fig. 3a). In addition to single channel bodies with lateral accretion strata, stacked channel bodies also occur, which suggests that successive river channels tended to re-occupy pre-existing drainage pathways. In some places, fragments of transported trunks occur immediately above the channel-bounding erosion surfaces (Fig. 3b). They are generally reduced to small pieces with moderate to well-rounded surfaces. Set thickness is commonly on the order of 1 m or less, but ranges up to 10 m; cross-bedding is of several types (trough to tabular) and scales.

The interchannel areas, which are of greater extent than the channels, were undoubtedly substrates for the flora. The overbank fines (lithofacies B) exhibit an array of post-depositional changes, such as a strong, climatically induced overprint with a degree of palaeosol formation (Fig. 3e). Sometimes large fossil trunks with attached branches have been found in an almost autochthonous position, embedded in pale brown to greenish, horizontally bedded overbank siltstones (Fig. 3d). Associated fine-grained clastic sediments usually lack compression fossils. All remains, even small fern pinnae and rhachises, are preserved three-dimensionally (Fig. 5b–d). Nevertheless, as is typical of well-drained settings with patchy vegetation, little organic matter has been incorporated into the sediment, and even that was likely to have been oxidized. Although drab-haloed root traces are frequently recognized (Fig. 3e), disturbance by roots was minor, and the surface may have been exposed occasionally to deflation, reworking and reddening.

One particular type of source for fossils is found at the top of the alluvial cycles and is

Fig. 5. (**a**) Detail of Fig. 3a. Lithofacies C: fossil-rich lacustrine and sheet flood deposits of the cycle tops, often enriched in silcrete nodules. Hammer is 28 cm long. (**b**) Pinna of psaroniaceous tree fern showing sporangia densely filled with spores. SNB 430; scale bar equals 500 μm. (**c**) Pinna of psaroniaceous tree fern and other three-dimensionally preserved plant remains. Note the well-sorted, highly mature, fine-grained sandstone. SNB 430; scale bar equals 2 mm. (**d**) Isolated, three-dimensionally preserved pinna of a psaroniaceous tree fern. The sandstone matrix was partially removed by sand blasting. Dernbach collection; scale bar equals 10 mm. (**e**) *Sphenophyllum* axis in transverse section showing the central triarch stele and remains of the cortex tissue. SNB 430; scale bar equals 500 μm. (**f**) Different plant remains three-dimensionally preserved in fine-grained silicified sandstone. MfNC K5253; scale bar equals 5 mm. (**g**) Silicified horizon almost exclusively consisting of small bivalves. MfNC F13726; scale bar equals 5 mm. (**h**) Silicified, cylindrical structured body in transverse section showing some kind of undulated concentric growth, which could be interpreted as stromatolite. MfNC K5254; scale bar equals 10 mm.

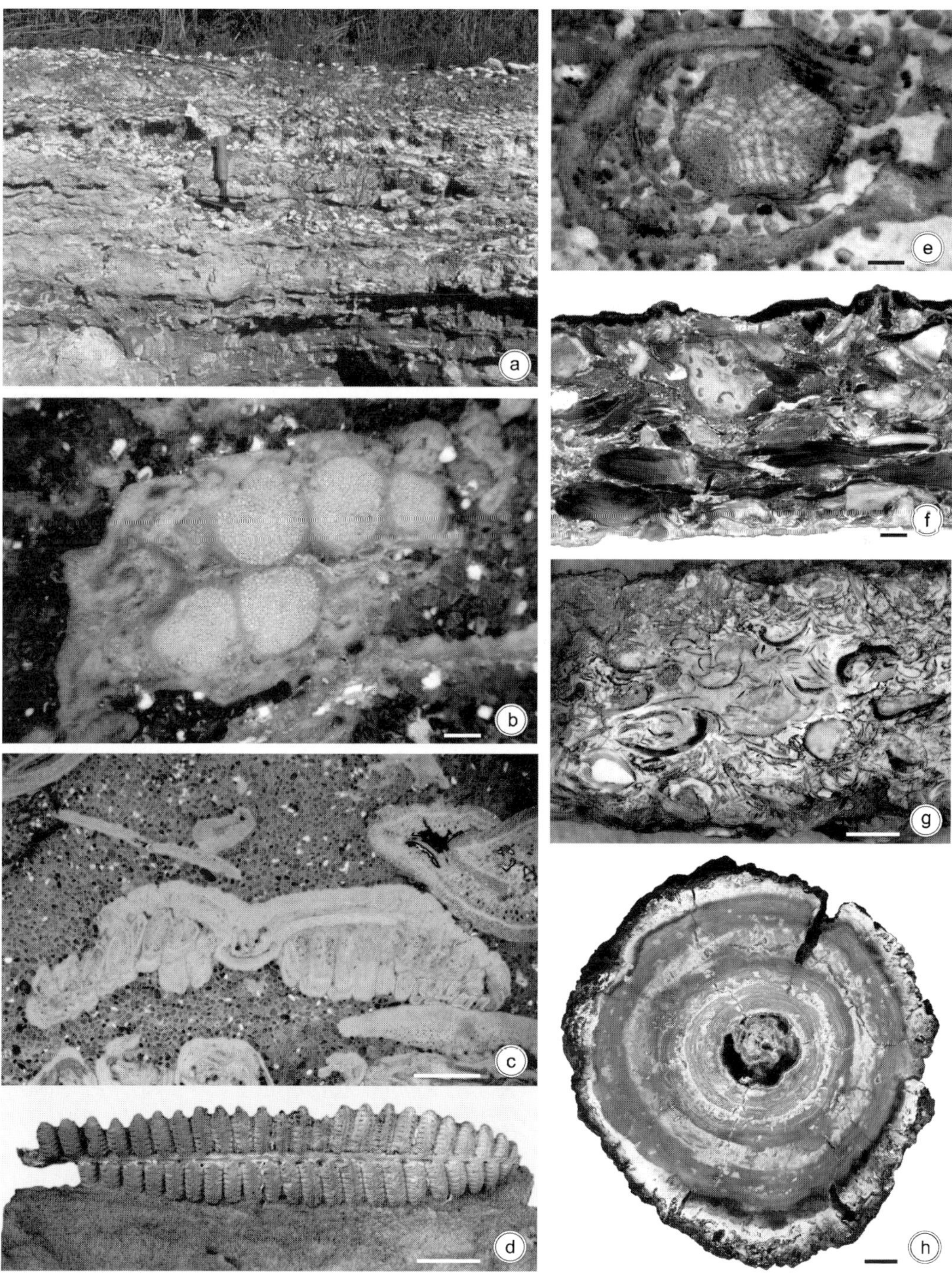
a
b
c
d
e
f
g
h

Table 1. *Comparison of fossil taxa recognized at Chemnitz and northern Tocantins localities*

Fossil remains	Chemnitz (Leukersdorf Formation)	northern Tocantins (Pedra de Fogo/Motuca Formations)
Ferns	*Psaronius tenius* Stenzel 1907 *Psaronius simplex* Unger 1847 *Psaronius musaeformis* Corda 1845 *Psaronius cottae* Corda 1845 *Psaronius gutbierii* Corda 1845 *Psaronius chemnitzensis* Corda 1845 *Psaronius ungerii* Corda 1845 *Psaronius plicatus* Stenzel 1907 *Psaronius scolecolithus* Unger 1847 *Psaronius pusillus* Unger 1847 *Psaronius asterolithus* Cotta 1832 emend. Stenzel 1854 *Psaronius infarctus* Unger 1847 *Psaronius quadrangulus* Stenzel 1907 *Psaronius pictus* Stenzel 1907 *Psaronius cinctus* Stenzel 1907 *Psaronius klugei* Stenzel 1907 *Psaronius haidingerii* Stenzel 1854 *Psaronius punctatus* Stenzel 1907 *Psaronius helmintholithus* Cotta 1832 emend. Stenzel 1854 *Psaronius spissus* Stenzel 1907 *Psaronius weberii* Sterzel 1887 *Psaronius coalescens* Stenzel 1907 *Scolecopteris arborescens* (Schlotheim 1820) Stur 1883 *Scolecopteris candolleana* (Brongniart 1833) Stur 1883 *Scolecopteris* sp. *Pecopteris mentiens* Sterzel 1880 *Remia pinnatifida* (Gutbier 1835) Knight 1985 emend. Kerp *et al.* 1991 *Lobatopteris geinitzii* (Gutbier 1849) Wagner 1983 *Nemejcopteris feminaeformis* (Schlotheim ex Sterzel 1893) Barthel 1968 *Grammatopteris baldaufii* (Beck 1920) Hirmer 1927 *Asterochlaena laxa* Stenzel 1889 *Asterochlaena ramosa* (Cotta 1832) Stenzel 1889 *Zygopteris primaria* (Cotta 1832) Corda 1845 *Ankyropteris brongniartii* (Renault 1869) Mickle 1980 *Anachoropteris pulchra* Corda 1845 *Tubicaulis solenites* Cotta 1832 *Tubicaulis* cf. *berthieri* Bertrand & Bertrand 1911	*Psaronius brasiliensis* (Brongniart 1874) Herbst 1985 *Psaronius sinuosus* Herbst 1991 *Tietea singularis* (Solms-Laubach 1913) Herbst 1999 *Tietea derbyi* Herbst 1992 *Grammatopteris freitasii* Rössler & Galtier 2002 *Dernbachia brasiliensis* Rössler & Galtier 2002 *Botryopteris nollii* Rössler & Galtier 2003 *? Tubicaulis* sp. *Stipitopteris* sp. *Pecopteris* sp. (3 different types) Sterile and fertile marattiaceous pinnules
Sphenophytes	*Calamites multiramis* Weiss 1884 *Calamitea striata* Cotta 1832 *Calamites* sp. (group of *Calamites alternans* Germar / *Calamites varians* Sternberg var. *insignis* Weiss) *Arthropitys bistriata* (Cotta 1832) Goeppert 1864/65 *Arthropitys ezonata* (Goeppert 1864/65) Rössler & Noll 2006 *Calamostachys tuberculata* (Sternberg 1825) Weiss 1884 *Dicalamophyllum altendorfense* Sterzel 1880 *Dicalamophyllum validum* Florin 1939 *Metacalamostachys dumasii* (Zeiller 1892) Barthel 1989 *Calamophyllites* sp. *Annularia spinulosa* (Sternberg 1821) Barthel 2000 *Annularia spicata* (Gutbier 1849) Schimper 1869 *Annularia carinata* (Gutbier 1849) Barthel 1976	*Arthropitys* sp. (A 1–5, B 1–3 Rössler & Noll 2002) *Sphenophyllum* sp.

Table 1. *Continued*

Fossil remains	**Chemnitz** (Leukersdorf Formation)	**northern Tocantins** (Pedra de Fogo/Motuca Formations)
Lycophytes	*Sigillaria* cf. *brardii* BRONGNIART 1828	undetermined cortex layers showing leaf cushions/traces
Seed ferns	*Medullosa stellata* COTTA 1832	
	Medullosa leuckartii GOEPPERT & STENZEL 1881	
	Medullosa porosa COTTA 1832	
	Medullosa solmsii SCHENK 1889	
	Myeloxylon elegans (COTTA 1832) BRONGNIART 1828	
	Taeniopteris jejunata GR. EURY 1877	
	Taeniopteris abnormis GUTBIER 1835	
	Dicksonites pluckenetii (SCHLOTHEIM 1820) STERZEL 1881	
	Sphenopteris sp.	
	Callistophyton sp., diarch roots with secondary growth	
	Neurocallipteris planchardii (ZEILLER) CLEAL, SHUTE & ZODROW 1995	
	Neuropteris sp.*Alethopteris schneideri* (STERZEL 1881) BARTHEL 1976	
	Cyclopteris sp.	
	Neuropteris cordato-ovata WEIß (1869) KRINGS & KERP 2000	
	Autunia naumannii (GUTBIER 1849) KERP 1988	
	Peltaspermum sp.	
Gymnosperms Incertae sedis	*Noeggerathia zamitoides* STERZEL 1918	*Dadoxylon* sp. (3 different types)
	Rhabdocarpus disciformis STERNBERG var. *laevis* WEISS	Diarch gymnospermous roots with secondary growth
Conifers	*Walchia piniformis* STERNBERG 1825	
	Ernestiodendron filiciformis (STERNBERG 1825) FLORIN 1934	
	Gomphostrobus bifidus (GEINITZ 1873) POTONIÉ 1891	
	Walchiopremnon sp.	
	Tylodendron saxonicum WEISS	
	Dadoxylon saxonicum (REICHENBACH 1836) GOEPPERT emend. FRENTZEN 1931	
	Dadoxylon sp.	
Cordaitaleans	*Cordaixylon brandlingii* (LINDLEY & HUTTON 1831) FELIX 1882	
	Cordaixylon sp.	
	Artisia sp.	
	Cordaicladus sp.	
	Broad-leaved cordaits of the group: *Cordaites principalis* - *C. borassifolius* - *C. palmaeformis*	
	Cardiocarpus reniformis GEINITZ 1858	
Algae and others	Algae (Charophyte gyrogonites)	? Stromatolite structures encrusting (?) plant axes
Animal remains	*Diadectes* sp. =*Phanerosaurus naumanni* VON MEYER 1859-61 cf. *Onchiodon* sp.	*Prionosuchus plummeri* PRICE 1948
	Isolated branchiosaur remains	*Brazilichthys macrognathus* COX & HUTCHINSON 1991
	Different types of coprolites (e.g. probable of oribatid mites)	Fragments of ctenacanth and xenacanth sharks, edestid holocephalians, and dipnoans, fish scales
	Ostracods	Coprolites
	Gastropods	Bivalves
	Infaunal ichnia of the *Scoyenia*-type	Ostracods
	Borings of various shape and size in petrified wood	Borings

attributed to lithofacies C. It consists of horizontally bedded sandstone sheets (Figs 5a, f), partially associated with silica nodules of irregular size/shape and silicified shell layers (Fig. 5g). Both the fossil content and the whole suite of sedimentary structures suggest a dynamic lacustrine environment, which may have been temporarily flooded and partially desiccated, but was largely integrated into an extensive drainage system. If the silica nodules are interpreted as silcretes, they suggest a locally to regionally warm humid setting in areas of very mature soil development (Milnes & Thiry 1992), or point to evaporation of water from the silica solution during warm-arid times (Walther 1993). Silcretes are normally formed at or below the surface in river valleys. In such cases, silica is mobilized in solution and re-precipitated locally. Precipitation may sometimes occur where upward-moving, silica-rich solutions meet downward-percolating water rich in salts, particularly sodium salts (Smale 1973), and, as such, are likely to be found in association with lake sediments.

The existence of ancient lakes and ponds as part of a large drainage network is confirmed by, for example, horizons of densely stacked freshwater bivalves (Fig. 5g), which are sometimes recognized in life position and with their conches preserved, and by other aquatic organisms, such as fishes or amphibians (Cox & Hutchinson 1991). Sometimes cylindrically structured bodies of 5–15 cm diameter have been found, which show undulatory concentric zonation. These are interpreted as possible stromatolites (Fig. 5h), although the type of core (plant structures?) remains open to interpretation. The rock matrix of the weakly horizontally bedded sandstone bodies consists mainly of highly siliceous, mature and well-sorted quartz arenite (Fig. 5c). Silicification is not confined to the enclosed plant fragments (but may have started there). Parts of the surrounding porous sediment are also silicified. The pattern of silicification is irregular to patchy, and has caused platy aggregations of fossil remains in a sandy matrix. The quartz grains are cemented by quartz cement, which has grown around original grains to make a very compact quartz mosaic. In these cases, isolated plant fragments, sometimes oriented – such as last order pinnae, small shoots and woody chips – exhibit an outstanding type of three-dimensional preservation (Figs 5b–f). This material offers an interesting source for future research. Sometimes, among these silicified sandstones, transitions to pure cherts have been recognized (Rössler & Noll 2002).

Fossil record and comparisons between both petrified forests

The fossil-bearing Permian sequence of Chemnitz enables the study of late Palaeozoic floral elements because of the following advantages:

(1) the sites of plant growth and burial are largely identical;
(2) the fossil-bearing horizon can be attributed to a single volcanic event (Fischer 1990);
(3) modern day volcanic events are suitable analogues for comparison to the processes that occurred at Chemnitz (Hoblitt *et al.* 1981; Wiatt 1981) and help to interpret the taphonomic conditions;
(4) Chemnitz is the type locality of several widely distributed Permian fossil plant genera, and finds from here have been researched for more than 260 years;
(5) one of the largest and scientifically most valuable collections of Permian petrified plants is permanently available for comparison and future research.

The deposits from the northern Tocantins region also provide many exceptional large specimens. As at Chemnitz, this site is preserved for continued scientific study. Furthermore, the excellent preservation of the Tocantins fossils provides a basis for future whole plant investigation. This is necessary for serious examination of the variability of several characters, such as the internal organization of the trunks as a basis for more reliable taxonomy.

Dominance patterns, growth habits and ecologies of ferns

The first impression is that there is a striking similarity between the Chemnitz and Tocantins localities: the floras are characterized by the abundance of marattialean tree ferns that seem to have become the dominant trees of tropical lowland forests in the basal Permian, continuing the dominance pattern of tree ferns known from the Late Carboniferous (Pfefferkorn & Thomson 1982; DiMichele & Phillips 1994). Also noteworthy is the diversity of zygopterid and filicalean ferns that are represented by several organ genera (e.g. *Grammatopteris, Ankyropteris, Zygopteris, Tubicaulis, Asterochlaena, Anachoropteris, Botryopteris* and *Dernbachia*), mostly representing small trees or scrambling, climbing and epiphytic forms. These ferns had evolved a remarkable array of growth architectures (DiMichele & Phillips 2002; Rössler & Galtier 2003), paralleling those seen, for example, in modern ferns

(Christ 1910; Jones 1987). Although all the aforementioned taxa appear to have colonized the wet lowlands, primarily in clastic-substrate floodplain environments, they may have also ventured into different ecological roles; they represent trees of different size and habit, scramblers, ground-rooted climbers, epiphytes and ground-covering plants.

Until a few years ago our knowledge of fossil plants from the Permian of Brazil was restricted to several species of psaroniaceous tree ferns, such as *Psaronius brasiliensis* (Derby 1913; Beurlen 1970; Petri & Fulfaro 1983; Herbst 1985). Solm-Laubach (1913) recognized a new taxon of marattialean tree ferns, which he named *Tietea singularis*. This fossil fern seems to be restricted to the Southern Hemisphere and dominates the fossiliferous exposures in Brazil by ratios of up to 90% (Fig. 3c). Recently, Herbst (1992, 1999) added to the list of species within these genera descriptions of *Tietea derbyi* and *Psaronius sinuosus,* both representing median to upper trunk portions. Unfortunately, the basal parts of these tree ferns, which bore dense mantles of adventitious roots, have rarely been collected to date. The material offered for sale at various places creates the misleading impression that the Gondwanan tree ferns must have been smaller and less anatomically complex than their relatives in Euramerica. Recent studies provide convincing evidence that this was not the case (Rössler & Noll 2002). There is now reason to believe that fern axes measuring as much as 1 m across at their base were by no means uncommon. As far as branching is concerned, there is an obvious dominance of the so-called 'four-rowed' type, which includes both *Tietea singularis* and *Psaronius brasiliensis*. The first discoveries of monomeristelic petiole remains (organ-genus *Stipitopteris* belonging to *Psaronius*) and different permineralized pinnae now complement the trunks of these tree ferns (Fig. 5b–d), although remains of the polymeristelic *Tietea* petioles are lacking. Contrary to the otherwise rich flora in general, petioles that belong to *Psaronius* are still lacking from the Chemnitz site, despite more than 260 years of collecting and study.

Closer examination shows that the Tocantins flora was by no means as uniform as long claimed in the literature. Several new tree ferns were recently discovered (Rössler & Galtier 2002*a*, *b*). These include *Grammatopteris freitasii* (Fig. 6c), evidence of a very rare genus, which had previously included only two species (*G. rigollotii* and *G. baldaufii*) and had only been known from specimens found in Autun/France and Chemnitz/Germany. It is referred to an *incertae sedis* group of filicalean ferns, because the reproductive organs are not known. The adaptation of the *Grammatopteris* plant to the tree-fern habit is comparable to that known in living tree-ferns and in the Carboniferous–Permian *Psaronius*. However, in *Psaronius*, the innermost roots were bounded by an interstitial tissue that did not exist in *Grammatopteris*. The large mantles of aerial adventitious roots in *Grammatopteris* taper and become smaller as they rise upwards, and they are entirely lacking in the upper portion of the trunk, where persistent petiole bases surround the trunk. The roots are relatively small and diarch in construction, which makes them comparable with those of several other fossil and recent fern families. New discoveries have made it possible to study different growth phases, ranging from small examples, measuring just 4 cm in diameter, to larger specimens whose basal transverse sections measure nearly 40 cm across.

The new *G. freitasii* became known in much greater detail than the previously described taxa, particularly with regard to its ontogenetic variability. The finds show a variety of ontogenetic stages and provide different sections from the base of these trunks upwards to their tops (Rössler & Galtier 2002*a*). *G. freitasii* specimens confirm the existence of a diagnostic feature represented by the genus *Grammatopteris* and considerably extend our knowledge of the geographic distribution of this type of fern. The trunks possess an exarch solid protostele surrounded by a dense mantle of petiole bases, below which the trunk is completely clothed in a root mantle (Fig. 6c). The stem cortex contains several types of sclerotic tissue. Leaf-trace vascular tissue develops from oval-shaped to tangentially elongated, typically bar-shaped strands. Some characters, such as the variability of the whole vascular architecture of the plant, the petiole cortex tissues and the departure of pinnae, have recently been studied for the first time (Rössler & Galtier 2002*a*). According to Miller (1971), ferns similar to *Grammatopteris* are thought to be of considerable importance or a link to the origin and early development of the Osmundaceae, the oldest-known extant fern family, although none of the currently known fossil fern species can definitely be considered the actual progenitor. Therefore, new finds that could improve our knowledge of the early evolution of extant ferns are of special interest. However, both sterile and fertile fronds of *G. freitasii* have yet to be found.

That the marginal root zone of tree fern trunks may have provided a frequently used substrate supporting the development of epiphytic or climbing plants is not only exhibited by a variety of extant plants, but is also confirmed for

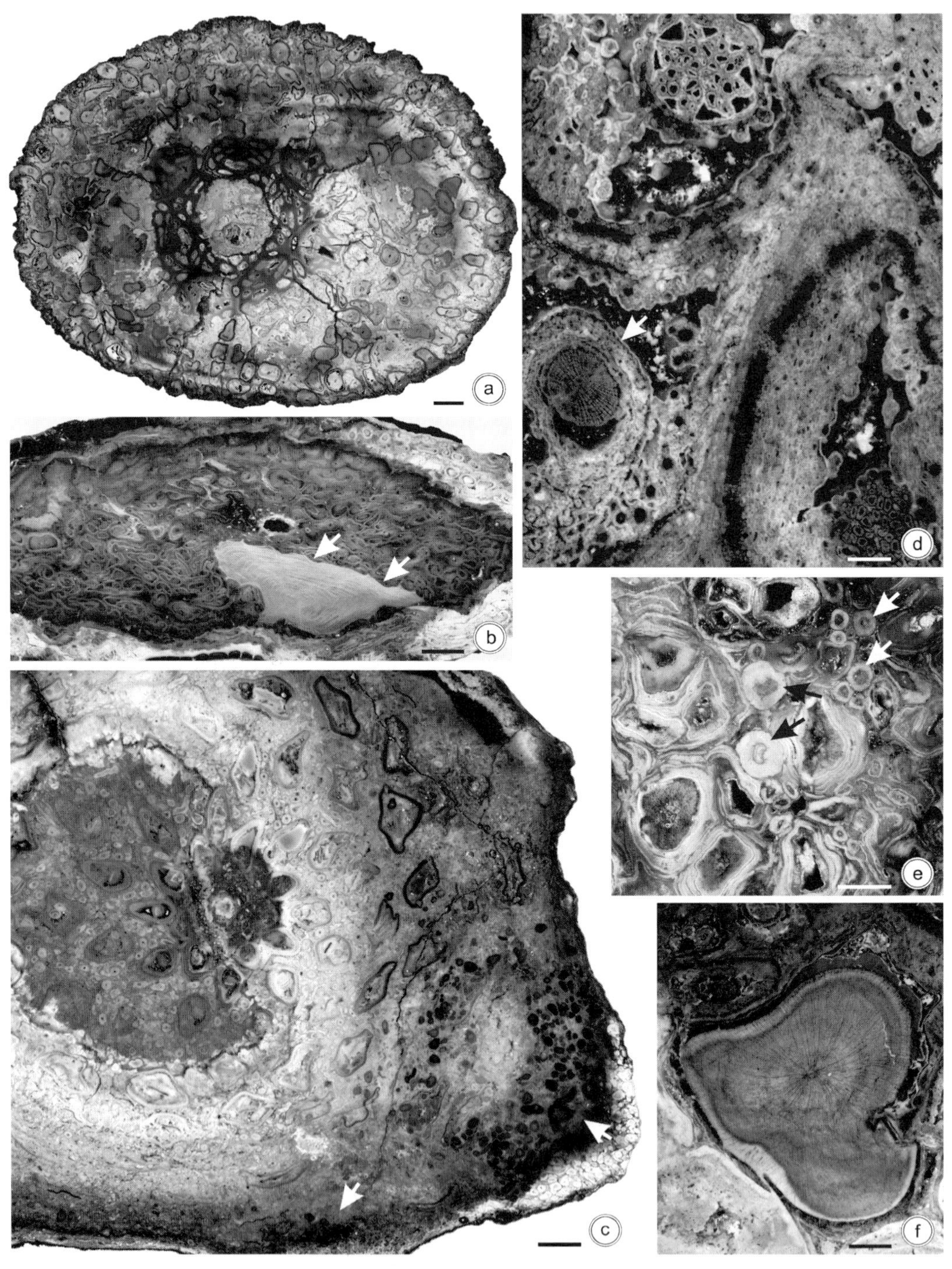
a
b
c
d
e
f

ancient equivalents by rich fossil evidence. Based on many examples from the Upper Carboniferous, Rothwell (1996) underlined the importance of tree-fern trunks that provided a unique environment for both lianas and epiphytes. Rössler (2000) reported different plants or plant organs that became trapped among free aerial roots of large basal *Psaronius* trunks from the Permian petrified forest of Chemnitz. Among them are climbing axes of *Ankyropteris brongniartii*, abundant small *Tubicaulis* cf. *berthierii* epiphytes (Fig. 7j), different isolated *Anachoropteris pulchra* foliar members, *Callistophyton* roots and two *Dadoxylon* specimens representing one *Amyelon* type root of *Cordaites,* and one conifer stem. The massive tree-fern root mantles protected the growth of different developing plants and may have improved the possibility of preservation of smaller plants and plant organs in particular. Bertrand (1909) was first drawn to epiphytic *Tubicaulis* specimens after studying the Chemnitz material. Later, Bertrand & Bertrand (1911) established *T. berthieri* for a single specimen from the Permian of Autun, France. Sahni (1931) reported epiphytic specimens of *Tubicaulis* from Chemnitz and confirmed the noteworthy frequency of these plants by mentioning one specimen with seven *Tubicaulis* axes. Although it seems obvious that the *Tubicaulis* axes were frequent epiphytes on *Psaronius* trunks, it still remains unclear which sterile or fertile frond type belongs to *Tubicaulis* ferns and what the entire plant looked like. The latter discoveries at Chemnitz stimulated similar observations on the base of large tree-fern root mantles at other localities, such as northern Tocantins.

Botryopteris nollii, which was the first-described botryopterid fern from the Southern Hemisphere (Rössler & Galtier 2003), also grew in close association with tree ferns. It has been found rooted within the marginal trunk base of both *Grammatopteris freitasii* and *Psaronius brasiliensis* ferns (Fig. 6c, e). However, variable growth form is one of the most conspicuous characters of *B. nollii.* This characteristic may underline the striking growth form diversity seen in the whole family Botryopteridaceae (Holmes 1989; Galtier & Phillips 1996), which may have contributed to a wide array of ecological adaptations. Surprisingly, additional specimens of the same species from Tocantins comprise false trunks composed of foliar members of different orders and associated adventitious roots (Rössler & Galtier 2003). This results from the formation of closely spaced, spirally arranged foliar members of different order on upright stems and of the repetitive development of shoots from foliar-borne buds inside the trunk. Multiseriate multicellular trichomes and adventitious roots complete the compound trunks and fill the empty space between cauline and foliar members.

In contrast to other species, such as *B. cratis,* the trichomes of *B. nollii* are not restricted to the shoot apex or primordial region. They represent a special feature of the adult stems and could have provided efficient protection against herbivory. This construction is similar to that previously described for *B. forensis* from the Upper Carboniferous coal balls (Phillips 1974; Rothwell 1991). Furthermore, *B. nollii* from Tocantins represents one of the largest and also geologically youngest botryopterids known to date. *Botryopteris nollii* is characterized by nearly circular protostelic stems and stipes/rachises that show a dorsiventral ω-shaped vasculature. Rössler & Galtier (2003) analysed the vegetative morphology, the three-dimensionally preserved architecture of the vascular system and specific histological details as observed in several series of sections. Based on its morphology, growth habit and habitat, it may be concluded that *B. nollii* depended on high humidity settings created by either the dense root mantles of tree fern trunks or the formation of a false trunk of its own. *Botryopteris* is one of the most common and completely known filicalean fern genera (Galtier & Phillips 1977), which ranges from the Lower Carboniferous to the Permian and was most abundant in Upper Carboniferous equatorial wetland forests. However, only a few species are relatively completely understood as whole plants, including cauline parts, foliar members, pinnules and reproductive organs. Since evidence of the persistance of *Botryopteris* into the Permian is rare, the finds from northern Tocantins State are of special interest.

Fig. 6. (**a**) Basal trunk portion of *Dernbachia brasiliensis* in transverse section. Note the formation of a root mantle. MfNC K5173; scale bar equals 10 mm. (**b**) Climbing gymnosperm (?pteridosperm) among the vascular bundles of a *Tietea singularis* trunk. MfNC K5256; scale bar equals 10 mm. (**c**) *Grammatopteris freitasii* trunk in transverse section showing *Botryopteris nollii* epiphytes (arrows) at the trunk periphery. MfNC K4969b; scale bar equals 10 mm. (**d**) Juvenile gymnosperm root (diarch woody stele and cortex) among the aerial roots of a *Tietea singularis* trunk. MfNC K5255b; scale bar equals 500 μm. (**e**) Three first-order foliar members (black arrows) and several second order foliar members (white arrows) of *Botryopteris nollii* among aerial roots of *Psaronius brasiliensis*. MfNC K5257; scale bar equals 5 mm. (**f**) Mature gymnosperm root among the aerial roots of a *Tietea singularis* trunk. MfNC K5255a; scale bar equals 2 mm.

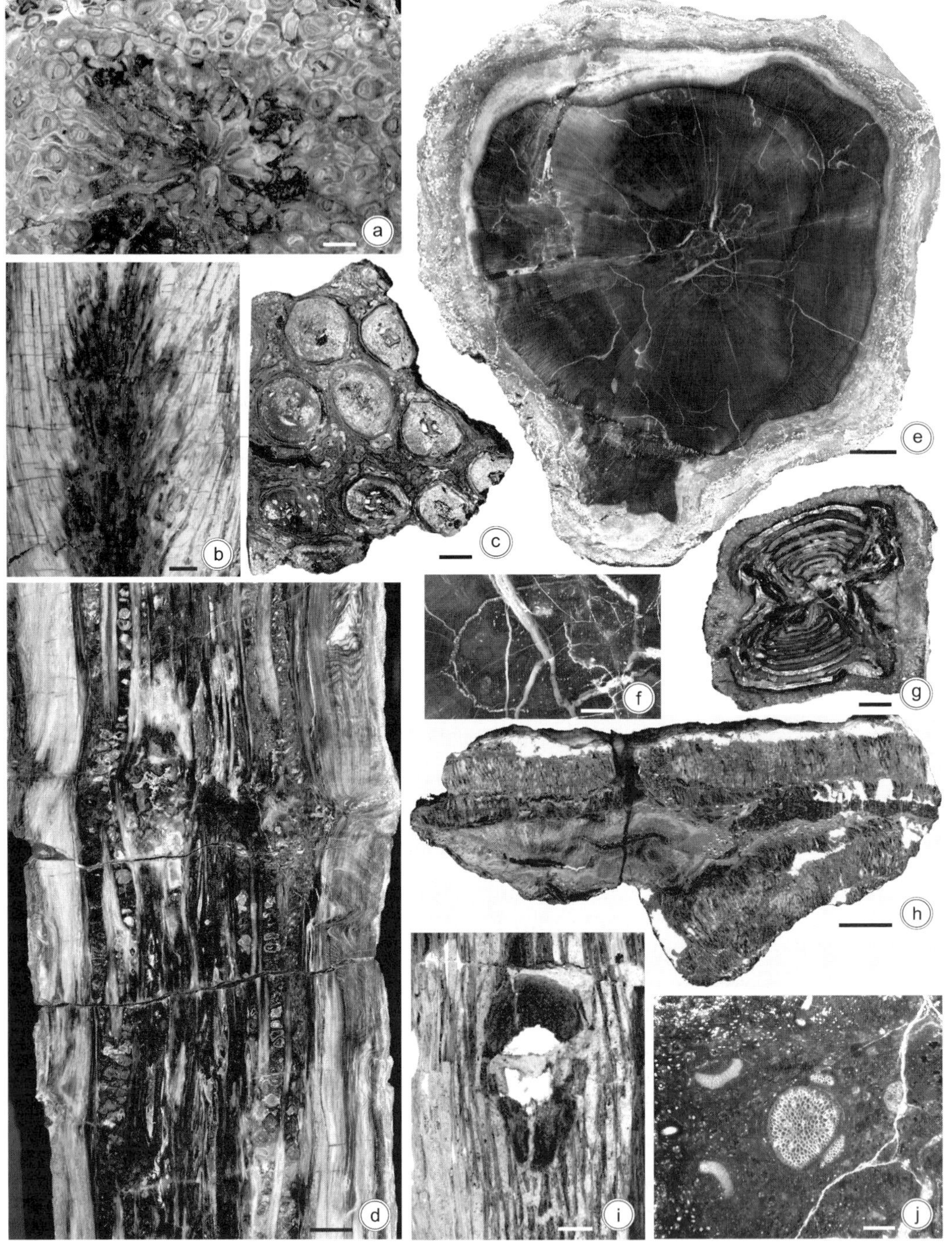
a
b
c
d
e
f
g
h
i
j

Another newly discovered fern is *Dernbachia brasiliensis* (Fig. 6a). Although detailed anatomical characters of ferns or fern-like plants are reported for a large number of late Palaeozoic permineralized taxa, and have been used in phylogenetic analyses, few taxa exhibit such a striking combination of stelar and foliar characters as this new genus (Rössler & Galtier 2002*b*). It shows the combination of a highly specialized stem, which has one of the largest actinosteles known in the plant record, and an evolutionarily advanced type of dorsiventral foliar anatomy, which we know in similar form in several extant ferns. The growth habit of *D. brasiliensis* can be reconstructed as a small-sized tree fern. The trunks are 60–170 mm in diameter and possess a large actinostele surrounded by a relatively narrow parenchymatous stem cortex extending into the petiole bases and traversed by adventitious aerial roots (Fig. 6a). Leaf traces rapidly develop from an oval xylem trace over a ω-shaped strand to a pair of nearly circular strands. Despite the large number of late Palaeozoic ferns now known, there are only a few forms that show an actinostelic stem.

Comparative investigations suggest a relatively close relationship between *Dernbachia* with *Asterochlaena* from the Permian petrified forest of Chemnitz (Fig. 7a, b). Both provide impressive examples of mosaic evolution and a combination of different types of steles and petioles among Early Permian ferns. The general phyllotaxy of *D. brasiliensis* and *A. laxa* is similar. The relatively large petioles of *Dernbachia* are arranged in pseudowhorls, leading to a variable number of orthostichies as well as to two opposite sets of parastichies. Like *Dernbachia, Asterochlaena* is very rare and appears to have a limited geographic occurrence, being restricted to the Permian of Chemnitz and one more locality near Nová Paka in the Křkonoše-Piedmont Basin, Czech Republic. Both species, *A. ramosa* and *A. laxa,* are known from Chemnitz. Like *Dernbachia, A. laxa* is a small tree-fern stem with a central actinostele, which is surrounded by a mantle of up to 50 spirally arranged, rather loose semi-erect persistent petioles and adventitious roots (Fig. 7a). Although *Asterochlaena* was described more than a century ago, its phylogenetic position, like that of *Dernbachia*, is still not convincingly established due to the lack of information on fertile organs. Nevertheless, details of stelar and foliar anatomy permit *Asterochlaena* to be assigned to the biseriate zygopterid ferns (Rössler & Galtier 2002*b*). Both taxa, *A. laxa* and *D. brasiliensis,* show a number of similar anatomical features: protoxylem tracheids form a narrow band in the middle of the cauline xylem arms; metaxylem tracheid walls show simple scalariform thickening/pitting; leaf traces separate from the stele as oval-shaped bipolar strands and develop into tangentially enlarged strands with slight adaxial curvature; and pinnae depart in two lateral rows, skewed slightly to the abaxial side of the petiole. The most important differences concern the stelar and petiole trace organization, which justifies the separation at the generic level.

Fig. 7. (**a**) *Asterochlaena laxa,* part of the stem showing the central actinostele and surrounding parenchymatous stem cortex traversed by outgoing leaf traces (slightly C-shaped clepsydroid vascular strand of adaxial curvature) and adventitious roots (mainly horizontal course). MfNC K5228; scale bar equals 5 mm. (**b**) *Asterochlaena laxa*, longitudinal section of the actinostelic stem showing attachment of surrounding petioles. Güldner collection; scale bar equals 5 mm. (**c**) *Zygopteris primaria* foliar members showing H-shaped vascular strands. MfNC K5230; scale bar equals 10 mm. (**d**) *Medullosa stellata*, longitudinal section of a stem showing internal vascular organisation and outgoing branch traces (arrow). MfNC K4019; scale bar equals 20 mm. (**e**) *Medullosa stellata* f. *lignosa*, stem with tuffitic cover in transverse section. Note the narrow pith (arrow) and the huge mantle of maxoxylic wood. MfNC K4496; scale bar equals 20 mm. (**f**) Detail of Fig. 8e showing small vascular bundles among the pith parenchyma. Scale bar equals 5 mm. (**g**) *Psaronius simplex*, stem in transverse section. MfNC K2514; scale bar equals 10 mm. (**h**) *Sigillaria* cf. *brardii* in transverse section showing the xylem core surrounded by cortex tissue. MfNC K5260a; scale bar equals 10 mm. (**i**) *Sigillaria* cf. *brardii*, outer surface of the trunk showing cortex tissue and one leaf trace. MfNC K5260c; scale bar equals 2 mm. (**j**) *Tubicaulis* cf. *berthierii*, small protostelic stem surrounded by C-shaped vascular traces. MfNC K231a; scale bar equals 2 mm.

Still enigmatic and puzzling: gymnosperms

Edaphic variation in wetland environments may have appeared through the distribution of subdominant floral elements, such as arborescent sphenopsids and different growth forms of gymnosperms. In addition to the overwhelming amount of psaroniaceous tree-ferns, *Arthropitys*-type calamite trees occur in association with pteridosperms and gymnosperms of uncertain affinity. Most of the gymnospermous stems from northern Tocantins are still under investigation. Initial results suggest overlapping combinations of character states, such as branching patterns, leaf-trace geometry, xylem/ray anatomy and

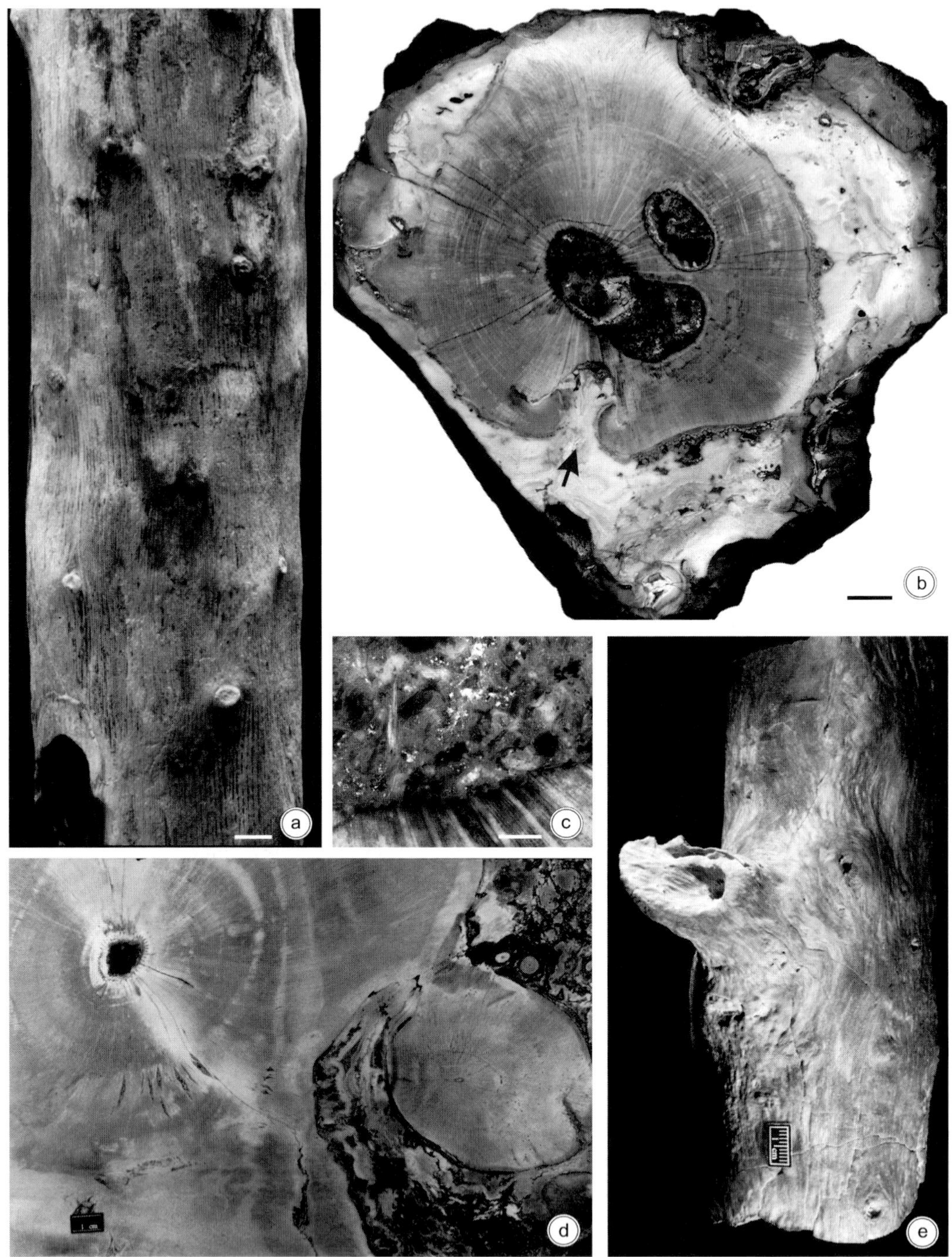
a
b
c
d
e
1 cm

existence of sclerotic nests or secretory ducts within the pith parenchyma. Such characteristics have been used to distinguish different gymnosperm groups in Euramerica, like cordaitaleans and conifers (Noll & Wilde 2002). This seems to give the new finds from Tocantins a somewhat exotic appearance, or sheds new light on the evolution of both morphological and anatomical characters among gymnosperms of the Northern and Southern Hemispheres.

In the light of the new discoveries from Chemnitz presented by Rössler (2000), which show *Callistophyton* roots growing on *Psaronius* trunks, the observation of Zeiller (1888, pl. VIII, fig. 1), regarding the doubtful connection of *Psaronius* stems and *Dicksonites sterzelii* fronds, may be regarded as solved: *Callistophyton*/*Dicksonites* was not in organic connection with *Psaronius*, but was a climber with a slender flexuous stem on *Psaronius* trunks, as it was previously drawn but not explained in detail by Laveine (1989). This interpretation does not contradict the proposed whole plant reconstruction provided by Galtier & Béthoux (2002), which shows the plant growing as a small scrambler in dense stands among sphenophytes, although their compression material did not show any roots. But, it seems that differences exist among *Dicksonites* specimens, especially regarding the frond architecture, which suggest the existence of several species. In addition to bipartite fronds described from the Late Carboniferous (see Galtier & Béthoux 2002), we also know of rather robust monopodial fronds from the Early Permian of Lauterecken, Saar–Nahe Basin, Germany (Rössler 2001, p. 154, fig. 386).

The Tocantins permineralized material shows unique diarch gymnosperm roots with a conspicuous amount of secondary growth, which have been found trapped within the marginal root mantle of tree ferns, such as *Grammatopteris freitasii* and *Tietea singularis* (Fig. 6d, f). These roots are to a certain degree comparable with, but unfortunately less well preserved than, the *Callistophyton* roots repeatedly identified within the large *Psaronius weberii* specimen from the Permian of Chemnitz (Rössler 2000). Some tree ferns additionally show climbing gymnosperms (possible pteridosperm axes) among their aerial roots (Noll *et al.* 2004) or between their stelar vascular bundles (Fig. 6b). As shown by Kerp & Krings (1998) and Krings *et al.* (2003), there are a number of potential climbing or scrambling seed ferns recently identified in late Palaeozoic wetland ecosystems, which they have reconstructed in more detail than has been attempted previously.

The Medullosales, an order of plants prominently featured in all palaeobotanical text books, is based on material first described from Chemnitz. The most impressive specimens are large permineralized trunks that were collected at the end of the nineteenth century, including specimens that show attached leaf bases of the *Myeloxylon* type. However, this group of seed ferns, which is so characteristic of the Chemnitz petrified forest (Weber & Sterzel 1896), remains unrecorded in Tocantins. Even in Chemnitz, medullosan seed ferns remain enigmatic. They include the largest seed ferns presently known and show a unique combination, optimization and arrangement of stelar types, combined with sophisticated growth forms and reproductive strategies. Medullosan seed ferns adapted well to gradual environmental changes during the late Palaeozoic until they became extinct at the end of the Permian. As new approaches demonstrate, they represent more than the classical, often-repeated, arboreal growth forms. In addition to the woody trees represented by the generotype *M. stellata*, and the *M. stellata* f. *lignosa*-type (Fig. 7e, 7f), there are frond-base supported taxa, such as *M. leuckartii* and most of the Carboniferous taxa, including non-self supporting, liana-like stems, which required some kind of mechanical support during growth (Dunn *et al.* 2003). Stems from the Carboniferous of England (*M. anglica*) and North America (*M. noei*), which are frequently offered as representatives of medullosan seed ferns and are included in biomechanical and phylogenetic analysis, are completely different in both their internal vascular organization and growth architecture. They have only a few things in common with the classical *M. stellata* trees with thick mantles of secondary xylem, on which this genus was originally based (Cotta 1832). Unfortunately, the type species has never been incorporated in any reconstruction. New preparations made on huge specimens from Chemnitz largely

Fig. 8. (**a**) Outer surface of a *Arthropitys*-type calamite showing branch traces, which alternate in successive nodes. MfNC K4552; scale bar equals 10 mm.
(**b**) *Arthropitys*-type calamite in transverse section showing borings and pathogenic reaction of the wood (arrow). MfNC K5259; scale bar equals 10 mm.
(**c**) Detail of Fig. 7b showing boring cavity filled with small woody fragments. Scale bar equals 2 mm.
(**d**) Huge basal portion of an *Arthropitys*-type calamite in transverse section with attached woody roots of different size. MfNC K5258; scale bar equals 10 mm.
(**e**) Terminal portion of an *Arthropitys*-type calamite. Note the branch traces and one permanent woody branch. MfNC K5266; scale bar equals 10 mm.

question the idea of a medium-sized tree carrying a crown of large compound leaves. Branch traces recognized on the surface of woody trunks (Fig. 7d) instead point to a growth form similar to arborescent coniferophytes.

Outstanding specimens of arborescent sphenopsids

The calamite remains found in northern Tocantins are still treated in open nomenclature (Rössler & Noll 2002), but new observations call into question the usual classical calamite reconstructions. Among late Palaeozoic plants, the calamite trees were highly successful and flourished in different tropical wetland environments. These unique, tubular, persisting trees are among the most interesting, but at the same time also the most puzzling, late Palaeozoic plants. The fossil record of late Palaeozoic arboreal sphenopsids, which extends over continents from Euramerica to Cathaysia and to Gondwana, is exceptionally rich and diverse in terms of preservation. The pith casts of hollow stems, and coalified compressions of the leafy shoots and strobili, are among the most common fossils found in Upper Carboniferous and Lower Permian strata. Moreover, they persisted for more than 60 Ma from the early Namurian (Gerrienne *et al.* 1999) to the Late Permian (Wang *et al.* 2003).

Calamite trees combine a number of developmental and morphological peculiarities, such as arborescence and aerial growth linked to an underground system of rhizomes, hollow stems combined with extensive wood development and a nodal branch architecture. In the literature, calamite trees are often uniformly reconstructed on the basis of their widely distributed pith cast preservations (Daviero & Lecoustre 2000). However, distinct characters necessary for reconstructing the growth architecture are often absent on pith casts. Recently at Tocantins, some large, almost complete stems were found (Figs 4d, 8a), which indicate different types of branching and a high variation of branching patterns, far more than previously thought. Fossil sphenopsids from Chemnitz confirm that few branches seem to have been permanent (Fig. 8e). The plants show considerable secondary xylary growth that, as expected, starts external to the stele, but without leaving any indication of its existence on the pith surface. Therefore, major branches that may have influenced plant biomechanics in particular are currently not illustrated in most reconstructions. Other branches, without secondary xylem and probably grown and abscised seasonally, are in fact seen on the pith cast, but they are not adequate for recognizing the growth architecture of the plants as a whole. Some kind of climatic cyclicity is suggested by growth rings found on several *Arthropitys* stems, both from Chemnitz and Tocantins.

From Chemnitz a huge, up to 60 cm diameter, calamite trunk base was recently reported (Fig. 4b), which represents the largest anatomically preserved calamite specimen known to date (Rössler & Noll 2006). The exceptional trunk attributed to *Arthropitys ezonata* was found nearly *in situ*, still standing upright and embedded in coarse-grained pyroclastics. This species is characterized by its rather homogeneous, loose wood with no clearly distinct interfascicular rays and fascicular wedges, but with high proportions (approximately 50%) of parenchyma in the secondary xylem. It has been suggested recently (Rössler & Noll 2006) that some, if not all, of the Permian calamite trees may have survived short seasonal episodes of dryness, during which they reduced water uptake and shed their herbaceous branches. This inference is supported by the large-diameter, long-living trunks. Although maximum stem diameters in calamite trees of about 1 m are frequently discussed in the literature, the largest published stem actually observed with its permineralized tissues until now measured only about 30 cm (Andrews & Agashe 1965).

Recently, an *Arthropitys*-type calamite up to 40 cm in diameter was found from Tocantins that exhibits two primary bodies (i.e. a possible branching basal stem) and woody adventitious roots of variable size measuring from 1 to 8 cm across (Fig. 8d). This specimen points to calamite trees that may have become more and more independent from the underground rhizome during their ontogeny, a growth form that has already been suggested for some Permian calamite species (Barthel 2004). Future research will investigate whether single calamite trees could have lived without their rhizome, at least from a later growth stage onwards, when the plants were supported by stem secondary xylem and adventitious roots. The woody roots of different size may have anchored the upright stems in the sediment. The biomechanical properties of the axes, including the enormous weight and diverse types of side-branches, need to be evaluated to see if such a growth form is feasible.

In addition to those interesting ecological aspects, all characters used for calamite systematics need to be re-evaluated with regard to their taxonomic significance. As stated several times by previous authors, many characters, such as frequently used measurement ratios or parenchyma and xylem anatomical features, show considerable variability, which seems to be due to both ontogenetic change and references to edaphical/ecological variation. Large-diameter woody trunks also point to an extended individual age of the Permian calamite plants. In addition, the high percentage of parenchyma may represent 'succulence' to a certain degree. Both characters could be interpreted as adaptations to ensure reproduction rate and survival in a more dynamic or disturbed environment, compared with the climax vegetation of the Carboniferous coastal tropics (DiMichele & Phillips 1994). With regard to future plant reconstructions, one should be careful to avoid over-interpreting the value of single preservation states, such as pith cast compressions. Only the consideration of all aspects of the primary and secondary body, branching and ontogenetic development will reveal the plant's growth habit as a whole and lead to more consistent systematics.

The second view: distinguished plant-plant and plant-animal interactions

Re-examination of several thin sections of the Chemnitz collection resulted in multiple instances of evidence for arthropod activity. This is demonstrated by small oval faecal pellets of two significant size orders. In addition to the major vegetation elements, *Psaronius* tree ferns and related epiphytes/vines/scramblers are affected in particular by herbivory. I would like to suggest that arthropods not only used rotted plant parts as a habitat and for coprolite storage, but also used erect, still-living plants. This is suggested by taphonomic characters, such as plant orientation or delicate preservation of sensitive tissues. The general preference for soft tissues, such as the thin-walled cortical parenchyma of the *Psaronius* root mantles or phloem of climbing and epiphytic ferns or pteridosperms by the herbivores, was confirmed by Rössler (2000). Labandeira *et al.* (1997) reviewed the important role of oribatid mite detritivory affecting all major plant taxa, including lycopsids, sphenopsids, ferns, pteridosperms and cordaites during a narrow 17 Ma interval from the Euramerican tropic coal-swamp forests. Recent investigations could extend this evidence into more restricted Permian-age environments as well as to other plants. From Tocantins, a number of plant–plant and plant–animal interactions have been reported (Noll *et al.* 2004), such as diarch fern roots growing inside a scrambling gymnosperm (the latter growing close to the stele of *Psaronius*), a *Psaronius* root growing inside the pith of a gymnosperm, undetermined gymnosperm roots growing among adventitious aerial roots of *Grammatopteris* or at the periphery of *Tietea singularis* trunks (Fig. 6d, f), and *Sphenophyllum* axes growing inside the borings of a calamite or within the marginal root mantle of *Grammatopteris*. Both the penetration and displacement of soft tissues and the use of pre-existing cavities or borings within dense wood were recognized. This is illustrated by one calamite specimen that shows both borings later filled with plant debris (Fig. 8c) and pathogenic reaction of the outer secondary body (Fig. 8b). Some climbing/scrambling gymnosperms have been observed among the vascular tissue of marattialean tree fern stems (Fig. 6b; Noll *et al.* 2004, p. 34, figs 8–9).

Conclusions

The Permian is recognized as an unprecedented period of revolutionary geological and biological change on Earth. Pangaea was formed by collisional processes that greatly facilitated the widespread migration of land-based organisms. Intensive volcanism both shaped the Earth's surface and allowed the detailed preservation of fossil remains, which has enabled the recognition and reconstruction of ancient biotopes and food chains. Gymnosperms continued their successful development and began to dominate as the first mammal-like reptiles appeared. A wide range of coeval, edaphic heterogeneous environments and habitats developed, such as wet, fern-rich forests, extensive clastic-substrate flood plains, conifer-dominated forests on elevated slopes, seasonally dry basins, dunes and salty deserts (Ziegler 1990; Rössler 2003).

An outstanding example of the Permian fossil record is offered by so-called petrified forests, which are unique documents from the geological past with great potential for palaeobotanical research. Not least, they form a splendid cultural heritage worth saving for future generations. Both the Chemnitz and Tocantins localities characterized in this contribution provide outstanding three-dimensionally preserved fossils, which have been collected, scientifically studied and exhibited for many years. At both localities the ferns have an extensive fossil record, including excellent examples of both anatomy and gross morphology, although nearly all fossils are fragmentary remains, preserving only a portion of the plant body, often with no clear connection between reproductive and vegetative organs.

Both localities represent tree-fern-dominated, warm-humid, seasonally influenced wetland forests of the Lower Permian, although the taphonomic framework responsible for the preservation of floral elements is quite different. As demonstrated at Chemnitz, volcanic processes can capture and preserve an entire community. They provide us with exceptional conditions that permit study of preservation processes. Even though the preservational types they create are outstanding, the distribution of fossil plants is highly patchy, and instructive organ connections remain rather rare. At Chemnitz, most permineralized trunks have been found in a single layer. In the northern Tocantins region, modern erosion exposes an abundance of fossil trunks, which are preserved in several horizons of a multi-storey alluvial sequence. In both cases rapid cover and subsequent silicification have prevented the decay of the plants and have led to well-preserved fossils that permit comprehensive investigations of branching morphology, internal anatomy, taphonomy and palaeoecology.

One of the most striking characters at both localities is the presence of exceptional, large remains. Besides Chemnitz and Tocantins, no other Permian location has provided such a spectrum of almost complete trunks. As a consequence, the knowledge emerging from on-going research may improve the view of both ontogenetic and ecological variability of major Permian plants. Ecological variability, in particular, seems to be one of the most under appreciated facets of Permian palaeobotany. As several authors have stated, the large number of species, especially in *Psaronius*, represent growth stages rather than true species. This remains to be investigated among other plant groups.

Most of the classical and published specimens from Chemnitz have been distributed to collections around the world for over 200 years, thus complicating comprehensive reviews of several taxa. This may be one of the reasons that revisions of the genera *Psaronius* or *Medullosa* are still lacking. Otherwise, even today, new finds are adding to the picture of the Chemnitz ecosystem. Most recently, the first lycopsid, referred to as *Sigillaria* cf. *brardii* (Fig. 7h, 7i), was found at a temporary excavation in Chemnitz. Similarly, modern finds from Tocantins offer the opportunity to characterize fossil taxa more fully and to establish or support whole plant concepts more than was previously possible with specimens from other places.

One of the most striking features of both localities is the abundance of marattialean ferns. Whereas *Psaronius* is present in both the Northern and Southern Hemispheres, *Tietea* and *Tuvichapteris* remained restricted to Gondwana. The geographical origin of marattialean ferns appears to be tropical and Euramerican, based on their appearance during the latest Early Carboniferous in Europe (Gerrienne *et al.* 1999). Among *Psaronius* species, the distichous type (Fig. 7g) is normally regarded as primordial and evolutionary rather primitive. Although present in Chemnitz, this type is lacking in Tocantins. During the Early Permian, psaroniaceous ferns may have started rapid and massive distribution to the south, which culminated in the occurrence of marattialean ferns in the Triassic of Antarctica (Delevoryas *et al.* 1992).

Given that there is some degree of floristic similarity between Chemnitz and Tocantins, I suggest that, during the Early Permian, there appeared a broad phytogeographic connection, and plants could extend their ranges (at least in one direction) without strong barriers due to the geographic situation made by the formation of Pangaea (Ziegler 1990). Similar climatic conditions in mid-Europe and northern Tocantins may be explained by the comparable distances of both regions from the palaeo-equator, which enabled plant growth within similar megafacies belts and, therefore, in similar palaeo-latitudinal settings. The lack of many typical Gondwanan forms at Tocantins, and the presence of widely distributed Lower Permian floral elements in Tocantins co-existing with those typically found in Europe, is not unexpected since there is growing doubt about the conciseness of the classical floral realms. The new finds of *Grammatopteris, Botryopteris* and *Dernbachia* in northern Tocantins considerably extended our knowledge about the scarce record of anatomically preserved Permian fern groups of the Southern Hemisphere, which were represented by some members of the Filicales, such as *Skaaripteris,* or Osmundales, like *Palaeosmunda* and *Guairea* and, finally the common Psaroniales, including *Psaronius, Tietea* and *Tuvichapteris.* The discoveries of exceptional specimens of *Arthropitys*-type calamite trunks at the Chemnitz and Tocantins localities shed new light on arborescent sphenopsids and allow evaluation of the classical whole-plant reconstructions.

The comprehensive collection in the Chemnitz Museum of Natural History could serve as a source of unique material illustrating the Permian period in particular (Rössler 2002). Although new discoveries can be compared with this, many questions remain and new finds or ideas are needed.

I am much obliged to R. Noll (Tiefenthal), whose useful discussion, various help and efforts with the preparation of many specimens cannot be overestimated. Many thanks are due to M. Barthel (Berlin), and J. Galtier

(Montpellier), for their stimulating and helpful comments on the subject. Both for the suggestion to write this paper and for useful comments to the manuscript, I thank J. Schneider (Freiberg). D. Dias-Brito (Rio Claro) is acknowledged for providing additional literature and useful comments on the manuscript. J. Dunlop (Berlin), kindly improved the English. E. Potievsky (Chemnitz), E. Müller (Cunnersdorf), J. Urban (Hainichen), and H. Bieser, (Carlsberg) are acknowledged for technical assistance. S. G. Lucas, L. Tanner and W. A. DiMichele considerably improved the paper as reviewers.

References

ANDREWS, H. N. & AGASHE, S. N. 1965. Some exceptionally large calamite stems. *Phytomorphology*, **15**, 103–108.

ARAÚJO, V. A. 2001. Programa de Levantamentos Básicos do Brasil. PLGB. Araguaína. Folha SB 22-2D. *Estados do Tocantins e do Pará. Escala 1:250.000.* CPRM. DIEDIG, Departamento Brasilia.

BARBOSA, O. & GOMES. F. A. 1957. Carvão Mineral na Bacia Tocantins-Araguaia. *Divisio de Geologia & Mineralogia, Bio de Janeiro, Boletin*, **174**.

BARTHEL, M. 1976. Die Rotliegendflora Sachsens. *Abhandlungen des Staatlichen Museums für Mineralogie und Geologie zu Dresden*, **24**, 1–190.

BARTHEL, M. 2004. Die Rotliegendflora des Thüringer Waldes. Teil 2: Calamiten und Lepidophyten. *Naturhistorisches Museum Schleusinges, Veröffentlichungen Schloss Bertholdsburg*, **19**, 19–48.

BERNER, R. A. & KOTHAVALA, Z. 2001. GEOCARB III: a revised model for atmospheric CO_2 over Phanerozoic time. *American Journal of Science*, **301**, 182–204.

BERTRAND, P. 1909. Etudes sur la fronde Zygoptéridées. PhD thesis, University of Lille.

BERTRAND, C. & BERTRAND, P. 1911. Le *Tubicaulis berthieri* nov. sp. *Bulletin de la Société d'Histoire Naturelle, Autun*, **24**, 43–92.

BEURLEN, K. 1970. *Geologie von Brasilien. Beiträge zur Regionalen Geologie der Erde.* Gebrüder Borntraeger, Berlin and Stuttgart, **9**.

BRONGNIART, A. 1872. Notice sur le *Psaronius brasiliensis*. *Bulletin de la Societé Botanique de France, Series 5*, **19**, 3–10.

CHRIST, H. 1910. *Die Geographie der Farne*. Gustav Fischer, Jena.

COTTA, B. 1832. *Die Dendrolithen in Bezug auf ihren inneren Bau.* Arnoldische Buchhandlung, Leipzig and Dresden.

COX, C. B. & HUTCHINSON, P. 1991. Fishes and amphibians from the Late Permian Pedra de Fogo Formation of northern Brazil. *Palaeontology*, **34**, 561–573.

DAVIERO, V. & LECOUSTRE, R. 2000. Computer simulation of sphenopsid architecture. Part II. *Calamites multiramis* Weiss, as an example of Late Palaeozoic arborescent sphenopsids. *Review of Palaeobotany and Palynology*, **109**, 135–148.

DELEVORYAS, T., TAYLOR, T. N. & TAYLOR, E. L. 1992. A marattialean fern from the Triassic of Antarctica. *Review of Palaeobotany and Palynology*, **74**, 101–107.

DERBY, O. A. 1913. Observations on the stem structure of *Psaronius brasiliensis*. *American Journal of Science*, **36**, 489–497.

DERBY, O. A. 1915. Illustrations of the stem structure of *Tietea singularis*. *American Journal of Science*, **39**, 251–260.

DERNBACH, U. (ed.) 1996. *Petrified Forests.* D'oro-Verlag, Heppenheim.

DIAS-BRITO, D. & CASTRO, J. C. 2005. *Caracterização Geológica a Paleontológica do Monumento Natural das Árvores Fossilizadas do Tocantins*. Universidade Estadual Paulista, UNESP, Rio Claro, Relatorio Final.

DIMICHELE, W. A. & PHILLIPS, T. L. 1994. Paleobotanical and paleoecological constraints on models of peat formation in the Late Carboniferous of Euramerica. *Palaeogeography, Palaeoclimatology, Palaeoecology*, **106**, 39–90.

DIMICHELE, W. A. & PHILLIPS, T. L. 2002. The ecology of Paleozoic ferns. *Review of Palaeobotany and Palynology*, **119**, 143–159.

DINO, R., ANTONIOLI, L. & BRAZ, S-M. N. 2002. Palynological data from the Trisidela Member of upper Pedra de Fogo Formation ('Upper Permian') of the Parnaíba Basin, northeastern Brazil. *Revista Brasileira de Paleontologia*, **3**, 24–35.

DÖRING, H., FISCHER, F. & RÖSSLER, R. 1999. Sporostratigraphische Korrelation des Rotliegend im Erzgebirge-Becken mit dem Permprofil des Donezk-Beckens. *Veröffentlichungen des Museums für Naturkunde Chemnitz*, **22**, 29–56.

DUNN, M. T., KRINGS, M., MAPES, G., ROTHWELL, G. W., MAPES, R. H. & SUN, K. 2003. *Medullosa steinii* sp. nov., a seed fern vine from the Upper Mississippian. *Review of Palaeobotany and Palynology*, **124**, 307–324.

EULENBERGER, S., TUNGER, B. & FISCHER, F. 1995. Neue Erkenntnisse zur Geologie des Zeisigwaldes bei Chemnitz. *Veröffentlichungen des Museums für Naturkunde Chemnitz*, **18**, 25–34.

FALCON-LANG, H. J. & BASHFORTH, A. R. 2004. Pennsylvanian uplands were forested by giant cordaitalean trees. *Geology*, **32**, 5, 417–420.

FISCHER, F. 1990. Lithologie und Genese des Zeisigwald-Tuffs (Rotliegendes, Vorerzgebirgs-Senke). *Veröffentlichungen des Museums für Naturkunde Chemnitz*, **14**, 61–74.

GALTIER, J. & BÉTHOUX, O. 2002. Morphology and growth habit of *Dicksonites pluckenetii* from the Upper Carboniferous of Graissessac (France). *Géobios*, **35**, 525–535.

GALTIER, J. & PHILLIPS, T. L. 1977. Morphology and evolution of *Botryopteris* – a Carboniferous age fern. 2. Observations on Stephanian species from Grand' Croix, France. *Palaeontographica, B*, **164**, 1–32.

GALTIER, J. & PHILLIPS, T. L. 1996. Structure and evolutionary significance of Palaeozoic ferns. *In*: CAMUS, J. M., GIBBY, M. & JOHNS, R. J. (eds) *Pteridology in Perspective*. Royal Botanic Gardens, Kew, 417–433.

GERRIENNE, P., FAIRON-DEMARET, M. & GALTIER, J. 1999. A Namurian A (Silesian) permineralized flora from the Carrière du Lion at Enighoul (Belgium). *Review of Palaeobotany and Palynology*, **107**, 1–15.

HERBST, R. 1985. Nueva descripcion de *Psaronius arrojadoi* (Pelourde) (Marattiales) del Pérmico de Brasil. *Ameghiniana*, **21**, 243–58.

HERBST, R. 1992. Studies on Psaroniaceae. III. *Tietea derbyi* n. sp., from the Permian of Brazil. *In:* SCHAARSCHMIDT, F. (ed.) *International Symposium on Palaeobotany: Anatomical Investigations of Plant Fossils*. Courier Forschungsinstitut Senckenberg, **147**, 155–161.

HERBST, R. 1999. Studies on Psaroniaceae. IV. Two species of *Psaronius* from Araguaina, State of Tocantins, Brazil. *Facena*, **15**, 9–17.

HOBLITT, R. P., MILLER, C. D. & VALLANCE, J. W. 1981. Origin and stratigraphy of the deposit produced by the May 18 directed blast. *In*: LIPMAN, P. W. & MULLINEAUX, D. R. (eds) *The 1980 Eruptions of Mount St Helens, Washington.* U.S. Geological Survey Professional Paper, **1250**, 401–419.

HOLMES, J. 1989. Anomalous branching patterns in some fossil Filicales: implications in the evolution of the megaphyll and the lateral branch, habit and growth pattern. *Plant Systematics and Evolution*, **165**, 137–158.

JONES, D. L. 1987. *Encyclopaedia of Ferns*. Timber Press, Portland.

KERP, H. 1996. Post-Variscian Late Palaeozoic Northern Hemisphere gymnosperms: the onset to the Mesozoic. *Review of Palaeobotany and Palynology*, **90**, 263–285.

KERP, H. 2000. The modernization of landscapes during the Late Paleozoic–Early Mesozoic. *In*: GASTALDO, R. A. & DIMICHELE, W. A. (eds) *Phanerozoic Terrestrial Ecosystems*. Paleontological Society Papers, **2000**(6), 79–113.

KERP, H. & KRINGS, M. 1998. Climbing and scrambling growth habits: common life strategies among Late Carboniferous seed ferns. *Compte Rendu de l'Académie des Sciences, Paris, Série IIa*, **326**, 583–588.

KRINGS, M., KERP, H., TAYLOR, T. N. & TAYLOR, E. L. 2003. How Paleozoic vines and lianas got off the ground: on scrambling and climbing Carboniferous-Early Permian pteridosperms. *Botanical Review*, **69**, 204–224.

LABANDEIRA, C. C., PHILLIPS, T. L. & NORTON, R. A. 1997. Oribatid mites and the decomposition of plant tissues in Paleozoic coal-swamp forests. *Palaios*, **12**, 319–353.

LAVEINE, J.-P. 1989. *Guide paléobotanique dans le terrain houiller Sarro-Lorrain. Documents des Houillères du Bassin de Lorraine.* Merlebach, France.

LIMA, E. A. M. & LEITE, J. F. 1978. Projeto Estudo Global dos Recursos Minerais da Bacia Sedimentar do Parnaíba – Integração Geológica-Metalogenética. *Relatório Técnico DNPM, CPRM, Recife*, **1**, 124–132.

MILLER, C. N., 1971. Evolution of the fern family Osmundaceae based on anatomical studies. *Contributions from the Museum of Paleontology, University of Michigan*, **28**, 105–169.

MILNES, A. R. & THIRY, M. 1992. Silcretes. *In*: MARTINI, I. P. & CHESWORTH, W. (eds) *Weathering, Soils and Paleosols*. Elsevier, Amsterdam, 349–377.

NASDALA, L., GÖTZE, J., PIDGEON, R. T., KEMPE, U. & SEIFERT, T. 1998. Constraining a SHRIMP U-Pb age: micro-scale characterization of zircons from Saxonian Rotliegend rhyolites. *Contributions to Mineralogy and Petrology*, **132**, 300-306.

NOLL, R. & WILDE, V. 2002. Conifers from the 'uplands': petrified wood from central Germany. *In*: DERNBACH, U. & TIDWELL, W. D. I. (eds) *Secrets of Petrified Plants: Fascination from Millions of Years*. D'Oro-Verlag, Heppenheim, 88–103.

NOLL, R., RÖSSLER, R. & ROJKO, R. 2004. Neue permische Pflanzen und deren ungewöhnliche Wuchsorte. *Veröffentlichungen des Museums für Naturkunde Chemnitz*, **27**, 29–38.

PELOURDE, F. 1912. Observations sur le *Psaronius brasiliensis*. *Annales des Sciences Naturelles, Botanique, Serie 9*, **16**, 337–352.

PETRI, S. & FULFARO, V. J. 1983. *Geologia do Brasil (Fanerozóico)*. University of Sao Paulo, Ciencias Naturales, **9**.

PFEFFERKORN, H. W. & THOMSON, M. 1982. Changes in dominance patterns in Upper Carboniferous plant-fossil assemblages. *Geology*, **10**, 641–644.

PHILLIPS, T. L. 1974. Evolution of vegetative morphology in coenopterid ferns. *Annals of the Missouri Botanical Garden*, **61**, 427–461.

PINTO, C. P. & SAD, J. H. G. 1986. Revisão da Estartigrafia da Formação Pedra de Fogo, borda sudoeste da Bacia do Parnaíba. *Anais do 34th. Congresso Brásileiro Geologia, Goiânia*, **1**, 346–358.

PLUMMER, F. B. 1948. Estados de Piauí e Maranhão (Geologia). *Relatoria Annual do Conselho Nacional de Petroleo, Rio de Janeiro*, **1946**, 87–134.

PRICE, L. J. 1948. Um anfibio labirintodonte da Formação Pedra de Fogo, Estrado do Maranhão. *Divisão de Geologia e Mineralogia, Rio de Janeiro, Boletim*, **124**, 1–32.

ROSCHER, M. & SCHNEIDER, J. W. 2005. An annotated correlation chart for continental Late Pennsylvanian and Permian basins and the marine scale. *In*: LUCAS, S. G. & ZEIGLER, K. E. (eds) *The Nonmarine Permian*. New Mexico Museum of Natural History and Science Bulletin, **30**, 282–291.

RÖSSLER, R. 2000. The late Palaeozoic tree fern *Psaronius:* an ecosystem unto itself. *Review of Palaeobotany and Palynology*, **108**, 55–74.

RÖSSLER, R. (ed.) 2001. *Der versteinerte Wald von Chemnitz. Katalog zur Ausstellung Sterzeleanum.* Museum für Naturkunde, Chemnitz.

RÖSSLER, R. 2002. Between precious inheritance and immediate experience – Palaeobotanical research from the Permian of Chemnitz, Gemany. *In*: DERNBACH, U. & TIDWELL, W. D. I. (eds) *Secrets of Petrified Plants: Fascination from Millions of Years*, D'Oro, Heppenheim, 104–119.

RÖSSLER, R. 2003. Das Perm: Farnwälder, Glutwolken und Salzwüsten. *Biologie in unserer Zeit*, **33**(4), 244–251.

RÖSSLER, R. & BARTHEL, M. 1998. Rotliegend taphocoenoses preservation favoured by rhyolitic explosive volcanism. *Freiberger Forschungshefte, Hefte C*, **474**, 6, 59–101.

RÖSSLER, R. & GALTIER J. 2002*a*. First *Grammatopteris* tree ferns from the Southern Hemisphere: new insights in the evolution of the Osmundaceae from the Permian of Brazil. *Review of Palaeobotany and Palynology*, **121**, 205–230.

RÖSSLER, R. & GALTIER J. 2002*b*. *Dernbachia brasiliensis* gen. nov. et sp. nov.: a new small tree fern from the Permian of NE-Brazil. *Review of Palaeobotany and Palynology*, **122**, 239–263.

RÖSSLER, R. & GALTIER, J. 2003. The first evidence of the fern *Botryopteris* from the Permian of the Southern Hemisphere reflecting growth form diversity. *Review of Palaeobotany and Palynology*, **127**, 99–124.

RÖSSLER, R. & NOLL, R. 2002. Der permische versteinerte Wald von Araguaina/Brasilien: Geologie, Taphonomie und Fossilführung. *Veröffentlichungen des Museums für Naturkunde Chemnitz*, **25**, 5–44.

RÖSSLER, R. & NOLL, R. 2006. Sphenopsids of the Permian. 1: The largest known anatomically preserved calamite, an exceptional find from the petrified forest of Chemnitz, Germany. *Review of Palaeobotany and Palynology*.

ROTHWELL, G. W. 1991. *Botryopteris forensis* (Botryopteridaceae), a trunk epiphyte of the tree fern *Psaronius*. *American Journal of Botany*, **78**, 782–788.

ROTHWELL, G. W. 1996. Pteridophytic evolution: an often underappreciated phytological success story. *Review of Palaeobotany and Palynology*, **90**, 209–222.

SAHNI, B. 1931. On certain fossil epiphytic ferns found on stems of the Palaeozoic tree-fern *Psaronius*. *Proceedings of the 18th Indian Sciences Congress, Nagpur*, 270.

SCHNEIDER, J. 1989. Basic problems of biogeography and biostratigraphy of the Upper Carboniferous and Rotliegendes. *Acta Musei Reginaehradecensis, Serie A, Scientiae Naturales*, **23**, 31–44.

SCHNEIDER, J. W. 1996. Xenacanth teeth: a key for taxonomy and biostratigraphy. *Modern Geology*, **20**, 321–340.

SCHNEIDER, J., RÖSSLER, R. & GAITZSCH, B. 1995. Timelines of Late Variscan Volcanism. *Zentralblatt für Geologie und Paläontologie, Teil I*, **1994**(5/6), 477–490.

SMALE, D. 1973. Silcretes and associated silica diagenesis in southern Africa and Australia. *Journal of Sedimentary Petrology*, **43**, 1077–1089.

SPRENGEL, A. 1828. *Commentatio de Psarolithis Ligni Fossilis Genere*. Halle.

SOLM-LAUBACH, H. 1913. *Tietea singularis*. Ein neuer Pteridinenstamm aus Brasilien. *Zeitschrift für Botanik, Jena*, **5**, 673–700.

STERZEL, J. T. 1875. Die fossilen Pflanzen des Rothliegenden von Chemnitz in der Geschichte der Palaeontologie. *Berichte der Naturwissenschaftlichen Gesellschaft zu Chemnitz*, **5**, 71–243.

STERZEL, J. T. 1918. Die organischen Reste des Kulms und des Rotliegenden der Gegend von Chemnitz. *Abhandlungen der Königlich Sächsischen Gesellschaft der Wissenschaften, Mathematisch-physikalische Klasse*, **35**(5), 205–315.

WANG, S.-J., LI, S.-S., HILTON, J. & GALTIER, J. 2003. A new species of the sphenopsid stem *Arthropitys* from Late Permian volcaniclastic sediments of China. *Review of Palaeobotany and Palynology*, **126**, 65–81.

WALTHER, H. B. 1993. *Silcretes in Germany and Australia*. PhD thesis, Freiberg University of Mining and Technology.

WEBER, O. & STERZEL, J. T. 1896. Beiträge zur Kenntnis der Medulloseae. *Berichte der Naturwissenschaftlichen Gesellschaft zu Chemnitz*, **13**, 44–143.

WERNEBURG, R. 1993. Ein Eryopide (Amphibia) aus dem Rotliegend (Unterperm) des Erzgebirge-Beckens (Sachsen). *Freiberger Forschungshefte Hefte C*, **450**(1), 151–160.

WERNEBURG, R. & SCHNEIDER, J. 1996. The Permian temnospondyl amphibians of India. *In*: MILNER, A. R. (ed.) *Studies on Carboniferous and Permian Vertebrates*. Special Papers in Palaeontology, **52**, 105–128.

WIATT, R. B. 1981. Devastating pyroclastic density flow and attendant air fall of May 18: stratigraphy and sedimentology of deposits. *In*: LIPMAN, P. W. & MULLINEAUX, D. R. (eds) *The 1980 eruptions of Mount St. Helens, Washington*. U.S. Geological Survey, Professional Paper, **1250**, 439–458.

WITKE, K., GÖTZE, J., RÖSSLER, R., DIETRICH, D. & MARX, G. 2004. Raman and cathodoluminescence spectroscopic investigations on Permian fossil wood from Chemnitz: a contribution to the study of the permineralisation process. *Spectrochimica Acta A*, **60**, 2903–2912.

ZEILLER, R. 1888. *Études sur le terrain houiller de Commentry. Flore Fossile.* Premiere Partie. Saint-Étienne.

ZIEGLER, A. M. 1990. Phytogeographic patterns and continental configurations during the Permian Period. *In*: MCKERROW, W. S. & SCOTESE, C. R. (eds) *Palaeozoic Palaeogeography and Biogeography. Geological Society, London, Memoir*, **12**, 363–379.

Global Permian tetrapod biostratigraphy and biochronology

SPENCER G. LUCAS

New Mexico Museum of Natural History and Science, 1801 Mountain Road N.W., Albuquerque, New Mexico 87104-1375, USA (e-mail: spencer.lucas@state.nm.us)

Abstract: The most extensive Permian tetrapod (amphibian and reptile) fossil records from the western United States (New Mexico–Texas) and South Africa provide the basis for definition of 10 land-vertebrate faunachrons that encompass Permian time. These are (in ascending order): the Coyotean, Seymouran, Mitchellcreekian, Redtankian, Littlecrotonian, Kapteinskraalian, Gamkan, Hoedemakeran, Steilkransian and Platbergian. These faunachrons provide a biochronological framework with which to determine and discuss the age relationships of Permian tetrapod faunas. Their correlation to the marine time scale and its numerical calibrations indicate that the Coyotean is a relatively long time interval of about 20 Ma, whereas most of the other faunachrons are much shorter, about 1–2 Ma long each. The Platbergian may also be relatively long, 14 Ma, although this is not certain. This suggests slow rates of terrestrial tetrapod faunal turnover during most of the Early Permian and late Middle to Late Permian, but more rapid rates of turnover during the latest Early and most of the Middle Permian, especially during the explosive initial diversification of therapsids.

Permian tetrapod (amphibian and reptile) fossils are widely distributed (Fig. 1) and have long provided a basis for non-marine biostratigraphy and biochronology (see reviews by Lucas 1998*a*, 2002, 2004). Here, I document a formal global Permian tetrapod biochronology that recognizes 10 time intervals (land-vertebrate faunachrons) (Lucas 2005*d*). This biochronology is based on the body-fossil record of tetrapods and provides a tetrapod-based time scale that can be used to determine and discuss the temporal relationships of Permian tetrapod assemblages. It can also be cross-correlated with reasonable precision to the standard global chronostratigraphical scale for the Permian, which is based on marine biostratigraphy. This correlation, which can be numerically calibrated, indicates a wide range of evolutionary turnover rates of tetrapods during the Permian.

Abbreviations and terminology

Biostratigraphy documents the distribution of fossils in strata, whereas biochronology is concerned with the temporal distribution of taxa. The following abbreviations are used in the text: FAD, first appearance datum (a biochronological event); HO, highest occurrence (a biostratigraphic datum); LAD, last appearance datum (a biochronological event); LVF, land-vertebrate faunachron; LO, lowest occurrence (a biostratigraphic datum); SGCS, standard global chronostratigraphic scale, which is the global time scale based on marine biostratigraphy (Wardlaw *et al.* 2004).

In this paper, pelycosaurian-grade (primitive basal) synapsids are simply referred to as pelycosaurs. Advanced basal synapsids are referred to as therapsids, and the term 'reptile' is used instead of 'amniote'.

Problems and procedures

Most of the problems with developing a Palaeozoic (Late Devonian, Carboniferous and Permian) tetrapod biostratigraphy and biochronology reduce to one problem: the rarity or lack of good Palaeozoic tetrapod index fossils. Good index fossils are easily identified, abundant and have a broad geographical (facies) range but a short stratigraphical (temporal) range. Few, if any, Devonian–Carboniferous tetrapod genera or species meet these criteria (Carroll 1979; Lucas 2000). Only in the Permian do some Palaeozoic tetrapod taxa qualify as good index fossils, and most of these are of Middle to Late Permian age. Indeed, as tetrapod abundance, diversity and breadth of geographic distribution (globalization) increases through the Palaeozoic, the ability to use tetrapods in biostratigraphy and biochronology increases (Fig. 2).

Much of the published discussion of Palaeozoic tetrapod distribution has focused on ecological or taphonomic controls of their distribution. For example, many workers have stressed the differences between Palaeozoic tetrapod records in coal-bearing strata ('coal measures') and red beds (e.g. Rayner 1971), noting that it is difficult

From: LUCAS, S. G., CASSINIS, G. & SCHNEIDER, J. W. (eds) 2006. *Non-Marine Permian Biostratigraphy and Biochronology.* Geological Society, London, Special Publications, **265**, 65–93.
0305-8719/06/$15.00

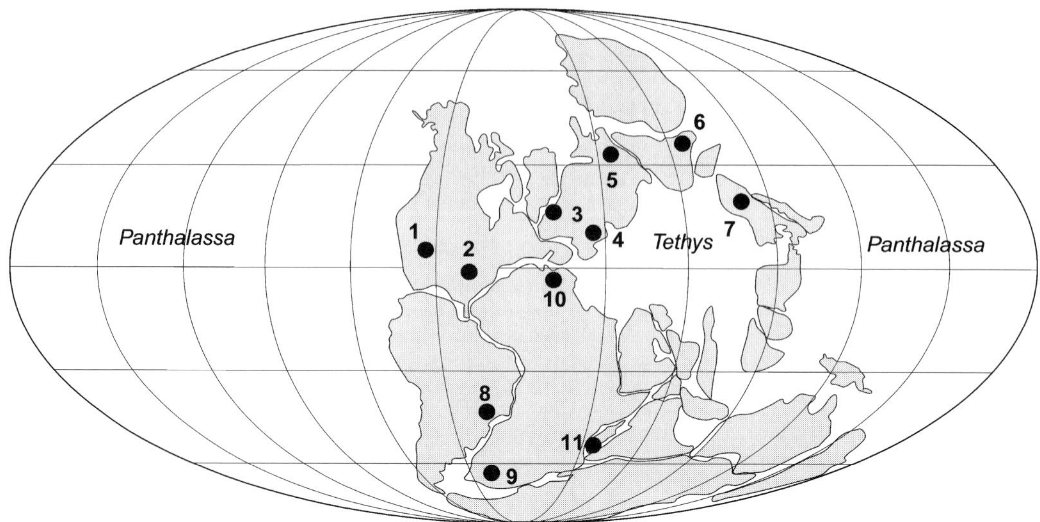

Fig. 1. Map of Permian Pangaea showing principal tetrapod localities. 1, western USA; 2, eastern USA (Dunkard); 3, Scotland; 4, western Europe (Rotliegend); 5, Russian Urals; 6, Junggur Basin, China; 7, Ordos Basin, China; 8, Paraná Basin, Brazil; 9, Karoo Basin, South Africa; 10, Morocco; 11, southern Madagascar.

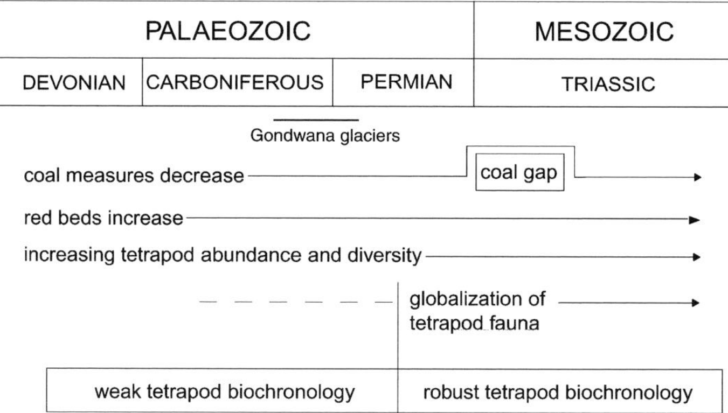

Fig. 2. Some parameters of the Devonian-Triassic world and an evaluation of the relative strength of tetrapod biochronology during that time interval.

to correlate between these two broadly conceived lithofacies. Other discussions explore in detail the sedimentological and taphonomical context of Palaeozoic tetrapod localities, indicating (or at least implying) that factors such as changes in climate or other environmental factors and taphonomic bias are the primary controls of Palaeozoic tetrapod distribution, not the actual temporal ranges of the tetrapod taxa themselves (e.g. Olson 1962; Olson & Vaughn 1970; Carroll 1979, 1997; Milner *et al.* 1986; Eberth *et al.* 2000). While these arguments have their merits – taphonomical and palaeoenvironmental factors do control the distributions of some tetrapod taxa – all Palaeozoic tetrapod taxa had distinct stratigraphic (temporal) ranges that make them of potential use in biostratigraphy (biochronology). This is particularly the case for some Permian tetrapod taxa, and this allows development of a global biostratigraphy and biochronology based on the record of Permian tetrapods (Lucas 2002, 2005*d*).

The subdivision of Permian time documented here is a biochronological scheme of 10 land-vertebrate faunachrons (Fig. 3). Lucas (1998*b*) explained the conceptual and methodological basis of such a scheme. The beginning of each LVF is defined by the FAD of a tetrapod genus,

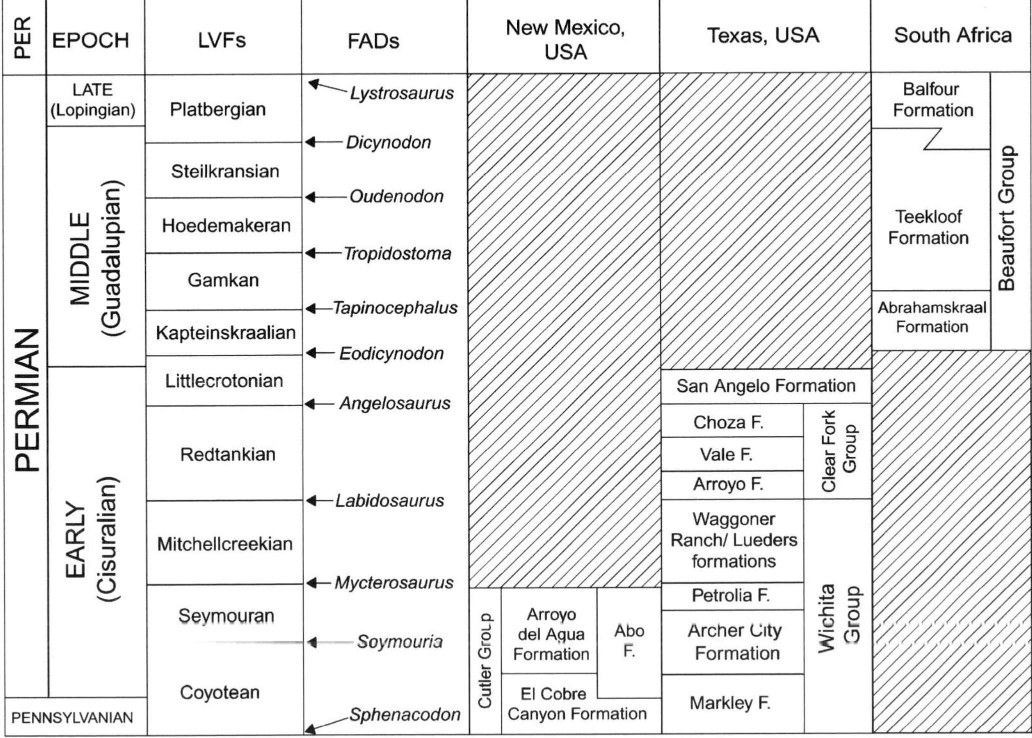

Fig. 3. The composite standard used to create a global tetrapod biochronology for the Permian.

so the end of an LVF is defined by the beginning of the succeeding LVF. The temporal succession of the FADs that define the beginnings of the LVFs is well established and allows all of Permian time to be encompassed by the LVFs. The primary basis for characterization of each LVF is a characteristic tetrapod assemblage that well represents the tetrapod fauna of the time interval. Robust index fossils of each LVF are temporally restricted, common, widespread and easily identified, so they do not include rare taxa restricted to a LVF, usually as a single record. The LVFs proposed here are the formalization (with some modification) of the biochronological scheme first proposed informally by Lucas (2002) and a detailed documentation of this scheme as presented by Lucas (2005*d*).

In this paper, the genus is the operational taxonomic unit for biostratigraphy. This is because most species-level taxa of Permian tetrapods are meaningless for correlation, as they are usually based on a single specimen or a local assemblage of well-preserved material and cannot be recognized at multiple localities (Williston 1915; Romer 1928). However, some species of Permian tetrapod genera (such as *Seymouria* and *Bolosaurus*) are of use in correlation, and taxonomic revisions of some other genera (such as *Eryops* and *Dimetrodon*) should produce species-level taxa of value to biostratigraphy. For example, Werneburg (1989) has also argued that species lineages (chronoclines) provide a more precise biostratigraphy than genus-based correlations. I agree with him in principle, but am unable to construct meaningful species lineages for most of the Permian tetrapod genera that are of value to a global biochronology. Nevertheless, where possible I do discuss species-level distinctions that are of use to correlation and that may be further developed to refine the biochronological framework proposed here.

The record of Permian tetrapods

The most extensive records of Permian tetrapods are from the western United States, western Europe, the Russian Urals, northern China and South Africa (Fig. 1). Relatively recent reviews of part or all of the Permian tetrapod record include Anderson & Cruickshank (1978), Olson (1989*a*), Milner (1990, 1993), Olson & Chudinov (1992),

Rubidge (1995*a*), Ivakhnenko *et al.* (1997), Berman *et al.* (1997), Lucas (1998*a*, 2002, 2004) and Rubidge & Sidor (2001). As extensive and long studied as the Permian record of tetrapods is, it is not without biases and imperfections. Most significant is the virtual geographic restriction of Early Permian tetrapods to the United States and western Europe, and the global gap in part of the Middle Permian tetrapod fossil record (Lucas 2004).

Note also that the biochronological framework proposed here has limitations. Thus, it cannot correlate endemic assemblages, such as that described from the Moradi Formation in Niger (Sidor *et al.* 2005), or isolated records, such as the discosauriscid seymouriamorph record from Tadjikistan (Ivakhnenko 1981).

Previous studies

Previous attempts to delineate a global tetrapod biostratigraphy or biochronology of the Permian are few (Fig. 4). Romer (1966, 1973) divided the Permian tetrapod record into three 'stages' that represent a three-fold subdivision of the Permian into Early (= Cisuralian on the SGCS), Middle (= part of the Guadalupian on the SGCS) and Late (= latter part of Guadalupian and Lopingian on the SGCS). As Romer noted, 'cotylosaurs' (his sense) and pelycosaurs dominate the early stage, best known from North America and the western European Rotliegend. The intermediate stage is dominated by therapsids and known mostly from South Africa and the Russian Urals. The third stage (Romer also called it the 'final phase') is dominated by advanced therapsids known mostly from South Africa.

Romer, thus, was decades ahead of the marine biostratigraphers in recognizing a more logical division of the Permian into three time intervals (or Epochs) instead of two. Indeed, it is also amazing that some of the Early Permian tetrapod biochronology documented here was already presaged by a remarkable and little utilized article by Romer (1928). In this article, Romer identified five stratigraphic zones (0–4) based on tetrapod fossils from the Texas Lower Permian section, and his zones nearly correspond to the faunachrons used here. Thus, Romer's zone 0

Romer (1966, 1973)	Anderson & Cruickshank (1978)		Cooper (1982)	Efremov (1937, 1952) Russia	Ivakhnenko *et al.* (1997) Russia			Rubidge *et al.* (1995) South Africa
final stage (Upper Permian)	endothiodontid/ dicynodontid empire		*Lystrosaurus* Zone	Zone IV	*Scutosaurus* Superzone	*Archosaurus rossicus* Zone		*Dicynodon* Assemblage Zone
						Scutosaurus karpinskii Zone	*Chroniosuchus paradoxus* Subzone	
							Jarilinus mirabilis Subzone	
		17	*Dicynodon* Zone			*Proelginia permiana* Zone	*Chroniosaurus levis* Subzone	
							Chroniosaurus dongusensis Subzone	
		16	*Cistecephalus* Zone			*Deltavjatia vjatkensis* Zone		*Cistecephalus* Assemblage Zone
intermediate stage (Middle Permian)	tapinocephalid empire			"Zone III"				
		15	*Robertia* Zone	Zone II	*Titanophoneus* Superzone	*Ulemosaurus svijagensis* Zone		*Tropidostoma* Assemblage Zone
								Pristerognathus Assemblage Zone
		14	*Venyukovia* Zone			*Estemmenosuchus uralensis* Zone		*Tapinocephalus* Assemblage Zone
						Parabradysaurus silantjevi Zone		
		13	*Otsheria* Zone	Zone I		*Clamorosaurus nocturnus* Zone		
		12	*Dimacrodon* Zone					*Eodicynodon* Assemblage Zone
		11						
early stage (Lower Permian)	edaphosaurid empire	10						
		9						
		8						
		7						
		6						
		5						
		4						
		3						
		2						
		1						

Fig. 4. Previously proposed schemes of Permian tetrapod biostratigraphy and/or biochronology.

approximates to the Coyotean, his zones 1 and 2 approximate to the Seymouran, his zone 3 is the Mitchellcreekian and his zone 4 is the Redtankian.

Anderson & Cruickshank (1978) recognized the same broad global divisions as Romer (1966, 1973), but they recast them as 'empires' (essentially the same concept as the chronofaunas of Olson 1952). Anderson & Cruickshank (1978, charts 2.1–2.2) also listed 17 Permian tetrapod zones, but did not explicitly define them (Fig. 4). However, from their chart 2.1, it is clear that zones 1–12 are based on the classical Texas stratigraphic units (1 = Pueblo Formation; 2 = Moran Formation; 3 = Putnam Formation; 4 = Admiral Formation; 5 = Belle Plains Formation; 6 = Clyde Formation; 7 = Lueders Formation; 8 = Arroyo Formation; 9 = Vale Formation; 10 = Choza Formation; 11 = San Angelo Formation; 12 = Flowerpot Formation), while zones 13 and 14 are equivalent to the Russian zones proposed by Efremov (1937) (13 = Zone I; 14 = Zones II and III), and the youngest zones are those of the South African Karoo Basin (15 = *Tapinocephalus* Zone; 16 = *Cistecephalus* Zone; 17 = *Daptocephalus* [= *Dicynodon*] Zone).

Cooper (1982) published a Middle to Late Permian tetrapod biostratigraphy very similar to zones 11–17 of Anderson & Cruickshank (1978), but with different terminology (Fig. 4). Thus, Cooper's (1982) *Dimacrodon* Zone is based on the vertebrate fossil assemblages of the San Angelo and Flowerpot formations of Texas (Olson 1962), while his *Otsheria* Zone in equivalent to Russian Zone I, his *Venyukovia* Zone to Russian Zone II, and his *Robertia* Zone to the *Tapinocephalus* Zone. Cooper (1982) assigned the *Lystrosaurus* Zone to the Permian, although most workers consider it (or most of it) to be Triassic.

Lucas (2002) proposed an informal Permian tetrapod biochronology that consisted of 10 faunachrons, labeled A–J. The LVFs doumented here are a formalization of this scheme with some modifications that correct errors in Lucas (2002) and reflect a more detailed understanding of the temporal distribution of Permian tetrapods (also see Lucas 2004). This article thus provides detailed documentation of the scheme formalized in brief by Lucas (2005*d*).

There have been no other explicit attempts to develop a global biostratigraphy or biochronology of Permian tetrapods, although correlation charts of the Permian tetrapod record are numerous (e.g. Romer 1966; Anderson 1981; Cheng 1981; Olson 1989*a*; Olson & Chudinov 1992). Nevertheless, two regional schemes of Permian tetrapod biostratigraphy have been extremely important (Fig. 4). Efremov (e.g. 1937, 1940) proposed a succession of four tetrapod 'zones' for the Middle to Upper Permian of the Russian Urals. Olson (1957) provided a useful English-language review of this record (also see Olson 1962; Olson & Chudinov 1992). Zone I is also called the Ocherian dinocephalian complex, and Zone II is the Isheevan dinocephalian complex. 'Zone III' lacks tetrapods, and Zone IV has been called the pareiasaurian faunal complex.

Ivakhnenko *et al.* (1997; also see Golubev, 1998, 2005) recently recast the Russian Permian tetrapod record in a new biostratigraphical scheme. They recognized two 'superzones': a *Titanophoneus* 'Superzone' equivalent to Zones I and II of Efremov (1937), and a *Scutosaurus* 'Superzone' equivalent to Zone IV of Efremov (Fig. 4). These 'superzones' are divided into eight zones, largely based on the stratigraphical ranges of dinocephalians. Furthermore, two of the zones in the *Scutosaurus* Superzone are divided into subzones based on the succession of chroniosuchian temnospondyls (Golubev 1998). The recent summary of the Permian tetrapod record in the southeastern Urals (Tverdokhlebov *et al.* 2005) does not modify the existing scheme of tetrapod biostratigraphy in the Russian section.

Rubidge *et al.* (1995) reviewed the evolution of the biostratigraphic understanding of the South African succession of Permian tetrapod assemblage zones originally proposed by Broom (1906, 1907, 1909) and Watson (1914*a*, *b*) and later elaborated by Kitching (1977) and Keyser & Smith (1977–1978). The current succession recognizes six assemblage zones of Mid- to Late Permian age (Fig. 4).

Composite standard

The most extensive Lower Permian tetrapod record is from the western United States, especially from Texas, Oklahoma and New Mexico (e.g. Olson 1967; Simpson 1979; Hook 1989; Berman 1993). The New Mexican and Texas records are used here to construct the Early Permian tetrapod biochronology of five LVFs (Fig. 3).

Fossil vertebrates have been collected from the non-marine Permian red beds in north-central Texas since the 1870s (see historical review by Craddock & Hook 1989). The collected vertebrate fossils have been published on extensively by E. D. Cope, E. C. Case, S. W. Williston, A. S. Romer and E. C. Olson, among others, and they provide the basis for much of what is known about the Early Permian evolution of tetrapods.

The lithostratigraphical nomenclature long applied to the Texas section was that of Plummer

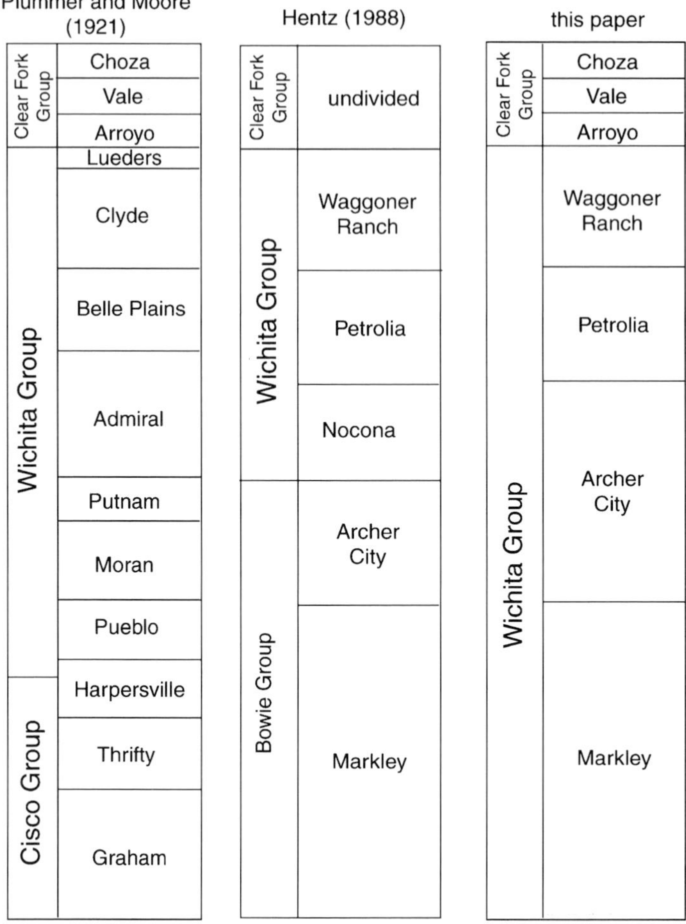

Fig. 5. Comparative lithostratigraphic nomenclature of the Lower Permian section in north-central Texas.

& Moore (1921), who named a series of rock formations of primarily marine origin, assigning them to the Cisco and Wichita groups of Pennsylvanian to Early Permian age (Fig. 5). Vertebrate palaeontologists, especially Romer (1928, 1935, 1958, 1974), readily placed vertebrate fossil localities in the Texas section into this lithostratigraphy. However, the vertebrate-fossil-bearing localities are mostly non-marine red beds split by thin marine limestone/shale horizons that correlate to, but are lithologically distinct from, the formations named by Plummer & Moore (1921).

Thus, Hentz (1988) justifiably created a new lithostratigraphical nomenclature for the Texas Permian red beds (Figs 5–6). Nevertheless, based on my field observations, two modifications need be made to his nomenclature.

(1) There is no significant (mappable) lithologic difference between the Archer City and Nocona formations (Hentz & Brown 1987; Hentz 1988; Hentz, pers. comm. 2001), so they should not be regarded as separate lithostratigraphical units. I thus abandon the term Nocona Formation and include all strata in this interval in the Archer City Formation (Fig. 5).
(2) Similarly, there is no lithological basis for separating the Bowie Group of Hentz from the Wichita Group, so I abandon Bowie Group and extend the base of the Wichita Group downward to the base of the Markley Formation (Fig. 5).

These changes modify the lithostratigraphy of the Texas Lower Permian red beds proposed by Hentz so that all formations are mappable units, and groups are associated mappable units, in accordance with accepted stratigraphic practice (NACSN 1983; Owen 1987).

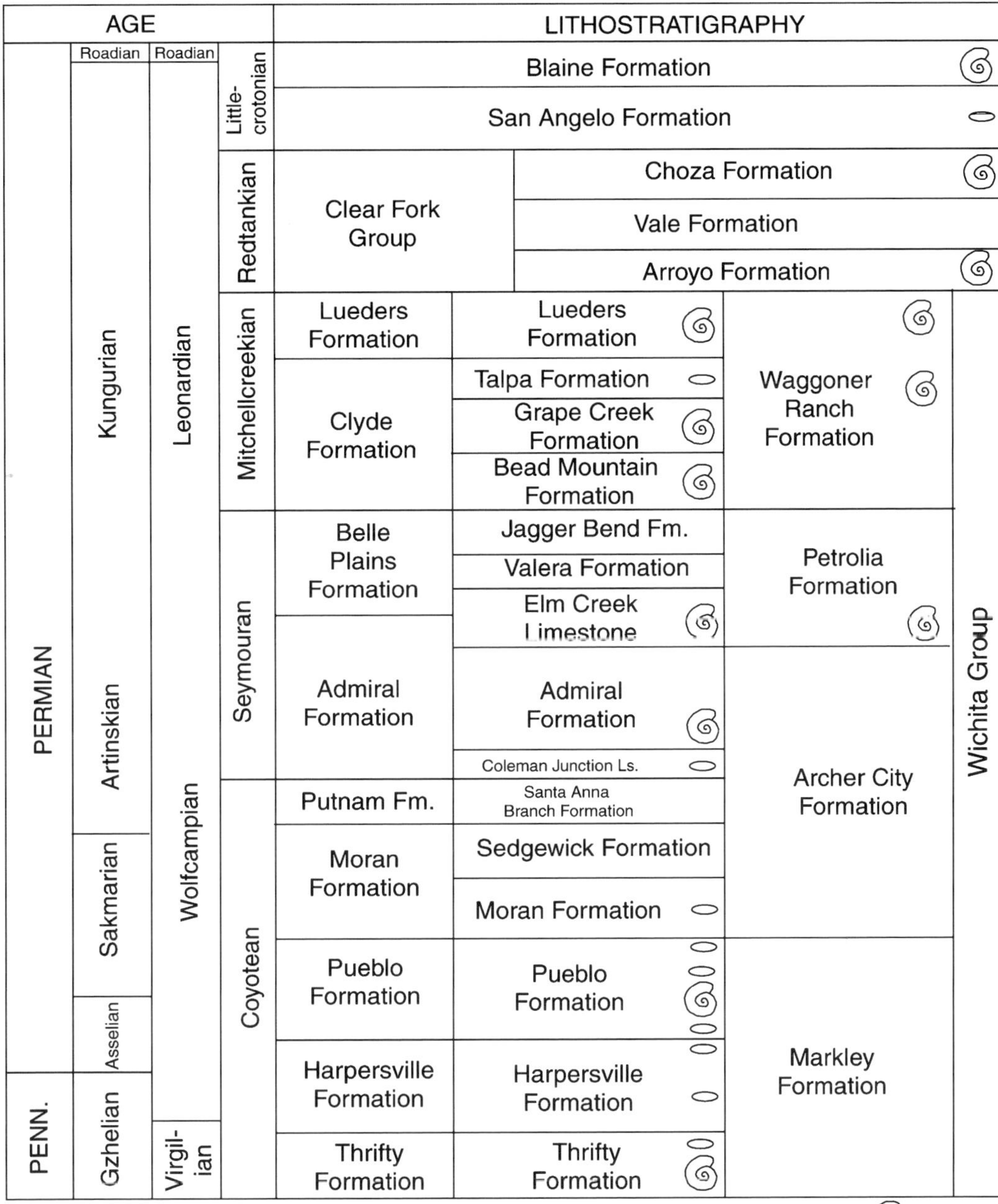

Fig. 6. Summary diagram of the Lower Permian section in north-central Texas showing cross-correlation of vertebrate biochronology and marine biostratigraphy. Formations on right of diagram are the tetrapod-bearing units.

The Texas Lower Permian red-bed section represents fluvial deposition on a broad coastal plain between a Permian seaway to the west and a series of ancestral Rocky Mountain uplifts (Ouachita, Arbuckle and Wichita) to the east and northeast (e.g. Brown 1973; Hentz 1988, 1989). The non-marine red beds intertongue with, and are laterally equivalent to, marine strata, allowing cross-correlation of non-marine and marine biostratigraphies (Fig. 6). This means it is possible to correlate directly a tetrapod biostratigraphy developed in the Texas red beds with a marine biostratigraphy based largely on fusulinids and ammonites (e.g. Böse 1917; Plummer & Moore 1921; Roth 1930; Dunbar & Skinner 1937; Plummer & Scott 1937; Henbest 1938; Lee *et al.* 1938; Miller & Furnish 1940; Skinner 1946; Miller & Youngquist 1947; Thompson 1954; Eardle 1960; Myers 1958, 1960, 1968; Kemp 1962; Kauffman & Roth 1966; Ross 1969; Vanderloop-Avery & Nestell 1984) and for which some conodont data are becoming available (Walsh & Barrick 2002; Wardlaw 2005).

The Texas section thus provides an excellent basis for Early Permian tetrapod biostratigraphy, and this biostratigraphy can be readily correlated to marine biostratigraphy (Fig. 6). However, this section has a glaring weakness in lacking an extensive record of tetrapods across the Pennsylvanian–Permian boundary. To remedy this, I have included the Pennsylvanian–Permian boundary record of tetrapods in north-central New Mexico (Rio Arriba County) to form a composite standard of New Mexico–Texas for the oldest Permian tetrapod faunachrons. The basis for this is as follows.

1. An extensive Upper Pennsylvanian tetrapod assemblage is known from the El Cobre Canyon Formation in the Cañon del Cobre of Rio Arriba County, New Mexico (Fig. 7). Co-occurring palynomorphs, megafossil plants and some of the tetrapod taxa themselves (such as *Desmatodon* and *Limnoscelis*) indicate an Upper Pennsylvanian age. (Vaughn 1963; Fracasso 1980; Hunt & Lucas 1992; Berman 1993; DiMichele & Chaney 2005; Lucas & Krainer 2005; Lucas *et al.* 2005*b*). This is the characteristic tetrapod assemblage of the Cobrean LVF of Lucas *et al.* (2005*b*).
2. Stratigraphically above this assemblage is a tetrapod assemblage that includes the LO of *Sphenacodon*. This assemblage, best known from the Arroyo del Agua area near Coyote in Rio Arriba County (Berman 1993; Lucas *et al.* 2005*c*), crosses the Virgilian–Wolfcampian boundary, so by the current time scale it is of latest Pennsylvanian to earliest Permian age. It has long been correlated (on a vertebrate palaeontological basis) to the lower part of the Wichita Group (Markley and lower Archer City formations) in Texas (e.g. Langston 1953; Romer 1960).
3. Stratigraphically higher, the LO of *Seymouria* is in the Arroyo del Agua Formation of the Cutler Group. It can be correlated to the *Seymouria*-bearing interval of the Wichita Group in Texas (Lucas *et al.* 2005*c*).

Thus, the New Mexican record superposes tetrapod assemblages that are entirely latest Pennsylvanian, cross the Pennsylvanian–Permian boundary and are of Early Permian age (Fig. 7). When combined with the Texas record, the tetrapod succession encompasses the entire Early Permian (Fig. 3).

The Middle to Upper Permian tetrapod fossil record and its biostratigraphy in the Karoo Basin of South Africa has long provided the classic succession of Middle to Late Permian tetrapod assemblages (Fig. 8). Karoo tetrapod fossils were discovered in 1838 and have been extensively studied and published on since the 1850s. Reviews by Rubidge (1995*b*, 2005), Smith & Keyser (1995*a–d*) and Kitching (1995) recognize six successive assemblage zones based on tetrapods. Here, I recast five of the South African assemblage zones as biochronological units (LVFs), using the FAD of a widespread and characteristic tetrapod taxon to define the beginning of each faunachron. This provides five LVFs for most of Middle and Late Permian time (Fig. 3). Thus, the New Mexico–Texas and South African tetrapod records provide a composite standard by which Permian tetrapod biochronology is defined (Fig. 3). However, few data are now available that allow the South African Middle to Late Permian tetrapod record to be cross-correlated to the SGCS.

In the Ural foreland basin the Russian succession of Middle to Late Permian tetrapod assemblages broadly correlates to the Karoo succession and has two advantages: not only can the lowermost (Kazanian) Russian tetrapods be tied to marine biostratigraphy, but the llawara magnetostratigraphic event (see below) has been identified in the Russian Tatarian, which provides another way of correlating the Russian section to the SGCS. Unfortunately, prior to the LO of *Dicynodon* in the Russian section (just above the Illawara event: Lozovsky *et al.* 2001), virtually all of its genus-level taxa are endemic and thus of limited biostratigraphical value. Rare exceptions include the parareptiles *Belebey* (also known in China) and *Macroleter* (reported from Oklahoma), but these provide only a limited basis for correlation. This is why long-standing attempts to correlate the Russian tetrapod assemblages to coeval assemblages in Gondwana (especially in the South African Karoo) have largely been based on assessments of the stage of evolution, usually expressed as family-level correlations (e.g. Rubidge 2005), not on low-level (genus or species) taxonomic identity, and thus

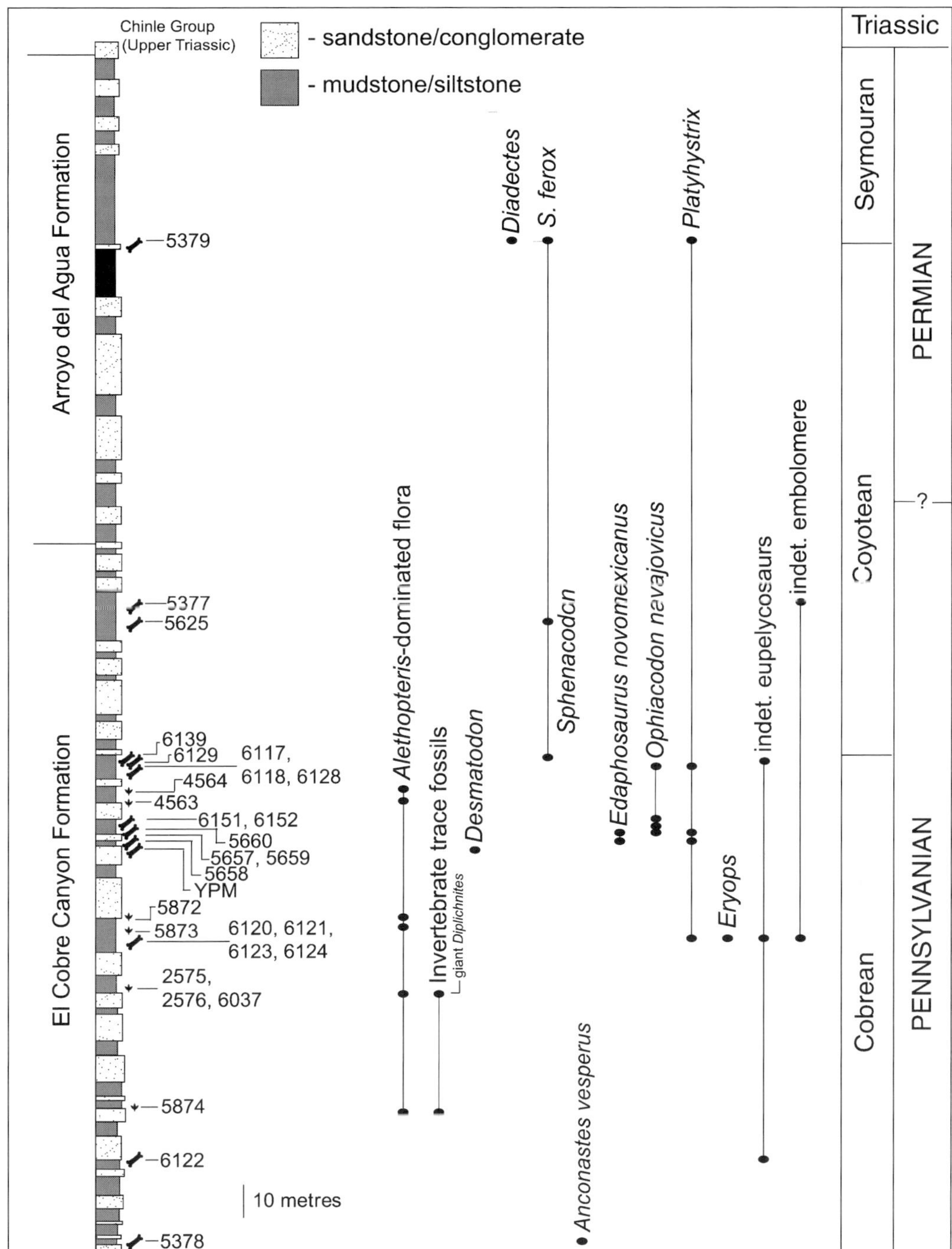

Fig. 7. Composite stratigraphic section of Cutler Group strata in El Cobre Canyon (Cañon del Cobre), northern New Mexico, and the distribution of vertebrate fossils and their ages (after Lucas *et al.* 2005*b*).

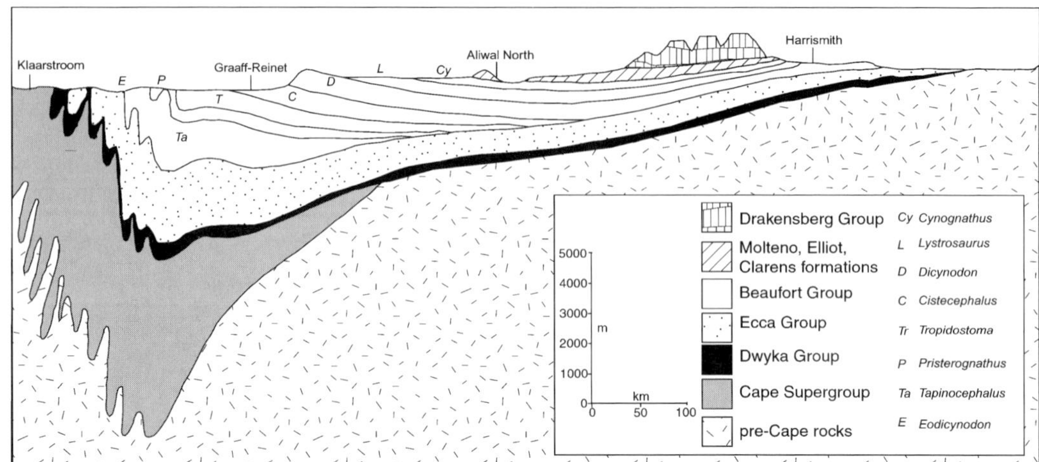

Fig. 8. Cross section showing distribution of vertebrate assemblage zones in the South African Karoo Basin (after Rubidge *et al.* 1995).

inherently imprecise and, to some, remain debatable. Therefore, I use the South African record as a more robust standard for Middle–Upper Permian vertebrate biostratigraphy than can be provided by the Russian record.

Land-vertebrate faunachrons

Coyotean LVF

Definition
The oldest interval of Permian time based on tetrapods is the Coyotean LVF. The name is for Coyote, New Mexico, near the many tetrapod bonebeds of Coyotean age in the upper part of the El Cobre Canyon Formation of the Cutler Group. Coyotean time begins with the FAD of the pelycosaur *Sphenacodon* and encompasses the Pennsylvanian–Permian boundary.

Index fossils
The eureptiles *Romeria* and *Protorothyris* are restricted to Coyotean time, as are the temnospondyls *Chenoprosopus, Edops, Neopteroplax, Neldasaurus* and *Brevidorsum* and the pelycosaur *Stereophallodon* (Fig. 9). However, none of these taxa are abundant or widespread enough to be robust index taxa.

Characteristic assemblage
The characteristic Coyotean assemblage is from the upper part of the El Cobre Canyon Formation (Cutler Group) in the Arroyo del Agua area of Rio Arriba County, New Mexico (Berman, 1993; Lucas *et al.* 2005*c*). It includes the temnospondyls *Eryops, Chenoprosopus, Zatrachys, Platyhystrix, Broiliellus* and *Ecolsonia*, the microsaur *Stegotretus*, an embolomere, a lepospondyl, the diadectomorphs *Diadectes* and *Tseajaia*, the parareptile *Bolosaurus*, the eureptile *Rhiodenticulatus*, the araeoscelid *Zarcasaurus* and the pelycosaurs *Sphenacodon, Aerosaurus, Edaphosaurus, Oedaleops* and *Ophiacodon.* The temnospondyl and eureptile components of the Coyotean are distinct from those of the Seymouran.

Principal correlatives
In Texas, the tetrapod assemblage from the Markley and lower part of the Archer City formations of the Wichita Group is of Coyotean age. This assemblage includes diverse temnospondyls (e.g. *Eryops, Edops, Neldasaurus, Zatrachys* and *Trimerorhachis*), a few microsaurs and nectrideans, anthracosaurs (*Archeria*), the diadectomorph *Diadectes*, the eureptiles *Protorothyris* and *Romeria* and diverse pelycosaurs (especially *Dimetrodon, Edaphosaurus* and *Stereophallodon*) (e.g. Hook, 1989 and references cited therein).

In the Arizona–Utah borderland (principally Monument Valley), the Halgaito Formation (Cutler Group) yields a Coyotean tetrapod assemblage that includes *Diplocaulus, Phlegethonia*?, a trimerorhachid, *Eryops, Platyhystrix, Archeria*, a limnoscelid, *Limnoscelis, Ophiacodon, Edaphosaurus, Sphenacodon* and an araeoscelid? (Vaughn 1962, 1964, 1965, 1966*a*, *b*, 1973; Frede *et al.* 1993; Sumida *et al.* 1999*a*, *b*). Sumida *et al.* (1999*b*) identified a single vertebra from the Halgaito Formation as *Seymouria*?, but this genus-level identification has been abandoned (S. Sumida, pers. comm. 2006).

taxa	Coyotean	Seymouran	Mitchellcreekian	Redtankian	Littlecrotonian
Acheloma				●	
Angelosaurus					●
Araeoscelis		●	●	●	
Archeria	●	●	●		
Aspidosaurus				●	
Bolosaurus	●	●	●		
Brachydectes	●		●		
Brevidorsum	●				
Broiliellus	●	●		●	
Cacops			●	●	
Captorhinus	●	●	●	●	
Carrolla		●			
Casea				●	
Caseoides					●
Chenoprosopus	●				
Cotylorhynchus				●	●
Crossotelos			●		
Ctenospondylus		●			
Cymatorhiza					●
Diadectes	●	●	●	●	
Dimetrodon	●	●	●	●	●
Diplocaulus	●	●	●	●	●
Ecolsonia	●				
Edaphosaurus	●	●	●	●	
Edops	●				
Eothyris		●			
Eryops	●	●	●	●	
Glaucosaurus			●		
Kahneria					●
Labidosaurikos			●	●	
Labidosaurus				●	
Lupeosaurus	●	●			
Macroleter					●
Mycterosaurus			●		
Neldasaurus	●				
Neopteroplax	●				
Ophiacodon	●	●	●		
Pantylus	●		●		
Pariotichus		●			
Parioxys	●	●			
Platyhystrix	●	●			
Protocaptorhinus		●	●		
Protorothyris	●				
Romeria	●				
Rothianiscus					●
Secodontosaurus		●		●	
Seymouria		●	●	●	
Slaughenhopia					●
Sphenacodon	●	●			
Stereophallodon	●				
Tersomius	●	●	●		
Trematopsis				●	
Trimerorhachis	●	●	●	●	
Tseajaia	●	●			
Varanodon					●
Varanosaurus		●	●	●	
Zatrachys	●	●		●	

Fig. 9. Temporal ranges of selected genera of Early Permian tetrapods.

In southwestern Colorado, the upper part of the Cutler Formation yielded a Coyotean-age assemblage that includes *Eryops*, *Platyhystrix*, a seymouriid, *Diadectes*, a captorhinid?, a haptodontid and '*Mycterosaurus*' (unreliable identification) (Lewis & Vaughn 1965; Wideman *et al.* 2005).

In the Lucero uplift of central New Mexico, the Red Tanks Member of the Bursum Formation yielded *Eryops*, *Trimerorhachis*, cf. *Archeria, Diadectes, Edaphosaurus, Sphenacodon* and *Dimetrodon* (Harris *et al.* 2004) – an assemblage of Coyotean age.

In southern Oklahoma, the upper part of the Oscar Group (especially the Waurika 1 locality) yielded a tetrapod assemblage of Coyotean age that includes *Diplocaulus, Trimerorhachis, Eryops, Archeria, Pantylus, Ophiacodon, Dimetrodon* and *Edaphosaurus* (Olson 1967; Simpson 1979).

In Brown County, Kansas, the Robinson locality in the upper Virgilian Soldier Creek Shale Member of the Bern Limestone yields a lysorophid, *Diplocaulus, Cricotus*, a trimerorhachid and cf. *Platyhystrix* (Foreman & Martin 1988). This assemblage may be of Coyotean age.

The Indian Cave Sandstone in Nemaha County, Nebraska, has yielded *Ophiderpeton, Phlegethontia, Captorhinus, Denderpetron* and a pelycosaur (Foreman & Martin 1988) – an assemblage that may be of Coyotean age. In Richardson County, Nebraska, the Eskridge Formation yields *Acroplous, Brachydectes*, a trimerorhachid, a microsaur, a diadectid and an edaphosaurid (Huttenlocker *et al.* 2005) and may also be of Coyotean age.

Tetrapods from the Washington Formation of the Dunkard Group in the west Virginia–Ohio–Pennsylvania borderland of the eastern United States (Moran 1952; Romer 1952; Olson 1975) include *Edops* (a Coyotean index taxon), as well as *Trimerorhachis, Diadectes, Edaphosaurus* and *Dimetrodon*, and are reasonably assigned a Coyotean age.

In Europe, some of the Rotliegend tetrapod assemblages (e.g. the lower *Protriton* and Gottlob horizons in the Thuringian forest) dominated by branchiosaurs (e.g. Boy 1993; Werneburg 1989, 2001) are apparently of Coyotean age, but the lack of shared taxa makes a direct tetrapod-based correlation impossible. The correlation, instead, must be based on other evidence which indicates that some of the Rotliegend tetrapod assemblages are of late Virgilian to middle Wolfcampian age (e.g. Roscher & Schneider 2005), which means they correlate to the Coyotean. The latest version of the Rotliegend amphibian zonation (Werneburg & Schneider 2006) recognizes nine zones based on species chronoclines that provide correlations in the Czech Republic, Germany, France, Poland and Italy. This is a provincial biostratigraphy in the Rotliegend extensional basins of Europe in which amphibian zones 3–9 appear to overlap Coyotean time as here defined.

Steyer (2000) critiqued Werneburg's biostratigraphy by arguing that taphonomic and palaeoecological factors have more control over amphibian distributions than actual temporal ranges, and by critiquing the species-chronocline method of taxonomy. However, Steyer's assertions about palaeoecology and taphonomy are largely undocumented, and the species-chronocline method is the preferred method used in the micropalaeontological taxonomy of the fusulinids and conodonts, the two biostratigraphic workhorses of the Permian SGCS. In principle, an extensive record of European amphibians should be amenable to such methods.

Comments

Lucas (2002) defined an informal faunachron A that is, in part, equivalent to the Coyotean. However, in 2002, I used the FAD of *Eryops* to define the beginning of Coyotean time, which clearly predates the FAD of *Sphenacodon* (e.g. Vaughn 1958; Harris *et al.* 2004; Lucas *et al.* 2005*b*). By using the FAD of *Sphenacodon* to define the beginning of the Coyotean, its beginning is close to (but almost certainly precedes) the Pennsylvanian–Permian boundary, making the Coyotean a shorter time interval than faunachron A of Lucas (2002).

Note that the distribution of tetrapod taxa in the New Mexico and Texas sections indicates that the Coyotean is equivalent to part of the Virgilian and much of the Wolfcampian (Fig. 6). Thus, *Sphenacodon* has its LO in Virgilian strata of the Bursum Formation in central New Mexico (Harris *et al.* 2004), and Coyotean tetrapods are found throughout the Markley and lower part of the Archer City formations in Texas, which means that the Coyotean encompasses most of Wolfcampian time.

Seymouran LVF

Definition

The Seymouran LVF is the time interval between the Coyotean and Mitchellcreekian LVFs. The name is for the town of Seymour, Baylor County, Texas, near the characteristic assemblage of the Seymouran, which is from the upper part of the Archer City Formation. The Seymouran LVF begins with the FAD of the seymouriamorph *Seymouria*.

Index fossils
The microsaurs *Carrolla* and *Pariotichus*, and the pelycosaurs *Ctenospondylus* and *Eothyris*, are restricted to the Seymouran. However, none of these taxa are abundant or widespread enough to be robust index taxa. The FADs of *Protocaptorhinus, Seymouria, Varanosaurus, Secodontosaurus* and *Araeoscelis*, and the LADs of *Parioxys, Platyhystrix* and *Sphenacodon*, also distinguish the Seymouran LVF (Fig. 9).

The species *Seymouria sanjuanensis* may be the best index taxon of the Seymouran, as it is known from Utah, New Mexico and Germany. The younger species, *S. baylorensis*, is late Seymouran through Redtankian, and *S. grandis* is from the Redtankian (Olson 1980).

Characteristic assemblage
The characteristic assemblage of the Seymouran is from the upper Archer City ('Nocona') and Petrolia formations of the Wichita Group in Texas. This assemblage has temnospondyls similar to those of the Coyotean (but without *Edops* and *Neldasaurus*), the microsaurs *Carrolla* and *Pantylus*, a few nectrideans, the anthracosaur *Archeria, Diadectes* and *Seymouria*, the eureptile *Protocaptorhinus*, diverse pelycosaurs (including the LOs of *Secodontosaurus* and *Varanosaurus*), the diapsid *Araeoscelis* and the parareptile *Bolosaurus* (see Hook 1989 and references cited therein).

Principal correlatives
In the Arizona–Utah borderland, the Organ Rock Shale (Cutler Group) yields a Seymouran tetrapod assemblage that includes *Seymouria, Eryops*, a trimerorhachid, a zatrachyid, the diadectomorphs *Tseajaia* and *Diadectes, Sphenacodon, Ophiacodon, Dimetrodon* and *Ctenospondylus* (Vaughn 1964, 1966*a*, *b*, 1973; Sumida *et al.* 1999*a*, *b*). The underlying Cedar Mesa Sandstone yields *Eryops* and *Sphenacodon* and could be either Coyotean or Seymouran in age.

In the Chama Basin of northern New Mexico, the superposition of Coyotean and Seymouran tetrapod assemblages is documented in the Arroyo del Agua Formation of the Cutler Group, where a Seymouran-age assemblage of *Seymouria, Sphenacodon, Diadectes, Platyhystrix* and an eryopid is stratigraphically above the characteristic Coyotean tetrapod assemblage (Lucas & Krainer 2005; Lucas *et al.* 2005*c*).

In northern Oklahoma, the Wellington Formation (especially the Perry and Orlando localities) yields an extensive tetrapod assemblage of Seymouran age that includes *Trimerorhachis, Zatrachys, Seymouria?, Brachydectes* (= *Lysorophus* of Wellstead 1991), *Eryops, Diplocaulus, Broiliellus, Diadectes, Archeria, Ophiacodon, Captorhinus, Edaphosaurus* and *Dimetrodon* (Olson 1967; Simpson 1979).

In Kansas, various localities in the upper Council Grove Group (especially those in the Speiser Shale) yield a probable Seymouran-age assemblage that includes *Brachydectes, Diplocaulus, Trimerorhachis* and *Euryodus* (Foreman & Martin 1988).

The Greene Formation of the Dunkard Group (localities principally in western Ohio) overlies the Washington Formation and yields *Brachydectes, Trimerorhachis, Eryops, Edaphosaurus* and *Ctenospondylus* (Berman & Berman 1975; Berman 1978). *Ctenospondylus* is also known from the 'Belle Plains Formation' (Petrolia Formation) in Texas and the Organ Rock Shale, both of which are Seymouran-age records, and this suggests a Seymouran age for the *Ctenospondylus* occurrence in the Green Formation. However, based primarily on chrondrichthyans, Lund (1975) correlated the Greene Formation to the lower Clear Fork Group of Texas, which suggests a Redtankian age.

Lower Permian red beds on Prince Edward Island in eastern Canada yield *Eryops, Seymouria, Diadectes* and a pelycosaur (Langston 1963; Spalding 1993), an assemblage of probable Seymouran age.

Berman & Martens (1993), Sumida *et al.* (1996, 1998), Berman *et al.* (2000, 2001, 2004) and Sumida *et al.* (2004), among others, documented tetrapods from the Tambach Formation of the Upper Rotliegend in Germany (also see Eberth *et al.* 2000), which include the trematopid *Tambachia, Seymouria,* the eureptile *Thuringothyris*, the diadectomorphs *Diadectes* and *Orobates*, the bolosaurid *Eudibamus*, a varanopid, a caseid and the pelycosaur *Dimetrodon*. This assemblage is of Seymouran age.

Comments
The Seymouran as used here is essentially the same as faunachron B of Lucas (2002).There is a substantial turn-over in the eureptile and pelycosaur components of the tetrapod fauna between the Coyotean and Seymouran (e.g. Romer & Price 1940, Clark & Carroll 1973; Heaton 1979; Hook 1989). Correlation of the Texas section indicates that the Seymouran straddles the Wolfcampian–Leonardian boundary (Fig. 6).

Mitchellcreekian LVF

Definition
The Mitchellcreekian LVF is the time interval between the Seymouran and Redtankian LVFs. The name is for Mitchell Creek near Lake Kemp

in Baylor County, Texas, which is near the characteristic tetrapod assemblage of the Mitchellcreekian in the Waggoner Ranch and Lueders formations. The Mitchellcreekian begins with the FAD of the pelycosaur *Mycterosaurus*.

Index fossils
The varanopid pelycosaur *Mycterosaurus* is not a common taxon, but it is known from Oklahoma, Texas and Ohio and is restricted to the Mitchellcreekian (Berman & Reisz [1982] note that its record in the Cutler Formation of Colorado [Lewis & Vaughn 1965] can be discounted). The nectridean *Crossotelos* and the pelycosaur *Glaucosaurus* are also restricted to Mitchellcreekian time but are not robust index taxa. The FAD of *Cacops* and the LADs of *Archeria, Bolosaurus, Brachydectes, Ophiacodon Protocaptorhinus, Pantylus* and *Varanosaurus* also distinguish the Mitchellcreekian (Fig. 9).

Characteristic assemblage
The characteristic Mitchellcreekian tetrapod assemblage is from the Waggoner Ranch and Lueders formations in Texas. The characteristic assemblage has only a few temnospondyls (except for abundant armoured dissorophids), gymnarthrids, *Brachydectes*, an aïstopod, abundant nectrideans (especially *Diplocaulus*), *Archeria* and *Diadectes*, eureptiles similar to those of Seymouran age, and diverse pelycosaurs (see Hook 1989, and references cited therein).

Principal correlatives
Tetrapod assemblages of Mitchellcreekian age are currently known only from Texas and Oklahoma (Olson 1967; Simpson 1979; Hook 1989; Burkhalter & May 2002). In southern Oklahoma, the tetrapod assemblage from the lower–middle Garber Formation (especially the South Grandfield and Northeast Frederick sites) is of Mitchellcreekian age and includes *Trimerorhachis, Tersomius, Brachydectes, Diplocaulus, Archeria, Diadectes, Captorhinus, Labidosaurikos, Ophiacodon, Dimetrodon* and *Araeoscelis*. The Richards Spur locality (a fissure-fill in Ordovician limestone) may also be of Mitchellcreekian age and includes *Phlegethontia, Doleserpeton, Cacops, Tersomius, Seymouria*, diverse gymnarthrids, *Captorhinus, Mycterosaurus, Bolosaurus* and a caseid. Indeed, the bolosaurid from Richards Spur, *Bolosaurus grandis*, is larger and more derived than the Coyotean–Seymouran bolosaurid, *B. striatus*, so they may be an ancestor-descendent lineage of biostratigraphic value (Lucas *et al.* 2005*a*).

Comments
The Mitchellcreekian as used here is faunachron C of Lucas (2002). However, Lucas (2002) used the FAD of '*Lysorophus*' (= *Brachydectes*) to define the beginning of his faunachron C because this corresponds to its LO in the Texas section. But, *Brachydectes* has older, Coyotean records outside of Texas (see above). The Mitchellcreekian is of Leonardian age (Fig. 6).

Redtankian LVF

Definition
The Redtankian LVF is the time interval between the Mitchellcreekian and Littlecrotonian LVFs. The name is for Red Tank, north of Seymour in Baylor County, Texas, near the characteristic tetrapod assemblage of the Redtankian, which is from the Clear Fork Group. The FAD of the eureptile *Labidosaurus* defines the beginning of the Redtankian.

Index fossils
Labidosaurus is an index fossil of Redtankian time but is rare. *Aspidosaurus, Casea, Acheloma* and *Trematopsis* are also restricted to the Redtankian, but they are not robust index taxa. The FADs of *Labidosaurikos* and *Cotylorhynchus* help to define the Redtankian, as do the LADs of *Araeoscelis, Broiliellus, Cacops, Captorhinus, Diadectes, Edaphosaurus, Eryops, Seymouria* and *Varanosaurus* (Fig. 9).

Characteristic assemblage
The characteristic Redtankian assemblage is from the Clear Fork Group (Arroyo, Vale and Choza formations) or Clear Fork Formation (where the three constituent formations are not distinct mappable units: Nelson *et al.* 2001) of Texas. It includes abundant *Brachydectes, Trimerorhachis* and *Diplocaulus*, as well as *Eryops, Trematops, Cacops, Trematopsis*, diverse dissorophids (including *Broiliellus, Aspidosaurus* and *Dissorophus*), diverse eureptiles (especially *Captorhinus, 'Captorhinikos', Captorhinoides* and *Labidosaurus*), *Seymouria, Diadectes*, diverse pelycosaurs (including *Casea, Dimetrodon, Varanosaurus, Secodontosaurus* and *Edaphosaurus*) and *Araeoscelis* (Olson 1952, 1954, 1958, 1989*b*; Olson & Mead 1982; Murry & Johnson 1987; Berman & Lucas, 2003).

Principal correlatives
Tetrapod assemblages of Redtankian age are currently known from Texas and Oklahoma. In Oklahoma, the LO of *Labidosaurus* is in the upper Garber Formation. The overlying Hennessey Group also yields a Redtankian assemblage that includes *Trematops, Tersomius, Trimerorhachis, Peroneodon, Brachydectes, Eryops, Captorhinus, Cotylorhynchus, Dimetrodon* and *Ophiacodon* (Olson 1967; Simpson 1979).

Comments
The Redtankian as used here is the same as faunachron D of Lucas (2002). Its characteristic assemblage is the classic Clear Fork Group chronofauna of Texas (Olson 1952) but, as will be discussed later, this chronofauna was of relatively short duration in geological time. The Redtankian is equivalent to part of the Leonardian (Fig. 6).

Littlecrotonian LVF

Definition
The Littlecrotonian LVF is the time interval between the Redtankian and Kapteinskraalian LVFs. The LVF derives its name from Little Croton Creek in Knox County, Texas, near the characteristic tetrapod assemblage in the San Angelo Formation. The FAD of the caseid pelycosaur *Angelosaurus* defines the beginning of the Littlecrotonian.

Index fossils
Most of the tetrapod genera of Littlecrotonian age are restricted to the time interval (Fig. 9), but only the 'microsaur' *Cymatorhiza*, the eureptile *Rothianiscus* (although it includes specimens that pertain to *Labidosaurikos*: Sumida, pers. comm. 2006) and the pelycosaur *Angelosaurus* are widely distributed in Texas–Oklahoma and thus may be relatively robust index taxa.

Characteristic assemblage
The youngest North American Leonardian tetrapod assemblage, from the San Angelo Formation of Texas (Olson & Beerbower 1953), is characteristic of this time interval. It is from localities in Knox, Foard and Hardeman counties in north-central Texas and includes the captorhinid *Rothianiscus,* the caseid pelycosaurs *Caseoides Cotylorhynchus* and *Angelosaurus*, the sphenacodontids *Steppesaurus* and *Tappensaurus* and the putative therapsid *Dimacrodon*. Olson (1962) later added these taxa to the San Angelo tetrapod assemblage: the temnospondyl *Slaugenhopia*, the captorhinid *Kahneria,* the sphenacodont *Dimetrodon,* the caseid *Caseopsis* and the 'therapsids' *Knoxosaurus, Gorgodon, Eosyodon, Driveria* and *Mastersonia*. Olson (1962) also reassigned *Tappenosaurus* and *Steppesaurus*, along with *Dimacrodon,* to the Therapsida. Abundant and diverse caseids are characteristic of Littlecrotonian time. However, all the 'therapsid' taxa from this assemblage have been re-evaluated and deemed to be based on fragmentary pelycosaur fossils (Parrish *et al.* 1986; Sidor & Hopson 1995).

Principal correlatives
The Flowerpot Formation of Texas and the Chickasha Formation of Oklahoma yield tetrapod assemblages of Littlecrotonian age (Olson 1962, 1965, 1967; Lucas 2004). Olson & Barghusen (1962) described vertebrate fossils from two localities in the Flowerpot Formation in Kingfisher County, Oklahoma, that yield the 'microsaur' *Cymatorhiza, Rothianiscus, Cotylorhynchus* and *Angelosaurus.*

Strata of the Chickasha Formation, which are laterally equivalent to the middle part of the Flowerpot Formation, yielded vertebrate fossils from about 20 localities, mostly in Blaine and Kingfisher counties, Oklahoma (Olson 1965). A single locality in McClain County, Oklahoma, also yielded unidentified bone from the Duncan Sandstone (Olson 1965). The Chickasha assemblage includes *Cymatorhiza*, the amphibians *Nannospondylus* and *Fayella, Rothianiscus, Cotylorhynchus, Angelosaurus* and the varanopid *Varanodon*. Olson (1972) subsequently added the nectridean *Diplocaulus* to this assemblage, and also described the supposed therapsid (actually a pelycosaur) *Watongia* (Olson 1974). Olson's (1980) *Seymouria agilis* from the Chickasha Formation assemblage has been reassigned to the parareptile *Macroleter*, a genus previously known only from Russia (Reisz & Laurin 2001).

Because Littlecrotonian time lasts until the beginning of the Kapteinskraalian, the gap between the Texas–Oklahoma assemblages just discussed and the oldest Kapteinskraalian assemblage ('Olson's gap') is of Littlecrotonian age (Fig. 3). The only tetrapod assemblage that may be in this gap is the Inta assemblage from the Pechora Basin in Russia. This assemblage is essentially an endemic amphibian fauna that resembles North American Early Permian amphibians in its stage of evolution, but cannot be otherwise correlated based on tetrapod biostratigraphy alone (Lucas 2004).

Comments
Olson (1962; and also Efremov 1956 and Olson & Chudinov 1992) consistently correlated the tetrapod assemblage of the San Angelo and Flowerpot formations with the oldest Middle Permian therapsid-bearing assemblages in Russia. This correlation was not based on shared low-level taxa (genera and species) but on the supposed abundance of therapsids in the Texas faunas and the presence of 'counterparts' (equivalent evolutionary grades) among the Texan and Russian amphibians and caseids. Recognition that all the San Angelo 'therapsid' fossils are actually pelycosaurs undermines this correlation and suggests that the therapsid-dominated faunas that are the oldest Permian assemblages in Russia and South Africa postdate the youngest North American Permian faunas

(e.g. Sidor & Hopson 1995). Marine biostratigraphy supports this, so there is a hiatus in the Permian tetrapod record ('Olson's gap') equivalent to part of Roadian time (Lucas 2004). Recently, Lozovsky (2005) has argued against this, but his arguments have been answered by Lucas (2005*a*).

Kapteinskraalian LVF

Definition

The Kapteinskraalian LVF is the time interval between the Littlecrotonian and Gamkan LVFs. The LVF derives its name from the Kapteinskraal River in South Africa, the type section of the *Eodicynodon* assemblage zone. The beginning of the Kapteinskraalian LVF is the FAD of the therapsid *Eodicynodon*.

Index fossils

Most of the tetrapod taxa of the characteristic Kapteinskraalian assemblage are limited to the LVF (Fig. 10), but lack a proven broad distribution (they are endemic to either South Africa or Russia) that would identify them as robust index taxa. The most primitive anomodonts (e.g. *Eodicynodon*, *Otsheria* and *Patronomodon*) and dinocephalians (e.g. *Australosyodon*, *Tapinocaninus*) are indexes of the Kapteinskraalian, but no genus-level taxon is widespread. The first therapsids appear during Kapteinskraalian time.

Characteristic assemblage

The characteristic Kapteinskraalian assemblage is from the lower Abrahamskraal Formation, Beaufort Group, South Africa (Rubidge 1995*b*). The characteristic tetrapod assemblage is the *Eodicynodon* Assemblage Zone and includes temnospondyls, a gorgonopsian, the therocephalians *Glanosuchus* and *Alopecodon*, the anomodont *Patranomodon*, the dicynodont *Eodicynodon* and the dinocephalians *Tapinocaninus* and *Australosyodon* (Rubidge 1995*b*, 2005 and references cited therein).

Principal correlatives

The oldest Russian tetrapod assemblages of Kazanian age (Russian Zone I: Ocher assemblage and part of Mezen assemblages) yield basal anteosaurid dinocephalians and anomodonts

taxa	Kapteins-kraalian	Gamkan	Hoedemakeran	Steilkraansian	Platbergian
Alopecodon	●	●			
Aulacephalodon				●	
Bradysaurus		●			
Cistecephlaus				●	
Dicynodon					●
Diictodon		●	●	●	●
Elliotsmithia		●			
Embrithosaurus		●			
Emydops		●	●	●	●
Endothiodon		●	●	●	
Eodicynodon	●				
Eunotosaurus		●			
Gorgonops		●	●	●	
Ictidosuchoides		●	●	●	●
Kingoria			●	●	●
Lycaenops			●	●	●
Otsheria	●				
Oudenodon				●	●
Pareiasaurus		●	●	●	●
Patranomodon	●				
Pelanomodon					●
Pristerodon		●	●	●	●
Pristerognathus		●			
Rhachiocephalus			●	●	
Rhinesuchus		●	●	●	●
Robertia		●			
Tapinocephalus		●			
Theriognathus					●
Tropidostoma			●		
Youngina			●	●	●

Fig. 10. Temporal ranges of selected genera of Mid- to Late Permian tetrapods.

(see Golubev 1998, 2005 for summaries). They predate the beginning of the Gamkan LVF and are therefore of Late Kapteinskraalian age.

Comments
The Kapteinskraalian as used here is faunachron F of Lucas (2002). The characteristic assemblage, the *Eodicynodon* assemblage zone in the Karoo Basin, is thought to be of Kazanian (Wordian) age, and older than the Russian Zone I and II assemblages (Rubidge & Hopson 1990; Lucas 2004; Rubidge 2005), but direct correlation with the SGCS is difficult. For many years, and by some today (Benton *et al.* 2004, fig. 1; Golubev 2005), Russian Zone II was thought to be the oldest therapsid fauna, but taxa from the *Eodicynodon* Assemblage Zone in South Africa are among the most primitive members of their groups; this is especially true of anomodonts and tapinocephaline dinocephalians (Rubidge 1993; Rubidge & Hopson 1996; Modesto *et al.* 1999, 2002, 2003; Modesto & Rubidge 2000; Modesto & Rybczynski 2000; Battail 2000; Rubidge & Sidor 2001). Thus, based on stage-of-evolution, the *Eodicynodon* Assemblage Zone is thought to be the oldest Mid-Permian tetrapod assemblage with therapsids. Given that no Early Permian tetrapod assemblage yields *bona fide* therapsids, it seems unlikely that the *Eodicynodon* assemblage zone is of Early Permian age. But, just how old it is in the Mid-Permian remains uncertain.

Gamkan LVF

Definition
The Gamkan LVF is the time interval between the Kapteinskraalian and Hoedemakeran LVFs. The name of the LVF is for the Gamka River, which adjoins the type locality of the *Tapinocephalus* assemblage zone. The beginning of the Gamkan LVF is the FAD of the dinocephalian *Tapinocephalus*.

Index fossils
Tapinocephalus, various other dinocephalians, *Eunotosaurus, Bradysaurus, Elliotsmithia, Pristerognathus* and *Robertia* are some of the better-known taxa restricted to the Gamkan but they are not robust index fossils. The FADs of *Diictodon, Endothiodon, Gorgonops, Ictidosuchoides, Pristerodon, Rhinesuchus* and *Emydops* and the LAD of *Alopecodon* help to define the Gamkan (Fig. 10). The Gamkan is the time of highest dinocephalian diversity.

Characteristic assemblage
The characteristic Gamkan tetrapod assemblage is from the upper Abrahamskraal and lower Teekloof formations, Beaufort Group, South Africa. It combines those of the *Tapinocephalus* Assemblage Zone of Smith & Keyser (1995*a*) and the *Pristerognathus* Assemblage Zone of Smith & Keyser (1995*b*). It thus includes the temnospondyl *Rhinesuchus*, pareiasaurs (especially *Bradysaurus*), the pelycosaur *Elliotsmithia*, diverse dinocephalians (especially *Tapinocephalus*), the anomodont *Galeops*, dicynodonts (especially *Diictodon*), two biarmosuchiana, several gorgonopsians and therocephalians (see Smith & Keyser 1995*a*, *b* and references cited therein).

Principal correlatives
In Zimbabwe, the Madumabisa Mudstones have yielded diverse dinocephalians (anteosaurids, tapinocephalids and *Criocephalosaurus*) (Boonstra 1946; Lepper *et al.* 2000) of probable Gamkan age.

The *Endothiodon* record in the K5 interval of the Ruhuhu Formation in the Ruhuhu depression in Tanzania may be of Gamkan age (Cox 1964; Gay & Cruickshank 1999).

Zone II (Isheevo) of the Russian Permian (Ivakhnenko *et al.* 1997) has long been correlated to the South African *Tapinocephalus* zone (e.g. Chudinov 1975) based on shared evolutionary counterparts in biarmosuchians, anteosaurid and tapinocephalid dinocephalians and anomodonts and therefore is of Gamkan age.

In the Ordos Basin of northern China, the Xidagou Formation yields the temnospondyl *Anakamacops*, an *Intasuchus*-like temnospondyl, the anthracosaurs *Ingentidens* and *Phratochronis*, the bolosaur *Belebey* (also known from Russian Zone II), a captorhinid, the dinocephalians *Sinophoneus* and *Stenocybus* and the anomodont *Biseridens*. This is the *Biseridens* assemblage of probable Gamkan age (Lucas 2005*b*).

In the Paraná Basin of southern Brazil, the Posto Queimado and Aceguá tetrapod assemblages include diverse dinocephalians and *Pareiasaurus* and are of probable Gamkan age (Araújo 1985; Barbarena *et al.* 1985*b*; Lee 1997; Langer *et al.* 1998; Langer 2000; Cisneros *et al.* 2005).

Recently described tetrapods from intraformational conglomerates of the Buena Vista Formation in northeastern Uruguay (on the southern flank of the Paraná Basin) include the procolophonoid *Pintosaurus*, a supposed varanopid pelycosaur (though I doubt this identification) and a temnospondyl, and may be a single biostratigraphic assemblage of Gamkan age (Marsicano *et al.* 2000; Piñeiro *et al.* 2003, 2004). I base this very tentative conclusion largely on the fact that the Buena Vista Formation is homotaxial to the Sango do Cabral Formation of Brazil and that the youngest varanopids are of Gamkan age (Modesto *et al.* 2001).

Comments
The Gamkan as used here is faunachron G of Lucas (2002). The fauna of the *Pristerognathus* Assemblage Zone is a depauperate subset of the *Tapinocephalus* Assemblage Zone, which I have not treated as distinctive, so it is considered to be of late Gamkan age. Boonstra (1969) divided his *Tapinocephalus* zone into three assemblages, lower, middle and upper (= *Pristerognathus* assemblage zone), so the potential exists for subdivision of the Gamkan LVF.

Hoedemakeran LVF

Definition
The Hoedemakeran LVF is the time interval between the Gamkan and Steilkransian LVFs. The name is for the Hoedemaker River in South Africa, near the type locality of the *Tropidostoma* assemblage zone. The Hoedemakeran LVF begins with the FAD of the dicynodont *Tropidostoma*.

Index fossils
Tropidostoma is an index taxon of Hoedemakeran time. The FADs of *Cistecephalus, Kingoria, Lycaenops, Rhachiocephalus* and *Youngina* help to identify the Hoedemakeran (Fig. 10).

Characteristic assemblage
The characteristic Hoedemakeran assemblage is from the middle Teekloof Formation, Beaufort Group, South Africa. The characteristic tetrapod assemblage is much of the *Tropidostoma* Assemblage Zone (below the LO of *Cistecephalus*) of Smith & Keyser (1995*c*) and includes the temnospondyl *Rhinesuchus*, the pareiasaur *Pareiasaurus*, therocephalians (but no scylacosaurids), gorgonopsians and numerous dicynodonts, especially *Diictodon, Pristerodon, Tropidostoma* and *Endothiodon*.

Principal correlatives
The lower part of the Kawinga Formation in the Ruhuhu depression of Tanzania yields a temnospondyl?, a pareiasaur?, *Endothiodon?, Rhachiocephalus, Pristerodon*? and *Pachytegos*? (Gay & Cruickshank 1999) and may be of Hoedemakeran age.

In the Paraná Basin of southern Brazil, the Serra do Cadeao locality yielded rhinesuchids and *Endothiodon* and is probably of Hoedemakeran age (Barbarena & Araújo 1975; Barbarena & Dias 1998; Barbarena *et al.* 1985*a*, *b*; Barbarena 1998; Cisneros *et al.* 2005).

In Russia, all or part of the *Proelginia permiana* assemblage zone (*sensu* Golubev 2005) may be of Hoedemakeran age. The tetrapod assemblage includes chroniosuchids, 'procolophonids,' pareiasaurs, burnetiids, gorgonopids, dicynodonts and cynodonts.

Comments
The Hoedemakeran as used here is faunachron H of Lucas (2002). There is a substantial turnover of pareiasaurs at the beginning of the Hoedemakeran, and a very diverse dicynodont fauna characterizes this LVF.

Steilkransian LVF

Definition
The Steilkransian LVF is the time interval between the Hoedemakeran and Platbergian LVFs. The name is for the Steilkrans farm in South Africa, which is the type locality of the *Cistecephalus* assemblage zone. The Steilkransian LVF begins with the FAD of the dicynodont *Cistecephalus*.

Index fossils
Aulacephalodon is an index taxon of the Steilkransian. The FAD of *Oudenodon* and the LADs of *Cistecephalus*, *Endothiodon, Gorgonops* and *Rhachiocephalus* also help to identify the Steilkransian (Fig. 10).

Characteristic assemblage
The characteristic Steilkransian tetrapod assemblage is from the Upper Teekloof Formation, Beaufort Group, South Africa. The characteristic tetrapod assemblage thus combines the uppermost *Tropidostoma* Assemblage Zone and the *Cistecephalus* Assemblage Zone (Smith & Keyser 1995*c*, *d*) and includes the temnospondyl *Rhinesuchus*, captorhinids, therocephalians, a biarmosuchian and dicynodonts, especially *Diictodon, Cistecephalus, Emydops, Aulacephalodon* and *Oudenodon*.

Principal correlatives
Cistecephalus occurs in the Madumabisa Mudstone Formation of the Luangwa Valley in Zambia, and the closely related *Kawingasaurus* is present in the Kawinga Formation in the Ruhuhu depression of Tanzania (Gay & Cruickshank 1999), so these are records of probable Steilkransian age.

In Malawi, the Chiweta beds are coal-bearing strata of the Karoo Supergroup that yield *Endothiodon*, *Oudenodon* and a new biarmosuchian and are of probable Steilkransian age (Haughton 1926; Jacobs *et al.* 2005).

In southern Madagascar, south of the Isalo massif, the lower Sakamena Formation yields *Oudenodon*, *Rhinesuchus* and various endemic reptiles (Piveteau 1926; Mazin & King, 1991), and is probably of Steilkransian age.

In the Pranhita–Godavari Valley of India, the Kundaram Formation yields a captorhinid and the dicynodonts *Endothiodon, Pristerodon, Emydops, Cistecephalus* and *Oudenodon*, an assemblage of Steilkransian age (Ray 1999, 2001).

In northern China, the *Shihtienfenia* assemblage from the Shihezi Formation in Henana and the Sunjiagou Formation in Shanxi yields the temnospondyl *Bystrowiana* and various pareiasaurs, especially *Shihtienfenia* (Lucas 2005*b*). The pareiasaurs are most similar to characteristic Steilkransian pareiasaurs such as *Scutosaurus, Pareiasaurus* and *Pareiasuchus*, which suggests a tentative correlation (Lucas 2005*b*).

Comments
The Steilkransian as used here is faunachron I of Lucas (2002). Its boundaries are marked by significant evolutionary turn-over in pareiasaurs, gorgonopsians and therocephalians.

Platbergian LVF

Definition
The Platbergian LVF is the time interval between the Steilkransian and Lootsbergian LVFs (see Lucas 1998*b* for definition of the Lootsbergian). The name is for Platberg in South Africa, which is the type locality of the *Dicynodon* assemblage zone. The Platbergian LVF begins with the FAD of the dicynodont *Dicynodon*.

Index fossils
Dicynodon is the key index taxon of Platbergian time. *Pelanomodon* and *Theriognathus* are also restricted to Platbergian time. The LADs of *Oudenodon*, *Aulacephalodon* and a variety of tetrapod taxa that become extinct at or just before the Permo-Triassic boundary also help to identify the Platbergian (Fig. 10).

Characteristic assemblage
The characteristic tetrapod assemblage is from the uppermost Teekloof and the Balfour formations, Beaufort Group, South Africa. The characteristic assemblage combines the uppermost *Cistecephalus* Assemblage Zone and the *Dicynodon* Assemblage Zone (Kitching 1995), a tetrapod assemblage dominated by dicynodonts (especially *Dicynodon, Diictodon* and *Pelanomodon*) with some biarmosuchians, diverse gorgonopsians and therocephalians (especially *Theriognathus*) and cynodonts (especially *Procynosuchus*).

Principal correlatives
The broad distribution of *Dicynodon* establishes the Platbergian as the most widely recognizable (correlateable) of the Permian LVFs. Tetrapod assemblages of Platbergian age are:

(1) Karoo Basin, South Africa, where specimens of *Dicynodon* first occur in the upper *Cistecephalus* Assemblage Zone and are the dominant tetrapod fossils in the *Dicynodon* Assemblage Zone of the Teekloof and Balfour formations (Kitching 1995);
(2) part of the Kawinga Formation in the Ruhuhu Valley of Tanzania (Haughton 1932; Gay & Cruickshank 1999; Maisch & Gebauer 2005);
(3) 'Horizon 5' of Boonstra in the Luangwa Valley, 4.8–6.4 km north of Nt'awere, Zambia (King & Jenkins, 1997);
(4) Cutties Hillock Quarry, Elgin, Scotland (Newton 1893; King 1988) in the Cutties Hillock Sandstone Formation (Benton & Walker 1985);
(5) the Hopeman Sandstone at Clashbach Quarry, Scotland (Clark 1999);
(6) various localities of the Upper Sokolki assemblage and Vyatskyan assemblage of the Russian Upper Tatarian (Amalitzky 1922; Sushkin 1926; Ivakhnenko *et al.* 1997; Kurkin 1999; Kalandadze & Kurkin 2000; Golubev 2000; Lucas 2005*b*);
(7) Quanzijie, Wutonggou and Guodikeng formations in the Junggur and Turpan basins, Xinjiang Province, China (Lucas 1998*a*, 2001, 2005*a*);
(8) Sunan Formation, Gansu and Naobaogou Formation, Nei Monggol, both Ordos Basin, China (Lucas 1998*a*, 2001, 2005*a*; Li *et al.* 2000);
(9) north of the Mekong River in the Luang–Prabang area of Laos (Battail *et al.* 1995; Battail 1997).

Comments
The Platbergian as used here is faunachron J of Lucas (2002). *Dicynodon* is a long recognized and extensively studied Permian dicynodont (King 1988). Nevertheless, the amount and significance of variation in the genus has never been fully documented and analysed, so that the species-level taxonomy of *Dicynodon* has remained open to discussion (Cluver & Hotton 1981; King 1988).

Recently, Angielczyk & Kurkin (2003) advocated a cladistic approach to the species-level taxonomy of *Dicynodon* that purports to split it into several genera that correspond to terminal nodes on a cladogram. Lucas & Kondrashov

(2004) referred to such an approach as 'cladotaxonomy', and defined a cladotaxon as a low-level taxon (genus or species) that corresponds to a clade in a cladistic analysis. Lucas (2005*c*) critiqued the cladotaxonomy of *Dicynodon* as basically typological, over-split, of little biological significance and premature.

With regard to alpha taxonomy, taxonomic identity should be demonstrated by morphological similarity analysed within the context of population variation. Such an analysis will produce species-level taxa of potential biological significance that can be organized into genera. This is preferable to the typology inherent to cladotaxonomy, which will recognize several genera in what was formerly *Dicynodon* based only on their perceived cladistic relationships. However, having said this, there still needs to be an extensive overhaul of the taxonomy of the genus *Dicynodon* to better assess its utility and the utility of its species in Permian biostratigraphy.

Cross-correlations

The Pennsylvanian–Permian boundary as currently defined falls within the Wolfcampian Stage, so the boundary is within the Coyotean LVF (Fig. 11). This is because the LO of *Sphenacodon* is Late Virgilian (Harris *et al.* 2004).

Sumida *et al.* (1999*a*, *b*) assigned a Late Pennsylvanian age to the Coyotean tetrapod assemblage of the Halgaito Formation of the Cutler Group in the Arizona–Utah borderland. They based this age assignment on Baars (1995, pp. 39–40), who stated that the mixed marine–non-marine strata of the Elephant Canyon Formation, the supposed lateral equivalent of the Halgaito Formation, is mostly of Late Pennsylvanian age. However, a review of the age data on and unresolved debate over the Elephant Canyon Formation (e.g. Welsh 1958; Baars 1962, 1987, 1991; Loope *et al.* 1990; Condon 1997) reveals a much more complex picture. Thus, whether or not the Elephant Canyon Formation is a valid lithostratigraphical unit is uncertain and, according to Condon (1997), the Halgaito Formation only correlates to the uppermost Elephant Canyon Formation (but see Baars 1987 for a different correlation). The Elephant Canyon Formation yields three temporally successive fusulinid assemblages: *Triticites*-dominated (Virgilian), *Schwagerina*-dominated (probably Bursum age, which is now latest Pennsylvanian) and *Pseudoschwagerina*-dominated (earliest Permian). Clearly, the Halgaito Formation and its tetrapod assemblage are close in age to the Pennsylvanian–Permian boundary, but it is not clear whether they are entirely Pennsylvanian or entirely Early Permian.

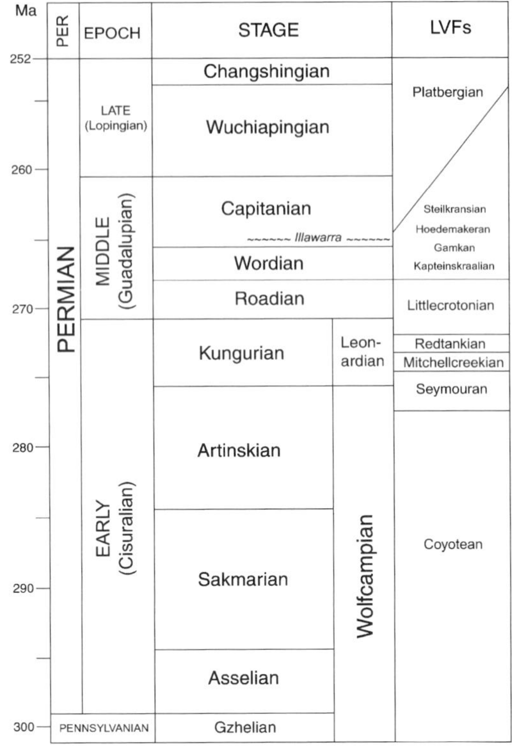

Fig. 11. Cross-correlation of the tetrapod biochronology proposed here to the SGCS of Wardlaw *et al.* (2004).

The Early Permian tetrapod record is restricted to North America and western Europe, so the biochronological scheme of Early Permian faunachrons has no current applicability outside of a Euramerican palaeoprovince. For most of the Coyotean, however, which was during the Gondwana glaciation, it is unlikely that any tetrapods lived in Gondwana.

Recent re-correlation of the North American Early Permian marine stages (Wolfcampian and Leonardian) to the standard Russian Cisuralian stages indicates that the Leonardian is only equivalent to the Kungurian, so the Wolfcampian is much longer than the Leonardian (Wardlaw *et al.* 2004). Numerical calibration of this part of the Permian time scale is imprecise, being based largely on interpolation between a cluster of radioisotopic ages near the Carboniferous–Permian boundary, an Artinskian U–Pb age from Russia of 280.3 ± 2.5 and the U–Pb age of the Capitanian base of

265.3 ± 0.2 Ma (Wardlaw *et al.* 2004). If the graphic correlation based on these numbers is used, then the Wolfcampian is about 23 Ma long (~ 276–299 Ma ago), and the Leonardian is only about 6 Ma long. This indicates that the Coyotean is 15–20 Ma long, whereas Seymouran time is closer to 5 Ma long. The three remaining Early Permian LVFs encompass less than 2 Ma each (Fig. 11). This suggests very little evolutionary turn-over in the tetrapod fauna during Coyotean time (a true chronofauna) followed by substantially higher faunal turn-over rates in the late Wolfcampian–Leonardian that may be related to the drier and more seasonal climates of the late Early Permian (e.g. Olson & Vaughn 1970).

At the Early-Middle Permian boundary, the basis for the LVFs shifts from North America to South Africa. I advocate recognition of a global gap between the youngest North American Permian tetrapods (San Angelo Formation and equivalents) and the oldest, therapsid-bearing faunas, those of Russian Zone I and the *Eodicynodon* Assemblage Zone of South Africa (Lucas 2004). Thus, Lucas (2004) explained in detail why the youngest North American Permian tetrapod assemblages (from the San Angelo, Flowerpot and Chickasha formations of Texas–Oklahoma) are late Leonardian in age. In brief, this is because intercalated marine strata yield Leonardian fusulinids, and overlying strata at the base of the Blaine Formation yield ammonoids of late Leonardian age (Fig. 6).

I have also accepted the argument (see above) that the *Eodicynodon* Assemblage Zone is probably the oldest therapsid-bearing assemblage, because it contains the most primitive therapsids. Therefore, I define the beginning of the Kapteinskraalian by the FAD of *Eodicynodon*, and consider the San Angelo assemblage to be older, and characteristic of the Littlecrotonian LVF (Fig. 11). Thus, the gap in the tetrapod record is equivalent to the younger part of the Littlecrotonian, which is part or all of the Roadian of the SGCS.

In the Russian Tatarian, the Illawara event is just below the LO of *Dicynodon*, which is approximately at the Urzhumian–Severodvinskian boundary, so this is ~ 265 Ma (Lozovsky *et al.* 2001; Menning 2001). If the LO of *Dicynodon* elsewhere is approximately synchronous (within resolution) with the Russian LO, then the Platbergian (= *Dicynodon* biochron of Lucas 1997) is very long, spanning about 14 Ma (~ 252–268 Ma ago). Furthermore, this indicates that the four Middle Permian LVFs, which are no older than Wordian, represent about 4 Ma of Permian time (Fig. 11). Faunal turn-over rates would thus have been extremely high during the Wordian, with explosive diversifications of dinocephalians and therapsids.

Nevertheless, a very long Platbergian necessitates substantial differences in sedimentation rates in the Karoo basin succession. Thus, the *Tapinocephalus* zone has a maximum thickness of 2000 m, whereas the other zones are 200–600 m thick. If the 500–m thick *Dicynodon* assemblage zone in the Karoo is 14 Ma long, then the underlying tetrapod zones, with a combined maximum thickness of about 3500 m, are squeezed into an interval about 4 Ma long. The average sedimentation rates would thus be about 36 mm/1000 years during the Platbergian, and 875 mm/1000 years for the Kapteinskraalian–Steilkransian, average rates of sedimentation that are well within the range of average rates for fluvial systems (Schindel 1980, 1982; Sadler 1981). However, whether or not such drastic changes in sedimentation rates are possible in the Karoo section needs to be addressed.

It is also possible that the LO of *Dicynodon* in the Russian section is much older than its LO in the Karoo Basin, with its LO in the Karoo being the result of immigration. Finally, there is the problem of the taxonomy of *Dicynodon* discussed above. What is called *Dicynodon* at its LO in the Russian section may not be the same taxon at its LO in the Karoo section. At present, I lack the data to resolve the problems posed by cross-correlation of the Platbergian to the SGCS, so, on Figure 11, I show the Platbergian base as a diagonal line that covers the range of possibilities. This is an important problem that needs resolution.

Traditionally, the Permian-Triassic boundary has been placed at the FAD of the dicynodont *Lystrosaurus*. However, it is likely that the FAD of *Lystrosaurus* is actually latest Permian (Lucas 1998*b*; Hancox *et al.* 2002; Retallack *et al.* 2003). Therefore, the boundary is within the Lootsbergian LVF of Lucas (1998*b*), which immediately follows the Platbergian of this paper. Thus, like the Carboniferous–Permian boundary, the Permian-Triassic boundary does not correspond to an LVF boundary.

I am grateful to T. Hentz and A. Milner for unpublished information. Collaboration in the field and the museum with D. Berman, D. Chaney, S. Harris, A. Henrici and K. Krainer influenced the content of this paper. Reviews by D. Berman, S. Harris, S. Modesto, B. Rubidge, S. Sumida and R. Werneburg corrected many shortcomings in the manuscript and are gratefully acknowledged.

References

AMALITZKY, V. 1922. Diagnoses of the new forms of vertebrates and plants from the Upper Permian of North Dvina. *Bulletin of the Academy of Science, St Petersburg*, **16**, 329–340.

ANDERSON, J. M. 1981. World Permo-Triassic correlations: their biostratigraphic basis. *In*: CRESSWELL, N. N. & VELLA, P. (eds) *Gondwana Five*. A. A. Balkema, Rotterdam, 3–10.

ANDERSON, J. M. & CRUICKSHANK, A. R. I. 1978. The biostratigraphy of the Permian and the Triassic. Part 5. A review of the classification and distribution of Permo-Triassic tetrapods. *Palaeontologia Africana*, **21**, 15–44.

ANGIELCZYK, K. D. & KURKIN, A. A. 2003. Has the utility of *Dicynodon* for Late Permian terrestrial biostratigraphy been overstated? *Geology*, **31**, 363–366.

ARAÚJO, D. C. 1985. Sobre *Pareiasaurus americanus* sp. nov., do Permiano Superior do Rio Grande do Sul, Brasil. I: Diagnose específica. *Anais da Academie Brasileira de Ciências*, **57**, 63–66.

BAARS, D. L. 1962. Permian System of Colorado Plateau. *American Association of Petroleum Geologists Bulletin*, **46**, 149–218.

BAARS, D. L. 1987. The Elephant Canyon Formation revisited. *Four Corners Geological Society Guidebook*, **10**, 81–90.

BAARS, D. L. 1991. The Elephant Canyon Formation – for the last time. *The Mountain Geologist*, **28**, 1–2.

BAARS, D. L. 1995. *Navajo County*. University of New Mexico Press, Albuquerque.

BARBARENA, M. C. 1998, *Australerpeton cosgriffi* n. g., n. sp.: a Late Permian rhinesuchoid amphibian from Brazil. *Anais da Academie Brasileira de Ciências*, **70**, 125–137.

BARBARENA, M. C. & ARAÚJO, D. C. 1975. Tetrapodos fossils de Sudamerica y deriva continental. I *Congreso Argentiono de Paleontologi i Bioestratigrafia Actas*, **1**, 497–504.

BARBARENA, M. C. & DIAS, E. V. 1998. On the presence of a short-snouted rhinesuchoid amphibian in the Rio do Rasto Formation (Late Permian of Paraná Basin, Brazil). *Anais da Academie Brasileira de Ciências*, **70**, 465–468.

BARBARENA, M. C., ARAÚJO, D. C. & LAVINA, E. L. 1985*a*. Late Permian and Triassic tetrapods of southern Brazil. *National Geographic Research*, **1**, 5–20.

BARBARENA, M. C., ARAÚJO, D. C., LAVINA, E. L. & AZEVEDO, S. A. K. 1985*b*. O estado atual do conheciemento sobre os tetrápodes Permianos e Triássico do Brasil meridional. *Coletânea de Trabalhos Paleontologicos, Sério Geologico*, **27**, 21–28.

BATTAIL, B. 1997. Les genres *Dicynodon* et *Lystrosaurus* (Therapsida, Dicynodontia) en Eurasie: une mise au point. *Géobios, Memoire Speciale*, **20**, 39–48.

BATTAIL, B. 2000. A comparison of Late Permian Gondwanan and Laurasian amniote faunas. *Journal of African Earth Sciences*, **31**, 165–174.

BATTAIL, B., DEJAX, J., RICHIR, P., TAQUET, P. & VE'RAN, M. 1995. New data on the continental Upper Permian in the area of Luang-Prabang, Laos. *Geological Survey of Vietnam Journal of Geology, Series B*, **5–6**, 11–15.

BENTON, M. J. & WALKER, A. D. 1985. Palaeoecology, taphonomy and dating of Permo-Triassic reptiles from Elgin, north-east Scotland. *Palaeontology*, **28**, 207–234.

BENTON, M. J., TVERDOKHLEBOV, V. P. & SURKOV, M. V. 2004. Ecosystem remodeling among vertebrates at the Permian–Triassic boundary in Russia. *Nature*, **432**, 97–100.

BERMAN, D. S. 1978. *Ctenospondylus ninehevensis*, a new species (Reptilia, Pelycosauria) from the Lower Permian Dunkard Group of Ohio. *Annals of Carnegie Museum*, **47**, 493–514.

BERMAN, D. S. 1993. Lower Permian vertebrate localities of New Mexico and their assemblages. *In*: LUCAS, S. G. & ZIDEK, J. (eds) *Vertebrate Palaeontology in New Mexico*. New Mexico Museum of Natural History and Science Bulletin, **2**, 11–21.

BERMAN, D. S. & BERMAN, S. L. 1975. *Broiliellus hektotopos* sp. nov. (Temnospondyli: Amphibia) Washington Formation, Dunkard Group, Ohio. *In*: BARLOW, J. A. & BURKHAMMER, S. (eds) *Proceedings of the first I. C. White memorial symposium 'The age of the Dunkard'*. West Virginia Geological and Economic Survey, Morgantown, 69–78.

BERMAN, D. S & LUCAS, S. G. 2003. *Aspidosaurus binasser* (Amphibia, Temnospondyli): a new species of Dissorophidae from the Lower Permian of Texas. *Annals of Carnegie Museum*, **72**, 241–262.

BERMAN, D. S. & MARTENS, T. 1993. First occurrence of *Seymouria* (Amphibia: Batrachosauria) in the Lower Permian Rotliegend of central Germany. *Annals of Carnegie Museum*, **62**, 63–79.

BERMAN, D. S. & REISZ, R. R. 1982. Restudy of *Mycterosaurus longiceps* (Reptilia, Pelycosauria) from the Lower Permian of Texas. *Annals of Carnegie Museum*, **51**, 423–453.

BERMAN, D. S., REISZ, R. R. & EBERTH, D. A. 1987. *Seymouria sanjuanensis* (Amphiba, Batrachosauria) from the Lower Permian Cutler Formation of north-central New Mexico and the occurrence of sexual dimorphism in that genus questioned. *Canadian Journal of Earth Sciences*, **24**, 1769–1784.

BERMAN, D. S., SUMIDA, S. S. & LOMBARD, R. E. 1997. Biogeography of primitive amniotes. *In*: SUMIDA, S. S. & MARTIN, K. L. M. (eds) *Amniote Origins*. Academic Press, San Diego, 85–139.

BERMAN, D. S., HENRICI, A. C., SUMIDA, S. S. & MARTENS, T. 2000. Redescription of *Seymouria sanjuanensis* (Seymouriamorpha) from the Lower Permian of Germany based on complete, mature specimens with a discussion of paleoecology of the Bromacker locality assemblage. *Journal of Vertebrate Paleontology*, **20**, 253–268.

BERMAN, D. S., REISZ, R. R., MARTENS, T. & HENRICI, A. C. 2001. A new species of *Dimetrodon* (Synapsida: Sphenacodontidae) from the Lower Permian records first occurrence of genus outside of North America. *Canadian Journal of Earth Sciences*, **38**, 803–812.

BERMAN, D. S, HENRICI, A. C., KISSEL, R. A., SUMIDA, S. S. & MARTENS, T. 2004. A new diadectid

(Diadectomorpha), *Orobates pabsti*, from the Early Permian of central Germany. *Bulletin of the Carnegie Museum of Natural History*, **35**, 1–36.

BÖSE, E. 1917. *The Permo-Carboniferous Ammonoids of the Glass Mountains, West Texas, and Their Stratigraphical Significance*. University of Texas Bulletin, **1762**.

BOONSTRA, L. D. 1946. Report on some fossil reptiles from Gunyanka's Kraal, Busi Valley. *Proceedings and Transactions of the Rhodesian Scientific Association*, **41**, 46–49.

BOONSTRA, L. D. 1969. The fauna of the *Tapinocephalus* Zone (Beaufort Beds of the Karoo). *Annals of the South African Museum*, **56**(1), 1–73.

BOY, J. A. 1993. Synopsis of the tetrapods from the Rotliegend (Lower Permian) in the Saar–Nahe Basin (SW Germany). *Pollichia*, **29**, 155–169.

BROOM, R. 1906. On the Permian and Triassic faunas of South Africa. *Geological Magazine*, **5**, 29–30.

BROOM, R. 1907. On the geological horizons of the vertebrate genera of the Karoo formation. *Records of the Albany Museum*, **2**, 156–163.

BROOM, R. 1909. An attempt to determine the horizons of the fossil vertebrates of the Karoo. *Annals of the South African Museum*, **7**, 285–289.

BROWN, L. F., JR. 1973. Cisco depositional systems in north-central Texas. *Bureau of Economic Geology University of Texas at Austin, Guidebook*, **14**, 57–73.

BURKHALTER, R. J. & MAY, W. J. 2002. Lower Permian vertebrates from southwest Oklahoma. *Oklahoma Geological Survey, Open-file Report*, **10–2002**, 75–80.

CARROLL, R. L. 1979, Problems in the use of terrestrial vertebrates for zoning of the Carboniferous. *In*: SUTHERLAND, P. K. & MANGER, W. L. (eds) *Neuvième Congrès International de Stratigraphie et de Géologie du Carbonifère, Comptes Rendus. 2: Biostratigraphy*. Southern Illinois University Press, Carbondale, 135–147.

CARROLL, R. L. 1997. Limits to knowledge of the fossil record. *Zoology*, **100**, 221–231.

CHENG, Z. 1981. Permo-Triassic continental deposits and vertebrate faunas of China. *In*: CRESSWELL, N. N. & VELLA, P. (eds) *Gondwana Five*. A. A. Balkema, Rotterdam, 65–70.

CHUDINOV, P. K. 1975. New facts about the fauna of the Upper Permian of the U.S.S.R. *Journal of Geology*, **73**, 117–130.

CISNEROS, J. C., ABDALA, F. & MALABARBA, M. C. 2005. Pareiasaurids from the Rio do Rasto Formation, southern Brazil: biostratigraphic implications for Permian faunas of the Paraná Basin. *Revista Brasiliera de Paleontologia*, **8**, 13–24.

CLARK, J. & CARROLL, R. L. 1973. Romeriid reptiles from the Lower Permian. *Harvard University Museum of Comparative Zoology Bulletin*, **144**, 353–407.

CLARK, N. D. L. 1999. The Elgin marvel. *Open University Geological Society Journal*, **20**, 16–18.

CLUVER, M. A. & HOTTON, N. 1981. The genera *Dicynodon* and *Diictodon* and their bearing on the classification of the Dicynodontia. *Annals of the South African Museum*, **83**, 99–146.

CONDON, S. M. 1997. *Geology of the Pennsylvanian and Permian Cutler Group and Permian Kaibab Limestone in the Paradox Basin, Southeastern Utah and southwestern Colorado*. U. S. Geological Survey, Professional Paper, **2000**-P.

COOPER, M. R. 1982. A Mid-Permian to earliest Jurassic tetrapod biostratigraphy and its significance. *Arnoldia Zimbabwe*, **9**, 77–103.

COX, C. B. 1964. On the palate, dentition and classification of the fossil reptile *Endothiodon* and related genera. *American Museum Novitates*, **2171**, 1–25.

CRADDOCK, K. W. & HOOK, R. W. 1989. An overview of vertebrate collecting in the Permian System of north-central Texas. *In*: HOOK, R. W. (ed.) *Permo-Carboniferous Vertebrate Paleontology, Lithostratigraphy and Depositional Environments of North-central Texas*. Society of Vertebrate Paleontology, Austin, 40–46.

DIMICHELE, W. A. & CHANEY, D. S. 2005. Pennsylvanian–Permian fossil floras from the Cutler Group, Cañon del Cobre and Arroyo del Agua areas in northern New Mexico. *In*: LUCAS, S. G., ZEIGLER, K. E. & SPIELMAN, J. A. (eds) *The Permian of Central New Mexico*. New Mexico Museum of Natural History and Science Bulletin, **31**, 26–33.

DUNBAR, C. O. & SKINNER, J. W. 1937. Permian Fusulinidae of Texas. *University of Texas Bulletin*, **3701**, 516–732.

EARDLE, D. H. 1960. Stratigraphy of Pennsylvanian and Lower Permian rocks in Brown and Coleman Counties, Texas. *U. S. Geological Survey, Professional Paper*, **315–D**, 55–78.

EBERTH, D. A., BERMAN, D. S., SUMIDA, S. S. & HOPF, H. 2000. Lower Permian terrestrial paleoenvironments and vertebrate paleoecology of the Tambach Basin (Thuringia, central Germany): the upland holy grail. *Palaios*, **15**, 293–313.

EFREMOV, I. A. 1937. On the stratigraphic subdivision of the continental Permian and Triassic of the USSR on the basis of the fauna of early Tetrapoda. *Doklady Akademiya Nauk SSSR*, **16**, 121–126.

EFREMOV, I. A. 1940. Preliminary description of new Permian and Triassic terrestrial vertebrates from the USSR. *Trudy Paleontologicheskaya Instituta*, **10**, 1–140.

EFREMOV, I. A. 1952. On the stratigraphy of the Permian red beds of the USSR based on terrestrial vertebrates. *Izvestiya Akademiya Nauk SSSR Geology Series*, **6**, 49–75.

EFREMOV, I. A. 1956. American elements in the fauna of the USSR. *Doklady Akademiya Nauk SSSR*, **111**, 1091–1094.

EWER, R. F. 1961. Anatomy of the anomodont *Daptocephalus leoniceps*. *Proceedings of the Zoological Society of London*, **136**, 375–402.

FOREMAN, B. C. & MARTIN, L. D. 1988. A review of Palaeozoic tetrapod localities of Kansas and Nebraska. *Kansas Geological Survey Guidebook*, **6**, 133–145.

FRACASSO, M. A. 1980. Age of the Permo-Carboniferous Cutler Formation vertebrate fauna from El Cobre Canyon, New Mexico. *Journal of Paleontology*, **54**, 1237–1244.

FREDE, S., SUMIDA, S. S. & BERMAN, D. S. 1993. New information on Early Permian vertebrates from the

Halgaito Tongue of the Cutler Formation of southeastern Utah. *Journal of Vertebrate Paleontology*, **13**, 36A.

GAY, S. A. & CRUICKSHANK, A. R. I. 1999. Biostratigraphy of the Permian tetrapod faunas from the Ruhuhu Valley, Tanzania. *Journal of African Earth Sciences*, **29**, 195–210.

GOLUBEV, V. K. 1998. Revision of the Late Permian chronosuchians (Amphibia, Anthracosauromorpha) from eastern Europe. *Paleontological Journal*, **32**, 390–401.

GOLUBEV, V. K. 2000. The faunal assemblages of Permian terrestrial vertebrates from eastern Europe. *Paleontological Journal*, **34**, 5211–5224.

GOLUBEV, V. K. 2005. Permian tetrapod stratigraphy. *In*: LUCAS, S. G. & ZIDEK, K. E. (eds) *The Nonmarine Permian.* New Mexico Museum of Natural History and Science Bulletin, **30**, 95–99.

HANCOX, P. J., BRANDT, D., REIMOLD, W. U., KOEBERL, C. & NEVELING, J. 2002. Permian-Triassic boundary in northwest Karoo Basin: current stratigraphic placement, implications for basin developmental models, and the search for evidence of an impact. *In*: KOEBERL, C. & MACLEOD, K. G. (eds) *Catastrophic Events and Mass Extinctions: Impacts and Beyond*, Geological Society of America, Special Paper, **356**, 429–444.

HARRIS, S. K., LUCAS, S. G., BERMAN, D. S & HENRICI, A. C. 2004. Vertebrate fossil assemblage from the Upper Pennsylvanian Red Tanks Member of the Bursum Formation, Lucero uplift, central New Mexico. *In*: LUCAS, S. G. & ZIEGLER, K. E. (eds) *Carboniferous–Permian Transition at Carriza Arrogo central New Mexico*. New Mexico Museum of Natural History and Science Bulletin, **25**, 267–283.

HAUGHTON, S. H. 1926. On Karroo vertebrates from Nyasaland. *Transactions of the Geological Society of South Africa*, **16**, 67–75.

HAUGHTON, S. H. 1932. On a collection of Karroo vertebrate fossils from Tanganyika Territory. *Quarterly Journal of the Geological Society, London*, **88**, 634–668.

HEATON, M. J. 1979. Cranial anatomy of primitive captorhinid reptiles from the Late Pennsylvanian and Early Permian, Oklahoma and Texas. *Oklahoma Geological Survey Bulletin*, **127**, 1–84.

HENBEST, L. G. 1938. Notes on the ranges of Fusulinidae in the Cisco Group (restricted) of the Brazos River region, north-central Texas. *University of Texas Publications*, **3801**, 237–247.

HENTZ, T. F. 1988. *Lithostratigraphy and Paleoenvironments of Upper Palaeozoic Continental Red Beds, North-Central Texas: Bowie (new) and Wichita (revised) Groups.* Bureau of Economic Geology, University of Texas at Austin, Report of Investigations, **170**.

HENTZ, T. F. 1989. Permo-Carboniferous lithostratigraphy of the vertebrate-bearing Bowie and Wichita Groups, north-central Texas. *In*: HOOK, R. W. (ed.) *Permo-Carboniferous Vertebrate Paleontology, Lithostratigraphy and Depositional Environments of North-central Texas.* Society of Vertebrate Paleontology, Austin, 1–21.

HENTZ, T. F. & BROWN, L. F. JR. 1987. *Wichita Falls–Lawton Sheet.* Geologic Atlas of Texas: Scale 1:250,000. Bureau of Economic Geology, University of Texas.

HOOK, R. W. 1989. Stratigraphic distribution of tetrapods in the Bowie and Wichita Groups, Permo-Carboniferous of north-central Texas. *In*: HOOK, R. W. (ed.) *Permo-Carboniferous Vertebrate Paleontology, Lithostratigraphy and Depositional Environments of North-central Texas.* Society of Vertebrate Paleontology, Austin, 47–53.

HUNT, A. P. & LUCAS, S. G. 1992. The paleoflora of the Lower Cutler Formation (Pennsylvanian, Desmoinesian?) *In*: El Cobre Canyon, New Mexico, and its biochronological significance. *New Mexico Geological Society, Guidebook*, **43**, 145–150.

HUTTENLOCKLER, A. K., PARDO, J. D. & SMALL, B. J. 2005. An earliest Permian nonmarine vertebrate assemblage from the Eskridge Formation, Nebraska. *In*: LUCAS, S. G. & ZEIGLER, K. E. (eds) *The Nonmarine Permian. New Mexico Museum of Natural History, Bulletin*, **30**, 133–143.

IVAKHNENKO, M. F. 1981. Discosauricidae from the Permian of Tadzhikistan. *Paleontological Journal*, **1981**, 90–102.

IVAKHNENKO, M. F., GOLUBYEV, V. K., GUBIN, YU. M., KALANDADZE, N. N., NOVIKOV, I. V., SENNIKOV, A. G. & RAUTIAN, A. S. 1997. *Permskiye I Triasovyye tetrapody vostochnoi Evropy* [Permian and Triassic Tetrapods of Eastern Europe]. GEOS, Moscow.

JACOBS, L. L., WINKLER, D. A., NEWMAN, J. D., GOMANI, E. M. & DEINO, A. 2005. Therapsids from the Permian Chiweta Beds and the age of the Karoo Supergroup in Malawi. *Palaeontologia Electronica*, **8**(1), 28A, 1–23.

KALANDADZE, N. N. & KURKIN, A. A. 2000. A new Permian dicynodont and the question of the origin of the Kannemeyeroidea. *Paleontological Journal*, **34**, 642–649.

KAUFFMAN, A. E. & ROTH, R. I. 1966. *Upper Pennsylvanian and Lower Permian Fusulinids from North-Central Texas.* Cushman Foundation for Foraminiferal Research, Special Publication, **8**, 1–49.

KEMP, A. H. 1962. The stratigraphic and geographic distribution of cephalopod genera in the Lower Permian of Baylor County, north-central Texas. *Journal of Paleontology*, **36**, 1124–1127.

KEYSER, A. W. & SMITH, R. M. H. 1977–1978. Vertebrate biozonation of the Beaufort Grup with special reference to the western Karoo Basin. *Annals of the Geological Survey, Republic of South Africa*, **12**, 1–35.

KING, G. M. 1988. Anomodontia. *Handbuch der Paleoherpetology*, **17C**.

KING, G. M. & JENKINS, I. 1997. The dicynodont *Lystrosaurus* from the Upper Permian of Zambia: evolutionary and stratigraphical implications. *Palaeontology*, **40**, 149–156.

KITCHING, J. W. 1977. *On the Distribution of the Karoo Vertebrate Fauna.* Bernard Price Institute for Palaeontological Research, Memoirs, **1**, 1–131.

KITCHING, J. W. 1995. Biostratigraphy of the *Dicynodon* Assemblage Zone. *In*: RUBIDGE, B. S. (ed.) *Biostratigraphy of the Beaufort Group (Karoo Supergroup). South African Committee for Stratigraphy, Biostratigraphic Series*, **1**, 29–34.

KURKIN, A. A. 1999. A new dicynodont from the Malaya Severnaya Dvina River excavations. *Paleontological Journal*, **33**, 297–301.

LANGER, M. C. 2000. The first record of dinocephalians in South America: Late Permian (Rio do Rasto Formation) of the Paraná Basin, Brazil. *Neues Jahrbuch für Geologie und Paläontologie, Abhandlungen*, **215**, 69–95.

LANGER, M. C., SCHULTZ, C. L. & LAVINA, E. L. 1998. On the first South American Anteosauroidea (Therapsida-Dinocephalia). *Ameghiniana*, **34**, 537.

LANGSTON, W. JR. 1953. *Permian Amphibians from New Mexico*. University of California, Publications in Geological Sciences, **29**, 349–416.

LANGSTON, W. JR. 1963. Fossil vertebrates and the Late Palaeozoic red beds of Prince Edward Island. *National Museum of Canada Bulletin*, **187**, 1–36.

LEE, M. S. Y. 1997. A taxonomic revision of pareiasaurian reptiles: implications for Permian terrestrial palaeoecology. *Modern Geology*, **21**, 231–298.

LEE, W., NICKELL, C. O., WILLIAMS, J. S. & HENBEST, L. G. 1938. *Stratigraphic and Paleontologic Studies of the Pennsylvanian and Permian Rocks in North-Central Texas*. University of Texas Bulletin **3801**.

LEPPER, J., RAATH, M. A. & RUBIDGE, B. S. 2000. A diverse dinocephalian fauna from Zimbabwe. *South African Journal of Science*, **96**, 403–405.

LEWIS, G. F. & VAUGHN, P. P. 1965. *Early Permian Vertebrates from the Cutler Formation of the Placerville area, Colorado*. U. S. Geological Survey, Professional Paper, **503-C**, 1–50.

LI, P., CHENG, Z. & LI, J. 2000. A new species of *Dicynodon* from Upper Permian of Sunan, Gansu, with remarks on related strata. *Vertebrata PalAsiatica*, **38**, 147–157.

LOOPE, D. B., SANDERSON, G. A. & VERVILLE, G. J. 1990. Abandonment of the name 'Elephant Canyon Formation' in southeastern Utah: physical and temporal implications. *The Mountain Geologist*, **27**, 119–130.

LOZOVSKY, V. A. 2005. Olson's gap or Olson's bridge: that is the question. *In*: LUCAS, S. G. & ZEIGLER, K. E. (eds) *The Nonmarine Permian*. New Mexico Museum of Natural History and Science Bulletin **30**, 179–184.

LOZOVSKY, V. R., KRASSILOV, V. A., AFONIN, S. A., BUROV, B. V. & YAROSHENKO, O. P. 2001. Transitional Permian–Triassic deposits in European Russia, and non-marine correlations. *In*: CASSINIS, G. (ed.) *Permian Continental Deposits of Europe. Regional Reports and Correlations.* Natura. Bresciana, Monografia, **25**, 301–310.

LUCAS, S. G. 1997. *Dicynodon* and Late Permian Pangaea. *In*: WANG, N. & REMANE, J. (eds) *Stratigraphy: Proceedings of the 30th International Geological Congress, Utrecht*. VSP International Science Publishers, **11**, 133–141.

LUCAS, S. G. 1998*a*. Toward a tetrapod biochronology of the Permian. *In*: LUCAS, S. G., ESTEP, J. W. & HOFFER, J. M. (eds) *Permian Stratigraphy and Paleaotology of the Robleds Mountains New Mexico.* New Mexico Museum of Natural History and Science Bulletin, **12**, 71–91.

LUCAS, S. G. 1998*b*. Global Triassic tetrapod biostratigraphy and biochronology. *Palaeogeography, Palaeoclimatology, Palaeoecology*, **143**, 347–384.

LUCAS, S. G. 2000. Toward a Carboniferous tetrapod biochronology. *Newsletter on Carboniferous Stratigraphy*, **19**, 20–22.

LUCAS, S. G. 2001. *Chinese fossil vertebrates*. Columbia University Press, New York.

LUCAS, S. G. 2002. Tetrapods and the subdivision of Permian time. *In*: HILLS, L.V., HENDERSON, C. M. & BAMBER, E. W. (eds) *Carboniferous and Permian of the World.* Canadian Society of Petroleum Geologists, Memoir **19**, 479–491.

LUCAS, S. G., 2004. A global hiatus in the Middle Permian tetrapod fossil record. *Stratigraphy*, **1**, 47–64.

LUCAS, S. G. 2005*a*. Olson's gap or Olson's bridge: an answer. *In*: LUCAS, S. G. & ZEIGLER, K. E. (eds) *The Nonmarine Permian.* New Mexico Museum of Natural History and Science Bulletin, **30**, 185–186.

LUCAS, S. G. 2005*b*. Age and correlation of Permian tetrapod assemblages from China. *In*: LUCAS, S. G. & ZEIGLER, K. E. (eds) *The Nonmarine Permian* New Mexico Museum of Natural History and Science Bulletin, **30**, 187–191.

LUCAS, S. G. 2005*c*. *Dicynodon* from the Upper Permian of Russia: biochronological significance. *In*: LUCAS, S. G. & ZEIGLER, K. E. (eds) *The Nonmarine Permian.* The New Mexico Museum of Natural History and Science Bulletin, **30**, 192–196.

LUCAS, S. G. 2005*d*. Permian tetrapod faunachrons. *In*: LUCAS, S. G. & ZEIGLER, K. E. (eds) *The Nonmarine Permian.* New Mexico Museum of Natural History and Science Bulletin, **30**, 197–201.

LUCAS, S. G. & KONDRASHOV, P. E. 2004. Early Eocene (Bumbanian) perissodactyls from Mongolia and their biochronological significance. *In*: LUCAS, S. G., ZEIGLER, K. E. & KONDRASHOV, P. E. (eds) *Paleogene Mammals.* New Mexico Museum of Natural History and Science Bulletin, **26**, 215–220.

LUCAS, S. G. & KRAINER, K. 2005. Cutler Group (Permo-Carboniferous) stratigraphy, Chama Basin, New Mexico. *In*: LUCAS, S. G., ZEIGLER, K. E. & SPIELMANN, J. A. (eds) *The Permian of Central New Mexico*. New Mexico Museum of Natural History and Science Bulletin, **31**, 90–100.

LUCAS, S. G., BERMAN, D. S, HENRICI, A. C. & HUNT, A. P. 2005*a*. *Bolosaurus* from the Lower Permian at Arroyo del Agua, New Mexico and its biostratigraphic significance. *In*: LUCAS, S. G., ZEIGLER, K. E. & SPIELMANN, J. A. (eds) *The Permian of Central New Mexico*. New Mexico Museum of Natural History and Science Bulletin, **31**, 125–127.

LUCAS, S. G., HARRIS, S. K., SPIELMANN, J. A., BERMAN, D. S & HENRICI, A. C. 2005*b*. Vertebrate biostratigraphy and biochronology of the Pennsylvanian–Permian Cutler Group, El Cobre Canyon, northern New Mexico. *In*: LUCAS, S. G., ZEIGLER, K. E. & SPIELMANN, J. A. (eds) *The Permian of Central New Mexico*. New Mexico Museum of Natural History and Science Bulletin, **31**, 128–139.

LUCAS, S. G., HARRIS, S. K. *et al.* 2005*c*. Early Permian vertebrate biostratigraphy at Arroyo Del Agua, Rio Arriba County, New Mexico. *In*: LUCAS, S. G., ZEIGLER, K. E. & SPIELMANN, J. A. (eds) *The Permian of Central New Mexico*. New Mexico Museum of Natural History and Science Bulletin, **31**, 163–169.

LUND, R. 1975. Vertebrate-fossil zonation and correlation of the Dunkard Basin. *In*: BARLOW, J. A. & BURKHAMMER, S. (eds) *Proceedings of the first I. C. White memorial symposium 'The age of the Dunkard'*. West Virginia Geological and Economic Survey, Morgantown, 171–178.

MAISCH, M. W. & GEBAUER, E. V. I. 2005. Reappraisal of *Geikia locusticeps* (Therapsida: Dicynodontia) from the Upper Permian of Tanzania. *Palaeontology*, **48**, 309–324.

MARSICANO, C., PEREA, D. & UBILLA, M. 2000. A new temnospondyl amphibian from the Lower Triassic of South America. *Alcheringa*, **24**, 119–123.

MAZIN, J. M. & KING, G. M. 1991. The first dicynodont from the Late Permian of Malagasy. *Palaeontology*, **34**, 837–842.

MENNING, M. 2001. A Permian time scale 2000 and correlation of marine and continental sequences using the Illawarra reversal (265 Ma). *In*: CASSINIS, G. (ed.) *Permian Continental Deposits of Europe. Regional Reports and Correlations.* Natura Bresciana, Monografia, **25**, 355–362.

MILLER, A. K. & FURNISH, W. M. 1940. *Permian Ammonoids of the Guadalupe Mountain Region and Adjacent Areas*. Geological Society of America, Special Paper, **26**.

MILLER, A. K. & YOUNGQUIST, W. 1947. *Lower Permian Cephalopods from the Texas Colorado River Valley*. University of Kansas Paleontological Contributions **2**, Mollusca, Article **1**, 1–15.

MILNER, A. R. 1990. The radiations of temnospondyl amphibians. *In*: TAYLOR P. D. & LARWOOD, G. P. (eds) *Major Evolutionary Radiations*. Clarendon Press, Oxford, 321–349.

MILNER, A. R. 1993. Biogeography of Palaeozoic tetrapods. *In*: LONG J. A. (ed.) *Palaeozoic Vertebrate Biostratigraphy and Biogeography*. Belhaven Press, London, 324–353.

MILNER, A. R., SMITHSON, T. R., MILNER, A. C., COATES, M. I. & ROLFE, W. D. I. 1986. The search for early tetrapods. *Modern Geology*, **10**, 1–28.

MODESTO, S. & RUBIDGE, B. 2000. A basal anomodont therapsid from the lower Beaufort Group, Upper Permian of South Africa. *Journal of Vertebrate Paleontology*, **20**, 515–521.

MODESTO, S. & RYBCZYNSKI, N. 2000. The amniote faunas of the Russian Permian: Implications for Late Permian terrestrial vertebrate biostratigraphy. *In*: BENTON, M. J., SHISHKIN, M. A., UNWIN, D. M. & KUROCHKIN, E. N. (eds) *The Age of Dinosaurs in Russia and Mongolia*. Cambridge University Press, Cambridge, 17–34.

MODESTO, S. P., RUBIDGE, B. S. & WELMAN, J. 1999. The most basal anomodont therapsid and the primacy of Gondwana in the evolution of the anomodonts. *Proceedings of the Royal Society of London, B*, **266**, 331–337.

MODESTO, S. P., SIDOR, C. A., RUBIDGE, B. S. & WELMAN, J. 2001. A second varanopseid skull from the Upper Permian of South Africa: implications for Late Permian 'pelycosaur' evolution. *Lethaia*, **34**, 249–259.

MODESTO, S. P., RUBIDGE, B. S. & WELMAN, J. 2002. A new dicynodont therapsid from the lowermost Beaufort Group, Upper Permian of South Africa. *Canadian Journal of Earth Sciences*, **39**, 1755–1765.

MODESTO, S. P., RUBIDGE, B. S. & WELMAN, J. 2003. A new basal dicynodont from the Upper Permian of South Africa. *Paleontology*, **46**, 211–223.

MORAN, W. E. 1952. Location and stratigraphy of known occurrences of fossil tetrapods in the Upper Pennsylvanian and Permian of Pennsylvania, west Virginia, and Ohio. *Annals of Carnegie Museum*, **33**, 1–44.

MURRY, P. A. & JOHNSON, G. D. 1987. Clear Fork vertebrates and environments from the Lower Permian of north-central Texas. *Texas Journal of Science*, **39**, 253–266.

MYERS, D. A. 1958. Stratigraphic distribution of some fusulinids from the Thrifty Formation, Upper Pennsylvanian, central Texas. *Journal of Paleontology*, **32**, 677–681.

MYERS, D. A. 1960. *Stratigraphic distribution of some Pennsylvanian Fusulinidae from Brown and Coleman Counties, Texas.* U. S. Geological Survey, Professional Paper, **315**-C.

MYERS, D. A. 1968. Schwagerina crassitectoria *Dunbar and Skinner, 1937, a fusulinid from the upper part of the Wichita Group, Lower Permian, Coleman County, Texas.* U. S. Geological Survey, Professional Paper, **600**-B, 133–139.

NACSN [North American Commission on Stratigraphic Nomenclature] 1983. North American stratigraphic code. *American Association of Petroleum Geologists Bulletin*, **67**, 841–875.

NELSON, W. J., HOOK, R. W. & TABOR, N. 2001. Clear Fork Group (Leonardian, Lower Permian) of north-central Texas. *Oklahoma Geological Survey Circular*, **104**, 167–169.

NEWTON, E. T. 1893. On some new reptiles from the Elgin Sandstones. *Philosophical Transactions of the Royal Society of London, Series B*, **184**, 431–503.

OLSON, E. C. 1952. The evolution of a Permian zvertebrate chronofauna. *Evolution*, **6**, 181–196.

OLSON, E. C. 1954. Notes on the stratigraphic and geographic ranges of certain genera of Permian vertebrates. *Journal of Geology*, **62**, 610–611.

OLSON, E. C. 1957. Catalogue of localities of Permian and Triassic terrestrial vertebrates of the territories of the U.S.S.R. *Journal of Geology*, **65**, 196–226.

OLSON, E. C. 1958. Fauna of the Vale and the Choza. 14. Summary, review and integration of the geology and the faunas. *Fieldiana Geology*, **10**, 397–448.

OLSON, E. C. 1962. Late Permian terrestrial vertebrates, U.S.A. and U.S.S.R. *Transactions of the American Philosophical Society, New Series*, **52**, 1–223.

OLSON, E. C. 1965. *Chickasha Vertebrates*. Oklahoma Geological Survey Circular, **70**.

OLSON, E. C. 1967. *Early Permian Vertebrates of Oklahoma*. Oklahoma Geological Survey Circular, **74**.

OLSON, E. C. 1972. *Diplocaulus parvus*, n. sp. (Amphibia, Nectridea) from the Chickasha Formation (Permian. Guadalupian) of Oklahoma. *Journal of Paleontology*, **46**, 656–659.

OLSON, E. C. 1974. On the source of the therapsids. *Annals of the South African Museum*, **64**, 27–46.

OLSON, E. C. 1975. Vertebrates and the biostratigraphic position of the Dunkard. *In*: BARLOW, J. A. & BURKHAMMER, S. (eds) *Proceedings of the first I. C. White memorial symposium 'The Age of the Dunkard'*. West Virginia Geological and Economic Survey, Morgantown, 155–165.

OLSON, E. C. 1980. The North America Seymouriidae. *In*: JACOBS, L. L. (ed.) *Aspects of Vertebrate History*. Museum of Northern Arizona Press, Flagstaff, 137–152.

OLSON, E. C. 1989*a*. Problems of Permo-Triassic terrestrial extinctions. *Historical Biology*, **2**, 17–35.

OLSON, E. C. 1989*b*. The Arroyo Formation (Leonardian: Lower Permian) and its vertebrate fossils. *Texas Memorial Museum Bulletin*, **35**, 1–25.

OLSON, E. H. & BARGHUSEN, H. 1962. Vertebrates from the Flowerpot Formation, Permian of Oklahoma. *Oklahoma Geological Survey Circular*, **59**, 5–48.

OLSON, E. C. & BEERBOWER, J. R. 1953. The San Angelo Formation, Permian of Texas, and its vertebrates. *Journal of Geology*, **61**, 389–423.

OLSON, E. C. & CHUDINOV, P. 1992. Upper Permian terrestrial vertebrates of the USA and Russia: 1991. *International Geology Review*, **34**, 1143–1160.

OLSON, E. C. & MEAD, J. G. 1982. The Vale Formation (Lower Permian) its vertebrates and paleoecology. *Texas Memorial Museum Bulletin*, **29**, 1–46.

OLSON, E. C. & VAUGHN, P. P. 1970. The changes of terrestrial vertebrates and climates during the Permian of North America. *Forma et Functio*, **3**, 113–138.

OWEN, D. E. 1987. Commentary: usage of stratigraphic terminology in papers, illustrations, and talks. *Journal of Sedimentary Petrology*, **57**, 363–372.

PARRISH, J. M., PARRISH, J. T. & ZIEGLER, A. M. 1986. Permian–Triassic paleogeography and paleoclimatology and implications for therapsid distribution. *In*: HOTTON, N. III, MACLEAN, P. D., ROTH, J. J. & ROTH, E. C. (eds) *The Ecology and Biology of Mammal-Like Reptiles*. Smithsonian Institution Press, Washington, DC, 109–131.

PIÑEIRO, G., VERDE, M., UBILLA, M. & FERIGOLO, J. 2003. First basal synapsids ('pelycosaurs') from the Upper Permian-lower Triassic of Uruguay, South America. *Journal of Paleontology*, **77**, 389–392.

PIÑEIRO, G., ROJAS, A. & UBILLA, M. 2004. A new procolophonoid (Reptilia, Parareptilia) from the Upper Permian of Uruguay. *Journal of Vertebrate Paleontology*, **24**, 814–821.

PIVETEAU, J. 1926. Paléontologie de Madagascar, XIII. Amphibiens et reptiles Permiens. *Annales de Paléontologie*, **15**, 1–128.

PLUMMER, F. B. & MOORE, R. C. 1921. Stratigraphy of the Pennsylvanian formations of north-central Texas. *University of Texas Bulletin*, **2132**, 1–237.

PLUMMER, F. B. & SCOTT, G. 1937. Upper Palaeozoic ammonites in Texas. *University of Texas Bulletin*, **3701**, 1–516.

RAY, S. 1999. Permian reptile fauna from the Kundaram Formation, Pranhita-Godavari Valley, India. *Journal of African Earth Sciences*, **29**, 211–218.

RAY, S. 2001. Small Permian dicynodonts from India. *Paleontological Research*, **5**, 177–191.

RAYNER, D. H. 1971. Data on the environment and preservation of late Palaeozoic tetrapods. *Proceedings of the Yorkshire Geological Society*, **38**, 437–495.

REISZ, R. R. 1986. Pelycosauria. *Handbook of Paleoherpetology*, **17A**, 1–102.

REISZ, R. R. & LAURIN, M. 2001. The reptile *Macroleter*, the first vertebrate evidence for correlation of Upper Permian continental strata of North America and Russia. *Geological Society of America Bulletin*, **113**, 1229–1233.

REISZ, R. R., BERMAN, D. S. & SCOTT, D. 1984. The anatomy and relationships of the Lower Permian reptile *Araeoscelis*. *Journal of Vertebrate Paleontology*, **4**, 57–67.

RETALLACK, G. J., SMITH, R. M. H. & WARD, P. D. 2003. Vertebrate extinction across Permian–Triassic boundary in Karoo Basin, South Africa. *Geological Society of America Bulletin*, **115**, 1133–1152.

ROMER, A. S. 1928. Vertebrate faunal horizons in the Texas Permo-Carboniferous red beds. *University of Texas Bulletin*, **2801**, 67–108.

ROMER, A. S. 1935, Early history of Texas redbeds vertebrates. *Geological Society of America Bulletin*, **46**, 1597–1658.

ROMER, A. S. 1952. Late Pennsylvanian and Early Permian vertebrates of the Pittsburgh–West Virginia region. *Annals of Carnegie Museum*, **33**, 47–113.

ROMER, A. S. 1958. The Texas Permian redbeds and their vertebrate fauna. *In*: WESTOLL, T. (ed.) *Studies on Fossil Vertebrates Essays Presented to D. M. S. Watson*. Athlone Press, London, 157–179.

ROMER, A. S. 1960. The vertebrate fauna of the New Mexico Permian. *New Mexico Geological Society, Guidebook*, **11**, 48–54.

ROMER, A. S. 1966. *Vertebrate Paleontology*. 3rd edition. University of Chicago Press, Chicago.

ROMER, A. S. 1973. Permian reptiles. *In*: HALLAM, A. (ed.) *Atlas of Palaeobiogeography*. Elsevier, Amsterdam, 159–167.

ROMER, A. S. 1974. The stratigraphy of the Permian Wichita redbeds of Texas. *Breviora*, **427**, 1–29

ROMER, A. S. & PRICE, L. I. 1940. *Review of the Pelycosauria*. Geological Society of America, Special Paper, **28**.

ROSCHER, M. & SCHNEIDER, J. W. 2005. An annotated correlation chart for continental Late Pennsylvanian and Permian basins and the marine scale. *In*: LUCAS, S. G. & ZEIGLER, K. E. (eds) *The Nonmarine Permian*. New Mexico Museum of Natural History and Science Bulletin, **30**, 282–291.

ROSS, C. A. 1969. Paleoecology of *Triticites* and *Dunbarinella* in the Upper Pennsylvanian strata of Texas. *Journal of Paleontology*, **43**, 298–311.

ROTH, R. 1930. New information on the base of the Permian in north-central Texas. *Journal of Paleontology*, **5**, 295.

RUBIDGE, B. S. 1993. New South African fossil links with earliest mammal-like reptile (therapsid) faunas from Russia. *South African Journal of Science*, **89**, 460–461.

RUBIDGE, B. S. (ed.) 1995*a*. *Biostratigraphy of the Beaufort Group (Karoo Supergroup)*. South African Committee for Stratigraphy, Biostratigraphic Series, **1**, 1–46.

RUBIDGE, B. S. 1995*b*. Biostratigraphy of the *Eodicynodon* Assemblage zone. *In*: RUBIDGE B. S. (ed.) *Biostratigraphy of the Beaufort Group (Karoo Supergroup)*. South African Committee for Stratigraphy, Biostratigraphic Series, **1**, 3–7.

RUBIDGE, B. S. 2005. Middle–Late Permian tetrapod faunas from the South African Karoo and their biogeographic significance. *In*: LUCAS, S. G., ZEIGLER, K. E. & SPIELMANN, J. A. (eds) *The Permian of Central New Mexico. New Mexico Museum of Natural History and Science, Bulletin*, **31**, 292–294.

RUBIDGE, B. S. & HOPSON, J. A. 1990. A new anomodont therapsid from South Africa and its bearing on the ancestry of Dicynodontia. *South African Journal of Science*, **86**, 43–45.

RUBIDGE, B. S. & HOPSON, J. A. 1996. A primitive anomodont therapsid from the base of the Beaufort Group (Upper Permian) of South Africa. *Zoological Journal of the Linnaean Society*, **117**, 115–139.

RUBIDGE, B. S. & SIDOR, C. A. 2001. Evolutionary patterns among Permo-Triassic therapsids. *Annual Review of Ecology and Systematics*, **32**, 449–480.

RUBIDGE, B. S., JOHNSON, M. R., KITCHING, J. W., SMITH, R. M. H., KEYSER, A. W. & GROENEWALD, G. H. 1995. An introduction to the biozonation of the Beaufort Group. *In*: RUBIDGE, B. S. (ed.) *Biostratigraphy of the Beaufort Group (Karoo Supergroup)*. South African Committee for Stratigraphy, Biostratigraphic Series, **1**, 1–2.

SADLER, P. M. 1981. Sediment accumulation rates and the completeness of stratigraphic sections. *Journal of Geology*, **89**, 569–584.

SCHINDEL, D. E. 1980. Microstratigraphic sampling and the limits of paleontologic resolution. *Paleobiology*, **6**, 408–426.

SCHINDEL, D. E. 1982. Resolution analysis: a new approach to the gaps in the fossil record. *Paleobiology*, **8**, 340–353.

SIDOR, C. A. & HOPSON, J. A. 1995. The taxonomic status of the Upper Permian eotheriodont therapsids of the San Angelo Formation (Guadalupian), Texas. *Journal of Vertebrate Paleontology*, **15**(3 supplement), 53A.

SIDOR, C. A., O'KEEFE, F. R. *et al.* 2005. Permian tetrapods from the Sahara show climate-controlled endemism. *Nature*, **434**, 886–889.

SIMPSON, L. C. 1973. Preliminary correlation of the Lower Permian of north Texas and Oklahoma. *Shale Shaker*, **24**, 68–72.

SIMPSON, L. C. 1979. Upper Gearyan and Lower Leonardian terrestrial vertebrate faunas of Oklahoma. *Oklahoma Geology Notes*, **39**, 3–21.

SKINNER, J. W. 1946. Correlation of Permian of west Texas and southeast New Mexico. *American Association of Petroleum Geologists Bulletin*, **30**, 1857–1874.

SMITH, R. M. H. & KEYSER, A. W. 1995*a*. Biostratigraphy of the *Tapinocephalus* Assemblage zone. *In*: RUBIDGE, B. S. (ed.) *Biostratigraphy of the Beaufort Group (Karoo Supergroup)*. South African Committee for Stratigraphy, Biostratigraphic Series, **1**, 8–12.

SMITH, R. M. H. & KEYSER, A. W. 1995*b*. Biostratigraphy of the *Pristerognathus* Assemblage zone. *In*: RUBIDGE, B. S. (ed.) *Biostratigraphy of the Beaufort Group (Karoo Supergroup)*. South African Committee for Stratigraphy, Biostratigraphic Series, **1**, 13–17.

SMITH, R. M. H. & KEYSER, A. W. 1995*c*. Biostratigraphy of the *Tropidostoma* Assemblage zone. *In*: RUBIDGE, B. S. (ed.) *Biostratigraphy of the Beaufort Group (Karoo Supergroup)*. South African Committee for Stratigraphy, Biostratigraphic Series, **1**, 18–22.

SMITH, R. M. H. & KEYSER, A. W. 1995*d*. Biostratigraphy of the *Cistecephalus* Assemblage zone. *In*: RUBIDGE, B. S. (ed.) *Biostratigraphy of the Beaufort Group (Karoo Supergroup)*. South African Committee for Stratigraphy, Biostratigraphic Series, **1**, 23–28.

SPALDING, D. A. E. 1993. *Bathygnathus*, Canada's first 'dinosaur'. *Modern Geology*, **18**, 247–255.

STEYER, S. 2000. Are European Palaeozoic amphibians good stratigraphic markers? *Bulletin de la Société Géologique de France*, **171**, 127–135.

SUMIDA, S. S., BERMAN, D. S., EBERTH, D. A. & HENRICI, A. C. 2004. A terrestrial vertebrate assemblage from the Late Palaeozoic of central Germany, and its bearing on Lower Permian palaeoenvironments. Fossils and Strata, **50**, 113–123.

SUMIDA, S. S., BERMAN, D. S. & MARTENS, T. 1996. Biostratigraphic correlations between the Lower Permian of North America and central Europe using the first record of an assemblage of terrestrial tetrapods from Germany. *PaleoBios*, **17**, 1–12.

SUMIDA, S. S., BERMAN, D. S. & MARTENS, T. 1998. A new trematopid amphibian from the Lower Permian of central Germany. *Palaeontology*, **41**, 605–629.

SUMIDA, S. S., WALLISER, J. B. D. & LOMBARD, R. E. 1999*a*. Late Palaeozoic amphibian-grade tetrapods of Utah. *In*: GILLETRE, D. (ed.) *Vertebrate Paleontology in Utah*. Utah Geological Survey, Miscellaneous Publication, **99–1**, 21–30.

SUMIDA, S. S., LOMBARD, R. E., BERMAN, D. S & HENRICI, A. C. 1999*b*. Late Palaeozoic amniotes and their near relatives from Utah and northeastern Arizona, with comments on the Permian–Pennsylvanian boundary in Utah and northern Arizona. *In*: GILLETRE, D. (ed.) *Vertebrate Paleontology in Utah*. Utah Geological Survey, Miscellaneous Publication, **99–1**, 31–43.

SUSHKIN, P. P. 1926. Notes on the pre-Jurassic Tetrapoda from Russia. 1. *Dicynodon Amalitzkii*. *Palaeontologia Hungarica*, **1**, 323–327.

THOMPSON, M. L. 1954. *American Wolfcampian Fusulinids*. University of Kansas, Paleontological Contributions 14, Protozoa 5.

TOERIEN, M. J. 1953. The evolution of the palate in South African Anomodontia and its classificatory significance. *Palaeontologia Africana*, **1**, 49–117.

TVERDOKHLEBOV, V. P., TVERDOKHLEBOVA, G. I., MINIKH, A. V., SURKOV, M. V. & BENTON, M. J. 2005. Upper Permian vertebrates and their sedimentological context in the south Urals, Russia. Earth-Science Reviews, **69**, 27–77.

VANDERLOOP-AVERY, M. L. & NESTELL, M. K. 1984. Carbonate facies of the Pueblo Formation–Lower Permian, eastern shelf; transgressive phylloid algal biostromes. *Society of Economic Paleontologists and Mineralogists Permian Basin Chapter, Publication*, 84–23, 34–62.

VAUGHN, P. P. 1958. On the geologic range of the labyrinthodont amphibian *Eryops*. *Journal of Paleontology*, **32**, 918–922.

VAUGHN, P. P. 1962, Vertebrates from the Halgaito Tongue of the Cutler Formation, Permian of San Juan County, Utah. *Journal of Paleontology*, **36**, 529–539.

VAUGHN, P. P. 1963, The age and locality of the Palaeozoic vertebrates from El Cobre Canyon, Rio Arriba County, New Mexico. *Journal of Paleontology*, **37**, 283–296.

VAUGHN, P. P. 1964. Vertebrates from the Organ Rock Shale of the Cutler Group, Permian of Monument Valley and vicinity, Utah and Arizona. *Journal of Paleontology*, **38**, 567–583.

VAUGHN, P. P. 1965. Vertebrates from the Organ Rock Shale of the Cutler Formation, Permian of Monument valley and vicinity, Utah and Arizona. *Journal of Paleontology*, **38**, 567–583.

VAUGHN, P. P. 1966*a*. Comparison of Early Permian vertebrate faunas of the Four Corners region and north-central Texas. *Los Angeles County Museum, Contributions in Science*, **105**, 1–13.

VAUGHN, P. P. 1966*b*. *Seymouria* from the Lower Permian of southeastern Utah, and possible sexual dimorphism in that genus. *Journal of Paleontology*, **40**, 603–612.

VAUGHN, P. P. 1973. Vertebrates from the Cutler Group of Monument Valley and vicinity, Utah and Arizona. *New Mexico Geological Society, Guidebook*, **24**, 99–105.

WALSH, T. R. & BARRICK, J. E. 2002. Conodonts from the Elm Creek Formation (Artinskian) of north-central Texas U. S. A.: fauna from a Permian intermittently restricted shallow shelf. *Geological Society of America, Abstracts with Programs*, **34**(2), A–26.

WARDLAW, B. R. 2005. Age assignment of the Pennsylvanian–Early Permian succession of north central Texas. *Permophiles*, **46**, 21–22.

WARDLAW, B. R., DAVYDOV, V. & GRADSTEIN, F. M. 2004. The Permian Period. *In*: GRADSTEIN, F. M., OGG, J. G. & SMITH, A. G. (eds) *A Geologic Time Scale 2004*: Cambridge University Press, Cambridge, 249–270.

WATSON, D. M. S. 1914*a*. The zones of the Beaufort Beds of the Karroo System in South Africa. *Geological Magazine, New Series*, **(6)** 1, 203–208.

WATSON, D. M. S. 1914*b*. On the nomenclature of the South African pareiasaurians. *Annals and Magazine of Natural History*, **14**, 98–102.

WELLSTEAD, C. F. 1991. Taxonomic revision of the Lysorophia, Permo-Carboniferous lepospondyl amphibians. *Bulletin of the American Museum of Natural History*, **209**, 1–90.

WELSH, J. E. 1958. Faunizones of the Pennsylvanian and Permian rocks in the Paradox basin. *Intermountain Association of Petroleum Geologists, Guidebook*, **9**, 153–162.

WERNEBURG, R. 1989. Labyrinthodontier (Amphibia) aus dem Oberkarbon und Unterperm Mitteleuropas Systematik, Phylogenie und Biostratigraphie. *Freiberger Forschungschrifte, Hefre C*, **436**, 7–57.

WERNEBURG, R. 1993. *Eryops* in the Thuringian Forest? *Pollichia*, **29**, 171–176.

WERNEBURG, R. 2001. Die Amphibien- und Reptilienfaunen aus dem Permokarbon des Thüringer Waldes. *Beitrage Geologie Thüringen Neue Folge*, **8**, 125–152.

WERNEBURG, R. & SCHNEIDER, J. W. 2006. Amphibian biostratigraphy of the European Permo-carboniferous. *In*: LUCAS, S. G., CASSINIS, G. & SCHNEIDER, J. W. (eds) *Non-marine Permian Biostratigraphy and Biochronology*. Geological Society, London, Special Publications, **265**, 201–215.

WIDEMAN, N. K., SUMIDA, S. S. & O'NEIL, M. 2005. A reassessment of the taxonomic status of the materials assigned to the early Permian tetrapod genera *Limnosceloides* and *Limnoscelops*. *In*: LUCAS, S. G. & ZEIGLER, K. E. (eds) *The Nonmarine Permian*. New Mexico Museum of Natural History and Science Bulletin, **30**, 358–362.

WILLISTON, S. W. 1915, New genera of Permian reptiles. *American Journal of Science*, **39**, 575–579.

Permo-Carboniferous climate: Early Pennsylvanian to Late Permian climate development of central Europe in a regional and global context

MARCO ROSCHER & JOERG W. SCHNEIDER

Institut für Geologie, TU Bergakademie Freiberg, B.v. Cotta-Strasse 2, 09596 Freiberg, Germany (e-mail: schneidj@geo.tu-freiberg.de)

Abstract: A well-justified stratigraphical correlation of continental successions and new palaeogeographic reconstruction of Pangaea reveal new insights into the northern Pangaean climate development influenced by palaeogeography, palaeotopography, glacio-eustatic sealevel changes and ocean currents. The overall Permo-Carboniferous aridization trend was interrupted by five wet phases. These are linked to the Gondwana icecap. The aridization and weakening of wet phases over time were not only caused by the drift of northern Pangaea to the arid climatic belt, but also by the successive closure of the Rheic Ocean, which caused the expansion of arid/semi-arid environments in the Lower/Middle Permian. The end of the Gondwana glaciation rearranged ocean circulation, leading to a cold, coast-parallel ocean current west of northern Pangaea, blocking moisture coming with westerly winds. The maximum of aridity was reached during the Roadian/Wordian. The Trans-Pangaean Mountain Belt was non-existent. Its single diachronous parts never exceeded an average elevation of 2000 m. The maximum elevation shifted during time from east to west. The Hercynian orogen never acted as an orographic east–west barrier, and the Inter-Tropical Convergence Zone was widely displaced, causing four seasons (dry summer/winter, wet spring/autumn) at the equator and a strong monsoon system.

The climate history of the European realm during the Late Carboniferous (Pennsylvanian) and Permian is stored in many solitary basins (Fig. 1) within the Hercynian orogen and the foreland basin. The story of Westphalian climate is well known because of the numerous investigations of the coal-bearing Variscan foredeep. The younger Westphalian is characterized by a slight aridization (Abbink & van Kronijnenburgvan Cittert 2003; Opluštil 2004), which was accompanied by an increase of seasonality. Nevertheless, the environment was strongly influenced by the ocean and epi-continental seas with multiple transgression events. The last extensive marine ingression was the Aegir/Mansfield Band (Westphalian B/C). The post-Westphalian climate development is more differentiated and not well known.

The Permo-Carboniferous climate of the central European realm is unquestionably dominated by an aridization trend (e.g. Chumakov & Zharkov 2002), which is not as simple as previously thought. Based on an improved stratigraphical correlation chart, several more humid phases are provable within the Late Carboniferous and the Early Permian. These so-called 'wet phases' are interpreted as a result of the waxing and waning of the Gondwana icecap. This is supported by the correlation of the continental European basins via isotopic ages to the Karoo Basin and adjacent areas. Further literature studies about this glaciation revealed numerous inconsistencies. Most important are the exact age, geographical position and size of the icecap. Nevertheless, the low-frequency cycles of this indubitably large glaciation are reflected in the continental sediments of northern Pangaea. This influence, and thus the strength of the wet phases, decreased during the Early Permian, although no decrease in the amplitude of waxing and waning of the icecap is reported. This discrepancy is solved by the new palaeogeographic reconstruction presented here of Late Carboniferous to Early Permian Pangaea (Fig. 2). The reduction of marine influence, which is essentially coupled to glacioeustatic sealevel, is based on the slow closure of the Rheic Ocean. In contradiction to the common palaeogeographic maps (e.g. Scotese 2001), this ocean persisted until the Middle to Late Permian. A second problem exists: the east–west barrier of the Trans-Pangaean Mountain Belt (Keller & Hatcher 1999) used in every modern (including altitudes) palaeoclimatic model, did not exist as a single orogen. Also, the height of individual mountain elements seems to be overestimated.

From: LUCAS, S. G., CASSINIS, G. & SCHNEIDER, J. W. (eds) 2006. *Non-Marine Permian Biostratigraphy and Biochronology*. Geological Society, London, Special Publications, **265**, 95–136.
0305-8719/06/$15.00

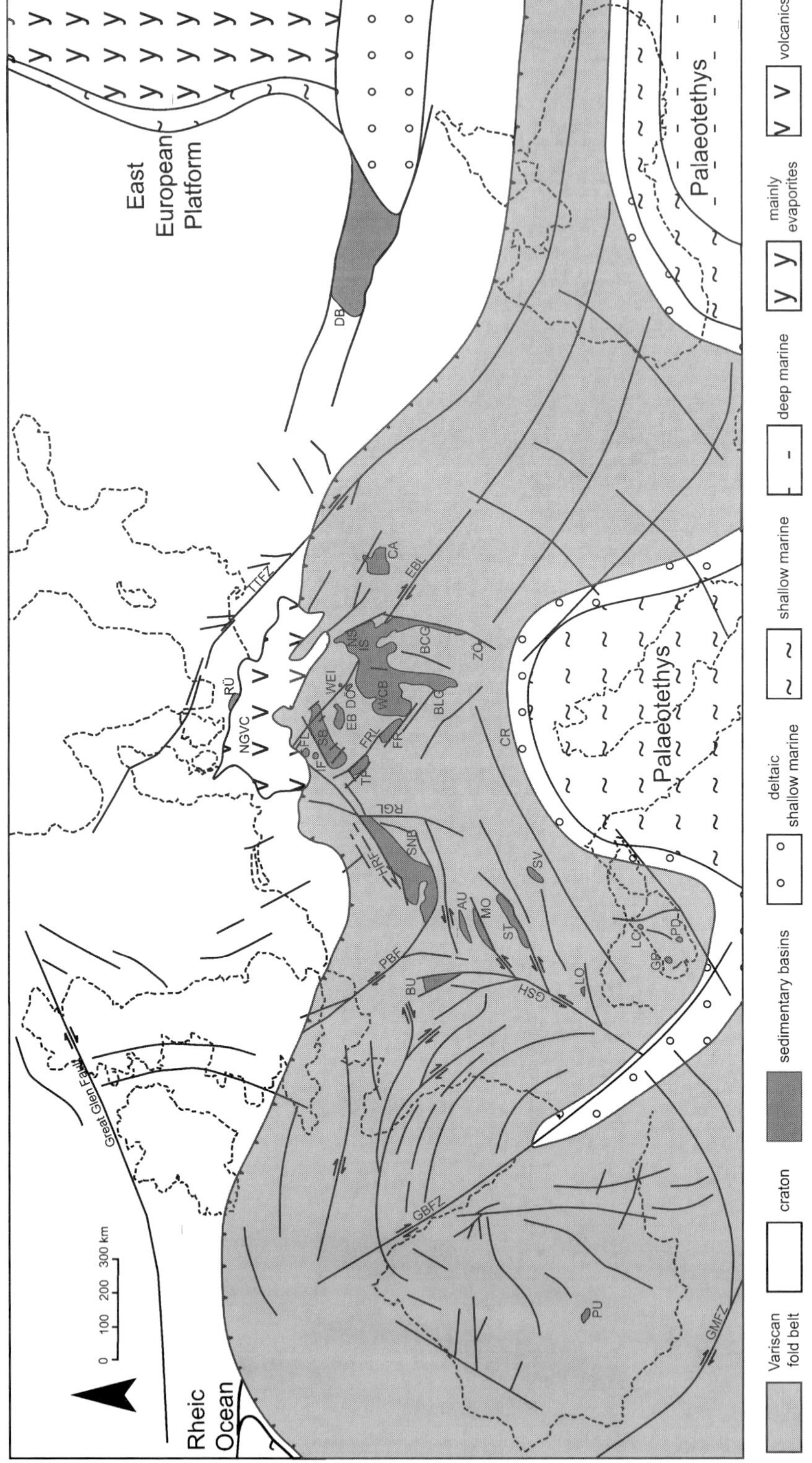
East European Platform
DB
Palaeotethys
Palaeotethys
TTFZ
RÜ
NGVC
CA
EBL
NS
IS
BCG
WEI
DO
EB
WCB
ZO
BLG
FL
SB
IF
FRL
FR
TP
CR
RGL
SNB
HRF
SV
AU
MO
ST
PBF
BU
GSH
LO
LC
PD
GR
Great Glen Fault
GBFZ
PU
GMFZ
Rheic Ocean
0 100 200 300 km
Variscan fold belt
craton
sedimentary basins
deltaic shallow marine
shallow marine
deep marine
mainly evaporites
volcanics

Fig. 2. Palaeogeographic map at 300 Ma, grid not arranged to the palaeo-equator (adapted from Kroner, in Schneider *et al.* 2006).

These problems lead to a different reconstruction of palaeoclimate and its forcing mechanisms. The model presented here is based on analogies and climate sensitive sediments.

The European basins and their sedimentological development

Each of the Late to post-Westphalian European basins has its own tectonic, sedimentological and climatic history. This paper describes climate signals based on our personal research and extensive literature studies. All signals that could give a hint of at least regional climate changes are displayed in Figures 15a, b and 16. For the comparability of these datasets, a well-based stratigraphy (Fig. 15a, b) is indispensable. This is a balanced combination of isotopic ages and litho-, bio-, event and tectono- stratigraphy. The international scale is adopted from Menning & German Stratigraphic Commission (2002) with some modifications, especially concerning the Carboniferous–Permian boundary, which is set here at 299 Ma, according to Ramezani *et al.* (2003). Linking times between the basins in the Late Carboniferous to early Lower Rotliegend has an estimated error of ±1 Ma. This relatively small inaccuracy for continental deposits was

Fig. 1. Geographic position of important Permo-Carboniferous basins, Northern and Southern Permian Basin are omitted. AU, Autun Basin; BLG, Blanice Graben; BCG, Boskovice Graben; BU, Bourbon l'Archambault Basin; CA, Carpathian Basin; CR, Carnic Alps; DB, Donetsk Basin; DÖ, Döhlen Basin; EB, Erzgebirge Basin; EBL, Elbe Lineament; FL, Flechting Block; FRL, Franconian Lineament; FR, Franconian Basin; GBFZ, Golf of Biscay Fracture Zone; GMFZ, Gibraltar Minas Fracture Zone; GP, Guardia Pisano Basin; GSH, Grand Sillon Houllier Fracture Zone; HRF, Hunsrück Fracture; IF, Ilfeld Basin; IS, Intra Sudetic Basin; KP, Krkonoše Piedmont Basin; LC, Lu Caparoni Basin; LO, Lodève Basin; MO, Montceau les Mines Basin; NGVC, North German Vulcanite Complex; NS, North Sudetic Basin; PBF, Pays de Bray Fracture; PD, Perdasdefogu Basin; PU, Puertollano Basin; RGL, Rhein Graben Lineament; SB, Saale Basin; SNB, Saar–Nahe Basin; ST, St. Etienne Basin; SV, Salvan-Dorénaz Basin; TF, Thuringian Forest Basin; TTFZ, Tornquist–Teysseyre Fracture Zone; WCB, Western and Central Bohemian Basins; WEI, Weissig Basin; ZÖ, Zöbingen.

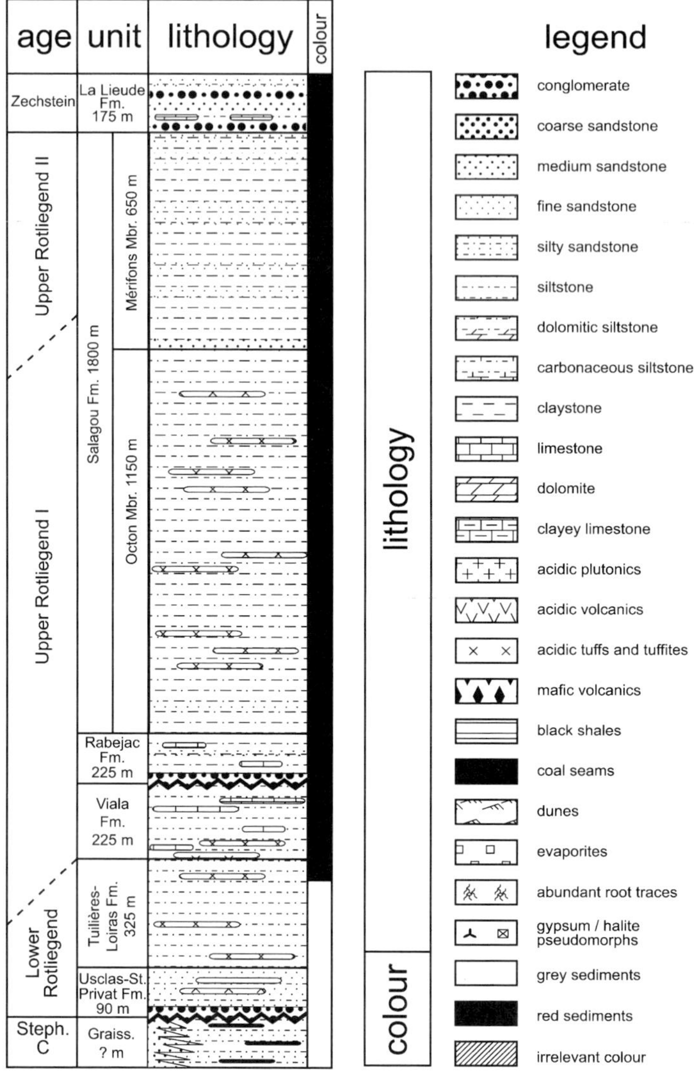

Fig. 3. General succession of the Lodève Basin.

reached by a combination of biostratigraphy (e.g. Schneider 1982; Hampe 1989; Werneburg 1999; Schneider *et al.* 2005*a*), geochronology (e.g. Hess & Lippolt 1989; Lippolt & Hess 1996; Goll & Lippolt 2001; Lützner *et al.* 2003; Lützner *et al.* 2006) and tectonostratigraphy (e.g. Stollhofen & Stanistreet 1994; Schneider *et al.* 1995*b*; Stollhofen *et al.* 1999), as well as lithostratigraphical correlations and comparison of sediment thicknesses (excluding volcanics) and sedimentary facies. For further information see Roscher & Schneider (2005). The connection to marine profiles, as well as to the international time scale, is based on isotopic dating and has a larger tolerance. All the climate indicative features discussed below are displayed in Figure 15a, b. By summarizing all these characteristics through space and time, at least several supra-regional climate events are obvious.

Lodève Basin

The sedimentation (Fig. 3) of the Lodève area (150 km^2) started in the Stephanian, unconformably above Hercynian structures, with grey clastics, several coal seams and lacustrine deposits in a terminal fan system (Gand *et al.* 2001). The overlying fluvial clastics and lacustrine black shales were deposited on an alluvial fan and by a floodplain system with eutrophic lakes in the basin centres. In the Upper Tuilières–Loiras Formation, the facies change from grey to red, as the environment changed from lacustrine to dominantly fluvial. The overlying late Autunian is characterized by a floodplain environment with sheetfloods and a braided river system with adjacent lakes. Above an erosional unconformity, fanglomeratic fan deposits follow. Further deposition took place in an alluvial plain/floodplain environment with periodically water-filled ponds grading into the playa sediments of the overlying Salagou Formation. Cycles of the Octon Member consist of 1-m-thick, massive, structure-less red-brown clayey siltstones (vertisols) and beige-coloured, 1-m-thick calcareous siltstones with characteristic desiccation cracks. Indications of a semi-arid to arid climate, such as vertisols (instead of calcisols) and desiccation crack horizons, became increasingly frequent. The occurrence of fossils is almost completely restricted to ephemeral channels that contain aquatic organisms adapted to dryness of seasonal or longer frequency (conchostracans, triopsids; cf. Gand *et al.* 1997*a*, *b*; Garric 2001).

Sedimentary cycles of the Mérifons Member consist of centimetre- to decimetre thick, often laminated, red-brown siltstone and grey-green, 1-cm-thick siltstone with calcareous cements of a distal fan environment. Sand patch fabric and vertisols, as well as dewatering structures, indicate, periodic wetting and drying (cf. Hardie *et al.* 1978). The La Lieude Formation starts quite abruptly with sheetflood and braided river conglomerates, as well as large-scale cross-bedded, pebbly channel sandstones. Frequent greenish-whitish colours of calcareous-cemented sandstone layers and calcic soils indicate a higher ground water level, causing reducing conditions. Calcisols, invertebrate burrows of *Scoyenia*-type, root-penetrated siltstones, tree vegetation and a modern tetrapod track association (Gand 1993; Gand *et al.* 2000) appear within the first conglomeratic levels.

The Graissessac Formation is dated by Bruguier *et al.* (2003) to 295.5 ± 5.1 Ma (U–Pb) and the macroflora is ascribed to the late Stephanian by Doubinger *et al.* (1995). The Sakmarian (*Melanerpeton pusillum–Melanerpeton gracile* zone to *Discosauriscus austriacus* zone see Werneburg 1996, Werneburg & Schneider 2006) Usclas-St Privat Formation and the Tuilières-Loiras Formation were deposited in a semi-humid climate with seasonal rainfall indicated by laminated (varved) lake sediments and an absolutely conifer-dominated macroflora (85%; Galtier, in Lopez *et al.* 2005). The red facies of the Viala Formation (289.3 ± 6.7 Ma (U–Pb) (Schneider *et al.* 2006) were deposited under semi-arid conditions, as indicated by desiccation cracks, xeromorphic calcisols, vertisols and rare pseudomorphs after gypsum and halite crystals. The Rabejac Formation points to seasonal to episodic heavy precipitation events in an overall semi-arid climate. The overlying Octon Member (284 ± 4 Ma U–Pb, Schneider *et al.* 2006) of the Salagou Formation is conspicuous by the absence of plant roots and invertebrate burrows, such as *Scoyenia*, pointing to a lowered ground water level. The maximum of aridity was reached in the Octon Member, which is supported by the geochemical investigations of Körner *et al.* (2003) and Schneider *et al.* (2006). The overlying Mérifons Member is characterized by a semi-arid climate with fluvial and fan deposits. It is dated by insects and conchostracans to a Kungurian to Roadian age (Gand *et al.* 1997*a*, *b*; Bethoux *et al.* 2002). The base of the La Lieude Formation is supposed to be next to the Illawara Reversal (Bachtadse, pers. comm.). These drastic changes in litho- and biofacies patterns from dry playa to wet alluvial plain environments are indications of a rapid increase in the rate of precipitation. This is supposed to be an affect of the Late Permian Bellerophon and Zechstein transgression (Schneider *et al.* 2006).

The overall trend of the climate development within this basin is expressed by an aridization trend from the Stephanian Graissessac Formation above the red Viala and Rabejac formations to the playa environment of the Octon and Mérifons Member (Salagou Formation). After that the trend reverses. This humidization is not strong enough to allow the development of extensive lacustrine or swamp deposits, but the sediments of the La Lieude Formation indicate much more precipitation since the Artinskian than the underlying units do.

Autun Basin

Sediments of this basin (Fig. 4) started on Variscan granitic and metamorphic basement with the deposition of the grey conglomerates and sandstones of Mont Pelé and the intercalated, folded Epinac coal seams. Above the angular unconformity are coarse clastics with intercalated coal seams that, towards the basin centre, grade into varved carbonaceous shales of a palustrine to lacustrine environment. The overlying deposits are characterized by bituminous shales, sandstones containing layers of siderite, and dolomitic horizons of a lacustrine to fluvial environment. The frequency of lacustrine horizons decreases in the Surmoulin Formation, which is built up of a large sequence of grey mudstones with rare sandy, cross-bedded

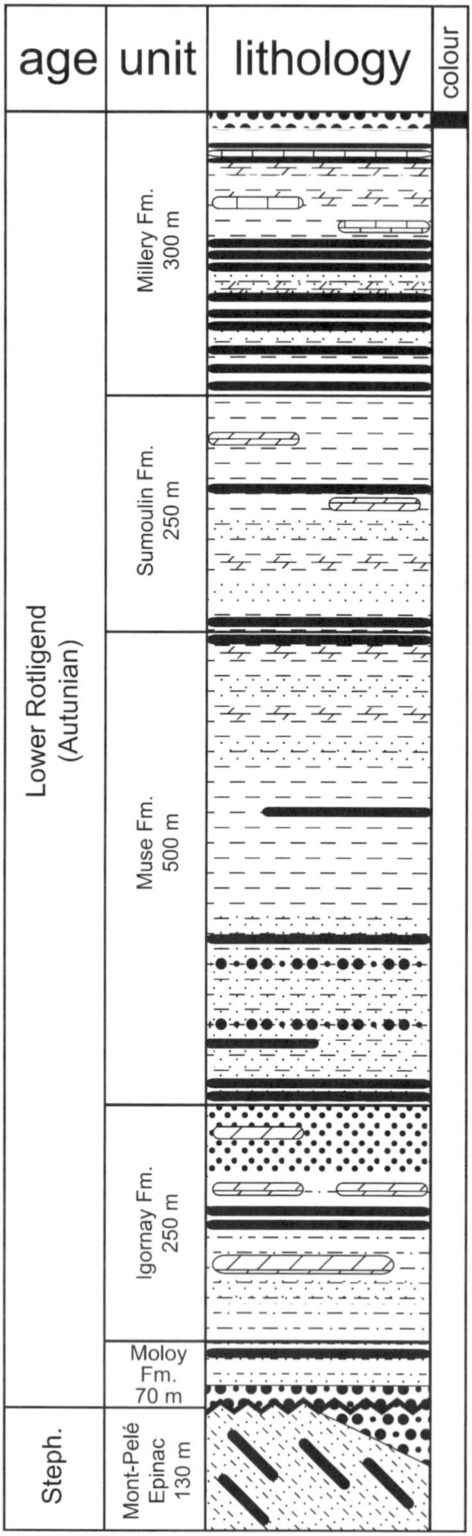

age
unit
lithology
colour
Lower Rotligend (Autunian)
Millery Fm. 300 m
Sumoulin Fm. 250 m
Muse Fm. 500 m
Igornay Fm. 250 m
Moloy Fm. 70 m
Steph.
Mont-Pelé Epinac 130 m

Fig. 4. General succession of the Autun Basin. For legend see Figure 3.

dolomitic intercalations (Chateauneuf & Pacaud 2001). The subsequent Millery Formation is mainly composed of varved, dolomitic, lacustrine mudstones. This series contains 10 lacustrine bituminous horizons that are terminated by the Margenne boghead coal. Towards the top and the basin border the deposits grade into red coarser clastics, which were formerly described as the Saxonian Curgy Formation.

The sequence from the unconformity at the base of the Moloy Formation to the top of the Millery Formation is defined as the stratotype of the Autunian. The stratotype is of poor quality because of problems with the lower and upper biostratigraphical and lithostratigraphical boundaries (Broutin *et al.* 1999). The Autunian stage is based on macro- and microfloras. Because of the problems with the strong reliance of plants on climate, the correlation presented here is based on fossil fauna. The Muse Formation is dated to the *Melanerpeton sembachense–Apateon dracyiensis* or *Apateon flagrifer flagrifer–Branchierpeton reinholdi* zone, early to middle Asselian/early Lower Rotliegend (Werneburg 1996, Werneburg & Schneider 2006) and *Syscioblatta dohrni–Sysciophlebia balteata* zone, early–middle Asselian/early Lower Rotliegend (Roscher & Schneider 2005). The Millery Formation contains teeth of *Bohemiacanthus* 'type Buxières', which is dated there (cf. page 101) to 289 ± 4 Ma (Pb/Pb) (Schneider *et al.* 2006) and the *Melanerpeton pusillum–M. gracile* zone, latest Lower Rotliegend (Werneburg & Schneider 2006).

This leads to various problems regarding the exact stratigraphical position of the Autunian. In brief, Broutin *et al.* (1999) published that the floras of the upper Goldlauter Formation of the Thuringian Forest Basin (cf. page 104) contain younger elements that do not occur in the Autunian of Autun, but the Goldlauter Formation, dated by amphibians, is older than the late Autunian Millery Formation. So, the dryness-adapted, modern Autunian floras, which do not occur at the stratotype, appear already in the mid-Autunian of the Thuringian Forest Basin. This indicates the sensitivity of plants to climate. The problems with the base of the Autunian are caused by the unconformity that separates the Autunian from the Stephanian as a structural unit (Chateauneuf & Pacaud 2001). This boundary could be used in the sense of a lithostratigraphical marker but not as a biostratigraphical one, which is necessary for stage definition. The base

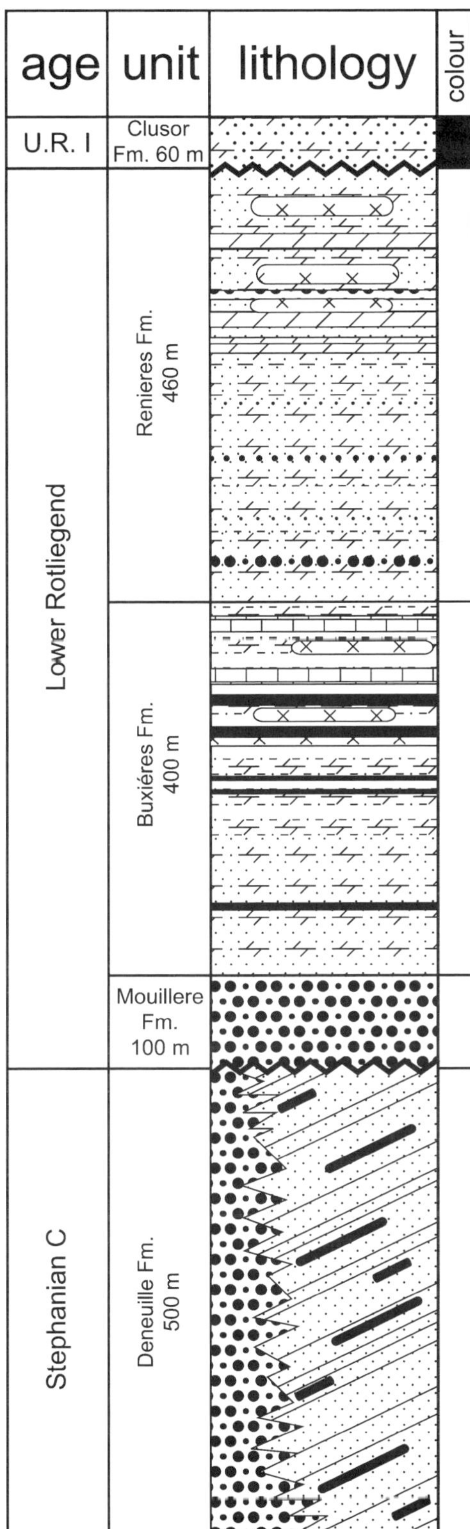

of the Autunian is not defined by any index fossil (Broutin *et al.* 1999). Therefore, this term should not be used in the sense of a biostratigraphical unit, but can be used as a lithostratigraphical unit comparable, to or even equivalent, to the Lower Rotliegend. For further information see Schneider (2001).

The climate evolution of this depositional area is characterized by an aridization trend from the humid coal seams of Epinac and Moloy to the semi-humid fluvial to lacustrine, seasonally laminated grey clastics of the Muse and Surmoulin formations, which lack larger bituminous and carbonaceous sequences, to the semi-arid coarse red clastics of the upper Millery Formation. This trend is interrupted by the extended humid to semi-humid lacustrine sequence in the lower and middle Millery Formation.

Bourbon l'Archambault Basin

The sedimentation (Fig. 5) in this basin (500 km^2) started on crystalline (granitic) basement of the Massif Central with the alluvial-dominated Deneuille Formation (Steyer *et al.* 2000) with thin coal seams of a fluvial and marginal palustrine environment. The next sedimentary cycle started with alluvial arkoses and conglomerates (Steyer *et al.* 2000). A vertical transition to coarse dolomitic sandstones of fluvial origin is observed. They grade into palustrine coarse sandstones that contain coal seams, lacustrine bituminous clay and siltstones, and dolomites with approximately 30 partial basin-wide correlatable tuff and tuffite horizons (Paquette & Feys 1989). The overlying Renière Formation (Debriette 1997) consists of the fluvial-dominated Renière A Member (conglomerates, arkoses and silt- and claystones; Paquette & Feys 1989) and the lacustrine to palustrine Renière B Member, which is defined by the first occurrence of dolomitic horizons (Debriette 1993). It contains several tuff layers (Paquette 1980). The overlying Clusor Formation (Debriette 1997) is dominated by red sandstones and siltstones. The humid to semi-humid Deneuille Formation is ascribed to the Stephanian (Paquette & Feys 1989). The Buxières Formation is dated by Werneburg (2003) to the *Melanerpeton gracile–Discosauriscus pulcherrimus* zone (late Lower Rotliegend, Sakmarian) and to the *Sysciophlebia alligans* zone or *Sysciophlebia* n. sp. B zone in the sense of Schneider & Werneburg (1993). The coal

Fig. 5. General succession of the Bourbon l'Archambault Basin. U. R. I, Upper Rotliegend I. For legend see Figure 3.

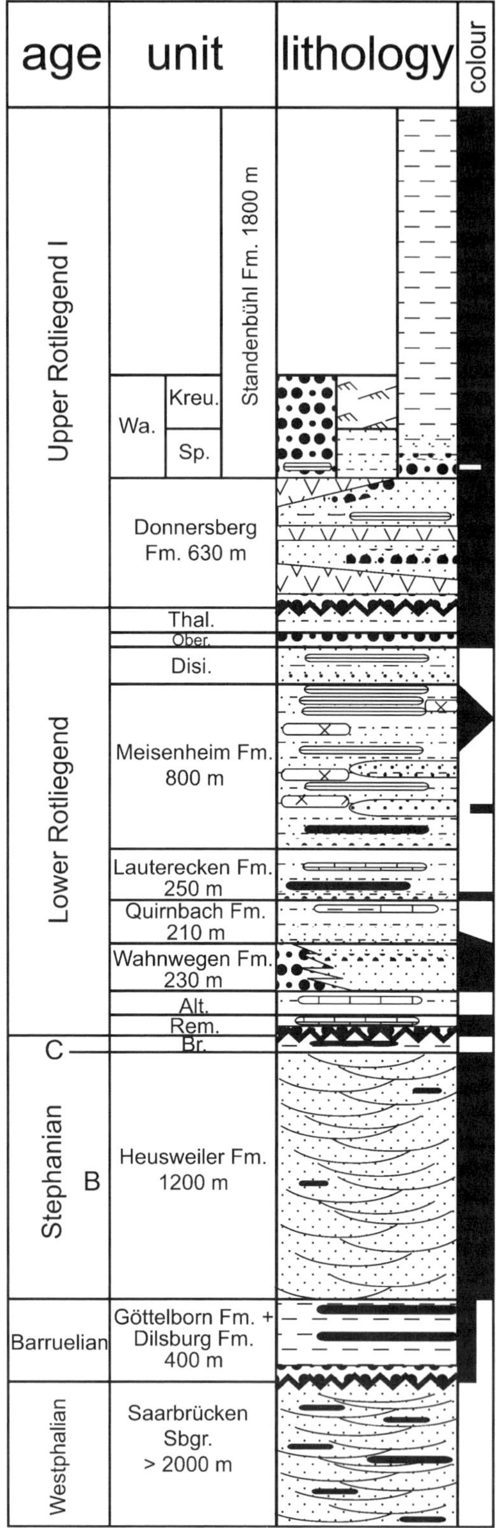

Fig. 6. General succession of the Saar–Nahe Basin. Br., Breitenbach Fm. 80 m; Rem., Remigiusberg Fm. 100 m; Alt., Altenglan Fm. 120 m; Disi., Disibodenberg Fm. 180 m; Ober., Oberkirchen Fm. 70 m; Thal., Thallichtenberg Fm. 120 m; Sp., Sponheim Fm. 240 m; Wa., Wadern Fm. 500 m; Kreu., Kreuznach Fm. 260 m. For legend see Figure 3.

seams of this formation show different palynological associations. The first is dominated by plants of a peat bog environment, whereas the second is dominated by a xerophilous floral association (Steyer *et al.* 2000). This indicates the development of local humid coal swamps within a semi-humid to semi-arid fluvial to lacustrine environment. The Clusor Formation is considered to be Saxonian (Upper Rotliegend) (Brulhet 1982). The climate development within this basin is characterized by an aridization trend from the coal-bearing Stephanian Deneuille Formation to the fluvial Renière Formation and the red beds of the Clusor Formation. This trend is interrupted by a short humidization event in the coal- and black shale-bearing, humid Buxières Formation.

Saar–Nahe Basin

The Westphalian to mid-Permian sediments of the *c.* 300 × 100 km Saar–Nahe Basin (Fig. 6) are subdivided into four subgroups. These are, from oldest to youngest, the Saarbrücken Subgroup, the Ottweiler Subgroup (Göttelborn–Breitenbach Formation), the Glan Subgroup (Remigiusberg–Thallichtenberg Formation) and the Nahe Subgroup (Donnersberg–Standenbühl Formation).

The whole Carboniferous section of this basin contains 140 named coal seams (120 are workable) that are concentrated in the Westphalian. The fluvio-lacustrine Saarbrücken Subgroup (about 3000 m thick: Müller & Konzan 1989) consists of rapid facies changes between grey clay, siltstones and sandstones with intercalated conglomerates and coal seams. Subsequent deposition started with a reddish conglomerate overlain by grey-green to reddish sandstones and claystones with several larger coal seams. This sequence is overlain by red, red-brown and violet-grey claystones, arkoses and conglomerates with various transitions. The environment is dominated by fluvial deposits (Boy 2003) with transitions at the top to lacustrine strata with coal seams.

The Remigiusberg Formation is characterized by alluvial red beds, fan conglomerates and fluvial to fluvio-lacustrine red clastics. Vertically, it grades into lacustrine grey to grey-green fine sediments with several intercalated bituminous limestones to black shales. The next cycle (Wahnwegen Formation) started with a prograding fan indicated by red, medium to coarse sandstones and conglomerates, and an environmental change from lacustrine to fluvial, again grading into grey fine clastics, with rare intercalations of lacustrine black shales and limestone horizons. Just above the basal conglomerate of the fluvio-lacustrine Lauterecken Formation, the up to 15-cm-thick, Odenbach carbonate coal seam spans an area of about 3300 km^2. Within the overlying Meisenheim Formation, approximately 40 tuff horizons occur in grey-brown to grey sandstones with sporadic regional coal seams. The fluvio-lacustrine middle part, with black shales and red sandstones, grade upward to lacustrine grey-brown to grey fine sandstones and pelites with subordinate red clastics (Stapf 1990). The overlying, fluvial-dominated red-grey conglomerates, arkoses and fine sandstones change with a fining-upward trend to red, yellow and grey sandstones and fine clastics. Above that, the Nahe Subgroup follows after a distinct hiatus (Stollhofen *et al.* 1999). It starts with the volcano-sedimentary Donnersberg Formation, consisting of red-grey conglomerates, arkoses and red to grey-green pelites that are strongly influenced by syn-sedimentary tectonics and volcanism (cf. Stollhofen 1994). The overall facies is described as fluvial red beds of a floodplain with meandering features and bituminous biolaminites of very small extent. The upper part of the Nahe Subgroup is represented in the west by playa-like pelites (indicated by the freshwater jellyfish *Medusina limnica*) of the Standenbühl Formation. The Wadern Formation in the northwest is dominated by red, alluvial-fan breccias and conglomerates. In the upper part, these fan deposits are interbedded with the red aeolian sandstone of the Kreuznach Formation. Importantly, in a fine clastic part of the Lower Wadern Formation, a last fish-containing lake horizon is intercalated.

Intensive investigations within this basin lead to a well-sustained stratigraphical correlation. The Westphalian Saarbrücken Subgroup is well dated with macro- and microfloras. Near the Carboniferous–Permian transition, the stratigraphical importance of floras decreases because not all European basins were in the same climatic belt at this time. Therefore, the stratigraphical correlation chart (Fig. 15a) is based on fauna and isotopic ages. The most important facts are summarised here (for further information see Roscher & Schneider 2005).

The Luisenthal Formation of the Saarbrücken Subgroup belongs to the *Archimylacris lubnensis* zone, Lower Westphalian D (Schneider *et al.* 2005*a*). The Dilsburg Formation is dated isotopically to 302.7 ± 0.6 Ma (Burger *et al.* 1997). The Stephanian A Göttelborn Formation belongs in the *Sysciophlebia* sp. A zone and the basal Stephanian B Heusweiler Formation in the *Spiloblattina pygmaea* zone (Schneider 1982; Schneider & Werneburg 1993; Schneider *et al.* 2005*a*). The Breitenbach Formation belongs to the *Branchierpeton saalensis–Apateon intermedius* zone (Werneburg 1996), *Bohemiacanthus* Ug zone (Schneider & Zajic 1994), *Sysciophlebia euglyptica–Syscioblatta dohrni* zone (Schneider & Werneburg 1993) and is geochronologically dated to 300 ± 2.4 Ma (Burger *et al.* 1997), which is Stephanian C. The Meisenheim Formation is dated to 297 ± 3.2 Ma (Königer 2000). The top of the Meisenheim Formation belongs to the *Melanerpeton pusillum–M. gracile* zone (Werneburg & Schneider 2006) and in the *Spiloblattina odernheimensis* zone (Schneider & Werneburg 1993; Schneider *et al.* 2005*a*). The youngest age is documented in the Wadern Formation, Sobernheim horizon, by the *Moravamylacris kukalovae* zone (Roscher & Schneider 2005).

The climatic development of this basin is marked by an obvious aridization, from the humid, coal-bearing grey sediments of the Saarbrücken Subgroup above the semi-humid wet red beds of the Heusweiler Formation, and the semi-humid to semi-arid red beds of the Wadern Formation to the semi-arid to arid playa sediments of the Standenbühl Formation. This overall trend is a multiphase one with several humid episodes in the Breitenbach, Altenglan, Disibodenberg and Wadern formations.

Thuringian Forest Basin

The Thuringian Forest Basin, an approximately 40 × 60 km NW–SE oriented depression, also called the SW Saale Basin, is one of the best, if not the best biostratigraphically investigated and correlated basin in the Variscan area (Schneider 1996; Lützner *et al.* 2003; Andreas *et al.* 2005). The basin is situated on the deeply eroded and peneplained Visean Thuringian Main Granite and the Ruhla Crystalline High of the inverted Mid-German Crystalline High (MGCH) at the outer border of the Variscan Orogen. This basin was at least partially connected to the Saale Basin by river systems (see below).

Sedimentation (Fig. 7) began with red (basin margin) and grey (basin centres) conglomerates and coarse arkosic sandstones followed by fluvial to lacustrine and palustrine fine clastics with fossiliferous lake horizons and thin coal seams of the Gehren Subgroup (Möhrenbach and

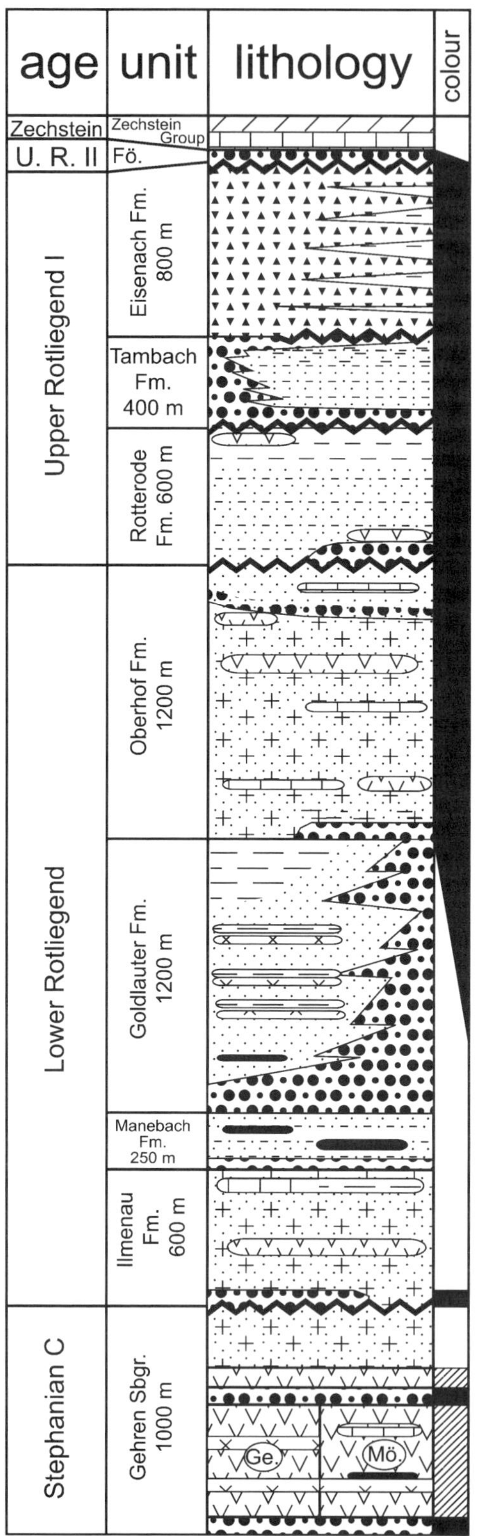

Fig. 7. General succession of the Thuringian Forest Basin. U. R. II, Upper Rotliegend II; Ge., Georgenthal Fm.; Mö., Möhrenbach Fm.; Fö., Förtha Fm. 20 m. For legend see Figure 3.

Georgenthal formations). Along intense fracture tectonic elements, the effusion of up to 1000 m of intermediate to acidic pyroclatics and lavas took place, and fluvial and muddy red beds that interfinger with laminated black shale lake deposits are intercalated. After a basin-wide erosional disconfomity, pyroclastics as well as rhyolites with minor intercalations of red-brown sediments were deposited. They grade into a fluvial-lacustrine grey facies with root-penetrated fine clastics and carbonaceous to sapropelitic, laminated lacustrine deposits. This is followed by nearly exclusively grey facies of fluvial deposits with a high ground water level. After a relief reactivation, red-brown alluvial fan conglomerates and fluvial sandstones grade into brownish to greenish fluvio-lacustrine and very finely laminated (varved) black shales of the depocentres intercalated by basin-wide tuff marker-beds. This sequence is overlain by interfingering red and grey facies with a high amount (up to 90%) of volcanics (acidic, intermediate and basaltic). Vertically it grades to sometimes playa-like red beds. The last perennial lake horizon of the Thuringian Basin is very widespread and grades laterally from calcareous, bituminous, varved black shales into red, varved, carbonateclay laminites (Schneider & Gebhardt 1993). After extensive erosion the exclusively red facies patterns show an alluvial fan to alluvial plain environment with temporary pools, root-horizons and sometimes playa-like (containing freshwater jellyfish *Medusina limnica*) clay- and siltstone of *Scoyenia*-facies.

With a shift of the depocentres to the north, again after a hiatus, the Tambach Formation follows. Facies patterns range from very coarse, matrix-supported wadi-fill conglomerates to proximal and distal debris-flow dominated alluvial fan clastics with fluvial, reworked aeolian sandstones, primarily accumulated as dunes on the top of the fans in the hinterland. *Scoyenia*-facies and complete bioturbation of *Planolites montanus*-type, indicative for wet red beds, is typical of these alluvial plain deposits. The flora consists of xerophilic walchians and cones of the drought-adapted *Calamites gigas*. Tambach is famous for complete articulated vertebrate skeletons, preserved in mud flows (Martens 1988; Berman & Martens 1993; Berman *et al.* 2000, Eberth *et al.* 2000). The fauna consist of reptiles and terrestrially adapted amphibians; fishes are missing. The conglomerates and monotonous red, silty, sandy to clayey haloturbated siltstones with abundant millimetre-sized gypsum crystal casts in the top of this succession were deposited on an apron of alluvial fans with predominantly sheet flood deposits that, towards the basin centre, interfinger with fine clastics of playa mudflats. Well-rounded coarse sand and granule grains (2–3 mm) in the alluvial fan fine clastics are conspicuous. The playa-jellyfish *Medusina limnica* is common in claystones; ephemeral pond deposits contain conchostracans, and newly discovered are leaves of *Taeniopteris* sp. (Voigt & Rössler 2004). The youngest Permian continental sediments, presumably deposited just before the marine Zechstein transgression, are badly sorted, indistinctly horizontally stratified, matrix-supported debris flow conglomerates. The primary red colour changed to grey some metres below the marine Zechstein conglomerate; even granite pebbles are completely leached to pale grey. The horizontal nodule layers are regarded as ground water calcretes because no plant root structures are present. Schneider (1996) has interpreted this leaching and the ground water calcretes as effects of the marine pre-Zechstein ingressions and the Zechstein transgression (Wuchiapingian) into the Southern Permian Basin (see below, page 108), which caused a maritime print onto the strong arid continental climate. Higher precipitation rates and changing ground water levels triggered this calcrete formation and the leaching, which are observed at many outcrops along the southern coastline of the Zechstein Sea. Marine reworked coarse clastics and the Kupferschiefer (Copper slate) form the base of the Zechstein.

A short summary of the stratigraphy follows (for further information see Roscher & Schneider 2005). The Stephanian C age of the Möhrenbach Formation (in the SE) and the Georgenthal Formation (in the NW) is confirmed by biostratigraphy (Schneider & Werneburg 1993; Werneburg 1996) and radiometric data (Roscher & Schneider 2005; Lützner *et al.* 2006). The environment indicates a humid to semi-humid climate. The basal Rotliegend Ilmenau Formation was deposited after an extensive erosional hiatus and is dated, together with the overlying Manebach Formation, by amphibians to the *Apateon dracyiensis–Melanerpeton sembachense* zone (Werneburg 1996; Werneburg & Schneider 2006). The sedimentary structures, coal seams, lake deposits and the floral remains of both formations indicate a humid climate. The following semi-humid to semi-arid Goldlauter Formation is biostratigraphically dated to the middle Lower Rotliegend by amphibians, insects and sharks and geochronologically as 288 ± 7 Ma (Lützner *et al.* 2003, 2006). Facies architecture of the late Lower Rotliegend Oberhof Formation indicates a semi-humid climate, which gradually changed to a semi-arid climate at the top. The semi-arid wet red-bed and playa sediments of the Rotterode Formation are ascribed to the *Moravamylacris kukalovae* zone (Roscher & Schneider 2005).

The stratigraphical position of the overlying Tambach Formation is defined by the *Lioestheria monticula* zone (Martens 1987) = *L. andreevi* zone (Holub & Kozur 1981); *Seymouria sanjuanensis*, latest Wolfcampian (Berman & Martens 1993); *Dimetrodon teutonis*, early Kungurian (Werneburg & Schneider 2006). However, the tetrapod footprint assemblage of the Tambach Sandstone comprising *Amphisauropus*, *Ichniotherium*, *Dimetropus*, *Varanopus*, and *Tambachichnium* (Voigt 2005) and suggests an age for the Tambach Formation not younger than late Artinskian. A Sakmarian age is more likely, because the typical Artinskian ichnotaxa such as *Erpetopus* (Choza Formation, Texas), *Hyloidichnus* (Hermit Shale, Grand Canyon; Rabejac Formation, Lodève Basin) and *Dromopus palmatus* (Rabejac Formation) are missing in the Tambach Formation (Voigt 2005; Voigt, pers. comm. 2005). Based on the litho- and biofacies, the climate conditions are attributed to semi-humid to semi-arid.

The exact age of the arid to semi-arid playa-like Eisenach Formation remains unclear. The Förtha Formation is of Capitanian to Abadehian age (*Lueckisporites virkkiae*, *Corisaccites*: Kozur 1988). The overlying Zechstein Group is dated by *Merrillina divergens* in the first cycle (Werra) carbonates to early Wuchiapingian (Bender & Stoppel 1965; Kozur 1994) and *Mesogondolella britannica* in the Kupferschiefer, Wuchiapingian (Legler *et al.* 2005). The 257.3 ± 1.6 Ma Re–Os isotopic age of the Kupferschiefer at the base of the Zechstein was published by Brauns *et al.* (2003).

The climatic development of the Thuringian Forest Basin is marked by an aridization from the humid, grey, coal-bearing volcaniclastics of the Gehren Subgroup above the semi-humid Goldlauter Formation to the semi-arid to arid playa sediments of the Eisenach Formation. This trend reversed in the Upper Permian with the influence of the extensive Zechstein transgression. The aridization in the Late Carboniferous and Early to Middle Permian is interrupted by some humidization events, namely in the Oberhof and Tambach formations.

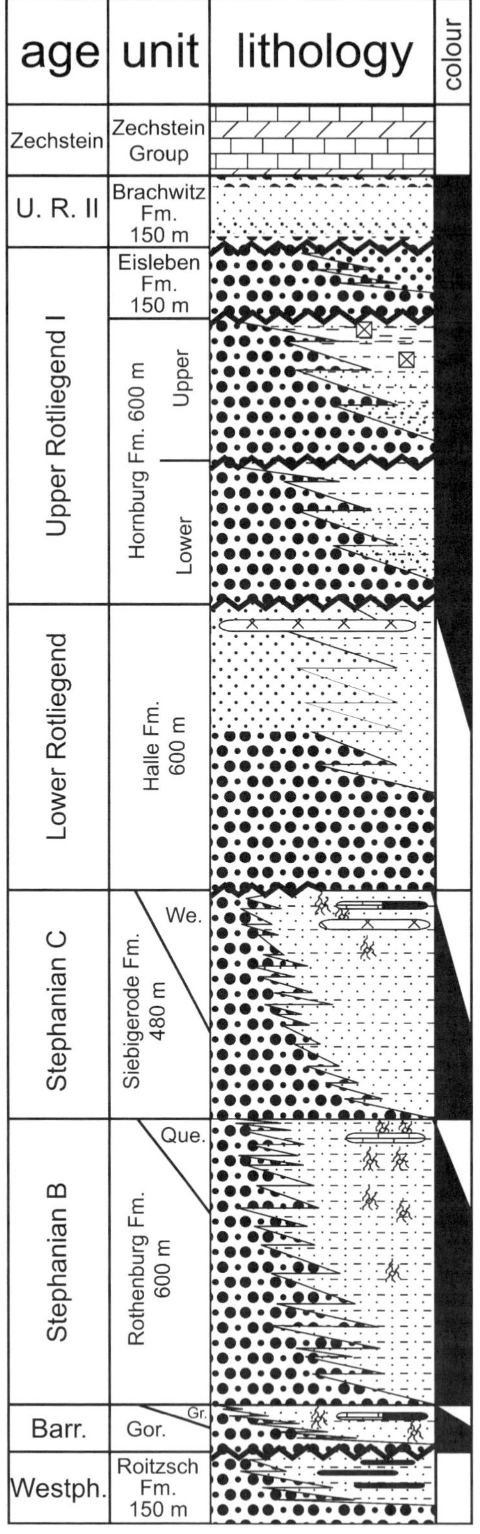

Fig. 8. General succession of the Saale Basin. U. R. II, Upper Rotliegend II; Gor., Gorenzen Fm. 60–100 m; Gr., Grillenberg Subfm. 20 m; Que., Querfurt Subfm. 150 m; We., Wettin Subfm. 300 m. For legend see Figure 3.

Saale Basin

The Saale Basin (*senso stricto*) is a continental basin of 150 × 90 km (Schneider *et al.* 2005*b*). The underlying Visean to Westphalian sediments (Steinbach & Kampe pers. comm. 2005) are known only from drill cores and belong to the southern border of the Variscan fore-deep (Gaitzsch 1998). The sediments of the Roitzsch Formation (Fig. 8), unconformable on Namurian A sediments, were deposited in a drainage system that started at the NE end of the Central Bohemian Basin (Opluštil & Pešek 1998) and ended in the Variscan foredeep (Gaitzsch *et al.* 1999). Above basal coarse clastics, conglomerate/rooted soil cycles with several coal seams, several decimetres to 1 m thick, and fine clastics follow. The Stephanian basin-fill rests disconformably on older sediments and Variscan metamorphics and consists mainly of wet red beds (*Scoyenia*-facies) of proximal to distal alluvial fan and alluvial plain environments of the Mansfeld Subgroup (Gorenzen Formation to Siebigerode Formation). Grey sediments of fluvial, lacustrine and palustrine facies, designated as subformations, are found in depocentres near the top of each megacycle. They contain several impure coal seams and carbonaceous horizons. The widely distributed wet red beds (*Scoyenia*-facies) comprise alluvial fan/sheetflood/braided river associations with immature to mature calcisoils and metre-thick calcretes and decimetre-thick greyish to reddish micritic limestones of ephemeral lakes. At the top, the change of channel geometry indicates a transition from braided to meandering rivers, and the grey facies is vertically and laterally more widespread and includes lacustrine limestones and black shales in the flood basins, as well as palustrine deposits of back-swamp environments that contain several workable coal seams.

The fish faunas of the Querfurt and the Wettin subformations indicate that the Saale Basin was interconnected with an extensive European drainage system that was destroyed at the beginning of the Rotliegend by volcano-tectonic reorganization of the basins (Schneider & Zajic 1994; Schneider *et al.* 2000). This event is indicated in the Saale Basin by the sudden deposition of the 'Kieselschiefer-Quarzit-Konglomerat' (Chert-Quartzite Conglomerate) at the base of the Halle Formation, which grades vertically into reddish sand and siltstones. After a long hiatus, the completely red alluvial fan to alluvial plain conglomerates and sandstones as well as occasional sheet flood sediments with interfingering siltstones and claystones were deposited. These fine clastics represent playa deposits characterized by *Medusina limnica*, millimetre-large halite hopper crystals and metre-deep desiccation cracks. Very noticeable features are aeolian deposits – dune sandstones (flooding of dunes during playa-lake formation resulted in strong deformation of these dune sandstones; Schneider & Gebhardt 1993) as well as dry sandflat sandstones with lag deposits of wind-transported, coarse, well-rounded grains. The latter may also have been re-deposited by flash floods, forming decimetre-thick sequences of bimodal, coarse-grained sandstones with fine-grained, well-sorted matrix of primary aeolian origin. This is capped by a sequence of red conglomerates, sandstones and siltstones of proximal to distal alluvial fans.

After a hiatus, the braided river, sheetflood and wet sandflat deposits of the Eisleben Formation, which is regarded as equivalent to the Dethlingen and Hannover formations of the Southern Permian Basin (Legler *et al.* 2005), were deposited. In the area of the Saale Basin, the Zechstein commonly starts with marine re-worked aeolian sands and above them the Kupferschiefer (Copper slate).

The stratigraphical correlation of this basin is based on the following data: the Wettin Subformation (Siebigerode Formation) belongs to the *Apateon intermedius–Branchierpeton saalensis* zone (Werneburg 1996; Werneburg & Schneider 2006), *Bohemiacanthus* Ug zone (Schneider & Zajic 1994), and *Sysciophlebia euglyptica–Syscioblatta dohrni* zone (Schneider & Werneburg 1993) and is geochronologically dated to 293 ± 2 Ma (Goll & Lippolt 2001). The overlying Halle Formation belongs to the *Apateon intermedius–Branchierpeton saalensis* zone (Werneburg 1996; Werneburg & Schneider 2006) and is dated by Breitkreuz & Kennedy (1999) as 297–301 ± 3 Ma. The conchostracans of the Upper Hornburg Formation belong to the *Lioestheria andreevi–Pseudestheria graciliformis–Palaeolimnadiopsis wilhelmsthalensis* assemblage zone (Hoffmann *et al.* 1989; Schneider *et al.* 2005*a*). The Eisleben Formation is lithostratigraphically correlated to the Hannover Formation of the North German Depression (see below).

The climate development of the Saale Basin demonstrates an aridization trend, starting with the humid, grey, coal-bearing sediments of the Gorenzen Formation, through the semi-humid, grey sediments of the Halle Formation, to the semi-arid to arid playa-like red clastics of the Upper Hornburg and Brachwitz formations. This trend is stopped at least by the sedimentation of the epi-continental Zechstein Sea and interrupted by the Stephanian C semi-humid to humid Siebigerode Formation.

North German/Polish Depression

The Central European Basin (CEB) is one of the largest basins on earth. It provides one of the most voluminous data sets for understanding the evolution of similar giant intracontinental accumulations of sediments (Fig. 9). The history of this basin started with formation of the Middle to Late Permian continental megaplaya/megasabkha system of the Southern Permian Basin (SPB). Early to Late Carboniferous sediments in the area of the SPB belong to the Variscan foredeep, which ceased during the Stephanian. Here, sediments from Late Westphalian (comparable to the Ruhr area: Hoth *et al.* 1990) up to the Zechstein base in the eastern part of northern Germany are considered. They comprise a grey, coal-bearing paralic facies (with coal seams up to 1.8 m thick) with a last extensive marine incursion (Aegir or Mansfield marine band) at the Westphalian B/C (Duckmantian/Bolsovian) boundary. After that, the facies changed from grey sediments to wet red beds. The uppermost thin root horizons and a thin coal seam are restricted to the basal Westphalian D.

The red, conglomeratic Mönchgut Formation is overlain by huge volcanic complexes, up to 3000 m thick, with small basins adjacent to and inside them. They are filled with red fan-conglomerates and variegated, red to grey alluvial plain sediments with laminated (varved) grey to reddish bituminous lacustrine limestones as well as tuff and tuffite horizons (Gaitzsch 1995*a*; Schneider *et al.* 1995*a*). Laminated lacustrine black shales, red alluvial plain to ephemeral lacustrine fine clastics, locally replaced by alluvial fan conglomerates, overly this sequence.

The SPB originated by thermal subsidence and the start of post-Variscan rifting processes linked with extrusions of upper mantle basalts (Gebhardt *et al.* 1991). The basin extended from England over the southern North Sea and northern Germany to Poland with a length of 2500 km and a width of 600 km. It was filled by about 2500 m of siliciclastics and evaporites of the Upper Rotliegend II, as well as 2000 m of siliciclastics, carbonates and evaporites during the Zechstein (Ziegler 1990*b*). The Upper Rotliegend II basin fill is dominated by desert sediments affected by an arid to semi-arid climate. Alluvial fans and dunes occur, especially at the southern basin margin, whereas saline lake deposits dominate in the centre. The sedimentation was controlled tectonically as well as climatically, as is reflected by the lithostratigraphical subdivision. Tectonically driven large-scale cycles are interpreted as formations, whereas the members are climatically governed cycles on a smaller scale (Gast 1991, 1993, 1995; Gaupp *et al.* 2000). Basaltic extrusions, which are well known from the Altmark region, the Soltau High and the Horn Graben, are signs of rifting as an important controlling parameter in the evolution of the basin (Gebhardt *et al.* 1991; Gebhardt 1994; Stemmerik *et al.* 2000). Short-term marine ingressions superpose the climatically driven, coarsening-upward cycles of the Hannover Formation (Gebhardt 1994; Legler 2005). The Zechstein Transgression was a sudden flooding event triggered by a sea level highstand and rifting in the Viking Central Graben system (Smith 1979; Glennie 1989). Eustatic sealevel highstands (0.8 to 1 Ma cycles observed in the Proto-Atlantic: Legler 2005) before and together with the Zechstein transgression, as well as the more or less coeval *Bellerophon* transgression, put a maritime print on the otherwise strongly continental climate of northern Pangaea.

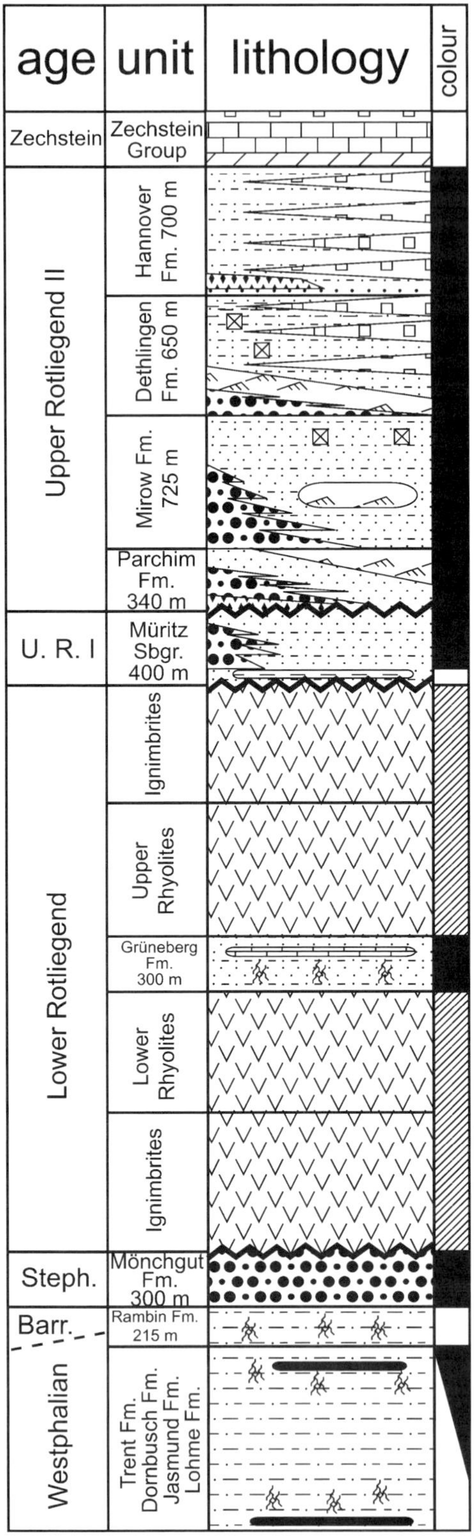

The stratigraphy of this basin is based on isotopic data and individual biostratigraphical constraints. The Lower Ignimbrite was dated by Breitkreuz & Kennedy (1999) as 302 ± 3 Ma. The sediment intercalations within the volcanics belong to the *Bohemiacanthus* Om–Ugo zone and the *Pseudestheria paupera* and *Pseudestheria palaeoniscorum* Zones, middle Lower Rotliegend (Gaitzsch 1995*b*, *c*). The Müritz Subgroup belongs to the *Lioestheria andreevi–Pseudestheria graciliformis–Palaeolimnadiopsis wilhelmsthalensis* assemblage zone, Upper Rotliegend I (Hoffmann *et al.* 1989; Schneider *et al.* 2005*a*). The age of the Parchim Formation is fixed by the Illawara Reversal in its basal portion (Menning 1995). In this basin, the conodont *Mesogondolella britannica* was found in the Kupferschiefer (Copper slate) and demonstrates a Wuchiapingian age (Legler *et al.* 2005).

The climatic development of this basin is marked by an overall aridization trend from the grey, coal-bearing Westphalian sediments, via the coaly wet red beds of the Mönchgut Formation, to the wet red beds of the Müritz Subgroup and the Upper Rotliegend II playa/sabkha red beds of the Elbe and Havel Subgroup. With the pre-Zechstein transgressions the environment

Fig. 9. General succession of the North German/Polish Depression. U. R. I, Upper Rotliegend I. For legend see Figure 3.

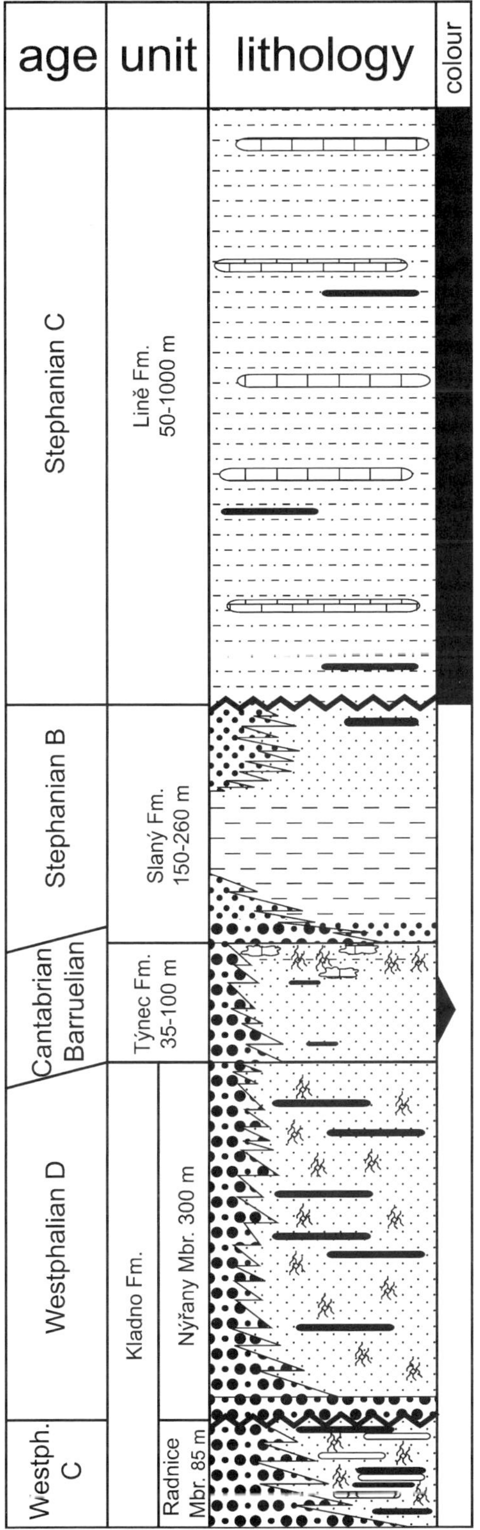

Fig. 10. General succession of the Western and Central Bohemian basins. For legend see Figure 3.

became more humid and, finally, marine with the Zechstein transgression.

Central and western Bohemian basins

The formerly connected basins (140 km × 25 km) of western and central Bohemia are (beginning from the west): the Plzeň Basin, Manětín Basin, Radnice Basin, Žihle Basin, Kladno-Rakovník Basin and Mšeno-Roudnice Basin. The basement consists of Late Proterozoic, weakly metamorphosed sediments and volcanic units with subordinate, medium to high grade metamorphic rocks (Pešek 2004). The basal sediments (Fig. 10) are characterized by fine- to medium-grained, grey, green or red-brown unbedded sediments, with matrix-supported conglomerates of an alluvial fan, to alluvial plain environment on uneven bedrock. These environments grade vertically to extensive peat bogs in a broad alluvial plain with many coal seams, often separated by numerous tuffitic layers. A conglomerate complex was deposited after a hiatus. After that the facies changed slowly, as indicated by the decrease of coal seams and sediment colour change to red-brown. Overlying are fluvio-lacustrine grey conglomerates to sandstones with rare, intercalated greenish or brownish red mudstones. The following biogenic laminated and varvite-like claystones are ascribed to a lacustrine environment of an extensive freshwater lake with a depth of several tens to some hundred metres. The stronger fluvial influence and shallowing of the environment is indicated by an increase of coarser sediments (grey to reddish violet sandstones) and culminates with the reappearance of coal seams, up to 1 m thick, that formed at the lakeside. After a depositional break, mostly red to deep red siltstones with a low coal content were deposited. The facies with several calcareous horizons is ascribed to an alluvial plain with a system of small, interconnected lakes.

The stratigraphy of these deposits is determined by various methods. The Radnice Member is dated by macroflora (Pešek 2004) and insects (*Archimylacris lubnensis* zone, Schneider *et al.* 2005*a*) to late Bolsovian to early Westphalian D, and has a high amount of coal, indicating a warm and humid climate. For the Nyřany Member, Werneburg (1989, 1996, 2001) indicated, by amphibians, the *Branchiosaurus salamandroides–Limnogyrinus elegans* zone (Westphalian D to Stephanian B). Climatically, it is in distinguishable from its precursor. The Týnec Formation is dated by macroplants to a Barruelian age (Pešek 2004), and the climate conditions were drier and possibly warmer. After a climate change to more humid conditions, the Slaný Formation was deposited during the early Stephanian B *Sysciophlebia grata* zone (Schneider 1982). During this time, finely laminated (varved) claystones formed within a freshwater lake of about 55 000 km² (Pešek 2004). The Líně Formation was dated to Stephanian C age by amphibians: *Apateon intermedius–Branchierpeton saalensis* to *Apateon dracyiensis–Melanerpeton sembachense* zone (Werneburg & Schneider 2006) and macroplants (Pešek 2004). The climate was drier than before, but cannot be regarded as pronouncedly arid (Pešek 2004) because of the interconnected lakes.

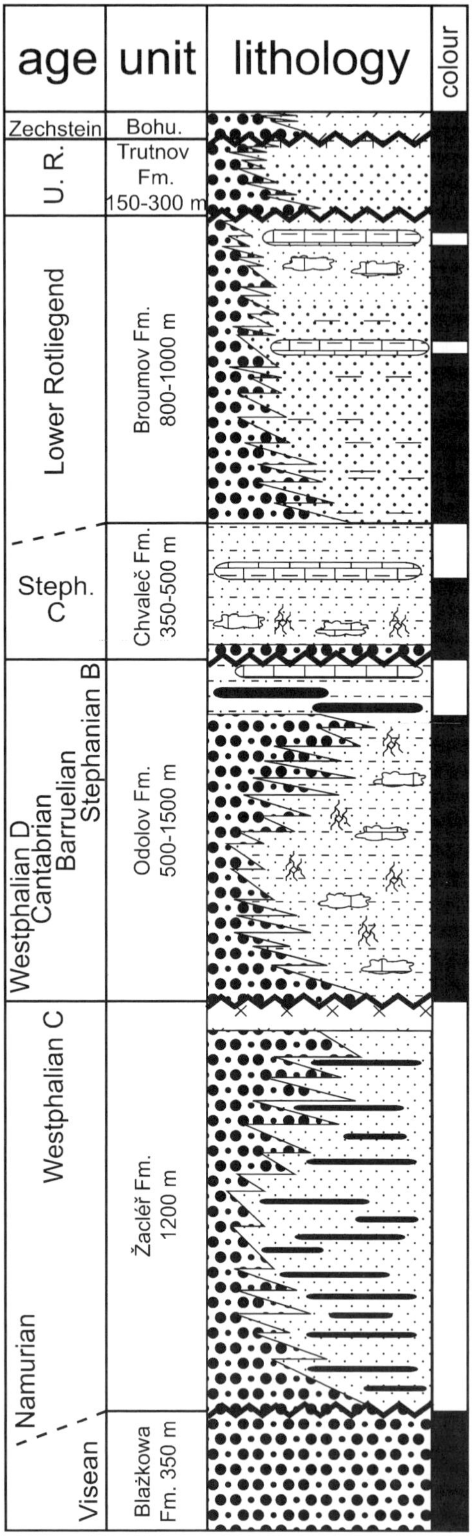

The climate development of all these basins shows an overall aridization starting from the warm-humid, coal-forming environments of the Kladno Formation (Westphalian D–?Cantabrian) to the alluvial plain red beds of the Líně Formation (Stephanian C). This trend was interrupted by the more humid climate, starting in the uppermost part of the Týnec Formation and culminating in the large lake development of the lower Slaný Formation (earliest Stephanian B). Within the upper Slaný Formation the aridization strengthened again.

Krkonoše Piedmont and Intra-Sudetic basins

The volcano-sedimentary filling of these basins covers the crystalline basement of the Krkonoše–Jizera complex and partially the Late Proterozoic and Ordovician/Silurian low-grade metamorphic rocks. The basal purple to brown-red conglomerates of the Intra-Sudetic Basin (ISB, 1800 km²; Fig. 11) fills up the palaeo-relief. The following sequence is characterized by cyclic fluvial and alluvial plain sediments, dominated by brownish to variegated conglomerates and sandstones with rare intercalations of pelites. Within these upward-fining cycles more than 60 coal seams were deposited. The final phase of the Žacléř Formation (ISB) deposition was accompanied by strong volcanic activity that produced the up to 100 m thick and laterally widespread Křenov rhyolithic tuff (309.0 ± 3.7 Ma: Lippolt *et al.* 1986). After a short break in sedimentation, the Odolov (ISB) and the Kumburk Formation of the Krkonoše Piedmont Basin (KPB, 1100 km²,

Fig. 11. General succession of the Intra-Sudetic Basin. U. R., Upper Rotliegend; Bohu., Bohuslavice Fm. 30–120 m. For legend see Figure 3.

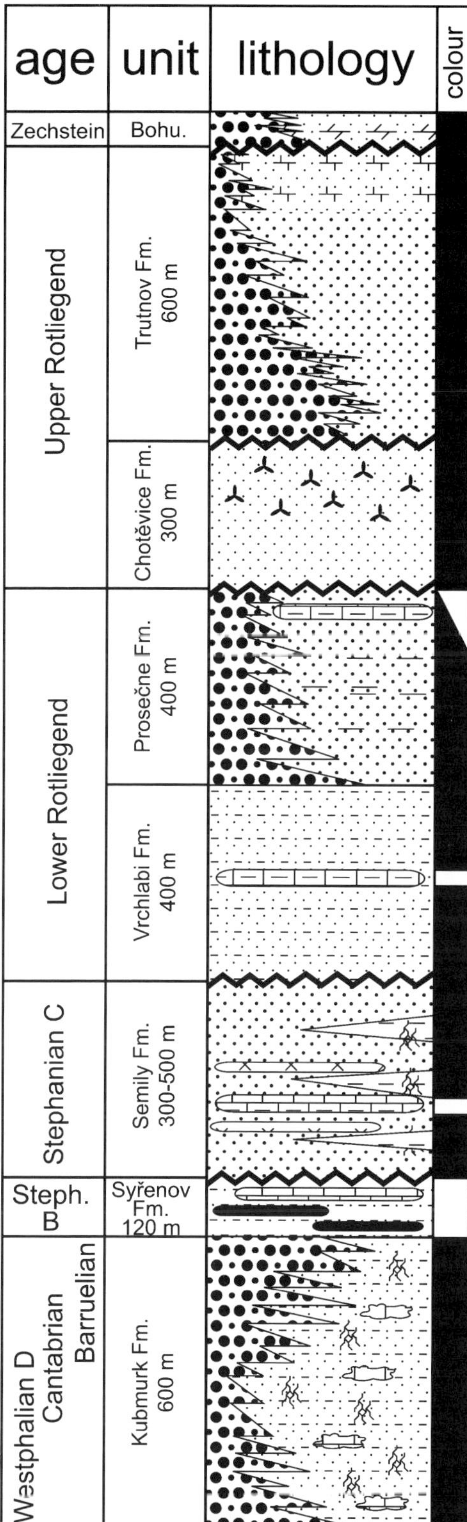

Fig. 12) began with mostly reddish pelites arkoses and conglomerates of an alluvial plain (Opluštil & Pešek 1998). Next to the top of the Odolov Formation (ISB) (= Syřenov Formation, KPB) the facies changes to grey alluvial plain sediments containing several coal seams and some lacustrine deposits, which were connected to the large lake of the western and central Bohemian basins (Slaný Formation: Pešek 2004). In the KPB, it is overlain by mostly red-coloured, coarse-grained sediments with occasional mudstone intercalations deposited on a broad braid plain with common pedogenic horizons and a widespread horizon of bituminous shales with thin lacustrine carbonate and volcaniclastic beds (Opluštil & Pešek 1998).

The Vrchlabi Formation (KPB) as well as the Chvaleč Formation (ISB) are separated from the underlying units by an unconformity. The basal conglomerate of the latter has no equivalent in the KPB. The overlying sediments are dominated by red-brown silty and clayey sediments with common palaeosols. Some layers of lacustrine-laminated (varved), bituminous, grey limestones with cherts are intercalated in this sequence of alluvial plain wet red beds. The overlying deposits are an assemblage of red-brown arkoses with intercalated conglomerates, thin pelites and limestones of ephemeral lakes. The environment is marked by the development from an alluvial plain with ephemeral ponds to lacustrine conditions and black-shale deposition. The upper part of the Broumov Formation (ISB) is characterized by brownish, playa-like alluvial plain fine clastics with common carbonate nodules. Within the erosively overlying Chotěvice Formation (KPB) the lithology changed from red-brown sandstones to brown-red claystones with abundant gypsum pseudomorphs as the environment changed from alluvial plain to a playa-like system. The trend continues with monotonous red-brown conglomerates, sand-, silt- and claystones, containing redeposited windblown grains of the overlying formations. The latest Permian Bohuslavice Formation consists of partially dolomitic conglomerates and sandstones, and was deposited by periodic streams and sheet floods.

The age of the units, shown in Figure 15b, is based of the following. The Blażkowa Formation is of late Viséan and/or early Namurian age (Pešek 2004), as determined by macroplants in this and the Late Namurian–Bolsovian Žacléř

Fig. 12. General succession of the Krkonoše Piedmont Basin. Bohu., Bohuslavice Fm. 30–120 m. For legend see Figure 3.

Formation. These units suggest a semi-humid warm climate. The earliest Westphalian D part of the Odolov Formation represents a warm-humid climate, indicated by the coal seams. The Kumburk Formation was deposited during late Westphalian D to Barruelian and is comparable to the semi-humid, wet red-bed environment of the lower Odolov Formation. The more humid conditions of the early Stephanian B (*Sooblatta stephanensis* Schneider 1983) Syřenov Formation correspond to the upper Odolov Formation. The Semily Formation indicates semi-arid conditions for the Stephanian C. The base of the Chvaleč Formation is characterized by wet red beds of a semi-arid climate. The appearance of lacustrine black-shale sediments in the upper part indicates a humid to semi-humid climate, as in the Lower Rotliegend Vrchlabi Formation. At the top of this formation a transition to more arid conditions is indicated. The Prosečné Formation is lithostratigraphically correlated to the Broumov Formation (ISB), which is determined by Werneburg (1996, 2001) to be in the *Melanerpeton pusillum–Melanerpeton gracile* zone.

The aridization trend, starting in the upper Vrchlabi Formation, continues in the lower part of the Prosečné and middle part of the Broumov Formation, but is reversed in the development of the Kalna and Ruprechtice Limestone Horizon. The re-occurrence of alluvial plain deposits in the upper Broumov Formation is consistent with an ongoing aridization to the playa system of the Chotěvice Formation. Similar to the sediments of other basins, with its arid features as well as its aeolian content, the Trutnov Formation indicates an Upper Rotliegend I to II age. The Bohuslavice Formation, which is, by analogy to the North Sudetic Basin, assigned to a Zechstein age, indicates a semi-arid climate.

The sedimentary successions of these basins show a well-established aridization trend from the humid to semi-humid Blażkowa Formation and Žacléř Formation, to the wet red beds of the Kumburk/Odolov Formation, through the semi-humid to semi-arid red beds of the Prosečné/Broumov Formation to the arid evaporitic playa environment of the Chotěvice Formation. Within this long geological record, several short-lived humidization events (reversals) are observable in the coal-bearing lower Odolov Formation, the coal-bearing Syřenov/upper Odolov Formation (humid) and the prominent lacustrine horizons in the lower Vrchlabi/upper Chvaleč Formation (semi-humid) and the upper Prosečné/middle Broumov Formation (semi-humid to semi-arid). Generally, each reversal of the aridization trend is weaker than its precursor. The aridization maximum is reached within the Chotěvice and Trutnov formations. After that a climatic reversal is observable, characterized by humidization up to the semi-arid climate of the Bohuslavice Formation.

Boskovice Graben

The N–S-striking elongated Boskovice Graben is about 100 km long and 3–10 km wide. Sedimentation (Fig. 13) began with the red-brown, clast-supported Balinka Conglomerate and cyclically intercalated sand- and siltstones on a crystalline basement. It grades to fluvio-lacustrine, grey sand-, silt- and claystones that are interrupted by several coal seams and subordinate lacustrine black shales. The overlying reddish alluvial plain deposits of the *Scoyenia*-facies are capped by the decimetre-thick Oslavany Conglomerate. Sedimentation continued, with cyclic red-brown and yellow-brown silt- and sandstones with intercalated arkoses to conglomerates and some thin grey successions. The environment is characterized by an alluvial plain to lacustrine facies that is strongly influenced by alluvial fan deposits. A facies change within the following deposits is only traceable by the intercalations of some larger complexes of bituminous, clay-rich limestones to carbonaceous black shales. They developed on an alluvial plain with several ephemeral pond and lakes and are concentrated in the lower and upper parts of the Letovice Formation.

The humid Rosice–Oslavany Formation is dated by macroplants to Stephanian C. The Padochov Formation was deposited subsequently. A major hiatus is marked by the Oslavany Conglomerate, which is thought to be the lithostratigraphical Rotliegend base. This is well supported by fossil insects in the Říčany Horizon (*Spiloblattina homigtalensis*, early Lower Rotliegend: Schneider 1980, 1982; Schneider & Werneburg 1993). The environment of this unit indicates a semi-humid climate in the lower part, developing to semi-arid conditions in the upper part. The Letovice Formation spans the Lower Rotliegend to the Upper Rotliegend I. The Zboněk–Svitavka Horizon belongs to the *Sysciophlebia alligans* zone (late Lower Rotliegend: Schneider & Werneburg 1993). The Bačov Horizon with the localities of Bačov, Obora and Sudice belongs to the *Moravamylacris kukalovae* zone (Schneider 1980) and the *Discosauriscus austriacus* zone (Werneburg 1996), which is early Upper Rotliegend I. Both horizons, the Zboněk–Svitavka Horizon in the lower part of the Letovice Formation, as well as the Míchov and Bačov Horizons in the uppermost part, are represented by a large sequence of grey clastics

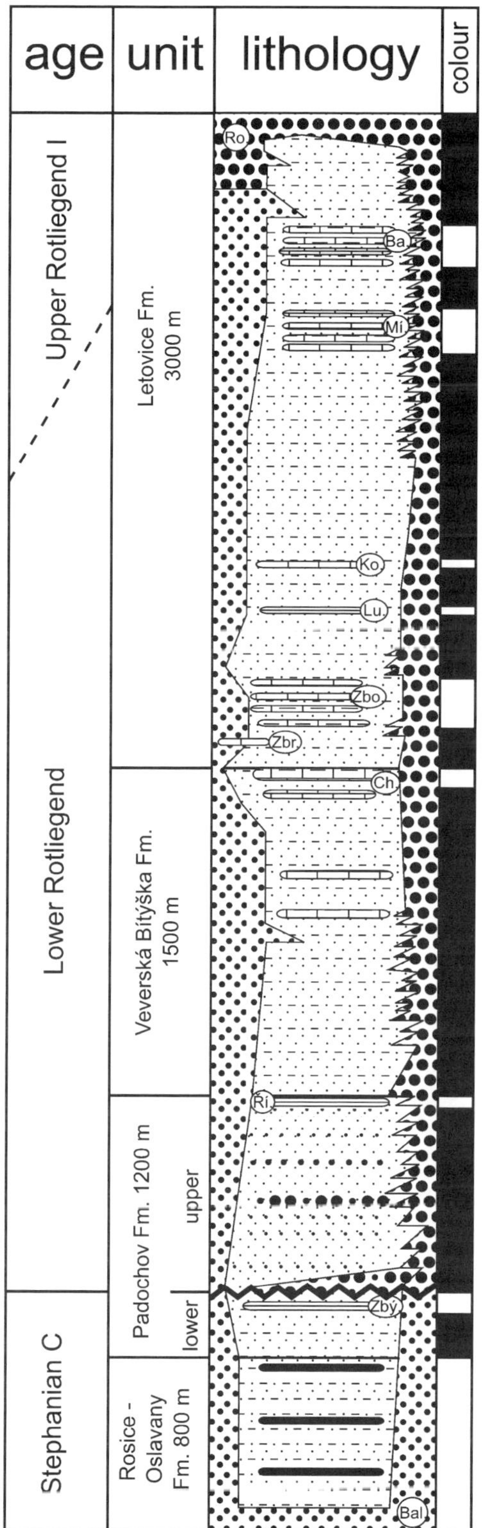

Fig. 13. General succession of the Boskovice Graben. Bal., Balinka Conglomerate; Zbý., Zbýšov Horizon; Ří., Říčany Horizon; Ch., Chudčice Horizon; Zbr., Zbraslavec Horizon; Zbo., Zboněk-Zvitávka Horizon; Lu., Lubě Horizon; Ko., Kochov Horizon; Mí., Míchov Horizon; Ba., Bačov Horizon; Ro., Rokytná Conglomerate. For legend see Figure 3.

with metre-thick lacustrine, bituminous, calcareous and extensively laminated black shales. Therefore, they indicate a semi-humid climate, whereas the red clastics in between show semi-arid conditions.

The palaeoclimate reconstruction of this basin shows a well-established aridization trend through the whole profile, from the humid Rosice–Oslavany Formation, through the semi-arid red beds of the Veverská Bítýška Formation, to the semi-arid alluvial plain sediments of the Letovice Formation. This trend is interrupted by two humid to semi-humid phases that produced large lake horizons in the lower and upper Letovice Formation.

Moroccan basins

In the Moroccan Meseta and High Atlas Mountains, late Palaeozoic sediments and volcanics crop out, which document the development of this area during the formation of the Mauretanide part of the Hercynian orogeny. In the Western Meseta, late Visean to Early Westphalian marine turbidite sequences mark the early stages of foreland basin development, which is interpreted as the southern extension of the European Variscan belt. During Stephanian and Permian times, pure continental intramontane basins developed (Saber *et al.* 1995; for details, see Hmich *et al.* 2006). Here only a short synthesis is presented.

Biostratigraphically well-dated grey sediments of earliest Stephanian B age in the Souss Basin (Fig. 14) are transitional between the Early Stephanian wet phase and the Middle Stephanian dry phase. In a first preliminary approximation, the Moroccan Souss Basin could be interpreted as being situated during this time in the southern subtropical summer-wet belt (biome 2 of Ziegler 1990*a*). The wet red beds (*Scoyenia* facies) of the Khenifra Basin (Fig. 14), based on macrofloras, are regarded as transitional Autunian to Saxonian (Broutin *et al.* 1998). This fits well into the Artinskian wet phase and is perhaps transitional into the Kungurian. The wet red beds (*Scoyenia* facies) of the Tourbihine Member, Ikakern Formation, in the Argana Basin (Fig. 14) are dated by pareiasaur remains, which are closely related to those from the Late

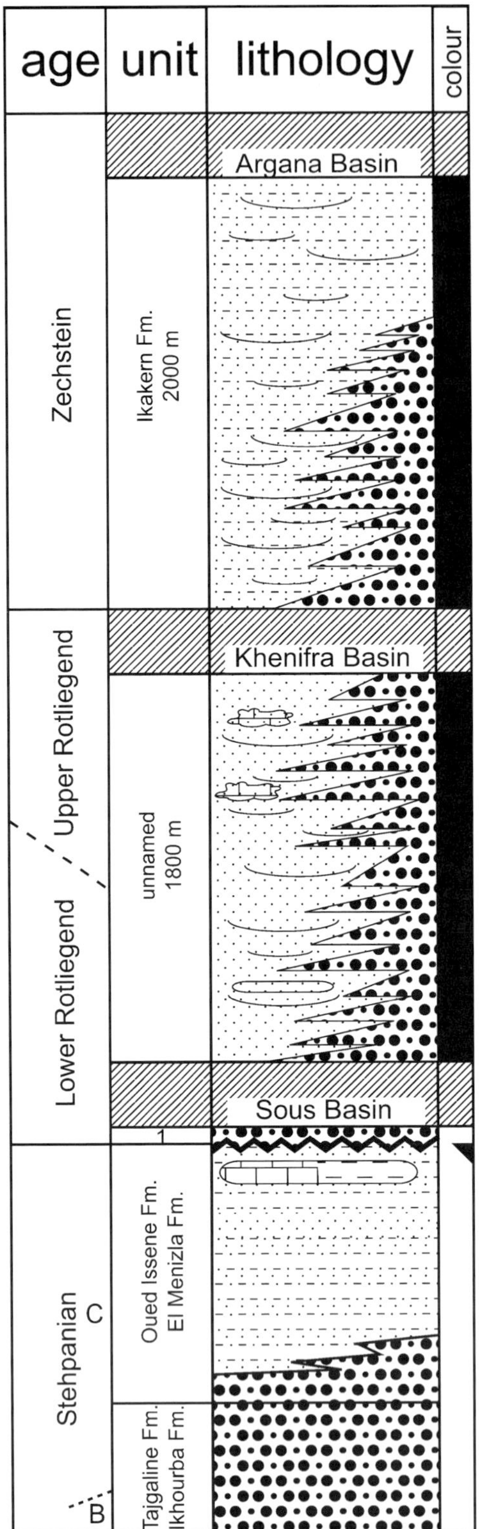

Fig. 14. General succession of the Moroccan basins. For legend see Figure 3.

Permian of Elgin, Scotland, and the Zechstein of Richelsdorf, Germany, as Middle Wuchiapingian (Jalil per. comm. 2002). Wet red beds similar to those in the Argana Basin are known in Europe, such as those from the La Lieude Formation, Late Permian of the Lodève Basin. They originated during the Wuchiapingian wet phase.

Climatic development of northern Pangaea

The most significant climate indicators mentioned above are shown in Figure 15a, b. Based on a well-justified stratigraphy it is obvious that the climate development of each basin shows more than local climate variations. Nevertheless, the climate of every basin was influenced by local features, such as subsidence, orographic barriers, slope gradient and exposition and vegetative cover. All these meso- to microclimates were local modifications of the inter-regional Euramerican climate. In addition to the undisputed aridization trend in northern Pangaea, several wet phases are correlatable between the basins. These phases are characterized by more humid conditions during a short period (2–3 Ma) in contrast to the over- and underlying units. Because of the general aridization trend, every single wet phase is weaker than its precursor. The wet phases are situated in the Barruelian, Stephanian C to early Lower Rotliegend, late Lower Rotliegend and Early Upper Rotliegend I. A further one phase the Westphalian C/D cannot be demonstrated by the dataset used here, but its occurrence is suggested because of the mechanism discussed below.

The major problem for the reconstruction of continental climate is the lack of complete successions. Therefore, a well-justified stratigraphy for many basins with incomplete successions is indispensable. All boundaries of the wet phases are fixed within continuous successions, but within different basins. For example, the lower boundary of the Stephanian C to early Lower Rotliegend wet phase can be defined only in the Saar–Nahe Basin, the Thuringian Forest Basin, the Saale Basin and the Intra-Sudetic Basin, whereas in the other sections (Fig. 15a, b) this boundary cannot be documented because the wet phase starts just above a discontinuity. These wet phases can be seen also in many basins with a smaller extent, for example the Ilfeld Basin and the Döhlen Basin (Fig. 15a).

The separation of climatic versus tectonic imprints in the sediment facies patterns is difficult, but within complete successions a drastic influence of tectonics can be ruled out. Within the late Lower Rotliegend and early Upper Rotliegend major lake horizons of wet phases are summarized. The combination of tectonic artifacts (closed system lakes) is ruled out by the distribution and migration of aquatic faunas (Schneider & Zajic 1994; Boy & Schindler 2000). The maximum of aridity was reached in the late Upper Rotliegend I to early Upper Rotliegend II. This is supported by geochemical and sedimentological investigations by Schneider *et al.* (2006). They place the maximum of aridity within the Octon Member of the Salagou Formation (Lodève Basin). This is also based on the sedimentary features of the Eisenach Formation of the Thuringian Forest Basin and the Chotěvice and Trutnov formations of the Krkonoše Piedmont Basin and Intra-Sudetic Basin. The last but most dramatic climate change in the central European area was caused by the Zechstein and *Bellerophon* transgression. The filling of the Northern and Southern Permian Basin, as well as the weaker pre-Zechstein transgressions, represent the source area of the moisture that influenced the arid climate. The effect of this large epi-continental sea is well recognized in the rapid change of facies patterns within the Late Permian of the Lodève Basin (La Lieude Formation).

The correlation of the wet phases to the global standard is difficult, because at the moment no profile with interfingering marine/continental sediments has been sufficiently studied. Therefore, no biostratigraphical link between the Permian continental sediments and the marine standard is possible, so all age determinations of the wet phases are given in 'Central European continental terms'. The correlation to the international scale presented here is based on various isotopic ages of different basins (for further information, see Roscher & Schneider 2005). The Barruelian wet phase is correlated to the middle to late Kasimovian, and the late Stephanian C to early Lower Rotliegend wet phase to the early Asselian. The late Lower Rotliegend wet phase is in the early Sakmarian and the early Upper Rotliegend I wet phase is in the early Artinskian. For the latter, the correlation is the weakest because its absolute age is not clearly defined by isotopic ages. To understand the mechanism responsible for the climatic variations we must look at the southern hemisphere.

Climate development of southern Pangaea (Karoo Basin)

When these more humid phases (wet phases) are plotted on a stratigraphical correlation chart, it is striking that they correlate with the deglaciation cycles of the Dwyka Group of the Karoo Basin (Fig. 15b & 16). The stratigraphy of the Dwyka Group is defined by isotopic ages published by Bangert *et al.* (1994, 1999), Stollhofen (1999), Stollhofen *et al.* (1999, 2000), and Rohn & Stollhofen (2000). Visser (1995, 1996, 1997) and Scheffler *et al.* (2003) dealt with the relative sea level during these times. The positioning of the transgression and regression cycles in relation to the lithological profile by both authors is the same. Differences occur with reference to absolute ages, potentially an artefact of different dating techniques and time scales.

The humidity increases towards the top of every deglaciation cycle. It is widely accepted that these cycles are of a deepening-upward (transgressive) nature. The sealevel fluctuations have a frequency of 5–7 Ma (Scheffler *et al.* 2003). This duration is not known from any orbital frequency. Therefore, the cause of these fluctuations cannot be explained at present. Nonetheless, the impact of the waxing and waning of the Gondwanan glaciation is traceable globally. The higher frequency glacio-eustatic fluctuations are represented in the marine-paralic Kansas cyclothems (Heckel 1999, 2002*a*, *b*). These high-frequency cycles are not traceable within continental sections because the age resolution for this kind of sediments is not high enough, but the lower frequency of 5–7 Ma is reflected on the continent by humid-arid alternations.

The Gondwana glaciation

The Gondwana Glaciation began around the Mid-Carboniferous boundary at 320 Ma (Bruckschen *et al.* 1999; Saltzman 2003) or just before it at 330–335 Ma (Holkerian–Asbian boundary) (Dickins 1996; Wright & Vanstone 2001; Wright 2003). Saltzman (2003) ascribed the onset of the large-scale Gondwana Glaciation to the closure of an equatorial seaway. The simultaneous reorganization of the global ocean currents supported the transportation of moisture to the southern continent. This equatorial seaway can be seen, for example, in the Western Meseta (Morocco), where the late Viséan to Early Westphalian marine turbidite sequences mark the early stages of foreland basin development (Abbou *et al.* 2001). Rapid isostatic uplift, after subduction under Laurussia (maximum of high pressure metamorphism at 340 Ma), of the continental Gondwanan crust caused the closure of the strait between the Palaeo-Tethys and the Rheic Ocean (Fig. 17). This supports the older Viséan Age for the beginning of the glaciation.

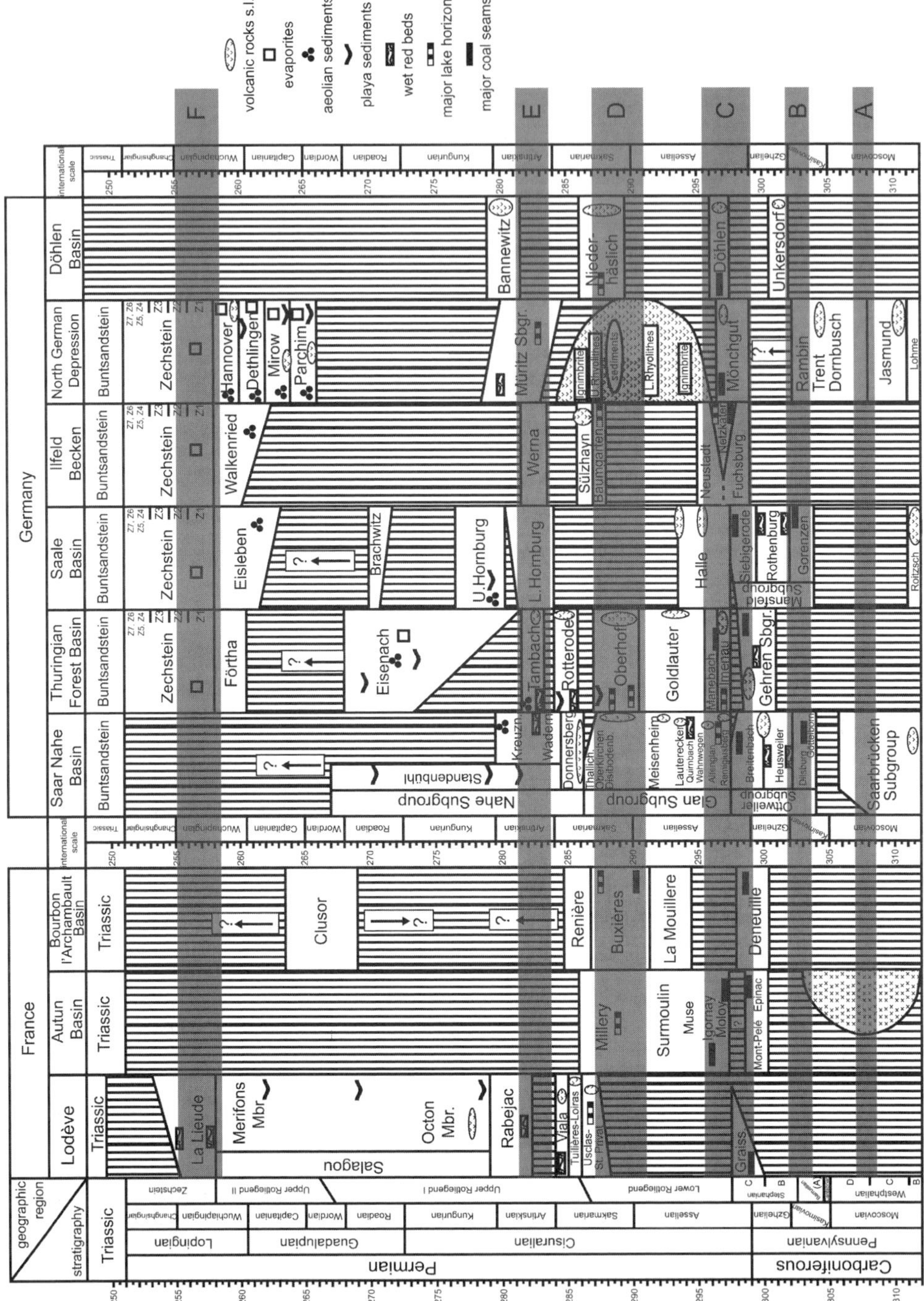

Fig. 15(a). Correlation chart of European Basins. A, assumed Westphalian C/D wet phase; B, Barruelian wet phase; C, Stephanian C to early Lower Rotliegend wet phase; D, late Lower Rotliegend wet phase; E, Early Upper Rotliegend I wet phase; F, Zechstein/Bellerophon Transgression.

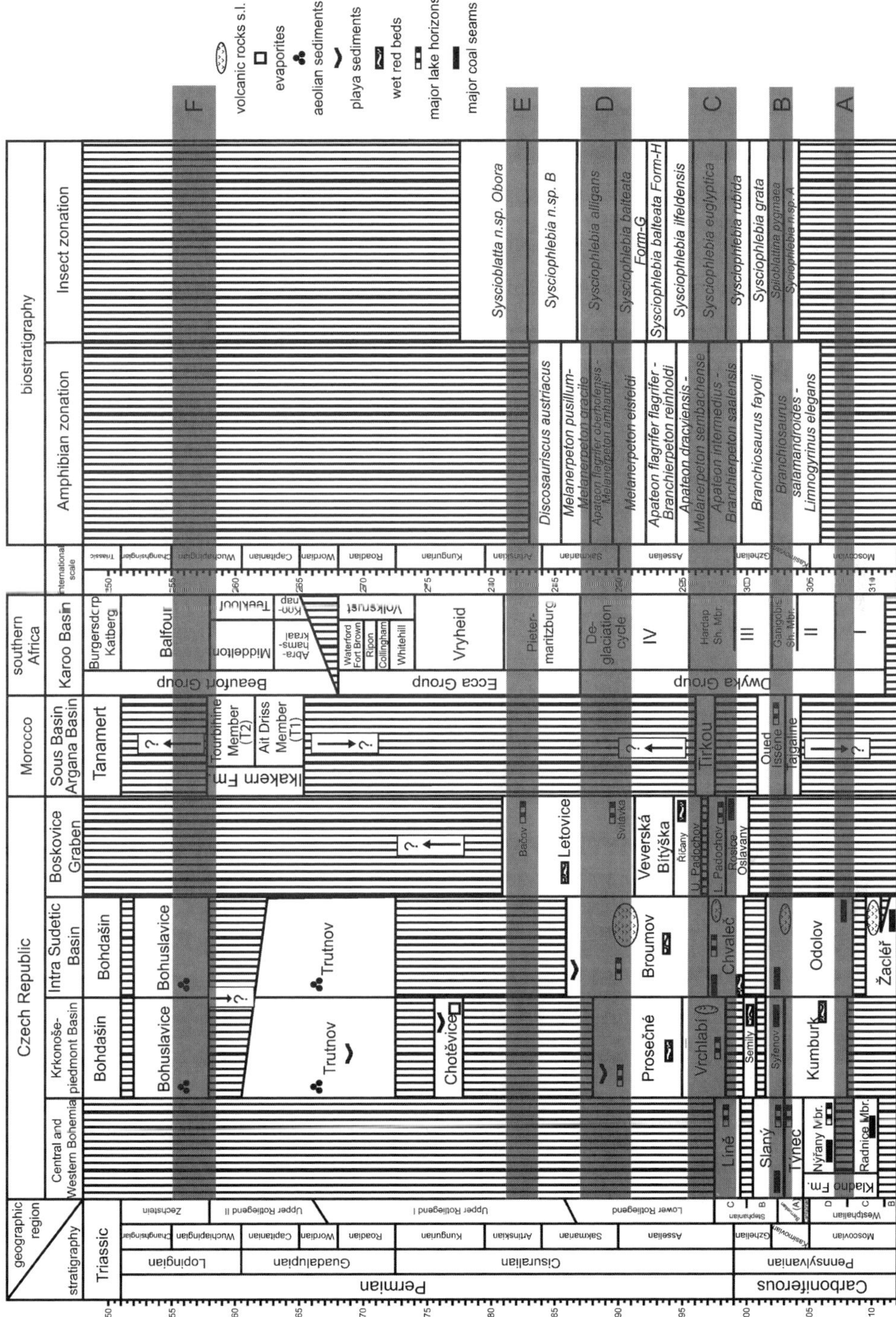

(**b**) Correlation chart of European Basins, the Karoo Basin and biostratigraphy. A, assumed Westphalian C/D wet phase; B, Barruelian wet phase; C, Stephanian C to early Lower Rotliegend wet phase; D, late Lower Rotliegend wet phase; E, Early Upper Rotliegend I wet phase; F, Zechstein/Bellerophon Transgression.

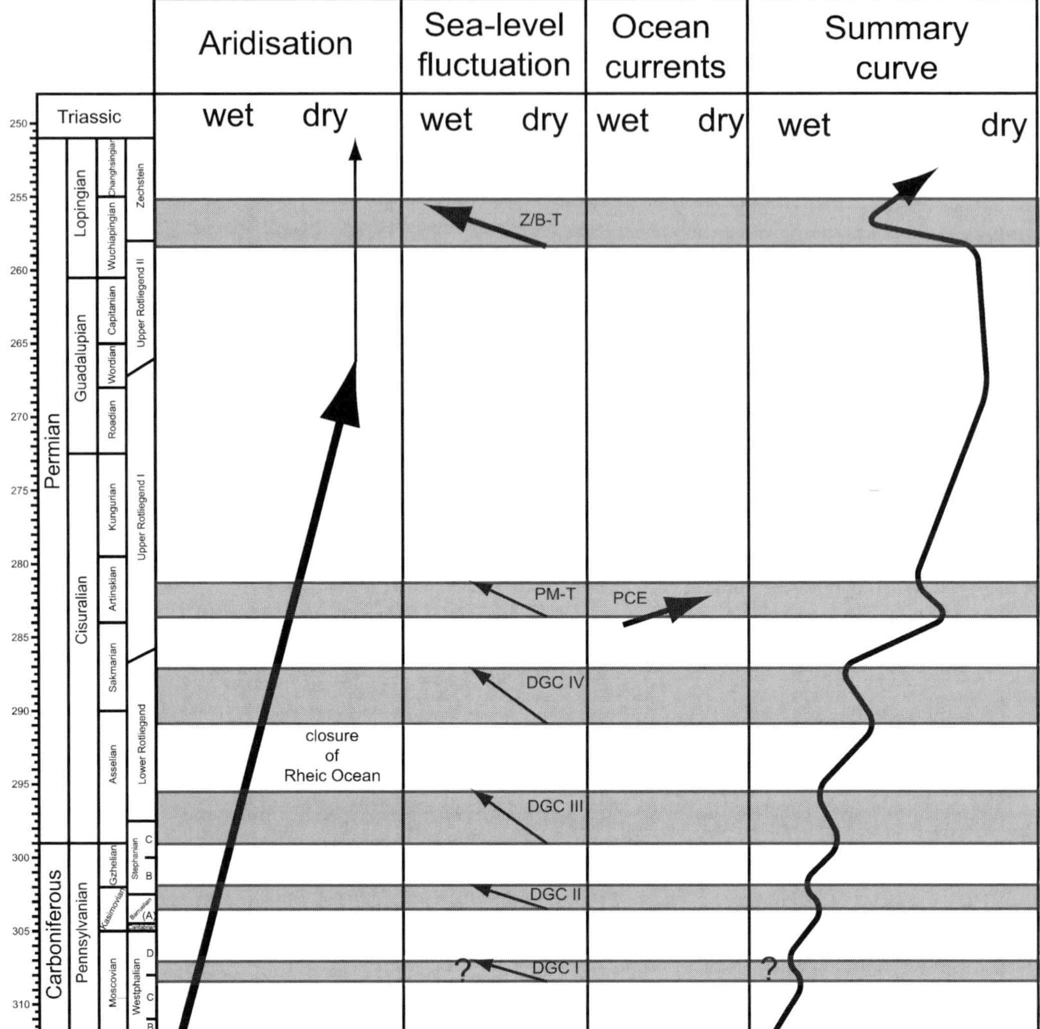

Fig. 16. Climate curve. Thickness of lines indicates the strength of the single process: DGC I–IV, Deglaciation Cycle 1–4 of the Gondwana Icecap; PM-T, Pietermaritzburg Transgression; Z/B-T, Zechstein/Bellerophon Transgression; PCE, Permian Chert Event.

The Gondwana Glaciation is recorded by glacial sediments in South Africa, South America, Australia, India and Antarctica. As in the Pleistocene, this icecap supported the expression of Milankovich cycles. This high cyclic sedimentation is described in Canada (Falcon-Lang 2003, 2004), as well as in the Mid-Continent Basin (e.g. Heckel 1999, 2002*a*, *b*). Within these fourth to fifth order cyclothems, glacio-eustatic sea-level changes of 60 to 200 m are documented (e.g. Crowell 1978, 1999; Veevers & Powell 1987; Heckel 1994; Isbell *et al.* 2003). An ice-volume difference of 35–115 million km^3 was calculated by Crowell & Baum (1991) between glacial and interglacial periods to produce eustatic sealevel fluctuations of this scale. This implies a huge icecap located over Antarctica, Australia, South America, India and South Africa, as proposed by Ziegler *et al.* (1997), Golonka (2000) and Scotese (2001).

In contradiction, Isbell *et al.* (2003) reported ice-free Trans-Antarctic Mountains during Carboniferous times and an ice marginal position for it in the Early Permian. Broutin *et al.* (1995) pointed out that the early Westphalian Al Khlata Formation of Oman was influenced only by alpine glaciers. Similar deposits of the same age are described from Queensland, Australia (Dickins 1996). The global maximum of glaciation is described by DiMichele *et al.* (1996) and Sano *et al.* (2003) for the late Westphalian. For the Paraná Basin in South America, dos Santos *et al.* (1996) published the ice maximum adjacent to the Permian–Carboniferous boundary, as Broutin *et al.* (1995) did for Oman. For the Early Permian, the glaciation is described as alpine-type for the Oman region (Angiolini *et al.* 2003*a*, *b*); in Australia glaciation occurred in five phases (Dickins 1996) and sediments of alpine glaciers in Queensland were deposited between non-glacial fluvial, lacustrine and shallow marine strata (Jones & Fielding 2004). For South Africa, Catuneanu (2004) published a general direction of ice movement from north to south.

To summarize, it remains unclear where and when this huge, continent-wide polar cap was established, which produced the estimated 60–200 m eustatic sealevel changes. A lower eustatic amplitude seems more realistic, so the ice-covered area was probably not as large as originally thought. Sedimentological research has shown that the amplitude of the glacio-eustatic fluctuations are on a scale of a maximum of 100 m for the major cyclothems, with incised valleys up to 30 m deep (Soreghan & Giles 1999; Feldman *et al.* 2005).

One of the most familiar and best-investigated profiles within the outcrop area of the glacial sediments is the Dwyka Group in the Karoo Basin, South Africa. The sedimentology and palaeontology of this unit have been published by Visser (1995, 1996, 1997). Isotopic ages were measured by Turner (1999) and Stollhofen *et al.* (2000), and geochemical investigations were carried out by Scheffler *et al.* (2003). The main features are the four deglaciation cycles with fining-upward sequences (Figs 15b, 16). Correlating this profile by isotopic ages (Fig. 15b) to the international time scale, the four eustatic maxima are in late Moscovian (Westphalian C/D, ~308 Ma), late Kasimovian (Barruelian, ~304 Ma), early Asselian (Stephanian C, ~298 Ma) and Asselian–Sakmarian transition (late Lower Rotliegend, ~289 Ma). The Gondwana glaciation ends at the beginning of the Sakmarian. This is concurrent with a worldwide classical *Eurydesma* transgression (Visser 1997). Sakmarian transgressive systems are also described in Australia (Dickins 1996) and Oman (Angiolini *et al.* 2003*a*, *b*). With the disappearance of the polar cap, the highly cyclical transgression-regression sequences within coastal and shallow marine sedimentary basins also stop (Beauchamp & Baud 2002).

The next eustatic transgression in southern Gondwana is in the late Artinskian (early Upper Rotliegend I, ~282 Ma) Pietermaritzburg transgression (Visser 1997). The cause for this flooding is not found in the icecap. After the Asselian, only seasonal and/or local glaciations are reported (e.g. Dickins 1996; Henderson 2003). All these Late Carboniferous and Early Permian transgressions occurred with a constant rhythm of about 5–7 Ma, ascribed to the Gondwana Glaciation. The Pietermaritzburg transgression, however, does not fit this pattern. Perhaps eustatic sealevel fluctuations with a low frequency of 5–7 Ma drove the glaciation, and only the higher order cycles are glacio-eustatic. The cause for the 5–7 Ma cyclicity is not known, and no processes with a comparable frequency have been described.

Palaeo-ocean currents and salinity

Oceanic and atmospheric circulations control the distribution of climatic belts. They are responsible for the transport of heat from the equator to the poles and vice versa. Near the equator (~30°S to ~30°N) the oceans carry more than 50% of the heat, and in higher latitudes the atmosphere plays the more important role (Von der Haar & Oort 1973).

The position, direction and relative temperature of the ocean currents with respect to their surroundings significantly influence the modern continental environment. Two types of large currents that impact continental climate should be mentioned as recent examples. The cold and upwelling currents of the Humboldt and Benguela Stream cause zones of high bio-production in the marine environment and coastal deserts on the land. These effects occur because the moist air, coming from the oceans with westerly winds, deposits rain over the cold ocean and does not reach the continent. The second example is the Gulf Stream in the northern Atlantic. This brings heat and moisture from the Gulf of Mexico to the northern high latitudes of Europe and is responsible for the mild climate in the United Kingdom and Norway. All these currents are part of a conveyor-belt-like heat pump that stretches around the whole world. The mechanism that keeps these circulations in motion changed during the past. Today, thermal and thermohaline circulation patterns produce cool oxygen- and nutrient-rich bottom water that generates areas of high organic productivity when upwelling. The opposite mode, haline circulation, is driven by the sinking of warm, high salinity water due to intense evaporation. The Mediterranean Sea is a small recent analogue (Talley 1996). The high salinity trail that leaves the strait of Gibraltar can be observed across the entire Atlantic (Worthington & Wright 1970). The warm saline bottom waters are dis- to anoxic and poor in nutrients. Their upwelling generates areas of lowered organic productivity and areas that are hostile to life (Kidder & Worsley 2004).

The reconstruction of palaeo-ocean current systems is mostly made by numerical models (Winguth *et al.* 2002). These calculations for an atmosphere/ocean simulation are based on many parameters, such as palaeogeography, seafloor spreading, palaeotopography, salinity, carbon dioxide level in the atmosphere, vegetation cover and position of ice. Given these variably known parameters, the models can only be tested by looking at the effects of the circulation on shallow marine and continental systems.

The model of Winguth *et al.* (2002) for the Middle Permian (Wordian) shows a cost-parallel, cold current at the western border of Euramerica. This is supported by the investigations of Beauchamp & Baud (2002), who described a cold stream along the west coast of North America that caused the Permian Chert Event (PCE). This event is marked by shallow-water chert deposition, which implies a cooler environment on the shelf than could be assumed by latitude. This model is supported by LePage *et al.* (2003), who discovered warm-temperate macrofloras within Kungurian cold water shelf carbonates in the Canadian High Arctic, which requires warm-temperate continental climates in the neighbourhood of cold oceanic conditions. For this part of the model (Winguth *et al.* 2002), the reconstructed ocean currents match the geologic record.

On the other side of the Panthalassic Ocean, there is a discrepancy between the model and biogeography. Shi & Grunt (2000) and Weldon & Shi (2003) showed that, during Roadian and Wordian time, cool water currents from Australia to Siberia, Mongolia and the Pamir Mountains are necessary to explain the distribution of coldwater adapted brachiopods (*Terrakea* Booker 1930). This is not explained by the numerical model of Winguth *et al.* (2002), although these authors used different values for the atmospheric carbon dioxide concentration. However, the variations of this parameter do not deliver satisfying reconstructions. Perhaps changing other parameters, such as salinity, topography, palaeogeography, could provide models to explain this geological situation.

Palaeogeography

For most recent numeric palaeoclimate models, there has been the question of using a Pangaea A or B reconstruction for calculation (Klein 1994; Kutzbach & Ziegler 1994; Fluteau *et al.* 2001; Gibbs *et al.* 2002; Berthelin *et al.* 2003). Pangaea A, the Wegnerian one, is not under discussion for the Triassic (Muttoni *et al.* 2003). But, for the Carboniferous and Early Permian, the Pangaea B reconstruction introduced by Irving (1977) better fits the palaeomagnetic data as well as the numeric climate models versus geological record (Fluteau *et al.* 2001). Nevertheless, the Pangaea B model has one master problem: how was it transformed to the younger Wegnerian Pangaea A?

Major strike-slip systems are proposed by many authors (e.g. Doblas *et al.* 1998; Muttoni *et al.* 2003; Youbi *et al.* 2003). But, these continent-wide mega-shear systems have not been demonstrated in the field. Also, the geometry of a strike-slip fault zone through the whole of Pangaea is difficult to imagine, because a stress field that produces only one major fault along the compressive Gondwana–Eurasia suture is not plausible. In the older literature (e.g. Arthaud & Matte 1977), there are descriptions of some large dextral transverse fault systems within Europe. The Tornquist–Teysseyre, Bay of Biscay, Gibraltar and South Atlas fracture zones are mentioned in many publications, such as Ziegler (1990*b*), Doblas *et al.* (1998) and Golonka (2000).

Kroner (in Schneider *et al.* 2006) plotted a Pangaea A (Wegnerian) reconstruction (255 Ma: map after Scotese 2001) on a true sphere with a recent Earth diameter (Fig. 2). The grid and Laurussia were fixed in position, and everything else except Gondwana and Laurussia was omitted. Now, the Ouachita, Alleghenian, Appalachian, Mauretanid and Variscan Orogen were stripped back along their major fault systems. During this, the four European fracture zones mentioned above became most important. The relative motion of Gondwana with respect to Laurussia was reconstructed by the combination of structural elements as fault zones, fold and thrust belts and magmatic activity. Because of the geometric constraints on the sphere, the complete closure of the Rheic Ocean requires two consecutive plate rotation processes, as already demonstrated by Wise (2004) for the deformation style of the Pennsylvania salient of the North American Appalachians. Therefore, the reconstruction of Pangaea 300 Ma ago (Fig. 2) looks somewhat different from that in use by most authors (Ziegler 1988, 1990*b*; Ziegler *et al.* 1997; Golonka 2000; Scotese 2001).

The post-Caledonian orogens along the Gondwana–Laurussia suture testify to collision processes taking place from the Early Devonian to the Late Permian. The existence of a continuous Trans-Pangaean Mountain Belt, with high altitudes reaching from the Variscides to the Ouachita Mountains (Keller & Hatcher 1999), as used for climate modelling by Kutzbach & Ziegler (1994), Fluteau *et al.* (2001) and Gibbs *et al.* (2002), requires contemporaneous orogenic processes along the whole plate boundary. Actually, the individual orogens give evidence of diachronous tectono-thermal events. http://dict.tu-chemnitz.de/dings.cgi?o=3001;count=50;service=de-en;query=auf. Particularly with regard to uplift and exhumation processes taking place in the Early to Late Carboniferous and Permian, large differences between the European Variscides, the North American Appalachian Orogen and the Ouachita Mountains exist.

The climax of high-pressure metamorphism during the central European Variscan Orogen occurred in the Early Carboniferous. Strong erosion and differential uplift accompanied the subsequent rapid exhumation and cooling of the high-grade metamorphic complexes. At the end of the Carboniferous, the orogenic crust of Europe equilibrated (Kroner *et al.* 2004). Finally, dextral transpression during the Permo-Carboniferous took place in local, small-scale tectono-thermal events (Gaitzsch 1998; Kroner & Hahn 2003; Capuzzo & Wetzel 2004). Late Carboniferous arc-continent collision led to the very low-grade fold and thrust belt of the Ouachita Mountains at the southern edge of Laurentia (Thomas 1989). There is no hint of large-scale uplift and erosion at a subsequent stage. The final closure of the Rheic Ocean during the Permian, caused by the clockwise rotation of Gondwana and contemporaneous shearing along the South Atlas and Gibraltar Fracture Zone (Fig. 2), created the Alleghenian–Mauretanides Orogen of North America and western Africa (Arthaud & Matte 1977; Ziegler 1982). Convergent tectonics, producing large-scale fold and thrust belts, started in the Late Carboniferous and ceased in the Late Permian. Tectonothermal events, uplift and exhumation processes continued during the whole time span (Hatcher *et al.* 1989). Large-scale uplift and erosion lasted until the Triassic.

As a consequence of the processes delineated above, a Pangaea-wide mountain chain during the Middle Permian is very doubtful. High altitudes should occur in Appalachian and west African mountain belts at this time. In comparison, a strong topography is ruled out in the region of the Ouachita Mountains and the European Variscides. Furthermore, the reconstruction of the late Palaeozoic plate movement of Gondwana relative to Laurussia (Figs 2 & 17) shows that there is no need for a strong collision between South America and Laurentia. Instead of this collision and orogen as published by many authors (e.g. Kutzbach & Ziegler 1994; Keller & Hatcher 1999; Fluteau *et al.* 2001), the remnant of the Rheic Ocean formed an embayment from the Panthalassia Ocean to the mid-European areas that persisted at least from Early to Middle Permian time. This oceanic embayment was situated in the low latitudes of Pangaea and would explain the marine Permian sediments described by Benison & Goldstein (2000) in the southwest of the Mid-Continent Basin (Utah, Colorado, west Wyoming), as well as the basin facies in the southwest of the Mid-Continent Basin. This deep shelf facies is not situated next to an orogen, as shown by Heckel (1999, 2002*a*, *b*).

Palaeotopography of Pangaea in European surroundings

The elevation of the Variscides, Mauretanides and Appalachians is of major importance for European and global climate during the Permo-Carboniferous. Different estimates exist for the altitude range of the Variscan mountain chain, some in excess of 5000 m (Kutzbach & Ziegler 1994; Becq-Giraudon *et al.* 1996; Fluteau *et al.* 2001; Gibbs *et al.* 2002). All numerical climate models that include topography generalize an equatorial mountain chain, as published by, for example, Keller & Hatcher (1999), with an average elevation of about 2000–2500 m, contrary to Gibbs *et al.* (2002), who assumed a maximum height of 1800 m. Based on our knowledge of the European realm (section 6) these hypotheses do not seem plausible.

The culmination of pressure-emphasized metamorphism during the Variscan continent–continent collision was during the Early Carboniferous (Mississippian) (Gaitzsch 1998; Gaitzsch *et al.* 1999; Bosse *et al.* 2000; Matte 2001; Medaris *et al.* 2003; Rodriguez *et al.* 2003; Kroner *et al.* 2004), at about 340–350 Ma. The following main uplift/exhumation processes documented a rapid cooling of the high-pressure domains by isothermal decompression (Kröner & Willner 1998, and references therein), so the main elevation should have been reached just after that. The uplift and tectonic exhumation of the high-grade

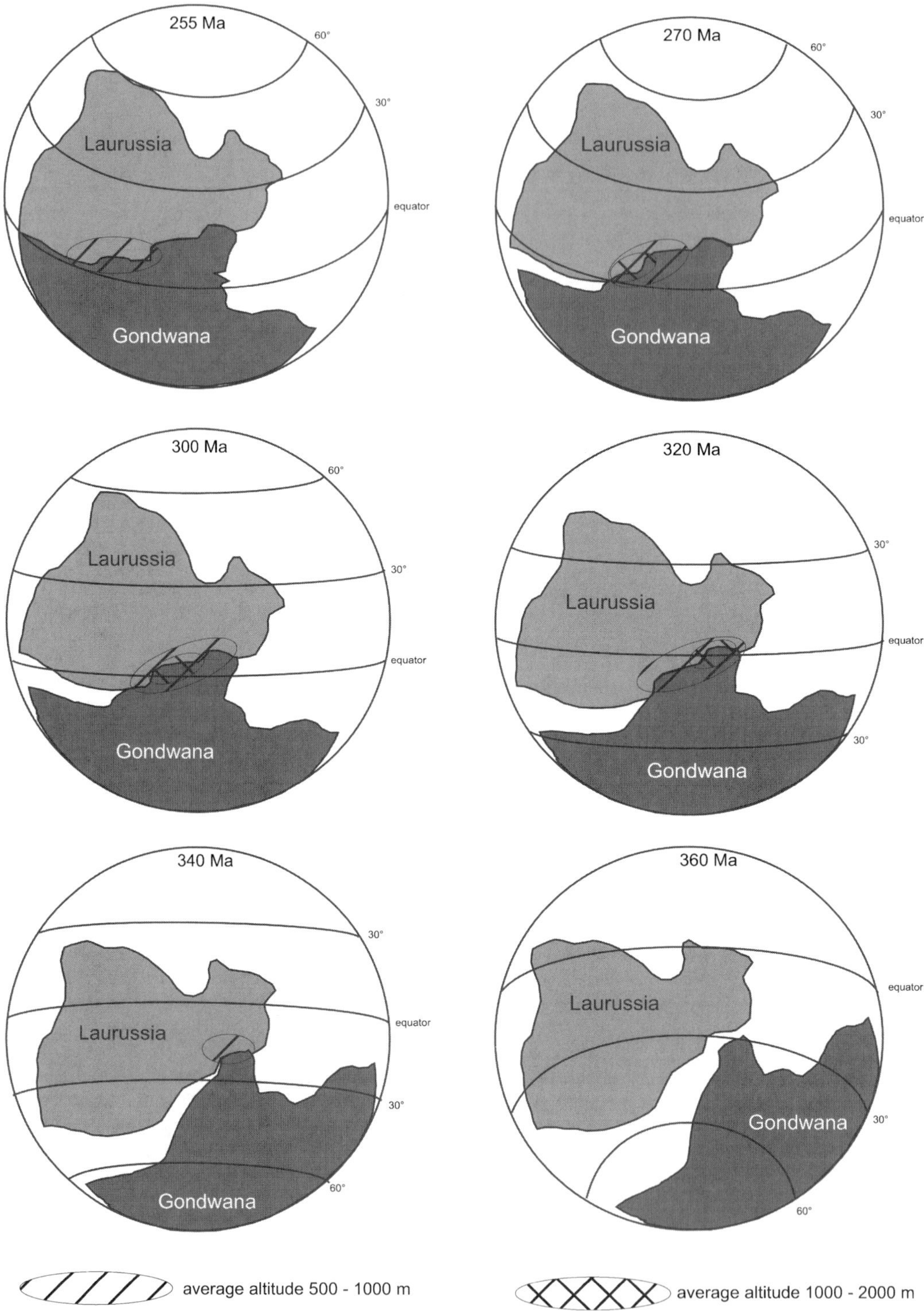

Fig. 17. Sketch map of the Pangaea formation and estimated palaeotopography.

metamorphics was simultaneous with the erosion, as is observed today in New Zealand (Simpson *et al.* 1994). When the orogen reached its maximum elevation, most of its roof was already eroded. The first undeformed sediments within the Variscan realm were described by Ahrendt *et al.* (2001) from the Erzgebirge Basin, which rests on deformed turbiditic flysch sediments. These sediments were dated by macroflora to the Early Carboniferous (e.g. Gothan 1949; Daber 1959). Palynological investigations by Bek (1997 pers. comm.) delivered a younger, Asbian age (early Late Viséan, V3b). Isotopic dating of a tuff yielded an age of 330 ± 4 Ma (Gehmlich *et al.* 2000), and Ahrendt *et al.* (2001) demonstrated the existence of Variscan detritus by dating white mica with a temperature of closure to ~330 Ma. Capuzzo & Wetzel (2004) reported a rapid removal of more than 5 km of upper Variscan crust in the intramontane Salvan-Dorénaz Basin during the Late Carboniferous. They justified their assumption by $^{39}Ar/^{40}Ar$ chronometry of white mica from subaerially cooled volcanics in contrast to ages derived from the sedimentary detritus above the volcanics. Intramontane basins, which may have existed, would have been cannibalized during the Late Carboniferous. Also, the Bohemian basins and the French basins of the Massif Central (cf. pages 98–102), which rest on Variscan roots, granites and high metamorphic complexes, started their evolution by at least Stephanian time. For example, the basement of the Blanice Graben consists of metamorphic rocks derived from a depth of 30–35 km (Opluštil 2004). Therefore, an attempt to explain high altitudes versus sedimentary basins is presented here.

Recent denudation rates of 3.0 ± 1.3 to 4.5 ± 1.7 mm/a (3.5 mm/a in average) for the western Himalayan mountain range, which is in a totally different climatic belt than the Variscides, were published by Garzanti *et al.* (2005). This would imply an erosion of 10 km within a time span of less than 3 Ma. This does not seem to be very realistic. To bring it down to earth, using an erosion rate of 0.5 mm/a, an orogen summit of 10 km would be eroded during 20 Ma. The only factor to account for the lowering of the erosion rate is that the linear calculation does not represent nature.

Thus, during most of the Carboniferous, the Variscan mountain range was situated next to the equator in a warm-humid tropical climate. This implies a high precipitation rate and consequently a high weathering and erosion rate. This allows a comparison to recent New Zealand, where extremely high exhumation rates are observed. The uplift rate for this region over the last 16 ka has been calculated at 13.7 mm/a (Simpson *et al.* 1994). However, the topography does not reflect this uplift because only some individual mountains exceed 3000 m. Comparing palaeo-erosion rates with recent ones is difficult because of the differences in the evolutionary level of the flora, but it gives a hint of truth. Also, few Palaeozoic plants had root systems comparable to modern plants.

The Thuringian Main Granite, which forms the basement of the Thuringian Forest Basin, underwent, by dextral transpression, an exhumation from 310–300 Ma by a denudation of 8–10 km of overlying rocks (Zeh *et al.* 1998). This underlines the hypothesis of fast erosion and levelling of the Variscan mountain chain. For the Pyrenees, Maurel *et al.* (2004) published ages of granite intrusions of 305 Ma. Using various isotopic systems, these authors estimated cooling rates of 30 °C/Ma. These rates decrease at 290 Ma to 1 °C/Ma. In comparison to Central Europe, it is the same duration of the orogen but the phase is displaced by about 15–20 Ma. This displacement can also be seen by the later maximum of anatexis in Central Iberia between 335–305 Ma (Montero *et al.* 2004).

However, a palaeotopographical reconstruction of the whole Hercynian orogen has not been established, and the juxtaposition of contemporaneous sedimentary basins and deeper crustal units excludes a Hercynian-wide, Himalayan-type mountain belt at any time of the orogeny. At least by the beginning of the Stephanian (~305 Ma), the Central European Variscan orogen was levelled to a hilly landscape. However, it cannot be ruled out that local transpressional horst systems produced alpine-type mountains of limited extent. Recent analogues are the Tatra Mountains in the Czech Republic (Sperner *et al.* 2002).

Generally, the clockwise rotation of Gondwana (Fig. 17) caused the migration of the maximum elevation through the continent. The first uplift occurred in the central east of Pangaea. This was continued in the Early to Middle Permian by the Appalachians and Mauretanides.

Seasonality

As shown above, laminated (varved) sediments of perennial lakes are known since at least the late Stephanian (Stephanian B of Bohemian basins: Pešek 2004) and can be traced up to the youngest perennial lakes (Oberhof Formation,

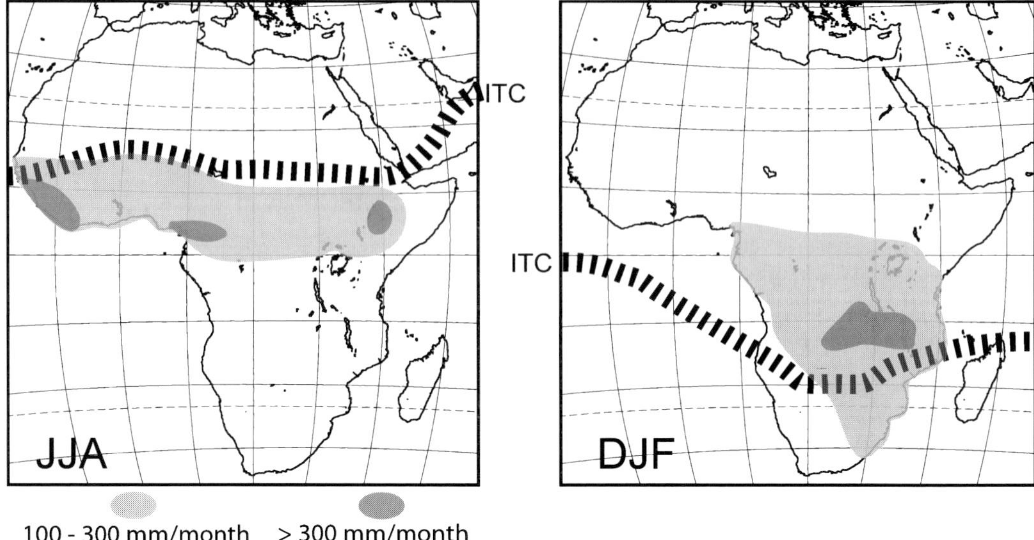

Fig. 18. Recent precipitation over Africa. JJA, northern summer; DJF, northern winter.

Buxières Formation, etc., see Fig. 15a, b) in the topmost Lower Rotliegend (Schneider 1996). They are interpreted as the result of annual changes of wet and dry seasons of strong monsoonal climate (Schäfer & Stamm 1989; Stapf 1989; Schneider & Gebhardt 1993; Clausing & Boy 2000). Further, Ziegler (1990*a*) showed by investigations on macroplants that, during the Permian, the biome of the ever-wet tropics disappeared in central and eastern Pangaea next to the equator, but not on the Cimmerian continents. Exceptions are reported only from Morocco, Spain and Texas, but, without references. The model proposed here is based on an increasing seasonality starting in the Stephanian and gaining strength in the Permian, which inhibits ever-wet biomes. The huge landmasses in the northern and southern mid-latitudes generated a strong monsoonal climate, as has already been modelled, for example by Parrish (1993) and Kutzbach & Ziegler (1994). During Permian time the zone of Inter-Tropical Convergence (ITC) was much more displaced over the continents, like today (recent values for Africa ~20°N/S: Zahn 1991; see Figure 18). In the Permian it reached latitudes of about 30°N and 30°S (Kutzbach & Ziegler 1994). The associated precipitation belt was not displaced as much as the ITC (Fig. 19).

The broad shift of the rainfall area is comparable to that of eastern Africa, but on a larger scale. Seleshi & Zanke (2004), for example, published data for Ethiopia; in the south at 5°N, the wet seasons are in the spring and autumn. This seasonal distribution is comparable to the equatorial climate in central Pangaea. Additionally, the biomes compare well. The recent arid to semi-arid savannahs of Ethiopia (Zahn 1991) may be a modern analogue for the Permian summer-wet tropical deciduous forests or savannahs (biome 2 *sensu* Ziegler 1990*a*). Northern summer (JJA) and winter (DJF) dry seasons are expected for the whole of central Pangaea (Fig. 19). This is caused by the relatively large distance to the ITC and its precipitation belt and the influence of trade winds. Spring and autumn wet seasons occur while the ITC is placed directly above the basins, which are discussed herein. All Permian climate models assume moisture for the precipitation over the CEBs associated with the trade winds from the Tethys. These passat winds rained out in front of the Trans-Pangaean Mountain Chain (Keller & Hatcher 1999). But, as shown above, these mountains were not as high as would be necessary to induce heavy rainfall accompanied with a complete moisture loss at the southern flank of this supposed mountain chain.

In the modern world, only a small part of the central African rain is achieved from the east. For ancient times, by analogy, the moisture of the ITC must have been obtained from the west, as it is today over Africa. The large distance to the Panthalassic Ocean is in contradiction to this

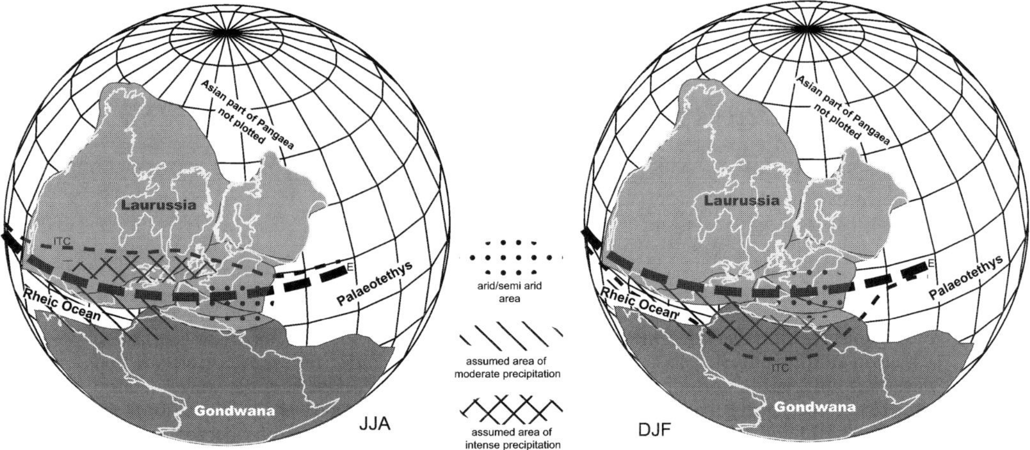

Fig. 19. Precipitation over Gondwana 300 Ma ago, grid not arranged to the palaeo-equator. E, equator; ITC, Inter-Tropical Convergence Zone; JJA, northern summer; DJF, northern winter.

hypothesis, if existing palaeogeographic reconstructions are utilized (e.g. Ziegler 1988, 1990*b*; Ziegler *et al.* 1997; Golonka 2000; Scotese 2001). The palaeogeographic reconstruction of Kroner (in Schneider *et al.* 2006), used here, does not create this problem. The supercontinent Pangaea receives its precipitation from the mega-ocean Panthalassia, with its embayment of the remnant Rheic Ocean. Compared to most previous climate models, the topography is assumed not to have been high enough to produce a continent-wide rain shadow (Fig. 17). Why did the first seasonal sediments occur during Stephanian B? The Variscan Orogen started to rise as an elevation near the mid-Carboniferous boundary. The emerging mountain chain was high enough to induce a significant effect on the distribution of rainfall. By the beginning of the Stephanian B, the morphology of the Variscan Mountains was no longer important for the climate evolution of central Pangaea. This coincides with the estimates made earlier for erosion versus exhumation rates.

Aridization

The pronounced seasonality was strengthened by development of more arid conditions in the continental interior. This aridization began in the Westphalian (Chevalier *et al.* 2003) and is interrupted by multiple wet phases. Gibbs *et al.* (2002) attributed the aridization of the northern part of Pangaea to the northward movement of the whole continent over about 10°–15°N away from the equator. But, this explains only the younger half (Middle to Late Permian) of a trend that began in the Carboniferous. If this continental movement was the only cause of the northern Gondwanan aridization, there should have been a humidization trend in the Carboniferous while moving from the south towards the equator. But, there is no indication of this.

Also, this movement should but did not leave a traceable record of a southward shifting of the ever-wet tropical belt. The appearance of more arid/continental conditions can also be interpreted as being due to a barrier to maritime influences of the Rheic Ocean remnant. The slow closure of the Rheic Ocean between North and South America during the Pennsylvanian and Permian had strong effects on the climate. It changed step by step from maritime humid/semi-humid climates to more continental arid/semi-arid conditions. Gradually, the large Carboniferous coal swamps disappeared towards the Permian. The aridization trend in the European basins is marked by the spreading of red-bed facies. Near the Sakmarian–Artinskian boundary, this effect becomes stronger (Fig. 16). For example, Hembree *et al.* (2004) described episodic, perhaps seasonal droughts in the late Wolfcampian (late Sakmarian–early Asselian) coastal plain of Kansas. They based their hypothesis on amphibian burrows and ephemeral ponds (Speiser Shale, Kansas). Similar phenomena are observed in the Lodève Basin (Körner *et al.* 2003; Schneider *et al.* 2006). The increasing pace of the aridization cannot be explained by a faster closure of the Rheic Ocean.

The breakdown of the Carboniferous coldhouse and the associated Gondwana glaciation by the Sakmarian caused a reorganization of the major ocean currents. This is expressed by the onset of the Permian Chert Event (Beauchamp & Baud 2002). The sedimentation of cherty sequences in a shallow marine environment is interpreted as a cold-water environment. Because this depositional style can be found along the northwestern Pangaea shelf, they attributed this to the presence of a cold longshore current. The model of Winguth *et al.* (2002) showed that the cold conditions could be explained simply by upwelling. This could have also been induced by the offshore transport of surface water due to the wind-driven Ekman transport (perpendicular to the right of the wind direction in the northern hemisphere) (Gibbs *et al.* 2002).

This current had strong effects on continental climate. This cold barrier to the west of the north-drifting continent inhibited the moisture, which came with the westerly winds to rain over the continent. Recent similarities can be found in SW Africa (Namib Desert) and the SW of South America (Atacama Desert), due to the Benguela and Humboldt Currents. These two phenomena, the slow closure of the Rheic Ocean since the Viséan, and the rapid change of the oceanic circulation patterns, are assumed to be the cause of the central Pangaea aridization. Studies of very young sediments in the marine successions next to the Congo River mouth (Schefuss *et al.* 2005) revealed correlations between monsoon strength and equatorial precipitation. The stronger the monsoon, the less it rains. They attributed this relationship to the effect of the monsoon winds on cloud distribution. A stronger wind with an east–west component transports more clouds away to the ocean. This causes less precipitation on the continent.

The gradual closure of the Rheic Ocean had two effects on the continental climate in Central Europe. First, the distance to the ocean in the west increased and, second, the monsoon strengthened. This effect of accelerated aridization can be observed in many basins in Central Europe (Figs 15a, b & 16). The post-Artinskian climate development in the Euramerican continental environment is not reconstructable as accurately as in the Late Carboniferous and Early Permian.

In most of the European basins, with the exception of the Saar–Nahe Basin and the Lodève Basin, sedimentation from the middle Cisuralian up to the late Guadalupian is interrupted by several hiatuses (Fig 15a, b). Only scattered deposits without sustainable biostratigraphy occur in most basins. The Standenbühl Formation of the Saar–Nahe Basin (Fig. 15a) is not well enough investigated. The only section with essentially continuous sedimentation is the Salagou Formation of the Lodève Basin. These red beds were well investigated by Körner *et al.* (2003) and Schneider *et al.* (2006) in a sedimentological and geochemical sense and for its climatic implications. Within this succession, the ongoing aridization is obvious, and the maximum is reached in the upper Octon Member (Fig. 15a). This unit is estimated to be of Roadian age. The following Mérifons Member seems to have been deposited under more humid conditions.

A process that could be responsible for this maximum of aridization is published by Shi & Grunt (2000) and Weldon & Shi (2003). They showed the necessity for a cool current along the Cimmerian continents to explain the distribution of cold-water-adapted brachiopods. The brachiopod *Terrakea* Booker 1930 from Australia is distributed in Siberia, Mongolia and the Pamirs during Roadian and Wordian time. This cold current could prevent moisture from the east reaching the continent. But, this current was next to the Cimmerian continents, and the Tethys was located to the west of it. So, this current is a hint of a global reorganization or modification of the oceanic currents. Until more sections of continental Middle Permian sediments are thoroughly investigated, the climatic development of the Central European region for this time span is based on estimates only. Probably, the climax of the Permian aridization produced conditions that did not allow normal sedimentation, due to the lack of the transport medium. Alternatively, nearly the entire European region was elevated above base level by doming (due to a superplume: Doblas *et al.* 1998; or crustal underplating: Bachmann & Hoffmann 1995). Most of the scattered Upper Rotliegend I deposits of Central Europe are characterized by angular to subangular components and a high content of primary aeolian sediments. This is interpreted to indicate a more or less exclusive physical weathering with a deficit of debris removal over large areas of Permian Europe.

Aeolian sediments

The appearance of aeolian sediments is connected to the previously mentioned aridization trend. The first occurrence of important amounts of wind-transported grains is observable in the early Upper Rotliegend I (Fig. 15a, b). The first dunes appear in the Upper Rotliegend I of the Hornburg and Kreuznach formations, but are more prominent in the Upper Rotliegend II

(Walkenried Formation, Ilfeld Basin; Northern and Southern Permian Basin: Glennie 1982).

Two features controlled the occurrence of aeolian sediments. The first was the aridity within northern Pangaea. As was shown earlier, the process of ongoing aridization produced large areas of arid to semi-arid environments. Therefore, a large amount of sediment was available. The second controlling fact was the velocity of the wind. In order to produce large dune systems, a strong and directionally constant wind system is necessary. Glennie (1982) showed that, in the Upper Rotliegend II of the Northern Permian Basin, mega-dunes occur. For the formation of such giant structures, comparable to the Pleistocene ones, high wind velocities are necessary. The high atmospheric velocities in the Pleistocene were caused by the compression of the climatic belts by the spreading out of the icecaps. In the Late Permian (late Guadalupian and Lopingian) no large glacial deposits are known. The climatic belts in the Late Permian were not compressed but spread out. This is supported by the investigations of Cúneo (1996), who showed that the Antarctic Continent was forested with cold temperate and humid *Glossopteris* vegetation. Ziegler (1990*a*) ascribed it to biome 6 (nemoral broadleaf–deciduous forests), which, in recent times, are typical in eastern and central Europe as well as in the eastern USA.

The discrepancy of the high velocity of the winds over northern Pangaea and the spreading out of the climatic belts is difficult to explain. Two facts should be considered. On the one hand, the Pangaean continent was a much larger connected land area than today. On the other hand, the ever-wet tropical biome disappeared and the equator was surrounded by desert to semi-desert regions. The development of this huge wasteland was caused by closure of the Rheic Ocean. The marine influence on the continental climate was retreated step by step until the final oceanic embayment, the Rheic Ocean remnant, was closed in the Late Permian (Fig. 17). This retreat of the ocean due to plate tectonics caused, in the Permian, a slow shift from marine to continental climate in northern Pangaea. Starting from the Lower Permian, the monsoon system strengthened so that larger dune systems could be established. With the further development of the exceptionally large and dry central Pangaean desert, the wind velocities were raised to values comparable to those of the Pleistocene

Conclusions

The Permo-Carboniferous climate was marked by an aridization trend (Chumakov & Zharkov 2002). This trend was interrupted by several wet phases in the Barruelian, Stephanian C to early Lower Rotliegend, late Lower Rotliegend and early Upper Rotliegend I. These more humid phases can be seen in all European and North African sedimentary basins (compare Figs 1 & 15a, b). All these wet phases are linked to the waxing and waning of the Gondwana icecap. This is supported by correlations with the South African Karoo Basin. The melting of the icecap caused a eustatic sealevel rise, which led to a more humid climate in North Pangaea. A further wet phase in the Westphalian C/D cannot be proven by the dataset used here, but its existence is presumed because of its origin due the waxing and waning of the Gondwanan icecap.

The wet phases themselves, as a part of the aridization trend, weakened over time. The aridization and the weakening of the wet phases were caused not only by the northward drift of the supercontinent and the shifting of North Pangaea to the arid climatic belt (which was modelled by Gibbs *et al.* 2002). A spreading-out of the arid/semi-arid belt in the Lower to Middle Permian is traceable by the disappearance of ever-wet tropical associations (Ziegler 1990*a*). This spreading-out can only be explained by using a new reconstruction of the configuration of Pangaea (cf. Figs 2 & 17). During the Late Carboniferous and the Early Permian, an oceanic embayment of the Panthalassic Ocean, a large remnant of the Rheic Ocean, existed between North and South America. This ocean was successively closed during the Permian. The retreat of this water body displaced the source area of moisture for central North Pangaea stepwise to the west. With increasing distance from the ocean, the aridization and the monsoon system strengthened. The stronger monsoon system transported more clouds to the offshore area (Schefuss *et al.* 2005). The result was less precipitation on the continent. Later, at the end of the Gondwana Glaciation in the latest Sakmarian, oceanic circulation was rearranged. This led to a cold coast parallel ocean current that induced chert sedimentation on the shallow shelf along the west coast of North Pangaea. The cold-water surface temperatures blocked moisture coming with the westerly winds.

The maximum of aridity was reached during Roadian and Wordian time. A chief premise of all these hypotheses is that the major part of the precipitation was sourced to the west. The Trans-Pangaean Mountain Belt was non-existent in the new palaeogeographic reconstruction (Figs 2 & 17). The altitude of the Hercynian system never exceeded an average of 2000 m, and the maximum elevation of the Trans-Pangaean orogeny shifted during time from the east to the west with the ongoing closure of the Rheic Ocean. The Hercynian Orogen never acted as a large orographic barrier blocking moisture and precipitation from the west. The Variscian Orogen was levelled down at least by the middle Stephanian (Stephanian B). The lack of an orographic east–west barrier led to a strong seasonality. The Inter-Tropical Zone of Convergence (ITC) was displaced over larger distances than today. This huge displacement of the associated precipitation area caused four seasons at the equator. The summer and winter were dry because the ITC was further north or south. Only the spring and autumn were marked by heavy precipitation. This environment was not suitable for ever-wet tropical ecosystems. These can only persist in areas where onshore winds bring moisture from the oceans to the continents. This was the case for Cathaysia and the regions around the Rheic Ocean (Morocco, Spain and Texas: Ziegler 1990*a*). The Permo-Carboniferous climate of North Pangaea is a result of many processes that originated in palaeogeography, palaeotopography, glacio-eustatic sealevel changes and ocean current organization (Fig. 16), which can only be separated by investigations based on well-justified stratigraphic correlation of the sedimentary basins (Fig. 15a, b).

We thank S. Voigt, B. Legler, O. Elicki and J. Fischer for critical discussions of the manuscript and various hints on this topic, and U. Kroner for support with the palaeogeographic model. The reviewers are acknowledged for their critical evaluation and improvement of the manuscript. This publication results partly from projects: Menning & Schneider Me 1134/5 'Devonian – Carboniferous – Permian Correlation Chart'; Schn 408/7 'Reference Profile Lodève', within the main project 'Evolution of the System Earth in the Late Palaeozoic: Clues from Sedimentary Geochemistry'; and Schn 408/10 'Permian Playa to Sabkha' within the main project SPP 1135 'Central European Basin System' of the Deutsche Forschungsgemeinschaft (DFG). It is a contribution to the tasks of the working group '*Marine – non-marine correlations*' of the Subcommission on Permian Stratigraphy of the IUGS.

References

ABBINK, O. A. & VAN KRONIJNENBURG-VAN CITTERT, J. H. A. 2003. A palaeoecological approach to the Pennsylvanian (Upper Carboniferous) palynology of the Netherlands. *XVth International Congress on Carboniferous and Permian Stratigraphy. Utrecht, 2003*, Abstracts, 3–4.

ABBOU, M. B., SOULA, J.-C. *et al.* 2001. Controle tectonique de la sedimentation dans le systeme de bassins d'avant-pays de la Meseta marocaine: controls of deposition by thrust-propagation folds in the Moroccan Meseta foreland basin system. *Comptes Rendus de l'Academie des Sciences, Series IIA*, **332**, 703–709.

AHRENDT, H., BÜTTNER, A., TISCHLER, M. & WEMMER, K. 2001. K/Ar dating on detrital micas and ε_{Nd} characteristics for provenance studies in the Saxothuringian Zone of the Variscides (Thuringia ans Saxony, Germany). *Zeitschrift der Deutschen Geologischen Gesellschaft*, **152**, 351–363.

ANDREAS, D., LÜTZNER, H. & SCHNEIDER, J. W. 2005, in press. Karbon des Thüringer Waldes – Kraichgau-Senke und Thüringisch-Fränkische Grabenzone. *In*: DEUTSCHE STRATIGRAPHISCHE KOMMISION & WREDE, V. (eds) *Stratigraphie von Deutschland. V. Das Oberkarbon (Ponnsy Ivanium) in Deutschland.* Courier Forschungsinstitut Senkenberg, **254**, 403–418.

ANGIOLINI, L., BALINI, M., GARZANTI, E., NICORA, A. & TINTORI, A. 2003*a*. Gondwanan deglaciation and opening of Neotethys: the Al Khlata and Saiwan Formations of Interior Oman. *Palaeogeography, Palaeoclimatology, Palaeoecology*, **196**, 99–123.

ANGIOLINI, L., BALINI, M., GARZANTI, E., NICORA, A., TINTORI, A., CRASQUIN, S. & MUTTONI, G. 2003*b*. Permian climatic and paleogeographic changes in Northern Gondwana: the Khuff Formation of Interior Oman. *Palaeogeography, Palaeoclimatology, Palaeoecology*, **191**, 269–300.

ARTHAUD, F. & MATTE, P. 1977. Late Paleozoic strike-slip faulting in southern Europe and northern Africa: Result of a right-lateral shear zone between the Appalachians and the Urals. *Geological Society of America Bulletin*, **88**, 1305–1320.

BACHMANN, G. H. & HOFFMANN, N. 1995. Entstehung des Mitteleuropäischen Rotliegend-Beckens. *Terra Nostra*, **7**, 12–15.

BANGERT, B., LORENZ, V. & ARMSTRONG, R. 1994. Bentonitic tuff horizons of the Permo-Carboniferous Dwyka Group in southern Africa; volcaniclastic deposits as ideal time markers. *Journal of African Earth Sciences*, **27**(1A), 18–19.

BANGERT, B., STOLLHOFEN, H., LORENZ, V. & ARMSTRONG, R. 1999. The geochronology and significance of ash-fall tuffs in the glaciogenic Carboniferous-Permian Dwyka Group of Namibia and South Africa. *Journal of African Earth Sciences*, **29**, 33–49.

BEAUCHAMP, B. & BAUD, A. 2002. Growth and demise of Permian biogenic chert along northwest Pangea: evidence for end-Permian collapse of thermohaline circulation. *Palaeogeography, Palaeoclimatology, Palaeoecology*, **184**, 37–63.

BECQ-GIRAUDON, J.-F., MONTENAT, C. & VAN DEN DRIESSCHE, J. 1996. Hercynian high-altitude phenomena in the French Massif Central: tectonic implications. *Palaeogeography, Palaeoclimatology, Palaeoecology*, **122**, 227–241.

BENDER, H. & STOPPEL, D. 1965. Perm-Conodonten. *Geologisches Jahrbuch*, **82**, 331–364.

BENISON, K. C. & GOLDSTEIN, R. H. 2000. Sedimentology of ancient saline pans: an example from the Permian Openche Shale, Williston Basin, North

Dakota, U.S.A. *Journal of Sedimentary Research*, **70**, 159–169.

BERMAN, D. S. & MARTENS, T. 1993. First occurrence of *Seymouria* (Amphibia; Batrachosauria) in the Lower Permian Rotliegend of central Germany. *Annals of Carnegie Museum*, **62**, 63–79.

BERMAN, D. S., HENRICI, A. C., SUMIDA, S. S. & MARTENS, T. 2000. Redescription of *Seymouria sanjuanensis* (Seymouriamorpha) from the Lower Permian of Germany based on complete, mature specimens with a discussion of paleoecology of the Bromacker locality assemblage. *Journal of Vertebrate Paleontology*, **20**, 253–268.

BERTHELIN, M., BROUTIN, J., KERP, H., CRASQUIN-SOLEAU, S., PLATEL, J.-P. & ROGER, J. 2003. The Oman Gharif mixed paleoflora: a useful tool for testing Permian Pangea reconstructions. *Palaeogeography, Palaeoclimatology, Palaeoecology*, **196**, 85–98.

BETHOUX, O., NEL, A., GAND, G., LAPEYRIE, J. & GALTIER, J. 2002. Discovery of the genus *Iasvia* Zalessky, 1934 in the Upper Permian of France (Lodeve Basin) (Orthoptera, Ensifera, Oedischiidae). *Géobios*, **35**, 293–302.

BOSSE, V., FERAUD, G., RUFFET, G., BALLÈVRE, M., PEUCAT, J.-J. & DE JONG, K. 2000. Late Devonian subduction and early-orogenic exhumation of eclogite-facies rocks from the Champtoceaux Complex (Variscan belt, France). *Geological Journal*, **35**, 297–325.

BOY, J. A. 2003. Exkursion 2: Paläoökologie permokarbonischer Seen. *Terra Nostra*, **5**, 188–215.

BOY, J. A. & SCHINDLER, T. 2000. Ökostratigraphische Bioevents im Grenzbereich Stephanium/Autunium (höchstes Karbon) des Saar–Nahe-Beckens (SW-Deutschland) und benachbarter Gebiete. *Neues Jahrbuch für Geologie und Paläontologie, Abhandlungen*, **216**, 89–152.

BRAUNS, C. M., PÄTZOLD, T. & HAACK, U. 2003. A Re-Os study bearing on the age of the Kupferschiefer black shale at Sangerhausen (Germany). *XVth International Congress on Carboniferous and Permian Stratigraphy. Utrecht, 2003*, Abstracts, 66.

BREITKREUZ, C. & KENNEDY, A. 1999. Magmatic flare-up at the Carboniferous/Permian boundary in the NE German Basin revealed by SHRIMP zircon ages. *Tectonophysics*, **302**, 307–326.

BROUTIN, J., AASSOUMI, H., EL WARTITI, M., FREYTET, P., KERP, H., QUESADA, C. & TOUTIN-MORIN, N. 1998. The Permian Basins of Tiddas, Bou Achouch and Khenifra (central Morocco). Biostratigraphic und palaeo-phytogeographic implications. *In*: CRASQUIN-SOLEAU, S. & BARRIER, É. (eds) *Peri-Tethys Memoir. 4: Epicratonic Basins of Peri-Tethyan Platforms*. Mémoires du Muséum National d'Histoire Naturelle, Paris, **179**, 257–278.

BROUTIN, J., CHATEAUNEUF, J. J., GALTIER, J. & RONCHI, A. 1999. L'Autunian d'Autun reste-t-il une référence pour les dépôts continentaux du Permien inférieur d'Europe? Apport des données paléobotaniques. *Géologie de la France*, **2**, 17–31.

BROUTIN, J., ROGER, J. *et al.* 1995. The Permian Pangea: phytogeographic implications of new paleontological discoveries in Oman (Arabian Peninsula). *Comptes Rendus – Academie des Sciences. Geoscience, Serie IIa*, **321**, 1069–1086.

BRUCKSCHEN, P., OESEMANN, S. & VEIZER, J. 1999. Isotope stratigraphy of European Carboniferous: proxy signals for ocean chemistry, climate and tectonics. *Chemical Geology*, **161**, 127–164.

BRUGUIER, O., BECQ-GIRAUDON, J. F., CHAMPENOIS, M., DELOULE, E., LUDDEN, J. & MANGIN, D. 2003. Application of *in situ* zircon geochronology and accessory phase chemistry to constraining basin development during post-collisional extension: a case study from the French Massif Central. *Chemical Geology*, **201**, 319–336.

BRULHET, J. 1982. *Le bassin de Bourbon l'Archambault, example des relations: tectonique–sedimentation–climat regissant la vie d'un bassin Autunien du nord du Massif Central*. Compagnie Génerale des Materiès Nucléaires, Velizy Villacoublay.

BURGER, K., HESS, J. C. & LIPPOLT, H. J. 1997. Tephrochronologie mit Kaolin-Kohlentonsteinen: Mittel zur Korrelation paralischer und limnischer Ablagerungen des Oberkarbons. *Geologisches Jahrbuch, A*, **147**, 3–39.

CAPUZZO, N. & WETZEL, A. 2004. Facies and basin architecture of the Late Carboniferous Salvan-Dorenaz continental basin (Western Alps, Switzerland/France). *Sedimentology*, **51**, 675–697.

CATUNEANU, O. 2004. Basement control on flexural profiles and the distribution of foreland facies: the Dwyka Group of the Karoo Basin, South Africa. *Geology*, **32**, 517–520.

CHATEAUNEUF, J. J. & PACAUD, G. 2001. *The Autun Permian Basin: the Autunian Stratotype. Guide-book of the Field Trip*. French Association of Permian Geologists.

CHEVALIER, F., DE LEEUW, K., BRACCINI, E. & ROS, J.-B. 2003. Sandy body reconstruction and paleogeographical mapping of Westphalian deposits (central offshore, the Netherlands). *XVth International Congress on Carboniferous and Permian Stratigraphy, Utrecht, 2003*, Abstracts 92–93.

CHUMAKOV, N. M. & ZHARKOV, M. A. 2002. Climate during Permian-Triassic Biosphere Reorganizations. Article 1: Climate of the Early Permian. *Stratigraphy and Geological Correlation*, **10**, 586–602.

CLAUSING, A. & BOY, J. A. 2000. Lamination and primary production in fossil lakes: relationship to palaeo climate in the Carboniferous–Permian transition. *In*: HARDT, M. B. (ed.) *Climates: Past, Present and Future*. Geological Society, London, Special Publications, **181**, 5–16.

CROWELL, J. C. 1978. Gondwana glaciation, cyclothemes, continental positioning, and climate change. *American Journal of Science*, **278**, 1345–1372.

CROWELL, J. C. 1999. Pre-Mesozoic ice ages. *Geological Society of America, Memoir*, **192**, 1–106.

CROWELL, J. C. & BAUM, S. K. 1991. Estimating Carboniferous sea-level fluctuations from Gondwana ice extent. *Geology*, **19**, 975–977.

CÚNEO, N. R. 1996. Permian phytogeography in Gondwana. *Palaeogeography, Palaeoclimatology, Palaeoecology*, **125**, 75–104.

DABER, R. 1959. Die Mittel-Visé-Flora der Tiefbohrungen von Doberlug-Kirchhain. *Geologie, Beiheft 26*, 1–83.

DEBRIETTE, P. 1993. Le bassin permien de Bourbon l'Archambault et le Sillon Houiller (Allier-France). *Extrait du Bulletin de la Société d'Histoire Naturelle d'Autun*, **144**, 34.

DEBRIETTE, P. 1997. *Bassins Stephano-Permiens du sud-ouest du Bourbonnais*. Livret-guide de la visite Association des Géologues du Permian à Buxièresles-Mines.

DICKINS, J. M. 1996. Problems of a Late Palaeozoic glaciation in Australia and subsequent climate in the Permian. *Palaeogeography, Palaeoclimatology, Palaeoecology*, **125**, 185–197.

DIMICHELE, W. A., PFEFFERKORN, H. W. & PHILLIPS, T. L. 1996. Persistence of Late Carboniferous tropical vegetation during glacially driven climatic and sea-level fluctuations. *Palaeogeography, Palaeoclimatology, Palaeoecology*, **125**, 105–128.

DOBLAS, M., OYARZUN, R. *et al.* 1998. Permo-Carboniferous volcanism in Europe and northwest Africa: a superplume exhaust valve in the centre of Pangaea? *Journal of African Earth Sciences*, **26**, 89–99.

DOS SANTOS, P. R., ROCHA-CAMPOS, A. C. & CANUTO, J. R. 1996. Patterns of late Palaeozoic deglaciation in the Paraná Basin, Brazil. *Palaeogeography, Palaeoclimatology, Palaeoecology*, **125**, 165–184.

DOUBINGER, J., VETTER, P., LANGIAUX, J., GALTIER, J. & BROUTIN, J. 1995. *La flore fossile du bassin houiller de Saint-Etienne*. Muséum National d'Histoire Naturelle, Paris, *Mémoires*, **164**.

EBERTH, D. A., BERMAN, D. S., SUMIDA, S. S. & HOPF, H. 2000. Lower Permian terrestrial paleoenvironments and vertebrate paleoecology of the Tambach Basin (Thuringia, central Germany): the upland holy grail. *Palaios*, **15**, 293–313.

FALCON-LANG, H. J. 2003. Response of Late Carboniferous tropical vegetation to transgressive-regressive rhythms at Joggins, Nova Scotia. *Journal of the Geological Society, London*, **160**, 643–648.

FALCON-LANG, H. J. 2004. Pennsylvanian tropical rain forests responded to glacial-interglacial rhythms. *Geology*, **32**, 689–692.

FELDMAN, H. R., FRANSEEN, E. K., JOECKEL, R. M. & HECKEL, P. H. 2005. Impact of longer-term modest climate shifts on architecture of high-frequency sequences (cyclothems), Pennsylvanian of Midcontinent U.S.A. *Journal of Sedimentary Research*, **75**, 350–368.

FLUTEAU, F., BESSE, J., BROUTIN, J. & RAMSTEIN, G. 2001. The Late Permian climate: what can be inferred from climate modelling concerning Pangea scenarios and Hercynian range altitude? *Palaeogeography, Palaeoclimatology, Palaeoecology*, **167**, 39–71.

GAITZSCH, B. 1995*a*. *Extramontane Senken im variscischen Finalstadium in Norddeutschland: Lithofaziesmuster, Tektonik und Beckenentwicklung*. PhD thesis, TU Bergakademie Freiberg.

GAITZSCH, B. 1995*b*. Zwischensedimente. *In*: PLEIN, E. (ed.) *Stratigraphie von Deutschland. I: Norddeutsches Rotliegendbecken. Rotliegend-Monographie. Teil II*. Courier Forschungs institut Senckenberg, **183**, 101–102.

GAITZSCH, B. 1995*c*. Grüneberg-Formation. *In*: PLEIN, E. (ed.) *Stratigraphie von Deutschland. I. Norddeutsches Rotliegendbecken. Rotliegend-Monographie. Teil II*. Courier Forschungs institut Senckenberg, **183**, 102–106.

GAITZSCH, B. 1998. Flysch und Frühmolassen im östlichen Saxothuringikum und im Saxolugikum. *Terra Nostra*, **98**, 49–52.

GAITZSCH, B., RÖSSLER, R., SCHNEIDER, J. W. & SCHRETZENMEYR, S. 1999. Neue Ergebnisse zur Verbreitung potentieller Muttergesteine im Karbon von Nord- und Mitteldeutschland. *Geologisches Jahrbuch, A*, **149**, 25–58.

GAND, G. 1993. La palichnofaune de vertébrés tétrapodes du bassin permien de Saint-Affrique (Aveyron): comparisons et conséquences stratigraphiques. *Géologie de la France*, **1**, 41–56.

GAND, G., GARRIC, J. & LAPEYRIE, J. 1997*a*. Biocénoses à Triopsidés (Crustacea), Branchiopoda) du Permien du bassin de Lodève (France). *Géobios*, **30**, 673–700.

GAND, G., LAPEYRIE, J., GARRIC, J., NEL, A., SCHNEIDER, J. & WALTER, H. 1997*b*. Découverte d'Arthropodes et de bivalves inédits dans le Permien continental (Lodèvois, France). *Comptes Rendus de l'Academie des Science, Paris, Série, IIa*, **325**, 891–898.

GAND, G., GARRIC, J., DEMATHIEU, G. & ELLENBERGER, P. 2000. La Palichnofaune de vertebres tetrapodes du Permien superieur du bassin de Lodeve (Languedoc-France). Tetrapod (vertebrate) ichnofauna from the Upper Permian of the Lodeve Basin, Languedoc, France. *Palaeovertebrata*, **29**, 1–82.

GAND, G., GALTIER, J., GARRIC, J., SCHNEIDER, J. W. & KÖRNER, F. 2001. *The Graissessac Carboniferous and Lodève Permian basins (Languedoc-France). Field Trip Guidebook*. 7 May 2001. International Meeting on The Stratigraphic and Structural Evolution of the Late Carboniferous to Triassic Continental and Marine Successions in Tuscany (Italy). Regional reports and general correlations. Siena, 30 April–7 May 2001.

GARRIC, J. 2001. Les rigoles fossilifères du Saxonien du bassin de Lodève. *Bulletin de la Societe d'Histoire Naturelle d'Autun*, **174**(2), 7–48.

GARZANTI, E., VEZZOLI, G., ANDO, S., PAPARELLA, P. & CLIFT, P. D. 2005. Petrology of Indus River sands: a key to interpret erosion history of the Western Himalayan Syntaxis. *Earth and Planetary Science Letters*, **229**, 287–302.

GAST, R. 1991. The perennial Rotliegend saline lake in NW-Germany. *Geologisches Jahrbuch, A*, **119**, 25–59.

GAST, R. 1993. Sequenzanalyse von äolischen Abfolgen im Rotliegenden und deren Verzahnung mit Küstensedimenten. *In*: MÜLLER, E. P. & PORTH, H. (eds) *Zur Geologie und Kohlenwasserstoff-Führung des Perm im Ostteil der Norddeutschen Senke*. Geologisches Jahrbuch, **131**, 117–139.

GAST, R. 1995. Sequenzstratigraphie. *In*: PLEIN, E. (ed.) *Stratigraphie von Deutschland. I. Norddeutsches Rotliegendbecken Rotliegend Monographien. Teil II.* Senkenbergische Naturforschende Gesellschaft, Frankfurt a.M., 47–54.

GAUPP, R., GAST, R. & FORSTER, C. 2000. Late Permian Playa Lake Deposits of the Southern Permian Basin (Central Europe). *In*: GIERLOWSKI-KORDESCH, E. H. & KELTS, K. R. (eds) *Lake Basins through Space and Time*. AAPG Studies in Geology, **46**, 75–86.

GEBHARDT, U. 1994. Zur Genese der Rotliegend-Salinare in der Norddeutschen Senke (Oberrotliegend II, Perm). *Freiberger Forschungsheft, Hefte C*, **452**, 3–22.

GEBHARDT, U., SCHNEIDER, J. & HOFFMANN, N. 1991. Modelle zur Stratigraphie und Beckenentwicklung im Rotliegenden der Norddeutschen Senke. *Geologisches Jahrbuch, A*, **127**, 405–427.

GEHMLICH, M., LINNEMANN, U., TICHOMIROWA, M., GAITZSCH, B., KRONER, U. & BOMBACH, K. 2000. Geochronologie oberdevonischer bis unterkarbonischer Magmatite der Thüringischen und Bayrischen Faziesreihe sowie variszischer Deckenkomplexe und der Frühmolasse von Borna-Hainichen (Saxothuringisches Terrane). *Zeitschrift der Deutschen Geologischen Gesellschaft*, **151**(4), 337–363.

GIBBS, M. T., MCALLISTER, P., KUTZBACH, J. E., ZIEGLER, A. M., BEHLING, P. J. & ROWLEY, D. B. 2002. Simulations of Permian climate and comparisons with climate-sensitive sediments. *Journal of Geology*, **110**, 33–55.

GLENNIE, K. W. 1982. Early Permian (Rotliegendes) Palaeowinds of the North Sea. *Sedimentary Geology*, **34**, 245–265.

GLENNIE, K. W. 1989. Some effects of the Late Permian Zechstein transgression in northern Europe. *In*: BOYLE, R. W., BROWN, A. C., JEFFERSON, C. W., JOWETT, E. C. & KIRKHAM, R. V. (eds) *Sediment-hosted Stratiform Copper Deposits*. Geological Association of Canada, Special Papers, **36**, 557–565.

GOLL, M. & LIPPOLT, H. J. 2001. Biotit-Geochronologie (^{40}Ar rad/K, ^{40}Ar rad/ ^{39}Ar K, ^{87}Sr rad/ ^{87}Rb) spät-variszischer Magmatite des Thüringer Waldes. *Neues Jahrbuch für Geologie und Paläontologie, Abhandlungen*, **222**, 353–405.

GOLONKA, J. 2000. *Cambrian-neogen: plate tectonic maps*. Habilitation Thesis, Uniwersytet Jagiellonski.

GOTHAN, W. 1949. *Die Unterkarbon-Flora der Dobrilugker Tiefbohrungen*. Abhandlungen der Geologischen Landesanstalt, Berlin, **217**.

HAMPE, O. 1989. Revision der Triodus-Arten (Chondrichthyes: Xenacanthida) aus dem saarpfälzischen Rotliegenden (Oberkarbon-Perm, SW-Deutschland) aufgrund ihrer Bezahnung. *Paläontologische Zeitschrift*, **63**, 79–101.

HARDIE, L. A., SMOOT, J. P. & EUGSTER, H. P. 1978. Saline lakes and their deposits: a sedimentological approach. *International Association of Sedimentologists, Special Publications*, **2**, 7–41.

HATCHER, D. H., THOMAS, W. A., GEISER, P. A., SNOKE, A. W., MOSHER, S. & WILTSCHKO, D. V. 1989. Alleghanian orogen. *In*: HATCHER, R. D. J., THOMAS, W. A. & VIELE, G. W. (eds) *The Appalachian-Ouachita Orogen in the United States*. Geological Society of America, Boulder, The Geology of North America, **F-2**. 233–318.

HECKEL, P. H. 1994. Evaluation of evidence for glacio-eustatic control over marine Pennsylvanian cyclothemes in North America and consideration of possible tectonic effects. *In*: DENNISON, J. M. & ETTERSOHN, F. R. (eds) *Tectonic and Eustatic Controls on Sedimentary Cycles*. Society for Sedimentary Geology, Concepts in Sedimentology and Paleontology, **4**, 65–87.

HECKEL, P. H. 1999. *Middle and Upper Pennsylvanian (Upper Carboniferous) Cyclotherm Succession in Midcontinent Basin, U.S.A.* Kansas Geological Survey, Calgary, Open-file Report 99–27, Field Trip **8**.

HECKEL, P. H. 2002*a*. Overview of Pennsylvanian cyclothems in Midcontinent North America and brief summary of those elsewhere in the world. *In*: HILLS, L. V., HENDERSON, C. M. & BAMBER, E. W. (eds) *Carboniferous and Permian of the World. XIV International Congress of the Carboniferous and Permian Proceedings*. Canadian Society of Petroleum Geologists, Memoir, **19**, 79–98.

HECKEL, P. H. 2002*b*. Genetic stratigraphy and conodont biostratigraphy of upper Desmoinesian-Missourian (Pennsylvanian) cyclothem succession in Midcontinent North America. *In*: HILLS, L. V., HENDERSON, C. M. & BAMBER, E. W. (eds) *Carboniferous and Permian of the World. XIV International Congress of the Carboniferous and Permian*. Canadian Society of Petroleum Geologists, Memoir, **19**, 99–119.

HEMBREE, D. I., MARTIN, L. D. & HASIOTIS, S. T. 2004. Amphibian burrows and ephemeral ponds of the Lower Permian Speiser Shale, Kansas: evidence for seasonality in the midcontinent. *Palaeogeography, Palaeoclimatology, Palaeoecology*, **203**, 127–152.

HENDERSON, C. M. 2003. Global correlations of the Kungurian Stage and its paleogeograpic implications for the assembly of Pangaea. *XVth International Congress on Carboniferous and Permian Stratigraphy. Utrecht, 2003*, Abstracts 208–210.

HESS, J. C. & LIPPOLT, H. J. 1989. Isotopic evidence for the stratigraphic position of the Saar–Nahe Rotliegende volcanism. Part III, Synthesis of results and geological implications. *Neues Jahrbuch für Geologie und Paläontologie, Abhandlungen*, **1989**, 553–559.

HMICH, D., SCHNEIDER, J. W., SABER, H., VOIGT, S., EL WARTITI, M. 2006. New continental Carboniferous and Permian faunas of Morocco: implications for biostratigraphy, palaeobiogeography and palaeoclimate. *In*: LUCAS, S. G., SCHNEIDER, J. W. & CASSINIS, G. (eds) *Non-Marine Permian Biostratigraphy and Biochronology*. Geological Society, London, Special Publications, **265**, 297–324.

HOFFMANN, N., KAMPS, H.-J. & SCHNEIDER, J. 1989. Neuerkenntnisse zur Biostratigraphie und Paläodynamik des Perms in der Norddeutschen Senke: ein Diskussionsbeitrag. *Zeitschrift für Angewandte Geologie*, **35**, 198–207.

HOLUB, V. & KOZUR, H. 1981. Revision einiger Conchostracen-Faunen des Rotliegend und biostratigraphische Auswertung der Conchostracen des Rotliegend. *Geologisch – Palaeontologische Mitteilungen Innsbruck*, **11**, 93–94.

HOTH, K., WEYER, D., KAHLERT, E. & DÖRING, H. 1990. Das Standardprofil des Oberkarbons von Vorpommern. *Zeitschrift für Angewandte Geologie*, **36**, 361–368.

IRVING, E. 1977. Drift of the major continental blocks since the Devonian. *Nature*, **270** (5635), 304–309.

ISBELL, J., LENAKER, P. A., ASKIN, R. A., MILLER, M. F. & BABCOCK, L. E. 2003. Reevaluation of the timing and extent of the late Paleozoic glaciation in Gondwana: role of the Transantarctic Mountains. *Geology*, **31**, 977–980.

JONES, A. T. & FIELDING, C. R. 2004. Sedimentological record of the late Paleozoic glaciation in Queensland, Australia. *Geology*, **32**, 153–156.

KELLER, G. R. & HATCHER, R. D. J. 1999. Some comparisons of the structure and evolution of the southern Appalachian–Ouachita orogen and portions of the Trans-European Suture Zone region. *Tectonophysics*, **314**, 43–68.

KIDDER, D. L. & WORSLEY, T. R. 2004. Causes and consequences of extreme Permo-Triassic warming to globally equable climate and relation to the Permo-Triassic extinction and recovery. *Palaeogeography, Palaeoclimatology, Palaeoecology*, **203**, 207–237.

KLEIN, G. D. 1994. *Pangea: Paleoclimate, Tectonics, and Sedimentation during Accretion, Zenith and Breakup of a Supercontinent*. Geological Society of America, Special Paper, **288**.

KÖNIGER, S. 2000. Verbreitung, Fazies und stratigraphische Bedeutung distaler Ascheentuffe der Glan-Gruppe im karbonisch-permischen Saar–Nahe-Becken (SW-Deutschland). *Mainzer Geowissenschaftliche Mitteilungen*, **29**, 97–132.

KÖRNER, F., SCHNEIDER, J. W., HOERNES, S., GAND, G. & KLEEBERG, R. 2003. Climate and continental sedimentation in the Permian of the Lodève Basin (southern France). *Bolletin della Societá Geologica Italiana*, **2**, 185–191.

KOZUR, H. 1988. The age of the Central European Rotliegendes. *Zeitschrift für Geologische Wissenschaften*, **16**, 907–915.

KOZUR, H. 1994. The correlation of the Zechstein with the marine standard. *Jahrbuch der Geologischen Bundesanstalt, Wien*, **137**, 85–103.

KRÖNER, A. & WILLNER, A. P. 1998. Time of formation and peak of Variscan HP-HT metamorphism of quartz-feldspar rocks in the central Erzgebirge, Saxony, Germany. *Contributions to Mineralogy and Petrology*, **132**, 1–20.

KRONER, U. & HAHN, T. 2003. Sedimentation, Deformation und Metamorphose im Saxothuringikum während der variszischen Orogenese: Die komplexe Entwicklung von Nord-Gondwana während kontinentaler Subduktion und schiefer Kollision. *Geologica Saxonica*, **48/49**, 133–146.

KRONER, U., LINNEMANN, U. & ROMER, R. L. 2004. Synthesis of the geological history of Saxo-Thuringia: from Cadomian accretion to Variscan collision. *Geologica Saxonica*, **48/49**, 9–11.

KUTZBACH, J. E. & ZIEGLER, A. M. 1994. Simulation of Late Permian climate and biomes with an atmosphere-ocean model; comparisons with observations. *In*: ALLEN, J. R. L., HOSKINS, B. J., SELLWOOD, B. W., SPICER, R. A. & VALDES, P. J. (eds) *Palaeoclimates and Their Modelling; With Special Reference to the Mesozoic Era*. Chapman & Hall, London, 119–132.

LEGLER, B. 2005. Marine Prä-Zechstein-Ingressionen in das Südliche Permbecken - Ein multidisziplinärer Ansatz. *In*: *DGMK/ÖGEW-Frühjahrstagung des Fachbereichs Aufsuchung und Gewinnung*. Deutsche Wissenschaftliche Gesellschaft für Erdöl, Erdgas und Kohle e.V., Celle, 127–135.

LEGLER, B., GEBHARDT, U. & SCHNEIDER, J. W. 2005. Late Permian non-marine-marine transitonal profiles in the central Southern Permian Basin, northern Germany. *International Journal of Earth Sciences. DOI: 10.1007/s00531-005-0002-5.*

LEPAGE, B. A., BEAUCHAMP, B., PFEFFERKORN, H. W. & UTTING, J. 2003. Late Early Permian plant fossils from the Canadian High Arctic: a rare palaeoenvironmental/climatic window in northwest Pangea. *Palaeogeography, Palaeoclimatology, Palaeoecology*, **191**, 345–372.

LIPPOLT, H. J. & HESS, J. C. 1996. Numerische Stratigraphie permokarbonischer Vulkanite Zentraleuropas. Teil II: Westharz. *Zeitschrift der Deutschen Geologischen Gesellschaft*, **147**(1), 1–9.

LIPPOLT, H. J., HESS, J. C., HOLUB, V. & PEŠEK, J. 1986. Correlation of Upper Carboniferous deposits in the Bohemian Massif (Czechoslovakia) and in the Ruhr District (FR Germany): evidence from Ar^{40}/Ar^{39} ages of tuff layers. *Zeitschrift der Deutschen Geologischen Gesellschaft*, **137**, 447–464.

LOPEZ, M., GAND, G., GARRIC, J. & GALTIER, J. 2005. *Playa Environments in the Lodève Permian Basin and the Triassic Cover (Languedoc-France). Excursion guide*. Association Sedimentologistes Français, Montpellier.

LÜTZNER, H., MÄDLER, J., ROMER, R. L. & SCHNEIDER, J. W. 2003. Improved stratigraphic and radiometric age data for the continental Permo-Carboniferous reference-section Thüringer-Wald, Germany. *International Congress on Carboniferous and Permian Stratigraphy. Utrecht, May, 2003, Abstracts*, 338–341.

LÜTZNER, H., LITTMANN, S., MÄDLER, J., ROMER, R. L. & SCHNEIDER, J. W. 2006. Radiometric and biostratigraphic data of the Permo-Carboniferous reference section Thüringer Wald. *XVth International Congress on Carboniferous and Permian Stratigraphy. Utrecht, 2003, Proceedings*, (in press).

MARTENS, T. 1987. Conchostracan zone: succession. *In*: LÜTZNER, H. (ed.) Sedimentary and Volcanic Rotliegendes of the Saale Depression. Symposium on Rotliegendes in Central Europe, May 1987, Excursion Guide, Zentralinstitut für die Physik der Erde, Potsdam.

MARTENS, T. 1988. First evidence of terrestrial tetrapods with North-American faunal elements in the red beds of Upper Rotliegendes (Lower Permian, Tambach beds) of the Thuringian Forest (G.D.R.):

first results. *Acta Musei Reginaehradecensis, Serie A, Scientiae Naturales*, **23**, 99–104.

MATTE, P. 2001. The Variscan collage and orogeny (480–290 Ma) and the tectonic definition of the Armorica microplate: a review. *Terra Nova*, **13**, 122–128.

MAUREL, O., RESPAUT, J.-P., MONIE, P., ARNAUD, N. & BRUNEL, M. 2004. U-Pb emplacement and ^{40}Ar/^{39}Ar cooling ages of the eastern Mont-Louis granite massif (Eastern Pyrenees, France). *Comptes Rendus des Geosciences*, **336**, 1091–1098.

MEDARIS, J. G., DUCEA, M., GHENT, E. & IANCU, V. 2003. Conditions and timing of high-pressure Variscan metamorphism in the South Carpathians, Romania. *Lithos*, **70**, 141–161.

MENNING, M. 1995. A numerical time scale for the Permian and Triassic periods: an integrated time analysis. *In*: SCHOLLE, P., PERYT, T. M. & SCHOLLE-ULMER, N. (eds) *Permian of the Northern Pangea., vol. 1*. Springer, Heidelberg, 77–97.

MENNING, M. & GERMAN STRATIGRAPHIC COMMISSION 2002. A geologic time scale 2002. *In*: GERMAN STRATIGRAPHIC COMMISSION (ed.) *Stratigraphic Table of Germany 2002*. Potsdam.

MONTERO, P., BEA, F., ZINGER, T. F., SCARROW, J. H., MOLINA, J. F. & WHITEHOUSE, M. 2004. 55 million years of continuous anatexis in central Iberia: single-zircon dating of the Peña Negra Complex. *Journal of the Geological Society, London*, **161**, 255–263.

MÜLLER, E. & KONZAN, H. P. 1989. *Geologische Übersicht. Erläuterungen zur Geologischen Karte des Saarlandes 1:50 000*. Geologisches Landesamt des Saarlandes, Saarbrücken, pp. 5–39.

MUTTONI, G., KENT, D. V., GARZANTI, E., BRACK, P., ABRAHAMSEN, N. & GAETANI, M. 2003. Early Permian Pangea 'B' to Late Permian Pangea 'A'. *Earth and Planetary Science Letters*, **215**, 379–394.

OPLUŠTIL, S. 2004. Late Carboniferous tectono-sedimentary evolution and related terrestrial biotic changes on the North Variscan and Appalachian forelands, and adjacent paralic and continental basins. *Geologica Balcanica*, **34**, 51–67.

OPLUŠTIL, S. & PEŠEK, J. 1998. Stratigraphy, palaeoclimatology and palaeogeography of the Late Palaeozoic continental deposits in the Czech Republic. *In*: CRASQUIN-SOLEAU, S., IZART, A., VASLET, D. & DE WEVER, P. (eds) *Peri-Tethys: Stratigraphic Correlations. Vol. 2*. Geodiversitas, **20**, 597–620.

PAQUETTE, Y. 1980. *Le Bassin Autunien de l'Aumance (Allier)*. PhD thesis, Universite de Dijon.

PAQUETTE, Y. & FEYS, R. 1989. Le bassin de Bourbon-l'Archambault (Aumance). *In*: CHATEAUNEUF, J.-J. AND FARJANEL, G. (eds) *Synthèse Géologigue des Bassins Permiens Français*. Memoire, de Bureau de Rechesches Géologiques et Minières, **128**, 43–54.

PARRISH, J. T. 1993. Climate of the supercontinent Pangea. *Journal of Geology*, **101**, 215–233.

PEŠEK, J. 2004. *Late Paleozoic Limnic Basins and Coal Deposits of the Czech Republic*. Musei Rerum Naturalium Bohemiae Occidentalis, Plzen, Geologica Editio Specialis, **1**.

RAMEZANI, J., DAVYDOV, V. I., NORTHRUP, C. J., BOWERING, S. A., CHUVASHOV, B. I. & SNYDER, W. S. 2003. Volcanic ashes in the Upper Paleozoic of the Southern Urals: opportunities for the high-precision calibration of the Upper Carboniferous – Cisuralian interval. *XVth International Congress on Carboniferous and Permian Stratigraphy. Utrecht, 2003*, Abstracts 431–432.

RODRIGUEZ, J., COSCA, M. A., GIL IBARGUCHI, J. I. & DALLMEYER, R. D. 2003. Strain partitioning and preservation of ^{40}Ar/^{39}Ar ages during Variscan exhumation of a subducted crust (Malpica–Tui complex, NW Spain). *Lithos*, **70**, 111–139.

ROHN, R. & STOLLHOFEN, H. 2000. The Permian age of the Passa Dois Group (Parana Basin, southern Brazil) re-affirmed. *31st International Geological Congress*, Abstract volume. Rio de Janeiro, August 1987.

ROSCHER, M. & SCHNEIDER, J. W. 2005. An annotated correlation chart for continental Late Pennsylvanian and Permian basins and the marine scale. *In*: LUCAS, S. G. & ZEIGTER, K. E. (eds) *The nonmarine Permian*, New Mexico Museum of Natural History and Science Bulletin, **30**, 282–291.

SABER, H., EL WARTITI, M., BROUTIN, J. & TOUTIN MORIN, N. 1995. L'intervalle Stephano-Permien (fin du cycle varisque au Maroc). *Gaia*, **11**, 57–71.

SALTZMAN, M. R. 2003. Late Paleozoic ice age: oceanic gateway or pCO_2. *Geology*, **31**, 151–154.

SANO, H., FUJII, S. & MATSUURA, F. 2003. Response to Middle to Late Carboniferous sea-level fall: mid-oceanic seamount-capping carbonates, Akiyoshi, Japan. *XVth International Congress on Carboniferous and Permian Stratigraphy. Utrecht, 2003*, Abstracts 464–465.

SCHÄFER, A. & STAMM, R. 1989. Lakustrine Sedimente im Permokarbon des Saar–Nahe-Beckens. *Zeitschrift der Deutschen Geologischen Gesellschaft*, **140**, 259–276.

SCHEFFLER, K., HOERNES, S. & SCHWARK, L. 2003. Global changes during Carboniferous–Permian glaciation of Gondwana: linking polar and equatorial climate evolution by geochemical proxies. *Geology*, **31**, 605–608.

SCHEFUSS, E., SCHOUTEN, S. & SCHNEIDER, R. R. 2005. Climatic controls on central African hydrology during the past 20,000 years. *Nature*. DOI: 10.1038/nature03945.

SCHNEIDER, J. W. 1980. Zur Entomofauna des Jungpaläozoikums der Boskovicer Furche (CSSR). Teil 1: Mylacridae (Insecta, Blattodea). *Freiberger Forschungsheft, Hefte C*, **357**, 43–55.

SCHNEIDER, J. W. 1982. Entwurf einer Zonengliederung für das euramerische Permokarbon mittels der Spiloblattinidae (Blattodea, Insecta). *Freiberger Forschungsheft, Hefte C*, **375**, 27–47.

SCHNEIDER, J. W. 1983. Die Blattodea (Insecta) des Paläozoikums. Teil 1: Systematik, Ökologie und Biostratigraphie. *Freiberger Forschungsheft, Hefte C*, **382**, 106–145.

SCHNEIDER, J. W. 1996. Biostratigraphie des kontinentalen Oberkarbon und Perm im Thüringer Wald, SW-Saale-Senke: Stand und Probleme. *Beiträge zur Geologie von Thüringen, Neue Folge*, **3**, 121–151.

SCHNEIDER, J. W. 2001. Rotliegendstratigraphie: Prinzipien und Probleme. *Beitrage zur Geologie von Thüringen*, **8**, 7–42.

SCHNEIDER, J. W. & GEBHARDT, U. 1993. Litho- und Biofaziesmuster in intra- und extramontanen Senken des Rotliegend (Perm, Nord- und Ostdeutschland). *Geologisches Jahrbuch, A*, **131**, 57–98.

SCHNEIDER, J. W. & WERNEBURG, R. 1993. Neue Spiloblattinidae (Insecta, Blattodea) aus dem Oberkarbon und Unterperm von Mitteleuropa sowie die Biostratigraphie des Rotliegend. *Naturhistorischen Museum Schloss Bertholdsburg, Schleusingen, Veröffentlichungen*, **7/8**, 31–52.

SCHNEIDER, J. W. & ZAJIC, J. 1994. Xenacanthodier (Pisces, Chondrichthyes) des mitteleuropäischen Oberkarbon und Perm: Revision der Originale zu GOLDFUSS 1847, BEYRICH 1848, KNER 1867 und FRITSCH 1879–1890. *Freiberger Forschungsheft, Hefte C*, **452**, 101–151.

SCHNEIDER, J. W., GEBHARDT, U. & GAITZSCH, B. 1995*a*. Müritz-Subgruppe. *In*: PLEIN, E. (ed.) *Stratigraphie von Deutschland. I. Norddeutsches Rotliegendbecken. Rotliegend-Monographie Teil II*. Courier Forschungsinstitut Senckenberg, **183**, 107–109.

SCHNEIDER, J. W., RÖßLER, R. & GAITZSCH, B. 1995*b*. Timelines of Late Variscan volcanism: a holostratigraphic synthesis. *Zentralblatt für Geologie und Paläontologie*, **1**, 477–490.

SCHNEIDER, J. W., HAMPE, O. & SOLER-GIJÓN, R. 2000. The Late Carboniferous and Permian: aquatic vertebrate zonation in Southern Spain and German basins. *In*: *Palaeozoic Vertebrate Biochronology and Global Marine/Non-Marine Correlation: Final Report of IGCP 328 (1991–1996)*. Courier Forschungsinstitut Senkenberg, **223**, 543–561.

SCHNEIDER, J. W., GORETZKI, J. & RÖSSLER, R. 2005*a*. Biostratigraphisch relevante nicht-marine Tiergruppen im Karbon der variscischen Vorsenke und der Innensenken. *In*: DEUTSCHE STRATIGRAPHISCHE KOMMISSION & WREDE, V. (eds) *Stratigraphie von Deutschland. V: Das Oberkarbon Pennsylvanium) in Deutschland.* Courier Forschungsinstitut Senkenberg, **254**, 103–118.

SCHNEIDER, J. W., RÖSSLER, R., GAITZSCH, B., GEBHARDT, U. & KAMPE, A. 2005*b*. Saale-Senke. *In*: DEUTSCHE STRATIGRAPHISCHE KOMMISSION & WREDE, V. (eds) *Stratigraphie von Deutschland. V: Das Oberkarbon Pennsylvanium) in Deutschland.* Courier Forschungsinstitut Senkenberg, **254**, 419–440.

SCHNEIDER, J. W., KÖRNER, F., ROSCHER, M. & KRONER, U. 2006. Permian climate development in the northern peri-Tethys area: the Lodève Basin, French Massif Central, compared in a European and global context. *Palaeogeography, Palaeoclimatology, Palaeoecology*, doi:10.1016/j.palaeo.2006.03.057.

SCOTESE, C. R. 2001. *Atlas of Earth History. 1. Paleogeography*, PALEOMAP Project, Arlington, Texas.

SELESHI, Y. & ZANKE, U. 2004. Recent changes in rainfall and rainy days in Ethiopia. *International Journal of Climatology*, **24**, 973–983.

SHI, G. R. & GRUNT, T. A. 2000. Permian Gondwana–Boreal antitropicality with special reference to brachiopod faunas. *Palaeogeography, Palaeoclimatology, Palaeoecology*, **155**, 239–263.

SIMPSON, G. D. H., COOPER, A. F. & NORRIS, R. J. 1994. Late Quaternary evolution of the Alpine Fault zone at Paringa, south Westland, New Zealand. *New Zealand Journal of Geology & Geophysics*, **37**, 49–58.

SMITH, D. B. 1979. Rapid marine transgressions and regressions of the Upper Permian Zechstein Sea. *Journal of the Geological Society, London*, **136**, 155–156.

SOREGHAN, G. S. & GILES, K. A. 1999. Amplitudes of Late Pennsylvanian glacioeustasy. *Geology*, **27**, 255–258.

SPERNER, B., RATSCHBACHER, L. & NEMČOK, M. 2002. Interplay between subduction retreat and lateral extrusions: tectonics of the Western Carpathians. *Tectonics*, **21(**6).

STAPF, K. R. G. 1989. Biogene fluvio-lakustrine Sedimentation im Rotliegend des permokarbonen Saar–Nahe-Beckens (SW-Deutschland). *FACIES*, **20**, 169–198.

STAPF, K. R. G. 1990. Einführung lithostratigraphischer Formationsnamen im Rotliegend des Saar–Nahe-Beckens (SW-Deutschland). *Mitteilungen der Pollichia*, **77**, 111–124.

STEMMERIK, L., INESON, J. R. & MITCHELL, J. G. 2000. Stratigraphy of the Rotliegend Group in the Danish part of the northern Permian Basin, North Sea. *Journal of the Geological Society, London*, **157**, 1127–1136.

STEYER, J. S., ESCUILLIE, F. *et al.* 2000. New data on the flora and fauna from the?uppermost Carboniferous-Lower Permian of Buxieres-les-Mines, Bourbon l'Archambault Basin (Allier, France). A preliminary report. *Bulletin de la Société Géologique de France*, **171**, 239–249.

STOLLHOFEN, H. 1994. Synvulkanische Sedimentation in einem fluviatilen Ablagerungsraum: Das basale 'Oberrotliegend' im permokarbonen Saar–Nahe-Becken. *Zeitschrift der Deutschen Geologischen Gesellschaft*, **145**, 343–378.

STOLLHOFEN, H. 1999. Karoo Synrift-Sedimentation und ihre tektonische Kontrolle am entstehenden Kontinentalrand Namibias. *Zeitschrift der Deutschen Geologischen Gesellschaft*, **149**, 519–632.

STOLLHOFEN, H. & STANISTREET, I. G. 1994. Interaction between bimodal volcanism, fluvial sedimentation and basin development in the Permo-Carboniferous Saar–Nahe Basin (south-west Germany). *Basin Research*, **6**, 245–267.

STOLLHOFEN, H., FROMMHERZ, B. & STANISTREET, I. G. 1999. Volcanic rocks as discriminants in evaluating tectonic versus climatic control on depositional sequences, Permo-Carboniferous continental Saar–Nahe Basin. *Journal of the Geological Society, London*, **156**, 801–808.

STOLLHOFEN, H., STANISTREET, I. G., BANGERT, B. & GRILL, H. 2000. Tuffs, tectonism and glacially related sea-level changes, Carboniferous–Permian, southern Namibia. *Palaeogeography, Palaeoclimatology, Palaeoecology*, **161**, 127–150.

TALLEY, L. D. 1996. North Atlantic circulation and variability, reviewed for the CNLS conference. *Physica D: Nonlinear Phenomena*, **98**, 625–646.

THOMAS, W. A. 1989. The Appalachian–Ouachita orogen beneath the Gulf Coastal Plain between the outcrops in the Appalachian and Ouachita Mountains. *In*: HATCHER, R. D. J., THOMAS, W. A. & VIELE, G. W. (eds), *The Appalachian–Ouachita Orogen in the United States*. The Geological Society of America, Boulder, The Geology of North America, **F-2**, 537–553.

TURNER, B. R. 1999. Tectonostratigraphical development of the Upper Karoo foreland basin: orogenic unloading versus thermally-induced Gondwana rifting. *Journal of African Earth Sciences*, **28**, 215–238.

VEEVERS, J. J. & POWELL, C. M. 1987. Late Paleozoic glacial episodes in Gondwanaland reflected in transgressive-regressive depositional sequences in Euramerica. *Geological Society of America Bulletin*, **98**, 475–487.

VISSER, J. N. J. 1995. Post-glacial Permian stratigraphy and geography of southern and central Africa: boundary conditions for climatic modelling. *Palaeogeography, Palaeoclimatology, Palaeoecology*, **118**, 213–243.

VISSER, J. N. J. 1996. Controls on Early Permian shelf deglaciation in the Karoo Basin of South Africa. *Palaeogeography, Palaeoclimatology, Palaeoecology*, **125**, 129–139.

VISSER, J. N. J. 1997. Deglaciation sequences in the Permo-Carboniferous Karoo and Kalahari basins of Southern Africa: a tool in the analysis of cyclic glaciomarine basin fills. *Sedimentology*, **44**, 507–521.

VOIGT, S. 2005. *Die Tetrapodenichnofauna des kontinentalen Oberkarbon und Perm im Thüringer Wald: Ichnotaxonomie, Paläoökologie und Biostratigraphy*. Cuvillier Verlag, Göttingen.

VOIGT, S. & RÖSSLER, R. 2004. Taeniopterid-type leave fragments: the first record of macrophytic remains from the Eisenach Formation (Rotliegend, Permian, Thuringian Forest). *Hallesches Jahrbuch für Geowissenschaften, B*, **18**, 27–38.

VON DER HAAR, T. H. & OORT, A. H. 1973. New estimate of annual poleward energy transport by Northern Hemisphere oceans. *Journal of Physical Oceanography*, **3**, 169–172.

WELDON, E. A. & SHI, G. R. 2003. Global distribution of *Terrakea* Booker, 1930 (Productidina, Brachiopoda): implications for Permian marine biogeography and Eurasia-Gondwana correlations. *XVth International Congress on Carboniferous and Permian Stratigraphy. Utrecht, 2003*, Abstracts 583.

WERNEBURG, R. 1989. Some notes to systematic, phylogeny and biostratigraphy of labyrinthodont amphibians from the Upper Carboniferous and Lower Permian in Central Europe. *Acta Musei Reginaehradecensis, Serie A, Scientriae Naturales*, **XXII (1989)**, 117–129.

WERNEBURG, R. 1996. Temnospondyle Amphibien aus dem Karbon Mitteldeutschlands. *Naturhistorischen Museums Schloss Bertholdsburg, Schleusingen, Veröffentlichungen*, **11**, 23–64.

WERNEBURG, R. 1999. Biostratigraphy and palaeobiogeography of the European Permo-Carboniferous using labyrinthodont amphibians. *In*: CASSINIS, G., CORTESOGNO, L., GAGGERO, L., PITTAU, P., RONCHI, A. & SARRIA, E. (eds) *Late Palaeozoic Continental Basins of Sardinia*. Field trip guidebook 15–18 September. International Field Conference on The Continental Permian of the Southern Alps and Sardinia (Italy). Regional reports and general correlations. Brescia, 15–25 September 1999. Pavia University. Abstracts, 27.

WERNEBURG, R. 2001. Die Amphibien- und Reptilien-Faunen im Permokarbon des Thüringer Waldes. *Beiträge zur Geologie von Thüringen, Neue Folge*, **8**, 125–152.

WERNEBURG, R. 2003. The branchiosaurid amphibians from the Lower Permian of Buxières-les-Mines, Bourbon l'Archambault Basin (Allier, France) and biostratigraphic significance. *Bulletin de la Societe Géologiques de France*, **174**, 343–349.

WERNEBURG, R. & SCHNEIDER, J. W. 2006. Amphibian biostratigraphy of the European Permo-Carboniferous. *In*: LUCAS, S. G., SCHNEIDER, J. W. & CASSINIS, G. (eds) *Non-Marine Permian Biostratigraphy and Biochronology*. Geological Society, London, Special Publications, **265**, 201–215.

WINGUTH, A. M. E., HEINZE, C. *et al.* 2002. Simulated warm polar currents during the Middle Permian. *Paleoceanography*, **17**, 18.

WISE, D. U. 2004. Pennsylvania salient of the Appalachians; a two-azimuth transport model based on new compilations of Piedmont data. *Geology*, **32**, 777–780.

WORTHINGTON, L. V. & WRIGHT, W. R. 1970. *North Atlantic Ocean Atlas. Vol. 4*. Woods Hole Oceanographic Institute, Massachusetts.

WRIGHT, P. 2003. The influence of glacio-eustasy, climate and sea-water chemistry changes, and biological events on the super-giant Permo-Carboniferous hydrocharbon reservoires of Kazakhstan. *XVth International Congress on Carboniferous and Permian Stratigraphy. Utrecht, 2003*, Abstracts 589.

WRIGHT, V. P. & VANSTONE, S. D. 2001. Onset of late Palaeozoic glacio-eustasy and the evolving climates of low latitude areas: a synthesis of current understanding. *Journal of the Geological Society, London*, **158**, 579–582.

YOUBI, N., BEN ABBOU, M. *et al.* 2003. The Permian volcanism of North Africa and its insertion in the geodynamic framework of the Central Pangean Permo-Carboniferous Province. *XVth International Congress on Carboniferous and Permian Stratigraphy. Utrecht, 2003*, Abstracts 592–593.

ZAHN, U. 1991. *Diercke Weltatlas. 2. Auflage*. Westermann Schulbuchverlag GmbH, Braunschweig.

ZEH, A., BRÄTZ, H., BANKWITZ, E. & BANKWITZ, P. 1998. Maximale Intrusionstiefe und Exhumierungsrate des Thüringer Hauptgranites, Mitteldeutsche Kristallinzone, sowie Hinweise zum regionalen Spannungsfeld während der Intrusion. *Terra Nostra*, **98**, 169–172.

ZIEGLER, A. M. 1990*a*. Phytogeographic patterns and continental configurations during the Permian

Period. *In*: McKerrow, W. S. & Scotese, C. R. (eds) *Palaeozoic Palaeogeography and Biogeography*. Geological Society, London, Memoir, **12**, 363–379.

Ziegler, A. M., Hulver, M. L. & Rowley, D. B. 1997. Permian world topography and climate. *In*: Martini, I. P. (ed.) *Late Glacial and Postglacial Environmental Changes: Pleistocene, Carboniferous-Permian, and Proterozoic*. Oxford University Press, Oxford, 111–146.

Ziegler, P. A. 1982. Faulting and graben formation in western and central Europe. *Philosophical Transactions of the Royal Society of London, A*, **305**, 113–143.

Ziegler, P. A. 1988. *Evolution of the Arctic-North Atlantic and the Western Tethys*. American Association of Petroleum Geologists, *Memoirs*, **43**.

Ziegler, P. A. 1990*b*. *Geological Atlas of Western and Central Europe*. Shell International Petroleum Maatschappij, The Hague.

Permian tetrapod ichnofacies

ADRIAN P. HUNT & SPENCER G. LUCAS

New Mexico Museum of Natural History, 1801 Mountain Road NW, Albuquerque, New Mexico 87104-1375, USA (e-mail: adrian.hunt@state.nm.us)

Abstract: Three fundamental terms in ichnology are:

(1) assemblage, which is equivalent to an assemblage of body fossils;
(2) ichnocoenosis, which is a trace fossil assemblage produced by a biological community that can be characterized by morphological criteria;
(3) ichnofacies, which refers to recurrent ichnocoenoses that represent a significant portion of Phanerozoic time.

There are five archetypal vertebrate ichnofacies for non-marine environments (*Chelichnus, Grallator, Carichnium, Batrachichnus, Characichichnos*) of which four are present in the Permian:

(1) *Chelichnus* ichnofacies – *Chelichnus* ichnocoenosis;
(2) *Batrachichnus* ichnofacies – *Batrachichnus* ichnocoenosis;
(3) *Brontopodus* ichnofacies – *Pachypes* ichnocoenosis;
(4) *Characichichnos* ichnofacies – *Serpentichnus* ichnocoenosis. The *Chelichnus* and *Characichichnos* ichnofacies occur throughout the Permian, the *Batrachichnus* ichnofacies is restricted to the Early Permian and the *Brontopodus* to the Middle to Late Permian. The *Batrachichnus* ichnocoenosis can be divided into the *Ichniotherium* sub-ichnocoenosis, *Amphisauropus* sub-ichnocoenosis and the *Dimetropus* sub ichnocoenosis, which represent a spectrum of non-marine environments from alluvial fan to tidal flat.

Permian tetrapod tracks are among the most abundant in the Phanerozoic track record, and they have been fundamental to the development of vertebrate ichnology (Hunt & Lucas 2005*c*). Permian tetrapod tracks were first described more than 175 years ago, so they were the first scientifically described tetrapod tracks (Grierson 1828; Sarjeant 1974; Pemberton & Gingras 2003). The largest Palaeozoic tracksites and track collections are from the Permian of the southwestern United States (e.g. papers in Lucas & Heckert 1995). In addition, the largest ichnofauna from any Phanerozoic aeolianite is from the Permian Coconino Sandstone of Arizona (e. g. Gilmore 1926, 1927, 1928).

Permian ichnofaunas have been extensively studied around the world during the past 175 years, with a recent renaissance sparked by the study of the large samples of tracks from the Robledo Mountains of southern New Mexico, USA. Permian tracks are currently known from five continents (Fig. 1). Permian ichnology has also had a significant impact on the concept of ichnofacies. The study of tetrapod ichnofacies was stimulated by Baird (1965), who noted that the differences between Permian ichnofaunas of the red beds of the American West and those of aeolianites were the result of facies differences.

The purpose of this paper is to describe tetrapod ichnofacies of the Permian. This necessitates a discussion of the philosophy and classification of ichnofacies and a brief review of Permian ichnofaunas. USNM refers to the United States National Museum, Washington.

Tetrapod ichnofacies

Seilacher (1964) recognized that there are recurrent associations of invertebrate ichnofossils, which he referred to as ichnofacies. Subsequently, he named six archetypal ichnofacies after typical ichnofossils (Seilacher 1964, 1967). Seilacher's work stimulated numerous studies of invertebrate ichnofacies (Hunt & Lucas 2006). Lockley *et al.* (1994) first discussed vertebrate ichnofacies in detail, and Hunt & Lucas (2006) have recently proposed a comprehensive scheme of archetypal tetrapod ichnofacies. The archetypal tetrapod ichnofacies parallel invertebrate archetypal ichnofacies and do not incorporate them (Hunt & Lucas 2006).

From: LUCAS, S. G., CASSINIS, G. & SCHNEIDER, J. W. (eds) 2006. *Non-Marine Permian Biostratigraphy and Biochronology*. Geological Society, London, Special Publications, **265**, 137–156.
0305-8719/06/$15.00

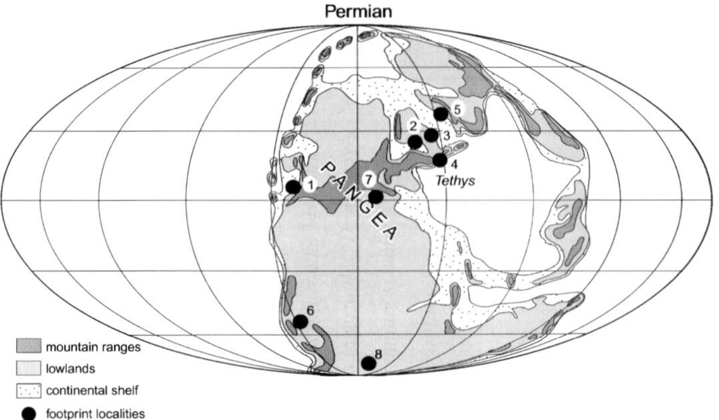

Fig. 1. Distribution of principal Permian tetrapod tracksites on Permian Pangaea. Locations are: 1, western United States; 2, France; 3, Germany; 4, Italy; 5, Russia; 6, Argentina; 7, Morocco; 8, South Africa.

Ichnocoenoses and ichnofacies

It is important to establish a clear nomenclature in any discussion of ichnofacies (Hunt & Lucas 2006). There is a large literature about invertebrate ichnology that discusses nomenclature (e.g. Bromley 1996; Keighley & Pickerill 2003), and vertebrate palaeontologists should be as consistent as possible in their terminology (Hunt & Lucas 2006).

The basic collective term for trace fossils is 'ichnoassemblage', which is equivalent to an assemblage of body fossils (Bromley 1996). This term implies nothing about the origin of the trace fossils. It is rare to be able to physically rework trace fossils (Bromley 1996; Hunt & Lockley 1997). However, trace fossil assemblages may not reflect actual communities because:

(1) time averaging caused by relatively slow deposition rates can result in structures produced by successive communities being superimposed in the same rock unit;
(2) ecological tiering can result in deep burrows penetrating into beds containing trace fossils of an entirely different community (Bromley 1996).

These two circumstances are obviously more relevant to invertebrate traces than to tetrapod footprints (Hunt & Lucas 2006).

The term 'ichnocoenosis' was originally proposed by Dvitashvili (1945) to represent the traces of a biological community. This term has subsequently been utilized in several different ways (e.g. Bromley 1996; Keighley & Pickerill 2003). There is now consensus that an ichnocoenosis can be defined as a trace fossil assemblage produced by a biological community that can be characterized by morphological criteria (independent of depositional environment or biological affinities) (e.g. Bromley 1996; McIlroy 2004). As ichnocoenoses represent traces of different communities, they have temporal and geographical ranges that represent the extent of the animal community.

Seilacher (1964, p. 303) introduced the term 'ichnofacies' for 'general trace associations, or types of ichnocoenoses, representing certain facies with a long geologic range'. Seilacher (1964) initially introduced four ichnofacies and subsequently added two more (Seilacher 1967). These high-level ichnofacies have been referred to as Seilacherian (Bromley 1996) or archetypal (Frey & Pemberton 1987); we prefer the latter term. There is consensus in the invertebrate ichnological literature that ichnofacies should refer to recurrent ichnocoenoses that represent a significant portion of Phanerozoic time.

Hunt & Lucas (2006) argued that all tetrapod ichnofacies defined prior to 2005 should be considered ichnocoenoses as they represent the traces of specific communities that lack the stratigraphical and geographical ranges that would warrant their consideration as archetypal ichnofacies. Therefore, to parallel invertebrate ichnology, Hunt & Lucas (2006) proposed five archetypal tetrapod ichnofacies (Table 1).

Two traditions in ichnology

Two distinct traditions can be identified in ichnology, which Hunt & Lucas (2003, 2004, 2005*b*, 2006) termed the ethological and the biotaxonomic. Invertebrate palaeontologists mostly use an ethological approach to ichnology by describing

Table 1. *Archetypal tetrapod ichnofacies (after Hunt & Lucas 2006)*

Archetypal Tetrapod Ichnofacies	Predominant trace fossil types	Constituent Permian ichnocoenoses	Inferred environment
Chelichnus	Low diversity ichnofaunas (less than 4 ichnogenera) of tetrapod tracks that have equant shape with subequal manual and pedal impressions and short digit impressions	*Chelichnus* (= *Laoporus*) (Early Permian; Lockley *et al.* 1994; Hunt & Lucas 2005*b*)	Aeolian crossbeds
Batrachichnus	Majority of tracks are of quadrupedal carnivores; medium-high diversity (4–8 ichnogenera)	*Batrachichnus* (Early Carboniferous–Early Permian; Hunt & Lucas, 2005): separable into sub-ichnocoenoses: (1) *Ichniotherium* – inland/distal alluvial fan settings characterized by the presence of *Ichniotherium* and a paucity of *Dimetropus;* (2) *Amphisauropus* – alluvial plain settings characterized by the presence of *Amphisauropus*; and (3) *Dimetropus* – coastal/tidal flat settings characterized by the relative abundance of *Batrachichnus* and *Dimetropus* (Hunt & Lucas 2005*b*)	Tidal flat-fluvial plain
Brontopodus	Majority of tracks are terrestrial herbivores with small quantity (generally > 10%) of terrestrial carbonate carnivore tracks; medium-high diversity (4–8 ichnogenera)	*Pachypes* (herein) *Ichniotherium* (herein)	Coastal plain, clastic or marine shoreline
Grallator	Medium-high diversity ichnofaunas (5–8 ichnogenera) with tracks (usually dominant) of tridactyl avian and non-avian theropods	None	Lacustrine margin
Characichichnos	Parallel scratch marks and fish swimming trails (*Undichna*)	*Serpentichnus* (herein)	Shallow lacustrine/aquatic

Note that Hunt & Lucas (2006) advocated italicizing the eponymous ichnogenus in ichnofacies (contra Bromley 1996).

and naming behavioural interactions between an organism and the substrate. In contrast, vertebrate palaeontologists have principally applied a biotaxonomical approach by attempting to relate tracks and traces to the taxonomy of the producer. In other words, vertebrate ichnologists treat vertebrate ichnotaxa as proxies of biotaxa. Palaeoentomologists are unusual among invertebrate ichnologists in utilizing a biotaxonomical approach.

Ichnofacies have been defined as associations of ichnotaxa recurrent in time and space. Hunt & Lucas (2003, 2004, 2005*b*, 2006) considered that there must be two different kinds of ichnofacies, one peculiar to each ichnological tradition. They created the terms 'ethoichnofacies' for invertebrate ichnology and 'biotaxonichnofacies' for tetrapod ichnology (Hunt & Lucas 2003, 2004, 2005*b*, 2006). These terms are useful to discriminate between the two kinds of ichnofacies. We assume, unless stated otherwise, that all vertebrate ichnofacies are biotaxonichnofacies and all invertebrate ichnofacies are ethoichnofacies. Ichnofacies based on traces of fossil arthropods, such as in palaeosols, could be biotaxonichnofacies.

Tetrapod ichnofacies

Baird (1965) was apparently the first to raise the issue of ichnofacies with regard to tetrapods when he noted that the differences between the Permian tetrapod ichnofaunas of red beds and aeolianites in the American West might be the result of facies differences. However, it was 30 years before Lockley *et al.* (1994) provided a cogent synthesis and discussion of tetrapod ichnofacies. The majority of relevant work over the last decade has concentrated on late Palaeozoic tetrapod ichnocoenoses (Hunt *et al.* 1995*d*, 2005*b*, *d*; Hunt & Lucas 1998*b*, 2003, 2004, 2005*b*, 2006; Hunt & Santucci 1998; Lockley & Meyer 2000; Melchor & Sarjeant 2004; Lockley 2006).

Hunt & Lucas (2006) defined five archetypal tetrapod ichnofacies for non-marine environments: *Chelichnus, Grallator, Brontopodus, Batrachichnus* and *Characichichnos* ichnofacies. Four of these ichnofacies are recognized in the Permian (*Chelichnus, Brontopodus, Batrachichnus, Characichichnos*) (Table 1).

Permian tetrapod track record

The global record of Permian tetrapod footprints encompasses localities in North America, South America, Europe, Russia and Africa (Fig. 1; Tables 2–3).

North America

North American Permian tetrapod ichnofaunas are all of Early Permian age. They are distributed from northeastern Canada to the southwestern United States (Table 2).

There are only two notable ichnofaunas in Canadian: northern Nova Scotia and Prince Edward Island (Mossman & Place 1989; Van Allen *et al.* 2005) (Fig. 2). There is an even more sparse record of Permian tracks in the eastern United States, with small assemblages in Ohio and West Virginia (Cotton *et al.* 1995).

The most extensive Permian ichnofaunas in North America occur in the southwestern United States. These ichnofaunas are arguably the most important Permian ichnofaunas in the world and are among the most significant in the Phanerozoic because of

(1) large sample sizes;
(2) direct correlation with the marine time scale;
(3) diverse extramorphological variants in individual ichnofaunas;
(4) large aeolian ichnofaunas.

The longest-studied ichnofaunas and the largest aeolian ichnofaunas in the southwest are from Arizona, whereas New Mexico has the most numerous localities and specimens from red beds (Haubold *et al.* 1995*a*; Hunt *et al.* 1995*b*, *c*, 2005*a*, *b*, *f*; Haubold 1996, 2000; Lucas *et al.* 2004; Lucas & Hunt 2006). The New Mexico localities are geographically widespread and encompass Wolfcampian environments, ranging from tidal flat to mountain front (Hunt *et al.* 1995*d*, 2005*d*). These assemblages are from the Earp Formation (Big Hatchet Mountains), the Robledo Mountains Formation of the Hueco Group (Robledo, Doña Ana and San Andres mountains), the Abo Formation (Caballo and Fra Cristobal mountains, Joyita Hills, Abo Pass) and the Sangre de Cristo Formation (Villanueva). Relative abundances of the ichnotaxa vary between sites, but *Dromopus* and *Batrachichnus* dominate, and co-occur with *Amphisauropus, Varanopus, Dimetropus, Hyloidichnus, Ichniotherium* and *Limnopus* (e.g. Haubold 2000; Haubold & Lucas 2001*a*; Lucas *et al.* 2001, 2004; Hunt *et al.* 2005*b*).

Texas tracksites have the most extensive record of Leonardian tracksites in North America. The best-known site is in the upper part of the Choza Formation at Castle Peak near Abilene, Texas (Haubold & Lucas 2001*b*). There is only a single Leonardian locality in Oklahoma in the lower part of the Hennessey Formation at Oklahoma City (Lucas & Suneson 2002).

Table 2. *Permian ichnofaunas, ichnofacies and ichnocoenoses of North America*

Location	Age	Stratigraphic unit	Selected ichnotaxa	Ichnofacies	Ichnocoenosis (Sub-ichnocoenosis)	Comments (and principal references)
Brule, Nova Scotia	?Early Permian (?Wolfcampian)	Cape John Formation	*Amphisauropus Varanopus Dimetropus*	*Batrachichnus*	*Batrachichnus (Amphisauropus)*	Van Allen *et al.* (2005)
Prince Edward Island	Early Permian (Leonardian)	Hillsborough Formation	*Amphisauropus*	*Batrachichnus*	*Batrachichnus (Amphisauropus)*	*Ichniotherium* was incorrectly identified from this site – specimens probably represent *Amphisauropus* or *Gilmoreichnus* (Mossman & Place 1989; Calder *et al.* 2004)
Southeast Ohio	Wolfcampian	Dunkard Group	*Dromopus Limnopus*	*Batrachichnus*	*Batrachichnus (?Dimetropus)*	(Cotton *et al.* 1995)
Northwest West Virginia	Wolfcampian	Waynesburg Sandstone	*Limnopus Dimetropus*	*Batrachichnus*	*Batrachichnus (?Dimetropus)*	Too small an ichnofauna to be definitive about sub-ichnofacies (Cotton *et al.* 1995)
Oklahoma City, Oklahoma	Leonardian	Hennessey Formation	*Amphisauropus cf. Dromopus*	*Batrachichnus*	*Batrachichnus (Amphisauropus)*	(Lucas & Suneson 2002)
Castle Peak, Texas	Leonardian	Choza Formation	*Varanopus Erpetopus Dromopus*	*Batrachichnus*	*Batrachichnus (?Amphisauropus)*	*?Amphisauropus* sub-ichnofacies based onpresence of *Varanopus.* (Haubold & Lucas 2001*b*, 2003)
Lake Kemp, Texas	Leonardian	Lueders Formation	*Dimetropus*	*Batrachichnus*	*Batrachichnus (?Dimetropus)*	Too small an ichnofauna to be definitive about sub-ichnofacies – no small tracks (Lucas & Hunt 2005)
Lake Kemp, Texas	Leonardian	Clear Fork Formation	*Batrachichnus cf. Amphisauropus*	*Batrachichnus*	*Batrachichnus (?Amphisauropus)*	(Lucas & Hunt 2005)
Sid McAdams locality, Texas	Leonardian	Vale Formation	*Dimetropus*	*Batrachichnus*	*Batrachichnus (?Dimetropus)*	Too small an ichnofauna to be definitive about sub-ichnofacies – no small tracks. (Lucas & Hunt 2005)
San Angelo, Texas	Leonardian	Blaine Formation	*aff. Amphisauropus aff. Chelichnus*	*?Batrachichnus*	*Batrachichnus (?Amphisauropus)*	No published detailed description (Lucas & Hunt 2005)

Table 2. *Continued*

Location	Age	Stratigraphic unit	Selected ichnotaxa	Ichnofacies	Ichnocoenosis (Sub-ichnocoenosis)	Comments (and principal references)
San Juan Mountains, Colorado	Wolfcampian	Cutler Formation	*Limnopus*	*Batrachichnus* (?*Dimetropus*)	*Batrachichnus*	Too small an ichnofauna to be definitive about sub-ichnofacies (Baird 1965)
Maroon Bells Wilderness Area, Colorado	Wolfcampian	Maroon	*Ichniotherium Varanopus Dimetropus Tambachichnium*	*Brontopodus*	*Ichniotherium*	(Voigt *et al.* 2005)
Monument Valley area, Arizona	Wolfcampian	Organ Rock Shale	*Dromopus Gilmoreichnus*	*Batrachichnus*	*Batrachichnus* (?*Dimetropus*)	Too small an ichnofauna to be definitive about sub-ichnofacies. (Haubold *et al.* 1995*a*)
Eastern Arizona	Leonardian	Schnebbly Hill Formation	*Dimetropus*	*Batrachichnus*	*Batrachichnus* (?*Dimetropus*)	Too small an ichnofauna to be definitive about sub-ichnofacies. (Hunt & Lucas 2005*a*; Hunt *et al.* 2005*e*, *f*)
Grand Canyon, Arizona	Wolfcampian	Hermit Shale	*Batrachichnus Ichniotherium*	*Batrachichnus*	*Batrachichnus* (*Ichniotherium*)	*Ichniotherium* is based upon USNM 11707 (Haubold 1971; Haubold *et al.* 1995*a*) (Hunt & Lucas 2005*a*; Hunt *et al.* 2005*e*, *f*)
Big Hatchet Mountains, New Mexico	Wolfcampian	Earp Formation	*Batrachichnus Dromopus*	*Batrachichnus*	*Batrachichnus* (?*Dimetropus*)	Too small an ichnofauna to be definitive about sub-ichnofacies (Lucas & Hunt 1995)
Robledo Mountains, Dona Ana Mountains, San Andres Mountains, New Mexico	Wolfcampian	Robledo Mountains Formation	*Batrachichnus Dromopus Dimetropus*	*Batrachichnus*	*Batrachichnus* (*Dimetropus*)	(Haubold *et al*, 1995*a*; Hunt *et al.* 1995*b*; Lucas *et al.* 1995*b*)
Robledo Mountains, New Mexico	Wolfcampian	Robledo Mountains Formation	*Serpentichnus Batrachichnus*	*Characichichnos*	*Serpentichnus*	(Braddy *et al.* 2003)

Table 2. *Continued*

Location	Age	Stratigraphic unit	Selected ichnotaxa	Ichnofacies	Ichnocoenosis (Sub-ichnocoenosis)	Comments (and principal references)
Caballo Mountains, San Andres Mountains, New Mexico	Wolfcampian	Abo Formation	*Batrachichnus Limnopus Dimetropus*	*Batrachichnus*	*Batrachichnus (Dimetropus)*	(Lucas *et al.* 2005*b*)
Fra Cristobal Mountains, New Mexico	Wolfcampian	Abo Formation	*Batrachichnus Dimetropus*	*Batrachichnus*	*Batrachichnus (Dimetropus)*	(Lucas *et al.* 2005*c*)
Joyita Hills, New Mexico	Wolfcampian	Abo Formation	*Dromopus Limnopus Dimetropus Amphisauropus*	*Batrachichnus*	*Batrachichnus (Amphisauropus)*	(Hunt *et al.* 1995*c*)
Abo Pass, New Mexico	Wolfcampian	Abo Formation	*Amphisauropus Batrachichnus Dromopus Varanopus*	*Batrachichnus*	*Batrachichnus (Amphisauropus)*	(Lucas *et al.* 2001)
Lucero uplift, Placitas, New Mexico	Wolfcampian	Abo Formation	*Batrachichnus Dromopus Dimetropus*	*Batrachichnus*	*Batrachichnus (Dimetropus)*	(Lucas *et al.* 2004)
Lucero uplift, New Mexico	Wolfcampian	DeChelly Sandstone	*Amphisauropus Dimetropus Limnopus*	*Batrachichnus*	*Batrachichnus (Amphisauropus)*	(Lucas *et al.* 2005*a*)
Villanueva, New Mexico	Wolfcampian	Sangre de Cristo Formation	*Limnopus Batrachichnus Ichniotherium*	*Batrachichnus*	*Batrachichnus (Ichniotherium)*	(Hunt *et al.* 1995*d*)
Lake Powell, Utah	Wolfcampian	Cedar Mesa Sandstone	*Chelichnus*	*Chelichnus*	*Chelichnus*	*Chelichnus ?grandis* was previously identified as cf. *Anomalopus* (Lockley & Madsen 1993)
San Juan River area, Utah	Wolfcampian	Cedar Mesa Sandstone	*Chelichnus*	*Chelichnus*	*Chelichnus*	(Loope 1984; Sumida *et al.* 1999)
Grand Canyon, Seligman area, Arizona	Leonardian	Coconino Sandstone	*Chelichnus*	*Chelichnus*	*Chelichnus*	(Hunt & Lucas 2005*a*; Hunt *et al.* 2005*e, f*)
Monument Valley area, Arizona	Leonardian	DeChelly Sandstone	*Chelichnus*	*Chelichnus*	*Chelichnus*	(Hunt & Lucas 2005*a*; Hunt *et al.* 2005*e, f*)
Front Range, Colorado	Leonardian	Lyons Sandstone	*Chelichnus*	*Chelichnus*	*Chelichnus*	(Toepelman & Rodeck 1936; Lockley & Hunt 1995)

Ages based on Lucas & Hunt (2006) or original sources.

Table 3. *Permian ichnofaunas, ichnofacies and ichnocoenoses outside North America*

Location	Age	Stratigraphic unit	Selected ichnotaxa	Ichnofacies	Ichnocoenosis Sub-ichnocoenosis	Comments (and principal references)
Lodève Basin, France	Sakmarian	'Autunian Group' (Usclas-St. Privat, Tuilieres-Loiras, Viala formations) F1–F3 formations	*Amphisauropus*, cf. *Ichniotherium* *Batrachichnus* *Varanopus*	*Batrachichnus*	*Batrachichnus* (?*Ichniotherium*)	Insufficient data to discriminate if separate *Amphisauropus* and *Ichniotherium* sub-ichnofacies present: Associations I and II of & Durand (2006) (Cassinis & Santi 2005; Gand & Durand 2006)
Lodève Basin, France	Artinskian	Rabejac, lower Salagou formation (F5) F4–F5 (lower)	*Varanopus Dromopus* *Dimetropus* *Hyloidichnus* *Batrachichnus*	*Batrachichnus*	*Batrachichnus* (*Dimetropus*)	Association III of Gand & Durand (2006) (Cassinis & Santi 2005; Gand & Durand 2006)
Lodève Basin, France	Wuchiapingian	Salagou Formation (upper) = La Lieude Formation	*Brontopus* *Merifontichnus* *Lunaepes Planipes*	*Brontopodus*	*Pachypes*	Associations IV of Gand & Durand (2006) 'La Lieude' level. (Cassinis & Santi 2005; Gand & Durand 2006)
Estérel and Bas-Argens basins, Provence, France	Artinskian-Kungurian	Bayonne Formation	*Varanopus Dromopus*	*Batrachichnus*	*Batrachichnus* (*Dimetropus*)	(Cassinis & Santi 2005; Gand & Durand 2006)
Estérel and Bas-Argens basins, Provence, France	Guadalupian	Les Pradinaux Formation	*Varanopus* *Batrachichnus* *Dromopus Hyloidichnus* *Lunaepes*	*Batrachichnus*	*Batrachichnus* (*Dimetropus*)	(Cassinis & Santi 2005; Gand & Durand 2006)
Estérel and Bas-Argens basins, Provence, France	Guadalupian	Le Mitan and Le Muy formations	*Dromopus Limnopus*	*Batrachichnus*	*Batrachichnus* (*Dimetropus*)	(Cassinis & Santi 2005; Gand & Durand 2006)
Estérel and Bas-Argens basins, Provence, France	Wuchiapingian	La Motte Formation	*Varanopus Dimetropus*	*Batrachichnus*	*Batrachichnus* (*Dimetropus*)	(Cassinis & Santi 2005; Gand & Durand 2006)
Toulon-Cuers basin, Provence, France	Guadalupian	St-Mandrier Formation	*Batrachichnus* *Dimetropus* *Varanopus Limnopus*	*Batrachichnus*	*Batrachichnus* (*Dimetropus*)	(Cassinis & Santi 2005; Gand & Durand 2006)
Cantabrian Mountains, Spain	Artinskian	Sagra Formation	*Hyloidichnus* cf. *Limnopus*	*Batrachichnus*	*Batrachichnus* (?*Dimetropus*)	(Cassinis & Santi 2005)

Table 3. *Continued*

Location	Age	Stratigraphic unit	Selected ichnotaxa	Ichnofacies	Ichnocoenosis Sub-ichnocoenosis	Comments (and principal references)
Orobic, Val Trompia basins, Italy	Artinskian	Collia Formation	*Amphisauropus Dromopus Varanopus Ichniotherium Batrachichnus*	*Batrachichnus*	*Batrachichnus* (*Ichniotherium*)	(Cassinis & Santi 2005)
Tregiovo basin, Italy	Artinskian-Kungurian	Tregiovo Formation	*Dromopus*	*Batrachichnus*	*Batrachichnus*	Too small an ichnofauna to be definitive about sub-ichnofacies. (Cassinis & Santi 2005)
Western Dolomites, Italy	Wuchiapingian	Val Gardena Formation	*Pachypes Dicynodontipus Rhynchosauroides*	*Brontopodus*	*Pachypes*	(Cassinis & Santi 2005)
Thuringian Forest, Germany	Asselian-Sakmarian	Goldlauter and Oberhof Formations	*Dromopus Ichniotherium Amphisauropus Batrachichnus Varanopus*	*Brontopodus*	*Ichniotherium*	(Haubold 1984; Voigt 2005)
Thuringian Forest, Germany	Artinskian	Tambach Sandstone	*Ichniotherium Varanopus Dimetropus Tambachichnium*	*Brontopodus*	*Ichniotherium*	(Voigt 2005)
Saar-Nahe, Wetterau basins, SW Germany	Asselian-Kungurian	Altenglan Formation to Nahe Group, Kusel and Lebach Groups	*Amphisauropus Batrachichnus*	*Batrachichnus*	*Batrachichnus* (*Amphisauropus*)	(Haubold & Stapf 1998)
Cornberg, Germany	Capitanian-Wuchapingian	Cornberger Sandstein	*Chelichnus*	*Chelichnus*	*Chelichnus*	(McKeever & Haubold 1996)
Krkonoše, Czech Republic	Gzhelian-Sakmarian	Semily, Vrchlabi, and Proseèné formations	*Batrachichnus Amphisauropus Ichniotherium*	*Batrachichnus*	*Batrachichnus* (*Ichniotherium*)	(Haubold 1984; Voigt 2004)
Poland	Carboniferous-?Artinskian	Slupiec Formation	*Amphisauropus Dimetropus Dromopus*	*Batrachichnus*	*Batrachichnus* (*Amphisauropus*)	(Haubold 1984)
Scotland	Wuchiapingian	Corncockle, Lochabriggs and Hopeman sandstones	*Chelichnus*	*Chelichnus*	*Chelichnus*	(McKeever & Haubold 1996)

Table 3. *Continued*

Location	Age	Stratigraphic unit	Selected ichnotaxa	Ichnofacies	Ichnocoenosis Sub-ichnocoenosis	Comments (and principal references)
Staffordshire, England	Lower Permian	Enville Group	*Ichniotherium cf. Amphisauropus Dromopus Batrachichnus*	*Batrachichnus*	*Batrachichnus* (*Ichniotherium*)	(Haubold & Serjeant 1973)
Russia, Caucasus	Early Permian	unnamed	cf. *Dromopus* cf. *Dimetropus*	*Batrachichnus*	*Batrachichnus*	Lucas *et al.* (1999)
Russia, Cisurals	Wuchiapingian	Severodvinskian Gorinzont	*Batrachichnus*	*Batrachichnus*	*Batrachichnus*	(Tverdokhlebov *et al.* 1997)
Russia, northern	Wuchiapingian	Severodvinskian Gorinzont	pareiasaur or dicynodont tracks	*Brontopodus*	*Pachypes*	(Gubin *et al.* 2001)
Karoo basin	Guadalupian-Lopingian	Beaufort Group	pareiasaurs, dicynodonts	*Brontopodus*	*Pachypes*	(Seeley 1904; Smith 1993)
Carapacha basin, Argentina	Artinskian	Carapacha Formation	*Batrachichnus*, *Hyloidichnus* cf. *Varanopus*	*Batrachichnus*	*Batrachichnus* (?*Amphisauropus*)	cf. *Amphisauropus* of Melchor & Serjeant (2004) is not diagnostic and could represent *Limnopus*. ?*Amphisauropus* sub-ichnofacies is based on presence of cf. *Varanopus*. (Melchor & Sarjeant 2004)
Carapacha basin, Argentina	Artinskian	Carapacha Formation	*Chararichnos* ?*Undichna*	*Characichichnos*	*Serpentichnus*	(Melchor & Sarjeant 2004)
Eastern Permian basin, Argentina	Guadalupian	Yacimiento Los Reyunas Formation	*Chelichnus*	*Chelichnus*	*Chelichnus*	(Melchor 2001)
Paganzo basin, Argentina	Early Permian	Patquía Formation	*cf. Dromopus cf. Gilmoreichnus*	?*Batrachichnus*	*Batrachichnus*?	No published detailed description (Caselli & Arcucci 1999)
Paraná basin, Brazil	?Guadalupian	Rio do Rastro Formation	Swimming trace	*Characichichnos*	? *Serpentichnus*	Too small an ichnofauna to be definitive about ichnocoenosis (Leonardi 1987)
Khenifra basin, Morocco	Kungurian	Unit B	cf. *Batrachichnus* *Dromopus Limnopus*	*Batrachichnus*	*Batrachichnus* (?*Dimetropus*)	(Hmich *et al.* 2006)

Table 3. *Continued*

Location	Age	Stratigraphic unit	Selected ichnotaxa	Ichnofacies	Ichnocoenosis Sub-ichnocoenosis	Comments (and principal references)
Tiddas basin, Morocco	Kungurian	'Upper Formation'	*Hyloidichnus*	*Batrachichnus* (?*Dimetropus*)	*Batrachichnus*	(Hmich *et al.* 2006)
Argana basin, Morocco	?middle Wuchiapingian	Ikakern Formation (Tourbihine Member)	*Synaptichnium* *Rhynchosauroides*	*Pachypes*	*Pachypes*	Hmich *et al.* (2006) consider the Tourbihine Member to be Late Permian (middle Wuchiapingian) based on undescribed pareiasaur remains, but the presence of *Synaptichnium* suggests that this unit may be partly Early Triassic in age. (Hmich *et al.* 2006)

Ages based on Lucas & Hunt (2006) or original sources.

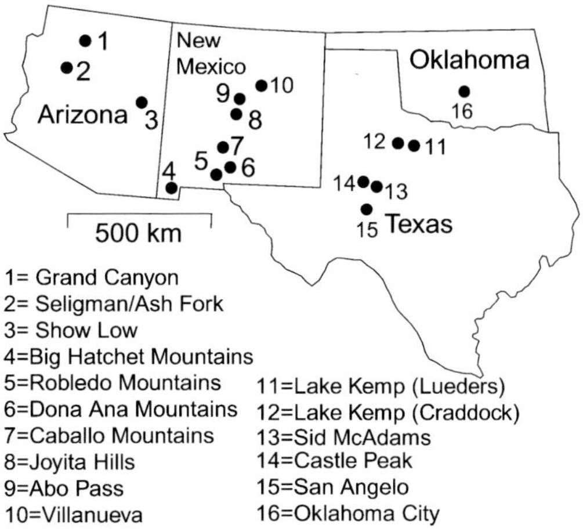

Fig. 2. Distribution of principal footprint localities in the western United States.

There are three areas in Colorado that yield tracks, in the San Juan Mountains (Cutler Group), Front Range (Lyons Sandstone) and the central part of the state (Maroon Formation) (Baird 1965; Lockley & Hunt 1995; Voigt *et al.* 2005). Localities in Utah are restricted to the southeastern portion of the state and generally yield only small ichnoassemblages (Loope 1984; Lockley & Madsen 1993; Sumida *et al.* 1999).

Europe

The European Permian tetrapod footprint record comes principally from three countries (Germany, France and Italy), although Lower Permian tracks are also known from the United Kingdom, Spain, Poland, and the Czech Republic (Table 3) (e.g. Haubold 1984; Haubold & Lucas 2001*a*; Cassinis & Santi 2005; Voigt 2005; Gand & Durand 2006).

The German record is from a series of Rotliegend basins and spans the late Carboniferous through the Wuchiapingian, although most tracks are Early Permian in age. The German record has a long history of study and was long considered the standard for the Early Permian. However, the German ichnofaunas, while numerous, are hampered by relatively poor age control and, with notable exceptions (e.g. Thuringian Forest Basin), small sample sizes. A similar record occurs in France, although these faunas include more ichnofaunas of Guadalupian and Wuchiapingian age (Lucas & Hunt 2006). Smaller Early Permian ichnofaunas occur in Poland, the Czech Republic and Spain (Table 3).

The Italian record includes both Early and Late Permian ichnofaunas, with the latter being the most significant in the world. The Upper Permian 'ichnoassociation' is best known from the Val Gardena and Bellerophon formations in the Bletterbach Gorge section in northern Italy (e.g. Conti *et al.* 1977; Cassinis & Santi 2005). The ichnogenera *Pachypes, Rhynchosauroides* and *Dicynodontipus* are characteristic.

Small assemblages of Permian tetrapod tracks have been reported from three areas in Russia, but there is great potential for future work, particularly in the Late Permian (Tverdokhlebov *et al.* 1997; Lucas *et al.* 1999; Gubin *et al.* 2001).

South America

The majority of Permian tracks from South America are derived from Argentina and have been described in recent years by Melchor and associates (e.g. Melchor & Poiré 1992; Melchor 1997; 2001; Melchor & Sarjeant 2004). The Argentinian track record is significant because it contains ichnofaunas from both aeolian and red-bed environments that are similar to those from Laurasia, indicating the global cosmopolitan nature of Early Permian ichnofaunas (e.g. Hunt & Lucas 1988*a*). There is a single locality of a swimming trace from Brazil (Leonardi 1987).

Africa

There is a substantial record of tetrapod footprints in Middle–Upper Permian strata in the Karoo Basin of South Africa (e.g. Seeley 1904; Smith 1993). These are primarily tracks of pareiasaurs and dicynodonts, but they have not been described in any detail. The demonstrated abundance of tracks, their age and fine preservation (e.g. Smith 1993) indicates that they need more in-depth study and that they will make an interesting comparison to the Upper Permian tracks from northern Italy. Recently, several localities of Early and Late Permian tracks have been reported from Morocco (Hmich *et al.* 2006).

Permian tetrapod ichnofacies and ichnocoenoses

Batrachichnus *ichnofacies*

Hunt & Lucas (2006) redefined the *Batrachichnus* ichnofacies to refer to medium-diversity ichnofaunas in which the majority of tracks are of quadrupedal carnivores with a moderate–high diversity (4–8 ichnogenera). This ichnofacies represents tidal flat–fluvial plain facies from the Devonian to the Middle Triassic.

The *Batrachichnus* ichnofacies encompasses one ichnocoenosis in the Permian, the *Batrachichnus* ichnocoenosis, which ranges from the Early Carboniferous to Early Permian, and is separable into sub-ichnocoenoses (Hunt & Lucas 2005*b*, 2006). The Early Permian of New Mexico has been pivotal in the development of concepts of tetrapod ichnofacies (Hunt & Lucas 2003, 2004, 2005*b*; Hunt *et al.* 2005*c*, *d*; Lucas *et al.* 2004; Lucas 2005). New Mexico provides an unrivalled spectrum of depositional environments (with tetrapod ichnofaunas) in the Early Permian, ranging from tidal flats to alluvial fans. It is thus possible to reconstruct an ecological transect from south to north, from tidal flat through distal alluvial fan (Fig. 3). This transect has been discussed in some detail over the last decade (Hunt *et al.* 1995*d*, 2005*d*; Hunt & Lucas 2003, 2004, 2005*b*; Lucas *et al.* 2004; Lucas 2005). The ichnofaunas from these ecosystems encompass the common Early Permian ichnotaxa *Batrachichnus, Limnopus, Amphisauropus, Dromopus, Dimetropus* and *Hyloidichnus.*

Hunt & Lucas (Hunt *et al.* 1995*d*, 2005*d*; Hunt & Lucas 2003, 2004, 2005*b*, 2006; Lucas *et al.* 2004; Lucas 2005) utilized the New Mexico record to distinguish three subdivisions of the *Batrachichnus* ichnocoenosis. There are existing terms for ichnological units smaller in scale than an ichnocoenosis. Bromley (e.g. 1990) has introduced two terms: 'association' and 'ichnoguild'. The tem 'ichnoguild' is not appropriate, as it is based on tiering and trophic structure (Bromley 1990). 'Association' could be used in this context, but we prefer the term 'sub-ichnocoenosis' to emphasize that these entities are intergradational sub-units of the ichnocoenosis. Thus we follow Hunt & Lucas (2006) in utilizing the term sub-ichnocoenosis, and we recognize three such units within the *Batrachichnus* ichnocoenosis (Fig. 3):

(1) *Ichniotherium* sub-ichnocoenosis – inland/distal alluvial fan settings characterized by common *Ichniotherium* and a paucity of *Dimetropus;*
(2) *Amphisauropus* sub-ichnocoenosis – alluvial plain settings characterized by the presence of *Amphisauropus* and rare *Ichniotherium* (Hunt *et al.* 2005*e*);
(3) *Dimetropus* sub-ichnocoenosis – coastal/tidal flat settings characterized by the relative abundance of *Batrachichnus* and *Dimetropus*.

Fig. 3. Palaeogeographic map of New Mexico during the Early Permian and north–south transect of Early Permian red beds in New Mexico showing distribution of possible tetrapod ichnofacies (after Hunt *et al.* 2005*d*; Lucas 2005).

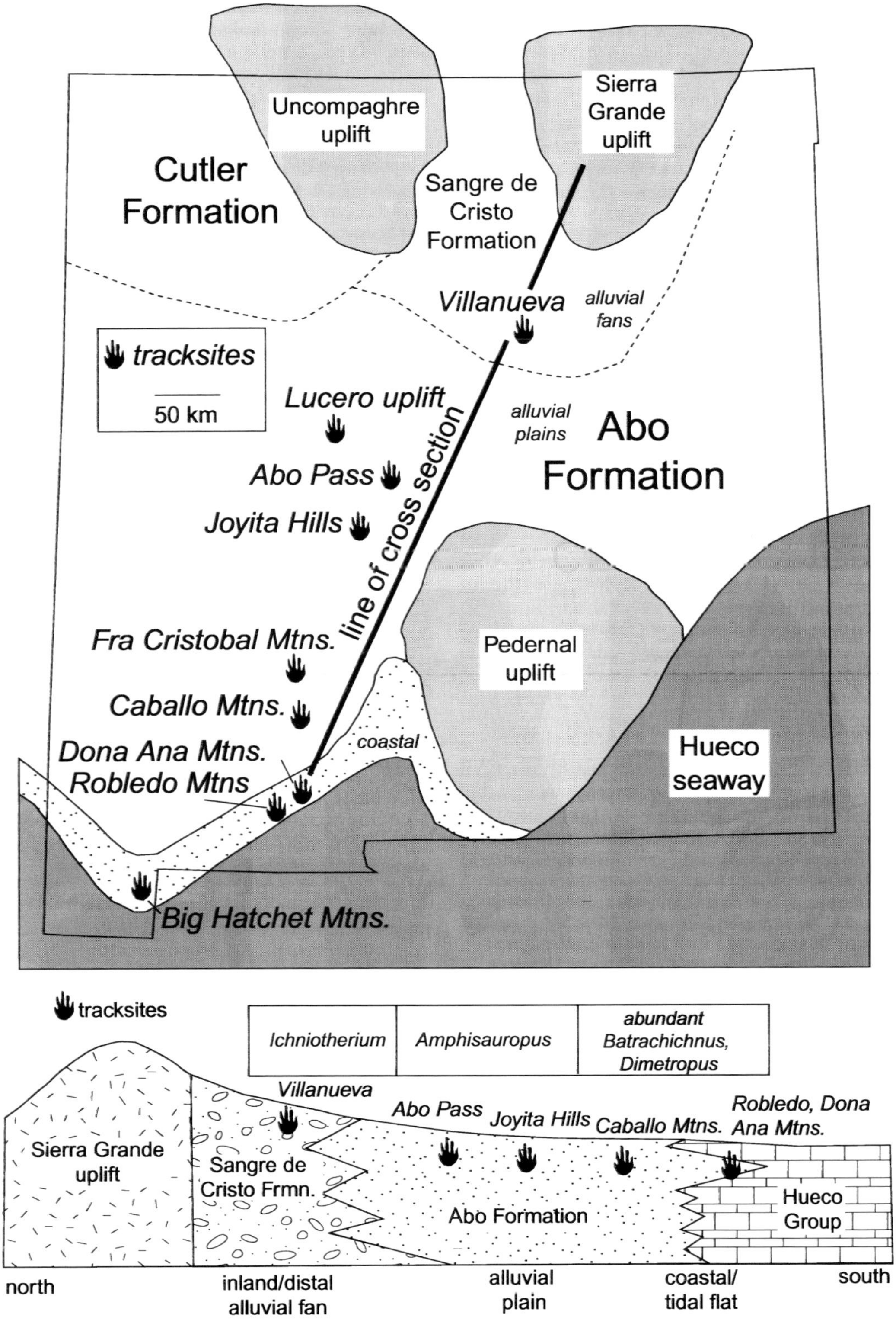

Uncompaghre uplift
Sierra Grande uplift
Cutler Formation
Sangre de Cristo Formation
Villanueva
alluvial fans
tracksites
50 km
Lucero uplift
alluvial plains
Abo Formation
Abo Pass
Joyita Hills
line of cross section
Fra Cristobal Mtns.
Pedernal uplift
Caballo Mtns.
coastal
Hueco seaway
Dona Ana Mtns.
Robledo Mtns
Big Hatchet Mtns.
tracksites
Ichniotherium
Amphisauropus
abundant Batrachichnus, Dimetropus
Villanueva
Abo Pass
Joyita Hills
Caballo Mtns.
Robledo, Dona Ana Mtns.
Sierra Grande uplift
Sangre de Cristo Frmn.
Abo Formation
Hueco Group
north
inland/distal alluvial fan
alluvial plain
coastal/ tidal flat
south

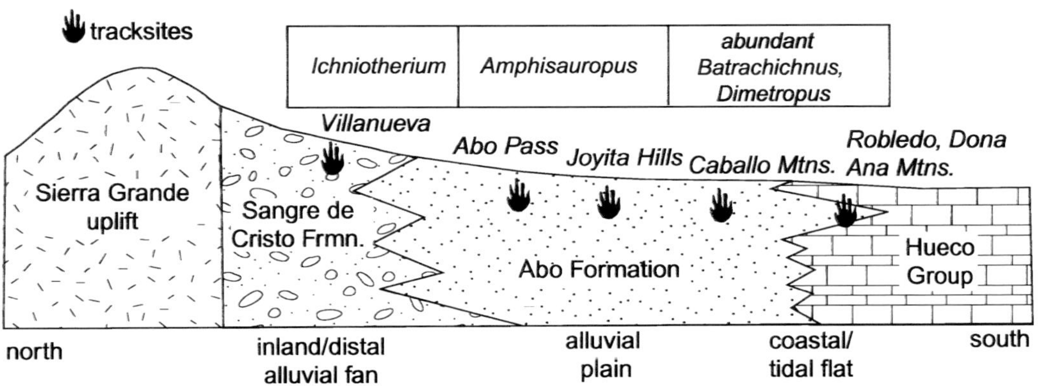

The New Mexico record also has potential for the study of tetrapod ethoichnofacies (Hunt *et al.* 2005*b*).

This trichotomy has utility in the southwestern United States, and the three sub-ichnocoenoses can be plotted through Arizona and Colorado (Fig. 4). *Ichniotherium* is only common at two tracksites in the *Batrachichnus* ichnocoenosis, both of which conform to the *Ichniotherium* sub-ichnocoenosis model in being far from the shoreline and closer to mountain fronts. These tracksites are the Villanueva tracksite in the Sangre de Cristo Formation, northern New Mexico (Hunt *et al.* 1995*e*) and the Hermit Formation tracksite in the Hermit Formation in Grand Canyon National Park, Arizona (Hunt & Santucci 1998; Hunt *et al.* 2005*d*). Further inland, *Ichniotherium* is the dominant tracktype, and ichnofaunas of this type are classified within the *Brontopodus* ichnofacies because they have a preponderance of tracks of herbivores (e.g. Maroon Bells Wilderness Area tracksite in the Maroon Formation, Colorado: Voigt *et al.* 2005).

These subdivisions of the *Batrachichnus* ichnocoenosis provide a worldwide basis for discriminating environments in other areas in Lower Permian strata (Tables 2–3). Many of the Early Permian ichnofaunas of Europe are from intermontane basins and, predictably, many represent the *Ichniotherium* or *Amphisauropus* sub-ichnocoenoses. *Tambachichnium* is potentially of use in defining or sub-dividing the *Ichniotherium* sub-ichnocoenosis, although there is need for a careful assessment of its relationship to *Varanopus* (cf. Voigt *et al.* 2005, figs 4 & 5).

There is potential for the recognition of other sub-ichnocoenoses. For example, it is possible that there is an *Erpetopus* sub-ichnocoenosis that could be discriminated for playa environments based on the Castle Peak locality in Texas (Haubold & Lucas 2001*b*, 2003). More study is needed of the distribution of *Erpetopus* and of its probable synonym *Camunipes*.

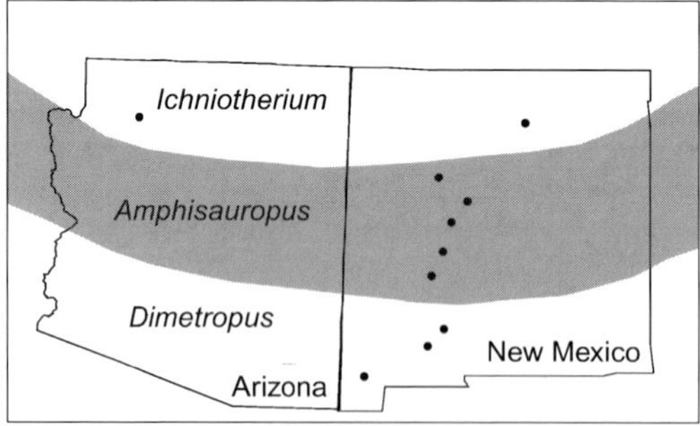

Fig. 4. Geographic distribution of sub-ichnocoenoses of the *Batrachichnus* ichnocoenosis (Hunt *et al.* 2005*d*).

Another example of a possibility for discriminating sub-ichnocoenoses lies with the ichnofaunas of southern New Mexico (Hunt *et al.* 2005*d*). The Robledo Mountains Formation of the Hueco Group represents a complex intercalation of siliclastic tidal flat (red beds) and shallow marine shelf (limestones and calcareous shales) deposits (Lucas *et al.* 1995*a*, *b*). The more productive localities are in the Robledo Mountains (e.g. Haubold *et al.* 1995*a*) and occur in fine sandstones, siltstones and mudstones that formed on tidal flats (Lucas *et al.* 1995*b*). The Doña Ana Mountains are located a few tens of kilometres to the northeast of the Robledo Mountains, and they also yield significant ichnofaunas from the Robledo Mountains Formation (Lucas *et al.* 1995*b*). However, the Robledo Mountains Formation in the Doña Ana Mountains is lithologically very different from that preserved in the Robledo Mountains. In the Doña Anas, the Robledo Mountains Formation:

(1) has a prevalence of reduced colours (green);
(2) contains large conglomeratic channels;
(3) contains carbonized plant material;
(4) has a higher mean grain size;
(5) has abundant ripple lamination;
(6) yields some bone material (Lucas *et al.* 1995*b*; Hunt *et al.* 1995*b*, 2005*d*). Pending more detailed sedimentological analylsis, our hypothesis is that the Robledo Mountains Formation in the Robledo Mountains represent a sediment-starved clastic, tidal-flat shoreline, and in the Doña Ana Mountains this unit represents a shoreline with appreciable clastic input and areas of reduction that we crudely refer to as deltaic. Thus, there is potential to examine ichnological changes along the Hueco shoreline between the Robledo Mountains and Doña Ana Mountains that may mirror changes in depositional environments. There are certainly taphonomic differences in tetrapod track preservation between the two areas that are in need of additional study, and this may lead to the recognition of new sub-ichnocoenoses (Hunt *et al.* 2005*d*).

Chelichnus *ichnofacies*

Hunt & Lucas (2006) proposed the *Chelichnus* archetypal ichnofacies for ichnofaunas that have a low diversity (less than four ichnogenera) of tetrapod tracks whose manual and pedal tracks are equant in shape, subequal in size and have short digit impressions. This ichnofacies is recurrent in dune faces in aeolian environments of Permian–Jurassic age, although other tetrapod ichnofacies can be present in aeolian environments. The *Chelichnus* ichnofacies encompasses one Permian ichnocoenosis (originally named as an ichnofacies): the *Chelichnus* (= *Laoporus*) ichnocoenosis (Lockley *et al.* 1994; Hunt & Lucas 2003, 2004, 2005*b*, 2006).

The *Chelichnus* ichnocoenosis is of Early to Late Permian age (Lucas & Hunt 2006). The largest sample sizes of this ichnocoenosis are in the southwestern United States (e.g. Gilmore 1926, 1927, 1928; Lockley *et al.* 1995). These ichnofaunas include the Cedar Mesa Sandstone in Utah, the Coconino and DeChelly sandstones in Arizona and the Lyons Sandstone in Colorado. These ichnofaunas range in age from Wolfcampian (Cedar Mesa) through early Leonardian (DeChelly) to late Leonardian (Coconino) (Lucas & Hunt 2006).

The longest-studied ichnofaunas of the *Chelichnus* ichnocoenosis are from the Corncockle, Lochabriggs and Hopeman sandstones of Scotland (Wuchapingian) and the Cornberger Sandstein of Germany (Capitanian–Wuchapingian) (e.g. Grierson 1828; Haubold 1984, 1996, 2000; McKeever & Haubold 1996; Haubold & Lucas 2001*a*). The most recently discovered ichnoassemblage of this ichnocoenosis is from the Guadalupian (Wordian?) Yacimiento Los Reyunas Formation of Argentina (Melchor 2001; Lucas & Hunt 2006).

The *Chelichnus* ichnocoenosis has a very low diversity and, with a few exceptions, it is monogeneric. McKeever & Haubold (1996) reduced the panoply of ichnotaxonomic names that had been proposed for Permian aeolian tracks to three species of *Chelichnus* (*C. gigas*, *C. duncani* and *C. bucklandi*). This scheme has received broad consensus and was rapidly applied in Europe, the United States and Argentina (Hunt & Lucas 1988*a*; Melchor 1997; Hunt & Santucci 1998). The three species of *Chelichnus*, as redefined by McKeever & Haubold (1996), are solely based on size, which is not an ideal criterion, but this ichnotaxonomy is of operational utility.

There are very few examples of morphotypes in this ichnocoenosis other than *Chelichnus*. This is especially notably given the very large sample sizes of some ichnofaunas (e.g. Lockley *et al.* 1995, fig. 5). Examples of other morphotypes include a lacertoid track from the DeChelly Sandstone (Haubold *et al.* 1995*b*) and an unidentified form from the Coconino Sandstone (Hunt & Santucci 2001), both from Arizona.

Chelichnus is conventionally considered to represent a caseid (e.g. Haubold 1984), but the trackmaker is not clearly understood. Ironically, the aeolianites of the *Chelichnus* ichnocoenosis preserve abundant small arthropod trails (e.g. Hunt & Lucas 2005*a*), but the tetrapod tracks do not preserve much detail. Tracks from the *Batrachichnus* ichnocoenosis have been identified as *Chelichnus*, notably from Provence, France (e.g. Gand & Durand 2006). However, it is not certain that this attribution is correct. It is interesting that Late Triassic to Early Jurassic aeolianites in North and South America preserve tracks with morphologies broadly similar to *Chelichnus* (Hunt & Lucas 2006). This morphology may represent an adaptation to locomotion on loose sand.

The DeChelly Sandstone is a classic aeolianite in Arizona and northwestern New Mexico that yields extensive tetrapod ichnofaunas in Arizona (Lockley *et al.* 1995; Morales & Haubold 1995; Hunt *et al.* 2005*d*, *f*). In the Lucero uplift of central New Mexico, the DeChelly (formerly Meseta Blanca Member of Yeso Formation: Lucas *et al.* 2005*c*) includes intercalated red beds in its lower portion. At Carrizo Arroyo in the Lucero uplift, these red beds yield tetrapod tracks that Lucas *et al.* (2005*c*) assigned to *Limnopus*, *Amphisauropus* and *Dimetropus*. This ichnofauna clearly pertains to the red-bed *Batrachichnus* ichnofacies (Hunt & Lucas 2005*b*). Thus, the DeChelly presents unique example of a stratigraphical unit of Palaeozoic age that grades laterally from aeolian to fluvial environments *and* yields ichnofaunas pertaining to the globally pervasive *Chelichnus* and *Batrachichnus* ichnocoenoses. This demonstrates that these two ichnocoenoses are contemporaneous and track changes in environments of deposition. There is a need for further study of the DeChelly to examine lateral changes in both ichnofaunas and ichnocoenoses.

Brontopodus *ichnofacies*

Hunt & Lucas (2006) proposed the *Brontopodus* ichnofacies for medium-diversity ichnofaunas in which the majority of tracks are of terrestrial herbivores, with a small quantity (generally >10%) of terrestrial carnivore tracks. This ichnofacies includes coastal plain-shoreline environments.

Intermontane ichnofaunas of the Early Permian are dominated by the tracks of diadectomorphs (*Ichniotherium*). We assign such ichnofaunas to the *Ichniotherium* ichnocoenosis. This ichnocoenosis is present in the Maroon Bells Wilderness Area tracksite in the Maroon

Formation, Colorado (Voigt *et al.* 2005) and the Tambach Sandstone and Goldlauter and Oberhof formations of the Thuringian Forest, Germany (Voigt 2005). In the southwestern United States, *Ichniotherium* becomes increasingly abundant in more inland ichnofaunas. In the *Batrachichnus* ichocoenosis, *Ichniotherium* is rare in the *Amphisauropus* sub-ichnocoenosis and more common in the *Ichniotherium* sub-ichnocoenosis. The prevalence of *Ichniotherium* in the more inland ichnocoenoses of the *Brotopodus* ichnofacies (*Ichnotherium* ichnocoenosis) continues this trend.

Middle to Late Permian ichnofaunas from water-laid facies are dominated by the tracks of dicynodonts and paraeiasaurs. We name this ichnocoenosis the *Pachypes* ichnocoenosis, with the most characteristic ichnofauna being that of the Val Gardena Formation in northern Italy (Conti *et al.* 1977; Cassinis & Santi 2005). The Early Triassic is characterized by the *Chirotherium* ichnocoenosis of the *Batrachichnus* ichnofacies (Hunt & Lucas 2006). The *Brontopodus* ichnofacies does not become dominant again until the Late Jurassic (Hunt & Lucas 2006).

Characichichnos *ichnofacies*

Hunt & Lucas (2006) named the *Characichichnos* ichnofacies for medium-diversity ichnofaunas in which the majority of tracks are swimming traces (parallel scratch marks) and fish swimming trails (*Undichna*). This ichnofacies represents shallow lacustrine (and tidal) environments. Several Permian ichnofaunas pertain to this ichnofacies (e.g. portions of ichnofaunas from the Carapacha Formation of Argentina: Melchor & Sarjeant 2004; portions of the ichnofaunas from the Robledo Mountains Formation of New Mexico, USA: Braddy *et al.* 2003).

We name the *Serpentichnus* ichnocoenosis for these ichnofaunas, which are characterized by the swimming traces of small temnospondyls (*Serpentichnus, Batrachichnus*) (e.g. Braddy *et al.* 2003). The *Serpentichnus* ichnocoenosis is far less common than any other ichnocoenosis in the Permian. This abundance partly represents the true rarity of subaqueous vertebrate traces. However, it may also reflect that fact that swimming traces are often overlooked by vertebrate ichnologists because they do not provide sufficient information about trackmakers (biotaxa). In addition, tetrapod ichnofaunas from a single stratigraphical unit are usually considered as a whole, whereas they may represent different environments and ichnofacies.

Discussion

The history of the study of tetrapod tracks is nearly as old as that of vertebrate fossils. However, throughout the 19th and much of the 20th centuries, few vertebrate palaeontologists devoted the majority of their careers to palaeoichnology, with a few notable exceptions (e.g. Hitchcock, Ellenberger, Leonardi, Haubold). The First International Symposium of Dinosaur Tracks and Traces at the New Mexico Museum of Natural History in 1986 (Gillette & Lockley 1989) served as a catalyst for a renaissance of tetrapod ichnology. In the subsequent decade, several books were devoted to vertebrate ichnology (e.g. Gillette & Lockley 1989; Thulborn 1990; Lockley 1991), and there were major advances in many areas, including ichnotaxonomy (e.g. Haubold *et al.* 1995*a*; Haubold 1996). In this time frame, Lockley *et al.* (1994) provided the first overview of tetrapod ichnofacies.

Hunt & Lucas (2006) erected an over-arching scheme to encompass and classify tetrapod ichnofacies through the Phanerozoic. This structure, which is applied in this paper, has utilities on several levels. It facilitates the discrimination of pervasive ichnocoenoses and ichnofacies and provides information on broad patterns of faunal evolution, environmental changes and the distribution of facies. On a medium scale, the recognition of distinct ichnofacies and ichnocoenoses within sedimentary basins provides information on local palaeoenvironmental successions. The recognition of distinct ichnocoenoses, and even ichnofacies, within a single stratigraphical unit provides the potential to aid in fine-scale palaeoenvironmental analysis and to dispel the pervasive, but usually unstated notion, that all tetrapod tracks in a formation represent a single ichno-assemblage and a single environment.

Permian tetrapod tracks have been pivotal in the development of tetrapod ichnofacies. Baird's (1965) insightful observation on Permian ichnofacies sparked interest in this topic, and the Wolfcampian ichnofaunas of New Mexico provided a basis for both discrimination of small-scale (sub-ichnocoenoses) and large-scale (archetypal ichnofacies) models. The synthetic study of tetrapod ichnofacies is still relatively new, but it provides great potential in future analyses of diverse subjects ranging from environmental interpretation to biog|eography.

We thank S. Voigt, L. Buatois and an anonymous reviewer for helpful comments on the manuscript.

References

BAIRD, D. 1965. Footprints from the Cutler Formation. *In*: LEWIS, G. E., VAUGHN, P. P. & BAIRD, D. (eds) *Early Permian Vertebrates from the Cutler Formation of the Placerville area, Colorado, with a section on footprints from the Cutler Formation.* U. S. Geological Survey, Professional Paper, **503C**, 47–50.

BRADDY, S. J., MORRISSEY, L. B. & YATES, A. M. 2003. Amphibian swimming traces from the Lower Permian of southern New Mexico. *Palaeontology*, **46**, 671–683.

BROMLEY, R. G. 1990. *Trace Fossils: Biology and Taphonomy.* Unwin Hyman, London.

BROMLEY, R. G. 1996. *Trace Fossils: Biology and Taphonomy.* Unwin Hyman, London [2nd edition].

CALDER, J. R., BAIRD, D. & URDANG, E. B. 2004. On the discovery of tetrapod trackways from Permo-Carboniferous redbeds of Prince Edward Island and their biostratigraphic significance. *Atlantic Geology*, **40**, 217–226.

CASELLI, A. & ARCUCCI, A. B. 1999. Huellas de tetrápodos en eolianitas de la Formación Patquía (Pérmico), La Rioja, Argentina. *1st Simposio Argentino del Paleozoico Superior, Abstracts*, 19–20.

CASSINIS, G. & SANTI, G. 2005. Permian tetrapod footprint assemblages from southern Europe and their stratigraphic implications. *In*: LUCAS S. G. & ZEIGLER, K. E. (eds) *The Nonmarine Permian.* New Mexico Museum of Natural History and Science Bulletin, **30**, 26–38.

CONTI, M. A., LEONARDI, G., MARIOTTI, N. & NICOSIA, U. 1977. *Tetrapod Footprints of the 'Val Gardena Sandstone' (north Italy). Their Paleontological, Stratigraphic and Paleoenvironmental Meaning.* Palaeontographica Italica, New Series, **40**, 91.

COTTON, W. D., HUNT, A. P. & COTTON, J. E. 1995. Paleozoic vertebrate tracksites in eastern North America. *In*: LUCAS, S. G. & HECKERT, A. B. (eds) *Early Permian Footprints and Facies.* New Mexico Museum of Natural History and Science, Bulletin, **6**, 189–211.

DVITASHVILI, L. S. 1945. Tsenozy zhivykh organizmov I organicheskikh ostatkov [Cenoeses of living organisms and organic remains]. *Akademiya Nauk Gruzin SSR*, **6**, 527–534. [In Russian]

FREY, R. W. & PEMBERTON, S. G. 1987. The *Psilonichnus* ichnofacies, and its relationship to adjacent marine and nonmarine ichnocoenoses along the Georgia coast. *Bulletin of Canadian Petroleum Geology*, **35**, 333–357.

GAND, G. & DURAND, M. 2006. Tetrapod footprint ichnoassociations from French Permian basins: comparisons with other Euramerican ichnofaunas. *In*: LUCAS, S. G., SCHNEIDER, J. W. & CASSINIS, G. (eds) *Non-Marine Permian Biostratigraphy and Biochronology.* Geological Society, London, Special Publications, **265**, 157–177.

GILLETTE, D. D. & LOCKLEY, M. G. (eds) 1989. *Dinosaur Tracks and Traces.* Cambridge University Press, Cambridge.

GILMORE, C. W. 1926. *Fossil Footprints from the Grand Canyon.* Smithsonian Miscellaneous Collections, **77**(9).

GILMORE, C. W. 1927. *Fossil Footprints from the Grand Canyon: Second Contribution.* Smithsonian Miscellaneous Collections, **80**(3).

GILMORE, C. W. 1928. *Fossil Footprints from the Grand Canyon: Third Contribution.* Smithsonian Miscellaneous Collections, **80**(8).

GRIERSON, J. 1828. On footsteps before the flood, in a specimen of red sandstone. *Edinburgh Journal of Science*, **8**, 130–134.

GUBIN, Y. M., BULANOV, V. V., GOLUBEV, V. K. & PETUKHOV, S. V. 2001. The first traces of big Permian reptiles in eastern Europe. *Journal of Vertebrate Paleontology*, **21**(3, supplement), 57A.

HAUBOLD, H. 1984. *Saurierfährten.* A. Ziemsen Verlag, Wittenberg Lutherstadt.

HAUBOLD, H. 1996. Ichnotaxonomie und Klassifikation von Tetrapodenfährten aus dem Perm. *Hallesches Jahrbuch für Geowissenschaften, Reihe B*, **18**, 23–88.

HAUBOLD, H. 2000. Tetrapodenfährten aus dem Perm–Kenntnisstand und Progress 2000. *Hallesches Jahrbuch für Geowissenschaften, Reihe B*, **22**, 1–16.

HAUBOLD, H. & LUCAS, S. G. 2001*a*. Early Permian tetrapod tracks – preservation, taxonomy, and Euramerican distribution. *In*: CASSINIS, G. (ed.) *Permian Continental Deposits of Europe. Regional Reports and Correlations. Natura Bresciana, Monografia*, **25**, 347–354.

HAUBOLD, H. & LUCAS, S. G. 2001*b*. Die Tetrapodenfährten der Choza Formation (Texas) und das Artinsk-Alter der Redbed-Ichnofaunen des Unteren Perm. *Hallesches Jahrbuch für Geowissenschaften, Reihe B*, **23**, 79–108.

HAUBOLD, H. & LUCAS, S. G. 2003. Tetrapod footprints of the Lower Permian Choza Formation at Castle Peak, Texas. *Paläontologische Zeitschrift*, **77**, 247–261.

HAUBOLD, H. & SARJEANT, W. A. S. 1973. Tetrapodenfährten aus den Keele und Enville Groups (Permokarbon: Stefan und Autun) von Shropshire und South Staffordshire, Grossbritannien. *Zeitschrift fur Geologische Wissenschaft*, **8**, 895–933.

HAUBOLD, H. & STAPF, H. 1998. The Early Permian tetrapod track assemblage of Nierstein, Standenbuhl Beds, Rotliegend, Saar-Nahe Basin, SW-Germany. *Hallesches Jahrbuch für Geowissenschaften, Reihe B*, **20**, 17–32.

HAUBOLD, H. HUNT, A. P., LUCAS, S. G. & LOCKLEY, M. G. 1995*a*. Wolfcampian (Early Permian) tracks from Arizona and New Mexico. *In*: LUCAS, S. G. & HECKERT, A. B. (eds) *Early Permian Footprints and Facies.* New Mexico Museum of Natural History and Science Bulletin, **6**, 135–166.

HAUBOLD, H., LOCKLEY, M. G., HUNT, A. P. & LUCAS, S. G. 1995*b*. Lacertoid footprints from Permian dune sandstones, Cornberg and DeChelly sandstones: *In*: LUCAS, S. G. & HECKERT, A. B. (eds) *Early Permian Footprints and Facies.* New Mexico Museum of Natural History and Science Bulletin, **6**, 235–244.

HMICH, D., SCHNEIDER, J. W., SABER, H., VOIGT, S. & EL WARTITI, M. 2006. New continental Carboniferous and Permian faunas of Morocco: implications for biostratigraphy, palaeobiogeography and palaeoclimate. *In*: LUCAS, S. G., CASSINIS, G. & SCHNEIDER, J. W. (eds) *Non-Marine Permian Biostratigraphy and Biochronology*. Geological Society, London, Special Publications, **265**, 297–324.

HUNT, A. P. & LOCKLEY, M. G. 1997. Taphonomy of vertebrate tracks: direct deposits into the fossil record. *Journal of Vertebrate Paleontology*, **17**(3, supplement), 54A.

HUNT, A. P. & LUCAS, S. G. 1998*a*. Implications of the cosmopolitanism of Permian tetrapod ichnofaunas. *In*: LUCAS, S. G., ESTEP, J. W. & HOFFER, J. M. (eds) *Permian Stratigraphy and Paleontology of the Robledo Monutains, New Mexico. New Mexico Museum of Natural History and Science Bulletin*, **12**, 55–57.

HUNT, A. P. & LUCAS, S. G. 1998*b*. Vertebrate ichnofaunas of New Mexico and their bearing on Lower Permian vertebrate ichnofacies. *In*: LUCAS, S. G., ESTEP, J. W. & HOFFER, J. M. (eds) *Permian Stratigraphy and Paleontology of the Robledo Monutains, New Mexico. New Mexico Museum of Natural History and Science Bulletin*, **12**, 63–65.

HUNT, A. P. & LUCAS, S. G. 2003. Tetrapod ichnofacies in the Paleozoic. *In*: MARTIN, A. J. (ed.) *Workshop on Permo-Carboniferous Ichnology, Program and Abstracts*, Alabama Museum of Natural History, Alabama Geological Survey, 8–11.

HUNT, A. P. & LUCAS, S. G. 2004. Tetrapod ichnofacies in the Paleozoic. *32nd International Geological Congress 2004, Florence, Abstracts*, **1**, 600.

HUNT, A. P. & LUCAS, S. G. 2005*a*. Nonmarine Permian track faunas from Arizona, USA. *In*: LUCAS, S. G. & ZEIGLER, K. E. (eds) *The Nonmarine Permian*. New Mexico Museum of Natural History and Science Bulletin, **30**, 128–131.

HUNT, A. P. & LUCAS, S. G. 2005*b*. Tetrapod ichnofacies and their utility in the Paleozoic. *In*: BUTA, R. J., RINDSBERG, A. K. & KOPASKA-MERKEL, D. C. (eds) *Pennsylvanian Footprints in the Black Warrior Basin of Alabama*. Alabama Paleontological Society, Monographs, **1**, 113–119.

HUNT, A. P. & LUCAS, S. G. 2005*c*. The significance of Permian tracks to the development of tetrapod ichnology. *In*: LUCAS, S. G. & ZEIGLER, K. E. (eds) *The Nonmarine Permian*. New Mexico Museum of Natural History and Science Bulletin, **30**, 127.

HUNT, A. P. & LUCAS, S. G. 2006. Tetrapod ichnofacies: a new paradigm. *Ichnos*. (In press)

HUNT, A. P. & SANTUCCI, V. L. 1998. Taxonomy and ichnofacies of Permian tetrapod tracks from Grand Canyon National Park, Arizona. *In*: SANTUCCI, V. L. & MCCLELLAND, L. (eds) *National Park Service Paleontological Research*. National Park Service Geological Resources Division, Technical Report NPS/NRGRD/GRDTR-98/01, 94–96.

HUNT, A. P. & SANTUCCI, V. L. 2001. An unusual tetrapod track morphology from the Permian Coconino Sandstone, Grand Canyon National Park, Arizona. *In*: SANTUCCI, V. L., SANTUCCI, V. L. & MCCLELLAND, L. (eds) *Proceedings of the 6th Fossil Resource Conference*. National Park Service Geological Resources Division, Technical Report NPS/NRGRD/GRDTR-01/01, 44–47.

HUNT, A. P., LUCAS, S. G., COTTON, W., COTTON, J. & LOCKLEY, M. G. 1995*a*. Early Permian vertebrate tracks from the Abo Formation, Socorro County, central New Mexico: a preliminary report. *In*: LUCAS, S. G. & HECKERT, A. B. (eds) *Early Permian Footprints and Facies*. New Mexico Museum of Natural History Bulletin, **6**, 263–268.

HUNT, A. P., LUCAS, S. G., HAUBOLD, H. & LOCKLEY, M. G. 1995*b*. Early Permian (Late Wolfcampian) tetrapod tracks from the Robledo Mountains, south-central New Mexico. *In*: LUCAS, S. G. & HECKERT, A. B. (eds) *Early Permian Footprints and Facies*. New Mexico Museum of Natural History Bulletin, **6**, 167–180.

HUNT, A. P., LUCAS, S. G. & LOCKLEY, M. G. 1995*c*. Paleozoic tracksites of the western United States. *In*: LUCAS, S. G. & HECKERT, A. B. (eds) *Early Permian Footprints and Facies*. New Mexico Museum of Natural History and Science Bulletin, **6**, 213–217.

HUNT, A. P., LUCAS, S. G., LOCKLEY, M. G., HAUBOLD, H. & BRADDY, S. 1995*d*. Tetrapod ichnofacies in Early Permian red beds of the American Southwest. *In*: LUCAS, S. G. & HECKERT, A. B. (eds) *Early Permian Footprints and Facies*. New Mexico Museum of Natural History and Science Bulletin, **6**, 295–301.

HUNT, A. P., LUCAS, S. G., SANTUCCI, V. L. & ELLIOTT, D. K. 2005*a*. Permian vertebrates of Arizona. *In*: HECKERT, A. B. & LUCAS, S. G. (eds) *Vertebrate Paleontology in Arizona*. New Mexico Museum of Natural History and Science Bulletin, **29**, 10–15.

HUNT, A. P., LUCAS, S. G. & SPIELMANN, J. A. 2005*b*. Early Permian tetrapod ethoichnofacies in New Mexico. *In*: LUCAS, S. G., ZEIGLER, K. E. & SPIELMANN, J. A. (eds) *The Permian of Central New Mexico*. New Mexico Museum of Natural History and Science Bulletin, **31**, 52–55.

HUNT, A. P., LUCAS, S. G. & SPIELMANN, J. A. 2005*c*. Early Permian tetrapod tracksites in New Mexico. *In*: LUCAS, S. G., ZEIGLER, K. E. & SPIELMANN, J. A. (eds) *The Permian of Central New Mexico*. New Mexico Museum of Natural History and Science Bulletin, **31**, 46–48.

HUNT, A. P., LUCAS, S. G. & SPIELMANN, J. A. 2005*d*. Paleoenvironmental transects and tetrapod biotaxonichnofacies in New Mexico. *In*: LUCAS, S. G., ZEIGLER, K. E. & SPIELMANN, J. A. (eds) *The Permian of Central New Mexico*. New Mexico Museum of Natural History and Science Bulletin, **31**, 49–51.

HUNT, A. P., LUCAS, S. G. & SPIELMANN, J. A. 2005*e*. The Permian ichnogenus *Ichniotherium cottae* from central New Mexico. *In*: LUCAS, S. G., ZEIGLER, K. E. & SPIELMANN, J. A. (eds) *The Permian of Central New Mexico*. New Mexico Museum of Natural History and Science Bulletin, **31**, 56–58.

HUNT, A. P., SANTUCCI, V. L. & LUCAS, S. G. 2005*f*. Vertebrate trace fossils from Arizona with special reference to tracks preserved in National Park Service units and note son the Phanerozoic distribution of fossil footprints. *In*: HECKERT, A. B. & LUCAS, S. G. (eds) *Vertebrate Paleontology in Arizona*. New

Mexico Museum of Natural History and Science Bulletin, **29**, 159–167.

KEIGHLEY, D. G. & PICKERILL, R. K. 2003. Ichnocoenoses from the Carboniferous of eastern Canada and their implications for the recognition of ichnofacies in nonmarine strata. *Atlantic Geology*, **39**, 1–22.

LEONARDI, G. 1987. The first tetrapod footprint in the Permian of Brazil. *Anais do X Congresso Brasileiro de Paleontologia, Rio de Janeiro*, **1**, 333–335.

LOCKLEY, M. G. 1991. *Tracking Dinosaurs: A New Look at An Ancient World*. Cambridge University Press, Cambridge.

LOCKLEY, M. G. 2006. A tale of two ichnologies: the different goals and missions of invertebrate and vertebrate ichnotaxonomy and how they relate in ichnofacies analysis. *Ichnos* (In press).

LOCKLEY, M. G. & HUNT, A. P. 1995. *Dinosaur Tracks and Other Fossil Footprints of the Western United States*. Columbia University Press, New York.

LOCKLEY. M. G. & MADSEN, J. H. JR. 1993. Early Permian vertebrate trackways from the Cedar Mesa Sandstone of eastern Utah. *Ichnos*, **2**, 147–153.

LOCKLEY, M. G., & MEYER, C. 2000. *Dinosaur Tracks and Other Fossil Footprints of Europe*. Columbia University Press, New York.

LOCKLEY, M. G., HUNT, A. P. & MEYER, C. 1994. Vertebrate tracks and the ichnofacies concept: implications for paleoecology and palichnostratigraphy. *In*: DONOVAN, S. (ed.) *The Paleobiology of Trace Fossils*. John Wiley, London. 241–268.

LOCKLEY, M. G., HUNT, A. P., HAUBOLD, H. & LUCAS, S. G. 1995. Fossil footprints in the DeChelly Sandstone of Arizona: with paleoecological observations on the ichnology of dune facies. *In*: LUCAS, S. G. & HECKERT, A. B. (eds) *Early Permian Footprints and Facies*. New Mexico Museum of Natural History Bulletin, **6**, 225–233.

LOOPE, D. B. 1984. Eolian origin of upper Paleozoic sandstones, southeastern Utah. *Journal of Sedimentary Petrology*, **54**, 563–580.

LUCAS, S. G. 2005. Tetrapod ichnofacies and Ichnotaxonomy: quo vadis? *Ichnos*, **12**, 157–162.

LUCAS, S. G. & HECKERT, A. B. (eds) 1995. *Early Permian Footprints and Facies*. *In*: LUCAS, S. G. & HECKERT, A. B. (eds) *Early Permian Footprints and Facies*. New Mexico Museum of Natural History and Science Bulletin, **6**.

LUCAS, S. G. & HUNT, A. P. 1995. Stratigraphy and paleontology of the Lower Permian Earp Formation. Big Hatchet Mountains, Hidalgo County, New Mexico. *In*: LUCAS, S. G. & HECKERT, A. B. (eds) *Early Permian Footprints and Facies*. New Mexico Museum of Natural History Bulletin, **6**, 287–294.

LUCAS, S. G. & HUNT, A. P. 2005. Permian tetrapod tracks from Texas. *In*: LUCAS, S. G. & ZEIGLER, K. E. (eds) *The Nonmarine Permian*. New Mexico Museum of Natural History and Science Bulletin, **30**, 202–206.

LUCAS, S. G. & HUNT, A. P. 2006. Permian tetrapod footprint: biostratigraphy and biochronology. *In*: LUCAS, S. G., CASSINIS, G. & SCHNEIDER, J. W. (eds) *Non-Permian Biostratigraphy and Biochronology*. Geological Society, London, Special Publications, **265**, 179–200.

LUCAS, S. G. & SUNESON, N. 2002. Amphibian and reptile tracks from the Hennessey Formation (Leonardian, Permian), Oklahoma County, Oklahoma. *Oklahoma Geology Notes*, **62**, 56–62.

LUCAS, S. G., HUNT, A. P. & HECKERT, A. B. 1995*a*. Preliminary report on paleontology of the Abo Formation, McLeod Hills, Sierra County, New Mexico. *In*: LUCAS, S. G. & HECKERT, A. B. (eds) *Early Permian Footprints and Facies*. New Mexico Museum of Natural History and Science Bulletin, **6**, 279–285.

LUCAS, S. G., HUNT, A. P., HECKERT, A. B. & HAUBOLD, H. 1995*b*. Vertebrate paleontology of the Robledo Mountains Member of the Hueco Formation, Doña Ana Mountains, New Mexico. *In*: LUCAS, S. G. & HECKERT, A. B. (eds) *Early Permian Footprints and Facies*. New Mexico Museum of Natural History and Science Bulletin, **6**, 269–275.

LUCAS, S. G., LERNER, A. J. & HAUBOLD, H. 2001. First record of *Amphisauropus* and *Varanopus* in the Lower Permian Abo Formation, central New Mexico. *Hallesches Jahrbuch für Geowissenschaften, Reihe B*, **23**, 69–78.

LUCAS, S. G., LERNER, A. J. & HUNT, A. P. 2004. Permian tetrapod footprints from the Lucero uplift, central New Mexico, and Permian footprint biostratigraphy. *In*: LUCAS, S. G. & ZIEGLER, K. E. (eds) *Carboniferous–Permian Transition at Carrizo Arroyo, Central New Mexico*. New Mexico Museum of Natural History and Science Bulletin, **25**, 291–300.

LUCAS, S. G., LOZOVSKY, V. R. & SHISHKIN, M. A. 1999. Tetrapod footprints from Early Permian redbeds of the northern Caucasus, Russia. *Ichnos*, **6**, 277–281.

LUCAS, S. G., MINTER, N. J., SPIELMANN, J. A., HUNT, A. P. & BRADDY, S. J. 2005*a*. Early Permian ichnofossil assemblage from the Fra Cristobal Mountains, southern New Mexico. *In*: LUCAS, S. G., ZEIGLER, K. E. & SPIELMANN, J. A. (eds) *The Permian of Central New Mexico*. New Mexico Museum of Natural History and Science Bulletin, **31**, 140–150.

LUCAS, S. G., MINTER, N. J., SPIELMANN, J. A. & SMITH, J. A. 2005*b*. Early Permian ichnofossils from the northern Caballo Mountains, Sierra County, New Mexico. *In*: LUCAS, S. G., ZEIGLER, K. E. & SPIELMANN, J. A. (eds) *The Permian of Central New Mexico*. New Mexico Museum of Natural History and Science Bulletin, **31**, 151–162.

LUCAS, S. G., SMITH, J. A. & HUNT, A. P. 2005*c*. Tetrapod tracks from the Lower Permian Yeso Group, central New Mexico. *In*: LUCAS, S. G., ZEIGLER, K. E. & SPIELMANN, J. A. (eds) *The Permian of Central New Mexico*. New Mexico Museum of Natural History and Science Bulletin, **31**, 121–124.

MCILROY, D. 2004. Some ichnological concepts, methodologies, applications and frontiers. *In*: MCILROY, D. (ed.) *The Application of Ichnology to Palaeoenvironmental and Stratigraphic Analysis*. Geological Society, London, Special Publications, **228**, 3–27.

McKeever, P. M. & Haubold, H. 1996. Reclassification of vertebrate trackways from the Permian of Scotland and related forms from Arizona and Germany. *Journal of Paleontology*, **70**, 1011–1022.

Melchor, R. N. 1997. Permian tetrapod ichnofaunas from Argentina: further evidence of redbed shallow lacustrine and aeolian vertebrate ichnofacies. *Workshop on the Ichnofacies and Ichnotaxonomy of the Terrestrial Permian, Halle, Abstracts and Papers*, 59–60.

Melchor, R. N. 2001. Permian tetrapod footprints from Argentina. *Hallesches Jahrbuch für Geowissenschaften, Reihe B*, **23**, 35–43.

Melchor, R. N. & Poiré, D. 1992. Sedimentological and paleoecological implications of invertebrate and tetrapod ichnocoenosis from a Permian fluvial/lacustrine sequence: Carapacha Formation, La Pampa Province, Argentina. *Cuarta Reunión Argentina de Sedimentología, Actas*, **III**, 247–257.

Melchor, R. N. & Sarjeant, W. A. S. 2004. Small amphibian and reptile footprints from the Permian Carapacha basin, Argentina. *Ichnos*, **11**, 57–78.

Morales, M. & Haubold, H. 1995. Tetrapod tracks from the Lower Permian DeChelly Sandstone of Arizona: systematic description. *In*: Lucas, S. G. & Heckert, A. B. (eds) *Early Permian Footprints and Facies*. New Mexico Museum of Natural History and Science Bulletin, **6**, 251–261.

Mossman, D. J. & Place, C. H. 1989. Early Permian fossil vertebrate footprints and their stratigraphic setting in megacyclic sequence II red beds, Prim Poiunt, Prince Edward Island. *Canadian Journal of Earth Sciences*, **26**, 591–605.

Pemberton, S. G. & Gingras, M. K. 2003. The Reverend Henry Duncan (1774–1846) and the discovery of the first fossil footprints. *Ichnos*, **10**, 69–75.

Sarjeant, W. A. S. 1974. A history and bibliography of the study of fossil vertebrate footprints in the British Isles. *Palaeogeography, Palaeoclimatology, Palaeoecology*, **16**, 265–378.

Seeley, H. G. 1904. Footprints of small fossil reptiles from the Karroo rocks of Cape Colony. *Annals and Magazine of Natural History, Series 7*, **14**, 287–289.

Seilacher, A. 1964. Biogenic sedimentary structures. *In*: Imbrie, J. & Newell, N. (eds) *Approaches to Paleoecology*. Wiley, New York, 296–316.

Seilacher, A. 1967. Bathymetry of trace fossils. *Marine Geology*, **5**, 413–428.

Smith, R. M. H. 1993. Sedimentology and ichnology of floodplain paleosurfaces in the Beaufort Group (Late Permian), Karoo sequence, South Africa. *Palaios*, **8**, 339–357.

Sumida, S. S., Lombard, R. E., Berman, D. S. & Henrici, A. C. 1999. Late Paleozoic amniotes and their near relatives from Utah and northeastern Arizona, with comments on the Permian-Pennsylvanian boundary in Utah and northeast Arizona. *Utah Geological Survey, Miscellaneous Publications*, **99-1**, 31–43.

Thulborn, T. 1990. *Dinosaur Tracks*. Chapman Hall, London.

Toepelman, W. C. & Rodeck, H. G. 1936. Footprints in late Paleozoic red beds near Boulder, Colorado. *Journal of Paleontology*, **10**, 660–662.

Tverdokhlebov, V. P., Tverdokhlebova, G. I., Benton, M. J. & Storrs, G. W. 1997. First record of footprints of terrestrial vertebrates from the Upper Permian of the Cis-Urals, Russia. *Palaeontology*, **40**, 157–166.

Van Allen, H. E. K., Calder, J. H. & Hunt, A. P. 2005. The trackway record of a tetrapod community in a walchian conifer forest from the Permo-Carboniferous of Nova Scotia. *In*: Lucas, S. G. & Zeigler, K. E. (eds) *The Nonmarine Permian*. New Mexico Museum of Natural History and Science Bulletin, **30**, 322–332.

Voigt, S. 2005. Die Tetrapodenichnofauna des kontinentalen Oberkarbon und Perm im Thüringer Wald: Ichnotaxonomie, paläoökologie und Biostratigraphie. Cuvillier Verlag, Göttingen.

Voigt, S., Small, B. J. & Sanders, F. 2005. A diverse terrestrial ichnofauna from the Maroon Formation (Pennsylvanian–Permian), Colorado: biostratigraphic and paleoecological significance. *In*: Lucas, S. G. & Ziegler, K. E. (eds) *The Nonmarine Permian*. New Mexico Museum of Natural History, Bulletin, **30**, 342–351.

Tetrapod footprint ichno-associations from French Permian basins. Comparisons with other Euramerican ichnofaunas

GEORGES GAND[1] & MARC DURAND[2]

[1]*Biogéosciences UMR 5561, Centre des Sciences de la Terre, Université de Bourgogne, 6 Boulevard Gabriel, 21000 Dijon, France (e-mail: georges.gand@wanadoo.fr)*

[2]*47 rue de Lavaux, 54520 Laxou, France*

Abstract: In order to take into account the studies of the European and American (USA) collections carried out by one of the authors, and of the recent nomenclatural revisions from new footprint discoveries, which have occurred during the last decade, the authors present a critical review of the French Permian palichnofauna. The distribution of the ichnospecies in the stratigraphy of the Lodève Basin, taken as a reference, is outlined. The ichno-associations are then compared with those of other French (Provence), European (Italy, Germany) and USA basins. Based on the ages of different ichnofossiliferous formations, three successive ichnofaunal units can be distinguished in the Permian of Europe. The first developed in the Cisuralian (Asselian to Kungurian). The second is found in the south of France in Kazanian to Lower Tatarian strata, equivalent to the Roadian–Wordian. The third and youngest, dated as Lopingian, is only found in Italy, in the Bolzano Basin. Because of sedimentary gaps, limited observations, sometimes erroneous determinations, and ichnospecies with great vertical distribution, it currently appears that footprints have a low utility for biochronological resolution. Nevertheless, they allow us to discriminate three time intervals in the Permian, as is also the case for skeletal remains.

There are 20 Permian basins in France (Fig. 1), but only a few of them have yielded tetrapod vertebrate tracks. The first were collected in 1903 by Delage (1912) near Neffiès, south of Lodève. After a long period during which no research was conducted, new observations were made in the Lodève Basin by Ellenberger & Ellenberger (1959) and by Heyler & Lessertisseur (1962). The latter two workers published a review of their finds, including descriptions of what were accepted as '14 new genera and 16 new species' (Heyler & Lessertisseur 1963).

After 1963, research focused on other areas, and a second review of the Permian palichnofauna of southern France, based mainly on footprints from the Lodève Basin, was published briefly by Ellenberger (1983*a*–*c*, 1984). In his short notes, the author presented and named 'almost 130' new ichnotypes but without description. Thirty-nine ichnogenera, in addition to the seven described earlier by Heyler & Lessertisseur (1963), characterized for Ellenberger (1983*a*–*c*) 'the marked individuality of the Lodève province as regards palaeontology, palaeogeography and palaeoecology'.

From 1980, in parallel with these works, Gand undertook a revision of French Permian footprints that involved prospecting in all the French basins and visiting several European fossil collections or sites: the Prague Museum in the Czech Republic, the museums of Halle, Gotha, Nierstein and the University of Mainz in Germany, the museums of Oxford, Cambridge, Birmingham, Nottingham, Keyworth and Manchester in England and those of Dumfries, Edinburgh and Elgin in Scotland. Field studies were also conducted in Scotland, Saarland, Thuringia, and Val Gardena in the Italian Dolomites. Reviews were published upon completion of these studies (Gand 1987; Châteauneuf & Gand 1989) that recognized only 14 ichnogenera and 22 ichnospecies; among them are almost all the ichnogenotypes defined by Haubold (1970, 1971, 1973). From the new French palichnological results, stratigraphical correlations were suggested by using the continental stages 'Autunian' and 'Saxonian' (Gand 1987; Gand & Haubold 1988).

Since Gand's work was presented, the data have been supplemented by new ichnogenus and ichnospecies descriptions from the Provence (Demathieu *et al.* 1992; Gand *et al.* 1995) and Lodève basins (Gand *et al.* 2000). If we include only footprints that represent the most faithful images of autopods, defined morphologically and morphometrically as ichnopopulations, the Permian French palichnofauna includes 21 ichnospecies accomodated in 15 ichnogenera.

From: LUCAS, S. G., CASSINIS, G. & SCHNEIDER, J. W. (eds) 2006. *Non-Marine Permian Biostratigraphy and Biochronology*. Geological Society, London, Special Publications, **265**, 157–177.
0305-8719/06/$15.00

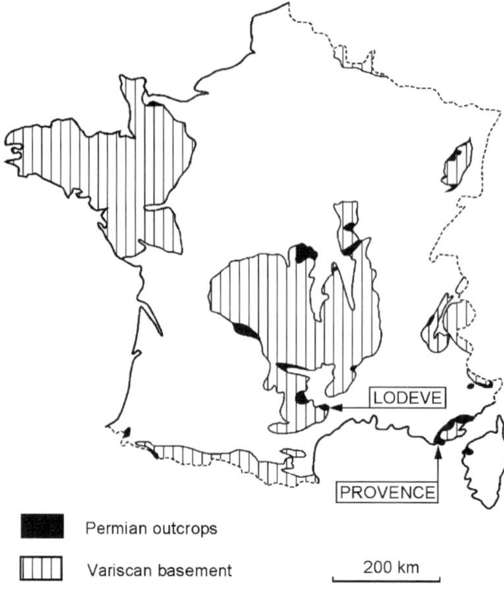

Fig. 1. Location of Permian basins in France.

This assessment is, of course, very distant from that of Haubold (2000), who reduced the southern French palichnologic content to four ichnogenera and eight ichnospecies. This is why we shall present the French footprints, discuss the nomenclatural changes of Haubold (1996, 1998*a*, 2000) and make comparisons between European and American ichnospecies. After that, we will discuss the stratigraphical aspects inferred from palichnological data.

The French palichnofauna

Footprints ascribable to temnospondyls

Branchiosauridea and/or *Micromelerpetondidea*

Batrachichnus salamandroides (Geinitz 1861) Haubold 1996 = *Anthichnium salamandroides* (Geinitz 1861) Haubold 1970; synonyms: *Nanipes minutus*, *Auxipes minor*, *Devipes caudatus*, *Crenipes abscurvus*, *Crenipes abrectus* Heyler & Lessertisseur 1963, and *Margennipes pansioti* Heyler 1984 (Figs 2a2,b3,c4&5,d4; 5a; 6a). These small prints of tetradactyl manus and pentadactyl pes were revised by Haubold (1970, 1971, 1973) and by Gand (1987, pp. 74). They are abundant with the same morphology in France, the Saar–Nahe (Germany) and the USA. They are attributed to small temnospondyls (Haubold 1970; Gand 1987).

While based on the morphology of *Batrachichnus* Woodworth 1900, and on its nomenclatural priority over *Anthichnium* Nopcsa 1923, Haubold used the binomials *B. delicatulus* for the American material (Haubold *et al.* 1995*a*) and *B. salamandroides* for the European traces (Haubold 1996). Because he inferred no significant differences between the two ichnospecies, *B. salamandroides* could only be used to indicate *Anthichnium* footprints.

Salichnium pectinatus Gand 1987 (Fig. 2c3) (= *Serripes pectinatus* Heyler & Lessertisseur 1963) and *Salichnium decessus* Gand 1987 (Fig. 2c1&2) (= *Acutipes decessus* and *Foliipes abscisus* Heyler & Lessertisseur 1963). These were allotted to microsaurians by Gand (1987). But for many of them, one can show that they represent slipped forms or undertracks of *Batrachichnus salamandroides* (Gand 1987; Haubold 1996) and *B. delicatulus* (Haubold *et al.* 1995*a*). So, they have no nomenclatural importance.

Eryopsidae

Limnopus (Permomegatherium) zeilleri (Delage 1912) Gand 1985; synonyms: *Permomegatherium zeilleri* Delage 1912, *Opisthopus ellenbergeri* Heyler & Lessertisseur 1963 (Figs 2b1&2,c6&7; 5p–r). This ichnospecies was described by Gand (1985, 1987), it is frequent in the Lower Permian lacustrine formation, and it is attributed to Eryopsidae. Many trackways were discovered in the uranium quarry worked by 'Cogema', where its morphological variability is quite clearly illustrated (Gand 1986, 1987). Dimensions and morphology of *L. zeilleri* resemble closely those of the American forms *Limnopus waynesburgensis* (Tilton 1931) Baird 1952 of the Lower Permian, and *L. littoralis* (Marsh 1894) Baird 1952 of the Upper Pennsylvanian.

Limnopus (= *Strictipes*) *regularis* (Heyler & Lessertisseur 1963) Haubold 1971; synonym: *Diversipes proclivis* Heyler & Lessertisseur 1963 (Fig. 2c8). We preserve this ichnotaxon that indicates centimetre traces with the digits definitively broader than those of *Batrachichnus salamandroides* (Gand 1987, p. 112). *Amphisauropus latus* (Fritsch 1901) Haubold 1970 (Fig. 2c10) is rare in France (Gand 1987, p.115).

Traces attributed to amniotes

Captorhinomorpha

Hyloidichnus major (Heyler & Lessertisseur 1963) Haubold 1971; synonyms: *Hyloidichnus* (= *Auxipes*) *major* (Heyler & Lessertisseur 1963) Haubold 1971, and *Garganolipes ballestrai* (Heyler & Montenat 1980) Gand 1987

(Figs 2d6&7,5h). The comparison between the French ichnospecies *H. major* and the holotype (n° 11518) *Hyloidichnus bifurcatus* Gilmore 1927 from the Hermit Shale Formation was studied at the National Museum in Washington, D.C., by G.G. It shows that there are few differences between these two ichnospecies. *H. major* is a very common footprint in the basins of south-eastern France (Gand 1987, 1993; Gand *et al.* 1995).

Varanopus curvidactylus Moodie 1929 (Figs 2d1, 5b&e), and *Microsauripus acutipes* Moodie 1929. Moodie (1929) described four ichnospecies from the northern slope of Castle Peak, upper Clear Fork Formation. Except for size, all have the same morphology with generally curvilinear fingers, a tetradactyl hand for *Erpetopus wilistoni, Microsauripus clarki* and *M. acutipes,* whereas it is pentadactyl for *V. curvidactylus*: ichnospecies reduced to only one specimen. In their revision, Haubold (1971) and Haubold & Lucas (2001*a*) keep only *E. willistoni* and *V. curvidactylus*. After having studied the American material in the Yale Peabody Museum, it appears more suitable to also preserve *M. acutipes* in this list because the material is abundant enough and well preserved. All these ichnospecies represent pentadactyl foot-hand couples with the curvilinear fingers slightly clawed.

Varanopus rigidus Gand 1989 (Figs 2c9, 5f&g, 6b). This ichnospecies is rare in the Lodève Basin, but more frequent in Provence where one finds it with *Hyloidichnus major*, often on the same level (Gand 1987, p. 163; Demathieu *et al.* 1992). The fingers are generally straight with short claws.

Subsequent to the definition of *V. rigidus*, the study of the footprint collection in the United States National Museum (USNM) of Washington, D. C., enabled us to more closely compare the French ichnospecies with *Hylopus hermitanus* Gilmore 1927, which was found with *Hyloidichnus bifurcatus* Gilmore 1927, at the same place; at the base of the Hermit Shale Formation. The analysis of the various characters of the traces and the trackways show that the French couples *V. rigidus* / *H. major* and American *H. hermitanus* / *H. bifurcatus* are similar. Thus, one could consider in the future that all of the French and American material represent the same taxa.

Currently, points of view diverge on the best taxonomic names to use. *V. rigidus*, partly, was integrated in *Varanopus curvidactylus* by Haubold & Lucas (2001*a*), which we contest. In addition, Haubold (1971) replaced *Hylopus* with *Gilmoreichnus,* while explaining later (Haubold *et al.* 1995*a*) that 'The ichnogenus *Gilmoreichnus* was introduced by Haubold (1971) because *Hylopus*, which was used by Gilmore (1927), should be restricted to certain Late Mississipian tracks'. This is a step that we do not accept. Otherwise, unlike Haubold (1971), Gand (1987) and Haubold *et al.* (1995*a*), we currently think that the *Gilmoreichnus (Hylopus) hermitanus* trackmakers are not small pelycosaurs but are captorhinomorphs, like the makers of *Hyloidichnus.*

Pelycosauria: Eupelycosauria

Dimetropus leisnerianus (Geinitz 1863) Haubold 1971 (Fig. 2c11,d2&3) is well represented in the Lodève Basin, in particular on two surfaces discovered and extended during the working of the Mas d'Alary quarry by 'Cogema'. One of them, known as C3, of 105 m^2, makes it possible to observe 16 trackways adding up 137 manus-pes pairs. The morphological variability of the ichnospecies is well illustrated by Gand (1986, 1987). The detailed study of *D. leisnerianus* was made by Gand (1987, p. 167).

Dimetropus nicolasi Gand & Haubold 1984 (Fig. 2b6,c12) is described in Gand & Haubold (1984) and Gand (1987, p. 178). Later discoveries showed that the *typus*-sample corresponds to undertracks of *Dimetropus leisnerianus. D. nicolasi* was used by Haubold *et al.* (1995*a*) to identify the American footprints.

Dimetropus (Gonfaronipes) latus (Heyler & Montenat 1980) Gand 1987 is restricted to one sample, which is not well preserved and has no significance at this time.

Pelycosauria: Edaphosauria

Ichniotherium cottae Pohlig 1885 is abundant in the Tambach Formation (Germany) where it occurs with *Dimetropus leisnerianus.* Many and large slabs with these two ichnospecies are preserved at the Gotha Museum where one can study their variability (Voigt & Haubold 2000; Voigt 2001). In the Lodève basin, we could identify *Ichniotherium* cf. *I. cottae* in three isolated cases (Fig. 2c14&15). The trackways observed on the C3 level as well as those of the Riviéral are swimming traces. Based on the various *I. cottae* morphotypes, they were allotted to this ichnogenus (Gand 1986, 1987 p. 186, 1989 p. 18), although some of these traces could also be slipped *Limnopus zeilleri.*

Caseomorpha, Therapsida or Therosauria ?

The following five ichnospecies are numbered 1–5 in order to cross-reference them with the work of Ellenberger (1983*a–c*). 1: *Merifontichnus thalerius* Gand *et al.* 2000 (Fig. 2e1), 2: *Lunaepes*

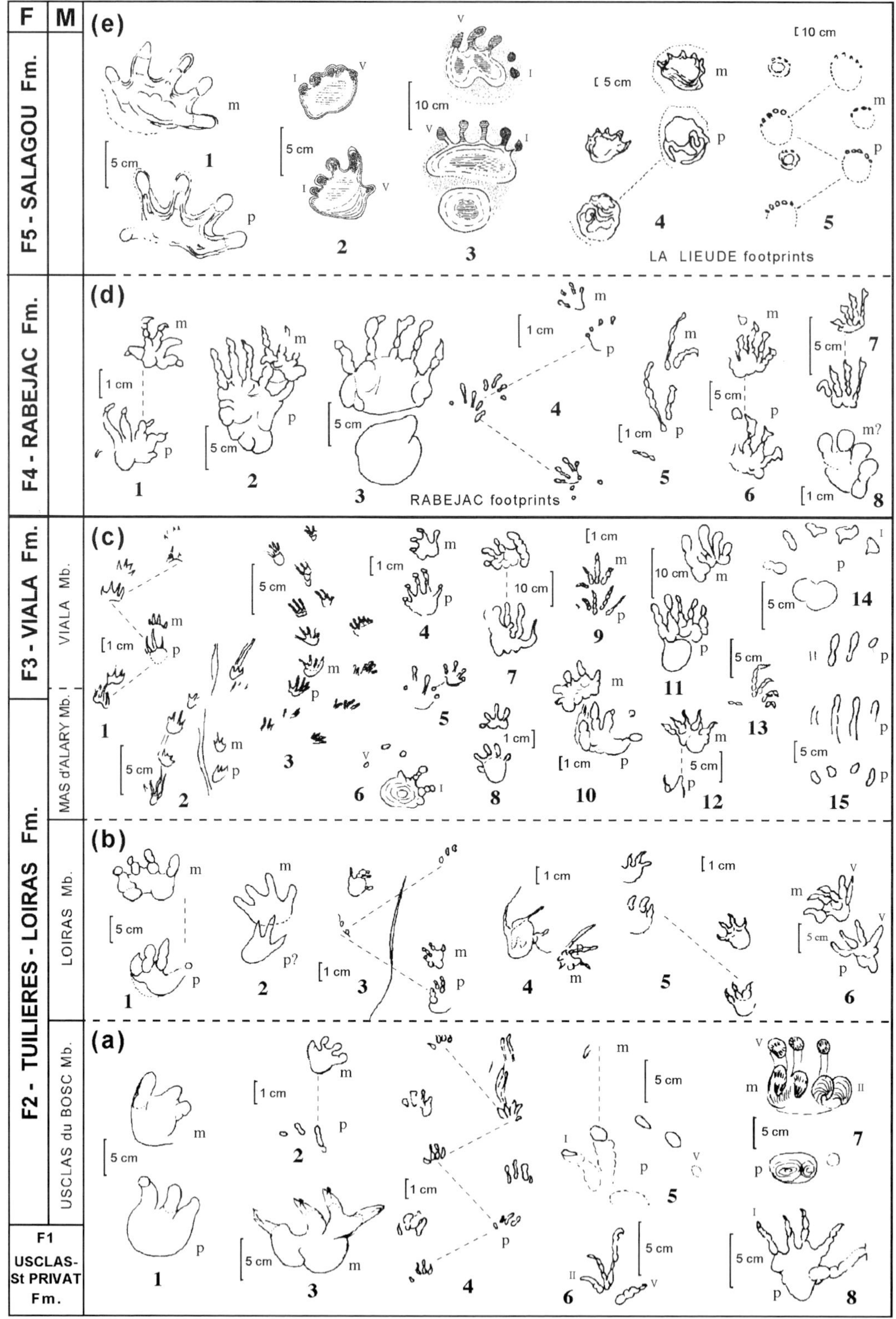
F
M
F5 - SALAGOU Fm.
F4 - RABEJAC Fm.
F3 - VIALA Fm.
VIALA Mb.
MAS d'ALARY Mb.
F2 - TUILIERES - LOIRAS Fm.
LOIRAS Mb.
USCLAS du BOSC Mb.
F1 USCLAS-St PRIVAT Fm.
(e)
(d)
(c)
(b)
(a)
LA LIEUDE footprints
RABEJAC footprints
5 cm
10 cm
1 cm
m
p
V
I
II

ollierorum Gand *et al.* 2000 (Fig. 2e2), 3: *Planipes brachydactylus* Gand *et al.* 2000 (Fig. 2e3), 4: *Brontopus giganteus* Heyler & Lessertisseur 1963 (Fig. 2e4), and 5: *Brontopus circagiganteus* Gand *et al.* 2000 (Fig. 2e5).

Ichnospecies 1–3 and 5 were named by Ellenberger (1983*a–c*) as 1: *Macrochelichnus thaleri*, 2: *Paranomodontipus ollieri*, 3: *Eocynodontipus antecursor* and *Pseudopithepus recurvidigitus* (written also *Pseudopithecopus recurvidigitus*), and 5: *Moschopopus enormis*, but these taxa were held to be invalid because they were given without any diagnosis and description as required by the International Code of Zoological Nomenclature (ICZN). This is the reason why, from the advice of the reviewer M.A. Conti, Gand *et al.* (2000) again described these footprints by modifying slightly the epithets of Ellenberger (1983*a,b,c*) in order to be in accordance with the recommendations of the ICZN. The ichnospecies *Lunaepes fragilis* Gand *et al.* 1995 (Fig. 6g) described in Provence is similar to *Lunaepes ollierorum*, which abounds in the Lodève Basin. Except *B. circagiganteus* and *B. giganteus* Heyler & Lessertisseur 1963, close to *Chelichnus titan* Jardine 1851–1853, attributable to a caseomorph, all the others are therapsid or therosaurian tracks (Gand *et al.* 2000).

Ichniotherium (*Cyclopus*) *aequalis* (Heyler & Lessertisseur 1963) Haubold 1971 is a trackway that is always unclear and debatable. *Planipes caudatus* Gand *et al.* 1995 (Fig. 6f) differs from *P. brachydactylus* in the presence of the tail trace. *Chelichnus incurvus* Gand *et al.* 1995 (Fig. 6c) is abundant in the St-Raphaël ichnofossiliferous site but was never found elsewhere.

Eosuchia and Araeoscelidia

Dromopus lacertoides (Geinitz 1861) Haubold 1971 (Fig. 2a6,b4,c13). Associated with *B. salamandroides* and *Limnopus zeilleri*, this ichnospecies is frequent in the playa facies. When it is reduced to manus-pes sets with only two fingers each, it is impossible to discriminate them from *Dromopus didactylus* (see above). *D. lacertoides* is very close to *D. agilis* Marsh 1894 from the Upper Carboniferous of the United States, which is also twinned with *Limnopus* (see above).

Dromopus didactylus (Moodie 1930) Gand & Haubold 1984, Gand 1987 (Figs 2d5, 5i–m). Moodie (1929) wrote about three ichnospecies with the same geological origin as *Varanopus curvidactylus*. They are *Varanopus palmatus*, *V. impressus* and *V. elrodi*, which were described on the basis of limited and fragmentary material. All have a lacertoid structure with the prevalence of two toes II–III or/and III–IV and the more subtle mark of I and V. The manus, smaller, has the same form. All these ichnospecies correspond to the manus-pes pairs, on small slabs probably coming from the same level. As they all have the same morphology, they can be assigned the same ichnospecies, *Varanopus palmatus* Moodie 1929, if it is admitted that the first ichnospecies described in a paper has priority. But, the holotype (n° 1241) is incomplete due to breakage, and *V. impressus* or *V. elrodi* are not appropriate because the toes are not fully printed. This is the reason why, following Sarjeant (1971), Gand & Haubold (1984) and Gand (1987) chosen the ichnospecies *Varanopus didactylus* Moodie 1930 to indicate these didactyl footprints, placing it in the ichnogenus *Dromopus* Marsh 1894, which has priority (Gand 1987, p. 204). The taxon *Dromopus didactylus* was accepted by Haubold in his revision of 1996 but he now prefers *Dromopus palmatus* (Haubold & Lucas 2001*a*). To validate this choice, a neotype would have to be created, which has not been done. Based on these different data, we prefer to use the name *Dromopus didactylus*.

Notice that this ichnospecies name was not well selected by Moodie because the didactyl aspect also appears in *Dromopus lacertoides,* which is rather localized in the basal part of the Lower Permian. Only the distance between the base of the IV toe and that of V makes it possible

Fig. 2. Main Permian footprints from the Lodève Basin. Stratigraphy: F = formations (Odin 1986); M = Autunian members from 'Cogema' (Laversanne 1976). Palichnofauna: (**a**) Usclas St-Privat Fm. and Tuilières-Loiras Fm. (Usclas du Bosc Mb.), 1–3, cf. *Limnopus*; 2, *Anthichnium salamandroides = Batrachichnus salamandroides*; 5, *Ichniotherium sp.*; 6, *Dromopus lacertoides*; 7, cf. *Ichniotherium*; 8, cf. *Dimetropus*. (**b**) Tuilières-Loiras Fm. (Loiras Mb.), 1 and 2, *Limnopus zeilleri*; 3, *Batrachichnus salamandroides*; 4, *Dromopus lacertoides*; 5, *'Gilmoreichnus brachydactylus' = B. salamandroides*; 6, *Dimetropus nicolasi*. (**c**) Tuilières-Loiras Fm. (Mas d'Alary Mb.) and Viala Fm., 1 and 2, *Salichnium decessus*; 3, *Salichnium pectinatus*; 4 and 5, *Anthichnium salamandroides = Batrachichnus salamandroides*; 6 and 7, *Limnopus (Permomegatherium) zeilleri*; 8, *Limnopus regularis*; 9, *Varanopus rigidus*; 10, *Amphisauropus latus*; 11, *Dimetropus leisnerianus*; 12, *Dimetropus nicolasi*; 13, *Dromopus lacertoides*; 14, *Ichniotherium* cf. *cottae*; 15, *Ichniotherium* cf. *cottae*. (**d**) Rabejac Fm., 1, *Varanopus curvidactylus*; 2 and 3, *Dimetropus leisnerianus*; 4, *Anthichnium salamandroides = Batrachichnus salamandroides*; 5, *Dromopus didactylus*; 6 and 7, *Hyloidichnus major*; 8, *Limnopus sp.* (**e**) Salagou Fm., 1, *Merifontichnus thalerius*; 2, *Lunaepes ollierorum*; 3, *Planipes brachydactylus*; 4, *Brontopus giganteus*; 5, *Brontopus circagiganteus*.

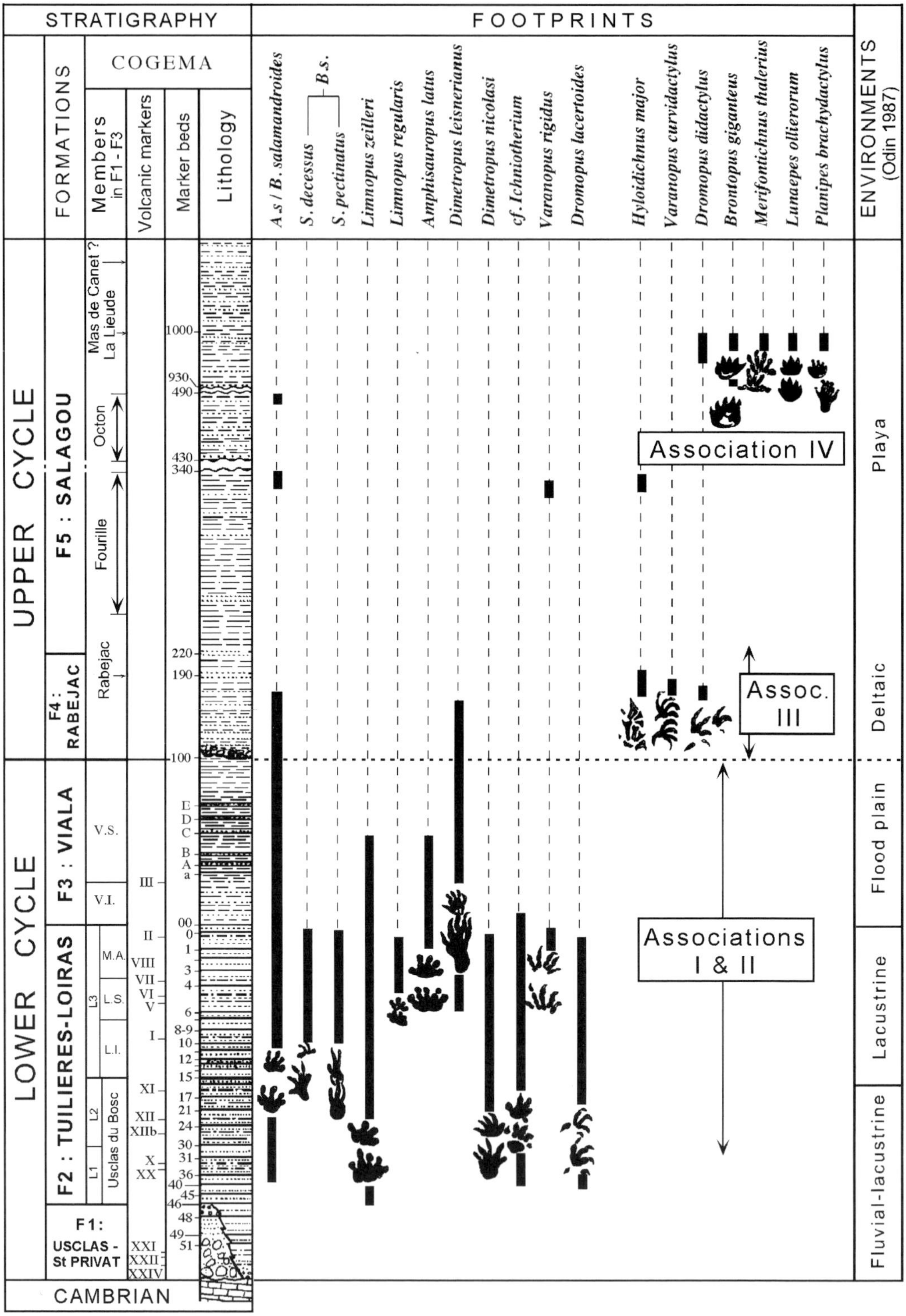

STRATIGRAPHY

FOOTPRINTS

ENVIRONMENTS (Odin 1987)

FORMATIONS

COGEMA

Members in F1 - F3

Volcanic markers

Marker beds

Lithology

A.s / B. salamandroides
S. decessus
B.s.
S. pectinatus
Limnopus zeilleri
Limnopus regularis
Amphisauropus latus
Dimetropus leisnerianus
Dimetropus nicolasi
cf. Ichniotherium
Varanopus rigidus
Dromopus lacertoides
Hyloidichnus major
Varanopus curvidactylus
Dromopus didactylus
Brontopus giganteus
Merifontichnus thalerius
Lunaepes ollierorum
Planipes brachydactylus

UPPER CYCLE

F5 : SALAGOU

Mas de Canet ?

La Lieude

Octon

Fourille

Rabejac

F4 : RABEJAC

1000

930

490

430

340

220

190

100

Association IV

Assoc. III

Playa

Deltaic

LOWER CYCLE

F3 : VIALA

F2 : TUILIERES-LOIRAS

F1: USCLAS - St PRIVAT

V.S.

V.I.

M.A.

L.S.

L.I.

L3

L2

L1

Usclas du Bosc

III

II

VIII

VII

VI

V

I

XI

XII

XIIb

X

XX

XXI

XXII

XXIV

E
D
C
B
A
a
00
0
1
3
4
6
8-9
10
12
15
17
21
24
30
31
36
40
45
46
48
49
51

Associations I & II

Flood plain

Lacustrine

Fluvial-lacustrine

CAMBRIAN

to differentiate the two ichnospecies. This requires complete prints, which are rather seldom observed.

Tambachichnium schmidti Müller 1954 or *Rhynchosauroides* Maidwell 1911. Many lacertoid footprints discovered in a disused quarry of St-Raphaël (Pradineaux Formation) were referred to *Tambachichnium schmidti* Müller 1954 by Gand *et al.* (1995) (Fig. 6e). This followed study by Gand in the Gotha Museum (Saxony, Germany) of the holotype and other lacertoid footprints, all coming from the Manebach Formation (Saale Basin). These last were represented by Pabst (1908, fig. 1, pl. XII).

The choice of *T. schmidti* was justified by the comparison of the manus traces, which have the same size and morphology in Saxony and France. Because he had only referred to the holotype, Haubold (1998*a*, p. 14) judged our determination as erroneous, which is far from being demonstrated. But, if the French lacertoid footprints are not *T. schmidti*, can they be referred to the ichnogenus *Rhynchosauroides*, which is known in the Val Gardena Sandstones and Bellerophon Formation (Italy), by two ichnospecies: *R. pallinii* Conti *et al.* 1977 and *R. palmatus* (Lull 1942)? The two sets of prints are close morphologically but are not of the same age: Artinskian for the German traces and Dzhulfian to Lower Dorashamian for those in Italy (Conti *et al.* 1986; Massari *et al.* 1999). Pending a statistical study, the answer remains open.

Protosauria

Pseudosynaptichnium esterelense Gand *et al.* 1995 was defined from a long trackway not seen elsewhere (Fig. 6h). Haubold (1998*a*, p. 14) suggested that these footprints are close to lacertoid traces that we have named *T. schmidti*, but we need more research to test this proposal.

At the end of this discussion, it appears that in each couple or triplet of taxa – *Batrachichnus salamandroides* / *B. delicatulus*, *Hyloidichnus bifurcatus* / *Hyloidichnus major*, *Erpetopus willistoni* / *Microsauripus acutipes* / *Varanopus curvidactylus*, *Hylopus hermitanus* = *Gilmoreichnus hermitanus* (*sensu* Haubold *et al.* 1995*a*) / *Varanopus rigidus*, and *Lunaepes ollierorum* / *Lunaepes fragilis* – the ichnospecies are so close that the first name could have nomenclatural priority.

Fig. 3. Vertical range of footprints in the Lodève Permian basin. Members: L.I., Lower Loiras; L.S., Upper Loiras; M.A., Mas d'Alary; V.I., Lower Viala; V.S., Upper Viala. Volcanic markers with roman numbers = 'cinérites' (dust-ash tuffs).

But, with respect to the stability of nomenclature, possibly it is better to continue using these binomials in a 'regional' sense. Anyway, before selecting the name having priority, it will be necessary to complete the morphological and morphometrical study by using modern methods (discriminant, Fourrier, Procrust analysis)

Most of the French taxa are ubiquitous in different formations of the Lodève, Provence, Saint-Affrique and Rodez (Sermels and Campagnac) basins. This is true for *Dromopus lacertoides*, *Dromopus didactylus*, *Limnopus zeilleri*, *Amphisauropus latus*, *Batrachichnus salamandroides*, *Hyloidichnus major*, *Dimetropus leisnerianus*, and *Varanopus curvidactylus* / *Microsauripus acutipes*.

Stratigraphical inferences

The reference Lodève Basin

The most complete Permian sedimentary section in France crops out in the Lodève Basin (Languedoc), and it also yields the greatest variety of non-marine palaeontological elements, distributed irregularly through the entire succession: palynomorphs and macroflora, invertebrate trace and body fossils (insects, crustaceans, etc.), tetrapod footprints (numerous) and bones (rather scarce). The series is composed of two sedimentary cycles separated by an unconformity (Odin 1986). The first cycle consists of three formations: F1–F3 ('Autunian Group'), and the second of two: F4 and F5 ('Saxonian group') (Figs 2&3).

The ichnoassociations

The strata containing footprints have been precisely located in the lithostratigraphy, which is well-known from uranium prospecting. This has formed the basis of a palichnostratigraphical scale in which several local ichnoassociations have been identified (Gand 1987; Châteauneuf & Gand 1989; Gand 1993) (Fig. 3). Within the first cycle (Usclas-St.Privat, Les Tuillères-Loiras and Viala formations), two ichnoassociations (I and II) were distinguished initially. If we take into consideration the equivalencies of *Salichnium decessus* and *S. pectinatus* (undertracks) with *Gilmoreichnus brachydactylus* (*sensu* Gand 1987) = *Batrachichnus salamandroides*, the interpretation of *D. nicolasi* as an undertrack corresponding to *Dimetropus leisnerianus,* and the uncertain presence of *Ichniotherium* (in any case rather rare), it can be seen that the first sedimentary cycle contains:

(1) Ichnotaxa confined there (*Limnopus zeilleri*, *Dromopus lacertoides*, *Amphisauropus latus* and *Ichniotherium*); therefore, they characterize this part of the sedimentary succession.
(2) Ichnotaxa that reach the basal part (Rabejac Formation) of the second cycle (*Batrachichnus salamandroides* and *Dimetropus leisnerianus*).

Above the unconformity, Association III (Rabejac) begins with the first appearance of *Varanopus curvidactylus* / *Microsauripus acutipes*, *Dromopus didactylus* and *Hyloidichnus major*. These two last ichnospecies, as well as *B. salamandroides*, are found up into a few levels of the F5 Salagou Formation, which is almost 2000 m thick. In the upper part of the Salagou Formation appears Association IV (La Lieude) with *Brontopus giganteus*, *B. circagiganteus*, *Planipes antecursor*, *Lunaepes ollierorum* and *Merifontichnus thalerius*. All are possibly therapsid traces except *Brontopus*, whose trackmaker could be a pelycosaurian (Gand *et al.* 2000).

Chronostratigraphy

The floristic content of the F1–F2 formations suggests, after comparison with that of the Autunian of Autun, an age ranging between the uppermost Ghzelian and the lower Sakmarian (Broutin *et al.* 1999). From the insects and conchostracans collected in the Salagou Formation, Nel & Schneider (in Gand *et al.* 1997) put forward 'a Kungurian to Tatarian age (Leonardian to Capitanian)' for this unit. Since that time, the distribution of various Odonatoptera (Nel *et al.* 1999) and Archaeorthoptera (Béthoux *et al.* 2001, 2002*a*,*b*, 2003*a*,*b*) suggested Artinskian, Ufimian and Kazanian ages. But, because of some undecided lithostratigraphical assignments and the fact that the stratigraphical range is not established for several taxa of reference, these age determinations are not consistent with one another. In a recent work, based on conchostracans, Schneider (in Gand *et al.* 2004*a*,*b*) modified slightly the age of the Salagou Formation, and included it between the Kungurian and the Changxingian.

Tentative correlations with others regions

The basins of Provence

The main stratigraphical marker through the Permian basins of Provence is the A7 Rhyolite (Fig. 4). In the central part of the Bas-Argens Basin (Le Muy area), it is composed of five successive flows. Radio-isotopic dating ($^{39}A/^{40}A$) gave an age of 272.5 $\pm$ 0. 3 Ma (Zheng *et al.* 1992), fitting well with the Artinskian–Kungurian boundary according to Jin *et al.* (1997).

Over a long time, footprints and tracks have been collected from formations overlying the A7 lava flow (Demathieu *et al.* 1992, Gand *et al.* 1995) (Fig. 5). *Batrachichnus salamandroides*, *Limnopus* sp., *Hyloidichnus major*, *Varanopus rigidus*, *Laoporus* sp., *Dimetropus* sp., *Dimetropus latus*, and cf. *Dromopus* were gathered mainly from the Pelitic Formation of the Luc Basin (i.e. La Motte Formation of the Bas-Argens Basin). *B. salamandroides* and *Limnopus zeilleri* came from the Muy and Mitan formations, and *Dromopus didactylus* from the Mitan Formation.

The Muy Formation yielded many plant remains (Visscher 1968; Germain 1968), among them coalified remnants, silicified woods (Vozenin-Serra *et al.* 1991) and various abundant palynomorphs. Such a floristic association, which is comparable to those found in the Zechstein (in the original sense), can be regarded as the youngest known currently in the French Permian (Broutin, in Toutin-Morin *et al.* 1994).

Other traces came from the Pradineaux Formation overlying the A7 Rhyolite. *B. salamandroides*, and *Varanopus curvidactylus* / *Microsauripes acutipes* were found in the Coulet-Redon quarry (Bas Argens). Numerous footprints were observed on a large slab belonging to the upper part near St-Raphaël town (Estérel): *B. salamandroides*, *Limnopus* sp., *Hyloidichnus major*, *Lunaepes fragilis*, *Planipes caudatus*, *Pseudosynaptichnium esterelense*, *Chelichnus incurvus*, and *Tambachichnium schmidti* (Fig. 4).

The Pradineaux Formation has been the subject of various attempts at dating. It includes the last important acidic flow (A11) that clearly predates the Illawarra Reversal, since it is cut by a fluorite-barite vein with adularia giving an isotopic age of 264 $\pm$ 2 Ma (Zheng *et al.* 1992). Near Agay, a thick unit of grey, fine-grained, lacustrine siliciclastics yields the oldest biostratigraphical elements known in that part of the series. They are mainly macrofloral remains, with very few palynomorphs, which allowed Visscher (1968) to propose an 'Early Thuringian' age, that is earliest Tatarian according to Menning (1994). Higher, the Pra Baucous lacustrine limestone beds yielded an ostracod association. According to their known stratigraphical occurrences in Russia, these species indicate an earliest Tatarian age, which can be correlated with the Early Midian of the Tethyan marine scale (Lethiers *et al.* 1993).

Recently, and for the first time, footprints have been discovered below the lava flow A7. *Varanopus curvidactylus* / *Microsauripus acutipes* and *Dromopus didactylus* were found in the

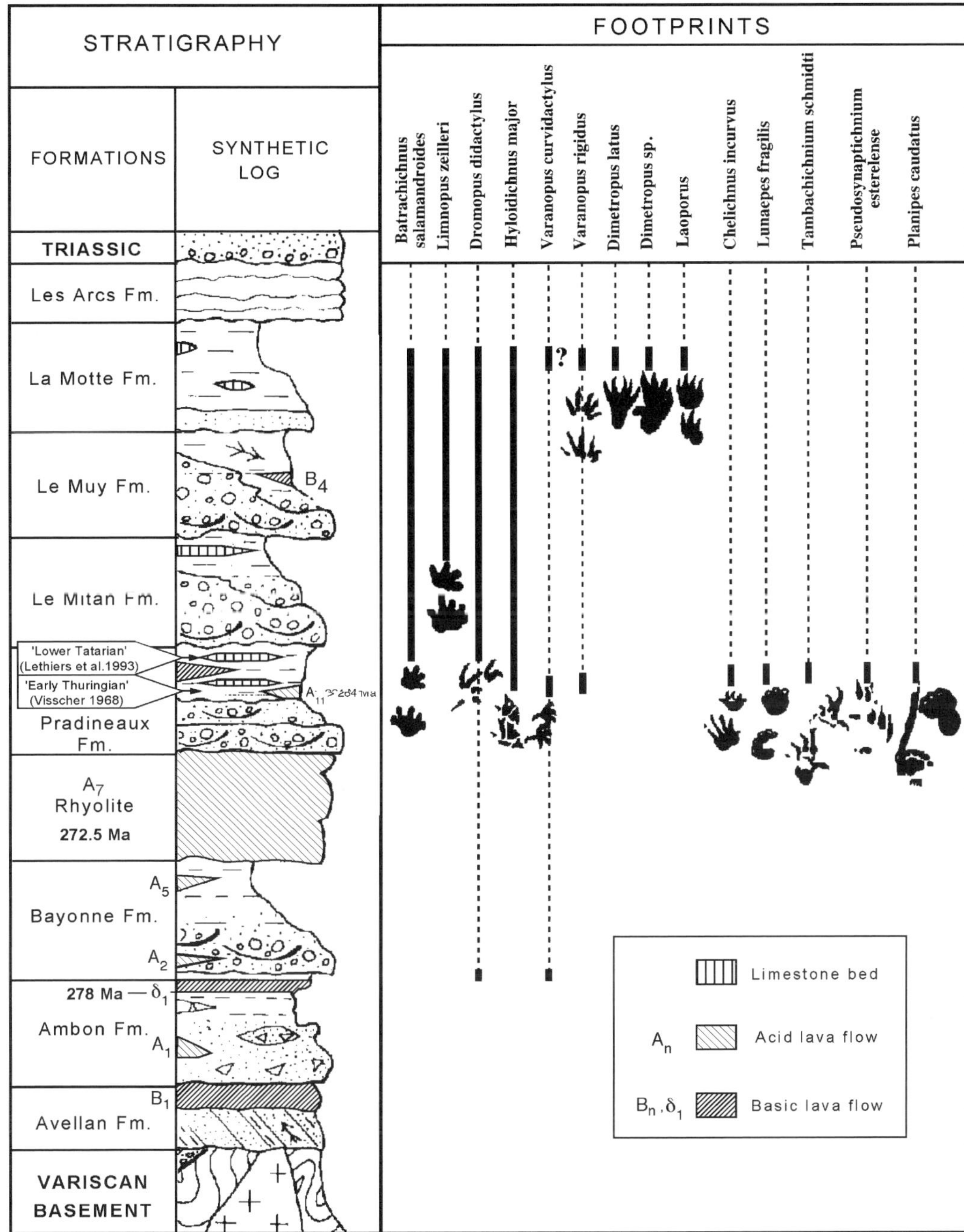

Fig. 4. Stratigraphical distribution of footprints in the Permian of Provence (Estérel–Bas Argens Basin).

Boson quarry in the Bayonne Formation (Durand *et al.* 2002). The ichnites are located above the δ 1 doleritic basalt, which represents the uppermost part of the Ambon Formation (Fig. 4). It was sampled for a $^{39}Ar/^{40}Ar$ dating, yielding an age of 278 ± 2 Ma (Zheng *et al.* 1992). According to Menning (1994) and Jin *et al.* (1997), this date is Artinskian.

As far as correlations are concerned, except for isolated ichnites such as *Laoporus* sp.,

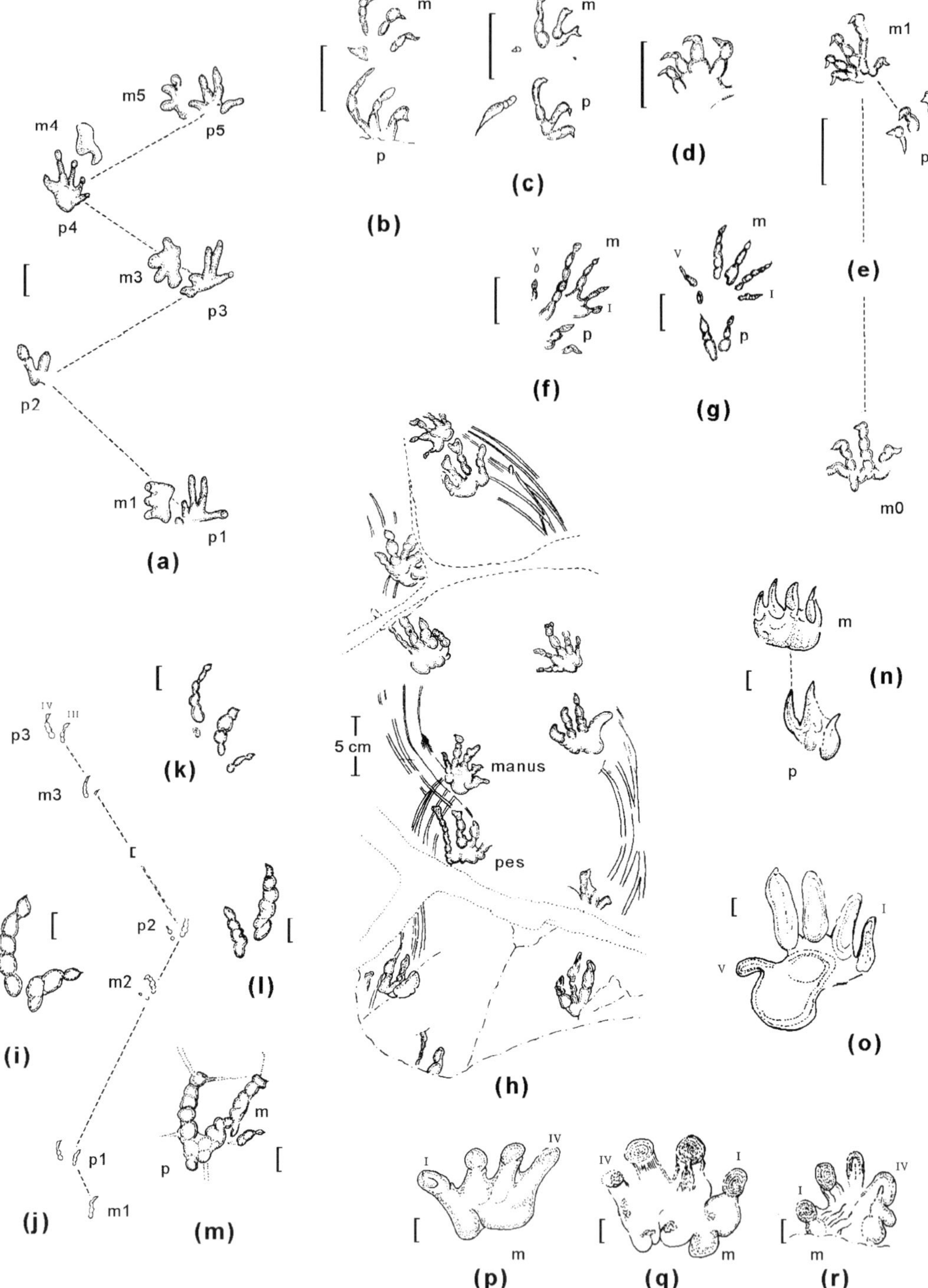

Fig. 5. Footprints from the Permian series of Provence. (**a**) *Batrachichnus salamandroides*, (**b–e**) *Varanopus curvidactylus*, (**f&g**) *Varanopus rigidus*, (**h**) *Hyloidichnus major*, (**i–m**) *Dromopus didactylus*, (**n**) *'Laoporus' sp.*, (**o**) *Dimetropus sp.*, (**p–r**) *Limnopus zeilleri*. Scale bar: 1 cm.

Dimetropus sp. and *Dimetropus latus*, the series of Provence contains traces also recorded from the Lodève Basin, but only in the second cycle (F4 and F5 formations). The Bayonne Formation yielded only *Varanopus curvidactylus* / *Microsauripus acutipes* and *Dromopus didactylus*, two taxa appearing in Association III. From common footprints, such as *Hyloidichnus major*, *Lunaepes* and *Planipes*, it is possible to correlate the formations located above the A7 Rhyolite with the Salagou Formation of Lodévois, containing Association IV. It should be noted that *Limnopus zeilleri* has a greater vertical distribution in the whole French Permian series than is observed in the Saint-Affrique and Lodève basins alone, ranking it close to that of *Batrachichnus salamandroides*.

Thus, the Permian series of Provence could be about the same age as that of Lodève: namely, between Artinskian and Lopingian. The Pradineaux Formation levels bearing *Lunaepes* and *Planipes* could be Tatarian with reference to the precise location of these ichnogenera in the Lodève Basin.

In a recent work, Haubold & Lucas (2001*a*), using the Late Artinskian age of the *Erpetopus willistoni*, *V. curvidactylus*, *Dromopus palmatus* association from the Choza Formation (Texas), assign the same age to the Pradineaux Formation, which contains the same ichnospecies. But, the recent discovery of *V. curvidactylus* / *Microsauripus acutipes* in the Bayonne Formation shows that the vertical range of the latter taxa is so large that it cannot be used as a good marker. So, the 'Artinskian' age of the Pradineaux Formation is not demonstrated by footprints. On the other hand, a review of all the other dating elements concludes that the Pradineaux Formation is of Wordian age (Durand, 2006).

The United States basins

The Permian palichnofauna of the United States has been the subject of an active revision in recent years (Haubold *et al.* 1995*a*,*b*; Haubold 1996; Haubold & Lucas 2001*a*,*b*, 2003), which leads to several correlations between the United States and Europe (Haubold 2000; Haubold & Lucas 2001*b*; Lucas 2002*a*,*b*).

In their work of 2001*a*, Haubold & Lucas refer to the Choza Formation of Texas and to the Robledo Mountains Formation of New Mexico, which have layers bearing footprints inserted in marine beds dated by fossils. The ichnofauna of the Choza Formation, with *Erpetopus willisti*, *Microsauripus acutipes*, *Varanopus curvidactylus*, and *Dromopus palmatus* (*sensu* Haubold & Lucas 2001*a*) = (*D. didactylus sensu* Gand & Haubold 1984), is the youngest, with a Late Artinskian age. This association is underlain by the Robledo Mountains Formation which, is dated as Early Artinskian. It comprises *Batrachichnus delicatulus*, *Dromopus agilis* / *lacertoides*, *Amphisauropus*, and *Dimetropus nicolasi*. As 'an identical composition and vertical succession are shown by the ichnofaunas of the red bed formations of the Lower Permian in Europe', the authors conclude 'The same geological age, Late Artinskian, is evident, e.g. for the Rabejac, Pradineaux–Mitan and Collio formations of some Permian basins in southern Europe'. Such a dating for the European formations can be accepted only if the vertical distribution of the ichnospecies of the Choza Formation does not exceed its stratigraphical limits. However, it is not the case since two of its ichnotaxa – *Varanopus curvidactylus* and *Dromopus palmatus* (*sensu* Haubold & Lucas 2001*a*) = *Varanopus palmatus* Moodie 1930 – appear in the Upper Asselian (Lucas 2002*a*).

When comparing the vertical range of the ichnospecies closest morphologically (*Batrachichnus delicatulus* ≈ *B. salamandroides*, *Dimetropus nicolasi* ≈ *D. leisnerianus*, *Gilmoreichnus hermitanus* ≈ *Varanopus rigidus*, *Hyloidichnus bifurcatus* ≈ *H. major*, and *Varanopus curvidactylus* ≈ *Microsauripus acutipes*), it is clear that there is not 'an identical composition and vertical succession' in every basin (Fig. 7), which prohibits any certainty concerning the correlations made from footprints. The results depend on the ichnospecies and the basin which are taken as reference. Thus, the comparison of the footprint vertical distributions of the United States basins (in Lucas 2002*a*,*b*) with that of the Provence basins (Fig. 4) would result in dating the American formations from lower Artinskian to Tatarian if one considers the *Varanopus curvidactylus* / *Microsauripus acutipes* distribution, and only equating Tatarian with *H. major* and *V. rigidus* (≈ *Gilmoreichnus hermitanus*).

Such a result underlines the importance of the palaeontological gaps within each basin, demonstrated in those of Lodève and Provence, and the great stratigraphical range of certain footprints: *H. major*, *Limnopus*, *B. salamandroides*, *V. curvidactylus* / *M. acutipes*, and *D. didactylus*.

Lombardy: Orobic and Trompia basins (northern Italy)

The palichnofauna of the Collio Formation was collected and studied at many sites in the Lombardy basins by Conti *et al.* (1977, 1991, 1997, 1999), Nicosia *et al.* (1999, 2000, 2001), Santi & Krieger (1999, 2001), Avanzini *et al.* (2001), Santi (2001, 2003), and Arduini *et al.* (2003). These last authors present a list by basin, including the

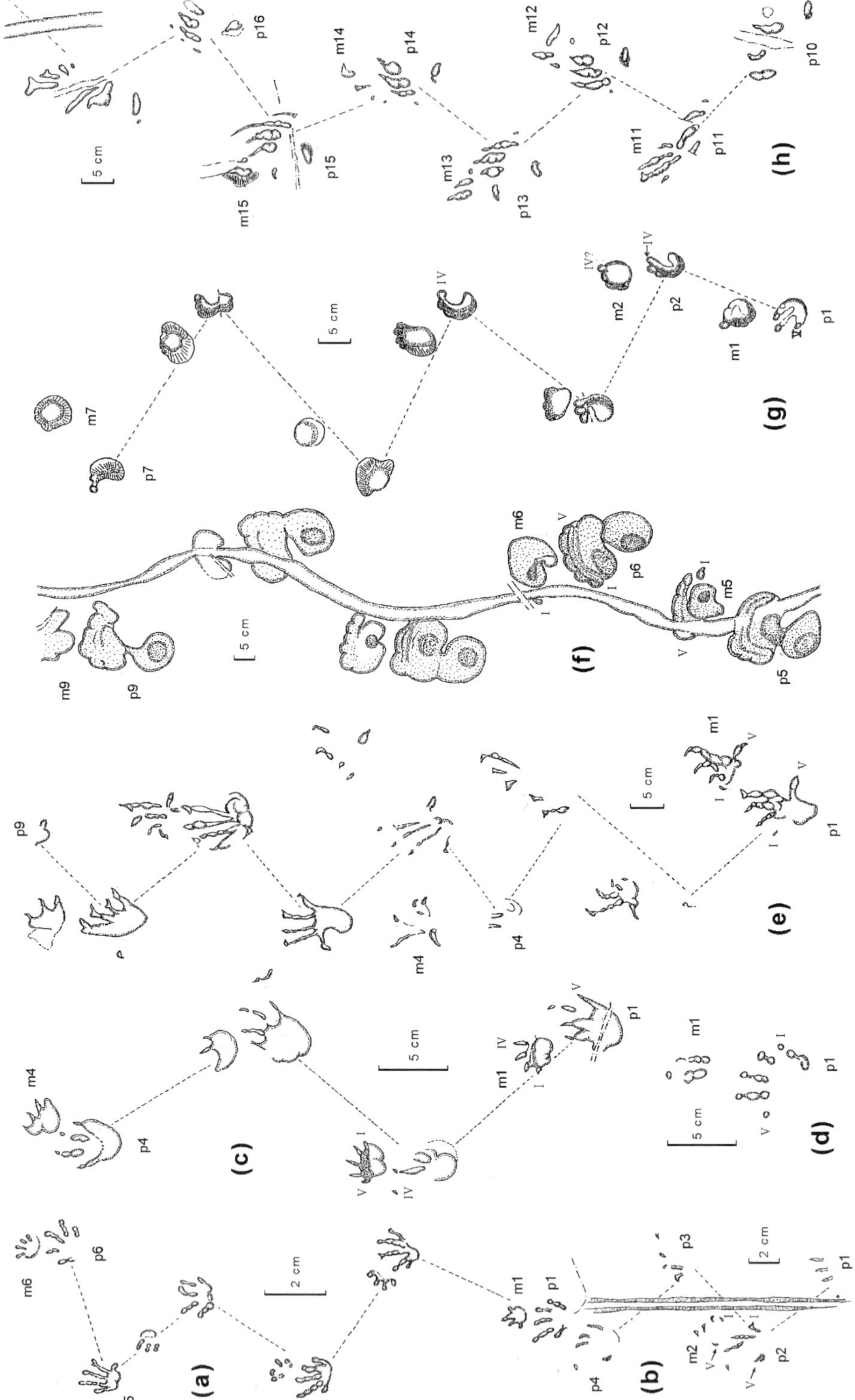
(a)
2 cm
m6
p6
p5
(b)
2 cm
m1
p1
p3
p4
m2
p2
p1
(c)
5 cm
m4
p4
m1
p1
(d)
5 cm
m1
p1
(e)
5 cm
p9
m4
p4
m1
p1
(f)
5 cm
m9
p9
m6
p6
m5
p5
(g)
5 cm
m7
p7
m2
p2
m1
p1
(h)
5 cm
m15
p15
p16
m14
p14
m13
p13
m12
p12
m11
p11
p10

following taxa: '*Batrachichnus*' *salamandroides*, *B.* sp., *Camunipes cassinisi*, *Amphisauropus latus*, *A. imminutus*, *V. curvidactylus*, *Dromopus lacertoides*, *D. didactylus* and *Ichniotherium cottae*.

After examination of the material in April 1999 in the Sapienza University (Rome), G.G. confirms the existence of *B. salamandroides*. While referring to its description and illustration in Ceoloni *et al.* (1987), *C. cassinisi* seems to be *Microsauripus acutipes*, but some footprints ascribed to this ichnogenus are clearly *B. salamandroides*. Thus, on a slab illustrated by Santi (2003, pl.Ic), the existence of digital scratches and the well marked metacarpal I results in recognizing a pentadactyl clawed hand, which is not the case for *M. acutipes*. Conversely, in Nicosia *et al.* (2000, p. 762), the '*A. imminutus*' trackway with a tetradactyl manus trace is not characteristic of the ichnogenus *Amphisauropus* in which the manus is pentadactyl.

All these ichnospecies are distributed in the Collio Formation, which rests on a volcanic member and is covered by the Verrucano. Isotopic studies make it possible to date it between 286/283 and 278/273 Ma (Cassinis *et al.* 1999). It is thus Artinskian based on the time scale of Menning (2001).

The Dolomites: Bolzano Basin (northern Italy)
Among the 24 taxa identified in the section of the Val Gardena Sandstone and Bellerophon Formation in the canyon of Butterloch-Bletterbach (Conti *et al.* 1977; Ceoloni *et al.* 1988), Avanzini *et al.* (2001) retained only the following ichnotaxa, which are common to other areas: Chirotheridae, *Rhynchosauroides* sp., *R.* aff. *palmatus*, *R. pallinii*, *Dicynodontipus* sp., *Ichniotherium* aff. *cottae*, *Ichniotherium accordii* and *Pachypes dolomiticus*. In this canyon, the Italian authors also identified: *Hyloidichnus tirolensis*, '*Chelichnus*' *tazelwuermi*, ?*Paradoxichnium radeinensis* and *Janusichnus bifrons*.

During the study of the material by G.G., in 1999, Umberto Nicosia specified that *Tridactylium* and *Phalangichnus perwangeri* had been invalidated, and that many footprints named *Dromopus*, *Phalangichnus*, *Varanopus curvidactylus* had been assigned to *Rhynchosauroides*. It is a fact that this ichnogenus is present and abundant in the Bolzano basin; and, we wondered whether the lacertoid trackways of St-Raphaël (France) could not comprise part of it. Another Italian ichnospecies, *Ichniotherium accordii*, suggests also a unification with *Merifontichnus thalerius* of the Lodève Basin. There is in both cases homopody and rather digitigrade manus traces, but the pes of *I. accordii* is generally plantigrade, whereas that of *M. thalerius* is digitigrade. More abundant material would thus be necessary to validate the assumption of a possible common ichnospecies between the Lodève and Bolzano basins.The comparison between *Pachypes dolomiticus* and *Brontopus giganteus*, both large round traces with indistinct digits, was made in Gand *et al.* (2000, p. 50). These ichnospecies differ in the size of the digits, which increase from the I to the IV in the first one, while the order is reversed in the second.

The association of the Bolzano Basin is well dated. It is indeed framed by levels with fusulinids and sporomorphs that make it possible to allot to it an age ranging from nearly 259 to 255 Ma (Avanzini *et al.* 2001), and is therefore Wuchiapingian.

The Thuringian Forest Basin (Germany)
Numerous footprints have been found in three regions in Germany: the Thuringian Forest, the Cornberg area, and the Saar–Nahe Basin.

In Thuringia, footprints were studied and named by Pabst (1908), and their nomenclature was re-examined by Haubold (1971, 1973, 1996, 1998*a*,*b*). The palichnofauna was collected in the Lower Rotliegend, for a long time referred to the Autunian and Lower Saxonian (Haubold 1984). In the 'Autunian' part, Haubold listed the following characteristic footprints: *Batrachichnus salamandroides*, *Gracilichnium jacobii*, *Jacobiichnus caudifer*, *Amphisauroides conrectus*, *A. discessus*, *Amphisauropus latus*, *A. imminutus*, *Hyloidichnus arnhardti*, *Varanopus microdactylus*, *Gilmoreichnus brachydactylus*, *G. kablikae*, *G. minimus*, *Dromopus lacertoides*, *Ichniotherium cottae* and *Dimetropus leisnerianus*. *Palmichnus tambachensis* and *Tambachichnium schmidti* are only found in the Tambach Formation of the 'Lower Saxonian'.

In his revision, Haubold (1998*a*) regards *P. tambachensis* as an extramorphologic form of *Varanopus microdactylus*. This latter ichnogenus is written between quotation marks; Gand (1987, p. 155) has suggested replacing it with *Hyloidichnus*. As in the French Lower Permian, *Foliipes*, *Serripes* and *Acutipes* (Heyler & Lessertisseur 1963), *Gracilichnium* and *Jacobiichnus* of the Thuringian Forest have only a regional interest because these taxa indicate slipped shapes of *B. salamandroides*. In our

Fig. 6. Trackways of the Saint-Sébastien slab (Pradineaux Formation, Estérel Basin, Provence). (**a**) *Batrachichnus salamandroides*, (**b**) *Varanopus rigidus*, (**c**) *Chelichnus incurvus*, (**d**) *Limnopus sp.*, (**e**) *Tambachichnium schmidti*, (**f**) *Planipes caudatus*, (**g**) *Lunaepes fragilis*, (**h**) *Pseudosynaptichnium esterelense*.

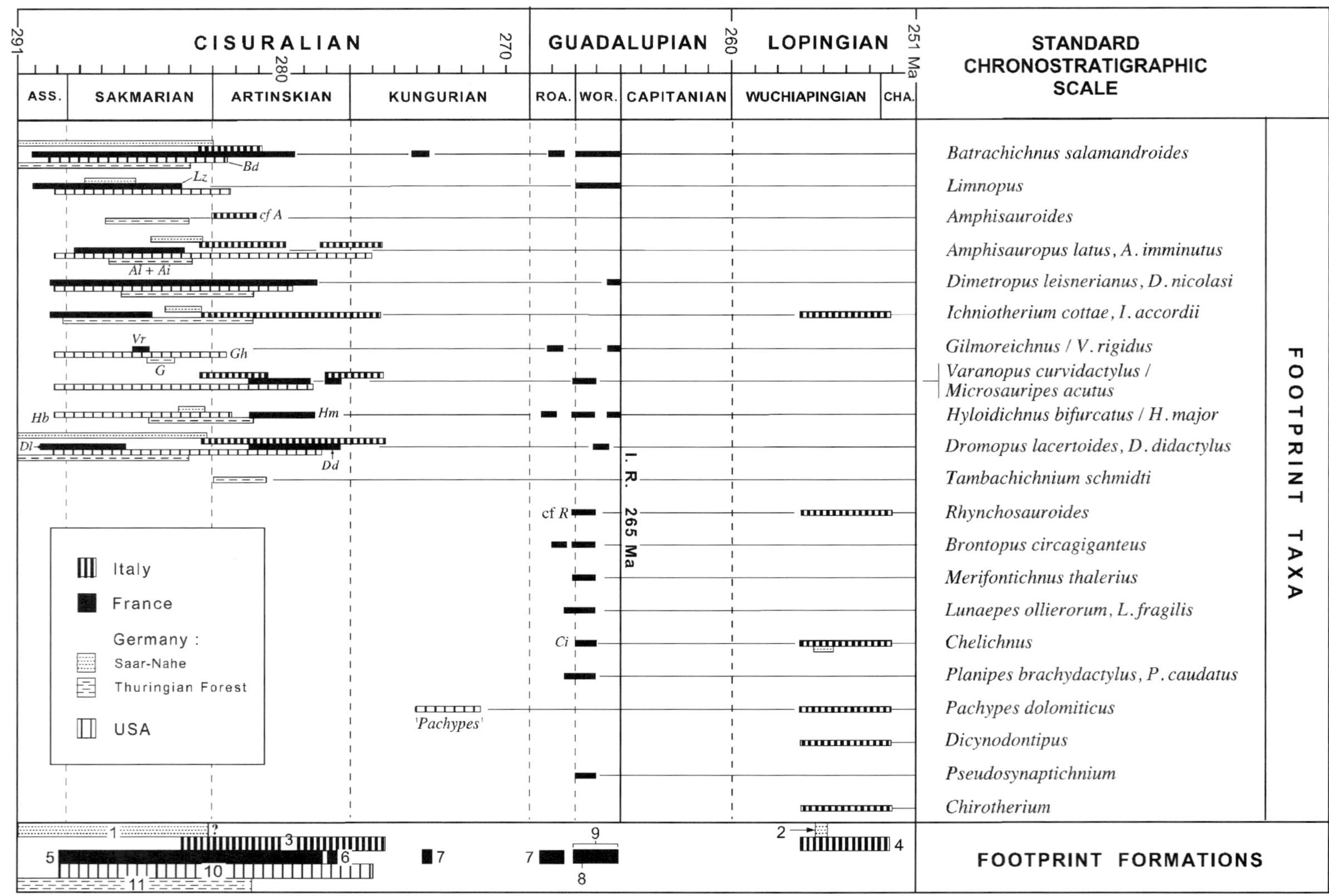
CISURALIAN
GUADALUPIAN
LOPINGIAN
STANDARD CHRONOSTRATIGRAPHIC SCALE
291
280
270
260
251 Ma
ASS.
SAKMARIAN
ARTINSKIAN
KUNGURIAN
ROA.
WOR.
CAPITANIAN
WUCHIAPINGIAN
CHA.
FOOTPRINT TAXA
Batrachichnus salamandroides
Limnopus
Amphisauroides
Amphisauropus latus, A. imminutus
Dimetropus leisnerianus, D. nicolasi
Ichniotherium cottae, I. accordii
Gilmoreichnus / V. rigidus
Varanopus curvidactylus / Microsauripes acutus
Hyloidichnus bifurcatus / H. major
Dromopus lacertoides, D. didactylus
Tambachichnium schmidti
Rhynchosauroides
Brontopus circagiganteus
Merifontichnus thalerius
Lunaepes ollierorum, L. fragilis
Chelichnus
Planipes brachydactylus, P. caudatus
Pachypes dolomiticus
Dicynodontipus
Pseudosynaptichnium
Chirotherium
Bd
Lz
cf A
Al + Ai
Vr
Gh
G
Hb
Hm
Dl
Dd
cf R
Ci
'Pachypes'
I. R. 265 Ma
Italy
France
Germany :
Saar-Nahe
Thuringian Forest
USA
FOOTPRINT FORMATIONS

opinion, the taxa *G. minimus*, *G. kablikae* and *Hyloidichnus arnhardti* are not characteristic enough to be preserved.

The vertical distribution of these ichnospecies, well localized stratigraphically, led Haubold (1984) to distinguish the Autunian from the Saxonian. But, given the impossibility of defining the Saxonian as a stage, his scale lost its chronostratigraphic utility. The 'Saxonian' of the Thuringian Forest Basin is restricted to the Tambach Formation, which produced, in addition to a beautiful ichnofauna (Pabst 1908), vertebrate remains that were studied by Sumida *et al.* (1996). From the bones, the footprint association is dated as Wolfcampian. Menning (1995) placed the Tambach Formation in the lower Artinskian; other subjacent formations (Manebach to Rotterode Formations) are dated from Asselian to Sakmarian. This age assignment is close to that which can be given based on the typical 'Autunian' palaeoflora, which is to say Asselian to early Sakmarian.

The Cornberg Sandstein (Hessian Depression, Germany)

The footprints at this locality were made on the surface of aeolian dune deposits. Schmidt (1959) studied them and distinguished several ichnospecies of *Phalangichnus*, *Akropus*, *Palmichnus*, *Chelichnus*, *Barypodus* and *Harpagichnus*, these last three ichnogenera being attributed to therapsids. Haubold (1971, 1984) synonymized the nomenclature a little, maintaining only *Laoporus* and *Chelichnus* at first, then *Chelichnus* alone in his last revision (Haubold 1996). The ichnogeneric differences introduced by Schmidt are due, for Morales and Haubold (1995), to 'certain preservational dunes facies influences, which are without taxonomic value'. This is, in our opinion, too simplifed. The problem was also discussed by Fichter (1994).

The Cornberg Sandstein Formation was dated as Tatarian by Menning (1995). In the stratigraphical scale, this formation is above the 'Illawara reversal', which indicates an age younger than 265 Ma.

The Saar–Nahe and Wetterau basin (Germany)

The Saar–Nahe palichnofaunas were collected by Boy, Fichter and Stapf, then studied with much meticulousness by Fichter (1976, 1982, 1983*a*,*b*, 1984) and Fichter & Kowalczyk (1983). From the study of the French Permian palichnofauna and that of the Thuringian Forest (Gand 1987), G.G. studied in 1994 the Fichter collection preserved at Mainz University and that of the Stapf family stored at the Nierstein Museum. Many traces were collected in the Nahe and Lebach Groups, and some others in the Kusel Group, which rests on the 'C Stephanian'. The ichnospecies are well located in lithostratigraphies established by Stapf (1992, 2003) and then by Boy & Fichter (1982, 1988). The last authors divided the Kusel Group into 7 formations (R–Lo3), the Lebach Group into 19 formations (Lo3–Lo10, D1–D2), and the upper Nahe Group into 8 formations (N1–N8). Based on the 'Autunian flora' (Kerp & Fichter 1985) and the varied vertebrate fauna, methodically described by Boy (1972, 1987*a*,*b*, 1988, 1989, 1993, 1995, 2003), the Kusel and Lebach Groups were assigned to the 'Autunian' and most of the Nahe Group (N2–N5 Formation) to the 'Saxonian'.

In the Kusel and Lebach groups, Fichter (1976, 1982, 1983*a*,*b*, 1984) identified all the ichnospecies recorded from the Thuringian Forest Formation, plus *Limnopus palatinus* and *Foliipes abscissus*, the latter also mentioned in France. Without entering into details, our determinations were different. We thus established the presence of *Batrachichnus salamandroides* (= *Saurichnites salamandroides* of Fichter) but also referred to *B. salamandroides* the specimens described by Fichter as *Foliipes abscissus, Jacobiichnus caudifer, Amphisauroides imminutus, Saurichnites incurvatus, Gilmoreichnus brachydactylus, G. kablikae, G. minimus, Varanopus microdactylus* and *Hyloidichnus arnhardti*. All these footprints show, indeed, at best, a tetradactyl manus without claw impressions. To determine *Gilmoreichnus*, for example, it is necessary to recognize pentadactyly and clawed footprints. The tail mark is not enough because it is irregular in the trackways of *B. salamandroides*. But, it is the true that, in this last ichnospecies,

Fig. 7. European and American stratigraphical ranges of footprint taxa. I.R., Illawarra Reversal. Stages: ASS, Asselian; ROA, Roadian; WOR, Wordian; CHA, Changhsingian. Footprint taxa: Bd, *Batrachichnus delicatulus*; Lz, *Limnopus zeilleri*; cf. A, cf. *Amphisauroides*; Al + AI, *Amphisauropus latus* and *A. imminutus*; Vr, *Varanopus rigidus*; Gh, *Gilmoreichnus hermitanus*; G, *Gilmoreichnus*; Hb, *Hyloidichnus bifurcatus*; Hm, *Hyloidichnus major*; Dl, *Dromopus lacertoides*; Dd, *Dromopus didactylus*; cf. R, cf. *Rhynchosauroides*, Ci = *Chelichnus incurvus*. Footprint formations: 1, Saar–Nahe basins, Kusel to Nahe Groups; 2, Cornberg Sandstone Fm.; 3, Lombardy basins, Collio Fm.; 4, Bolzano Basin, Val Gardena Sandstone and Bellerophon Fms; 5, Lodève Basin, Usclas-St-Privat to Rabejac Fms; 6, Estérel Basin, Bayonne Formation; 7, Lodève Basin, lower Salagou Fm.; 8, Salagou Fm., La Lieude Mb.; 9, Provence basins, Pradineaux to La Motte Fms; 10, USA basins; 11, Thuringian Forest Basin, Manebach to Tambach Fms.

distal digital scratches and a deep print of the metacarpal I may be mistaken for prints of the claws and digit I; thus, such extramorphological *B. salamandroides* could be assigned to *Gilmoreichnus*. *Amphisauroides* sp. with a tetradactyl manus and broad digits (Fichter 1983*a*, p. 56; 1983*b*, p. 25) are better assigned to *Limnopus palatinus*, but many *L. palatinus* are *B. salamandroides*. The presence of *Dromopus lacertoides, Amphisauropus latus* (= *Saurichnites intermedius* in Fichter 1983*a*) and *Limnopus zeilleri* is confirmed. These last ichnospecies were seen in the Stapf collection (Nierstein museum) and in that of Fichter (Mainz University), but *L. zeilleri* was named *Dimetropus leisnerianus* (Fichter 1976, p. 100; 1983*b*, p. 155) or *Ichniotherium cottae* (Fichter 1983*b*, p. 47). Actually, we never observed these two ichnotaxa in the Kusel and Lebach groups. Nevertheless, the sample PIM K 217 from the Wahnwegen Formation suggests the last one by its size and its plantigrady, but it is clearly lacertoid. Let us add to this list the presence of *Amphisauropus imminutus*, well characterized by only one trackway from the Stapf collection.

In the Nahe Group (N1–N8), which begins with a rhyolitic complex, red formations seldom produced complete traces. Fichter (1983*a,b*, 1984) identified in this group almost all the ichnospecies collected in the Kusel and Lebach groups, to which it is necessary to add *Phalangichnus alternans, Chelichnus, Laoporus* and *Anhomoiichnium*. We confirm *Dromopus lacertoides, Amphisauropus imminutus, A. latus* and *V. microdactylus*, whereas cf. *Chelichnus*, cf. *Laoporus, Phalangichnus* are doubtful, and *Anhomoiichnium* represents *D. didactylus*. *Varanopus curvidactylus* is a possible identification for only one specimen. Other rather common traces are represented only by the digital impressions with ends rounded like balls; they can be assigned to *Limnopus* or *A. latus*.

The ichnofauna discovered in the Wetterau (eastern extension of the Saar–Nahe Basin) comes from the Bleichenbach Formation. From the footprints, it was correlated 'with the upper part of the Nahe Group of the Saar–Nahe Basin'. Based on the figures of Fichter & Kowalczyk (1983), we recognized *B. salamandroides, Limnopus, A. latus* and *A. discessus*, but the footprints named *Dimetropus leisnerianus* and *G. kablikae* are not really determinable. *Ichniotherium cottae* does not seem characteristic of this ichnofauna, but the existence of this ichnotaxon is certain in the Sponheim (N4) Formation, since it is represented by a trackway preserved at Pollichia Natural History Museum of Bad Dürkheim. Certain *I. cottae* forms from the Wetterau suggest large *H. major* from the Lodève Basin (Gand 1987).

At the end of this revision, which was carried out by G.G. in 1994 (unpublished but communicated to Haubold and Fichter), the following ichnospecies were identified in the Saar–Nahe Basin: *Batrachichnus salamandroides, Limnopus zeilleri, L. palatinus, Amphisauropus latus, A. discessus, Varanopus microdactylus, Dromopus lacertoides, D. didactylus, Ichniotherium cottae* and cf. *H. major*. Such a result naturally modifies the palichnostratigraphy that was worked out by Boy & Fichter (1982) based on ichnospecies that we did not identify (see Fig. 7).

The majority of the footprints collected in the Kusel and Lebach Groups were associated with an 'Autunian' palaeoflora and a vertebrate fauna, both forming characteristic associations of this 'stage' (Boy 1987*a*). Following Broutin *et al.* (1999) it is possible to assign a latest Gzhelian to Early Sakmarian age to the footprint-bearing levels ranging between the base of the Kusel Group and the top of the N3 Formation of the Nahe Group. These results are in rather good agreement with isotopic data. On the one hand, just below the base of the Kusel Group, a tuff yielded a ^{40}A/^{39}A age of 300.3 ± 1.2 Ma (Burger *et al.* 1997), which can be compared with that of the Gzhelian–Asselian boundary (299.0 Ma) on the last geological time scale (Gradstein *et al.* 2004). On the other hand, the isotopic ages of volcanites at the base of the Nahe Group, in the Donnersberg N3 Formation, ranges from 298.7 ± 5.3 to 294.6 ± 4.3 Ma by the ^{40}Ar/^{39}Ar method, and from 292.1 ± 3.4 to 291 ± 1.3 Ma by the Rb/Sr method (Lippolt & Hess 1989).

Conclusions

At the end of these analyses, summarized in Figure 7, we can conclude that many ichnospecies, such as *B. salamandroides, Limnopus zeilleri, Dromopus lacertoides, D. didactylus*, and even *Amphisauropus latus*, have a great vertical distribution. They are thus not good biostratigraphical markers for inter-basin correlations. It is, of course, also the case for taxa whose geographical range is restricted, such as *Amphisauroides* and *Ichniotherium cottae*, which are not found in the United States and are little represented in France, or *Hyloidichnus*, which was not observed in the Lower Permian of Lombardy.

When considering the ichnospecies common to several basins, their vertical distribution differs so much that it is impossible to use them individually, nor even the ichnoassociations observed locally, in order to establish fine correlations between the footprint-bearing formations. It is nevertheless seen that the Lower Permian (Cisuralian) corresponds to a single 'ichnofaunal unit' (Avanzini *et al.* 2001; Haubold & Lucas 2001) characterized by some ichnotaxa: *Amphisauroides*, *Amphisauropus imminutus*, *A. latus* and *Dimetropus leisnerianus* / *D. nicolasi*.

After an important palichnological gap, corresponding at least partially with 'Olson's gap' (Lucas & Heckert 2001), a second 'ichnofaunal unit' can be recognized in the Lodève and Estérel (Provence) basins, in which one finds new ichnospecies. Many of them are attributable to therapsids, but one (*Brontopus giganteus*) is attributed to a late pelycosaurian or a pareiasaurian (Gand *et al.* 1995, 2000). The following ichnospecies seem to characterize this unit: *Lunaepes fragilis, L. ollierorum, Pachypes brachydactylus, P. caudatus, Merifontichnus thalerius, Brontopus giganteus, Chelichnus incurvus* and *Pseudosynaptichnium esterelense*. But it should be noticed that the French basins have in common only the ichnogenera *Lunaepes* and *Planipes*. Moreover, in these areas, the vertical distribution of most ichnospecies is not known conclusively: *B. giganteus* was recently discovered at a second level, 100 m lower than that of La Lieude. Since palaeontological and isotopic data make it possible to date the Estérel supra-A7 footprint association (≈ Lodève Association IV) as Wordian (Durand, 2006), the second Permian 'ichnofaunal unit' seems typical of the Middle Permian (Guadalupian).

It is then necessary to consider the Italian basin of Bolzano, where a third 'ichnofaunal unit', dated as Wuchiapingian, contains some ichnospecies with Triassic features: chirotheroid forms and *Rhynchosauroides*. It characterizes the Upper Permian (Lopingian). Thus, the French palaeoichnologic record allows us to discriminate three time intervals of Permian time in northern Pangaea, as do the skeletal remains (Lozovsky 2003).

The authors are very grateful to S. Lucas (Albuquerque Museum) and M. Lockley (University of Colorado at Denver) for their reviews of this paper, for their constructive remarks and suggestions and for their help concerning the English translation.

References

ARDUINI, P., KRIEGER, C., ROSSI, M. & SANTI, G. 2003. Early Permian vertebrate ichnoassociation in the Scioc Valley, Orobic Basin (Lombardy, Northern Italy). *Neues Jahrbuch für Geologie und Paläontologie, Monatshefte*, **7**, 385–399.

AVANZINI, M., CEOLONI, P. *et al.* 2001. Permian and Triassic tetrapod ichnofaunal units of Northern Italy: their potential contribution to continental biochronology. *In*: CASSINIS, G. (ed.) *Permian Continental Deposits of Europe and Other areas. Regional Reports and Correlations.* Natura Bresciana, Monografia, **25**, 89–107.

BAIRD, D. 1952. Revision of the Pennsylvanian and Permian footprints *Limnopus*, *Allopus* and *Baropus*. *Journal of Paleontology*, **26**, 832–840.

BÉTHOUX, O., NEL, A., GAND, G. & LAPEYRIE, J. 2001. *Surijoka lutevensis* nov. sp.: the first Glosselytrodea (Insecta) from the Upper Permian of France (Lodève Basin). *Géobios*, **34**, 405–413.

BÉTHOUX, O., NEL, A., GAND, G., LAPEYRIE, J. & GALTIER, J. 2002*a*. Discovery of the genus *Iasvia* Zalessky, 1934 in the Upper Permian of France (Lodève Basin) (Orthoptera: Ensifera: Oedischiidae). *Géobios*, **35**, 293–302.

BÉTHOUX, O., NEL, A., LAPEYRIE, J., GAND, G. & GALTIER, J. 2002*b*. *Raphogla rubra* n. gen., n. sp.: the oldest representative of the clade of modern Ensifera (Orthoptera: Tettigoniidea, Gryllidea). *European Journal of Entomology*, **99**, 111–116.

BÉTHOUX, O., NEL, A., GALTIER, J., LAPEYRIE, J. & GAND, G. 2003*a*. A new species of Tococladidae Carpenter, 1966 from the Permian of France (Insecta: Archaeorthoptera). *Géobios*, **36**, 275–283.

BÉTHOUX, O., NEL, A., LAPEYRIE, J., GAND, G. & GALTIER, J. 2003*b*. New Martynoviidae from the Permian of southern France (Lodève Basin) (Insecta: Palaeoptera : Diaphanopterodea). *Géobios*, **36**, 131–139.

BOY, J. A. 1972. Die Branchiosaurier (Amphibia) der saarpfalzischen Rotliegenden (Perm, SW-Deutschland. *Abhandlungen des Hessisches Landesamt für Bodenforschung*, **65**, 1–137.

BOY, J. A. 1987*a*. Die Tetrapoden-Lokalitäten des saarpfälzischen Rotliegenden (? Ober-Karbon–Unter-Perm; SW Deutschland) und die Biostratigraphie der Rotliegenden-Tetrapoden. *Mainzer Geowissenschaftlichen Mittelungen*, **16**, 31–65.

BOY, J. A. 1987*b*. Studien über die Branchiosauridae (Amphibia: Temnospondyli; Ober-Karbon-Unter-Perm). 2. Systematische Übersicht. *Neues Jahrbuch für Geologie und Paläontologie, Abhandlungen*, **174**, 175–104.

BOY, J. A. 1988. Über einige Vertreter der Eryopoidea (Amphibia: Temnospondyli) aus dem europäischen Rotliegend (? höchstes Karbon – Perm). 1. *Sclerocephalus. Paläontologische Zeitschrift*, **62**, 107–132.

BOY, J. A. 1989. Über einige Vertreter der Eryopoidea (Amphibia : Temnospondyli) aus dem europäischen Rotliegend (? höchstes Karbon–Perm). 2. *Acanthostomatops. Paläontologische Zeitschrift*, **63**, 133–151.

BOY, J. A. 1993. Synopsis of the tetrapods from the Rotliegend Permian (Lower Permian) in the Saar-Nahe Basin (SW-Germany). *In*: HEIDTKE, U. (ed.) *New Research on Permo-Carboniferous Faunas*. Pollichia-Bücher, Bad-Dürkheim, 155–169.

BOY, J. A. 1995. Über die Micromelerpetontidae (Amphibia: Temnospondili).1. Morphologie und Paläoökologie des *Micromelerpeton credneri* (Unter-Perm; SW-Deutschland). *Paläontologische Zeitschrift*, **69**, 429–457.

BOY, J. A. 2003. Paläoökologische Rekonstruktion von Wirbelttieren: Möglichkeiten und Grenzen. *Paläontologische Zeitschrift*, **77**, 123–152.

BOY, J. A. & FICHTER, J. 1982. Zur Stratigraphie des saarpfalzischen Rotliegenden (? Ober-Karbon–Unter-Perm; SW Deutschland). *Zeitschrift der Deutsche Geologische Gesellschaft*, **133**, 607–642.

BOY, J. A. & FICHTER, J. 1988. Zur Stratigraphie des höheren Rotliegenden im Saar-Nahe-Becken (Unter-Perm, SW-Deutschland) und seiner Korrelation mit anderen Gebieten. *Neues Jahrbuch für Geologie und Paläontologie, Abhandlungen*, **176**, 331–394.

BROUTIN, J., CHÂTEAUNEUF, J.-J., GALTIER, J. & RONCHI, A. 1999. L'Autunien d'Autun reste-t-il une référence pour les dépôts continentaux du Permien inférieur d'Europe? Apport des données paléobotaniques. *Géologie de la France*, **2**, 17–31.

BURGER, K., HESS, J. C. & LIPPOLT, H. J. 1997. Tephrochronologie mit Kaolin-Kohlentonsteinen: Mittel zur Korrelation paralischer und limnischer Ablagerungen des Oberkarbons. *Geologische Jahrbuch, A*, **147**, 1–39.

CASSINIS, G., CORTESOGNO, L., GAGGERO, L., MASSARI, F., NERI, C., NICOSIA, U. & PITTAU, P. (eds) 1999. *Stratigraphy and Facies of the Permian Deposits between Eastern Lombardy and the Western Dolomites. Field Trip Guidebook, 23–25 September 1999*. International Field Conference on 'The Continental Permian of the Southern Alps and Sardinia (Italy). Regional reports and general correlations', Brescia, 15–25 September 1999. Earth Science Department, Pavia University.

CEOLONI, P., CONTI, M. A., MARIOTTI, N., MIETTO, P. & NICOSIA, U. 1987. Tetrapod footprints from Collio Formation (Lombardy, northern Italy). *Memorie di Scienze Geologiche*, **39**, 213–233.

CEOLONI P., CONTI, M. A., MARIOTTI, N. & NICOSIA, U. 1988. New Late Permian tetrapod footprints from the southern Alps. *Memorie della Società Geologica Italiana*, **34**(1986), 45–65.

CHÂTEAUNEUF, J.-J. & GAND, G. 1989. Stratigraphie. *In*: CHÂTEAUNEUF, J.-J. & FARJANEL, G. (eds) *Synthèse Géologique des Bassins Permiens Français*. Mémoires du Bureau de Recherches Géologiques et Minières, **128**, 159–179.

CONTI, M. A., LEONARDI, G., MARIOTTI, N. & NICOSIA, U. 1977. Tetrapod footprints of the 'Val Gardena Sandstone' (north Italy). Their paleontological, stratigraphic and paleoenvironmental meaning. *Palaeontographia Italica*, **40**, 1–91.

CONTI, M. A., FONTANA, D. *et al.* 1986. Excursion 5: The Bletterbach-Butterloch section (Val Gardena Sandstone and Bellerophon Formation). *In*: S.G.I. and I.G.C.P. Project no.203, Proceedings of the Field Conference on Permian and Permian-Triassic boundary in the South-Alpine segment of the Western Tethys, and additional reports, Brescia, 4–12 July 1986. *Memorie della Società Geologica Italiana*, **34**, 91–110.

CONTI, M. A., MARIOTTI, N., MIETTO, P. & NICOSIA, U. 1991. Nuove ricerche sugli icnofossili della Formazione di Collio in Val Trompia (Brescia). *Natura Bresciana, Monografia*, **26**, 109–119.

CONTI, M. A., MARIOTTI, N., NICOSIA, U. & PITTAU, D. 1997. Selected bioevents succession in the continental Permian of the Southern Alps (Italy): improvements in intrabasinal and interregional correlations. *In*: DICKINS, J. M., YANG, Z. Y., YIN, H. F., LUCAS, S. Q. & ACHARYYA, S. T. (eds) *Late Palaeozoic and Early Mesozoic Circum-Pacific, Events and Their Global Correlation*. Cambridge University Press, Cambridge , 51–65.

CONTI, M. A., MARIOTTI, N., MANNI, R. & NICOSIA, U. 1999. Tetrapod footprints in the Southern Alps: an overview. *In*: CASSINIS, G., CORTEGOGNO, L., GAGGERO, L., MASSARI, F., NERI, C., NICOSIA, U. & PITTAU, P. (eds) *Stratigraphy and Facies of the Permian Deposits between Eastern Lombardy and the Western Dolomites; Field Trip Guidebook, 23–25 September 1999*. International Field Conference on 'The Continental Permian of the Southern Alps and Sardinia (Italy). Regional reports and general correlations', Brescia, 15–25 September 1999. Earth Science Department, Pavia University. 137–139.

DELAGE, A. 1912. *Empreintes de grands pieds de quadrupèdes de l'Hérault*. Mémoire de l'Académie des Sciences & Lettres, Montpellier.

DEMATHIEU, G., GAND, G. & TOUTIN-MORIN, N. 1992. La palichnofaune des bassins permiens provençaux. *Géobios*, **25**, 19–54.

DURAND, M. 2006. The problem of the transition from the Permian to the Tsiassie Series in southeastern France: comparison with other Peritethyan regions. *In*: LUCAS, S. G., CASSINIS, G. & SCHNEIDER, J. W. (eds) *Non-Permian Biostratigraphy and Biochronology*. Geological Society, London, Special Publications, **265**, 281–296.

DURAND, M., GAND, G. & CHÂTEAUNEUF, J.-J. 2002. *Les bassins permiens de Provence*. Livret-guide de la 16ème excursion annuelle de l'Association des Géologues du Permien, 9–11 Mai 2002.

ELLENBERGER, P. 1983*a*. *Sur l'intérêt paléontologique de la dalle à pistes de la Lieude (commune de Mérifons, Hérault, France)*. Bulletin de Liaison de la Société de Protection de la Nature du Languedoc Roussillon, Spécial, **1**.

ELLENBERGER, P. 1983*b*. Sur la zonation ichnologique du Permien moyen (Saxonien du bassin de Lodève, Hérault. *Comptes Rendus de l'Académie des Sciences, Paris, Séries II*, **297**, 553–558.

ELLENBERGER, P. 1983*c*. Sur la zonation ichnologique du Permien inférieur (Autunien du bassin de Lodève, Hérault. *Comptes Rendus de l'Académie des Sciences, Paris, Séries II*, **297**, 631–636.

ELLENBERGER, P. 1984. Données complémentaires sur la zonation ichnologique du Permien du Midi de la France: bassins de Lodève, Saint-Affrique et Rodez. *Comptes Rendus de l'Académie des Sciences, Paris, Séries II*, **299**, 581–586.

ELLENBERGER, F. & ELLENBERGER, P. 1959. Quelques pistes de vertébrés du Permien inférieur de Lodève. *Comptes Rendus de l'Académie des Sciences, Paris*, **248**, 437.

FICHTER, J. 1976. Tetrapodenfährten aus dem Unterrotliegenden (Autun, Unter-Perm) von Odernheim/Glan. *Mainzer Geowissenschaftlichen Mitteilungen*, **5**, 87–109.

FICHTER, J. 1982. Tetrapodenfährten aus dem Oberkarbon (Westfalium A und C) West-und Südwestdeutschlands. *Mainzer Geowissenschaftlichen Mitteilungen*, **11**, 33–77.

FICHTER, J. 1983*a*. Tetrapodenfährten aus dem saarpfälzischen Rotliegenden (?Ober- Karbon–Unter-Perm; Südwest-Deutschland) 1: Fährten der Gattungen *Saurichnites*, *Limnopus*, *Amphisauroides*, *Protritonichnites*, *Gilmoreichnus*, *Hyloidichnus* und *Jacobiichnus*. *Mainzer Geowissenschaftlichen Mitteilungen*, **12**, 9–121.

FICHTER, J. 1983*b*. Tetrapodenfährten aus dem saarpfälzischen Rotliegenden (? Ober- Karbon–Unter-Perm; Südwest-Deutschland) 2: Färthen der Gattungen: *Foliipes*, *Varanopus*, *Ichniotherium*, *Dimetropus*, *Palmichnus*, *Phalangichnus*, cf. *Chelichnus*, cf. *Laoporus* und *Anhomoiichnium*. *Mainzer Naturwissenschaftliches Archiv*, **21**, 125–186.

FICHTER, J. 1984. Neue Tetrapodenfährten aus den saarpfälzischen Standenbühl-Schichten (Unter Perm; SW Deutschland). *Mainzer Naturwissenschaftliches Archiv*, **22**, 211–229.

FICHTER, J. 1994. Permische Saurierfährten. Ein Diskussionsbeitrag zu der Bearbeitungsproblematik der Tetrapodenfährten des Cornberger Sandsteins (Perm, Deutschland) und des Coconino Sandsteins (Perm, USA). *Philippia*, **7**, 61–82.

FICHTER, J. & KOWALCZYCK, G. 1983. Tetrapodenfährten aus dem Rotliegenden der Wetterau und ihre stratigraphische Auswertung. *Mainzer Geowissenschaftliche Mitteilungen*, **12**, 123–158.

FRITSCH, A. 1901. *Fauna der Gaskohle und der Kalksteine der Permformation Böhmens*. F. Řivnáč, Prague, **4**(3), 93–97.

GAND, G. 1985. Significations paléobiologique et stratigraphique de *Limnopus zeilleri* dans la partie nord du bassin de St-Affrique. *Géobios*, **18**, 215–227.

GAND, G. 1986. Interprétations paléontologique et paléoécologique de quatre niveaux à traces de vertébrés observés dans l'Autunien du Lodèvois (Hérault). *Géologie de la France*, **2**, 155–176.

GAND, G. 1987. *Les traces de Vertébrés Tétrapodes du Permien français: Paléontologie, stratigraphie, paléoenvironnements*. Thesis, Bourgogne University.

GAND, G. 1989. *Varanopus rigidus*: une nouvelle ichnoespèce de vertébré tétrapode du Permien français attribuable à des Captorhinomorphes ou à des Procolophonoidés. *Géobios*, **22**, 227–291.

GAND, G. 1993. La palichnofaune de Vertébrés tétrapodes du bassin permien de St-Affrique (Aveyron): comparaisons et conséquences stratigraphiques. *Géologie de la France*, **3–4**, 41–56.

GAND, G. & HAUBOLD, H. 1984. Traces de Vertébrés du Permien du bassin de Saint-Affrique (Description, datation, comparaison avec celles du bassin de Lodève). *Géologie Méditerranéenne*, **11**, 321–348.

GAND, G. & HAUBOLD, H. 1988. Permian tetrapod footprints in central Europe, stratigraphical and paleontological aspects. *Zeitschrift für Geologische Wissenschaften*, **16**, 885–894.

GAND, G., DEMATHIEU, G. & BALLESTRA, F. 1995. La palichnofaune de vertébrés tétrapodes du Permien supérieur de l'Estérel (Provence, France). *Palaeontographica, Abteilung A*, **235**, 97–139.

GAND, G., LAPEYRIE, J., GARRIC, J., NEL, A., SCHNEIDER, J. & WALTER, H. 1997. Découverte d'Arthropodes et de bivalves inédits dans le Permien continental (Lodévois, France). *Comptes Rendus de l'Académie des Sciences, Paris, Série IIA*, **325**, 891–898.

GAND, G., GARRIC, J., DEMATHIEU, G. & ELLENBERGER, P. 2000. La palichnofaune de vertébrés tétrapodes du Permien Supérieur du bassin de Lodève (Languedoc-France). *Palaeovertebrata*, **29**, 1–82.

GAND, G., GALTIER, J., GARRIC, J., SCHNEIDER, J., KÖRNER, F. & BÉTHOUX, O. 2004*a*. Les bassins carbonifère de Graissessac et permien de Lodève (Languedoc-France): Livret-guide de l'excursion n° 3 du 7 mai 2001. *Bulletin de la Société d'Histoire Naturelle d'Autun*, **185**, 7–40.

GAND, G., GALTIER, J., GARRIC, J., SCHNEIDER, J., KÖRNER, F. & BÉTHOUX, O. 2004*b*. Les bassins carbonifère de Graissessac et permien de Lodève (Languedoc-France): Livret-guide de l'excursion n° 3 du 7 mai 2001. *Bulletin de la Société d'Histoire Naturelle d'Autun*, **186**, 9–40.

GEINITZ, H. B. 1861. *Dyas I. Dyas oder die Zechsteinformation und des Rotliegunde. vol. I: Die Animalischen ueberreste der Dyas*. William Engelmann, Leipzig.

GEINITZ, H. B. 1863. Beiträge zur Kenntnis der organischen Überreste in der Dyas (oder permischen Formation zum Theil) and Über den Namen Dyas. *Neues Jahrbuch für Mineralogie Geologie und Paläontologie*, 385–389.

GERMAIN, D. 1968. Au sujet des bois silicifiés du Permien supérieur du Muy (massif de l'Estérel, Var, France). *Bulletin de la Société Belge de Paléontologie et d'Hydrologie*, **57**, 203–215.

GILMORE, C. W. 1927. *Fossil Footprints from the Grand Canyon: Second Contribution*. Smithsonian Miscellaneous Collections, **80**(3).

GRADSTEIN, F. M., OGG, J. G., SMITH, A. G., BLEEKER, W. & LOURENS, L. J. 2004. A new geologic time scale with special reference to Precambrian and Neogene. *Episodes*, **27**, 83–100.

HAUBOLD, H. 1970. Versuch einer Revision der Amphibien-Fährten des Karbon und Perm. *Freiberger Forschungshefte, Hefte C*, **260**, 83–109.

HAUBOLD, H. 1971. Ichnia Amphibiorum et Reptiliorum fossilium. *Handbuch der Palaeoherpetology*, **18**.

HAUBOLD, H. 1973. Die Tetrapodenfährten aus dem Perm Europas. *Freiberger Forschungshefte, Hefte C*, **285**, 5–55.

HAUBOLD, H. 1984. *Saurierfährten*. Neue Brehm Bücherei, Wittenberg Lutherstadt.

HAUBOLD, H. 1996. Ichnotaxonomie und Klassifikation von Tetrapodenfährten aus den Perm. *Hallesches Jahrbuch für Geowissenschaften, B*, **18**, 23–88.

HAUBOLD, H. 1998*a*. The Early Permian tetrapod ichnofauna of Tambach, the changing concepts in ichnotaxonomy. *Hallesches Jahrbuch für Geowissenschaften Reihe B*, **20**, 1–16.

HAUBOLD, H. 1998*b*. The Early Permian tetrapod track assemblage of Nierstein, Standenbühl Beds, Rotliegend, Saar-Nahe Basin, SW-Germany. *Hallesches Jahrbuch für Geowissenschaften Reihe B*, **20**, 17–32.

HAUBOLD, H. 2000. Tetrapoden aus Perm. Kenntnisstand und Progress 2000. *Hallesches Jahrbuch für Geowissenschaften Reihe B*, **22**, 1–16.

HAUBOLD, H. & LUCAS, S. G. 2001*a*. The tetrapod footprints of the Choza Formation (Texas) and the Artinskian age of the Lower Permian ichnofaunas. *Hallesches Jahrbuch für Geowissenschaften Reihe B*, **23**, 79–108.

HAUBOLD, H. & LUCAS, S. G. 2001*b*. Early Permian tetrapod tracks: preservation, taxonomy, and Euramerican distribution. *In*: CASSINIS, G. (ed.) *Permian Continental Deposits of Europe and Other areas. Regional Reports and Correlations*. Natura Bresciana, Monografia **25**, 347–357.

HAUBOLD, H. & LUCAS, S. G. 2003. Tetrapod footprints of the Lower Permian Choza Formation at Castle Peak, Texas. *Paläontologische Zeitschrift*, **77**, 247–261.

HAUBOLD, H., HUNT, A. P., LUCAS, S. G. & LOCKLEY, M. G. 1995*a*. Wolfcampian (Early Permian) vertebrates tracks from Arizona and New Mexico. *In*: LUCAS, S. G. & HECKERT, A. B. (eds) *Early Permian Footprints and Facies*. New Mexico Museum of Natural History and Science Bulletin, **6**, 135–165.

HAUBOLD, H., LOCKLEY, M. G., HUNT, A. P. & LUCAS, S. G. 1995*b*. Lacertoid footprints from Permian Dune Sandstones, Cornberg and DeChelly Sandstones. *In*: LUCAS, S. G. & HECKERT, A. B. (eds) *Early Permian Footprints and Facies*. New Mexico Museum of Natural History and Science Bulletin, **6**, 235–244.

HEYLER, D. 1984. *Margennipes pansioti*, ng, nsp: Empreinte d'un vertébré tétrapode dans l'Autunien supérieur du bassin d'Autun. *Bulletin de la Société d'Histoire Naturelle, d'Autun*, **111**, 27–36.

HEYLER, D. & LESSERTISSEUR, J. 1962. Remarques sur les allures des Tétrapodes paléozoïques d'après les pistes du Permien de Lodève. *Colloques du Centre National de la Recherche Scientifique, Paris*, **104**, 123–134.

HEYLER, D. & LESSERTISSEUR, J. 1963. Pistes de Tétrapodes permiens dans la région de Lodève (Hérault). *Mémoire du Museum National d'Histoire Naturelle, Paris*, **11**, 125–221.

HEYLER, D. & MONTENAT, C. 1980. Traces de pas de vertébrés du Permien du Var. Intérêt biostratigraphique. *Bulletin du Museum National d'Histoire Naturelle, Paris, c*, **2**, 407–451.

JARDINE, W. 1851. *The ichnology of Annandale or illustrations of footmarks impressed on the New Red Sandstone of Corncockle Muir*. Privately published, Edinburgh.

JIN, Y. G., WARDLAW, B. R., GLENISTER, B. F. & KOTLYAR, G. V. 1997. Permian chronostratigraphic subdivisions. *Episodes*, **20**, 9–15.

KERP, H. & FICHTER, J. 1985. Die Makrofloren des Saarpfälzischen Rotliegenden (? Ober-Karbon – Unter-Perm; SW-Deutschland). *Mainzer Geowissenschaftliche Mitteilungen*, **14**, 159–286.

LAVERSANNE, J. 1976. *Sédimentation et minéralisation du Permien de Lodève (Hérault)*. PhD thesis, Paris-Sud University.

LETHIERS, F., DAMOTTE, R. & SANFOURCHE, J. 1993. Premières données sur les ostracodes du Permien continental de l'Estérel (SE de la France): s`à'--`-```ð¨-`-````2176`-```ðI`-`•paléoécologie. *Géologie Méditerranéenne*, **20**, 109–125.

LIPPOLT, H. J. & HESS, C. 1989. Isotopic evidence for the stratigraphic position of the Saar-Nahe Rotliegende volcanism. III: Synthesis of results and geological implications. *Neues Jahrbuch für Geologie und Paläontologie, Monatshefte*, **9**, 553–559.

LOZOVSKY, V. R. 2003. Correlation of the continental Permian of northern Pangea: a review. *Bolletino della Società Geologica Italiana, Volume speciale*, **2**, 239–244.

LUCAS, S. G. 2002*a*. North American Permian tetrapod footprint biostratigraphy. *Permophiles*, **40**, 22–24.

LUCAS, S. G. 2002*b*. Global Permian tetrapod footprint biostratigraphy and biochronology. *Permophiles*, **41**, 30–34.

LUCAS, S. G. & HECKERT, A. B. 2001. Olson"s gap: A global hiatus in the record of middle Permian tetrapods. *Journal of Vertebrate Paleontology*, **21**(3, supplement), 75.

LULL, R. S. 1942. Chugwater footprints from Wyoming. *American Journal of Science*, **240**, 500–504.

MAIDWELL, F. T. 1911. Note on footprints from the Keuper of Runcorn Hill. *Liverpool Geological Society Proceedings*, **11**, 140–152.

MARSH, O. C. 1894. Footprints of vertebrates in the Coal Measures of Kansas. *American Journal of Science*, **3**, 81–84.

MASSARI, F., NERI, C. *et al.* 1999. Excursion 3: The Bletterbach-Butterloch section (Val Gardena Sandstone and Bellerophon Formation). *In*: CASSINIS, G., CORTESOGNO, L., GAGGERO, L., MASSARI, F., NERI, C., NICOSIA, U. & PITTAU, P. (eds) *Stratigraphy and Facies of the Permian Deposits between Eastern Lombardy and the Western Dolomites. Field Trip Guidebook, 23–25 September 1999*. International Field Conference on 'The Continental Permian of the Southern Alps and Sardinia (Italy). Regional reports and general correlations; Brescia, 15–25 September 1999. Earth Science Department, Pavia University, 111–134.

MENNING, M. 1994. A numerical time scale for the Permian and Triassic periods: An integrated time analysis. *In*: SCHOLLE, P. A., PERYT, T. M & ULMER-SCHOLLE, D. S. (eds) *The Permian of Northern Pangea, Vol. 1*. Springer Verlag, Berlin, 77–97.

MENNING, M. 2001. A Permian time scale 2000 and correlation of marine and continental sequences using the Illawarra Reversal (265 Ma). *In*: CASSINIS, G. (ed.) *Permian Continental Deposits of Europe and Other Areas. Regional Reports and Correlations*. Natura Bresciana, Monografia, **25**, 355–362.

MOODIE, R. L. 1929. Vertebrate footprints from the red beds of Texas. I. *American Journal of Science*, **5**, 352–368.

MOODIE, R. L. 1930. Vertebrate footprints from the red beds of Texas. II. *American Journal of Science*, **38**, 548–565.

MORALES, M. & HAUBOLD, H. 1995. Tetrapod tracks from the Lower Permian DeChelly Sandstone of Arizona: systematic description. *In*: LUCAS, S. G. & HECKERT, A. B. (eds) *Early Permian Footprints and Facies*. New Mexico Museum of Natural History and Science Bulletin, **6**, 251–261.

MÜLLER, A. H. 1954. Zur Ichnologie und Stratonomie des Oberrotliegenden von Tambach, Thüringen. *Paläontologische Zeischrift*, **28**, 189–203.

NEL, A., GAND, G., GARRIC, J. & LAPEYRIE, J. 1999. The first recorded Protozygopterian insects from the Upper Permian of France. *Palaeontology*, **42**, 83–97.

NICOSIA, U., RONCHI, A. & SANTI, G. 1999. Permian tetrapod footprints from W Orobic basin (Northern Italy) with correlation. *In*: CASSINIS, G., CORTESOGNO, L., GAGGERO, L., PITTAU, P., RONCHI, A. & SARRIA, E. (eds) *Stratigraphy and Facies of the Permian Deposits between Eastern Lombardy and the Western Dolomites: Field Trip Guidebook, 23–25 September 1999*. International Field Conference on 'The Continental Permian of the Southern Alps and Sardinia (Italy): Regional Reports and General Correlations'. September 1999, Brescia, Abstracts, 28–30.

NICOSIA, N., RONCHI, A. & SANTI, G. 2000. Permian tetrapod footprints from Gerola Valley (Lombardy, Northern Italy). *Géobios*, **33**, 753–768.

NICOSIA, N., RONCHI, A. & SANTI, G. 2001. Tetrapod footprints from the Lower Permian of Western Orobic Basin (N. Italy). *In*: CASSINIS, G. (ed.) *Permian Continental Deposits of Europe and Other Areas. Regional Reports and Correlations*. Natura Bresciana, Monografia, **25**, 45–50.

NOPSCA, F. 1923. *Die Familien der Reptilien*. Fortschritte der Geologie und Paläontologie, Berlin, **2**.

ODIN, B. 1986. *Les formations permiennes, Autunien supérieur à Thuringien, du 'bassin' de Lodève (Hérault, France): stratigraphie, minéralogie, paléoenvironnements, corrélations*. PhD thesis, Aix-Marseille University.

PABST, W. 1908. Die Tierfährten in dem Rotliegenden Deutchlands. *Nova Acta Leopoldina*, **89**, 320–481.

POHLIG, H. 1885. Saurierfährten im unteren Rotliegenden von Friedrichroda. *Verhandlungen der Naturhistorischen Vereins der Preussischen Rheinlande und Westfalens*, **42**, 285–286.

SANTI, G. 2001. Tetrapod ichnology of Early Permian deposits in the Orobic basin (Southern Alps, Northern Italy). International Meeting on Stratigraphy and Structural Evolution on the Late Carboniferous to Triassic continental and marine successions in Tuscany (Italy). Regional Reports and General Correlation. Siena, 30 April–7 May 2001. Abstract volume, 59–60.

SANTI, G. 2003. Early Permian tetrapod footprints from the Orobic Basin (Southern Alps–Northern Italy). Data, problems, hypotheses. *Bolletino della Società Geologica Italiana, Volume Speciale*, **2**, 59–66.

SANTI, G. O. & KRIEGER, C. 1999. Nuove impronte di tetrapodi permiani dalla valle dell'Inferno (Lombardia, Italia Settentrionale). Osservazioni sistematiche e biochronologiche. *Geologia Insubrica*, **4**, 27–33.

SANTI, G. O. & KRIEGER, C. 2001. Lower Permian tetrapod footprints from Brembana Valley–Orobic Basin–(Lombardy, northern Italy). *Revue de Paléobiologie*, **20**, 45–68.

SARJEANT, W. A. S. 1971. Vertebrate tracks from the Permian of Castle Peak, Texas. *Texas Journal of Science*, **22**, 344–366.

SCHÄFER, A. & KORSCH, R. J. 1998. Formation and sediment fill of the Saar–Nahe Basin (Permo-Carboniferous, Germany). *Zeitschrift der Deutsche Geologische Gesellschaft*, **149**, 233–269.

SCHMIDT, H. 1959. Die Cornberger Färhten im Rahmen der Vierfüssler-Entwicklung. *Abhandlungen des Hessisches Landesamt für Bodenforschung*, **28**, 137 p.

STAPF, K. 1992. *The Rotliegend (Continental Permian) of the Saar–Nahe Basin (SW-Germany)*. Institut für Geowissenschaften of the Johannes Gutenberg-Universität, Mainz.

STAPF, K. 2003. Paläoböden im Rotliegend am Nordwestrand des Saar-Nahe-Beckens (SW-Deutschland). *Mitteilungen der Pollichia*, **90**, 61–120.

SUMIDA, S. S., BERMAN, D. S. & MARTENS, T. 1996. Biostratigraphic correlations between the Lower Permian of North America and central Europe using the first record of an assemblage of terrestrial tetrapods from Germany. *PaleoBios*, **17**, 1–12.

TILTON, E. 1931. Permian vertebrates tracks in West Virginia. *Geological Society of America, Bulletin*, **42**, 547–566.

TOUTIN, N. 1980. Le Permien continental de la Provence orientale France. Science thesis, Nice University.

TOUTIN-MORIN, N., BONIJOLY, D. *et al.* 1994. *Notice explicative: Carte géologique de la. France à 1150 000, feuille Fréjus-Cannes, n° 1024*. Bureau de Rechesche Geologique Miniere, Orléans.

VISSCHER, H. 1968. On the Thuringian age of the Upper Palaeozoic sedimentary and volcanic deposits of the Estérel, southern France. *Review of Palaeobotany and Palynology*, **6**, 71–83.

VOIGT, S. 2001. Variation and preservation of *Ichniotherium* in the Tambach Sandstone (Rotliegend, Thuringia). *In*: CASSINIS, G. (ed.) *Permian Continental Deposits of Europe and Other Areas. Regional Reports and Correlations*. Natura Bresciana, Monografia, **25**, 221–225.

VOIGT, S. & HAUBOLD, H. 2000. Analyse zur Variabilität der Tetrapodenfährte *Ichniotherium cottae* aus dem Tambacher Sandstein (Rotliegend, Unter-Perm, Thüringen). *Hallesches Jahrbuch für Geowissenschaften, B*, **22**, 17–58.

VOZENIN-SERRA, C., BROUTIN J. and TOUTIN-MORIN, N. 1991. Bois permiens du Sud Ouest de l'Espagne et du Sud-Est de la France. Implications pour la taxonomie des Gymnospermes paléozoïques et la phylogénie des Ginkgophytes. *Palaeontographica, Abteilung B*, **221**, 1–26.

WOODWORTH, J. B. 1900. Vertebrate footprints on Carboniferous shales of Plainville, Massachusetts. *Bulletin of the Geological Society of America*, **11**, 449–454.

ZHENG, J. S., MERMET, J.-F., TOUTIN-MORIN, N., HANES, J., GONDOLO, A., MORIN, R. & FÉRAUD, G. 1992. Datation ^{40}Ar-^{39}Ar du magmatisme et de filons minéralisés permiens en Provence orientale France. *Geodinamica Acta*, **5 (1991/1992)**, 203–215.

Permian tetrapod footprints: biostratigraphy and biochronology

SPENCER G. LUCAS & ADRIAN P. HUNT

New Mexico Museum of Natural History, 1801 Mountain Rd. NW, Albuquerque, New Mexico 87104-1375, USA

Abstract: Permian tetrapod footprints are known from localities in North America, South America, Europe and Africa. These footprints comprise four ichnofacies, the *Chelichnus* ichnofacies from aeolianites and the *Batrachichnus, Brontopodus* and *Characichnos* ichnofacies from water-laid (mostly red-bed) strata. Permian track assemblages of the *Chelichnus* ichnofacies are of uniform ichnogeneric composition and low diversity, range in age from Early to Late Permian, and thus are of no biostratigraphic significance. Footprints of the *Batrachichnus* and *Brontopodus* ichnofacies represent two biostratigraphically distinct assemblages: (1) Early Permian assemblages characterized by *Amphisauropus, Batrachichnus, Dimetropus, Dromopus, Hyloidichnus, Limnopus* and *Varanopus*; and (2) Middle to Late Permian assemblages characterized by *Brontopus, Dicynodontipus, Lunaepes, Pachypes, Planipes*, and/or *Rhynchosauroides*. Few Permian footprint assemblages are demonstrably of Middle Permian (Guadalupian) age, and there is a global gap in the footprint record equivalent to at least Roadian time. Permian tetrapod footprints represent only two biostratigraphically distinct assemblages, an Early Permian pelycosaur assemblage and a Middle to Late Permian therapsid assemblage. Therefore, footprints provide a global Permian biochronology of only two time intervals, much less than the ten time intervals that can be distinguished with tetrapod body fossils.

The global record of Permian tetrapod footprints encompasses localities in North America, South America, Europe and Africa (Fig. 1). Permian tetrapod footprints can be assigned to four ichnofacies, an aeolian *Chelichnus* ichnofacies and water-laid (red-bed) *Batrachichnus, Brontopodus* and *Characichnos* ichnofacies (Hunt & Lucas 2006). Various biostratigraphical schemes employing tetrapod footprints have been proposed, particularly for the Early Permian *Batrachichnus* ichnofacies, especially in Europe. For the purposes of a global Permian tetrapod footprint biostratigraphy, the operational taxonomic unit is the ichnogenus, as almost all ichnospecies are variants confined to a single locality and thus of little biostratigraphical value. Here, we rely primarily on the ichnotaxonomy of Haubold (1996, 2000) and McKeever & Haubold (1996) to review the biostratigraphical distribution of Permian tetrapod footprints to argue that on a global basis they only discriminate two intervals of Permian time.

Ichnotaxonomy

Biostratigraphy and biochronology are strongly dependent on taxonomy. This is because index taxa – those used to indicate age equivalence (correlation) – must be taxa with a well-founded and agreed taxonomy. Disagreements about correlations are often based on disagreements about taxonomy that undermine the identification of index taxa.

Prior to the mid-1990s, about 150 ichnogenera of Permian tetrapod footprints had been named (most of them of Early Permian age) (Haubold 2000). Many of these ichnogenera (and their ichnospecies) were based on small samples that appeared to demonstrate distinctive footprint structures and therefore seemed to justify the naming of many ichnotaxa. However, in 1994, the discoveries of Jerry MacDonald, an oustanding amateur footprint collector, in the Lower Permian strata of southern New Mexico, United States, became available for study (see articles in Lucas & Heckert 1995). MacDonald's collection consisted of more than 2000 slabs with footprints from a mega-tracksite in the Robledo Mountains of southern New Mexico, and localities in the field included many more. Most importantly, large surfaces were available for study that showed many footfalls of individual animals (trackways) in different substrate and gait conditions.

Peabody (1948) articulated much of the basis of the methodology that we (with Haubold and others) have employed to interpret the footprint variation in this huge sample. Like almost all other vertebrate ichnologists, we regard vertebrate ichnotaxa as proxies of biotaxa. In

From: LUCAS, S. G., CASSINIS, G. & SCHNEIDER, J. W. (eds) 2006. *Non-Marine Permian Biostratigraphy and Biochronology.* Geological Society, London, Special Publications, **265,** 179–200.
0305-8719/06/$15.00

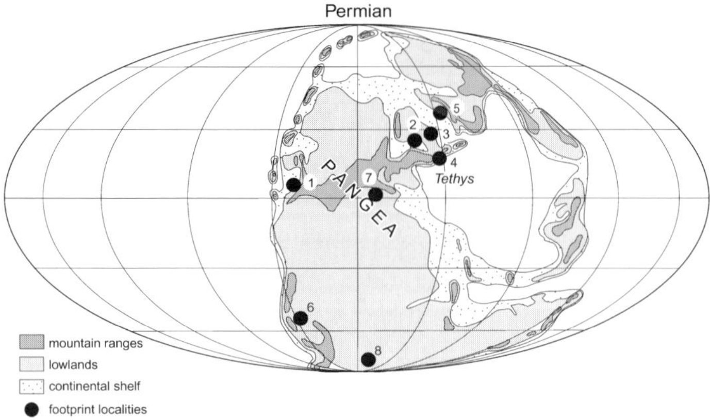

Fig. 1. Distribution of principal Permian tetrapod tracksites on Permian Pangea. Locations are: 1, western United States; 2, France; 3, Germany; 4, Italy; 5, Russia; 6, Argentina; 7, Morocco; 8, South Africa.

other words, we want different tetrapod ichnotaxa to correspond to discrete biotaxa, although we realize that the biotaxon equivalent to an ichnogenus is likely to be a family, suborder or order. To achieve this, we examine the various footprints, looking for what might be considered optimal tracks with a structure that best reflects the actual foot morphology. This optimal track structure is identified by matching the track to a presumed trackmaker. We consider other track structures that do not reflect the actual foot structure as suboptimal and regard them as extramorphological variants (Peabody's term). In other words, we judge the reason why the suboptimal tracks do not match the foot structure to be the result of differences in substrate, gait or other factors (especially taphonomy) that prevented an optimal footprint from being preserved. By looking at a wide range of variation in individual trackways and across multiple trackways, we separated what we concluded are many extramorphological variants from the optimal tracks (Fig. 2). The result has been the elimination of numerous ichnogenera that were demonstrably based on extramorphological variants of a valid ichnogenus based on optimal track morphology. The most striking example is the small temnospondyl track ichnogenus *Batrachichnus*, which includes tracks that have been called *Anthichnium, Crenipes, Dromillopus, Nanopus, Salichnium, Saurichnites* and many others (Haubold 1996; Haubold & Lucas 2001*a*).

Lucas (2005*a*) called this approach to tetrapod footprint ichnotaxonomy the 'fusion method' because it eliminates many names applied to extramorphological variants and recognizes as valid only one name based on an optimal track structure and its extramorphological variants (it thus fuses many names into one). The result, in popular parlance, has been ichnotaxonomic 'lumping' of the many Permian tetrapod footprint ichnogenera into a much smaller number of ichnogenera (e.g. Haubold 1996, 2000; McKeever & Haubold 1996; Voigt 2005). An easy measure of this is to compare Schult (1995), who recognized 23 ichnogenera in the Robledo Mountains megatracksite by attaching names to many extramorphological variants, with Hunt *et al.* (1995), who, using the fusion method, recognized only seven ichnogenera in the same sample.

One result of the sweeping ichnotaxonomic revisions of Haubold, Hunt and Lucas is to recognize that there is one tetrapod footprint assemblage (ichnofauna) in Lower Permian water-laid (usually red-bed) strata in the United States, Canada, Argentina, Germany, France, Italy, Russia and some other places in Europe (Hunt & Lucas 1998). In other words, the tetrapod footprints in Early Permian red beds are a single assemblage of broad, uniform composition. The following ichnogenera dominate: *Amphisauropus, Batrachichnus, Dimetropus, Dromopus, Hyloidichnus, Ichniotherium, Limnopus* and *Varanopus*. This assemblage is composed mostly of the tracks of temnospondyls, diadectomorphs, seymouriamorphs, procolophonids and pelycosaurs (Table 1). The North American record demonstrates that most (if not all) of these ichnogenera have long stratigraphical ranges through most or all of Early Permian time (Haubold & Lucas 2001*a*, *b*, 2003; Lucas 2002*b*). Furthermore, at the Robledo Mountains mega-tracksite in southern New Mexico, almost

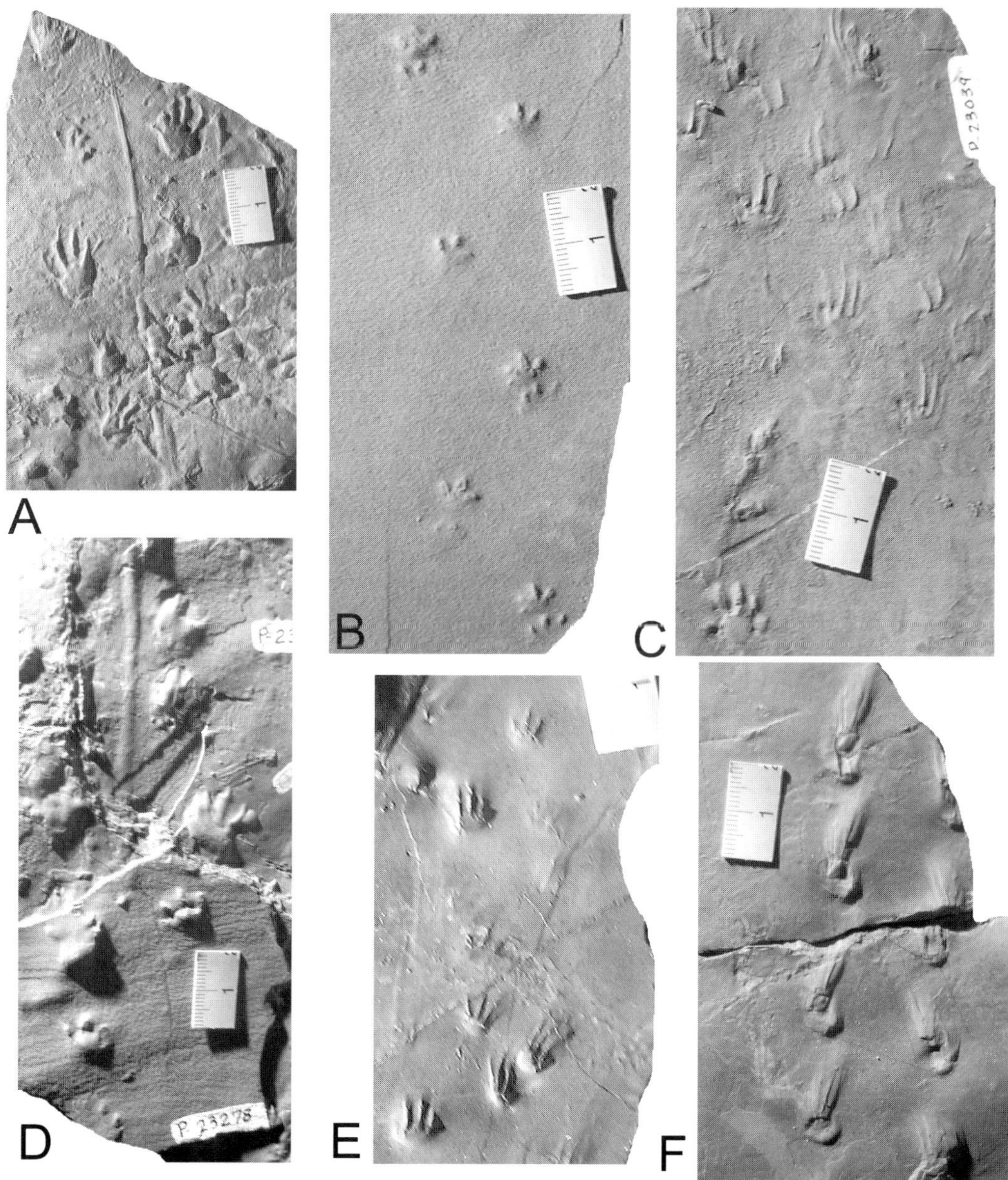

Fig. 2. Examples of small temnospondyl tracks from the Lower Permian Robledo Mountains megatracksite in southern New Mexico. All specimens are assigned to *Batrachichnus delicatulus* (Lull) and are in the collection of the New Mexico Museum of Natural History (NMMNH). They demonstrate nearly optimal footprint structure (A) and a variety of suboptimal extramorphological variants. The specimen in D is particularly interesting because it shows near optimal footprint structures with a median tail or body drag on the underside of one bedding plane (above), and the underside of a lower bedding plane (below) with underprints and no median drag. A, NMMNH P-23001; B, NMMNH P-23174; C, NMMNH P-29039-040; D, NMMNH P-23277-78; E, NMMNH P-23952; F, NMMNH P-23432.

Table 1. *Common Permian tetrapod footprint ichnogenera and the inferred trackmakers (based largely on Haubold 1996)*

Ichnofacies	Ichnogenera	Trackmaker
Batrachichnus (Early Permian)	*Amphisauropus*	seymouriamorph
	Batrachichnus	temnospondyl
	Dimetropus	pelycosaur
	Dromopus	araeoscelid
	Erpetopus	captorhinomorph
	Hyloidichnus	captorhinomorph
	Ichniotherium	diadectomorph
	Limnopus	temnsopondyl
	Tambachichnium	araeoscelid
	Varanopus	captorhinomorph
Chelichnus	*Chelichnus*	pelycosaur
Batrachichnus (Mid- to Late Permian)	*Brontopus*	therapsid
	Dicynodontipus	therapsid
	Lunaepes	therapsid
	Merifontichnus	therapsid
	Pachypes	pareiasaur
	Rhynchosauroides	eosuchian

all of these ichnogenera co-occur in a single, narrow stratigraphical interval. This suggests that local biostratigraphical zonations based on these ichnotaxa, especially those proposed in Germany and France, are not of global applicability and may also be of questionable utility, even at the local or regional scale.

A similar, broad-based ichnotaxonomic review of tracks of the *Chelichnus* ichnofacies has greatly simplified ichnotaxonomy, reducing ichnogeneric diversity to simply *Chelichnus* (Morales & Haubold 1995; McKeever & Haubold 1996). There has not, however, been a similar broad ichnotaxonomic revision of the Middle to Late Permian footprints attributed to pareiasaurs and therapsids. Because of this, we use the current names, although we are skeptical of the validity of some of them. Thus, our purpose here is not to revise ichnotaxonomy, so we list ichnogenera as reported by what we consider the most reliable source. Ichnogeneric names that we have placed in quotation marks are those we consider to be questionable identifications.

Ichnofacies

Permian tetrapod footprints have previously been assigned to two principal ichnofacies: an aeolian *Chelichnus* ichnofacies and a water-laid (red-bed) *Batrachichnus* ichnofacies (Hunt & Lucas 2005*b*, 2006; Hunt *et al.* 2005*c*, *d*). Hunt & Lucas (2006) have also assigned Mid- to Late Permian track assemblages from water-laid strata to the *Brontopodus* and *Characichnos* ichnofacies. Swanson & Carlson (2002) described Early Permian tetrapod footprints from dolomitic strata in Oklahoma and suggested that they may represent another, little known ichnofacies, but we regard this footprint assemblage as a poorly preserved example of the *Batrachichnus* ichnofacies.

The *Batrachichnus* ichnofacies encompasses ichnoassemblages in which the majority of tracks are of quadrupedal carnivores with a moderate to high diversity (four to eight ichnogenera). This ichnofacies represents tidal flat through fluvial plain environments from the Devonian to the Middle Triassic. The *Batrachichnus* ichnofacies encompasses one previously named ichnocoenosis, originally named as an ichnofacies: *Batrachichnus* from the Early Carboniferous to Early Permian, which is separable into sub-ichnocoenoses:

(1) inland/distal alluvial fan settings characterized by an abundance of *Ichniotherium* and a paucity of *Dimetropus*, the *Ichniotherium* sub-ichnocoenosis;
(2) alluvial plain settings characterized by the presence of *Amphisauropus*, the *Amphisauropus* sub-ichnocoenosis;
(3) coastal/tidal flat settings characterized by the relative abundance of *Batrachichnus* and *Dimetropus*, the *Dimetropus* sub-ichnocoenosis (Hunt & Lucas 2005*b*, 2006) (Fig. 3).

The *Chelichnus* ichnofacies encompasses ichnofaunas that have a low diversity (less than five ichnogenera) of tetrapod tracks in which manus and pes tracks are subequal in size and equant in shape, with short digit impressions. This ichnofacies is recurrent in dune faces of aeolian environments, and it extends from the Early Permian to the Early Jurassic. The *Chelichnus* ichnofacies encompasses two named ichnocoenoses (originally named as ichnofacies). These are the *Chelichnus* (= *Laoporus*) ichnocoenosis of Early Permian age (Lockley *et al.* 1994; Hunt & Lucas 2005*b*) and the *Brasilichnium* ichnocoenose of Late Triassic to Early Jurassic age (Lockley *et al.* 1994, 2004; Schultz-Pittman *et al.* 1996).

The *Brontopodus* ichnofacies encompasses medium diversity ichnoassemblages in which the majority of tracks are of terrestrial herbivores with a small quantity (generally

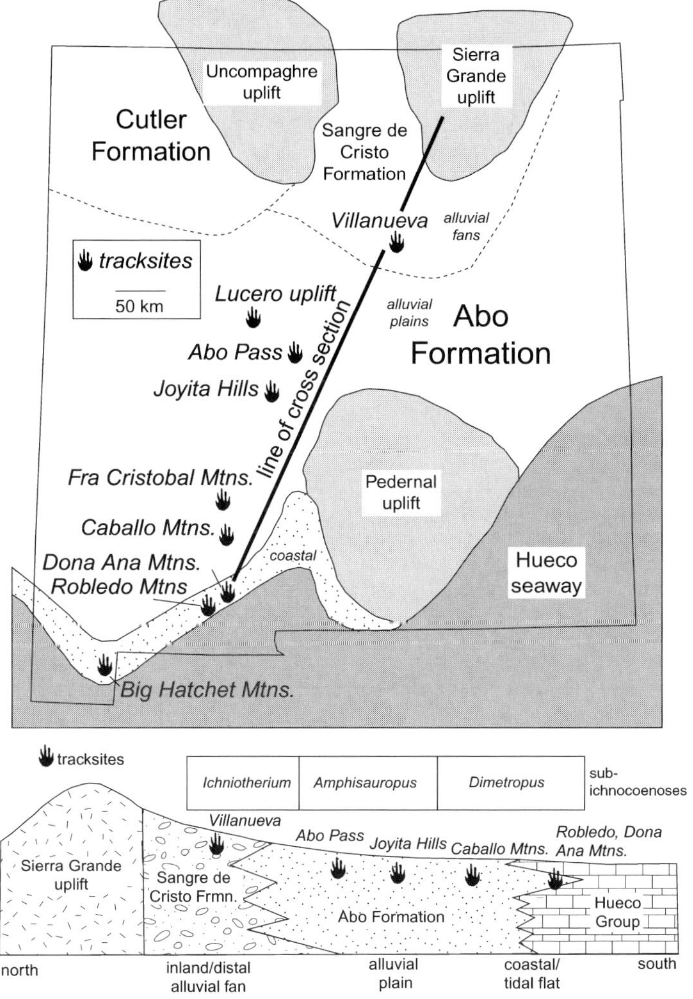

Fig. 3. Palaeogeographic map of New Mexico during the Early Permian and north–south transect of Early Permian red beds in New Mexico showing distribution of possible tetrapod sub-ichnocoenoses (from Lucas 2005*a*).

> 10%) of terrestrial carnivore tracks. This ichnofacies includes coastal plain-marine shore line environments and some lacustrine shorelines and ranges from Middle Permian to Recent in age (Hunt & Lucas 2006). It includes the *Pachypes* ichnocoenosis of Middle to Late Permian age.

The *Characichnos* ichnofacies of Hunt & Lucas (2006) encompasses medium diversity ichnofaunas in which the majority of tracks are swimming traces (parallel scratch marks) and fish swimming trails (*Undichna*). This ichnofacies represents shallow lacustrine (and tidal) environments.

Footprint-based biostratigraphy is often confined to a given ichnofacies largely because each ichnofacies has its own ichnotaxonomy. Thus, although the same trackmakers may have made tracks in different lithofacies, the tracks are so different morphologically that they receive different ichnotaxonomic names. For this reason, we do not attempt footprint-based correlations between ichnoassemblages of the temporally overlapping aeolian *Chelichnus* and the water-laid *Batrachichnus* and *Brontopodus* ichnofacies. The *Characichnos* ichnofacies consists of swimming traces and is of no biostratigraphic significance

Permian footprint distribution in space and time

North America

United States

In North America, tetrapod footprints of Permian age are found primarily in the western United States in Arizona, New Mexico and Texas, and some important sites are also known in the adjoining states of Utah, Colorado and Oklahoma (Fig. 4).

The *Chelichnus* ichnofacies in North America is best known from aeolian strata of the Coconino Sandstone in Arizona, although some other Permian aeolianites also yield tracks in Arizona, Utah, Colorado and New Mexico (Hunt & Lucas 2005*a*). Lull (1918) and Gilmore (1926, 1927, 1928) first described the Coconino tracks from the Grand Canyon of Arizona (where the widely used name *Laoporus* was introduced, although it is now recognized as a junior subjective synonym of *Chelichnus*), and Middleton *et al.* (1990) and Hunt *et al.* (2005*a*) provide a recent summary. The Coconino Sandstone is of late Leonardian age (Fig. 5). Note that it is directly overlain by marine strata of the late Leonardian Kaibab Formation (Hopkins 1990), and that the Coconino is homotaxial to the Leonardian Glorieta Sandstone of New Mexico and the San Angelo Formation of Texas (Middleton *et al.* 1990). In effect, Coconino dune fields were landward of the shorelines and coastal plains that deposited the Glorieta and San Angelo sediments during late Leonardian time.

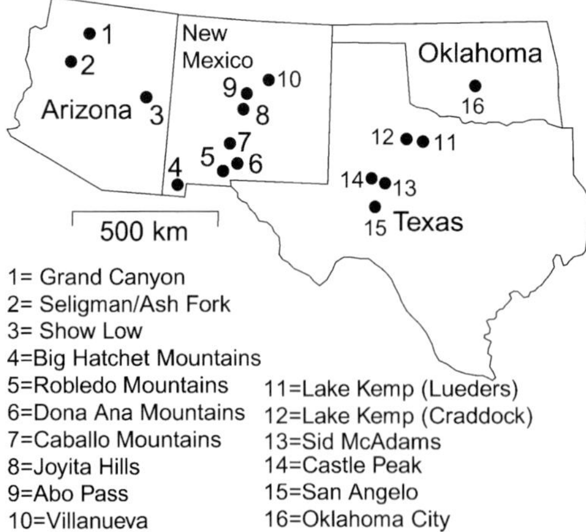

Fig. 4. Distribution of principal Permian tracksites in the western United States.

In the Lake Powell area of Utah, *Chelichnus* is known from the Wolfcampian Cedar Mesa Sandstone (Loope 1984; Lockley & Madsen 1993; Hunt & Lucas 2006), and in the Colorado Front Range it is known from the Leonardian Lyons Sandstone (e.g. Lockley & Hunt 1995). In Arizona, the early Leonardian DeChelly Sandstone (Blakey & Knepp 1989) yields *Chelichnus* and *Dromopus* (McKee 1934; Lockley *et al.* 1994, 1995; Morales & Haubold 1995; Haubold *et al.* 1995*b*). However, in New Mexico, mixed aeolian-fluvial facies of the DeChelly Sandstone yield *Amphisauropus, Dimetropus* and *Limnopus* (Lucas *et al.* 2005*c*).

In North America, the *Batrachichnus* ichnofacies is best understood in New Mexico, where numerous and extensive red-bed track assemblages of Early Permian age are known (see articles in Lucas & Heckert 1995; Lucas *et al.* 1998; Lucas *et al.* 2004; Hunt *et al.* 2005*b*) (Fig. 3). These assemblages are from red-bed ichnofacies of the Earp Formation (Big Hatchet Mountains), the Robledo Mountains Formation of the Hueco Group (Robledo, Doña Ana and San Andres Mountains), the Abo Formation (Caballo and Fra Cristobal Mountains, Joyita Hills, Abo Pass) and the Sangre de Cristo Formation (Villanueva) (Fig. 3). Relative abundances of the ichnotaxa vary between sites, but *Dromopus* and *Batrachichnus* dominate the ichnoassemblages, and co-occur primarily with *Dimetropus, Hyloidichnus* and *Limnopus* (e.g. Haubold 2000; Haubold & Lucas 2001*a*; Lucas *et al.* 2005*a*, *c*). Lucas *et al.* (2001) reported *Amphisauropus* and *Varanopus* from the Abo Pass tracksite, which is stratigraphically low in the Abo Formation (Fig. 5). *Ichniotherium* is present in some of the 'inland' assemblages in central and northern New Mexico (Hunt *et al.* 2005*e*).

Tracksites in the Sangre de Cristo and Abo formations are of Wolfcampian age, but a complete precise correlation and stratigraphical ordering of these sites has not yet been completed. Nevertheless, tracksites in the Robledo Mountains Formation in southern New Mexico are close in age to the Wolfcampian–Leonardian boundary (Kietzke & Lucas 1995; Lucas *et al.* 1995), whereas tracksites in the Caballo and Fra Cristobal Mountains and at Abo Pass are stratigraphically low in the Abo Formation and thus are of Mid-Wolfcampian age (Lucas *et al.* 2001, 2005*a*, *b*).

Various ichnogenera do vary stratigraphically. For example, *Batrachichnus* dominates tracksites stratigraphically low in the Abo Formation (Lucas *et al.* 2005*a*, *b*), whereas it is co-dominant with *Dromopus* at stratigraphically

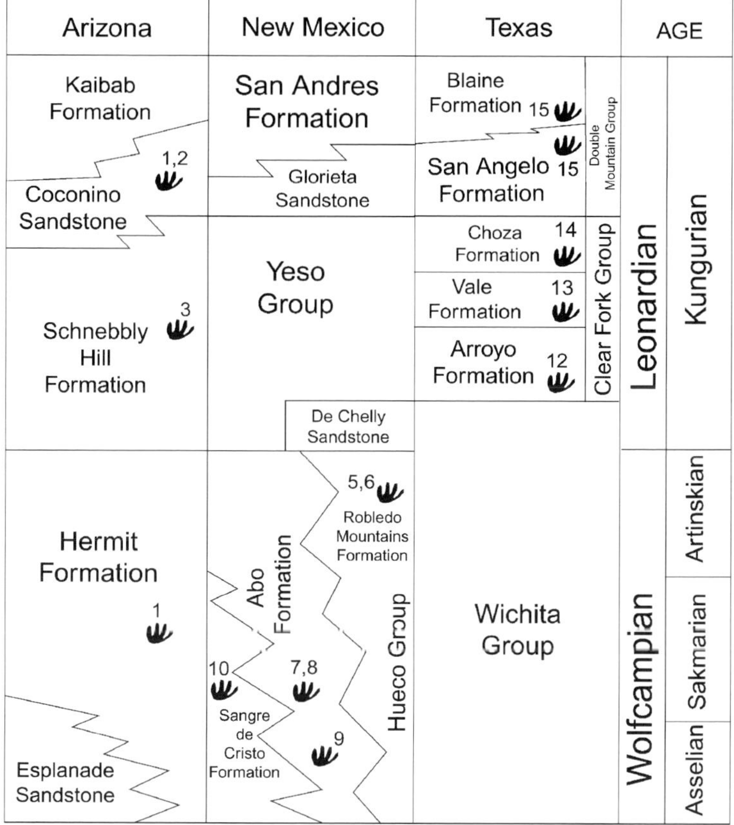

Fig. 5. Correlation of principal North American Permian tracksites.

higher sites. However, ichnogeneric composition does not vary significantly through the Abo Formation. The New Mexican red-bed track record thus encompasses most or all of Wolfcampian time and belongs to a single biostratigraphic assemblage.

In Arizona, the Wolfcampian Hermit Formation (Shale) (Blakey & Knepp 1989) yields tetrapod tracks assigned to the ichnogenera *Batrachichnus, Hyloidichnus, Ichniotherium* and *Limnopus* (Haubold *et al.* 1995*a*; Hunt & Santucci 1998; Hunt & Lucas 2005*a*; Hunt *et al.* 2005*a*). The stratigraphically higher Wolfcampian Organ Rock Shale in Monument Valley, Arizona, yields *Dromopus* and '*Gilmoreichnus*' (Vaughn 1964; Haubold *et al.* 1995*a*). In central Colorado, the Wolfcampian Maroon Formation yields *Dimetropus*, *Ichniotherium*, *Tambachichnium*, and *Varanopus* (Voigt *et al.* 2005). Also, in the San Juan Mountains of Colorado, Wolfcampian strata of the Cutler Formation yield *Limnopus* (Baird 1965).

Much less is known of Leonardian-age tracks in North America. A single specimen of *Dimetropus* is known from the Leonardian Schnebbly Hill Formation near Show Low in Arizona (Haubold *et al.* 1995*a*). A locality in the lower part of the Hennessey Formation at Oklahoma City yields *Amphisauropus* and possible *Dromopus* (Lucas & Suneson 2002). In Texas, *Dimetropus* is known from the Leonardian Vale Formation at the Sid McAdams locality in Taylor County (Olson & Mead 1982; Lucas & Hunt 2005), and Dalquest (1963) reported large amphibian tracks (*Limnopus*?) from the Leonardian Lueders Formation near Lake Kemp in Baylor County.

The classic North American Leonardian tracksite is in the upper part of the Choza Formation at Castle Peak near Abilene, Texas (Moodie 1929, 1930). Haubold & Lucas (2001*b*, 2003) revised the ichnotaxonomy at Castle Peak, and it comprises *Erpetopus, Varanopus* and *Dromopus*. We have recently collected a tracksite in the Arroyo Formation at Lake Kemp in Baylor County, Texas, and *Batrachichnus* dominates this assemblage, with fewer numbers of *Dromopus* and possible *Amphisauropus* (Lucas & Hunt 2005). Indeed, it might be tempting to suggest that *Erpetopus* and abundant *Varanopus* are characteristic of the Leonardian, although too few Leonardian age tracksites are known to confirm this. Furthermore, the Castle Peak and Lake Kemp tracksites are in playa and mudflat deposits of a broad, low relief coastal plain, quite different from the Wolfcampian tracksites in New Mexico, which come from strata that represent both inland floodplains (Sangre de Cristo and Abo formations) and coastal tidal flats (Robledo Mountains and Earp formations). Thus, the differences now perceived between Wolfcampian and Leonardian tetrapod tracks may be due to facies differences and not temporally significant.

The stratigraphically highest Permian tetrapod footprints from North America are in the San Angelo and Blaine formations at San Angelo, Tom Green County, Texas. Pittman *et al.* (1996) and Lucas & Hunt (2005) provided preliminary data on these tracks, which are large, indistinct tracks, possibly of a caseid pelycosaur, and rare *Amphisauropus*. The San Angelo and Blaine are late Leonardian in age (Fig. 5), and these youngest North American Permian tracks mirror the abundance of caseid pelycosaurs seen in the San Angelo Formation body fossil fauna (e.g. Olson 1962). It is also interesting that the common Coconino (correlative to the San Angelo) ichnogenus *Chelichnus* has been thought by some to be a caseid track, so this may provide a link between the *Chelichnus* and *Batrachichnus* ichnofacies.

In the eastern United States, tetrapod footprints are known from the Wolfcampian interval of the Dunkard Group in southeastern Ohio (*Dromopus* and *Limnopus*: Haubold 1971; Cotton *et al.* 1995) and in northwestern West Virginia (Waynesburg Sandstone, *Dimetropus* and *Limnopus*: Tilton 1926, 1931; Romer & Price 1940; Baird 1952).

It is extremely important that much of the Permian tetrapod footprint record in the western United States can be cross-correlated with marine biostratigraphy with great confidence (e.g. Lucas 2002*b*; Haubold & Lucas 2003). Thus, intercalated or bracketing marine strata in southern New Mexico and Texas contain biostratigraphical indicators (fusulinaceans, conodonts and/or ammonoids) that allow the track record to be readily correlated to the North American provincial stages Wolfcampian and Leonardian (Fig. 5).

Canada

Early Permian track records of the *Batrachichnus* ichnofacies are known in Canada from Nova Scotia and Prince Edward Island. At Brule in Nova Scotia, red beds of the Cape John Formation (Pictou Group) yield an extensive assemblage that comprises *Amphisauropus, Dimetropus, Dromopus, '?Gilmoreichnus', Limnopus* and *Varanopus* (Van Allen *et al.* 2005). Fossil plants indicate an age of Stephanian–Autunian for the Cape John Formation.

Red beds of the Hillsborough Formation on Prince Edward Island yield assemblages that comprise *Amphisauropus, 'Gilmoreichnus', 'Ichniotherium'* and *Varanopus* (Mossman & Place 1989; Calder *et al.* 2004). Based on associated fossil plants, these are of late Autunian age. The Canadian record thus encompasses characteristic ichnogenera of the *Batrachichnus* ichnofacies in Lower Permian strata.

Europe

The European Permian tetrapod footprint record (Fig. 6) comes principally from three countries – Germany, France and Italy – although Lower Permian tracks are also known from the United Kingdom, Spain, Poland, and the Czech Republic (e.g. Haubold 1973, 1984). Most of these records, including those from the United Kingdom, Spain, Poland and the Czech Republic, are of characteristic ichnogenera of the *Batrachichnus* ichnofacies of Early Permian age (e.g. Haubold 1970, 1971, 1973; Haubold & Sarjeant 1973; Cassinis & Santi 2005) and are not reviewed here.

Recently described footprints from the Tumlin Sandstone in Poland of supposed Late Permian age (Ptaszyński & Niedźwiedzki 2004) are actually of Early Triassic age (Racki 2005). In Scotland, footprints of the *Chelichnus* ichnofacies are known from the Corncockle, Hopeman and Lochabriggs formations (e.g. McKeever & Haubold 1996). These units predate the Zechstein transgression and are probably of late Capitanian or early Wuchiapingian age.

Here, we focus on the three track records – from Germany, France and Italy – of greatest importance to building a Permian footprint biostratigraphy and biochronology. An important aspect of the European track record is how poorly most of it can be correlated to the standard global stratigraphic scale (SGCS). In general, age control is based on fossil plants, and we consider it imprecise and questionable in places.

Germany

In Germany, the most extensive Permian track records are from the Thuringian and the

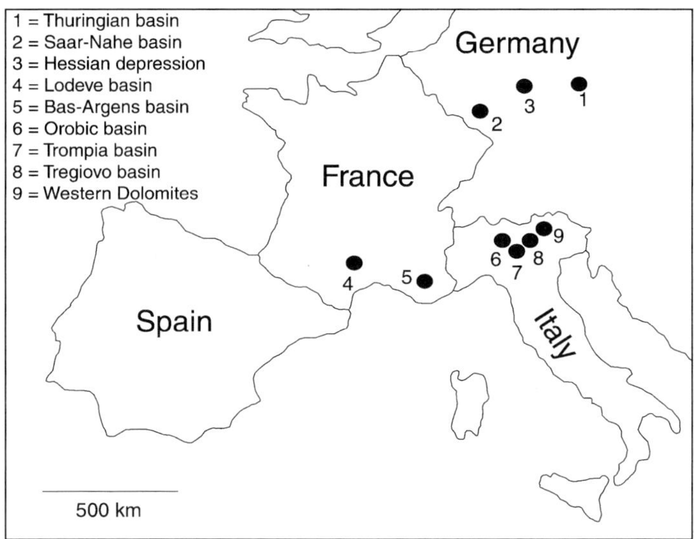

Fig. 6. Distribution of principal Permian tracksites in western Europe.

Saar–Nahe basins (Fig. 6). The Thuringian record, which is all assigned to the *Batrachichnus* ichnofacies, is of some historical significance, as one of the first known records of Permian footprints. Voigt (2005) has recently revised this record, and his revision indicates an essentially consistent tetrapod footprint assemblage from the Georgenthal through the Tambach formations, that is, from Late Pennsylvanian to Artinskian time (Fig. 7). This assemblage consists of *Amphisauropus, Batrachichnus, Dimetropus, Dromopus* and *Ichniotherium*. In Thuringia, the LO (lowest occurrence) of *Varanopus* is in the Oberhof Formation, whereas the LO of *Tambachichnium* is in the Goldlauter Formation. The Georgenthal-Tambach interval is assigned to the Gzhelian–Artinskian based on cockroach, selachian and amphibian biostratigraphy as well as Ar/Ar ages which indicate that the Oberhauf and Goldlauter formations are about 287–288 Ma (Roscher & Schneider 2005). The Thuringian record thus parallels the North American record by indicating an essentially uniform ichnoassemblage characteristic of the *Batrachichnus* ichnofacies from Late Pennsylvanian through most (or all) of the Early Permian.

In the Saar–Nahe Basin, tracks of the *Batrachichnus* ichnofacies are well known from the Glan and Nahe subgroups and have been extensively described by Fichter (1976, 1982, 1983*a*, *b*, 1984). The ichnogenera present are essentially the same as those in the Thuringian Basin, and include *Amphisauropus, Batrachichnus, Hyloidichnus* and *Varanopus*.

Boy & Fichter (1988*a*, *b*) used the footprint record from the Saar–Nahe Basin as the principal basis for recognition of six successive tetrapod footprint zones that spanned the Permian (Fig. 8). These are the (in ascending order) *Protritonichnites lacertoides, Saurichnites incurvatus, Varanopus microdactylus, Anhomoiichnium, Harpagichnus* and *Rhynchosauroides* zones. Boy & Fichter (1988*b*, p. 882) claimed that 'the biostratigraphic zonation of tetrapod tracks is not based on ecological and local climatic changes . . . but on large-scale faunal interchange across wide areas of Pangea'. Nevertheless, the biostratigraphical zonation of Boy & Fichter has been invalidated by taxonomic revision and further understanding of the stratigraphical distribution of Permian tetrapod footprint ichnogenera. Thus, their *Proitonichnites* is *Dromopus*, and what they termed *Anhomoiichnium* includes tracks now termed *Dromopus* and *Batrachichnus* (Haubold 1996). *Saurichnites incurvatus* of Boy & Fichter also is *Batrachichnus* (Haubold 1996). The zones are thus based on *Dromopus, Batrachichnus* and *Varanopus*, ichnogenera that routinely co-occur and have long stratigraphical ranges in the North American and the Thuringian Lower Permian sections. Furthermore, '*Harpagichnus*' (= *Chelichnus*) is the dominant ichnogenus of the *Chelichnus* ichnofacies and is found in Permian aeolianites

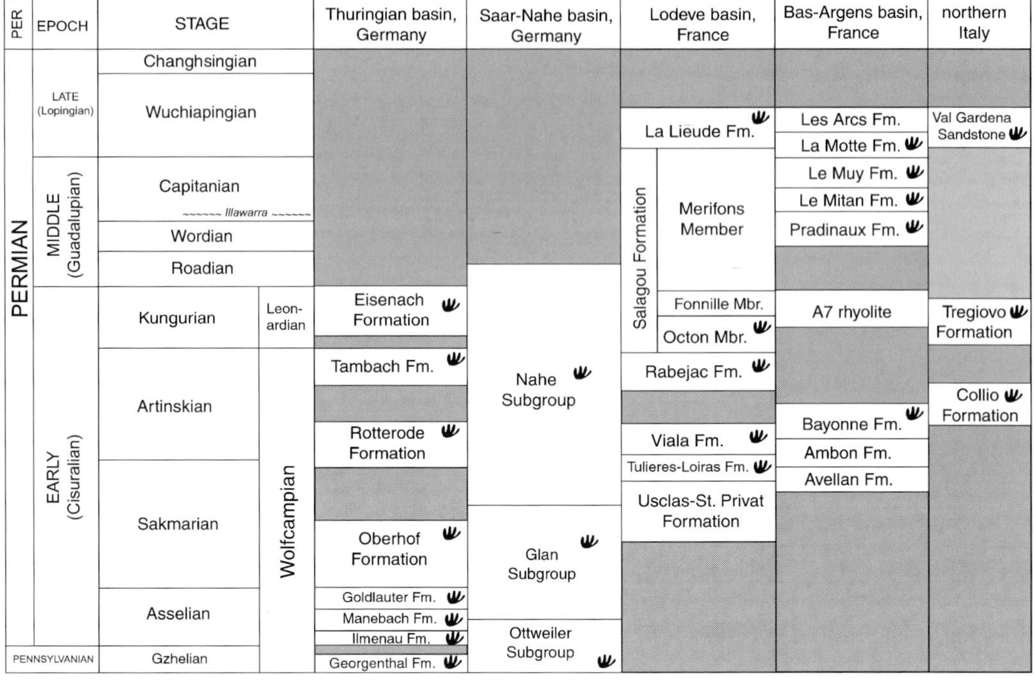

Fig. 7. Correlation of principal European Permian tracksites.

regardless of their precise age. The only zone that may be of value is the *Rhynchosauroides* zone, a tetrapod ichnogenus that has its lowest occurrence in the Upper Permian.

In the Hessian depression of Germany (Fig. 6), aeolianites of the Cornberg Sandstein yield an assemblage of *Chelichnus* (e.g. Schmidt 1959; Haubold 1996). The Cornberg Sandstein post-dates the Capitanian Illawarra reversal and is stratigraphically below the base of the Zechstein. This means it is either of late Capitanian or early Wuchiapingian age (Menning 1995; Roscher & Schneider 2005).

France

An extensive and well-studied Permian red-bed tetrapod footprint record is known from southern France, especially from the Lodève and Bas-Argens basins (e.g. Ellenberger 1983*a*, *b*, 1984; Gand 1987, 1993; Gand & Haubold 1988). Other French basins (e.g. Saint-Affrique) yield typical ichnogenera of the *Batrachichnus* ichnofacies from Lower Permian strata (e.g. Gand 1987, 1993) and are not reviewed here.

The Lodève Basin has the most stratigraphically extensive track record that encompass four 'associations' from a stratigraphical section up to 800 m thick (e.g. Heyler & Lessertisseur 1962, 1963; Ellenberger 1983*a*, *b*; Gand 1987; Gand & Durand 2006). The lower two associations are from the Tuilieres-Loiras and Viala formations. Ichnogenera from these strata are assigned to *Amphisauropus, Batrachichnus, Dimetropus*, cf. *Ichniotherium, Limnopus, 'Salichnium'* and *Varanopus*. All of these ichnogenera are found in the Tuilieres-Loiras Formation ('Association I'), and a subset of these ichnogenera (*Amphisauropus, Batrachichnus, Dimetropus* and *Limnopus*) comes from the overlying Viala Formation ('Association II'). Megafossil plants from the Tuilieres-Loras and Viala formations are generally considered Autunian (Broutin *et al.* 1999).

'Association III' is from the Rabejac Formation, which unconformably overlies the Viala Formation. The tracks are assigned to *Batrachichnus, Dimetropus, Dromopus, Hyloidichnus* and *Varanopus*. They thus do not differ substantially from the underlying track assemblages. Indeed, the Rabejac Formation is also of Early Permian age, given that the overlying Octon Member of the Salagou Formation is no younger than Kungurian.

In the overlying Salagou Formation, a few tracks are present in the Fonnile and Octon members: *Batrachichnus, Hyloidichnus* and

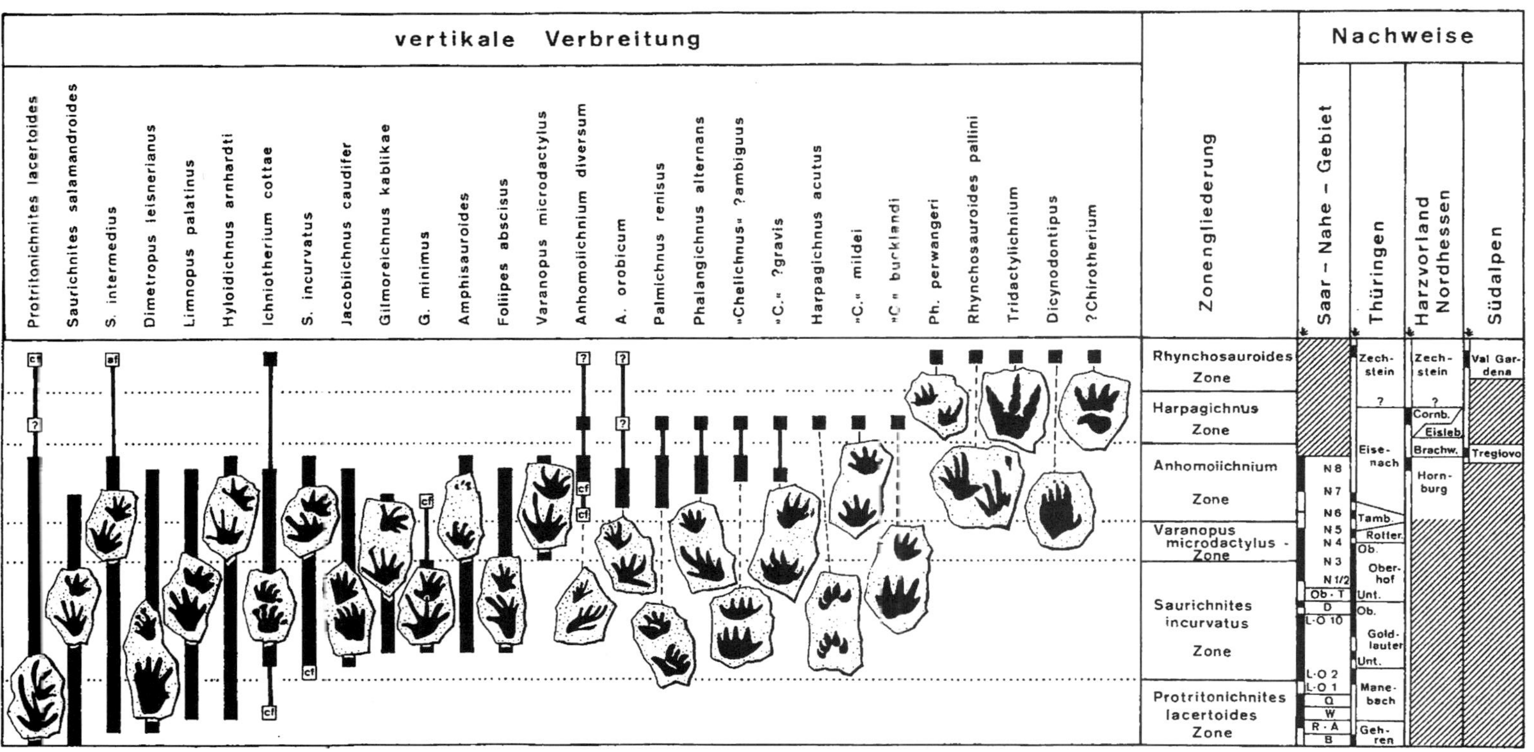

Fig. 3. Global Permian footprint bicstratigraphy of Boy & Fichter (1988b).

Varanopus. Conchostracans and insects, as well as a U/Pb age of 284±4 Ma from the Octon Member suggest it is either Artinskian or Kungurian in age (Gand *et al.* 1997; Bethoux *et al.* 2002; Roscher & Schneider 2005).

In contrast, the track assemblage of the La Lieude Formation (or member of the Salagou Formation of some authors), which is stratigraphically higher, above the Merifons Member of the Salagou Formation (Fig. 7), is quite distinctive. It encompasses the ichnogenera *Brontopus, Dromopus, Lunaepes, Merifontichnus*, and *Planipes*, which are mostly the tracks of therapsids. Roscher & Schneider (2005) assign the La Lieude Formation a Wuchiapingian age, linking it to the Zechstein and Bellerophon transgressions. Indeed, the footprint assemblage of the La Lieude Formation has much in common with that of the Wuchiapingian Val Gardena Sandstone in Italy (see below), so we assign it a Wuchiapingian age (Fig. 7).

The other biostratigraphically important French track record comes from the Bas-Argens Basin in southeastern Provence (Gand *et al.* 1995). Low in this section, a few tracks (*Dromopus*, *Varanopus*) are known from the Bayonne Formation, which is of probable Artinskian or Kungurian age. The overlying Pradinaux Formation yields a much more extensive track assemblage of the ichnogenera *'Chelichnus', Hyloidichnus, Lunaepes, Planipes, Pseudosynaptichnium, Tambachichnium* and *Varanopus*. The stratigraphically highest track assemblage is from the La Motte Formation: *Batrachichnus, Dromopus, 'Dimetropus', Hyloidichnus, 'Laoporus', Limnopus* and *Varanopus*.

The key age indicator in this succession is the so-called 'A7 rhyolite', which is unconformably overlain by the Pradinaux Formation. The latest and most reliable age estimate for the rhyolite is an Ar/Ar age of 272.5±0.3 Ma (Zheng *et al.* 1992), which is Late Kungurian on the standard global chronostratigraphic scale. Durand (2006) reviews age indicators for the Pradinaux Formation, which are megafossil plants, palynomorphs and ostracodes, to conclude that a Wordian age is most likely, although our reading of the age data indicates it could be younger. The overlying LeMuy Formation appears on a palaeobotanical basis to be of Zechstein (Wuchiapingian?) age.

Thus, in France, the Lower Permian strata produce track assemblages dominated by *Amphisauropus, Batrachichnus, Dimetropus, Dromopus, Hyloidichnus, Limnopus* and *Varanopus*. A stratigraphically much higher level in the Lodève Basin at La Lieude yields therapsid tracks, among others, and compares well with the Upper Permian ichnoassociation from Italy (see below). The Pradinaux Formation footprint assemblage may be of Wordian or slightly younger age, but an Artinskian age, as claimed by Haubold & Lucas (2003) seems highly unlikely. This means that the oldest Permian footprint assemblage with therapsid tracks is the Pradinaux assemblage.

Italy

In the Southern Alps of northern Italy (Fig. 6), Permian tetrapod tracks are found at two disparate stratigraphic intervals. The lower interval encompasses the Collio and Tregiovo formations in the Orobic, Val Trompia and Tregiovo basins. The younger interval is the Val Gardena (Gröden) Sandstone in the Western Dolomites. Avanzini *et al.* (2001) and Cassinis & Santi (2005) provide the most recent reviews of these assemblages in their stratigraphic context.

The Collio Formation in the Orobic basin yields *Amphisauropus, Batrachichnus, Dromopus* and *Varanopus* as well as '*?Camunipes*' and *?Ichniotherium* (Nicosia *et al.* 2000; Santi & Krieger 2001). In the Val Trompia Basin, the Collio Formation tracks have been assigned to *Amphisauropus, Batrachichnus, '?Camunipes', Dromopus, Ichniotherium* and *Varanopus* (Geinitz 1869; Curioni 1870; Berruti 1969; Ceoloni *et al.* 1987; Conti *et al.* 1991; Avanzini *et al.* 2001). Megafossil plant and palynomorph data indicate that the Collio Formation is of Early Permian age, either Artinskian or Kungurian. Furthermore, in the Val Trompia Basin, Schaltegger & Brack (1999) reported U–Pb zircon ages of 283±1 Ma and 280.5±2 Ma for the rhyolitic ignimbrites that bracket the Collio Formation. These are Artinskian ages on the current time scale (Fig. 7).

In the Tregiovo Basin, *Dromopus* has been reported from stratigraphically low in the Tregiovo Formation. Age assignments for the Tregiovo Formation based on megafossil plants and palynomorphs range from Kungurian to Ufimian. Thus, Italian ichnologists perceive the Tregiovo Formation tracks to be slightly younger than those of the Collio Formation, but to still represent one 'ichnoassociation' (e.g. Conti *et al.* 1997; Avanzini *et al.* 2001). We agree, and conclude the data best support an Artinskian–Kungurian age for this ichnoassociation.

In the Western Dolomites, an extensive track assemblage of the *Brontopodus* ichnofacies is known from the Val Gardena Sandstone (principal locality is Bletterbach Gorge) (Leonardi & Nicosia 1973; Leonardi *et al.* 1975; Conti *et al.* 1977, 1991; Nicosia *et al.* 2001). The principal ichnogenera documented are *Pachypes, Dicynodontipus, Rhynchosauroides* and *Varanopus*. The Val Gardena Sandstone interfingers with and is overlain by the marine Bellerophon Formation, which is of Wuchiapingian age (Ceoloni *et al.* 1988). This is a rare European example where a direct correlation of the tracks to the SGCS is possible. Thus, the upper 'ichnoassociation' of the Italian section (Avanzini *et al.* 2001) is fundamentally different from the lower 'ichnoassociation' in having tracks of therapsids and pareiasaurs. There is also a substantial temporal gap between the two ichnoassociations, equal to at least the entire Guadalupian (e.g. Cassinis *et al.* 2002; Lucas 2002*b*).

Russia

Despite the extensive outcrop area of non-marine Permian strata in Russia, few tetrapod track records have been documented. Lucas *et al.* (1999) reported a handful of tetrapod footprints (assigned to cf. *Dromopus* and cf. *Dimetropus*) from Early Permian red beds of the Caucasus. Tverdokhlebov *et al.* (1997) described red-bed tracks assigned to *Batrachichnus* from the Upper Tatarian of Russia, and Gubin *et al.* (2001) mentioned apparent pareiasaur tracks, also from the Upper Tatarian.

South America

Brazil

Leonardi (1987, 1994) reported tetrapod swimming traces (*Characichnos* ichnofacies) from the Rio do Rastro Formation at Tonetti in Paraná State, Brazil. This record, which Leonardi (1994, p. 46) correctly termed 'unclassifiable', is of Mid- or Late Permian age (the age of the Rio do Rastro Formation: Cisneros *et al.* 2005), but is of no biostratigraphical significance at present

Argentina

Melchor (2001) described Permian tetrapod footprints from Argentina in the Carapacha Basin (*Batrachichnus* ichnofacies tracks assigned to *Batrachichnus, Hyloidichnus* and 'cf. *Gilmoreichnus*') and the eastern Permian basin (*Chelichnus* ichnofacies tracks assigned to *Chelichnus*). Melchor (2001) suggested these records are of Late Permian age, but both records are more probably older. The Argentinian track record is significant because it suggests the presence in southern Gondwana during the Early to early Middle Permian of some of the characteristic ichnogenera of the *Batrachichnus* and *Chelichnus* ichnofacies.

In the Carapacha Basin of La Pampa Province, tetrapod footprints of the *Batrachichnus* and *Characichnos* ichnofacies are found in the Urre-Lauquen Member of the Carapacha Formation (Melchor 2001; Melchor & Sarjeant 2004). These have been assigned to *Batrachichnus, Hyloidichnus*, cf. *Amphisauropus* and cf. *Varanopus* and also include swimming traces assigned to *Characichnos*. Melchor (2001) and Melchor & Sarjeant (2004) claimed that the associated palaeoflora indicates an 'early Late Permian age', which means Kazanian on the time scale that they used. However, this palaeoflora lacks any tie to a marine time scale and, as Melchor & Cesari (1997, p. 628) stated, it '*could* have been deposited during the Late Permian' (our italics). Indeed, this is the 'Golondrinian' palaeoflora of Archangelsky & Cúneo (1984), which is younger than their 'Lubeckian' palaeoflora of Argentina. The Lubeckian palaeoflora has some direct ties to marine biostratigraphy that indicate it ranges in age from about Asselian to Sakmarian. However, the Golondrinian palaeoflora lacks such ties and is thought to begin in the Artinskian with an uncertain upper age limit (Archangelsky & Cúneo 1984).

In the eastern Permian basin of Mendoza Province, footprints of *Chelichnus* are present in the Areniscas Atigradas Member of the Yacimiento Los Reyunos Formation (Melchor 2001). A tuff below the tracks has been $^{40}Ar/^{39}Ar$ dated at ~ 266 Ma, although the scatter of single crystal ages from the tuff ranges from 263 to 269 Ma (Melchor 2000). Based on the age, Melchor (2001) concluded that the tracks are no older than Wordian (Kazanian). However, given the scatter of single crystal ages they could be as old as Roadian. If these tracks actually are Roadian or Wordian (they could be younger), they are one of the few known Middle Permian track records (Fig. 9).

Africa

Morocco

One Pennsylvanian and three Permian footprint records have been documented from Morocco. Hmich *et al.* (2006) report *Batrachichnus* and *Dromopus* from the El Menizla Formation of the Ida Ouzal Sub-Basin of the Souss Basin. Based on cockroach biostratigraphy, they assign this record a Stephanian B (late Kasimovian/middle Gzhelian) age.

Hmich *et al.* (2006) also document *Limnopus*, cf. *Batrachichnus* and *Dromopus* from 'unit B' in the Khenifra Basin. Based on the palaeoflora (Broutin *et al.* 1998), this occurrence is assigned a Kungurian (Autunian) age.

The 'upper formation' in the Tiddas Basin yielded tetrapod tracks assigned to '*Amphisauroides*', '*Gilmoreichnus*' and *Hyloidichnus* (El Wartiti *et al.* 1986; Broutin *et al.* 1987; Larhrib

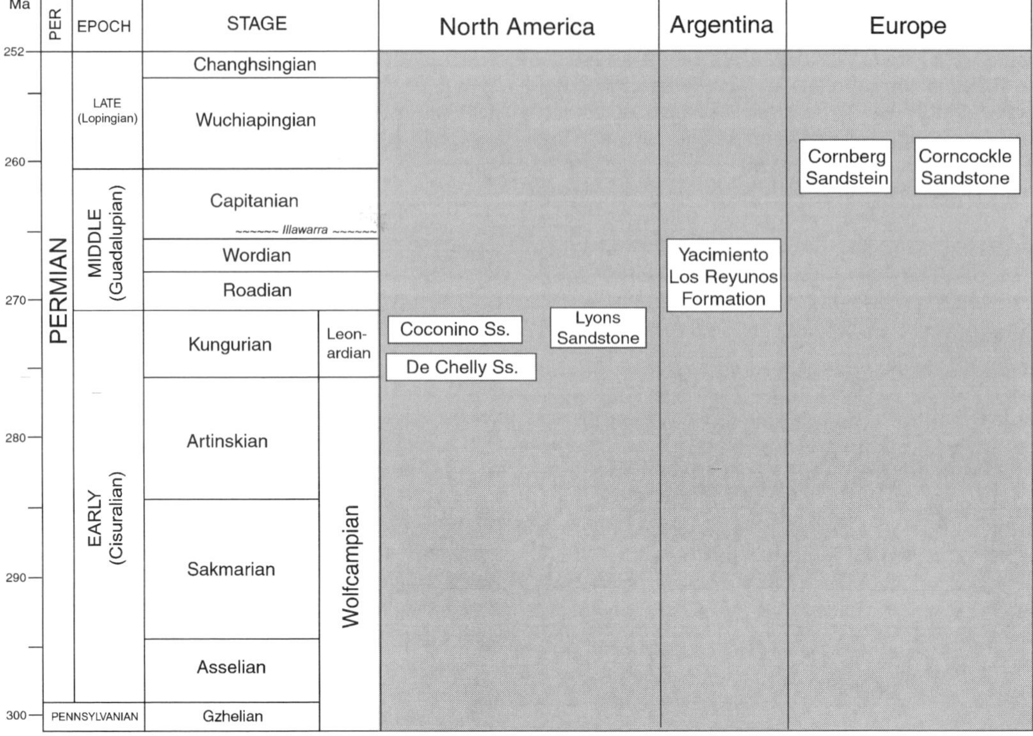

Fig. 9. Global correlation of selected Permian tetrapod tracksites of the *Chelichnus* ichnofacies.

1996). However, only the record of *Hyloidichnus* can be confirmed (Hmich *et al.* 2006). This record is also assigned a Kungurian age based on palaeoflora (Broutin *et al.* 1998).

Tracks assigned to *Synaptichnium* and *Rhynchosauroides* have been reported from the Tourbihine Member (T2) of the Ikakern Formation in the Argana Basin (Jones 1975; Hmich *et al.* 2006). Tetrapod body fossils from this unit have been assigned a Kazanian age (Jalil & Dutuit 1996), but Hmich *et al.* (2006) correlate this record to the Wuchiapingian 'wet phase', which essentially equates it to the Val Gardena Sandstone in northern Italy.

Thus, the Moroccan record indicates typical *Batrachichnus* ichnofacies in Upper Pennsylvanian to Lower Permian strata. A much younger ichnoassemblage is present in strata of probable Wuchiapingian age.

South Africa

A substantial record of tetrapod footprints apparently is present but largely undocumented in Mid- Upper Permian strata in the Karoo Basin of South Africa (e.g. Seeley 1904; Smith 1993; Ward 2004). These are primarily tracks of pareiasaurs, dinocephalians and dicynodonts and should be further studied and compared to the Middle and Upper Permian tracks from France and northern Italy, which they resemble. The track record from South Africa, once documented, should fill much of the Middle Permian global gap in the tetrapod footprint record.

Correlations

Chelichnus *ichnofacies*

The Permian *Chelichnus* ichnofacies is of the same ichnogeneric composition at all sites. The fact that Permian units of disparate ages, such as the Coconino and DeChelly formations in the United States, the Corncockle and Lochabriggs sandstones in Scotland, the Cornberg Sandstein in Germany and the Yacimiento Los Reyunos Formation in Argentina, have similar tetrapod ichnofossils is a reflection of shared ichnofacies, not of precise age equivalence (Fig. 9). Tetrapod footprints of the aeolian ichnofacies are thus of no biostratigraphical value as presently understood.

The aeolian trackmakers may have been some of the same animals as the red-bed trackmakers, and indeed one ichnogenus, *Dromopus*, is found in both ichnofacies. Furthermore, lithofacies transitional between aeolian and fluvial of the DeChelly Sandstone in central New Mexico yield typical *Batrachichnus* ichnofacies tracks, such as *Amphisauropus*, *Dimetropus* and *Limnopus* (Lucas *et al.* 2005*a*). But, in general, the aeolian track assemblages cannot be directly compared and correlated with the red-bed tracks: the tracks of both ichnofacies are too different in morphology.

Early Permian Batrachichnus *ichnofacies*

Tetrapod footprints of the Early Permian *Batrachichnus* ichnofacies are of broad, uniform composition, and ichnodiversity is much higher than in the *Chelichnus* ichnofacies. The following ichnogenera dominate: *Amphisauropus, Batrachichnus, Dimetropus, Dromopus, Hyloidichnus, Ichniotherium, Limnopus* and *Varanopus*. This assemblage is the tracks of temnospondyls, diadectomorphs, seymouriamorphs, captorhinomorphs and pelycosaurs (Table 1). The North American record demonstrates that most (if not all) of these ichnogenera have long stratigraphical ranges through most or all of Wolfcampian and Leonardian time (Haubold & Lucas 2001*a*, *b*; Lucas 2002*b*). Furthermore, at the Robledo Mountains megatracksite in southern New Mexico, all of these ichnogenera (except *Varanopus* and *Ichniotherium*) co-occur in a single, short stratigraphical interval. This suggests that local biostratigraphical zonations based on these ichnotaxa, especially those proposed in Germany and France, are not of global applicability and may also be of questionable local or regional utility. Thus, the Early Permian *Batrachichnus* ichnofacies yields a single biostratigraphical assemblage of tetrapod footprints found in the United States, Canada, Argentina, Germany, France, Italy, Russia and some other places in Europe (Fig. 10).

Middle to Late Permian Brontopodus *ichnofacies*

The Middle to Late Permian record of tetrapod footprints in water-laid facies is less extensive than but shows significant differences from the Early Permian record. This is a record dominated by the tracks of therapsids. Pareiasaur (*Pachypes*) and eosuchian tracks (*Rhynchosauroides*) also are diagnostic of this record. It is best known from Italy and France, and South African and Russian records demonstrate a broad distribution of this biostratigraphical assemblage. Its oldest occurrence appears to be Wordian, but most records are younger, of Capitanian–Wuchiapingian age.

Global gap

There is a stratigraphical gap in the global Permian tetrapod footprint record. This is the gap between the youngest Early Permian track records, which are as young as Kungurian, and the oldest well-documented Late Permian records, which are no older than Wordian. This gap, approximately equivalent to the Roadian, is approximately the same duration as the corresponding mid-Permian gap in the tetrapod body fossil record, which also approximately equals Roadian time (Lucas 2001, 2002*c*, 2004).

There are only a few described footprint assemblages that may fill this gap. The Pradinaux Formation assemblage in France is the key assemblage, as it documents the LO of therapsid tracks. We now accept the Pradinaux Formation assemblage as tentatively of Wordian age. Older age assignments (e.g. Haubold & Lucas 2003) seem unlikely, but an age as young as Wuchiapingian cannot be ruled out.

Global biostratigraphy and biochronology

An important question to ask of the Permian footprint record is how many useful biostratigraphic datum points can be identified? On a global basis, we believe there are only two: (1) the highest occurrence (HO) of pelycosaur tracks; and (2) the LO of therapsid tracks. Thus, we see no important biostratigraphical datum points within the Lower Permian record, as it consists of tracksites that yield the standard Early Permian ichnogenera that form a single, Lower Permian biostratigraphical assemblage that actually occurs in the Pennsylvanian as well. The HO of pelycosaur tracks is in assemblages that are no younger than Kungurian on the SGCS. Therefore, note that we reject the identification as *Dimetropus* by Demathieu *et al.* (1992) of some tracks from Middle Permian strata in the French Bas-Argens Basin.

The LO of therapsid tracks appears to be in the Pradinaux Formation of the Bas-Argens Basin in France. If this unit is of Wordian age, not younger, then the LO of therapsids in the track and body fossil record is essentially synchronous, or Wordian (Lucas 2004).

If we construct a global biochronology based on tetrapod footprints, it contains only two time intervals (Fig. 10). Lucas (2002*b*) recognized these same intervals, but believed the gap

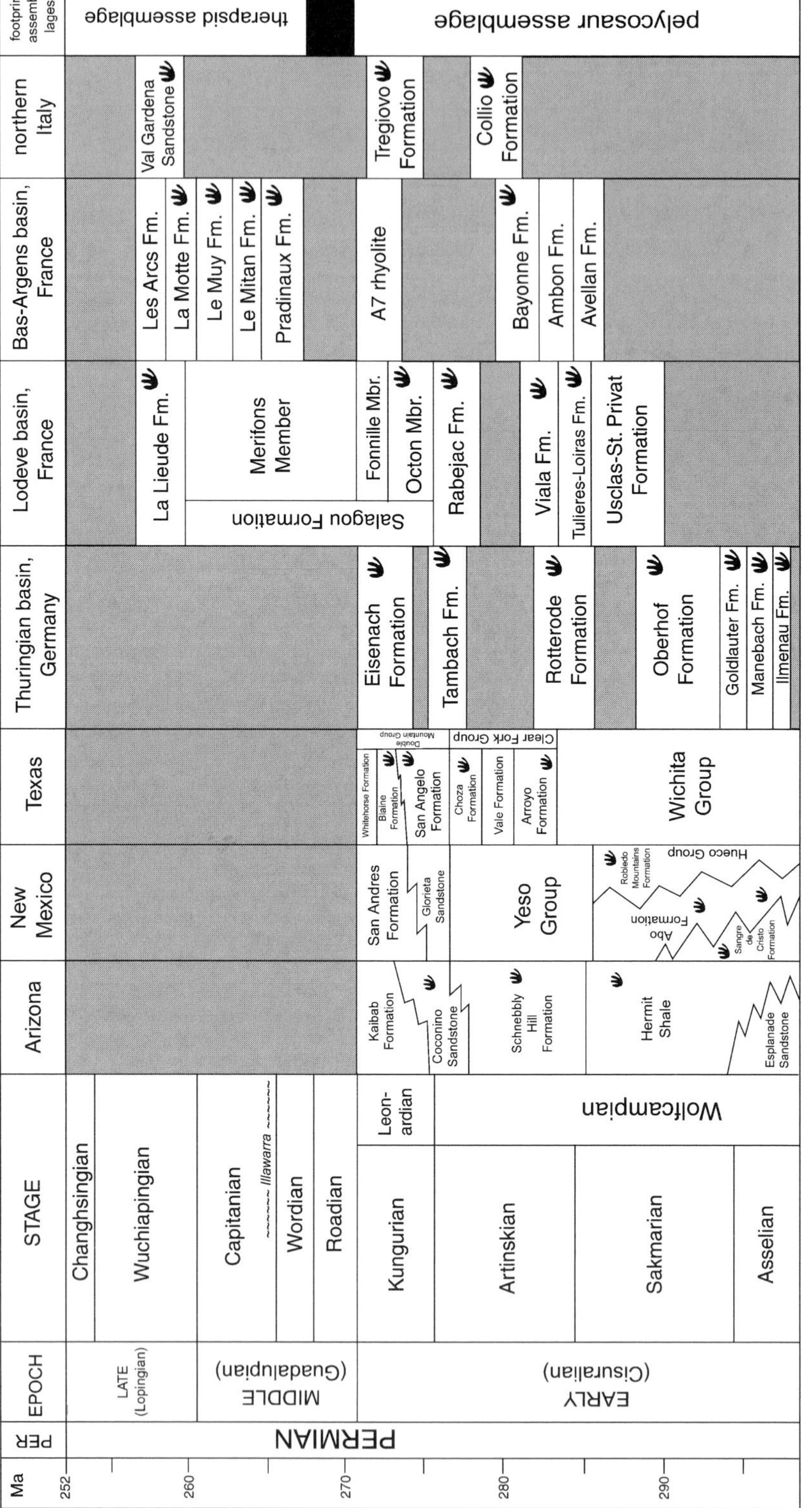

Fig. 10. Global correlation of selected Permian tetrapod tracksites of the *Batrachichnus* and *Brontopodus* ichnofacies.

between them to be longer than we indicate here. He named these two intervals the *Dromopus* and *Rhynchosauroides* biochrons, and noted that the *Dromopus* biochron has a temporal range of Pennsylvanian through Early Permian, and *Rhynchosauroides* has a temporal range of Late Permian through Late Triassic. However, this needs to be modified, as *Dromopus* does have records in the Middle and possibly Late Permian. Furthermore, *Rhynchosauroides* has its LO in Wuchiapingian strata, much younger than the LO of therapsid tracks. Therefore, we propose to identify a global Permian footprint biostratigraphy as consisting of a Lower Permian pelycosaur assemblage and a Middle–Upper Permian therapsid assemblage. Tetrapod footprints thus only discriminate two intervals of Permian time (Fig. 10).

In contrast, tetrapod body fossils can be used to discriminate about ten intervals of Permian time (Lucas 2002*a*, 2005*b*, 2006). Therefore, the tetrapod track record only resolves Permian time about 20% as well as does the tetrapod body fossil record. It thus represents an excellent example of the low biochronological resolution provided by tetrapod footprints (Lucas 1998).

We have benefited immensely in our studies of Permian tetrapod footprints from the collaboration and advice of J. Calder, H. Haubold, A. Lerner, M. Lockley and J. MacDonald. Reviews by J. Calder, M. Lockley, S. Voigt and an anonymous reviewer improved the content and clarity of the manuscript.

References

Archangelsky, S. & Cúneo, R. 1984. Zonación del Pérmico continental argentinoe sobre la base de sus plantas fósiles. *Memorias del III Congreso Latinoamericano de Paleontología y III Congreso de Exploración de Hidrocarbures*, **5**, 417–425.

Avanzini, M., Ceoloni, P. *et al.* 2001. Permian and Triassic tetrapod ichnofaunal units of northern Italy: their potential contribution to continental biochronology. *In*: Cassinis, G. (ed.) *Permian Continental Deposits of Europe. Legional Reports and Correlations.* Natura Bresciana, Monografia, **25**, 89–107.

Baird, D. 1952. Revision of the Pennsylvanian and Permian footprints *Limnopus*, *Allopus*, and *Baropus*. *Journal of Paleontology*, **25**, 832–840.

Baird, D. 1965. Footprints from the Cutler Formation. *U. S. Geological Survey, Professional Papers*, **503-C**, 47–50.

Berruti, G. 1969. Osservazioni biostratigrafiche sulle formazioni continentali pre-quaternarie delle valli Trompia e Sabbia. *Natura Bresciana, Monografia*, **6**, 3–32.

Béthoux, O., Nel, A., Gand, G., Lapeyrie, J. & Galtier, J. 2002. Discovery of the genus *Iasvia* Zalessky, 1934 in the Upper Permian of France (Lodève Basin) (Orthoptera, Ensifera, Oedischiidae). *Géobios*, **35**, 293–302.

Blakey, R. C. & Knepp, R. 1989. Pennsylvanian and Permian geology of Arizona. *Arizona Geological Society Digest*, **17**, 313–347.

Boy, J. A. & Fichter, J. 1988*a*. Zur Stratigraphie des höheren Rotliegend im Saar-Nahe-Becken (Unter-Perm; SW-Deutschland) und seiner Korrelation mit anderen Gebieten. *Neues Jahrbuch für Geologie und Paläontologie, Abhandlungen*, **176**, 331–394.

Boy, J. A. & Fichter, J. 1988*b*. Ist die stratigraphische Verbreitung der Tetrapodenfährten im Rotliegend ökologisch beeinflusst? *Zeitschrift für Geologische Wissenschaft*, **16**, 877–883.

Broutin, J., El Wartiti, M., Freytet, P., Heyler, D., Larhrib, M. & Morel, J.-L. 1987. Nouvelles découvertes paléontologiques dans le basin détritique carbonaté permien de Tiddas (Maroc central). *Comptes Rendus de l'Académie des Sciences, Paris, Série II*, **305**, 143–148.

Broutin, J., Aassoumi, H., El Wartiti, M., Freytet, P., Kerp, H., Quesada, C. & Toutin-Morin, N. 1998. The Permian basins of Tiddas, Bou Achouch and Khenifra (central Morocco). Biostratigraphic und palaeophytogeographic implications. *Mémoires du Muséum National d'Histoire Naturelle, Paris*, **179**, 257 278.

Broutin, J., Châteauneuf, J. J., Galtier, J. & Ronchi, A. 1999. L'Autunien d'Autun reste-t-il une référence pour les dépôts continentaux du Permien inférieur d'Europe? Apport des données paléobotaniques. *Géologie de la France*, **2**, 17–31.

Calder, J. R., Baird, D. & Urdang, E. B. 2004. On the discovery of tetrapod trackways from Permo-Carboniferous red beds of Prince Edward Island and their biostratigraphic significance. *Atlantic Geology*, **40**, 217–226.

Cassinis, G. & Santi, G. 2005. Permian tetrapod footprint assemblages from southern Europe and their stratigraphic implications. *In*: Lucas, S. G. & Zeigler, K. E. (eds) *The Nonmarine Permian.* New Mexico Museum of Natural History and Science Bulletin, **30**, 26–38.

Cassinis, G., Nicosia, U., Lozovsky, V. R. & Gubin, Y. M. 2002. A view on the Permian continental stratigraphy of the Southern Alps, Italy, and general correlation with the Permian of Russia. *Permophiles*, **40**, 4–16.

Ceoloni, P., Conti, M. A., Mariotti, N., Mietto, P. & Nicosia, U. 1987. Tetrapod footprints from Collio Formation (Lombardy, Northern Italy). *Memorie di Scienze Geologiche, Padova*, **39**, 213–233.

Ceoloni, P., Conti, M. A., Mariotti, N. & Nicosia, U. 1988. New Late Permian tetrapod footprints from Southern Alps. *Memorie della Società Geologica Italiana*, **34**, 45–65.

Cisneros, J. C., Abdala, F. & Malabarba, M. C. 2005. Pareiasaurids from the Rio do Rasto Formation, southern Brazil: biostratigraphic implications for Permian faunas of the Paraná Basin. *Revista Brasiliera de Paleontologia*, **8**,13–24.

Conti, M. A., Leonardi, G., Mariotti, N. & Nicosia, U. 1977. Tetrapod footprints of the "Val

Gardena Sandstone" (north Italy). Their paleontological, stratigraphic and paleoenvironmental meaning. *Palaeontographica Italica, New Series*, **40**, 1–91.

Conti, M. A., Mariotti, N., Mietto, P. & Nicosia, U. 1991. Nuove ricerche sugli icnofossili della Formazione di Collio in Val Trompia (Brescia). *Natura Bresciana*, **26**, *Monografia*, 109–119.

Conti, M. A., Mariotti, N., Nicosia, U. & Pittau, P. 1997. Succession of selected bioevents in the continental Permian of the Southern Alps (Italy): improvements of intrabasinal and interregional correlations. *In*: Dickins, J. M., Yang, Z. Y., Yin, H. F., Lucas, S. G. & Acharyya, S. K. (eds) *Late Palaeozoic and Early Mesozoic Circum-Pacific Events and Their Global Correlation. Cambridge University Press*, World and Regional Geology Series, **10**, 51–65.

Cotton, W. D., Hunt, A. P. & Cotton, J. E. 1995. Paleozoic vertebrate tracksites in eastern North America. *In*: Lucas, S. G. & Heckert, A. B. (eds) *Early Permian Footprints and Facies*. New Mexico Museum of Natural History and Science Bulletin, **6**, 189–211.

Curioni, G. 1870. *Osservazioni geologiche sulla Val Trompia*. Regio Istituto Lombardo di Scienze Lettere Arti, Memorie, Serie 3.

Dalquest, W. W. 1963. Large amphibian ichnites from the Permian of Texas. *Texas Journal of Science*, **15**, 220–224.

Demathieu, G., Gand, G. & Toutin-Morin, N. 1992. La palichnofaune des bassins Permiens provençaux. *Géobios*, **25**, 19–54.

Durand, M. 2006. The problem of the transition from the Permian to the Triassic Series in southeast France: comparison with other Peritethyan regions. *In*: Lucas, S. G., Cassinis, G. & Schneider, J. W. (eds) *Non-Marine Permian Biostratigraphy and Biochronology*. Geological Society, London, Special Publications, **265**, 281–296.

Ellenberger, P. 1983*a*. Sur la zonation ichnologique du Permien moyen (Saxonien) du bassin de Lodève (Hérault). *Compte Rendus Académie Science, Paris, Série II*, **297**, 553–558.

Ellenberger, P. 1983*b*. Données complémentaires ur la zonation ichnologique du Permien du Midi de la France (bassins de Lodève, Saint-Affrique et Rodez). *Compte Rendus Académie Science, Paris, Série II*, **299**, 581–586.

Ellenberger, P. 1984. Sur la zonation ichnologique du Permien moyen (Saxonien) du bassin de Lodève (Hérault). *Compte Rendus Académie Science, Paris, Série II*, **297**, 553–558.

El Wartiti, M., Broutin J. & Freytet, P. 1986. Premières découvertes paléontologiques dans les séries rouges carbonatées permiennes du bassin de Tiddas (Maroc Central). *Comptes Rendus de l'Académie des Sciences, Paris, Série II*, **303**, 263–268.

Fichter, J. 1976. Tetrapodenfährten aus dem Unterrotliegenden (Autun, Unter-Perm) von Odernheim/Glan. *Mainzer Geowissenschaftlichen Mitteilungen*, **5**, 87–109.

Fichter, J. 1982. Tetrapodenfährten aus dem Oberkarbon (Westfalium A und C) West-und Südwestdeutschlands. *Mainzer Geowissenschaftlichen Mitteilungen*, **11**, 33–77.

Fichter, J. 1983*a*. Tetrapodenfährten aus dem saarpfälzischen Rotliegenden (?Ober- Karbon–Unter-Perm; Südwest-Deutschland). 1: Fährten der Gattungen *Saurichnites*, *Limnopus*, *Amphisauroides*, *Protritonichnites*, *Gilmoreichnus*, *Hyloidichnus* und *Jacobiichnus*. *Mainzer Geowissenschaftlichen Mitteilungen*, **12**, 9–121.

Fichter, J. 1983*b*. Tetrapodenfährten aus dem saarpfälzischen Rotliegenden (?Ober- Karbon–Unter-Perm; Südwest-Deutschland). 2: Färhten der Gattungen: *Foliipes*, *Varanopus*, *Ichniotherium*, *Dimetropus*, *Palmichnus*, *Phalangichnus*, cf. *Chelichnus*, cf. *Laoporus* und *Anhomoiichnium*. *Mainzer Naturwissenschaftliches Archiv*, **21**, 125–186.

Fichter, J. 1984. Neue Tetrapodenfährten aus den saarpfälzischen Standenbühl-Schichten (Unter-Perm; SW-Deutschland). *Mainzer Naturwissenschaftlichesn Archiv*, **22**, 211–229.

Gand, G. 1987. *Les traces de vertébrés tétrapodes du Permien français*. Thesis, Université de Bourgogne.

Gand, G. 1993. La palichnofaune de vertébrés tétrapodes du bassin permien de Saint-Affrique (Aveyron): comparaisons et conséquences stratigraphiques. *Géologie de la France*, **1**, 41–56.

Gand, G. & Durand, M. 2006. Tetrapod footprint ichnoassociations from French Permian basins: comparisons with other Euramerican ichnofaunas. *In*: Lucas, S. G., Cassinis, G. & Schneider, J. W. (eds) *Non-Marine Permian Biostratigraphy and Biochronology*. Geological Society, London, Special Publications, **265**, 157–177.

Gand, G. & Haubold, H. 1988, Permian tetrapod footprints in central Europe, stratigraphical and palaeontological aspects. *Zeitschrift fur Geologische Wissenschaft*, **16**, 885–894.

Gand, G., Demathieu, G. & Ballestra, F. 1995. La palichnofaune de vertébrés tétrapodes du Permien supérieur de l'Estérel (Provence, France). *Palaeontographica, Abteilung A*, **235**, 97–139.

Gand, G., Lapeyrie, J., Garric, J., Nel, A., Schneider, J. & Walter, H. 1997. Découverte d'arthropodes et de bivalves inédits dans le Permien continental (Lodévois, France). *Comptes Rendus de l'Académie des Sciences, Paris, Série I*, **325**, 891–898.

Geinitz, H. B. 1869. Über fossile Pflanzenreste aus der Dyas von Val Trompia. *Neues Jahrbuch für Mineralogie, Geologie und Paläontologie*, **1869**, 456–461.

Gilmore, C. W. 1926. *Fossil Footprints from the Grand Canyon*. Smithsonian Miscellaneous Collections, **77**(9).

Gilmore, C. W. 1927. *Fossil Footprints from the Grand Canyon II*. Smithsonian Miscellaneous Collections, **80**(3).

Gilmore, C. W. 1928. *Fossil Footprints from the Grand Canyon III*. Smithsonian Miscellaneous Collections, **80**(8).

Gubin, Y. M., Bulanov, V. V., Golubev, V. K. & Petukhov, S. V. 2001. The first traces of big

Permian reptiles in eastern Europe. *Journal of Vertebrate Paleontology*, **21**(3, supplement), 57A.

HAUBOLD, H. 1970. Versuch einer Revision der Amphibienfährten des Karbon und Perm. *Freiberger Forschungshefte, Hefte C*, **260**, 83–110.

HAUBOLD, H. 1971. Ichnia amphibiorum et reptiliorum fossilium. *Handbuch der Paleoherpetologie*, **18**, 1–124.

HAUBOLD, H. 1973. Die Tetrapodenfährten aus dem Perm Europas. *Freiberger Forschungshefte, Hefte C*, **285**, 5–55.

HAUBOLD, H. 1984. *Saurierfährten*. A. Ziemsen Verlag, Wittenberg Lutherstadt.

HAUBOLD, H. 1996. Ichnotaxonomie und Klassifikation von Tetrapodenfährten aus dem Perm. *Hallesches Jahrbuch für Geowissenschaften, Reihe B*, **18**, 23–88.

HAUBOLD, H. 2000. Tetrapodenfährten aus dem Perm—Kenntnisstand und Progress 2000. *Hallesches Jahrbuch für Geowissenschaften, Reihe B*, **22**, 1–16.

HAUBOLD, H. & LUCAS, S. G. 2001*a*. Early Permian tetrapod tracks: preservation, taxonomy, and Euramerican distribution. *In*: CASSINIS, G. (ed.) *Permian Continental Deposits of Europe Regional Reports and Correlations*. Natura Bresciana, Monografia, **25**, 347–354

HAUBOLD, H. & LUCAS, S. G. 2001*b*. Die Tetrapodenfährten der Choza Formation (Texas) und das Artinsk-Alter der Redbed-Ichnofaunen des Unteren Perm. *Hallesches Jahrbuch für Geowissenschaften, Reihe B*, **23**, 79–108.

HAUBOLD, H. & LUCAS, S. G. 2003. Tetrapod footprints of the Lower Permian Choza Formation at Castle Peak, Texas. *Paläontologische Zeitschrift*, **77**, 247–261.

HAUBOLD, H. & SARJEANT, W. A. S. 1973. Tetrapodenfährten aus den Keele und Enville Groups (Permokarbon: Stefan und Autun) von Shropshire und South Staffordshire, Grossbritannien. *Zeitschrift für Geologische Wissenschaft*, **8**, 895–933.

HAUBOLD, H. HUNT, A. P., LUCAS, S. G. & LOCKLEY, M. G. 1995*a*. Wolfcampian (Early Permian) tracks from Arizona and New Mexico. *In*: LUCAS, S. G. & HECKERT, A. B. (eds) *Early Permian Footprints and Facies*. New Mexico Museum of Natural History and Science Bulletin, **6**, 135–166.

HAUBOLD, H., LOCKLEY, M. G., HUNT, A. P. & LUCAS, S. G. 1995*b*. Lacertoid footprints from Permian dune sandstones, Cornberg and DeChelly sandstones. *In*: LUCAS, S. G. & HECKERT, A. B. (eds) *Early Permian Footprints and Facies*. New Mexico Museum of Natural History and Science Bulletin, **6**, 235–244.

HEYLER, D. & LESSERTISSEUR, J. 1962. Remarques sur les allures des tétrapodes paléozoîques d'après les pistes du Permian de Lodève. Colloque CNRS, Paris 1961.

HEYLER, D. & LESSERTISSEUR, J. 1963. *Piste de tétrapodes permien dans la région de Lodève (Hérault)*. Mémoires du Musée d'Histoire Naturelle, Paris, Série C, **XI**.

HMICH, D., SCHNEIDER, J. W., SABER, H. & VOIGT, S. 2006. New continental Carboniferous and Permian faunas of Morocco: implications for biostratigraphy, palaeobiogeography and palaeoclimate. *In*: LUCAS, S. G., CASSINIS, G. & SCHNEIDER, J. W. (eds) *Non-Marine Permian Biostratigraphy and Biochronology*. Geological Society, London, Special Publications, **265**, 297–324.

HOPKINS, R. L. 1990, Kaibab Formation. In: BEUS, S. S. & MORALES, M. (eds) *Grand Canyon Geology*. Oxford University Press, New York, 225–245.

HUNT, A. P. & LUCAS, S. G. 1998. Implications of the cosmopolitanism of Permian tetrapod ichnofaunas. *In*: LUCAS, S. G., ESTEP, J. W. & HAFFER, J. M. (eds) *Permian Stratigraphy and Paleontology of the Robledo Mountains, New Mexico*. New Mexico Museum of Natural History and Science Bulletin, **12**, 55–57.

HUNT, A. P. & LUCAS, S. G. 2005*a*. Nonmarine Permian track faunas from Arizona, USA: ichnotaxonomy and ichnofacies. *In*: LUCAS, S. G. & ZEIGLER, K. E. (eds) *The Nonmarine Permian*. New Mexico Museum of Natural History and Science Bulletin, **30**, 128–131.

HUNT, A. P. & LUCAS, S. G. 2005*b*. Tetrapod ichnofacies and their utility in the Paleozoic. *In*: BUTA, R. J., RINDSBERG, A. K. & KOPASKA-MERKEL, D. C. (eds) *Pennsylvanian Footprints in the Black Warrior Basin of Alabama*. *Alabama Paleontological Society*, Monographs, **1**, 113–119.

HUNT, A. P. & LUCAS, S. G. 2006. Tetrapod ichnofacies: A new paradigm. *Ichnos*.

HUNT, A. P. & SANTUCCI, V. 1998. *Taxonomy and ichnofacies of Permian tetrapod tracks from Grand Canyon National Park, Arizona*. National Park Service Geological Resources Division Technical Report NPS/NRGRD/GRDTR-98/01, 94–96.

HUNT, A. P., LUCAS, S. G. & LOCKLEY, M. G. 1995. Paleozoic tracksites of the western United States. *In*: LUCAS, S. G. & HECKERT, A. B. (eds) *Early Permian Footprints and Facies*. New Mexico Museum of Natural History and Science Bulletin. **6**, 213–217.

HUNT, A. P., LUCAS, S. G., SANTUCCI, V. L. & ELLIOTT, D. K. 2005*a*. Permian vertebrates of Arizona. *In*: HECKERT, A. B. & LUCAS, S. G. (eds) *Vertelorate Paleontology in Arizona*. New Mexico Museum of Natural History and Science Bulletin, **29**, 10–15.

HUNT, A. P., LUCAS, S. G. & SPIELMANN, J. A. 2005*b*. Early Permian tetrapod tracksites in New Mexico. *In*: LUCAS, S. G., ZEIGLER, K. E. & SPIELMANN, J. A. (eds) *The Permian of Central New Mexico*. New Mexico Museum of Natural History and Science Bulletin, **31**, 46–47.

HUNT, A. P., LUCAS, S. G. & SPIELMANN, J. A. 2005*c*. Paleoenvironmental transects and tetrapod biotaxonichnofacies in the early Permian of the southwestern United States. *In*: LUCAS, S. G., ZEIGLER, K. E. & SPIELMANN, J. A. (eds) *The Permian of Central New Mexico*. New Mexico Museum of Natural History and Science Bulletin, **31**, 49–51.

HUNT, A. P., LUCAS, S. G. & SPIELMANN, J. A. 2005*d*. Early Permian tetrapod ethoichnofacies in New Mexico. *In*: LUCAS, S. G., ZEIGLER, K. E. & SPIELMANN, J. A. (eds) *The Permian of Central New Mexico*. New Mexico Museum of Natural History and Science Bulletin, **31**, 52–55.

HUNT, A. P., LUCAS, S. G. & SPIELMANN, J. A. 2005*e*. The Permian tetrapod ichnogenus *Ichniotherium cottae* from central New Mexico. *In*: LUCAS, S. E., ZEIGLER, K. E. & SPIELMANN, J. A. (eds) *The Permian of Central New Mexico.* New Mexico Museum of Natural History and Science Bulletin, **31**, 56–58.

JALIL, N. E. & DUTUIT, J. M. 1996. Permian captorhinid reptiles from the Argana Formation, Morocco. *Palaeontology*, **39**, 907–918.

JONES, D. F. 1975. *Stratigraphy, Environments of Deposition, Petrology, Age, and Provenance, Basal Red Beds of the Argana Valley, Western High Atlas Mountains, Morocco.* MSC thesis, New Mexico Institute of Mining and Technology, Socorro.

KIETZKE, K. K. & LUCAS, S. G. 1995. Some microfossils from the Robledo Mountains Member of the Hueco Formation, Doña Ana County, New Mexico. *In*: LUCAS, S. G. & HECKERT, A. B. (eds) *Early Permian Footprints and Facies.* New Mexico Museum of Natural History and Science Bulletin, **6**, 57–62.

LARHRIB, M. 1996. Flore fossile et séquences des formations rouges fluviatiles du basin autunien de Tiddas-Sebt Aït Ikkou (NW du Maroc central). *In*: MEDINA, F. (ed.) *Le Permien et le Trias du Maroc: État des Connaissances.* Editions Pumag, Marrakech, 19–29.

LEONARDI, G. 1987. The first tetrapod footprint in the Permian of Brazil. *Anais do X Congresso Brasileiro de Paleontologia, Rio de Janeiro*, **1**, 333–335.

LEONARDI, G. 1994. *Annotated Atlas of South America Tetrapod Footprints* (Devonian to Holocene). Companhia de Pesquisas de Recursos Minerais, Brasilia.

LEONARDI, G. & NICOSIA, U. 1973. Stegocephaloid footprint in the Middle Permian sandstone (Groedener Sandsteine) of the Western Dolomites. *Annali dell' Università di Ferrara, Nuova Serie*, **9**, 1116–1249.

LEONARDI, P., CONTI, M. A., LEONARDI, G., MARIOTTI, N. & NICOSIA, U. 1975. *Pachypes dolomiticus* n. gen. n. sp.; pareiasaur footprint from the 'Val Gardena Sandstone' (Middle Permian) in the western Dolomites (N. Italy). *Rendiconti dell'Accademia Nazionale dei Lincei, Classe Scienze, Matematiche, Fisiche e Naturali, Serie 8*, **57**, 221–232.

LOCKLEY, M. G. & HUNT, A. P. 1995. *Dinosaur Tracks and Other Fossil Footprints of the Western United States.* Columbia University Press, New York.

LOCKLEY, M. G. & MADSEN, J. H., JR. 1993. Early Permian vertebrate trackways from the Cedar Mesa Sandstone of eastern Utah: Evidence of predator-prey interaction. *Ichnos*, **2**, 147–153.

LOCKLEY, M. G., HUNT, A. P. & MEYER, C. 1994. Vertebrate tracks and the ichnofacies concept: Implications for paleoecology and palichnostratigraphy. *In*: DONOVAN, S. K. (ed.) *Trace Fossils.* John Wiley, New York, 241–268.

LOCKLEY, M. G., HUNT, A. P., HAUBOLD, H. & LUCAS, S. G. 1995. Fossil footprints in the DeChelly Sandstone of Arizona: with paleoecological observations on the ichnology of dune facies. *In*: LUCAS, S. G. & HECKERT, A. B. (eds) *Early Permian Footprints and Facies.* Bulletin of New Mexico Museum of Natural History and Science, **6**, 225–233.

LOCKLEY, M. G., LUCAS, S. G., HUNT, A. P. & GASTON, R. 2004. Ichnofaunas from the Triassic-Jurassic boundary sequences of the Gateway area, western Colorado: implications for faunal composition and correlations with other areas. *Ichnos*, **11**, 89–102.

LOOPE, D. B. 1984. Eolian origin of upper Paleozoic sandstones, southeastern Utah. *Journal of Sedimentary Petrology*, **54**, 563–580.

LUCAS, S. G. 1998. Toward a tetrapod biochronology of the Permian. *In*: LUCAS, S. G., ESTEP, J. W. & HOFFER, J. M. (eds) *Permian Stratigraphy of the Robledo Mountains, New Mexico.* New Mexico Museum of Natural History and Science Bulletin, **12**, 71–91.

LUCAS, S. G. 2001. A global hiatus in the Middle Permian tetrapod fossil record. *Permophiles*, **38**, 24–27.

LUCAS, S. G. 2002*a*. Tetrapods and the subdivision of Permian time. *In:* HILLS, L. V. & BAMBER, E. W. (eds) *Carboniferous and Permian of the World.* Canadian Society of Petroleum Geologists, Memoir, **19**, 479–491.

LUCAS, S. G. 2002*b*. North American Permian tetrapod footprint biostratigraphy. *Permophiles*, **40**, 22–24.

LUCAS, S. G. 2002*c*. The reptile *Macroleter*: First vertebrate evidence for correlation of Upper Permian continental strata of North America and Russia discussion. *Geological Society of America Bulletin*, **114**, 1174–1177.

LUCAS, S. G. 2004. A global hiatus in the Middle Permian tetrapod fossil record. *Stratigraphy*, **1**, 47–64.

LUCAS, S. G. 2005*a*. Tetrapod ichnofacies and Ichnotaxonomy: quo vadis? *Ichnos*, **12**, 157–162.

LUCAS, S. G. 2005*b*. Permian tetrapod faunachrons. *In*: LUCAS, S. G. & ZEIGLER, K. E. (eds) *The Nonmarine Permian.* New Mexico Museum of Natural History and Science Bulletin, **30**, 197–201.

LUCAS, S. G. 2006. Global Permian tetrapod biostratigraphy and biochronology. *In:* LUCAS, S. G., CASSINIS, G. & SCHNEIDER, J. W. (eds) *Non-Marine Permian Biostratigraphy and Biochronology*. Geological Society, London, Special Publications, **265**, 65–93.

LUCAS, S. G. & HECKERT, A. B. (eds) 1995. *Early Permian Footprints and Facies.* New Mexico Museum of Natural History and Science Bulletin, **6**.

LUCAS, S. G. & HUNT, A. P. 2005. Permian tetrapod tracks from Texas. *In:* LUCAS, S. G. & ZEIGLER, K. E. (eds) *The Nonmarine Permian.* New Mexico Museum of Natural History and Science Bulletin, **30**, 202–206.

LUCAS, S. G. & SUNESON, N. 2002. Amphibian and reptile tracks from the Hennessey Formation (Leonardian, Permian), Oklahoma County, Oklahoma. *Oklahoma Geology Notes*, **62**, 56–62.

LUCAS, S. G., ANDERSON, O. J., HECKERT, A. B. & HUNT, A. P. 1995. Geology of Early Permian tracksites, Robledo Mountains, south-central New Mexico. *In:* LUCAS, S. G. & HECKERT, A. B. (eds) *Early Permian Footprints and Facies.* New Mexico Museum of Natural History and Science Bulletin, **6**, 13–32.

LUCAS, S. G., ESTEP, J. W. & HOFFER, J. M. (eds) 1998. *Permian Stratigraphy and Paleontology of the Robledo Mountains, New Mexico*. New Mexico Museum of Natural History and Science Bulletin, **12**.

LUCAS, S. G., LOZOVSKY, V. R. & SHISHKIN, M. A. 1999. Tetrapod footprints from Early Permian red beds of the northern Caucasus, Russia. *Ichnos*, **6**, 277–281.

LUCAS, S. G., LERNER, A. J. & HAUBOLD, H. 2001. First record of *Amphisauropus* and *Varanopus* in the Lower Permian Abo Formation, central New Mexico. *Hallesches Jahrbuch für Geowissenschaften, Reihe B*, **23**, 69–78.

LUCAS, S. G., LERNER, A. J. & HUNT, A. P. 2004. Permian tetrapod footprints from the Lucero uplift, central New Mexico, and Permian footprint biostratigraphy. *In:* LUCAS, G. S. & ZEIGLER, K. E. (eds) *Carboniferous—Permian Transition of Carrizo Arroyo, Central New Mexico*. New Mexico Museum of Natural History and Science Bulletin, **25**, 291–300.

LUCAS, S. G., MINTER, N. J., SPIELMANN, J. A., HUNT, A. P. & BRADDY, S. J. 2005*a*. Early Permian ichnofossil assemblage from the Fra Cristobal Mountains, southern New Mexico. *In*: LUCAS, S. G., ZEIGLER, K. E. & SPIELMANN, J. A. (eds) *The New Mexico*. New Mexico Museum of Natural History and Science Bulletin, **31**, 140–150.

LUCAS, S. G., MINTER, N. J., SPIELMANN, J. A., SMITH, J. A. & BRADDY, S. J. 2005*b*. Early Permian ichnofossils from the northern Caballo Mountains, Sierra County, New Mexico. *In*: LUCAS, S. G., ZEIGLER, K. E. & SPIELMANN, J. A. (eds) *The Permian of Central New Mexico*. New Mexico Museum of Natural History and Science Bulletin, **31**, 151–162.

LUCAS, S. G., SMITH, J. A. & HUNT, A. P. 2005*c*. Tetrapod tracks from the Lower Permian Yeso Group, central New Mexico. *In*: LUCAS, S. G., ZEIGLER, K. E. & SPIELMANN, J. A. (eds) *The Permian of Central New Mexico*. New Mexico Museum of Natural History and Science Bulletin, **31**, 121–124.

LULL, R. S. 1918. Fossil footprints from the Grand Canyon of the Colorado. *American Journal of Science, Series 4*, **45**, 337–346.

MCKEE, E. D. 1934. An investigation of the light-colored, cross-bedded sandstones of Canyon De Chelly, Arizona. *American Journal of Science, Series 5*, **28**, 81–84.

MCKEEVER, P. M. & HAUBOLD, H. 1996. Reclassification of vertebrate trackways from the Permian of Scotland and related forms from Arizona and Germany. *Journal of Paleontology*, **70**, 1011–1022.

MELCHOR, R. N. 2000. Stratigraphic and biostratigraphic consequences of a new $^{40}Ar/^{39}Ar$ date for the base of the Cochico Group (Permian), eastern Permian basin, San Rafael, Mendoza, Argentina. *Ameghiniana*, **37**, 271–282.

MELCHOR, R. N. 2001. Permian tetrapod footprints from Argentina. *Hallesches Jahrbuch für Geowissenschaften, Reihe B*, **23**, 35–43.

MELCHOR, R. N. & CESARI, S. 1997. Permian floras from Carapacha basin, La Pampa Province, Argentina. Description and importance. *Géobios*, **30**, 607–633.

MELCHOR, R. N. & SARJEANT, W. A. S. 2004. Small amphibian and reptile footprints from the Permian Carapacha basin, Argentina. *Ichnos*, **11**, 57–78.

MENNING, M. 1995. A numerical time scale for the Permian and Triassic Periods. An integrated time analysis. *In*: SCHOLLE, P., PERYT, T. M. & ULMER-SCHOLLE, S. (eds) *Permian of the Northern Continents*, Vol. 1. Springer Verlag, Berlin, 77–97.

MIDDLETON, L. T., ELLIOTT, D. K. & MORALES, M. 1990. Coconino Sandstone. *In*: BEUS, S. S. & MORALES, M. (eds.) *Grand Canyon Geology*. Oxford University Press, New York, 183–202.

MOODIE, R. L. 1929. Vertebrate footprints from the red beds of Texas. *American Journal of Science*, **97**, 352–368.

MOODIE, R. L. 1930. Vertebrate footprints from the red beds of Texas. II. *Journal of Geology*, **38**, 548–565.

MORALES, M. & HAUBOLD, H. 1995. Tetrapod tracks from the Lower Permian DeChelly sandstone of Arizona: systematic description. *In*: LUCAS, S. G. & HECKERT, A. B. (eds) Early Permian Footprints and Facies. New Mexico Museum of Natural History and Science Bulletin, **6**, 251–261.

MOSSMAN, D. J. & PLACE, C. H. 1989. Early Permian fossil vertebrate footprints and their stratigraphic significance in megacyclic sequence II red beds, Prim Point, Prince Edward Island. *Canadian Journal of Earth Sciences*, **26**, 591–605.

NICOSIA, U., RONCHI, A. & SANTI, G. 2000. Permian tetrapod footprints from W Orobic Basin (Northern Italy). Biochronological and evolutionary remarks. *Géobios*, **33**, 753–768.

NICOSIA, U., SACCHI, E. & SPEZZAMONTE, M. 2001. New palaeontological data for the Val Gardena Sandstone. *In*: CASSINIS, G. (ed.) *Permian Continental Deposits of Europe Regional Reports and Correlations*. Natura Bresciana, Monografia, **25**, 83–88.

OLSON, E. C. 1962. *Late Permian Terrestrial vertebrates, U.S.A. and U.S.S.R.* Transactions of the American Philosophical Society, New Series, **52**, 1–223.

OLSON, E. C. & MEAD, J. 1982. The Vale Formation (Lower Permian), its vertebrates and paleoecology. *Texas Memorial Museum Bulletin*, **29**, 1–46.

PEABODY, F. 1948. Reptile and amphibian trackways from the Moenkopi Formation of Arizona and Utah. *University of California, Bulletin of Geological Sciences*, **27**, 295–468.

PITTMAN, J. G., SCHULTZ-PITTMAN, R., LOCKLEY, M. G. & WESTGATE, J. W. 1996. Two Permian footprint localities at San Angelo, Texas. *Journal of Vertebrate Paleontology*, **16**(3, supplement), 58A.

PTASZYŃSKI, T. & NIEDŹWIEDZKI, G. 2004. Late Permian vertebrate tracks from the Tumlin Sandstone, Holy Cross Mountains, Poland. *Acta Palaeontologica Polonica*, **49**, 289–320.

RACKI, G. 2005. 'Late Permian' vertebrate tracks from the Tumlin Sandstone of Poland: a commentary on some major implications. *Acta Palaeontologica Polonica*, **50**, 394–396.

ROMER, A. S. & PRICE, L. I. 1940. *Review of Pelycosauria*. Geological Society of America, Special Papers, **28**.

ROSCHER, M. & SCHNEIDER, J. W. 2005. An annotated correlation chart for continental Late Pennsylvanian and Permian basins with the marine scale. *In*: LUCAS, S. G. & ZEIGLER, K. E. (eds) *The Nonmarine Permian*. New Mexico Museum of Natural History and Science Bulletin, **30**, 282–291.

SANTI, G. & KRIEGER, C. 2001. Lower Permian tetrapod footprints from Brembana Valley – Orobic Basin – (Lombardy, Northern Italy). *Revue de Paléobiologie, Genève*, **20**, 45–68.

SCHALTEGGER, U. & BRACK, P. 1999. Radiometric age constraints on the formation of the Collio Basin (Brescian Prealps). *In*: CASSINIS, G., CORTESOGNO, L., GAGGERO, L., MASSARI, F., NERI, C., NICOSIA, U. & PITTAU, P. (coord.) *Stratigraphy and Facies of the Permian Deposits between Eastern Lombardy and the Western Dolomites: Field Trip Guidebook 23–25 September 1999*. International Field Conference on 'The Continental Permian of the Southern Alps and Sardinia (Italy). Regional reports and general correlations', Brescia, 15–25 September 1999. 71.

SCHMIDT, H. 1959. Die Cornberger Fährten im Rahmen der Vierfüsser Entwicklung. *Abhandlungen des Hessischen Landesamtes für Bodenforschung*, **28**, 1–137.

SCHULT, M. 1995. Vertebrate trackways from the Robledo Mountains Member of the Hueco Formation, south-central New Mexico. *In*: LUCAS, S. G. & HECKERT, A. B. (eds) *Early Permian Footprints and Facies*. New Mexico Museum of Natural History and Science Bulletin, **6**, 115–126.

SCHULTZ-PITTMAN, R. J., LOCKLEY, M. G. & GASTON, R. 1996. First reports of synapsid tracks from the Wingate and Moenave Formations, Colorado Plateau region. *Bulletin of the Museum of Northern Arizona*, **60**, 271–274.

SEELEY, H. G. 1904. Footprints of small fossil reptiles from the Karroo rocks of Cape Colony. *Annals and Magazine of Natural History, Series 7*, **14**, 287–289.

SMITH, R. M. H. 1993. Sedimentology and ichnology of floodplain paleosurfaces in the Beaufort Group (Late Permian), Karoo sequence, South Africa. *Palaios*, **8**, 339–357.

SWANSON, B. A. & CARLSON, K. J. 2002. Walk, wade, or swim? Vertebrate traces on an Early Permian lakeshore. *Palaios*, **17**, 123–133.

TILTON, J. L. 1926. Permian vertebrates from West Virginia. *Geological Society of America Bulletin*, **37**, 385–396.

TILTON, J. L. 1931. Permian vertebrate tracks in West Virginia. *Geological Society of America Bulletin*, **42**, 547–556.

TVERDOKHLEBOV, V. P., TVERDOKHLEBOVA, G. I., BENTON, M. J. & STORRS, G. W. 1997. First record of footprints of terrestrial vertebrates from the Upper Permian of the Cis-Urals, Russia. *Palaeontology*, **40**, 157–166.

VAN ALLEN, H. E. K., CALDER, J. H. & HUNT, A. P. 2005. The trackway record of a tetrapod community in a Walchian conifer forest from the Permo-Carboniferous of Nova Scotia. *In*: LUCAS, S. G. & ZEIGLER, K. E. (eds) *The Nonmarine Permian*. New Mexico Museum of Natural History and Science Bulletin, **30**, 322–332.

VAUGHN, P. P. 1964. A downslope trackway in the DeChelly Sandstone, Permian of Monument Valley. *Plateau*, **36**, 25–28.

VOIGT, S. 2005. Die Tetrapodenichnofauna des kontinentalen Oberkarbon und Perm im Thüringer Wald: Ichnotaxonomie, Paläoökologie und Biostratigraphie. Cuvillier Verlag, Göttingen.

VOIGT, S., SMALL, B. J. & SANDERS, F. 2005. A diverse terrestrial ichnofauna from the Maroon Formation (Pennsylvanian-Permian), Colorado: biostratigraphic and paleoecological significance. *In:* LUCAS, S. G. & ZEIGLER, K. E. (eds) *The Nonmarine Permian*. New Mexico Museum of Natural History and Science Bulletin, **30**, 342–351.

WARD, P. D. 2004. *Gorgon: Paleontology, Obsession, and the Greatest Catastrophe in Earth's History*. Viking, New York.

ZHENG, J. S., MERMET, J.-F., TOUTIN-MORIN, N., HANES, J., GONDOLO, A., MORIN, R. & FERAUD, G. 1992. Datation ^{40}Ar–^{39}Ar du magmatisme et de filons minéralisés permiens en Provence orientale (France). *Geodinamica Acta*, **5**, 203–215.

Amphibian biostratigraphy of the European Permo-Carboniferous

RALF WERNEBURG[1] & JOERG W. SCHNEIDER[2]

[1]*Naturhistorisches Museum Schloss Bertholdsburg, Burgstrasse 6, D-98553 Schleusingen, Germany (e-mail: museum.schleusingen@gmx.de)*

[2]*TU Bergakademie Freiberg, Cottastrasse 2, D-09596 Freiberg, Germany (e-mail: schneidj@geo.tu-freiberg.de)*

Abstract: A revised amphibian zonation for the European Pennsylvanian and Cisuralian (Upper Carboniferous to Lower Permian) with nine amphibian zones is presented. The index fossils belong to species-chronoclines with two or three closely related species. The time resolution of these amphibian zones is about 1.5–3.0 Ma. Biostratigraphical correlations with amphibian zones are applicable to 16 basins in the Czech Republic, Poland, France, Italy and Germany. The biostratigraphical potential of other tetrapods is discussed.

Tetrapod biostratigraphy or biochronology of the non-marine Permian and Carboniferous is an important and difficult task for palaeoherpetologists. Most of the tetrapod zonations for the Permian are reviewed by Lucas (1998, 2004). They were developed in North America for the Upper Carboniferous (Pennsylvanian) and Lower Permian (Cisuralian) as well as in South Africa, Russia and China for the Middle and Upper Permian (Guadalupian, Lopingian).

In Europe, tetrapod biostratigraphy was established for the Upper Carboniferous and Permian with tetrapod footprints and amphibians. Tetrapod footprint zonations for several basins in Europe were developed by Haubold (1980, 1984, 2000), Holub & Kozur (1981), Boy & Fichter (1982, 1988), Gand & Haubold (1988) and Kozur (1989). The general problem of tetrapod footprint biostratigraphy is that the tetrapod ichnospecies normally correspond to families or, in the best case, only genera in the osteological system. Therefore, tetrapod ichnospecies reflect larger steps in tetrapod phylogeny and consequently only represent longer time intervals (Lucas 1998, 2002).

Biostratigraphical zonations using osteological species of aquatic or semi-aquatic amphibians were presented by Boy (1987) and Werneburg (1989*a*, *b*, 1996). These amphibians belong to the dissorophoid Branchiosauridae and Micromelerpetontidae, to the stem-stereospondyl Archegosauriformes and to the seymouriamorph Discosauriscidae.

Basics of amphibian zonation

The newt-like families Branchiosauridae, Micromelerpetontidae and Discosauriscidae are ideal index fossils for biostratigraphy. They are common and have a wide distribution in space, but a narrow temporal range. Thousands of specimens are known of the branchiosaurs *Apateon dracyiensis* (Fig. 1a, b), *Apateon flagrifer* and *Apateon pedestris,* the discosauriscid *Discosauriscus austriacus*, as well as hundreds of individuals of the branchiosaurs *Melanerpeton sembachense, M. tenerum* and *Schoenfelderpeton prescheri.* In the European Permo-Carboniferous, up to four genera are now known with nearly 20 species and subspecies of the Branchiosauridae, four genera with 10 species of the Micromelerpetontidae and two genera with three species of the Discosauriscidae. At least 10 species of these three families are widespread and known from two or more basins in Europe.

The history of research of these three amphibian families spans as much as 130 years. Depending on the facies architectures and the degree of investigations, branchiosaurs can show a high frequency of occurrences in vertical sections. In the profile of the Thuringian Forest Basin, for example, about 11 successive levels with branchiosaurs are known, covering a time span from the Gzhelian up into Sakmarian, that is, 12 Ma (Werneburg 1989*a*, 2001*a*). The family Branchiosauridae exhibits a high rate of speciation with many short-lived species (but also with some long-lived species, such as *Apateon flagrifer* or *A. pedestris*).

The **taxonomic concept** of the Branchiosauridae and Micromelerpetontidae (Boy 1972 ff.; Werneburg 1986*a* ff.) as well as of the Discosauriscidae (Klembara & Meszároš 1992 ff.) has been extensively discussed, and a relatively wide consensus has been achieved. It is necessary to

From: LUCAS, S. G., CASSINIS, G. & SCHNEIDER, J. W. (eds) 2006. *Non-Marine Permian Biostratigraphy and Biochronology.* Geological Society, London, Special Publications, **265**, 201–215.

(a)

Fig. 1(a) & (b) The branchiosaur *Apateon dracyiensis* from the Lower Goldlauter Formation (Lower Rotliegend, Asselian) of Cabarz/Tabarz in the Thuringian Forest Basin (skull length about 8mm).

(b)

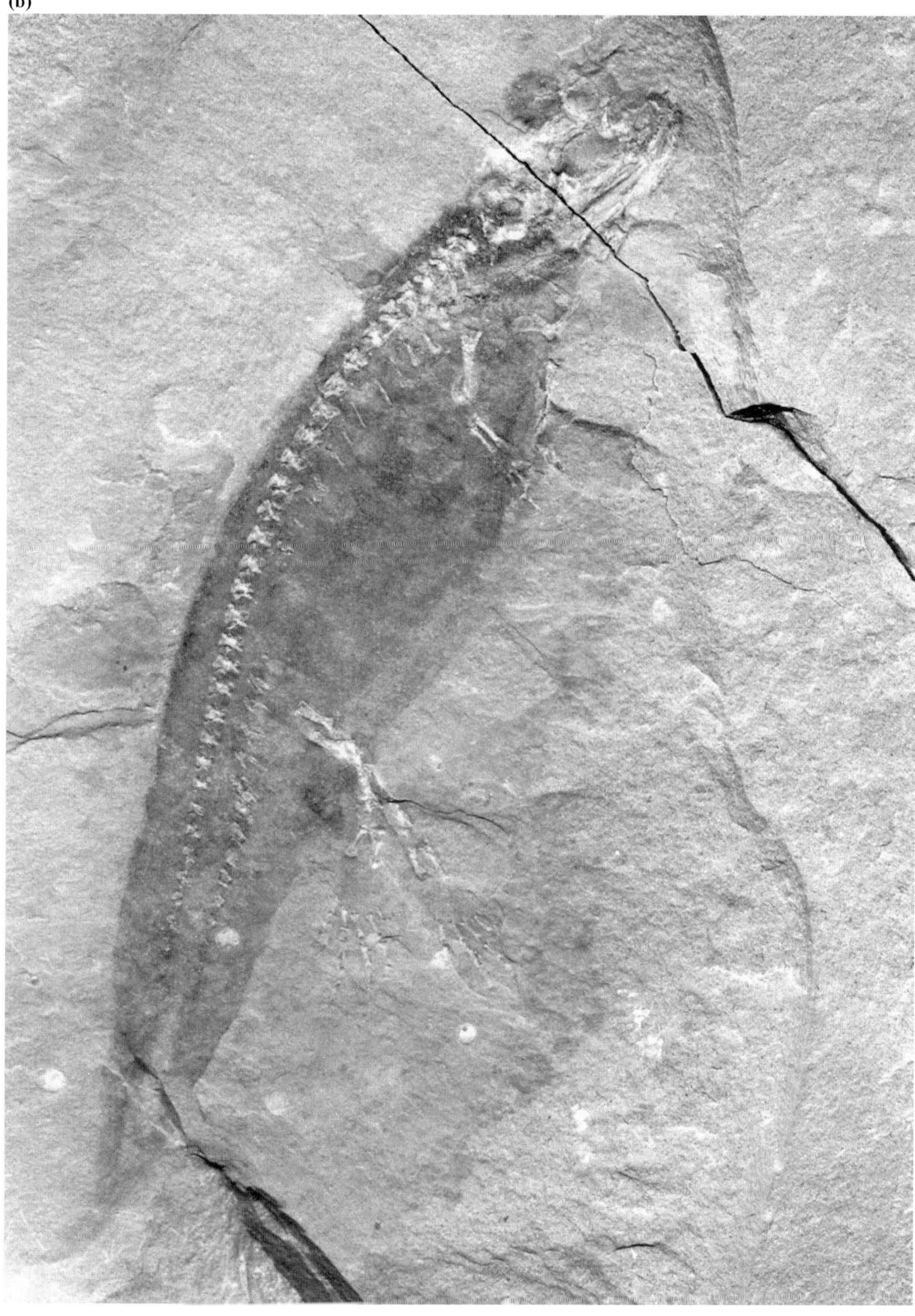

Fig. 1(b).

present some short remarks on the grammatically correct use of the *Apateon* species names. For a long time, the grammatical gender of the genus was not clear. The type species is *Apateon pedestris* Meyer, 1844. Werneburg (1988*a* ff.) has used the feminine gender for several species of *Apateon* (e.g. *A. flagrifera, A. caduca* and *A. umbrosa*). But, now it is clear that the genus *Apateon* has a masculine gender. Consequently, the species of *Apateon* have the following amended endings: *Apateon pedestris, A. caducus, A. flagrifer, A. kontheri, A. dracyiensis, A. intermedius* and *A. umbrosus*.

The palaeoecology of all three families is well known (e.g. Boy 2003; Boy & Sues 2000; Werneburg 2001*b*, 2002). Nearly all of their members are aquatic and neotenic. They lived in lakes of different sizes, and they are also known from grey and red facies deposits. It was possible to distinguish two ecomorphotypes, the pond-type and the stream-type, which are based on well-known forms based on extant larval salamanders (Werneburg 2002). The best fossil record is known from lacustrine deposits, but fine clastic intercalations in fluvial coarse clastics could also contain remains of those amphibians.

The phylogeny, especially of the branchiosaurs, is a more difficult problem. This amphibian family was able to adapt to the changing palaeogeographical and palaeoecological conditions in the European variscids through fast radiation and paedomorphosis (neoteny). The result is a complicated mosaic-pattern in their evolution. It is difficult to find clear-cut, uniquely derived characters for each species. A first draft of a phylogeny (Werneburg 1989*a*, *b*) was based on a detailed list of constitutive and diagnostic features as a basis for the definition for each taxon (Werneburg 1989*a*). Meanwhile, with more studies, this concept has been confirmed: for discosauriscids (Klembara 1997), *Micromelerpeton* (Boy 1995, 2000*a, b*) and *Melanerpeton* (Werneburg 1991).

The basic concepts of amphibian biostratigraphy presented here are lineages of two or three chronologically successive species. This is a pragmatic simplification for biostratigraphical application of the possibly more complicated phylogeny with assumed dichotomous and trichotomous speciation patterns. The species chronoclines consist of time-successive, closely related species of one evolutionary branch or lineage. Sometimes these chronoclines are supplied with directed feature trends.

The chronological sequence of species is often demonstrated by primary field data – by the successive occurrence of species in one lithostratigraphic profile. The consecutive range of species is determined by their more or less well-known First (FAD) and Last Appearance Dates (LAD). However, compared to marine sections, the patchy, discontinuous fossil record in continental deposits prevents the exact determination of FADs and LADs. So, it is easier to apply the First Occurrence Date (FOD; lowest occurrence) and the Last Occurrence Date (LOD; highest occurrence) for definition of the amphibian zones. Nevertheless, the increasing discoveries of amphibian sites (e.g. Ronchi & Tintori 1997; Werneburg 2003) and the cross-check with the insect zones (Schneider 1982; Schneider & Werneburg 1993) as well as isotopic ages (Lützner *et al.* 2003) are leading to increased improvements of the amphibian biostratigraphy.

Most of the fossil taxa are a mixture of biospecies and chronospecies. In the case of branchiosaurs we have some well-known communities of one lake with three different species, such as, *Apateon flagrifer, A. kontheri* and *Melanerpeton eisfeldi* from the Gottlob Lake in the Thuringian Forest Basin of Germany (Werneburg 1986*b*, 1988*a*). These are real biospecies with clear-cut features. Many of the other branchiosaurs may be chronospecies. Both kinds of species are used in the biostratigraphical concept of species chronoclines.

Amphibian zonation

The first amphibian biostratigraphy was established by Boy (1987) for the Lower Permian Rotliegend of the Saar–Nahe Basin. It is an assemblage zonation based on branchiosaurid and archegosauriform species known from this basin only. Boy (1987) proposed an additional biostratigraphic zonation for the middle European Rotliegend based on branchiosaurid relationship-groups:

- Zone of *Apateon pusillus* group (with uncertain relationship)
- Zone of *Apateon caducus* group (with *A. flagrifer*)
- Zone of *Apateon pedestris* (oldest one from Altenglan Formation).

This zonation is not really useful outside the Saar–Nahe Basin, because *Apateon pedestris* and *A. caducus* are known only from the Saar–Nahe Basin. Additionally, the basics of branchiosaurid taxonomy were not well enough advanced at that time to provide a sound basis for biostratigraphy.

Between 1986 and 1989, many branchiosaurid species were newly described or revised (Boy 1986, 1987; Werneburg 1986*a*, *b*, 1987, 1988*a*, *b*, *c*, 1989*a*, *b*, *c*). Most of these species are known from the Rotliegend section (uppermost Pennsylvanian to lower Cisuralian) of the Thuringian Forest Basin in Germany (Werneburg 2001*a*). From each formation of this section, branchiosaurids, micromelerpetontids and/or discosauriscids are known; altogether, 13 of the 23 known Thuringian amphibian species are present. The Thuringian Forest profile thus has the most complete succession of amphibian species in Europe. It was established as a reference section for the amphibian zonation by Werneburg (1989*a*, *b*, 1996). The amphibian succession of this profile covers the uppermost Stephanian C up to the uppermost Lower Rotliegend (upper Autunian or middle Sakmarian). It can be completed further down to Westphalian D time (Carboniferous, uppermost Moscovian) and up to lowermost Upper Rotliegend time (Saxonian, Sakmarian/Artinskian) by amphibian horizons from other European basins (see below). Most of the zone index fossils are not only recorded from the Thuringian Forest Basin but also from other basins in France, Italy (Sardinia), the Czech Republic, Poland and Germany. The revised version of the amphibian zonation is illustrated in Figure 2 and defined in the following synopsis.

Revised amphibian zonation

Branchiosaurus salamandroides – Limnogyrinus elegans zone

Definition: From the FODs of *Branchiosaurus salamandroides* and *Limnogyrinus elegans* up to the FOD of *Branchiosaurus fayoli.*

Occurrences: Nýřany Member, Kladno Formation, Westphalian D, Moskovian, of Nýřany and Tremošna, Plzeň Basin, Czech Republic.

Accompanying species: *Platyrhinops* cf. *P. lyelly, Scincosaurus crassus* et al., all from the Westphalian D of Nýřany.

Remarks: Some of the accompanying species, such as *P. lyelly*, are also known from the Westphalian D of Linton, Ohio, USA. This amphibian zone also provisionally includes the Cantabrian and Stephanian A (Kasimovian).

Branchiosaurus fayoli zone

Definition: From the FOD of *Branchiosaurus fayoli* up to the FOD of *Apateon intermedius* and *Branchierpeton saalensis.*

Occurrences:

- Stephanian B (Gzhelian) from Commentry, Commentry Basin, French Massif Central;
- ? Stephanian B (Gzhelian) from Montceau-les-Mines, Blanzy Basin, French Massif Central.

? Accompanying species: *Scincosaurus spinosus, Sauravus costei*, both from the Stephanian B of Montceau-les-Mines, Blanzy Basin.

Remarks: The FAD of *Branchiosaurus fayoli* is theoretically possible in the Cantabrian and Stephanian A (Barruelian), if amphibians are found in these beds.

Apateon intermedius – Branchierpeton saalensis zone

Definition: From the FODs of *Apateon intermedius* and *Branchierpeton saalensis* up to the LAD of *A. intermedius*, the FAD of *Apateon dracyiensis* and up to the FODs of *Melanerpeton sembachense* and *Limnogyrinus edani.*

Occurrences:

- Möhrenbach Formation (Stephanian C, Gzhelian) from Silbergrund and Moosbach, Thuringian Forest Basin;
- Wettin Member of the Siebigerode Formation (Stephanian C, Gzhelian) from Löbejün near Halle and the Halle Formation (lowermost Rotliegend, Gzhelian) from Petersberg near Halle, both Saale Basin;
- Netzkater Formation (lowermost Rotliegend, Gzhelian) of Neustadt, Ilfeld Basin, all Germany;
- Líně Formation (Stephanian C, Gzhelian) of Libechov, Roudnice Basin, Bohemia, Czech Republic.

Accompanying species: *Onchiodon manebachensis*, Stephanian C of Moosbach, Thuringian Forest Basin.

Remarks: The upper limit of this zone is well defined with the FAD of *Apateon dracyiensis.*

Apateon dracyiensis – Melanerpeton sembachense zone

Definition: From the FAD of *Apateon dracyiensis*, the FODs of *Melanerpeton sembachense* and *Limnogyrinus edani* up to the FODs of *Branchierpeton reinholdi* and *Apateon flagrifer flagrifer.*

Occurrences:

- Ilmenau Formation (earlier 'Lower Manebach Fm.', Lower Rotliegend, Gzhelian) from Sembachtal near Winterstein, Thuringian Forest Basin, Germany;
- ? Muse Formation (Lower Autunian or Rotliegend, Gzhelian/Asselian) from Dracy-St. Loup and Muse, Autun Basin, French Massif Central.

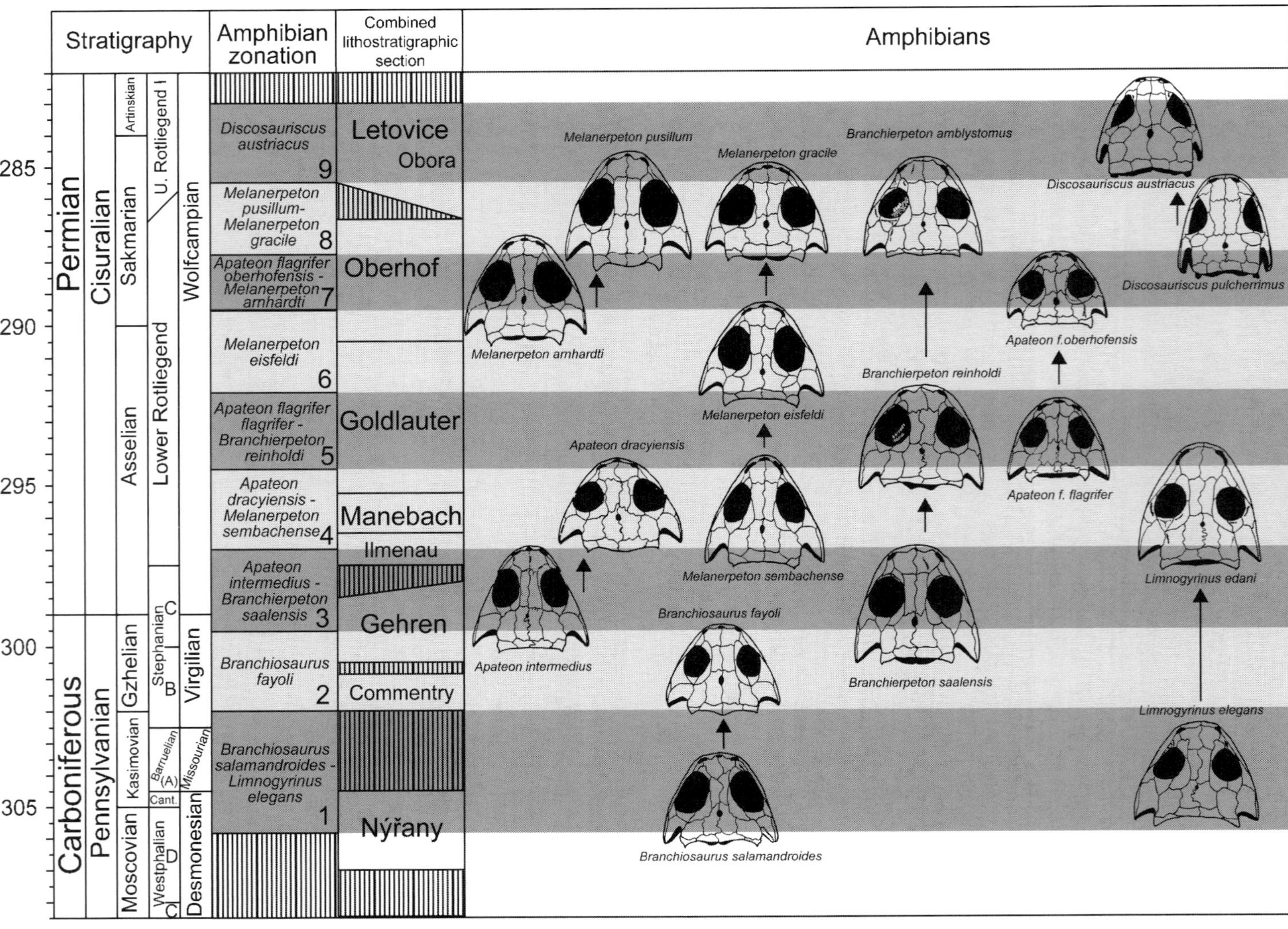

Stratigraphy
Amphibian zonation
Combined lithostratigraphic section
Amphibians
Permian
Carboniferous
Cisuralian
Pennsylvanian
Artinskian
Sakmarian
Asselian
Gzhelian
Kasimovian
Moscovian
U. Rotliegend I
Lower Rotliegend
Stephanian
C
B
Barruelian (A)
Cant.
Westphalian
D
C
Wolfcampian
Virgilian
Missourian
Desmonesian
Discosauriscus austriacus 9
Melanerpeton pusillum- Melanerpeton gracile 8
Apateon flagrifer oberhofensis - Melanerpeton arnhardti 7
Melanerpeton eisfeldi 6
Apateon flagrifer flagrifer - Branchierpeton reinholdi 5
Apateon dracyiensis - Melanerpeton sembachense 4
Apateon intermedius - Branchierpeton saalensis 3
Branchiosaurus fayoli 2
Branchiosaurus salamandroides - Limnogyrinus elegans 1
Letovice
Obora
Oberhof
Goldlauter
Manebach
Ilmenau
Gehren
Commentry
Nýřany
Melanerpeton pusillum
Melanerpeton gracile
Branchierpeton amblystomus
Discosauriscus austriacus
Discosauriscus pulcherrimus
Melanerpeton arnhardti
Apateon f.oberhofensis
Branchierpeton reinholdi
Melanerpeton eisfeldi
Apateon dracyiensis
Apateon f. flagrifer
Melanerpeton sembachense
Limnogyrinus edani
Branchiosaurus fayoli
Apateon intermedius
Branchierpeton saalensis
Limnogyrinus elegans
Branchiosaurus salamandroides
285
290
295
300
305

Accompanying species: *Onchiodon langenhani* from Sembachtal (Ilmenau Formation).

Remarks: The lower limit of this zone is well defined with the FAD of *Apateon dracyiensis.*

Apateon flagrifer flagrifer – Branchierpeton reinholdi zone

Definition: From the FODs of *Apateon flagrifer flagrifer* and *Branchierpeton reinholdi* up to the FOD of *Melanerpeton eisfeldi.*

Occurrences:

- Lower Goldlauter Formation (Lower Rotliegend, Asselian) from Sperbersbach, Pochwerksgrund, Untere Kniebreche and Cabarz/Tabarz, Thuringian Forest Basin, Germany;
- Börtewitz lake horizon, (Lower Rotliegend, Asselian) from Clennen and Börtewitz, NW-Saxony Basin, Germany;
- ? Muse Formation (Autunian or Lower Rotliegend, Gzhelian/Asselian) from Dracy-St. Loup and Muse, Autun Basin, French Massif Central.

Accompanying species: *Melanerpeton tenerum, Schoenfelderpeton prescheri* and *Apateon dracyiensis* from Börtewitz, Clennen, NW-Saxony Basin; *S. prescheri* and *A. dracyiensis* from Cabarz/Tabarz, Thuringian Forest Basin, all Germany.

Remarks: The correlation of the occurrences Börtewitz, Clennen and Cabarz/Tabarz is well founded on the common occurrence of *Branchierpeton reinholdi, Schoenfelderpeton prescheri* and *Apateon dracyiensis.* Following insect biostratigraphy, the Cabarz lake horizon clearly belongs to the Lower Goldlauter Formation of the Thuringian Forest Basin, Germany (Schneider & Werneburg 1993).

Melanerpeton eisfeldi – zone

Definition: From the FOD of *Melanerpeton eisfeldi* up to the LAD of *Apateon flagrifer flagrifer,* the FAD of *Apateon flagrifer oberhofensis* and the FOD of *Melanerpeton arnhardti.*

Fig. 2. Amphibian zonation for the European Permo-Carboniferous using next related species of species-chronorows from the aquatic families Branchiosauridae, Micromelerpetontidae and Discosauriscidae.

Occurrences:

- Upper Goldlauter Formation (Lower Rotliegend, Asselian) from Gottlob and Kesselgraben near Friedrichroda, Radelsgraben, Grenzwiese-Cabarz road and Oberschönau, all Thuringian Forest Basin, Germany;
- Rio su Luda-Formation (Asselian) from Is Alinus, Perdasdefogu Basin, Sardinia, Italy.

Accompanying species: *Apateon flagrifera flagrifera, Apateon kontheri, Onchiodon labyrinthicus,* all from the Gottlob near Friedrichroda, Thuringian Forest Basin.

Remarks: The upper limit of this zone is well defined with the FAD of *Apateon flagrifer oberhofensis.*

Apateon flagrifer oberhofensis – Melanerpeton arnhardti zone

Definition: From the FAD of *Apateon flagrifer oberhofensis* and the FOD of *Melanerpeton arnhardti* up to the LAD of *Melanerpeton arnhardti,* the FAD of *M. pusillum,* and the FODs of *M. gracile, Branchierpeton amblystomus* and *Discosauriscus pulcherrimus.*

Occurrence: Lower Oberhof Formation (Lower Rotliegend, Asselian/Sakmarian) from Lochbrunnen near Oberhof, Thuringian Forest Basin, Germany.

Accompanying species: *Onchiodon labyrinthicus* from Lochbrunnen near Oberhof, Thuringian Forest Basin.

Remarks: The lower limit of this zone is well defined by the FAD of *Apateon flagrifer oberhofensis.*

Melanerpeton pusillum – Melanerpeton gracile zone

Definition: From the FAD of *Melanerpeton pusillum,* the FODs of *Melanerpeton gracile, Branchierpeton amblystomus* and *Discosauriscus pulcherrimus* up to the FOD of *Discosauriscus austriacus* (Klembara 1997).

Occurrences:

- Upper Oberhof Formation (Lower Rotliegend, Sakmarian) from Im Grunde near Friedrichroda, Wintersbrunnen near Finsterbergen and Mösewegswiese near Tambach-Dietharz, Thuringian Forest Basin, Germany;
- Upper Niederhäslich Formation (Lower Rotliegend, Sakmarian) from Niederhäslich, Freital-Birkigt and Windberg near Freital, Dresden, Döhlen Basin, Germany;

- Olivětín Member of the Broumov Formation (Lower Rotliegend, Sakmarian) from Olivětín and Ruprechtice near Broumov, Intra-Sudetic Basin, Czech Republic;
- Upper Słupiec Formation (Lower Rotliegend, Sakmarian) from Ratno Dolne, Intra-Sudetic Basin, Poland;
- Upper Buxières Formation (Autunian or Lower Rotliegend, Sakmarian) from Buxières-les-Mines and La Queue d'Etang, Bourbon l'Archambault Basin, French Massif Central;
- Beds of Bert (Autunian or Lower Rotliegend, Sakmarian) from Bert, Blanzy Basin, French Massif Central;
- Uppermost Meisenheim Formation (L-0 10, Lower Rotliegend, Sakmarian) from Humberg, Wörsbach or Lebach, Saar–Nahe Basin, Germany.

Based on the occurrence of species, which typically accompany the zone species (see below), the following levels could be assigned to this zone, too, even though the zonespecies are missing at this time.

- ? Leukersdorf Formation (Lower/Upper Rotliegend, Sakmarian) from Chemnitz-Altendorf and Oberlungwitz, Erzgebirge Basin, Germany;
- ? Millery Formation (Autunian or Lower Rotliegend, Sakmarian) from Le Telots and Margenne, Autun Basin, French Massif Central;
- ? Usclas St. Privat Formation (Autunian or Lower Rotliegend, Sakmarian) from Usclas-St. Privat, Lodève Basin, southern French Massif Central.

Accompanying species:

- *Apateon flagrifer oberhofensis, Sclerocephalus jogischneideri* and *Onchiodon labyrinthicus* from the Thuringian Forest Basin;
- *Onchiodon labyrinthicus, Acanthostomatops vorax, Phanerosaurus pugnax, Batropetes fritschia* et al. from the Döhlen Basin;
- *Apateon pedestris, A. caducus, Melanerpeton humbergense, Micromelerpeton credneri, 'Cheliderpeton' latirostre, Archegosaurus decheni* and *Batropetes fritschia* from the Saar–Nahe Basin;
- *Apateon umbrosus, Onchiodon labyrinthicus* and *Cheliderpeton vranyi* from the Czech Intra-Sudetic Basin.

Remarks: The lower limit of this zone is well defined by the FAD of *Melanerpeton pusillum.*

Discosauriscus austriacus zone

Definition: From the FOD of *Discosauriscus austriacus* up to its unknown LAD.

Occurrences:

- Middle to upper Letovice Formation (upper Rotliegend, Sakmarian/Artinskian) from Bačov, Obora, Kochov, Drválovice et al. near Brno, Boscovice Graben, Czech Republic;
- ? Usclas St. Privat Formation (? upper 'Autunian' or Rotliegend, Sakmarian/ ?Artinskian) from Usclas-St. Privat, Lodève Basin, southern French Massif Central.

Accompanying species: *Discosauriscus pulcherrimus,* branchiosaurids and other amphibians from the Boskovice Graben.

Remarks: The upper limit of this zone is unknown.

Correlations

Most of the correlations are included under 'occurrences' in the revised amphibian zonation and are illustrated in Figure 3. At this point a few examples for the combined use of the amphibian zonation together with other methods are given. The insect wings of blattid cockroaches can be especially useful to solve problems of biostratigraphical correlation. The insect zonation of the Late Carboniferous and the Early Permian of Europe, North America and North Africa is based on species lineages with FADs (Schneider 1982; Schneider & Werneburg 1993, 2006; Schneider *et al.* 2003, 2004*a*,*b*). Other biostratigraphical methods are based on the teeth of xenacanth sharks (e.g. Schneider 1996; Schneider & Zajic 1994; Schneider *et al.* 2000) or on conchostracans (Martens 1983*a*, *b*, 1984). However, both of the latter methods have some limitations. The migration and distribution of xenacanth fishes is strongly restricted to interconnections of drainage systems and integrated lakes. Therefore, independent speciation processes could be contemporaneous in different, unconnected river and lake systems, producing different patterns of teeth. Conchostracans have a very high migration potential, because of wind distribution of their minute, drought-resistant eggs. Unfortunately, the distribution in time of the different species is not well enough known now for the deduction of certain FODs and LODs. Last, but not least, isotopic methods produce absolute age data of differing significance (cf. Lützner *et al.* 2003).

Fig. 3. Biostratigraphical correlations of selected formations and basins from the Permo-Carboniferous of Germany, the Czech Republic, Poland, France and Italy using the amphibian zonation.

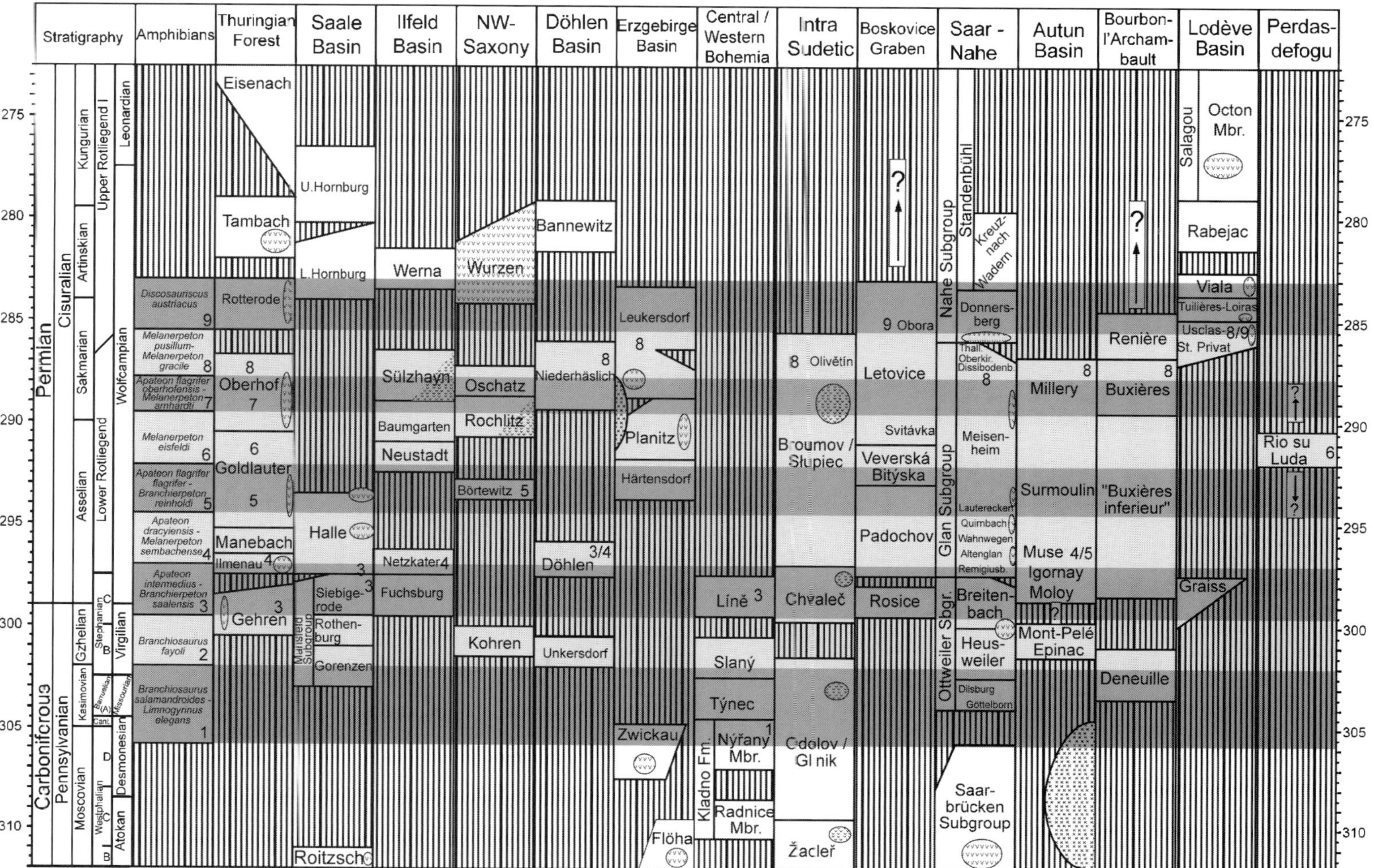

Stratigraphy
Amphibians
Thuringian Forest
Saale Basin
Ilfeld Basin
NW-Saxony
Döhlen Basin
Erzgebirge Basin
Central / Western Bohemia
Intra Sudetic
Boskovice Graben
Saar - Nahe
Autun Basin
Bourbon-l'Archambault
Lodève Basin
Perdasdefogu
Permian
Carboniferous
Cisuralian
Pennsylvanian
Kungurian
Artinskian
Sakmarian
Asselian
Gzhelian
Kasimovian
Moscovian
Upper Rotliegend I
Lower Rotliegend
Stephanian
Westphalian
Leonardian
Wolfcampian
Virgilian
Missourian
Desmonesian
Atokan
Discosauriscus austriacus 9
Melanerpeton pusillum - Melanerpeton gracile 8
Apateon flagrifer oberhofensis - Melanerpeton arnhardti 7
Melanerpeton eisfeldi 6
Apateon flagrifer flagrifer - Branchierpeton reinholdi 5
Apateon dracyiensis - Melanerpeton sembachense 4
Apateon intermedius - Branchierpeton saalensis 3
Branchiosaurus fayoli 2
Branchiosaurus salamandroides - Limnogynnus elegans 1
Eisenach
Tambach
Rotterode
Oberhof
Goldlauter
Manebach
Ilmenau
Gehren
U.Hornburg
L.Hornburg
Halle
Siebigerode
Rothenburg
Gorenzen
Mansfeld Subgroup
Roitzsch
Werna
Sülzhayn
Baumgarten
Neustadt
Netzkater
Fuchsburg
Wurzen
Oschatz
Rochlitz
Börtewitz 5
Kohren
Bannewitz
Niederhäslich
Döhlen
Unkersdorf
Leukersdorf
Planitz
Härtensdorf
Zwickau
Flöha
Líně 3
Slaný
Týnec
Kladno Fm.
Nýřany Mbr.
Radnice Mbr.
Olivětín
Broumov / Słupiec
Chvaleč
Odolov / Glinik
Žacléř
9 Obora
Letovice
Svitávka
Veverská Bitýska
Padochov
Rosice
Nahe Subgroup
Standenbühl
Kreuznach
Wadern
Donnersberg
Thall. Oberkir. Dissibodenb.
Meisenheim
Glan Subgroup
Lauterecken
Quirnbach
Wahnwegen
Altenglan
Remigiusb.
Ottweiler Sbgr.
Breitenbach
Heusweiler
Dilsburg
Göttelborn
Saarbrücken Subgroup
Millery
Surmoulin
Muse 4/5
Igornay
Moloy
Mont-Pelé
Epinac
Renière
Buxières
"Buxières inferieur"
Deneuille
Salagou
Octon Mbr.
Rabejac
Viala
Tuilières-Loiras
Usclas-St. Privat 8/9
Graiss
Rio su Luda 6

Example 1: Ilfeld Basin, Germany

Two index fossils are known from the Netzkater Formation of the small Ilfeld Basin in the south of the Hercynian Mountain:

- Based on blattid wings, the Netzkater Formation belongs to the *Sysciophlebia ilfeldensis* zone (Schneider 1982), which ranges from the lowermost Rotliegend (comparable with the early Halle or Ilmenau Formation) up to the Manebach Formation.
- Additionally, the branchiosaur *Apateon intermedius* was found in the Netzkater Formation, one of the zone species of the *A. intermedius – Branchierpeton saalensis* zone, which starts in the Stephanian C (Möhrenbach Formation) and goes up to the lowermost Rotliegend (basal Ilmenau or Halle Formation).

The combination of both independent methods gives an exact date for the Netzkater Formation of the Ilfeld Basin: Lowermost Rotliegend (comparable with the lower parts of the Ilmenau or Halle Formation; Gzhelian/Asselian transition).

Example 2: Bourbon l'Archambault Basin, Massif Central, France

Two excellent localities of fossil amphibians are known from this basin: Buxières-les-Mines and La Queue d'Etang near Bourbon l'Archambault. From both occurrences we know index fossils:

- The branchiosaur *Melanerpeton gracile* was described from the Upper Buxières-les-Mines Formation (Werneburg 2003). Therefore, these beds are part of the *Melanerpeton pusillum – M. gracile* zone, which indicates the topmost Lower Rotliegend (Autunian) to basal Upper Rotliegend (Saxonian), correlatable with the Upper Oberhof Formation up to Lower Letovice Formation. They are accompanied by large amphibians such as '*Cheliderpeton*' and *Onchiodon*.
- The insect wing of *Sysciophlebia* cf. *S. alligans* belongs to *Sysciophlebia alligans* and to the *Sysciophlebia* n. sp. B zone, which indicates an uppermost Lower Rotliegend age ('Autunian'), comparable with the Upper Oberhof Formation and basal Dissibodenberg Formation (Saar–Nahe Basin).
- The isotopic age of 288 ± 4 Ma ($^{207}Pb/^{206}Pb$ single zircon-evaporitization, TU Bergakademie Freiberg) of the tuff bed 'Lien Vert' in the Buxieres Formation corresponds to the 287 ± 2 Ma (Ar/Ar) of the Oberhof Formation (Lützner *et al.* 2003).

All three independent methods come to the same result: the Upper Buxières Formation correlates with the Upper Oberhof Formation/basal Dissibodenberg Formation (Sakmarian/Artinskian). The seymouriamorph *Discosauriscus pulcherrimus* (= *D. sacheti*) is accompanied by branchiosaurs and rare '*Cheliderpeton*' in the paper slates of La Queue d'Etang near Bourbon l'Archambault. This occurrence clearly belongs to the *Melanerpeton pusillum – M. gracile* zone and could be correlated with the Upper Buxières Formation of the 15-km-distant Buxières-les-Mines site.

Example 3: Perdasdefogu Basin, Sardinia, Italy

The first record of branchiosaurs from Sardinia was very unexpected (Ronchi & Tintori 1997), though the taxonomy of the branchiosaurs from the Is Alinus Lake of the Rio su Luda Formation is clear (Schneider *et al.* 2004*a*).

- Three branchiosaurid species lived in the Is Alinus-lake: *Melanerpeton eisfeldi, Apateon flagrifer* and *Apateon kontheri. M. eisfeldi* is the index species of the *M. eisfeldi* zone and indicates the species lineage from *Melanerpeton sembachense* up to *M. gracile* and thus a theoretical age from the Manebach up to the Lower Oberhof Formation in the reference section of the Thuringian Forest Basin. *Apateon flagrifer* is recorded from four amphibian zones: the *Apateon flagrifer flagrifer – Branchierpeton reihnholdi* zone (FOD in the Lower Goldlauter Formation) up to the *Melanerpeton pusillum – M. gracile* zone (extending up into the Upper Rotliegend). The coexistence of both species *M. eisfeldi* and *A. flagrifer* gives an age from the Lower Goldlauter up to the Lower Oberhof Formation for the Rio su Luda Formation of Sardinia. The third species, *Apateon kontheri*, completes the branchiosaurid fauna of the Is Alinus-lake, which is exactly the same fauna as the three species from the Gottlob Lake in the Thuringian Forest Basin (Upper Goldlauter Formation; Werneburg 1986*b*, 1988*a*).
- Insect wings are unknown from Sardinia up to now. The teeth of the xenacanthid shark *Bohemiacanthus* from the neighbouring Ortu Mannu section show a time range from the Manebach up to the Upper Goldlauter Formation (Schneider, in Freytet *et al.* 2002). However, in some French basins the biostratigraphical correlations with xenacanthid shark teeth give slightly older ages than indicated by amphibians and insects.

Table 1. *First Occurrence Date (FOD) of selected aquatic amphibians*

Time levels (FOD)	Genera	Families
Middle Sakmarian (Uppermost Lower Rotliegend, Upper Oberhof Formation	*Discosauriscus*	Discosauriscidae
Lower Asselian (Lowermost Lower Rotliegend, Ilmenau Formation)	*Melanerpeton*	–
Upper Gzhelian (Stephanian C, Gehren Subgroup, Möhrenbach Formation)	*Apateon,* *Branchierpeton*	–
Upper Moscovian (Westphalian D, Upper Kladno Formation)	*Branchiosaurus,* *Limnogyrinus*	Branchiosauridae, Micromelerpetontidae

Finally, it is possible to correlate the Rio su Luda Formation of the Perdasdefogu Basin in Sardinia with the Goldlauter Formation (perhaps up to the Lower Oberhof Formation) of the Thuringian Forest Basin with high confidence. Table 1 shows the prospects and limits for correlations with the FODs of genera and families used in the aquatic amphibian zonation.

Biostratigraphic potential of other tetrapods

Some aquatic, amphibious or terrestrial amphibian species and reptiles also have potential for biostratigraphical correlations. The species of the micromelerpetontid *Eimerisaurus* (Boy 2002*b*), *Micromelerpeton* with chrono-morphotypes (Boy 1995, 2002*a*), and *Archegosaurus* from the Rotliegend of the Saar–Nahe Basin (Germany) are very interesting, but they are known with certainty from this basin only. The taxonomies of '*Cheliderpeton*', *Sclerocephalus* as well as *Onchiodon* are in a state of flux. Different species of *Eryops* exist from the Upper Pennsylvanian up to the Kungurian (A.R. Milner, pers. comm. 2004). These species will be very useful for the biostratigraphy of the North American red beds. *Onchiodon* is very similar to *Eryops*, but it is not the same genus, which is necessary for biostratigraphical correlations. The biostratigraphical potential of the lepospondyl *Batropetes fritschia* (Rotliegend of the Saar–Nahe and Döhlen basins) as well as of the various species of *Osteocephalus, Scincosaurus* and *Sauravus* (Carroll *et al.* 1998) is probably much greater than based on current knowledge.

Sumida *et al.* (1996) have correlated the occurrence of the terrestrial amphibian assemblage with *Seymouria sanjuanensis, Diadectes* and 'cf. *Anconastes*' (= *Tambachia* Sumida *et al.* 1998) from the Tambach Formation (Upper Rotliegend) of the Thuringian Forest Basin with the 'earliest Permian Wolfcampian' of North America. However, the comparison of genera gives no precise correlations. The FOD and LOD of the species, *Seymouria sanjuanensis*, are known from the early to mid-Wolfcampian (Cutler Formation) of New Mexico (Berman *et al.* 1987) and from the late Wolfcampian (basal Organ Rock Shale) of Utah (Vaughn 1966). The next related species, *S. baylorensis*, is recorded from the late Wolfcampian (Nocona Formation, Admiral; Laurin 1996) or early Leonardian (Petrolia Formation, Belle Plains; Berman *et al.* 1987) as well as from the mid-Leonardian (Lower Clear Fork Group, Arroyo; Berman *et al.* 1987; Laurin 1996) of Texas. Both species could belong to a short species chronocline in the biostratigraphical sense. Therefore, the theoretical stratigraphical position of the Tambach Formation using *Seymouria sanjuanensis* reaches from the early to the late Wolfcampian of the North American time scale and from the Asselian up to the early Kungurian of the global time scale. That is a very long range and it includes the whole Lower Rotliegend and the early Upper Rotliegend of the Thuringian Forest section, that is, about 15 million years!

The diadectids have no biostratigraphical significance at present. *Diadectes absitus* and *Orobates pabsti* from the Tambach Formation of the Thuringian Forest Basin show large differences from the North American diadectids (Berman *et al.* 1998). The diadectids *Phanerosaurus naumanni* from the Leukersdorf Formation of the Erzgebirge Basin, and *Phanerosaurus pugnax* from the Niederhäslich Formation of the Döhlen Basin, are little known and they are not yet revised.

Terrestrial reptiles are relatively rare in the European Permian. The pelycosaurs are recorded with a little higher frequency. Some skeletons of *Haptodus* are known that belong together with the North American forms to the unique species *Haptodus baylei* (after Currie 1979). But this species has a range from the lower Gzhelian up to the Sakmarian. If no chronomorphical or specific differences exist,

Haptodus has no biostratigraphical significance. A similar situation is known from *Edaphosaurus* (Gzhelian–Sakmarian). *Dimetrodon* is recorded with *D. teutonis* from the Tambach Formation of the Thuringian Forest Basin (Berman *et al.* 2001). The genus *Dimetrodon* shows a chronomorphologic trend towards an overall increase in size in increasingly higher stratigraphic levels. *D. teutonis* is probably as primitive as *D. natalis,* with a similar small size from the Nocona Formation (late Wolfcampian, early Kungurian) of Texas (Berman *et al.* 2001). Up to now, *D. teutonis* gives the best indication for a late Wolfcampian to early Kungurian age of the Tambach Formation.

The caseid pelycosaurs are widespread in North America but they are concentrated in the middle and upper Leonardian. Sigogneau-Russell & Russell (1974) have described *Casea rutena* from the Upper Rotliegend (Saxonian) of the Rodez Basin in southern France. Sumida *et al.* (2001) have noted an undescribed caseid from the Tambach Formation of the Thuringian Forest Basin. A second caseid skeleton from Rodez is being published (R. Reisz, pers. comm. 2005). It is still uncertain whether or not the species *C. rutena* truly belongs to the genus *Casea* (large differences in skull features, robust proportions of the manus digits are more similar to *Cotylorhynchus*). Bones from a large caseid were discovered recently by Schneider and Körner in the Lodève Basin in the neighbourhood of Rodez. They are most comparable to *Cotylorhynchus*. If the taxonomy of caseids is clarified, this group will become very interesting for biostratigraphy.

The araeoscelid *Kadaliosaurus* from the Niederhäslich Formation (Sakmarian) of Niederhäslich in the Döhlen Basin is known from fragments. *Thuringothyris* from the Tambach Formation of the Thuringian Forest Basin is most closely related to *Paleothyris* from North America (Boy & Martens 1991), but there is no basis for biostratigraphical correlation as they are different genera. The same problem exists with the bolosaurid *Eudibamus* from the Tambach Formation (Berman *et al.* 2000).

Conclusions

The amphibian zonation of the European Permo-Carboniferous (Werneburg 1989*a*, *b*, 1996) is revised. The nine amphibian zones, together with 18 species and subspecies, are newly defined with their FAD or FOD. Species chronoclines that consist of two or three closely related species are the basic concept of this biochronology. These index fossils are mostly short-lived species with a clear taxonomy and palaeoecology. In this way, detailed correlations are possible.

Zone species are known from a number of basins in Europe. Thus far, correlations with amphibian zones are applicable to basins in the Czech Republic, Poland, France, Italy and Germany. The combination of amphibian zonation with insect zonation and isotopic ages promises the greatest success for biostratigraphical correlations in the European Permo-Carboniferous.

Unfortunately, most of the amphibian species from the Lower Permian of the Saar–Nahe Basin in Germany are virtually endemic and they can be used only for biostratigraphy in a single basin. A better understanding of the taxonomy and the stratigraphical distribution of the *Eryops* species could be very helpful for a biochronology of the North American red beds. All other amphibians and reptiles of the European Permo-Carboniferous are not now very suitable for detailed biostratigraphical correlations. Either they are not closely related species in the sense of a species chronocline, their taxonomy is unclear or the species has too great a stratigraphical range.

We thank the Deutsche Forschungsgemeinschaft (DFG) for their support of this study (WE 2833/3-1, SCHN 408/12-1). We thank A. Hunt (Albuquerque), S. G. Lucas (Albuquerque) and S. S. Sumida (San Bernardino) for their instructive reviews and linguistic improvements of the text. Best thanks to R. Schoch (Stuttgart for helpful discussion and to M. Roscher (Freiberg) for his support with computer graphics.

References

BERMAN, D. S., REISZ, R. R. & EBERTH, D. A. 1987. *Seymouria sanjuanensis* (Amphibia, Batrachosauria) from the Lower Permian Cutler Formation of north-central New Mexico and the occurrence of sexual dimorphism in that genus questioned. *Canadian Journal of Earth Sciences*, **24**, 1769–1784.

BERMAN, D. S., SUMIDA, S. S. & MARTENS, T. 1998. *Diadectes* (Diadectomorpha: Diadectidae) from the Early Permian of central Germany, with description of a new species. *Annals of Carnegie Museum*, **67**, 53–93.

BERMAN, D. S., REISZ, R. R., SCOTT, D., HENRICI, A. C., SUMIDA, S. S. & MARTENS, T. 2000. Early Permian bipedal reptile. *Science*, **290**(5493), 969–972.

BERMAN, D. S., REISZ, R. R., MARTENS, T. & HENRICI, A. C. 2001. A new species of *Dimetrodon* (Synapsida: Sphenacodontidae) from the Lower Permian of Germany records first occurrence of genus outside of North America. *Canadian Journal of Earth Sciences*, **38**, 803–812.

BOY, J. A. 1972. Die Branchiosaurier (Amphibia) des saarpfälzischen Rotliegenden (Perm, SW-Deutschland). *Hessisches Landesamt für Bodenforschung*, **65**, 1–137.

BOY, J. A. 1986. Studien über die Branchiosauridae (Amphibia: Temnospondyli). 1. Neue und wenig bekannte Arten aus dem mitteleuropäischen Rotliegenden (? Oberstes Karbon bis unteres Perm). *Paläontologische Zeitschrift*, **60**(1/2), 131–166.

BOY, J. A. 1987. Studien über die Branchiosauridae (Amphibia: Temnospondyli; Ober-Karbon-Unter-Perm). 2. Sytematische Übersicht. *Neues Jahrbuch für Geologie und Paläontologie, Abhandlungen*, **174**, 75–104.

BOY, J. A. 1995. Über die Micromelerpetontidae (Amphibia: Temnospondyli). 1. Morphologie und Paläoökologie des *Micromelerpeton credneri* (Unterperm; SW-Deutschland). *Paläontologische Zeitschrift*, **69**, 429–457.

BOY, J. A. 2002*a*. Über die Micromelerpetontidae (Amphibia: Temnospondyli). 2. *Micromelerpeton ulmetense* n. sp. und *Micromelerpeton(?) boyi* HEYLER. *Neues Jahrbuch für Geologie und Paläontologie, Abhandlungen*, **223**, 241–274.

BOY, J. A. 2002*b*. Über die Micromelerpetontidae (Amphibia: Temnospondyli). 3. *Eimerisaurus* n. g. *Neues Jahrbuch für Geologie und Paläontologie, Abhandlungen*, **225**, 425–452.

BOY, J. A. 2003. Paläontologische Rekonstruktion von Wirbeltieren: Möglichkeiten und Grenzen. *Paläontologische Zeitschrift*, **77**, 123–152.

BOY, J. A. & FICHTER, J. 1982. Zur Stratigraphie des saarpfälzischen Rotliegenden (?Oberkarbon-Unterperm; SW-Deutschland). *Zeitscrift der Deutschen Geologischen Gesellschaft*, **133**, 607–642.

BOY, J. A. & FICHTER, J. 1988. Zur Stratigraphie des höheren Rotliegend im Saar–Nahe–Becken (Unter-Perm; SW-Deutschland) und seiner Korrelation mit anderen Gebieten. *Neues Jahrbuch für Geologie und Paläontologie, Abhandlungen*, **176**, 331–394.

BOY, J. A. & MARTENS, T. 1991. A new captorhinomorph reptile from the Rotliegend of Thuringia (Lower Permian; eastern Germany). *Paläontologische Zeitschrift*, **659**(3/4), 363–389.

BOY, J. A. & SUES, H. D. 2000. Branchiosaurs: larvae, metamorphosis and heterochrony in temnospondyls and seymouriamorphs. *In*: HEATWOLE, H. & CARROLL, R. L. (eds) *Amphibian Biology, Palaeontology: the Evolutionary History of Amphibians.* Vol. 4. Surrey Beatty & Sons, Australia, 1150–1197.

CARROLL, R. L., BOSSY, K. A., MILNER, A. C., ANDREWS, S. M. & WELLSTEAD, C. F. 1998. Lepospondyli. *Handbuch der Paläoherpetologie*, **1**.

CURRIE, P. J. 1979. The osteology of haptodontine sphenacodonts (Reptilia: Pelycosauria). *Palaeontographica*, **4**(163), 130–168.

FREYTET, P., GALTIER, J., SCHNEIDER, J. W., RONCHI, A., TINTORI, A. & WERNEBURG, R. 2002. Early Permian biota from southeastern Sardinia (Ogliastra and Gerrei). *Rendiconti della Società Paleontologica Italiana*, **1**, 169–176.

GAND, G. & HAUBOLD, H. 1988, Permian tetrapod footprints in central Europe, Stratigraphical and palaeontological aspects. *Zeitschrift Geologischer Wissenschaft* **16**, 885–894.

HAUBOLD, H. 1980. Die biostratigraphische Gliederung des Rotliegenden (Permosiles) im mittleren Thüringer Wald. *Schriftenreihe Geologischer* Wissenschaften, **16**, 331–356.

HAUBOLD, H. 1984. Zyklen in der Evolution terrestrischer Vertebraten. *Zeitschrift Geologischer Wissenschaften*, **12**, 119–130.

HAUBOLD, H. 2000. Tetrapodenfährten aus dem Perm: Kenntnisstand und Progress 2000. *Hallesches Jahrbuch der Geowissenschaften, Reihe B*, **22**, 1–16.

HOLUB, V. & KOZUR, H. 1981. Revision einiger Tetrapodenfährten des Rotliegenden und biostratigraphische Auswertung. *Geologische – Paläontologische Mitteilungen, Innsbruck,* **11**(4), 149–193.

KLEMBARA, J. 1997. The cranial anatomy of *Discosauriscus* Kuhn, a seymouriamorph tetrapod from the Lower Permian of the Boskovice Furrow (Czech Republic). *Philosophical Transactions of the Royal Society of London, Series B*, **352**, 257–302.

KLEMBARA, J. & MESZÁROŠ, Š. 1992. New finds of *Discosauriscus austriacus* (Makowsky 1876) from the Lower Permian of Boskovice Furrow (Czecho-Slovakia). *Geologica Carpathica*, **43**, 305–312.

KOZUR, H. 1989. Biostratigraphic zonation in the Rotliegendes and their correlations. *Acta Musei Reginachradecensis, Serie Scientiae Naturales*, **22**, 15–30.

LAURIN, M. 1996. A Redescription of the Cranial Anatomy of *Seymouria baylorensis*, the best known Seymouriamorph (Vertebrata: Seymouriamorpha). *PaleoBios*, **17**, 1–16.

LUCAS, S. G. 1998. Toward a tetrapod biochronology of the Permian. *In*: LUCAS, S. G., ESTEP, J. W. & HOFFER, J. M. (eds) *Permian Stratigraphy and Paleontology of the Robledo Mountains, New Mexico.* New Mexico Museum of Natural History and Science Bulletin, **12**, 71–92.

LUCAS, S. G. 2002. Global Permian tetrapod footprint biostratigraphy and biochronology. *Permophiles*, **41**, 30–34.

LUCAS, S. G. 2004. A global hiatus in the Middle Permian tetrapod fossil record. *Stratigraphy*, **1**, 47–64.

LÜTZNER, H., MÄDLER, J., ROMER, R. L. & SCHNEIDER, J. W. 2003. Improved stratigraphic and radiometric age data for the continental Permo-Carboniferous reference-section Thüringer-Wald, Germany. *XVth International Congress on Carboniferous and Permian Stratigraphy, Utrecht, 2003, Abstracts*, 338–341.

MARTENS, T. 1983*a*. Zur Taxonomie und Biostratigraphie der Conchostraca (Phyllopoda, Crustacea) des Jungpaläozoikums der DDR Teil I. *Freiberger Forschungshefte, Hefte C*, **382**, 7–105.

MARTENS, T. 1983*b*. Zur Taxonomie und Biostratigraphie der Conchostraca (Phyllopoda, Crustacea) des Jungpaläozoikums der DDR. Teil II. *Freiberger Forschungshefte, Hefte C*, **384**, 24–48.

MARTENS, T. 1984. Zur Taxonomie und Biostratigraphie der Conchostraca (Phyllopoda, Crustacea) des Rotliegenden (oberstes Karbon bis Perm) im Saar-Nahe-Gebiet (BRD). *Freiberger Forschungshefte, Hefte C*, **391**, 35–57.

RONCHI, A. & TINTORI, A. 1997. First amphibian find in Early Permian from Sardinia (Italy). *Rivista Italiana di Paleontologia e Stratigrafia*, **103**, 29–38.

SCHNEIDER, J. 1982. Entwurf einer Zonengliederung für das euramerische Permokarbon mittels der Spiloblattinidae (Blattodea, Insecta). *Freiberger Forschungshefte, Hefte C*, **375**, 27–47.

SCHNEIDER, J. W. 1996. Xenacanth teeth: a key for taxonomy and biostratigraphy. *Modern Geology*, **20**, 321–340.

SCHNEIDER, J. & WERNEBURG, R. 1993. Neue Spiloblattinidae (Insecta, Blattodea) aus dem Oberkarbon und Unterperm von Mitteleuropa sowie die Biostratigraphie des Rotliegend. *Naturhistorisches Museum Schloss Bertholdsburg, Schleusingen Veröffentlichungen*, 7/8, 31–52.

SCHNEIDER, J. W. & WERNEBURG, R. 2006. Insect biostratigraphy of the Euramerican continental Late Pennsylvanian and Early Permian. *In*: LUCAS, S. G., CASSINIS, G. & SCHNEIDER, J. W. (eds) *Non-Marine Permian Biostratigraphy and Biochronology*. Geological Society, London, Special Publications, **265**, 325–336.

SCHNEIDER, J. W. & ZAJIC, J. 1994. Xenacanthiden (Pisces, Chondrichthyes) des mitteleuropäischen Oberkarbon und Perm. Revision der Originale zu Goldfuss 1847, Beyrich 1848, Kner 1867 und Fritsch 1879–1890. *Freiberger Forschungshefte*, **452**, 101–151.

SCHNEIDER, J. W., HAMPE, O. & SOLER-GIJÓN, R. 2000. The Late Carboniferous and Permian: Aquatic vertebrate zonation in southern Spain and German basins. *In*: BLIECK, A & TURNER, S. (eds) *Palaeozoic Vertebrate Chronology and Global Marine/Non-Marine Correlation.* Courier Forschungsinstitut Senckenberg, **223**, 543–561.

SCHNEIDER, J. W., WERNEBURG, R., LUCAS, S. G. & BETHOUX, O. 2003. Insect biochronozones – a powerful tool in the biostratigraphy of the Upper Carboniferous and the Permian. *Permophiles*, **42**, 11–13.

SCHNEIDER, J. W., EL WARTITI, M., HMICH, D., KERP, H., RONCHI, A., SABER, H. & WERNEBURG, R. 2004*a*. The continental Carboniferous and Permian of Morocco and Sardinia: links to south and north. *32nd International Geological Congress, Florence, 2004, Abstracts*, 748–749.

SCHNEIDER, J. W., LUCAS, S. G. & ROWLAND, J. M. 2004*b*. The Blattida (Insecta) fauna of Carrizo Arroyo, New Mexico: biostratigraphic link between marine and non-marine Pennsylvanian/Permian boundary profiles. *In*: LUCAS, S. G. & ZEIGLER, K. E. (eds) *Carboniferous-Permian Transition at Carrizo Arroyo, Central New Mexico.* New Mexico Museum of Natural History and Science Bulletin, **25**, 247–261.

SIGOGNEAU-RUSSELL, D. & RUSSELL, D. E. 1974. Étude du premier Caséidé (Reptilia, Pelycosauria) d'Europe occidentale. *Bulletin du Musee d'Histoire naturelle, Paris*, **230**, 145–214.

SUMIDA, S. S., BERMAN, D. S. & MARTENS, T. 1996. Biostratigraphic correlations between the Lower Permian of North America and Central Europe using the first record of an assemblage of terrestrial tetrapods from Germany. *PaleoBios*, **17**(2–4), 1–12.

SUMIDA, S. S., BERMAN, D. S. & MARTENS, T. 1998. A new trematopid amphibian from the Lower Permian of Central Germany. *Palaeontology*, **41**, 605–629.

SUMIDA, S. S., BERMAN, D. S., HENRICI, A. C. & MARTENS, T. 2001. The morphological diversification of taxa near the amphibian to amniote transition: new data from remarkably preserved fossils from the Lower Permian of central Germany. *Journal of Morphology*, **248**, 289.

VAUGHN, P. P. 1966. *Seymouria* from the Lower Permian of southeastern Utah, and possible sexual dimorphism in that genus. *Journal of Paleontology*, **40**, 603–612.

WERNEBURG, R. 1986*a*. Branchiosaurier aus dem Rotliegenden (Unterperm) der ČSSR. *Zeitschrift für Geologische Wissenschaften*, **14**, 673–686.

WERNEBURG, R. 1986*b*. Die Stegocephalen (Amphibia) der Goldlauterer Schichten (Unterrotliegendes, Perm) des Thüringer Waldes. Teil I: *Apateon flagrifer* (Whitt.). *Freiberger Forschungsheft, Hefte C*, **410**, 87–100.

WERNEBURG, R. 1987. Dissorophoiden (Amphibia, Rhachitomi) aus dem Westfal D (Oberkarbon) der ÈSSR. *Branchiosaurus salamandroides* Fritsch 1876. *Zeitschrift für Geologische Wissenschaften*, **15**, 681–690.

WERNEBURG, R. 1988*a*. Die Stegocephalen der Goldlauterer Schichten (Unterrotliegendes, Unterperm) des Thüringer Waldes, Teil II: *Apateon kontheri* n. sp., *Melanerpeton eisfeldi* n. sp. und andere. *Freiberger Forschungsheft, Hefte C*, **427**, 7–29.

WERNEBURG, R. 1988*b*. Die Amphibienfauna der Oberhöfer Schichten (Unterrotliegendes, Unterperm) des Thüringer Waldes. *Naturhistorisches Museum Schloss Bertholdsburg, Schleusingen, Veröffentlichungen*, **3**, 2–27.

WERNEBURG, R. 1988*c*. Die Stegocephalen (Amphibia) der Goldlauterer Schichten (Unterrotliegendes, Unterperm) des Thüringer Waldes. Teil III: *Apateon dracyiensis* (Boy), *Branchierpeton reinholdi* n. sp. und andere. *Veröffentlichungen Naturkundemuseum Erfurt*, **7**, 80–96.

WERNEBURG, R. 1989*a*. Labyrinthodontier (Amphibia) aus dem Oberkarbon und Unterperm Mitteleuropas: Systematik, Phylogenie und Biostratigraphie. *Freiberger Forschungshefte, Hefte C*, **436**, 7–57.

WERNEBURG, R. 1989*b*. Some notes to systematic, phylogeny and biostratigraphy of labyrinthodont amphibians from the Upper Carboniferous and Lower Permian in Central Europe. *Acta Musei Reginaehradecensis, Serie A, Sciential Naturales*, **XXII (1989)**, 117–129.

WERNEBURG, R. 1989*c*. Die Amphibienfauna der Manebacher Schichten (Unterrotliegendes, Unterperm) des Thüringer Waldes. *Naturhistorisches Museum Schloss Bertholdsburg, Schleusingen, Veröffentlichungen*, **4**, 55–68.

WERNEBURG, R. 1991. Die Branchiosaurier aus dem Unterrotliegend des Döhlener Beckens bei Dresden. *Naturhistorisches Museum Schloss Bertholdsburg, Schleusingen, Veröffentlichungen*, **6**, 75–99.

WERNEBURG, R. 1996. Temnospondyle Amphibien aus dem Karbon Mitteldeutschlands. *Naturhistorisches Museum Schloss Bertholdsburg, Schleusingen, Veröffentlichungen*, **11**, 23–64.

WERNEBURG, R. 2001*a*. Die Amphibien- und Reptilien-Faunen im Permokarbon des Thüringer Waldes. *Beiträge zur Geologie von Thüringen, Neue Folge*, **8**, 125–152.

WERNEBURG, R. 2001*b*. *Apateon dracyiensis* – eine frühe Pionierform der Branchiosaurier aus dem Europäischen Rotliegend. Teil 1: Morphologie. *Naturhistorisches Museum Schloss Bertholdsburg, Schleusingen, Veröffentlichungen*, **16**, 17–36.

WERNEBURG, R. 2002. *Apateon dracyiensis* – eine frühe Pionierform der Branchiosaurier aus dem Europäischen Rotliegend, Teil 2: Paläoökologie. *Naturhistorisches Museum Schloss Bertholdsburg, Schleusingen, Veröffentlichungen*, **17**, 17–32.

WERNEBURG, R. 2003. The branchiosaurid amphibians from the Lower Permian of Buxières-les-Mines, Bourbon l'Archambault Basin (Allier, France) and its biostratigraphic significance. *Bulletin de la Société Géologique de France*, **174**(4), 1–7.

Carboniferous–Permian actinopterygian fishes of the continental basins of the Bohemian Massif, Czech Republic: an overview

STANISLAV ŠTAMBERG

Regional Museum of Eastern Bohemia, Eliščino nábřeží 465, 500 01 Hradec Králové, Czech Republic (e-mail: s.stamberg@muzeumhk.cz)

Abstract: The actinopterygian fishes from the continental Westphalian to the Lower Permian basins in the Czech Republic are reviewed and compared with those of deposits of equivalent age in some central and western European Basins. Nine genera of actinopterygians belonging to eight families are known from the Westphalian–Stephanian sediments, and five genera belonging to four families are known from the Early Permian sediments of the Bohemian Massif. The new family Sceletophoridae is erected for the genus *Sceletophorus* from the Westphalian D. Very close relationships of the actinopterygian fauna between the Bohemian Massif and especially the basins of the French Massif Central are discussed.

Upper Carboniferous and Lower Permian freshwater sediments are widely distributed in the Bohemian Massif and are notable for their rich faunal content, particularly of actinopterygian fishes. Several basins in the Bohemian Massif were formed in intermontane depressions or within fault-bounded grabens. The most significant basins where Upper Carboniferous and Lower Permian actinopterygians can be studied are the Central and West Bohemian late Palaeozoic basins, the Intra-Sudetic Basin, the Krkonoše Piedmont Basin and the Boskovice Graben (Fig. 1). The scientific description of non-marine actinopterygians of this basins starts with Fritsch (1883–1901). Here, a review of the present knowledge is given for Westphalian, Stephanian and Permian times. The results obtained can be compared with the record in some other basins of central and western Europe, such as the Saar Basin, the Saale Basin, the basins of the French Massif Central and the Puertollano Basin in southern Spain.

Palaeogeographical situation

The Bohemian Massif was subjected to the Variscan orogeny, which is associated with the emplacement of magmatic rocks and the formation of huge mountain ranges, grabens and deep-reaching faults. Rapid enlargement of intramontane depositional basins during the late Westphalian and the Stephanian was marked by the deposition in lake deposits of large volumes of material eroded from the adjacent mountain ranges. The basins, of small size at first, merged into larger basins during the Stephanian; for example, the amalgamation of the many separate Central and West Bohemian late Palaeozoic basins into a lake of over 5000 km^2 took place during Stephanian B (Pešek *et al.* 2001). In contrast, filling of basins with sediments locally resulted in their shallowing and splitting into several sub-basins. Later, the Lower Permian sediments were deposited in several isolated basins, although the continuous character of some horizons indicates a vast, common depositional area. In the Boskovice Graben, the deepening of the depositional area can be observed together with the progress of deposition from the southern part of the basin, where Stephanian C sediments were deposited towards the north, where only Lower Permian sediments are known. Volcanic activity has been recorded in the Westphalian and Stephanian as well as in the Lower Permian. Based on my experience from individual palaeontological sites, it seems that volcanic activity strongly influenced life, especially in the Lower Permian lakes. Effusions of lava and falls of volcanic ashes in depositional areas resulted in interruptions of the development of faunal communities on a local to regional scale (Schneider & Zajíc 1984).

Climatic conditions

The Bohemian Massif was in the equatorial zone at the time of the Carboniferous/Permian transition. As evidenced by palaeomagnetic measurements (Krs & Pruner 1995), the Bohemian Massif was approximately between latitudes 2 and 3°N at these times. During the Late Carboniferous, a warm and humid climate alternated with more arid periods. Upper Carboniferous sediments

From: LUCAS, S. G., CASSINIS, G. & SCHNEIDER, J. W. (eds) 2006. *Non-Marine Permian Biostratigraphy and Biochronology*. Geological Society, London, Special Publications, **265**, 217–230.
0305-8719/06/$15.00

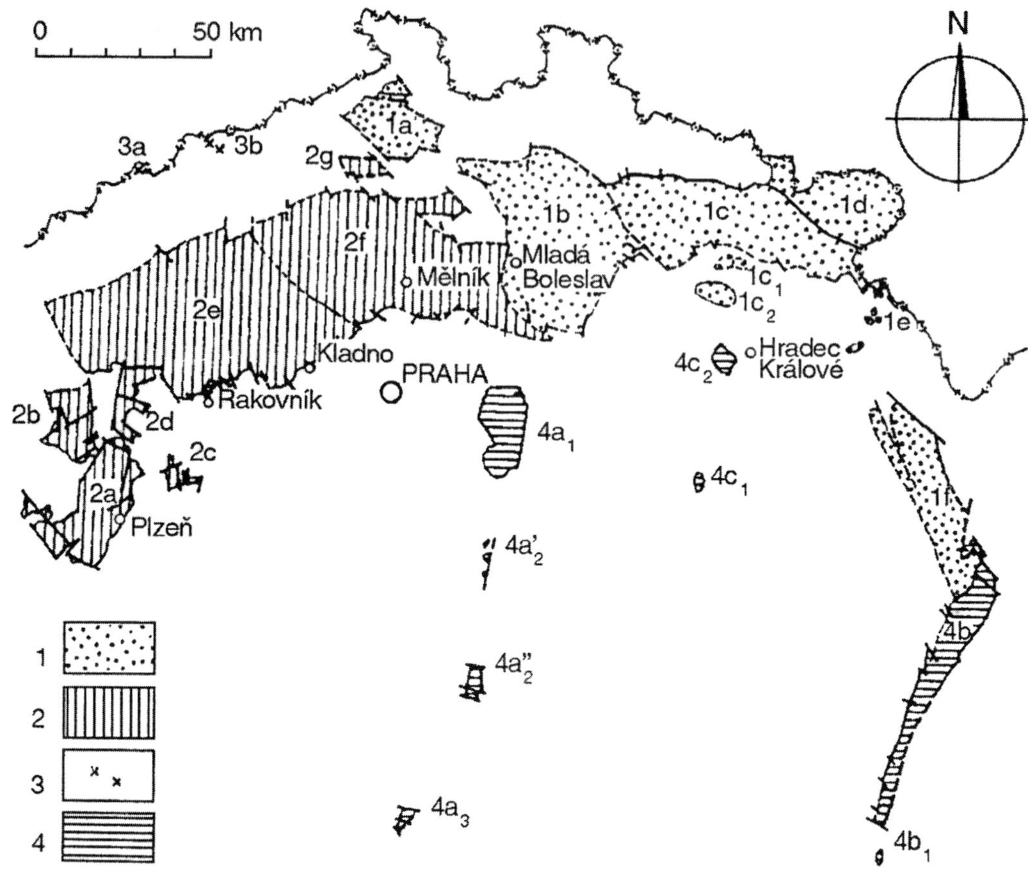

Fig. 1. Distribution of the limnic Permo-Carboniferous in the Czech Republic. After Pešek (2004). 1, Sudetic Late Palaeozoic Basins: a, Česká Kamenice; b, Mnichovo Hradiště; c, Krkonoše Piedmont; c_1, occurrence near Zvičina; c_2, occurrence on the Hořice crest; d, Intra-Sudetic; e, Permian occurrences in the Orlické Hory Mts.; f, Orlice. 2, Central and West Bohemian Late Palaeozoic Basins: a, Plzeň; b, Manětín; c, Radnice; d, Žihle; e, Kladno-Rakovník; f, Mšeno-Roudnice. 3, Krušné Hory Late Palaeozoic: a, occurrence near Brandov; b, occurrences between Moldava and Teplice v Čechách. 4, Late Palaeozoic in grabens: a, Blanice; a_1, northern (Český Brod region); a_2, central; a_2, Vlašim region; a_2, Tábor region; a_3, southern (České Budějovice region); b, Boskovice; b_1, occurrence near Miroslav; c, Jihlava; c_1, occurences in the Železné hory Mts; c_2, occurrences near Hradec Králové.

were deposited at the end of a period of relatively humid climate in a relatively stable, fluvial, lacustrine or possibly palustrine environment. A prominent increasing aridity of climate during the Permian (Pešek *et al.* 2001), however, resulted in the dominance of red sediments. Rather humid periods are represented by several well-defined intervals of grey-black claystones and limestones, frequently rich in dispersed organic substances (black shale facies) and rich in fauna. The arid character of the environment, however, induced relatively common drying of basins and frequent interruptions in the continuous development of organisms bound to the aquatic environment.

Actinopterygii of the Westphalian

The actinopterygians of the Westphalian (Westphalian D) are limited to the Central and West Bohemian late Palaeozoic basins (Table 1). The representatives of the families Haplolepidae Westoll, 1944, and Sceletophoridae, fam. nov. are distinguished.

The Haplolepidae Westoll, 1944, is represented by *Pyritocephalus sculptus* (Fritsch 1879). *Pyritocephalus sculptus* is a small fish, not exceeding 70 mm in total length, bearing several exceptional features. The most conspicuous is a large fenestration in the skull roof and characteristic formation of the dermal paired bones of the skull

Table 1. *Stratigraphy of Carboniferous units of the Central and West Bohemian basins (modified from Pešek 2004) with the occurrence of actinopterygians*

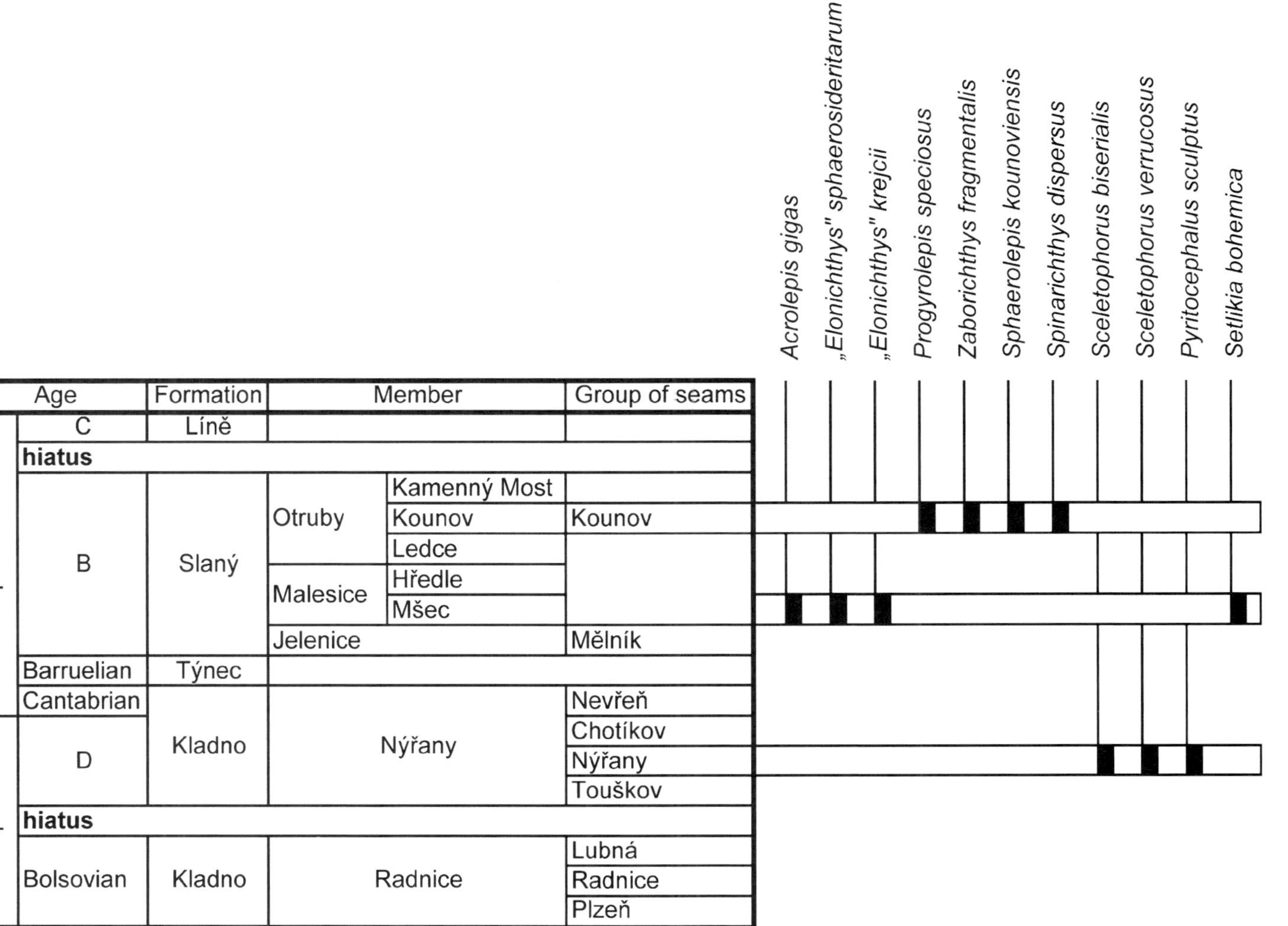

Age		Formation	Member		Group of seams
Stephanian	C	Líně			
	hiatus				
	B	Slaný	Otruby	Kamenný Most	
				Kounov	Kounov
				Ledce	
			Malesice	Hředle	
				Mšec	
			Jelenice		Mělník
	Barruelian	Týnec			
	Cantabrian	Kladno	Nýřany		Nevřeň
Westphalian	D				Chotíkov
					Nýřany
					Touškov
	hiatus				
	Bolsovian	Kladno	Radnice		Lubná
					Radnice
					Plzeň

roof (frontal, parietal, dermopterotic) coalesced to one unit. The maxilla has a large maxillary plate, and the jaws are without teeth. The series of one median and two pairs of lateral gulars between the lower jaw are present, but the branchiostegal rays are not developed. *Pyritocephalus sculptus* is very abundant in the Nýřany Member (Kladno Formation; Westphalian D).

The family Sceletophoridae, fam. nov., erected here, has the following diagnosis:

Body of fusiform shape. Pelvic fin placed closer to pectoral than to anal fin. Caudal fin with unequal lobes, but only moderately cleft. Fin rays of pectoral fin articulate from their base. Orbit large, lying well forward. Maxilla with large, square-shaped maxillary plate. Lower jaw stout. The upper and lower jaws bear robust, smooth, sharp-pointed teeth of equal size. Suspensorium nearly vertical. Opercular and subopercular of oblonged shape, small number of branchiostegal rays. Scales are rhombic.

The type genus of the family Sceletophoridae, fam. nov. is *Sceletophorus* Fritsch, 1894 and it is the single genus of the family (Fig. 2e). The genus *Sceletophorus* was formerly included in the family Trissolepididae Fritsch, 1895, together with the *Sphaerolepis* Frič, 1877 – the type genus of the family Trissolepididae (Štamberg 1991). *Sceletophorus* has several features similar to *Sphaerolepis* (large orbit, construction of the skull roof, type of dentition on the jaws, nearly vertical suspensorium), but several other important

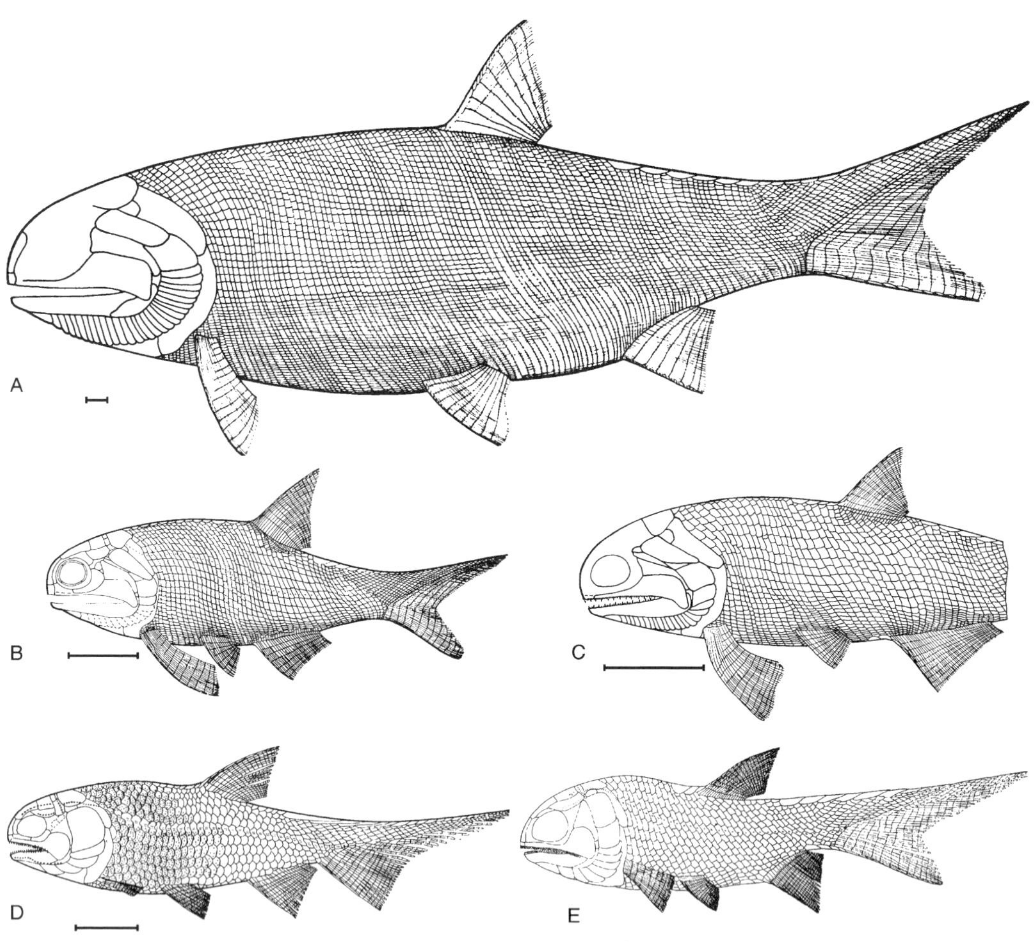

Fig. 2. Reconstruction of some Westphalian and Stephanian actinopterygians (from Štamberg 1991): (**A**) *Acrolepis gigas* (Frič, 1877); (**B**) '*Elonichthys*' *sphaerosideritarum* (Fritsch, 1895); (**C**) '*Elonichthys*' *krejcii* (Fritsch, 1895); (**D**) *Sphaerolepis kounoviensis* Frič, 1876; (**E**) *Sceletophorus biserialis* Fritsch, 1894. Scale bars represent 20 mm.

characters are well developed; the opercular and subopercular, greater number of branchiostegal rays, shape of the parasphenoid, caudal fin and moderately cleft, rhombic scales separate *Sceletophorus* from the Trissolepididae.

The new family Sceletophoridae includes *Sceletophorus biserialis* Fritsch 1894 and *Sceletophorus verrucosus* (Fritsch 1894) from the Westphalian D of the Central and West Bohemian late Palaeozoic basins. Both species include small specimens no more than 14 cm in total length. *Sceletophorus biserialis* is known from scores of skeletal remains.

Actinopterygii of the Stephanian

Stephanian sediments contain a relatively varied assemblage of actinopterygians (Tables 1–3). Their occurrence in the Bohemian Massif is limited to sediments of the Central and West Bohemian late Palaeozoic basins and, to a lesser extent, also of the Krkonoše Piedmont Basin and Intra-Sudetic Basin. Representatives of the families Pygopteridae Aldinger, 1937, Acrolepididae Aldinger, 1937, Elonichthyidae Aldinger, 1937, Trissolepididae Fritsch, 1895, Aeduellidae Heyler, 1969 and Igornichthyidae Heyler, 1977

Table 2. *Stratigraphy of the Czech part of the Intra-Sudetic Basin (modified from Pešek 2004) with the occurrence of actinopterygians*

Age		Formation	Member	Main Horizons	*Paramblypterus zeidleri*	*Paramblypterus* sp.	„*Amblypterus*" *vratislaviensis*	„*Amblypterus*" *kablikae*	*Sphaerolepis kounoviensis*
Triassic		Bohdašín							
Permian	Thuringian	Bohuslavice							
Permian	Saxonian	Trutnov							
Permian			local hiatus						
Permian	Autunian	Broumov	Martínkovice	Jetřichov					
Permian	Autunian	Broumov	Martínkovice	Hejtmánkovice					
Permian	Autunian	Broumov	Martínkovice	Vižňov					
Permian	Autunian	Broumov	Olivětín	Ruprechtice	■		■	■	
Permian	Autunian	Broumov	Nowa Ruda						
Permian	Autunian	Chvaleč	Bečkov	Bečkov					
Late Carboniferous	Stephanian C	Chvaleč	Verneřovice	Verneřovice					■
Late Carboniferous	Stephanian C	hiatus							
Late Carboniferous	Stephanian B	Odolov	Jívka	Radvanice Coal Seam					
Late Carboniferous	Barruelian	Odolov	Jívka	Bystré					
Late Carboniferous	Barruelian	Odolov	Jívka	Vítovy doly					
Late Carboniferous	Cantabrian	Odolov	Svatoňovice	Coal Seam					
Late Carboniferous	Westphalian D	Odolov	Svatoňovice						
Late Carboniferous	Bolsovian	Žacléř	Petrovice	Petrovice Coal Seam					
Late Carboniferous	Late Duckmantian	Žacléř	Prkenný důl-Žďárky	Buk or Strážovice Coal Seams					
Late Carboniferous	Early Duckmantian	Žacléř	Lampertice	Šverma mine Coal Seams					
Late Carboniferous	Late Langsettian	Žacléř	Lampertice	Šverma mine Coal Seams					
Late Carboniferous	Early Langsettian	Žacléř	Lampertice	Šverma mine Coal Seams					
Late Carboniferous	Late Namurian	Žacléř	Lampertice	Šverma mine Coal Seams					
Late Carboniferous	Middle Namurian	hiatus							
Early Carboniferous	Early Namurian	hiatus							
Early Carboniferous	Early Namurian	?							
Early Carboniferous	Viséan	Blažkov							

Table 3. *Stratigraphy of the Krkonoše Piedmont Basin (modified from Pešek 2004) with occurrence of actinopterygians*

Period	Age	Formation	Member	Main Horizons	*Paramblypterus rohani*	*Paramblypterus caudatus*	*Paramblypterus reussii*	*Paramblypterus zeidleri*	*Paramblypterus gelberti*	*Paramblypterus* sp.	„*Amblypterus*" *kablikae*	„*Amblypterus*" *feistmanteli*	*Amblypterus* sp.	*Aeduella* sp.	*Igornichthys* sp.	*Spinarichthys dispersus*	*Sphaerolepis kounoviensis*	*Elonichthys* sp.
Triassic	Anisian	Bohdašín	Upper															
	Scythian		Lower															
Permian	Thuringian	Bohuslavice																
	Saxonian	Trutnov	Suchovršice															
			Havlovice															
			Vlčice															
			Horní Město															
	angular discordance																	
	Autunian	Chotěvice																
		discordance																
		Prosečné	Upper	Kalná				■		■	■	■	■	■				
			Lower	Arcoses, Horní Branná = Mladé Buky														
		Vrchlabí	Čistá Sandstones	Kozinec														
				Háje														
			Stará Paka Sandstones	Rudník	■	■	■		■	■				■	■			
		discordance																
Late Carboniferous	Stephanian C	Semily	Upper	Štěpanice-Čikvásky = Ploužnice												■	■	■
			Middle															
			Lower															
	hiatus, erosion																	
	Stephanian B	Syřenov	Upper	Black claystone												■	■	■
			Lower	Syřenov														
	Barruelian	Kumburk	Štikov Arcoses															
	Cantabrian																	
	Westphalian D		Brusnice															

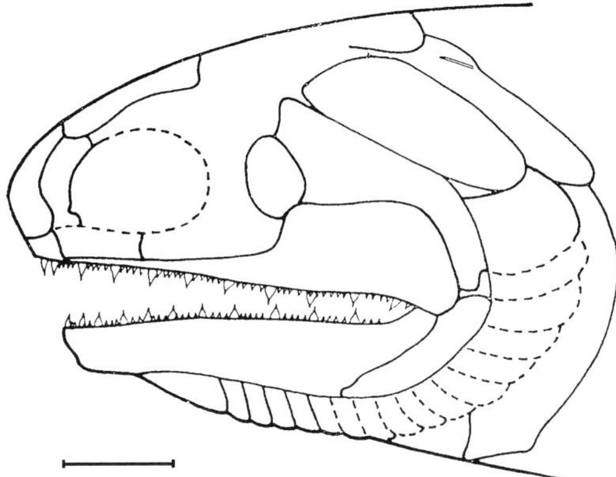

Fig. 3. Reconstruction of the head in lateral view of *Progyrolepis speciosus* (Frič, 1876) (from Štamberg 1991). Scale bar represents 20 mm.

have been described from the Stephanian sediments (Figs 2a–d, 3).

Among the family Pygopteridae, an important component of the faunas is the genus *Progyrolepis* (Fig. 3), namely the species *Progyrolepis speciosus* (Frič 1876). It is known from the Kounov Member of the Slaný Formation (Stephanian B, Central and West Bohemian late Palaeozoic basins). It is a predatory fish reaching 50–60 cm in length. Dermal bones of the head and the scales are markedly sculptured. Ten skeletal fragments have been reported, but isolated finds of its scales are more frequent. *Zaborichthys fragmentalis* Štamberg 1991, a predatory fish of the same type, is also placed within this family. It has been found in the Kounov Member of the Slaný Formation (Stephanian B, Central and West Bohemian late Palaeozoic basins). Scales of this species are equally sculptured and denticulated on their caudal margins. Only a single skeletal fragment with scales has been reported.

A prominent species of the Central and West Bohemian late Palaeozoic basins is *Acrolepis gigas* (Frič 1877). Only a single but articulated specimen has been found, reaching an impressive total length of 125 cm (Fig. 2a). This species belongs among the largest Permo-Carboniferous actinopterygians. Although an almost complete individual has been preserved, including some well-preserved dermal bones of the head, but dentition on the maxilla is not preserved. Its bones are markedly sculptured, and even relatively small scales on the trunk possess a characteristic sculpture. As indicated by comparison with other actinopterygians of the family Acrolepididae, it was the major predator. Apart from this whole individual, only several isolated scales have been found. This species occurs in the Mšec Member (Slaný Formation, Stephanian B), i.e., it dates from the period when the lake reached a minimum area of 5000 km^2, after several basins had merged.

The family Elonichthyidae is represented by the species *'Elonichthys' sphaerosideritarum* (Fritsch 1895) and *'Elonichthys' krejcii* (Fritsch 1895) from the Slaný Formation (Stephanian B) of the Central and West Bohemian late Palaeozoic basins. These two species were formerly (Štamberg 1991) referred to as *Watsonichthys sphaerosideritarum* and *Watsonichthys krejcii.* The study of the type material of the genus *Watsonichthys* and a comparison with *'Elonichthys' palatinus* described by Schindler (1993) clearly show that these two species are probably more closely related to *'Elonichthys' palatinus*, and can therefore be referred to as *'Elonichthys' sphaerosideritarum* (Fig. 2b) and *'Elonichthys' krejcii* (Fig. 2c). Both of these species belong to small predatory fishes not exceeding 15 cm in length. Their scales are also sculptured by ridges. Seven skeletal fragments of *'Elonichthys' sphaerosideritarum* and only the holotype of *'Elonichthys' krejcii* have been reported. Finds of their isolated scales are relatively common in the Slaný Formation (Stephanian B) and in the Líně Formation (Stephanian C, Central and West Bohemian late Palaeozoic basins) and the Syřenov Formation (Stephanian B) and Semily Formation (Stephanian C) of the Krkonoše Piedmont Basin.

The family Trissolepididae is represented by the species *Sphaerolepis kounoviensis* Frič, 1876, which bears a number of exceptional morphological features (Fig. 2d). This fish does not exceed 15 cm in total length, and its mouth is armed by numerous sharp teeth. It probably specialized in preying on invertebrates. Besides a number of advanced features, such as the shortened, square-shaped maxillary plate, more vertical suspensorium and a reduced number of branchiostegal rays, it possesses scales of cycloidal shape that are, moreover, sculptured with sharp, pointed tubercles similar to those of the modern *Perca fluviatilis.* This very common species is known from scores of skeletal remains and numerous finds of isolated scales. It occurs not only in the Central and West Bohemian late Palaeozoic basins but also in the Stephanian sediments of the Krkonoše Piedmont Basin and of the Intra-Sudetic Basin.

Another group of actinopterygian fishes, the Aeduellidae, has only recently been identified in the Permo-Carboniferous of the Bohemian Massif. Stephanian aeduellids are represented only by the species *Spinarichthys dispersus* (Fritsch 1894) from the Slaný Formation

(Stephanian B) of the Central and West Bohemian late Palaeozoic basins. *Spinarichthys* is a small fish reaching 8–10 cm in length. Its dentition consists of slender conical teeth similar to those known in *Aeduella blainvillei* (Agassiz 1833–43), a typical representative of the family initially described from the Early Permian of the French Massif Central, but also in *Paramblypterus* (Heyler 1969). The scales of the trunk are smooth and denticulated on their caudal margins. The related family Igornichthyidae is also represented by only a single species: *Setlikia bohemica* Štamberg & Zajíc, 1994, from the Slaný Formation (Stephanian B) of the Central and West Bohemian late Palaeozoic basins. *Setlikia* was an equally small fish, but it has been described only on the basis of preserved dermal bones of the skull roof. Skeletal remains of these two species are very rare: *Spinarichthys dispersus* is known from two skeletal fragments only, and *Setlikia bohemica* is known only from the holotype. On the other hand, isolated scales of *Spinarichthys dispersus* are abundant in the Slaný Formation and in the Líně Formation (Stephanian C) of the Central and West Bohemian late Palaeozoic basins and in the Syřenov Formation (Stephanian B) and Semily Formation (Stephanian C) of the Krkonoše Piedmont Basin (Štamberg 1989).

Actinopterygii of the Lower Permian

The Lower Permian sediments in the Bohemian Massif cover a larger area than the Westphalian and the Stephanian sediments. The stratal succession of the Central and West Bohemian late Palaeozoic basins is overlain by strata of Stephanian C age; younger beds are missing there. In contrast, Lower Permian sediments are extensive in the Krkonoše Piedmont Basin, the Intra-Sudetic Basin and the Boskovice Graben. Actinopterygii are present in all the above mentioned basins (Tables 2–4).

The actinopterygian faunas of Lower Permian age are dominated by fishes of the family Amblypteridae Romer, 1945. As shown by recent

Table 4. *Stratigraphy of the Boskovice Graben (modified from Zajíc & Štamberg 2004) with occurrence of actinopterygians*

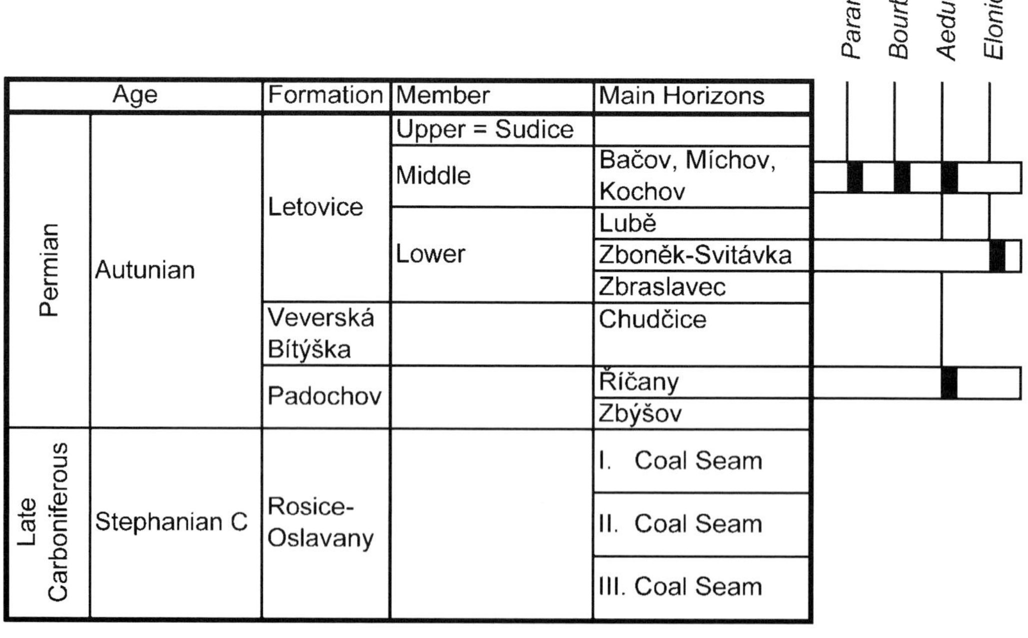

Age		Formation	Member	Main Horizons	*Paramblypterus* sp.	*Bourbonnella* sp.	*Aeduellidae*	*Elonichthyidae*
Permian	Autunian	Letovice	Upper = Sudice					
			Middle	Bačov, Míchov, Kochov	■	■	■	
			Lower	Lubě				
				Zboněk-Svitávka				■
				Zbraslavec				
		Veverská Bítýška		Chudčice				
		Padochov		Říčany			■	
				Zbýšov				
Late Carboniferous	Stephanian C	Rosice-Oslavany		I. Coal Seam				
				II. Coal Seam				
				III. Coal Seam				

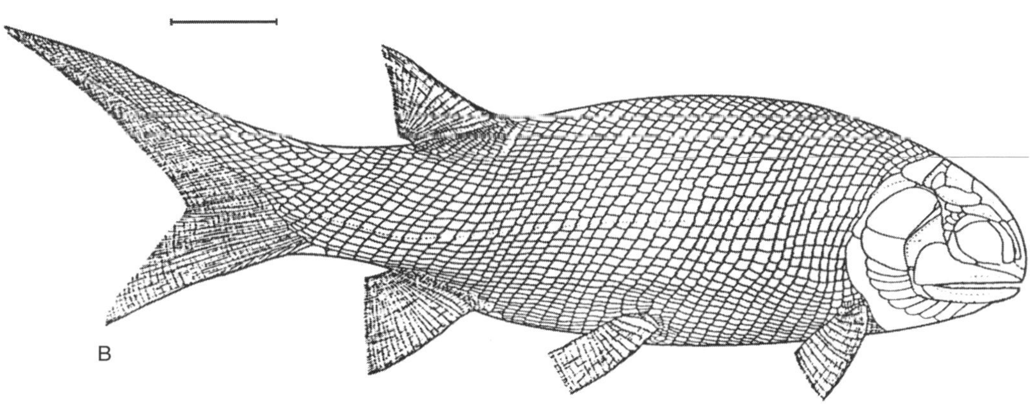

Fig. 4. *Paramblypterus* sp. from the Early Permian of the Boskovice Graben: (**A**) Museum Hradec Králové No 11024; (**B**) reconstruction of whole specimen in lateral view. Scale bar represents 20 mm.

investigations, fishes of the family Aeduellidae Heyler, 1969 were also a prominent component of the fauna in some areas, whereas the families Igornichthyidae, Heyler, 1977 and Elonichthyidae Aldinger, 1937 are relatively rare.

Amblypteridae are represented by the genera *Paramblypterus* and *Amblypterus.* The genus *Paramblypterus* (Fig. 4) encompasses the species *P. rohani* (Heckel 1861), *P. reussii* (Heckel 1861), *P. caudatus* (Heckel 1861) and *P. gelberti* (Goldfuss 1847) from the Rudník Horizon (Vrchlabí Formation, Early Permian) and *P. zeidleri* (Fritsch 1895) from the Kalná Horizon (Prosečné Formation, Early Permian) of the Krkonoše Piedmont Basin. Their dentition consists of numerous minute teeth embedded at the extremity of tubules arranged in a brush-like fashion. The scales of all species of the genus *Paramblypterus* are smooth, with no conspicuous ridges and with fine, concentrically arranged growth lines. The caudal margins of the scales are denticulated. Considering the similar trunk structure and the similar type of jaw dentition in the genus *Aeduella*, it can be deduced that the fishes of these two genera had similar feeding habits, living on minute aquatic arthropods or insects. The genus *Paramblypterus* is common in all basins.

The genus *Amblypterus* sp. (*sensu* Dietze 2000) has been also identified in the Kalná Horizon (Prosečné Formation) of the Krkonoše Piedmont Basin (Štamberg 2002), and the same genus probably includes the species *Amblypterus vratislaviensis* (Agassiz 1833–43), reported from the Olivětín Formation of the Intra-Sudetic Basin and based on many individuals. This species is about 15 cm long, and its scales and dentition are the same as those of the species of *Paramblypterus*. The species *Amblypterus kablikae* (Geinitz 1860) and *Amblypterus feistmanteli* Fritsch, 1895, as yet uncertain, are represented by several specimens from the Olivětín Formation of the Intra-Sudetic Basin and from the Prosečné Formation of the Krkonoše Piedmont Basin.

The Aeduellidae have only recently been identified in the Lower Permian of the Bohemian Massif (Štamberg 2002). They were found in the Vrchlabí Formation and Prosečné Formation of the Krkonoše Piedmont Basin, and in the Padochov Formation and Letovice Formation of the Boskovice Graben. Specimens belonging to *Aeduella* sp. (Prosečné Formation, Krkonoše Piedmont Basin) and *Bourbonnella* sp.(Letovice Formation, Boskovice Graben) bear typical aeduellid features, such as the shape of the opercular and subopercular, shape of the maxilla and only two branchiostegal rays. (Fig. 5).

Aeduellids showing morphological features of the family Aeduellidae but differing from the genera *Aeduella* and *Bourbonnella* were found in the southern part of the Boskovice Graben in the Říčany Horizon of the Padochov Formation. These very small fishes do not exceed 8–10 cm in length. Their scales are finely sculptured with concentrically arranged lines. Hundreds of individuals are present at a locality bearing a fitting name: the 'Fish Rock' (*Rybičková skála*) close to the village of Neslovice.

The third family encountered in the Lower Permian sediments are the Igornichthyidae with the genus *Igornichthys*. *Igornichthys* was a small predatory fish, 8–10 cm in length, that probably fed on insects. Its jaws were armed with relatively large, sharp teeth. The surfaces of the scales bear one or two sharply pointed tubercles. *Igornichthys* sp. is known from only two specimens from the Vrchlabí Formation of the Krkonoše Piedmont Basin.

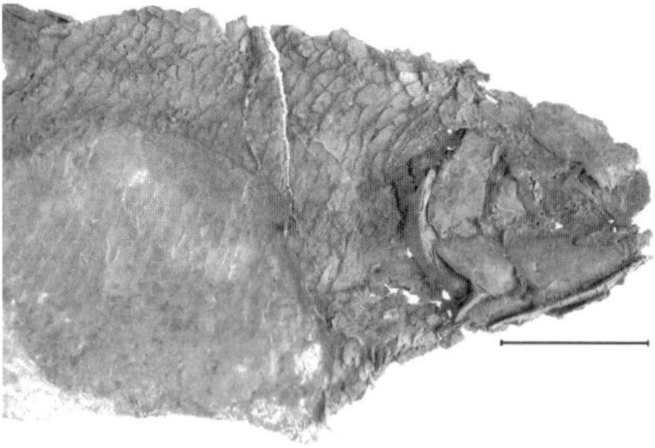

Fig. 5. The head and the oral region of the trunk of *Bourbonnella* sp. in lateral view, from the Early Permian of the Boskovice Graben. Museum Hradec Králové, No 63858. Scale bar represents 20 mm.

The representatives of the family Elonichthyidae (Aldinger 1937) have recently been discovered in the Lower Permian sediments of the Letovice Formation in the Boskovice Graben. Ten specimens, without more precise determination, are known. They are small predatory fishes ranging from 10 to 15 cm in total length. The scales are strongly sculptured by ridges that extend beyond the posterior scale margin to make a delicate denticulation. Lepidotrichs of the pectoral fin are not articulated at their base. The dentition on jaws consists of relatively large, sharply pointed teeth.

Discussion

The Westphalian and the Stephanian sediments of the Bohemian Massif yield eleven species of actinopterygians belonging to nine genera (Table 5). They constitute a relatively varied assemblage, including large predators (*Acrolepis gigas*, *Progyrolepis speciosus*, *Zaborichthys fragmentalis*); smaller predators ('*Elonichthys*' *sphaerosideritarum*, '*Elonichthys*' *krejcii*, *Sphaerolepis kounoviensis*, *Sceletophorus biserialis* and *Sceletophorus verrucosus*); small numbers of Aeduellidae (*Spinarichthys dispersus*) and Igornichthyidae (*Setlikia bohemica*), which fed on minute aquatic arthropods or insects; and the haplolepid fish *Pyritocephalus sculptus*, which probably fed largely on plankton. None of the recorded genera or species pass from the Stephanian to the Lower Permian. Similar phenomena have been discussed for freshwater sharks of the European basins in general by Schneider & Zajíc (1994) as well as by Boy & Schindler (2000) for aquatic vertebrates of the Saar–Nahe Basin. The faunal turnover at the Stephanian/Rotliegend ('Autunian') transition is regarded as result of the European volcanotectonical relief or basin reorganization (Schneider 1989; Schneider & Zajíc 1994; Schneider *et al.* 1995). This volcano-tectonical event caused the destruction of the formerly existing dewatering system and interbasinal connections. In some basins it caused the complete devastation of lake biotopes by long-lasting volcanic ash falls and by extensive lava flows. Obviously, the resettlement of the newly originating lakes took place from fluvial refuges or via newly originated drainage systems from the unaffected lake areas.

Table 5. *Actinopterygians of the continental basins of the Bohemian Massif*

Early Permian	Elonichthyidae	*'Elonichthyidae'*
	Amblypteridae	*Paramblypterus rohani* (Heckel 1861)
		Paramblypterus reussii (Heckel 1861)
		Paramblypterus caudatus (Heckel 1861)
		Paramblypterus gelberti (Goldfuss 1847)
		Paramblypterus zeidleri (Fritsch 1895)
		Paramblypterus sp.
		'Amblypterus' vratislaviensis (Agassiz 1833–43)
		'Amblypterus' kablikae (Geinitz 1860)
		'Amblypterus' feistmanteli Fritsch 1895
		Amblypterus sp.
	Aeduellidae	*Aeduella* sp.
		Bourbonnella sp.
	Igornichthyidae	*Igornichthys* sp.
Stephanian B–C	Pygopteridae	*Progyrolepis speciosus* (Frič 1876)
		Zaborichthys fragmentalis 'tamberg 1991
	Acrolepididae	*Acrolepis gigas* (Frič 1877)
	Elonichthyidae	*Elonichthys sphaerosideritarum* (Fritsch 1895)
		Elonichthys krejcii (Fritsch 1895)
	Trissolepididae	*Sphaerolepis kounoviensis* Frič 1876
	Aeduellidae	*Spinarichthys dispersus* (Fritsch 1894)
	Igornichthyidae	*Setlikia bohemica* 'tamberg & Zajíc 1994
Westphalian D	Sceletophoridae	*Sceletophorus biserialis* Fritsch 1894
		Sceletophorus verrucosus (Fritsch 1894)
	Haplolepidae	*Pyritocephalus sculptus* (Fritsch 1879)

After this event, the post-Stephanian Lower Permian fish faunas of the Czech basins are notable for their rich occurence of actinopterygians of the family Amblypteridae, namely *Paramblypterus rohani* and *Amblypterus vratislaviensis*. The substantial component of Early Permian actinopterygian fauna are Aeduellidae, namely the genera *Aeduella*, *Bourbonnella* and other as yet undetermined genera and species of the family. Amblypteridae and Aeduellidae probably fed on plankton and minute aquatic arthropods. It has to be noted that the number of individuals of the families, genera and species of Lower Permian actinopterygians will undergo major correction in the future as the author of this paper collected several thousand individuals or their fragments, which are now in the process of study and determination.

Focusing our attention on the actinopterygians from other basins of central and western Europe at the Carboniferous/Permian boundary, we find many common features but also differences. The Stephanian species *Acrolepis gigas* is exceptional and has no parallel in other basins, but another predator, *Progyrolepis,* is relatively abundant at the locality of Buxières-les-Mines (Poplin 1999). However, this lake horizon is of late Autunian (Asselian/Sakmarian) age (Werneburg 2003; Roscher & Schneider 2005). It has also been reported from the Stephanian C of the Puertollano Basin in Spain (Schneider *et al.* 2000; Forey & Young 1985). Actinopterygians of the *'Elonichthys' sphaerosideritarum* and *'Elonichthys' krejcii* type are probably represented by the species *Blanzychthys prorobisoni* Heyler & Poplin, 1994 in the basins of the French Massif Central (Montceau-les-Mines, Stephanian B: Heyler & Poplin 1994). A very closely related species, *'Elonichthys' palatinus,* is common in the Lower Permian sediments of the Saar Basin (Lauterecken to Odernheim Beds: Schindler 1993). *Watsonichthys eupterygius* from the Saar Basin, revised by Heyler (1976), can also be placed close to the group of the genus *Elonichthys* rather than to *Watsonichthys.* Actinopterygians belonging to the genus *Elonichthys* have also been reported from the Stephanian C of the Puertollano Basin (Forey & Young 1985).

Surprisingly, *Sphaerolepis kounoviensis*, prominent because of some of its characteristic bones and especially its scales, has not been recorded elsewhere with the exception of its occurrence in the Lower Permian sediments in northern Switzerland (Bürgin 1990). As has been pointed out, Aeduellidae of the Stephanian of the Bohemian Massif are represented by only two skeletal fragments of *Spinarichthys dispersus*; the genus *Aeduella*, however, is completely missing. In the Stephanian sediments of the basins of the French Massif Central, the genus *Aeduella* occures in the Monteau-les-Mines locality. The Aeduellidae now seem to have taken a much bigger share in the composition of the Permo-Carboniferous ichthyofauna than previously believed. Besides the basins of the French Massif Central, they are also present in the Stephanian Puertollano Basin, from where *Puertollanichthys ritchei* Forey & Young 1985, and probably also the genus *Bourbonnella*, have been reported. The latter genus was most probably found in New Mexico, United States, in the marine limey shale of the Atrasado Formation (Upper Pennsylvanian) as well (Gottfried 1987).

Aeduellidae became an important component of the faunas of continental basins, especially in the Lower Permian. Their relatively high abundance, with the typical species *Aeduella blainvillei* and several other genera (Heyler 1969), was reported from the basins of the French Massif Central. Their finds were later reported from the Lower Permian sediments of the Saar Basin (Heyler 1991) and from northern Switzerland (Bürgin 1990). In the Lower Permian of the Bohemian Massif, representatives of the genus *Aeduella* were found in the Krkonoše Piedmont Basin (Kalná Horizon of the Prosečné Formation) and of the genus *Bourbonnella* in the Boskovice Graben (Letovice Formation), and yet unspecified numerous specimens pertaining to the family Aeduellidae were found in the southern part of the Boskovice Graben (Padochov Formation). Their occurrences in the basins of the Bohemian Massif suggest that the Aeduellidae were abundant but restricted to certain horizons or to limited areas, being totally absent from the same horizons at other localities. Unlike aeduellids, paramblypterids are distributed throughout all fossiliferous horizons and all sites with fossils of Lower Permian vertebrate fauna. *Paramblypterus* can be taken as the most widely distributed genus in the Lower Permian fillings of the continental basins of the Bohemian Massif and other regions of Europe. No traces of this genus can be found in the Stephanian sediments of the Bohemian Massif; however, it is the most abundant actinopterygian genus in the Lower Permian.

In other regions of Europe, the situation is somewhat different. *Paramblypterus* has been reported from Stephanian sediments of the basins of the French Massif Central (Montceau-les-Mines) and Spain (Puertollano Basin). The genus then continues to develop, being very common also in the Early Permian, for example in Buxières-les-Mines and the Autun Basin in the French Massif Central, and the Thuringian Forest and Saar–Nahe Basin in Germany as well. In the Bohemian Massif, *Paramblypterus* occurs with several species and shares many features with the genus *Amblypterus* in the sense of the revision of Dietze (2000). The present state of knowledge does not allow us to distinguish between several thousand individuals collected from the Lower Permian sediments. To this date, individuals belonging to the genus have been clearly determined from the Lower Permian of the Krkonoše Piedmont Basin. The species *Amblypterus vratislaviensis* from the Intra-Sudetic Basin is also close to the genus *Amblypterus*.

The family Igornichthyidae, represented by the Stephanian species *Setlikia bohemica* and the Lower Permian genus *Igornichthys* in the Bohemian Massif, was much more common in the basins of the French Massif Central. Stephanian and Lower Permian sediments of France have yielded the species *Commentrya traquairi* Sauvage, 1888, *Igornella montcellensis* Heyler & Poplin, 1994, *Igornella comblei* Heyler, 1969, and *Igornichthys doubingeri* Heyler, 1969.

This brief comparison of the actinopterygian fauna of the Bohemian Massif in the time frame of latest Carboniferous Pennsylvanian to Lower Permian Middle Cisuralian with that from other coeval regions in Europe confirms that the basins of the French Massif Central are the closest parallel in their faunal composition. The numerous common features obviously combine many mutual differences. For example, while larger predatory actinopterygians are missing in the Lower Permian sediments of the Bohemian Massif, *Rhabdolepis* is present in the Saar Basin, and *Usclasichthys* and probably other genera are also present in the French basins. The presence of a haplolepid fish, *Blanzyhaplolepis beckaryae* Poplin, 1997, has been documented from Stephanian B of the Montceau-les-Mines Basin (Poplin 1997), while in the Bohemian Massif the haplolepid fish *Pyritocephalus sculptus* has been encountered in the Westphalian D only. It is necessary to say (Štamberg 1978) that species very closely related to *Pyritocephalus sculptus* are *Pyritocephalus lineatus* (Newberry 1856) from the Pennsylvanian of Linton (Ohio) and *Pyritocephalus gracilis* (Newberry & Worthen 1870) from the Pennsylvanian of Mazon Creek (Illinois). Also, *Pyritocephalus rudis* Westoll, 1944 from the Westphalian of Newsham (Northumberland) and *Pyritocephalus lowneyae* Huber, 1992 from the Pennsylvanian of New Mexico indicate an analogous environment in several regions in the Upper Carboniferous.

Relatively clearly defined fossiliferous horizons in the Lower Permian of the Bohemian Massif, with preserved mudcracks, ripples and other sole marks, are evidence of water-level fluctuations and periodic drying of basins. Studies at a number of sites, especially in the Lower Permian of the Boskovice Graben, show that the populations of actinopterygians and amphibians probably experienced giant booms under favourable conditions in the basin at sufficient water depth and oxygenation. The subsequent aridization of climate then caused a drop in water oxygenation and a drying of basins. Animals bound to aquatic environments were driven to mass extinction. The natural evolution of life in the Lower Permian basins of the Krkonoše Piedmont Basin was interrupted by volcanic activity, namely ash fall and lava effusions into the lake. This may explain why large individuals of actinopterygian fishes are absent from the Lower Permian deposits of the Bohemian Massif.

Mass occurrences of discosauriscid amphibians have been recorded in certain horizons of the Lower Permian fill of the Boskovice Graben. Actinopterygians must have formed a major component of their diet because discosauriscids are known to feed even on their weaker companions. These amphibians were probably replacing the missing larger predatory species of actinopterygians in the Lower Permian.

I thank C. Poplin (Muséum National d'Histoire Naturelle, Paris), S. G. Lucas (New Mexico Museum of Natural History and Science), J. Schneider and M. Röscher (Bergakademie Freiberg), R. Werneburg (Museum Schleusingen) and P. Forey (Natural History Museum, London) for their numerous critical comments and suggestions, which considerably improved the quality of this paper. This work was partly supported by the Grant Agency of the Czech Republic, Grant No 205/04/0941.

References

BOY, J. & SCHINDLER, T. 2000. Ökostratigraphische Bioevents im Grenzbereich Stephanium/Autunium (höchstes Karbon) des Saar-Nahe-Beckens (SW-Deutschland) und benachtbarter Gebiete. *Neues Jahrbuch für Geologie und Paläontologie, Abhandlung*, **216**, 89–152.

BÜRGIN, T. 1990. Palaeonisciden (Osteichthyes: Actinopterygii) aus dem Unteren Rotliegenden (Autunien) der Nordschweiz. *Eclogae Geologicae Helvetiae,* **83**, 813–827.

DIETZE, K. 2000. A revision of paramblypterid and amblypterid actinopterygians from Upper Carboniferous–Lower Permian lacustrine deposits of Central Europe. *Palaeontology*, **43**, 927–966.

FOREY, P. L. & YOUNG, V. T. 1985. Upper Stephanian fishes from the Puertollano Basin, Ciudad Real, Spain. *In*: LEMOS DE SOUZA, M. J. & WAGNER, R. H. (eds) *Papers on the Carboniferous Iberian Peninsula (Sedimentology, Stratigraphy, Paleontology, Tectonics and Geochronology).* Anais da Faculdade de Ciēncias do Porto, **64**(1983), Supplements, 233–244.

FRITSCH, A. 1883–1901. Fauna der Gaskohle und der Kalksteine der Permformation Böhmens. Band 1–4. Selbstverlag, Praha.

GOTTFRIED, M. D. 1987. A Pennsylvanian aeduelliform (Osteichthyes, Actinopterygii) from North America with comments on aeduelliform interrelationships. *Paläontologische Zeitschrift*, **61**, 141–148.

HEYLER, D. 1969. *Vertébrés de l'Autunien de France.* Cahiers de Paléontologie, Paris.

HEYLER, D. 1976. Sur le genre Amblypterus Agassiz (Actinoptérygien du Permien inférieur). *Bulletin de la Société d'Histoire Naturelle d'Autun*, **78**, 17–37.

HEYLER, D. 1991. Sur la présence de l'espèce *Aeduella blainvillei* (Pisces, Actinopterygii) dans le Permien Inférieur de la Sarre. *Bulletin de la Société d'Histoire Naturelle d'Autun*, **135**, 17–32.

HEYLER, D. & POPLIN, C. 1994. Les Poissons stéphaniens du Bassin de Montceau-les-Mines. *In*. POPLIN, C. & HEYLER, D. (eds) *Quand le Massif Central était sous l'Equateur: un écosystème carbonifère è Montceau-les-Mines*, Editions all CWRS Paris, 205–222.

KRS, M. & PRUNER, P. 1995. Palaeomagnetism and palaeogeography of the Variscan formations of the Bohemian Massif, comparison with other European regions. *Journal of the Czech Geological Society*, **40**, 3–36.

PEŠEK, J. 2004. Late Palaeozoic limnic basins and coal deposits of the Czech Republic. *Folia Musei Rerum Naturalium Bohemiae Occidentalis, Geologica*, (Special Edition) **1**.

PEŠEK, J., HOLUB, V., JAROS, J. *et al.* 2001. Geologie a ložiska svrchnopaleozoických limnických pánví České republiky. Český geologický ústav, Praha.

POPLIN, C. 1997. Le premier Haplolépiforme (Pisces, Actinopterygii) découvert en France (Carbonifère supérieur du bassin de Blanzy-Montceau, Massif Central). *Compte Rendus de l'Académie des Sciences, Serie IIa*, **324**, 59–66.

POPLIN, C. 1999. Un paléoniscoide (Pisces, Actinopterygii) de Buxières-les-Mines, témoin des affinités fauniques entre Massif Central et Boheme au passage Carbonifère-Permien. *Geodiversitas*, **21**, 147–155.

RÖSCHER, M. & SCHNEIDER, J. 2005. A Commented Correlation Chart for Continental Late Pennsylvanian and Permian Basins and the Marine Scale. *In*: LUCAS, S. G. & ZEIGTER, K. E. (eds) *The Nonmarine Permian.* New Mexico Museum of Natural History and Science, Butletin, **30**, 282–291.

SCHINDLER, T. 1993. '*Elonichthys*' *palatinus* n. sp., a new species of actinopterygians from the Lower Permian of the Saar-Nahe Basin (SW-Germany). *In*: HEIDTKE, U. (ed.) *New Research on Permo-Carboniferous Faunas.* Pollichia-Buch, Bad Durkheim, **29**, 67–81.

SCHNEIDER, J. 1989. Basic problems of biogeography and biostratigraphy of the Upper Carboniferous and rotliegendes. *Acta Musei Reginaehradecensis, Serie A, Scientiae Naturales*, **22**, 31–44.

SCHNEIDER, J. & ZAJÍC, J. 1994. Xenacanthiden (Pisces, Chondrichthyes) des mitteleuropäischen Oberkarbon und Perm – Revision der Originale zu Goldfuss 1847, Beyrich 1848, Kner 1867 und Fritsch 1879 – 1890. *Freiberger Forschungshefte, Hefte C*, **452**, 101–150.

SCHNEIDER, J., HAMPE, O. & SOLER-GIJÓN, R. 2000. The Late Carboniferous and Permian: aquatic vertebrate zonation in southern Spain and German basins. *In*: BLIECK, A. & TORNER, S. (eds) *Palaeozoic Vertechnite Chronology and Global Marine/Non-Marine Correlation: Final Report of IGCP (1991–1996)*. Courier Forschunginstitut Senckenberg, **223**, 543–561.

SCHNEIDER, J. W., RÖSSLER, R. & GAITZCH, B. 1995. Time lines of the Variscan volcanism – holostratigraphic synthesis. *Zentralblatt für Geologie und Paläontologie, Teil I*, **5/6**, 477–490.

ŠTAMBERG, S. 1978. New data on *Pyritocephalus sculptus* (Pisces) from the Carboniferous of the Plzeň Basin. Paleontological Conference '77', Charles University, Prague, 275–288.

ŠTAMBERG, S. 1989. Scales and their utilization for the determination of actinopterygian fishes (Actinopterygii) from Carboniferous basins of central Bohemia. *Journal of the Czech Geological Society*, **34**, 255–269.

ŠTAMBERG, S. 1991. Actinopterygians of the Central Bohemian Carboniferous Basins. *Acta Musei Nationalis Pragae, Series B*, **47**, 25–104.

ŠTAMBERG, S. 2002. Actinopterygian fishes from the new Lower Permian locality of the Krkonoše Piedmont Basin. *Journal of the Czech Geological Society*, **47**(3–4), 147–154.

WERNEBURG, R. 2003. The branchiosaurid amphibians from the Lower Permian of Buxières-les-Mines, Bourbon l'Archambault Basin (Allier, France) and its biostratigraphic significance. *Bulletin de la Société Géologique de France*, **174**, 1–7.

ZAJÍC, J. & ŠTAMBERG, S. 2004. Selected important fossiliferous horizons of the Boskovice Basin in the light of the new zoopaleontological data. *Acta Musei Reginaehradecensis, Series A, Scientiae Naturales*, **30**, 5–14.

Permian to Triassic sequences from selected continental areas of southwestern Europe

C. VIRGILI[1], G. CASSINIS[2] & J. BROUTIN[3]

[1]*Stratigraphy Department, Complutense University, Madrid, Spain (e-mail: carmina.virgili@gmail.com)*

[2]*Earth Science Department, Pavia University, via Ferrata 1, 27100 Pavia, Italy (e-mail: cassinis@unipv.it)*

[3]*UMR 5143 – Paléodiversité et Evolution des Embryophytes, Paris VI University, rue Cuvier 12, 75005 Paris, France (e-mail: jbroutin@snv.jussieu.fr)*

Abstract: This contribution is a tentative reconstruction of the still-debated geological history in the primarily continental domains now represented in various parts of southwestern Europe, between the end of the Variscan diastrophism and the beginning of the Alpine sedimentary evolution. Data and interpretations vary from one region of terrestrial rocks to another. Despite this, we have tried to highlight the most typical and significant geological features. From the Carboniferous to Triassic, palaeontological investigations of the macroflora, microflora and tetrapod footprints, as well as radiometric data, generally point out the presence of three main 'tectono-stratigraphic units' (TSUs), separated by marked unconformities and gaps of as yet uncertain duration. The most important geological episode generally started about the Early/Middle Permian boundary and later spanned discontinuously and intensely throughout Middle Permian (Guadalupian) time. It was characterized by specific tectonic, magmatic, thermal and basinal features, which could mark the presumed change suggested by some authors from a Pangaea B to a Pangaea A. In this context, it is worth mentioning that the unconformable Middle?–Upper Permian higher TSU in Spain consists of 'Buntsandstein'-type red beds, sometimes yielding a Thuringian flora; differently, in southern France, such as in the Lodève area, the Buntsandstein is Anisian and thus constitutes a later Triassic sequence, which rests unconformably above the as yet undefined (Mid–Late) Permian age assessment of the 'La Lieude fossil site'; in the Southern Alps, the 'Second tectono-sedimentary Cycle' emphasized from the recent literature, which is initially made up of the Verrucano Lombardo–Val Gardena Sandstone red clastics, is in part laterally and upwardly replaced, east of the Adige Valley, by the sulphate evaporite to shallow-marine Bellerophon Formation. It is thus represented by continental and marine sediments generally pertaining to Late Permian (post- 'Lower Tatarian') time and can be interpreted, in the light of the geological context of the region, as an Upper Permian and Lower Triassic TSU of a slightly younger numerical order (i.e. TSU 3 in place of TSU 2).

This paper deals with the Permian–Triassic sequences of Italy, France and Spain, which are essentially characterized by terrigenous and volcanic continental deposits (Fig. 1). Intrusive bodies are also widespread (southern and western Alps, Pyrenees, Sardinian–Corsican block, etc.). Marine sediments crop out in some Italian areas, where they are related to the most westerly branches of Palaeotethys. The extent and notable variations of the Permian–Triassic continental deposits preclude an exhaustive description here. In this context, the reader may examine other works of ours (e.g. Virgili 1989; Cassinis *et al.* 1992, 1995; Broutin *et al.* 1994; Cassinis 1996). Therefore, we stress only some selected areas generally involved in horst and graben structures. The following sections on Italy, France and Spain were compiled respectively by GC, JB and CV, while the conclusions are a joint effort.

Italy

Southern Alps

In the Permian Southern Alpine successions, two major, well-differentiated tectono-stratigraphic Units (TSUs), separated by a marked regional unconformity, are clearly evident (Fig. 2). They correspond to the Cycles 1 and 2 emphasized in the last decades by numerous authors (e.g. Italian IGCP-203 Group 1986; Cassinis *et al.* 2000*b*).

The Late Carboniferous to Early Permian lower unit, or TSU 1, up to about 2000 m thick, is

From: LUCAS, S. G., CASSINIS, G. & SCHNEIDER, J. W. (eds) 2006. *Non-Marine Permian Biostratigraphy and Biochronology*. Geological Society, London, Special Publications, **265**, 231–259.
0305-8719/06/$15.00

Fig. 1. Areas of southwestern Europe discussed in this paper (in grey).

generally made up of calc-alkaline acidic-to-intermediate volcanics and alluvial-to-lacustrine deposits (e.g. Collio and Tregiovo formations), both infilling intramontane, fault-bounded, transtensional subsiding basins isolated from each other by metamorphic and igneous structural highs (Fig. 2).

The boundary faults often coincide with long-lived features that were reactivated before and during the Alpine orogenesis (such as the Val Trompia, Valsugana, Giudicarie and Pusteria lines). Palaeontological investigations of the macroflora, microflora and tetrapod footprints (Tables 1, 2 & 4), as well as radiometric data (see Cassinis *et al.* 2002 and references therein) indicate that the aforementioned sedimentary and igneous succession probably started in the Late Carboniferous, but mainly developed during the Early Permian, and possibly persisted, at least locally (e.g. Tregiovo), until a slightly younger time, perhaps up to the 'Kazanian' (Conti *et al.* 1997; Pittau 1999*a*; Cassinis & Ronchi 2001). However, this presumed age assessment of the Tregiovo Formation, which is essentially based on palynological investigations, clashes with some radiometric dates of about 275 Ma (Bargossi *et al.* 2004), generally supporting a Kungurian age.

In the easternmost Carnia region, this lower TSU (Pontebba Supergroup) is dominated by Permo-Carboniferous shallow-marine sedimentary deposits, up to about 2000 m thick (Venturini 1990; Venturini, in Cassinis *et al.* 2000*a*). Biostratigraphical subdivisions are essentially based on fusulinids, which enable a zonal correlation with other parts of Tethys (Cassinis & Ronchi 2001).

The Mid? to Late Permian upper unit, which persisted up to the early Middle Triassic (Massari *et al.* 1994; Massari & Neri 1997), marks the onset of widespread erosion and cessation of volcanic activity in the whole Southern Alpine domain. The respective sediments are more widely distributed, although thinner (with a maximum thickness of about 600 m), than the older Lower Permian products; therefore, they form an almost continuous blanket that covers the basins of TSU 1 and the surrounding highs (Fig. 2).

Generally, the inception of this younger tectono-stratigraphical unit (TSU 2, corresponding to the 'second Cycle' or 'Cycle 2' of the literature) is characterized by the Verrucano–Val Gardena fluvial red beds. To the east of the Adige Valley, the latter formation is in part laterally and vertically replaced by the sulphate evaporite-to-shallow marine carbonate sediments of the Bellerophon Formation (Fig. 2). According to evidence from continental (macro- and microfloras, tetrapod footprints; Tables 3–4) and marine (foraminifers, algae, brachiopods, etc.; Fig. 3) fossils, this TSU 2 undoubtedly pertains, at least in part, to Late Permian time (e.g. Broglio Loriga *et al.* 1988; Massari *et al.* 1988, 1994; Massari & Neri 1997; Visscher *et al.* 1999; Pittau 1999*a*, *b*, 2001; Posenato & Prinoth 1999).

Recent studies on the brachiopods found in the so-called 'transitional beds' between the Bellerophon and Werfen formations of the eastern Southern Alps, which include bioclastic

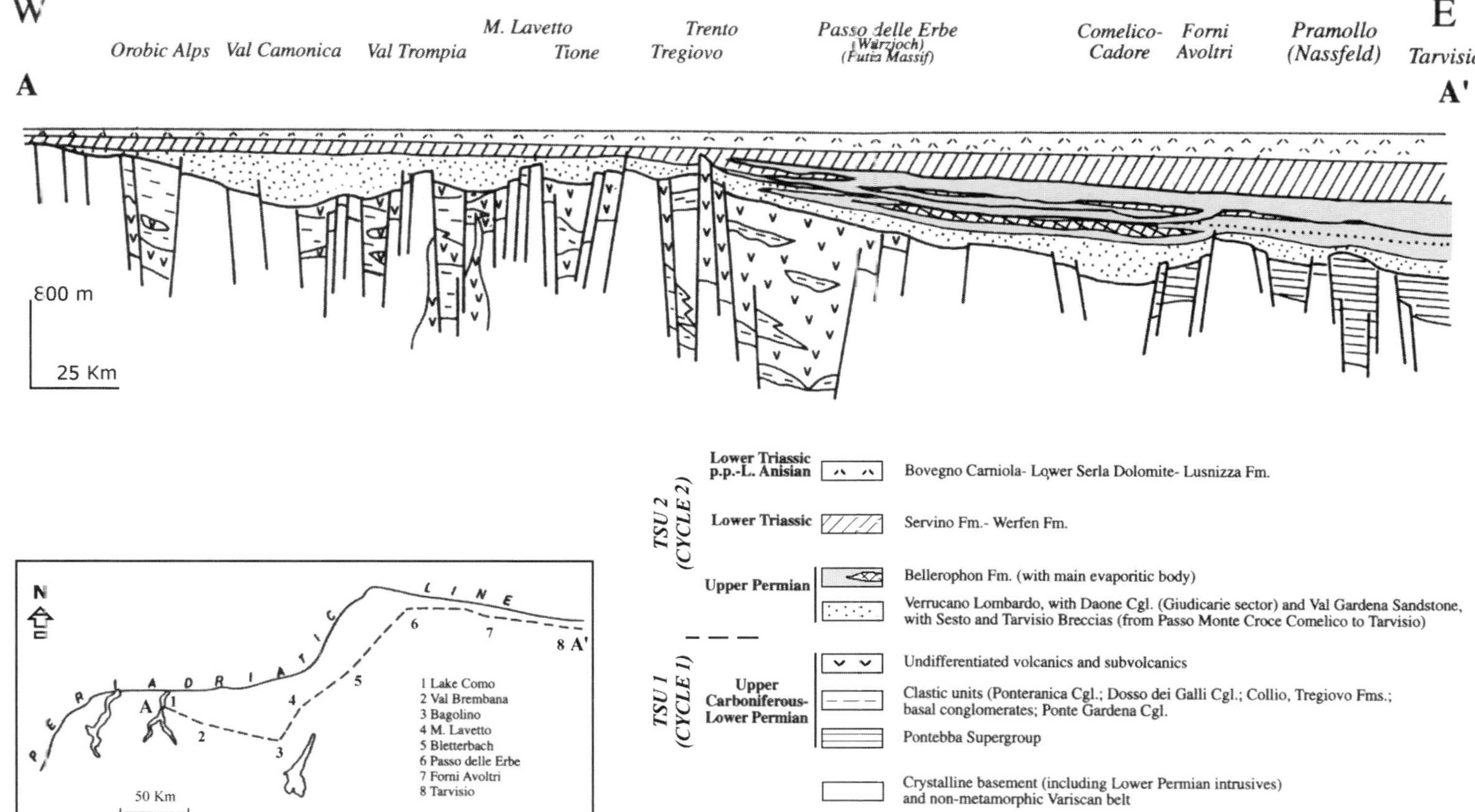

Fig. 2. Diagrammatic, simplified non-palinspastic cross-section (see trace on the inset map) through the Permian and the Lower Triassic of the central and eastern Southern Alps. Datum line: restored top of the Anisian succession, before the beginning of Middle Triassic tectonics (from Cassinis *et al.* 2002). TSU 1 and TSU 2, first and second 'Tectono-stratigraphic units' of the South-Alpine succession.

Table 1. *Macrofloras from some continental deposits of the central and eastern Southern Alps*

	WESTERN OROBIC BASIN	COLLIO BASIN	TREGIOVO BASIN TRENTINO-ALTO ADIGE	
EARLY PERMIAN	**Ponteranica Conglomerate** (Kerp *et al.*, 1996) *Cassinisia orobica* (Kerp *et al.*, 1996) conifer remains	**Collio Fm.** (Visscher *et al.*, 1999) *Sphenopteris suessii* *S. kukukiana* *S. patens* *"Sphenopteris"* cf. *interrupte-pinnata* *Hermitia* (al. *Walchia) geinitzii* *Walchiostrobus* sp. Unidentifiable conifer (and ginkgophyte?) remains	**Tregiovo Fm. - type locality** (Broutin, *pers. comm.*, 1990) *Ullmannia frumentaria* **Bolzano-Trento region** (Visscher *et al.*, 1999) *Lodevia* (al. *Callipteris)* cf. *nicklesii* *Lesleya* (al. *Taeniopteris) eckardtii* *Otovicia hypnoides* *Ullmannia frumentaria* *Walchia piniformis* *"Walchia" stricta* *Quadrocladus* (?) isolated *Ortiseia leonardii* leaves	EARLY PERMIAN - ? MIDDLE PERMIAN *p.p.*

(algae and foraminifers) grainstones and packstones, and are characterized by the presence of *Ombonia*, *Orthothetina* and *Janiceps*, also led Posenato (2001) to correlate a consistent part of the famous basal Tesero Member of the latter formation to the Late Changshingian, i.e. to place the beginning of 'the Werfen' at the top of the Permian. These results rule out the interpretation of Assereto *et al.* (1973), according to which a wide hiatus occurred between the '*Comelicania* beds' *sensu stricto* and the overlying, already recorded 'Tesero Horizon,' which has marked the base of the Werfen Formation since the pioneering work of Bosellini (1964).

As known, the Permian–Triassic (P/T) boundary is marked by the first appearance datum (FAD) of *Hindeodus parvus* (Yin 2000), after the recent ratification by the International Union of Geological Sciences of the Global Stratigraphic Section and Point (GSSP) at the base of Meishan section, southern China (Orchard 2001). In the Tesero section, the first occurrence of this conodont is 11 m above the Bellerophon/Werfen boundary (Nicora & Perri 1999), which is at least 8 m above the disappearance of the Permian-like components of the 'mixed fauna'. However, in the Bulla section, near Ortisei, *H. parvus* seems to occur at a lower level (1.3 m above the base of Werfen Formation: Farabegoli & Perri 1998), which is correlatable with a unit that, in the Tesero section, yielded the *Crurithyris* assemblage. We note that the Bulla section – characterized by relatively abundant conodonts – has recently been proposed by Perri & Farabegoli (2003) as the most reliable parastratotype section for the P/T boundary in the western Palaeotethys.

The palynomorph assemblages (Pittau 1999*b*, 2001) and the tetrapod footprints (e.g. Conti *et al.* 1986, 1999; Massari *et al.* 1988) highlighted in the Val Gardena Sandstone of the Butterloch/Bletterbach area and surroundings appear also to be in keeping with a Late Permian age. In this context, therefore, the Illawarra Reversal Event, which was recorded by Mauritsch & Becke (1983) and Dachroth (1988) in the Val Gardena red beds of Carnia and the western Dolomites respectively, and is placed by Menning (1995, 2001) at about 265 Ma in the 'Lower Tatarian', should be carefully reset.

The boundary between the two above-mentioned Permian TSUs is marked by an important regional unconformity associated with gaps of varying and as yet uncertain duration (Cassinis *et al.* 1999*a*, 2002; Cassinis & Ronchi 2001). However, from a general viewpoint, the palaeontological data and geological evidence suggest that the inception of this Upper Permian to Lower Triassic TSU is still open to debate.

Tuscany

In this region, the first northern consistent succession crops out in the Pisani Mountains (Fig. 4). The oldest sediments are represented by the S. Lorenzo Formation, as much as 300 m thick, which infills a fluvio-deltaic-lacustrine intramontane basin mainly composed of black silty shales grading into coarse-grained layers yielding Westphalian D (?) to Autunian plant fossils (Remy in Rau & Tongiorgi 1974). This formation is followed locally by the Asciano red breccias and conglomerates of undefined age. The presence of rhyolitic clasts in the overlying basal Verrucano suggests that volcanism occurred in Early Permian or slightly younger

Table 2. *Microfloras from the Collio and Tregiovo continental basins, central Southern Alps*

	COLLIO BASIN	TREGIOVO BASIN		
EARLY PERMIAN	**Collio Fm.** (Doubinger in Cassinis & Doubinger, 1991,1992) *Lundbladispora simoni* *Cordaitina* sp. *Potonieisporites* sp. *Nuskoisporites* sp. *Vestigisporites minutus* *Playfordiaspora crenulata* ALETE DISACCATES *Lueckisporites microgranulatus* *Flacisporites* cf. *zapfei* *Vittatina foveolata* *Vittatina costabilis* *Vittatina* sp. "A" Hochuli	**Tregiovo Fm.** (Pittau, 1999a) Pre-pollen and pollen grains -LOWER PART PSEUDOSACCATES *Endosporites* sp. MONOSACCATES *Cordaitina* *Florinites* *Densipollenites* *Nuskoisporites crenulata* *Potonieisporites* sp. *Crucisaccites* sp. *Vestigisporites* sp. *Vesicaspora schemeli* MONOLETE DISACCATES *Limitisporites* *Falcisporites schemeli* *Illinites* *Gardenasporites* spp. *G. leonardii* *G. oberrauchi* *Lueckisporites* *Alisporites* sp. TAENIATE DISACCATES *Distriatites insolitus* *Hamiapollenites* sp. *Protohaploxypinus perfectus* POLYPLICATES *Vittatina* spp. -MIDDLE PART Pre-pollen and pollen grains MONOSACCATES *Nuskoisporites* sp. *Cordaitina* *Florinites* *Densipollenites* *Samoilovichisaccites* sp. MONOLETE DISACCATES Common	ALETE DISACCATES Common TAENIATE DISACCATES POLYPLICATES *Vittatina* (rare) -UPPER PART Azonalete spores *Paludospora* *Psophosphaera* Pre-pollen and pollen grains MONOSACCATES *Nuskoisporites dulhuntyi* *N. klausii* *Paravesicaspora* *Cordaitina* (very rare) *Densipollenites* (very rare) *Florinites* (very rare) *Samoilovichisaccites* sp. MONOLETE DISACCATES *Falcisporites zapfei* *Gigantosporites aletoides* *G. hallstattensis* *Lueckisporites virrkiae* *Corissacites* sp. (? *alutas*) ALETE DISACCATES *Alisporites* *Platysaccus* *Scheuringipollenites* TAENIATE DISACCATES *Protodiploxypinus* *Striatoabieites multistriatus* *Strotersporites* *Sriatohaplopynites* POLYPLICATES *Vittatina costabilis* *V. subsaccata*	EARLY PERMIAN - ? MIDDLE PERMIAN *p.p.*

times, probably after the Asciano red beds were deposited, but before the Triassic Verrucano. Consequently, the aforementioned succession could represent, throughout the late Palaeozoic, a first tectono-stratigraphic unit (TSU 1) that is characterized, as in the continental Southern Alps, by the coexistence of sedimentary and volcanic products. On the contrary, the superjacent Tuscan Verrucano, the name of which is derived from the Pisani Mountains area and has generally been assigned to the Middle Triassic (Rau & Tongiorgi 1974; Rau *et al.* 1988), seems identifiable with a younger TSU (probably the third, i.e. prior to the marine sedimentation, based on correlation with other regions of western Europe located, e.g. in southern France).

South of the Arno River, in the surroundings of the village of Iano (near Volterra), the Late Variscan cover shows close affinities with the Permian South Alpine successions of Lombardy (Fig. 4). In particular, the Torri Formation and the overlying igneous products could be

Table 3. *Permian macrofloras and some significant palynomorph taxa from the Bletterbach–Butterloch section, western Dolomites*

WESTERN DOLOMITES: BLETTERBACH SECTION	
1. MACROFLORAS	2. MICROFLORAS most significant taxa
Val Gardena Sandstone (Visscher *et al.* 1999)	**Val Gardena Sandstone and Bellerophon Fm.** (Pittau 1999b)
Coniferous remains: *Ortiseia leonardii* *O. visscheri* *O. jonkeri* *Majonica alpina* *Dolomitia cittertiae* *Pseudovoltzia liebeana* *P. sjerpii*	MONOSACCATES *Paravesicaspora splendens* *Vesicaspora* spp.
Pteridosperm fragments *Peltaspermum martinsii*	MONOLETE DISACCATES *Gardenasporites* spp. *Limitosporites* spp. *Jugasporites* sp. *Falcisporites zapfei* *Lueckisporites* spp.
	ALETE DISACCATES *Klausipollenites schaubergeri* *Alisporites nuthallensis* *Platysaccus papilionis*
	TAENIATE DISACCATES *Lunatisporites* spp.

correlated with the lower TSU (TSU 1) of the Permian South-Alpine intramontane basins (Costantini *et al.* 1998; Lazzarotto *et al.* 2003). The Triassic fluvial Verrucano (TSU 3?), as in the Pisani Mountains area, lies unconformably on the Permian succession.

Eastward, between the Montagnola Senese and Mount Leoni, the Upper Palaeozoic of the Mid-Tuscan Ridge ('Monticiano-Roccastrada Unit') essentially pertains, from the literature, to Carboniferous marine deposits, such as the Carpineta and Farma formations, the Sant Antonio Limestone and the Spirifer Schists (Cocozza *et al.* 1987; Lazzarotto *et al.* 2003; Fig. 4); however, very recent palynological research (displaying *Lueckisporites virkkiae*, *Lunatisporites noviaulensis*, *Klausipollenites schaubergeri*, *Alisporites splendens*, *Striatopodocarpites* sp. etc.) would point to a Mid- or Late Permian age for at least the Farma Formation (Spina 2005). The Triassic clastic Verrucano again rests unconformably above this interval. Calcareous pebbles in its middle part (Quoio Formation) include *Meandrospira pusilla*, leading to a Scythian or earliest Anisian age assignment (Cocozza *et al.* 1975). The additional discovery of fusulinid-bearing clasts from the latest Carboniferous to the Early Permian (Engelbrecht 1984; Engelbrecht *et al.* 1989), allows new evaluations of the role played by marine sedimentation in southern Tuscany and nearby during Carboniferous–Triassic time.

Marine influences were also displayed in other Tuscan areas, such as in Elba Island (Fig. 4), where Bodechtel (1964) collected a sample with fusulinids referable to '*Parafusulina* sp.', which seems to belong to the Artinskian *Praeparafusulina lutigini–P. pseudojaponica* zones. Towards the east, in the Mount Amiata geothermal field, wells highlight a Palaeozoic metamorphic clastic sequence, including a carbonate horizon with Late Permian fusulinids, such as *Praeparafusulina* cf. *lutigini*, *Cancellina* sp., *Parafusulina* sp., *Pseudofusulina* sp., etc. According to Pasini (1991), this association pertains to the *Cancellina* Zone ('Kubergandinian').

Along the Tuscan 'Maremma', recent biostratigraphical research also pointed to a Late Permian or Early Triassic age for the Argentario and Poggio al Carpino sandstone formations (Cirilli *et al.* 2002; Lazzarotto *et al.* 2003). Therefore, on this basis, it is possible to date the unconformably overlying Civitella Marittima Formation (which is the oldest formation of the Verrucano Group) essentially as Mid-Triassic. According to the above age assessments, the Argentario–Poggio al Carpino sequences probably could be grouped into a second TSU (TSU 2).

Sardinia–Corsica block

During the late Palaeozoic, Sardinia and Corsica were characterized by a continental domain with a history resembling, in some respects, that deducible from the Carboniferous and Permian successions highlighted between southern Provence and eastern Spain. Both of the islands, in fact, according to palaeomagnetic and palaeofloristic data (Westphal *et al.* 1973; Olivet 1996; Broutin *et al.* 1994), occupied a position more or less close to this part of the European continent before their drift towards the present Tyrrhenian.

Sardinia

In Sardinia, the best known late- to post-Variscan continental deposits crop out to the

Table 4. *Permian ichnofaunas from some continental areas of the Southern Alps*

	W OROBIC BASIN	COLLIO BASIN	TREGIOVO BASIN	BLETTERBACH SECTION E
LATE PERMIAN				**Val Gardena Sandstone Fm.** *Pachypes dolomiticus* *Hyloidichnus tirolensis* *? Paradoxichnium radeinensis* *Ichniotherium accordii* *Rhynchosauroides pallinii* *R. palmatus* *Chirotherium* *"Chelichnus" tazelwurmi* *Dicynodontipus geinitzii* Gorgonopsian indet.
?			**Tregiovo Fm.** *"Dromopus" didactylus*	
EARLY PERMIAN	**Collio Fm.** *Amphisauropus latus* *A. imminutus* *Varanopus curvidactylus* *Dromopus lacertoides*	**Collio Fm.** *Batrachichnus* sp. *?Camunipes cassinisi* *Amphisauropus latus* *Ichniotherium cottae* *Dromopus lacertoides* *"Dromopus" didactylus*		

Orobic Basin (Ceoloni *et al.* 1987; Conti *et al.* 1999; Cassinis *et al.* 2000*b*; Santi 2001)
Collio Basin (Haubold & Katzung 1975; Ceoloni *et al.* 1987; Conti *et al.* 1991, 1997, 1999)
Tregiovo Basin (Conti *et al.* 1997, 1999)
Bletterbach-Butterloch section in the Western Dolomites (Conti *et al.* 1977, 1986, 1997, 1999; Ceoloni *et al.* 1988) (modified from Cassinis *et al.* 2002)

northwest in Nurra, approximately between Cala Viola and Mount Santa Giusta, and in the central-southeastern and southwestern sectors of the island, such as in Barbagia (Seui), Ogliastra (Perdasdefogu), Gerrei (Escalaplano), Sarcidano (Mulargia), Sulcis (Guardia Pisano) and Iglesiente (S. Giorgio) (Vardabasso 1950, 1966; Lombardi *et al.* 1974; Fontana *et al.* 1982, 2001; Cassinis *et al.* 1996, 1998, 1999*b*, 2003*a*, *b*; Cortesogno *et al.* 1998; Costamagna *et al.* 2000; Cassinis & Ronchi 2002; Pittau *et al.* 2002; Barca & Costamagna 2003). The succession can be generally subdivided into three TSUs (Figs 5–6). The lower (TSU 1) mainly consists of alluvial-lacustrine clastic deposits, rich in Autunian macro- and microfloral assemblages (Broutin *et al.* 1996; Ronchi *et al.* 1998), often associated with calc-alkaline acidic-to-intermediate volcanic products. Conglomerates generally occur below, lying unconformably on the Variscan metamorphic basement.

After this first TSU, the second (TSU 2) began with prevalent fluvial detritic red beds, marked again by a stratigraphical discontinuity. These unfossiliferous deposits possibly represent a succession corresponding to the 'Saxonian' and 'Thuringian' of the French authors (Cassinis & Ronchi 2002; Cassinis *et al.* 2003*b*), and thus interpreted as Permian in age. In some places, alkaline volcanics crop out, too.

The third TSU (TSU 3), or Sequence 3, consists of Lower to Middle Triassic red clastics (such as the so-called Cala Viola Sandstone in Nurra and the Escalaplano Formation in Gerrei, both biostratigraphically correlatable to the Germanic Buntsandstein (Pittau 1999*c*; Pittau in Costamagna *et al.* 2000), which begin in the former region with a thin, bright quartz-conglomerate ('Conglomerato del Porticciolo') resting unconformably on older Permian rocks, while in the latter area the Escalaplano Formation, which is characterized by varying

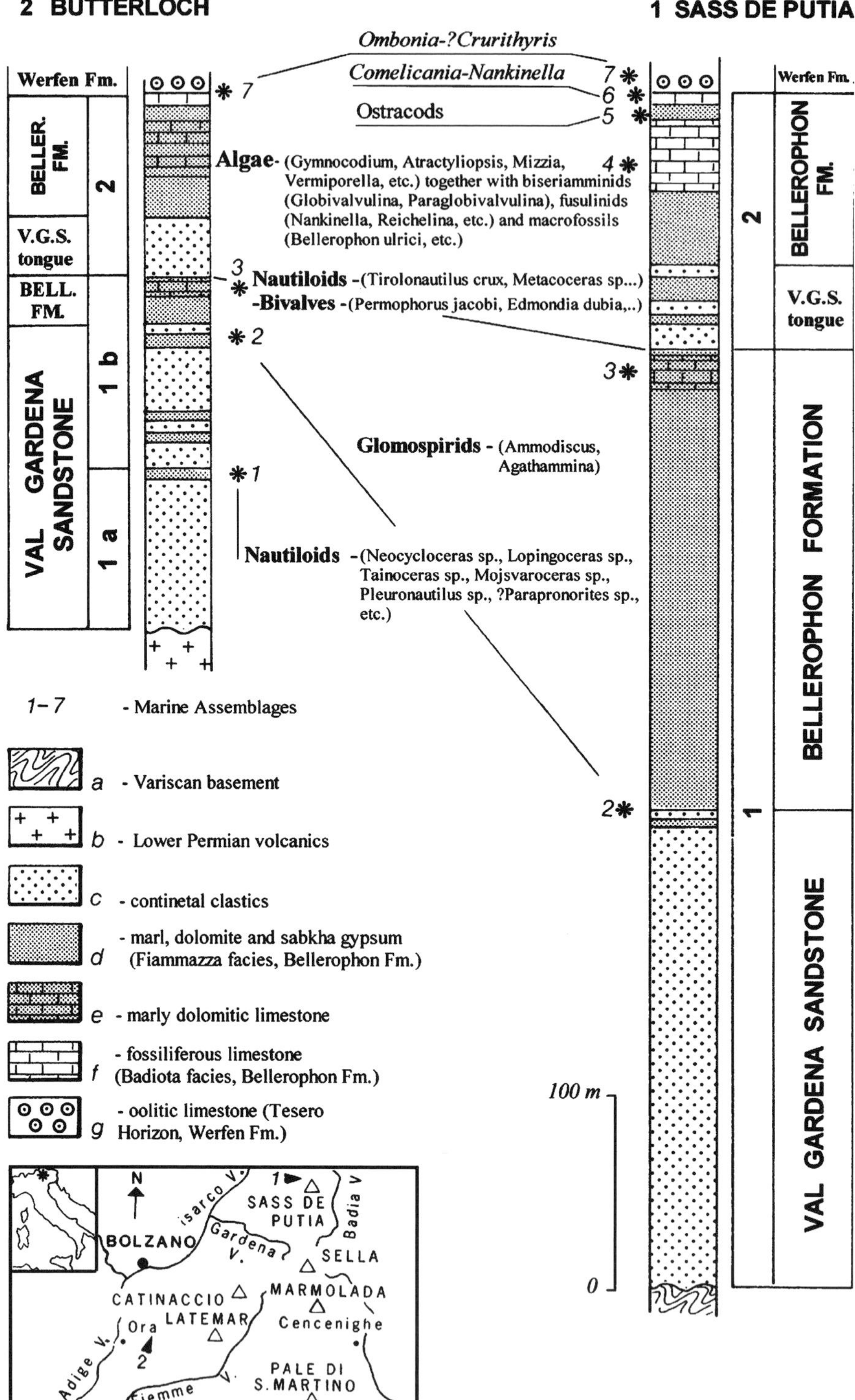

Fig. 3. Position and correlation of marine fossil assemblages in the Bletterbach-Butterloch and Sass de Putia sections, western Dolomites (modified from Broglio Loriga *et al.* 1988).

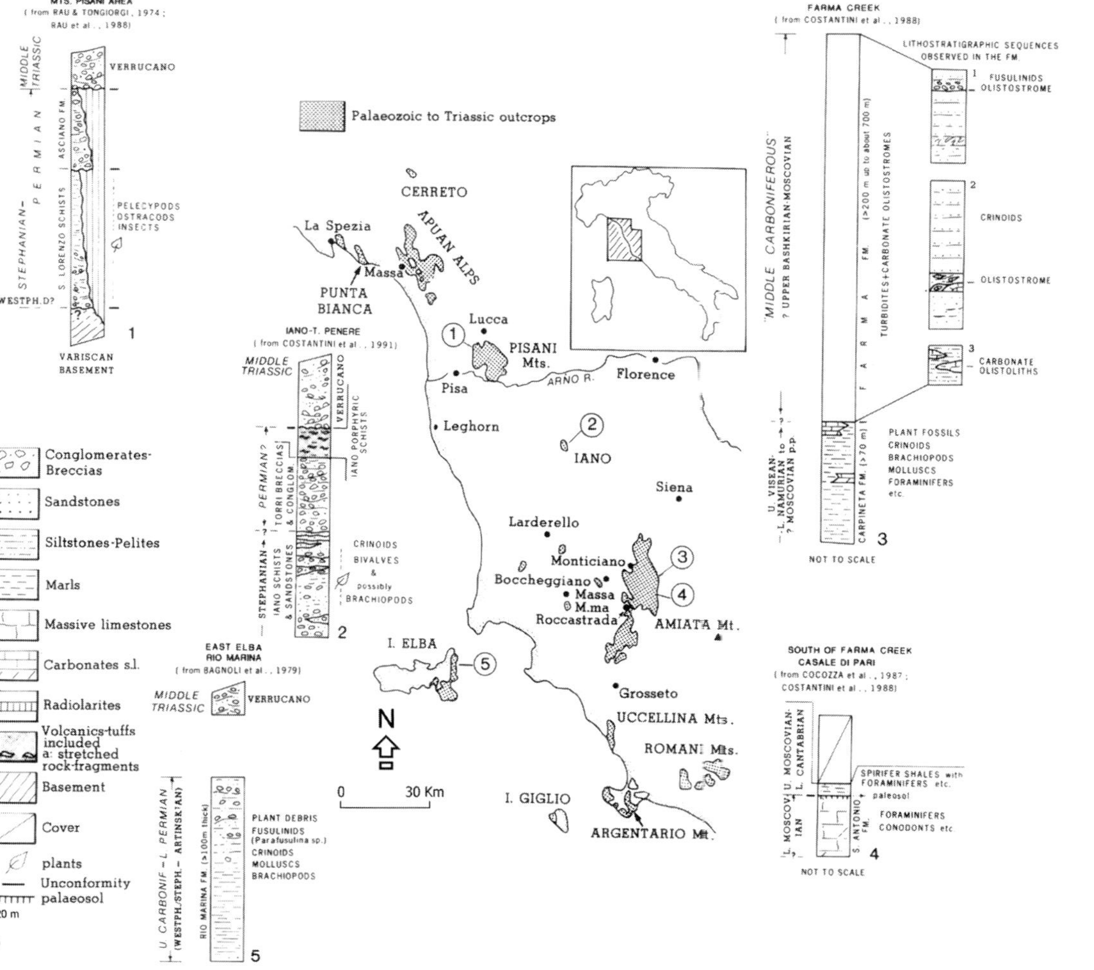

Fig. 4. Schematic Carboniferous–Triassic stratigraphic sections in Tuscany (data from the authors indicated above the columns; map from Elter & Pandeli 1990).

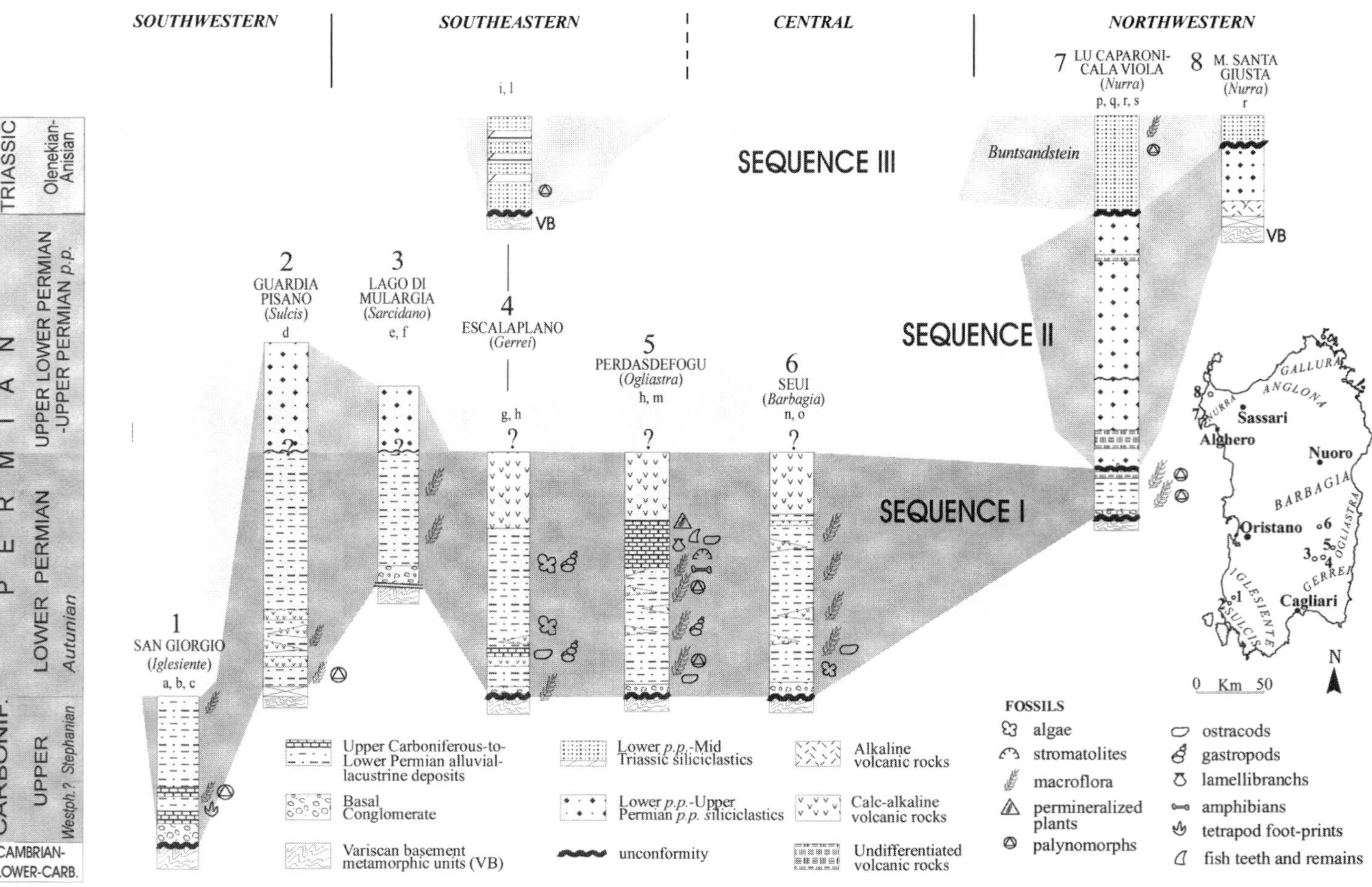
SOUTHWESTERN
SOUTHEASTERN
CENTRAL
NORTHWESTERN
TRIASSIC
Olenekian-Anisian
P E R M I A N
UPPER LOWER PERMIAN -UPPER PERMIAN p.p.
LOWER PERMIAN
Autunian
CARBONIF.
UPPER
Westph.? Stephanian
CAMBRIAN-LOWER-CARB.
1 SAN GIORGIO (Iglesiente) a, b, c
2 GUARDIA PISANO (Sulcis) d
3 LAGO DI MULARGIA (Sarcidano) e, f
4 ESCALAPLANO (Gerrei) g, h
i, l
5 PERDASDEFOGU (Ogliastra) h, m
6 SEUI (Barbagia) n, o
7 LU CAPARONI-CALA VIOLA (Nurra) p, q, r, s
8 M. SANTA GIUSTA (Nurra) r
VB
Buntsandstein
SEQUENCE III
SEQUENCE II
SEQUENCE I
Upper Carboniferous-to-Lower Permian alluvial-lacustrine deposits
Basal Conglomerate
Variscan basement metamorphic units (VB)
Lower p.p.-Mid Triassic siliciclastics
Lower p.p.-Upper Permian p.p. siliciclastics
unconformity
Alkaline volcanic rocks
Calc-alkaline volcanic rocks
Undifferentiated volcanic rocks
FOSSILS
algae
stromatolites
macroflora
permineralized plants
palynomorphs
ostracods
gastropods
lamellibranchs
amphibians
tetrapod foot-prints
fish teeth and remains
GALLURA
ANGLONA
NURRA
Sassari
Alghero
Nuoro
BARBAGIA
OGLIASTRA
Oristano
GERREI
IGLESIENTE
SULCIS
Cagliari
N
0 Km 50

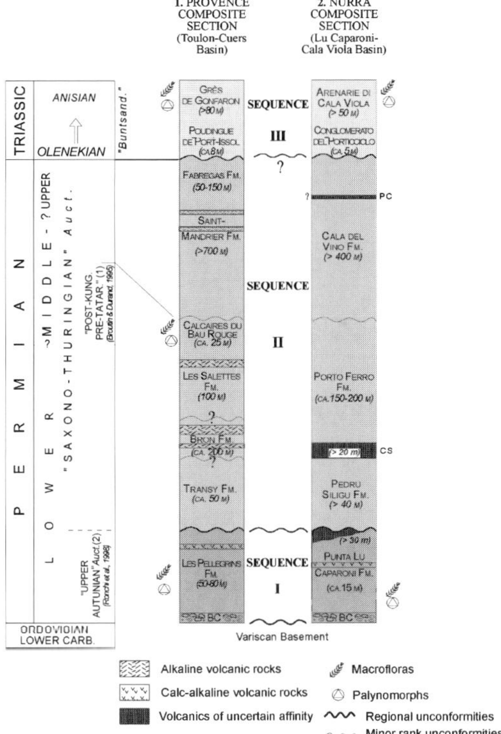

Fig. 6. Schematic and composite logs of the Permian and Triassic continental sequences of the Nurra (NW Sardinia) and Toulon (S Provence) basins, and, on the left, their potential correlation and dating. Columns are not to scale. BC, basal conglomerate; CS, Casa Satta volcanic rocks; PC, Ponte Crabolu volcanic rocks. In the time column, the dating (1) pertains to the Provence section, whereas (2) pertains to the Nurra section. (Modified from Cassinis *et al.* 2003*b*).

lithologies, directly overlies the Variscan metamorphic complex. The Ladinian marine Muschelkalk follows.

In this context, the 'post-Autunian' igneous activity of Sardinia mirrors the second alkaline magmatic cycle of French authors (e.g. Châteauneuf & Ferjanel 1989; Broutin *et al.* 1994; Deroin *et al.* 2001), which contrasts decidedly with that of the previous calc-alkaline Permo-Carboniferous cycle. In the northern and eastern sectors of the island, the emplacement of these alkaline products also occurred between the 'Autunian' and the Triassic Buntsandstein, that is, probably from the late Early Permian up to undefined Late Permian time. From recent research (Cassinis *et al.* 2003*b*), the remarkable similarities between the Permian–Triassic successions of the Nurra and Toulon basins (the latter in southern Provence) allow us to place them closely facing one another, and consequently to place the southern part of Sardinia near the eastern Pyrenees, and northwestern Corsica in front of the Estérel area, in Provence (Fig. 6).

Corsica

Corsica was dominated by very intense late Palaeozoic magmatism and represents the northern continuity of the coeval Sardinian igneous manifestation (Fig. 7). The French authors recognized the presence of two magmatic cycles (e.g. Vellutini 1977; Bonin 1988; Broutin *et al.* 1994; Deroin *et al.* 2001). The first, Late Carboniferous and Early Permian in age, up to about 280 Ma, was characterized by calc-alkaline volcanism and plutonism. In contrast, the upper magmatic cycle, essentially Middle to Late Permian in age, was represented by alkaline bimodal extrusive and intrusive products. As in other parts of southwestern Europe, this late-post Variscan magmatic succession includes some fluviolacustrine basin deposits, with macro- and microfloras ascribed to Carboniferous and/or Permian time (e.g. in Osani and Scandola).

Southern France

The best-known Permian and Triassic continental outcrops lie along the southern border of the Massif Central, namely, from south-west to south-east, in the Brive, Rodez, Saint-Affrique and Lodève basins (Fig. 8). Their stratigraphical successions also fit an informal subdivision into three TSUs. The Lodève Basin can be considered as 'a standard' for this group of Permian basins and is briefly commented on below.

Fig. 5. Schematic Upper Carboniferous to Triassic sections from the main continental basins of Sardinia, and their subdivision into three TSUs (or 'Sequences'). Location in the inset map; data from the references cited above the columns: **a**, Cocozza 1967; **b**, Del Rio 1973; **c**, Del Rio & Pittau 1999; **d**, Pittau *et al.* 1999*a*; **e**, Francavilla *et al.* 1977; **f**, Barca *et al.* 1995; **g**, Pecorini 1974; **h**, Ronchi 1997; **i**, Costamagna *et al.* 2000; **l**, Pittau *et al.* 1999*b*; **m**, Sarria 1987; **n**, Lauro *et al.* 1963; **o**, Sarria & Serri 1986; **p**, Gasperi & Gelmini 1980; **q**, Cassinis *et al.* 1996; **r**, Fontana *et al.* 2001; **s**, Cassinis *et al.* 2001; vertical distances are not time or thickness related. The above-mentioned 'Sequences', which are marked in the drawing by different patterns of grey, show on the left side the presumed age-assessment, according to a twofold subdivision of the Permian System (modified from Cassinis & Ronchi 2002).

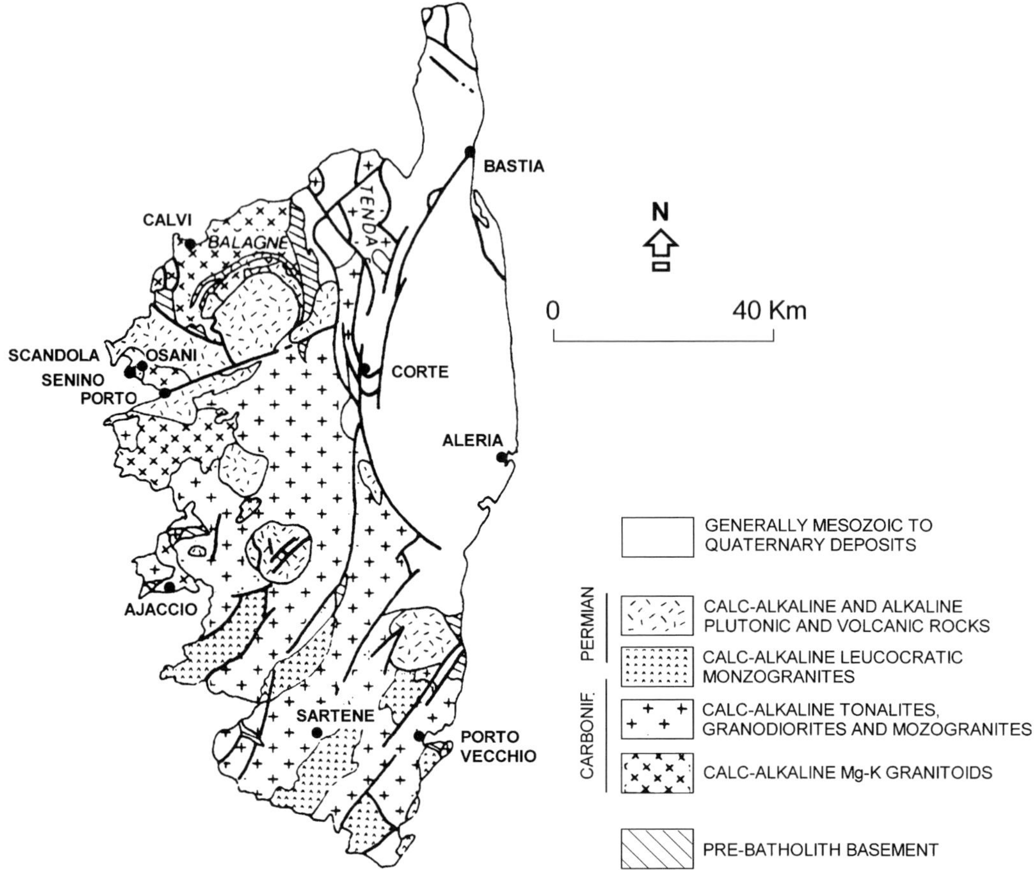

Fig. 7. Distribution and subdivision of the late Palaeozoic igneous rocks in Corsica (simplified from Ménot & Orsini 1990).

Lodève Basin

In the Lodève Basin the lower TSU (TSU 1), up to 800 m thick, rests unconformably on Stephanian or Cambrian rocks, and essentially consists of fluvio-lacustrine and fluvio-palustrine deposits (Fig. 9 and Table 5). From the base to the top, three lithostratigraphical units have been defined: the Usclas–St. Privat and Loiras formations ('grey-Autunian'), and the Tuilières Formation ('red-and-grey alternating Autunian') (Châteauneuf, in Broutin *et al.* 1992*a*). Whereas volcanic rocks are poorly represented in the Brive and Rodez Basins, rich ash layers occur mainly in the middle part of this lower succession of the Lodève Basin, possibly related to the coeval volcanism known from Sardinia, Provence and Corsica.

Macro- and microfloral assemblages pertain to the Early Permian (equivalent of the Millery Formation of the Autun Basin; Broutin *et al.* 1999), whereas vertebrate footprints point to a slightly older Permian age (equivalent of the Muse Formation of the Autun basin: Gand 1987). Both the 'grey' and 'alternating red-and-grey' 'Autunian' display a macroflora dominated by conifers (*Walchia* spp., *Culmitzschia* spp., *Otovicia hypnoides, Ernestiodendron filiciforme*) and callipterids (*Autunia conferta, Rhachiphyllum schenkii, Lodevia nicklesi, Dichophyllum flabellifera, Arnhardtia moureti*, etc.), associated with rare but stratigraphically relevant taxa, such as *Ginkgophyllum grassetii*, *Baiera* sp., *Ginkgoites lodevensis* and *Nephropsis* sp. The vertebrate footprint assemblage mainly includes *Antichnium salamandroides, Limnopus zeilleri, Ichniotherium cottae, Dromopus lacertoides* and *Dimetropus nicolasi*.

The Viala Formation follows and, after an unconformity, the alluvial-to-deltaic red conglomerates of the Rabejac Formation, which grades laterally and upwardly into playa siltstone

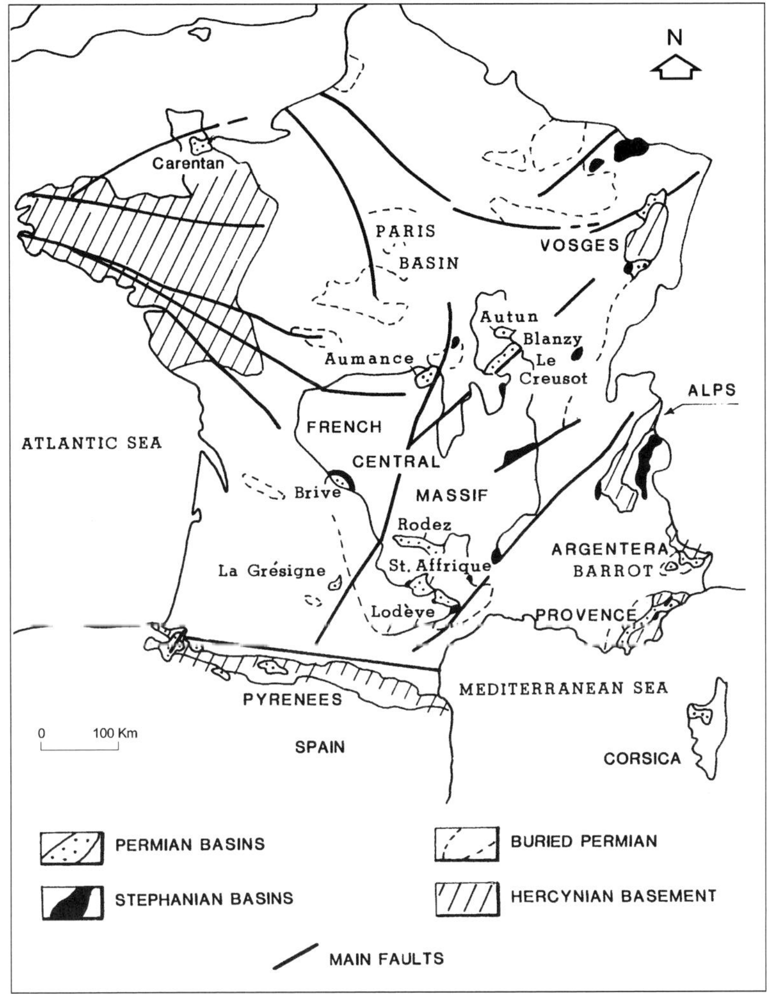

Fig. 8. Permian deposits of France.

sediments (Salagou Formation). The last two formations form the Permian upper TSU (TSU 2), locally over 2000 m thick. Therefore, the Permian ends with thousands of metres of fine-grained red beds, reflecting a very steady subsidence activity of the basin and still somewhat uncertain in age. As a matter of fact, the biostratigraphical data remain a little bit confused. According to Gand (1987), the above-mentioned upper stratigraphical succession includes three tetrapod ichnofauna assemblages.

For the Viala Formation, surprisingly, a big discrepancy was noted between the dating interpretations based on macrofloral and ichnofaunal elements on the one hand and microfloral data on the other hand. The vertebrate footprint association, which consists of *Salichnium decessus, S. pectinatus, Limnopus regularis, Gilmoreichnus brachydactylus, Varanopus rigidus, Amphisauropus latus* and *Dimetropus leisnerianus*, indicates a 'latest Autunian' age (Gand 1987) in agreement with the dating based on the macroflora (Broutin *et al.* 1992*a*). In contrast, Doubinger's interpretation (Doubinger *et al.* 1987) favours a younger Late Permian ('early Thuringian' age or, at least, clearly 'post-Autunian' age), based on the occurrence of markers such as *Gardenasporites heisseli, Hamiapollenites insolitus* and, above all, *Lueckisporites singhii* and *L. globosus*.

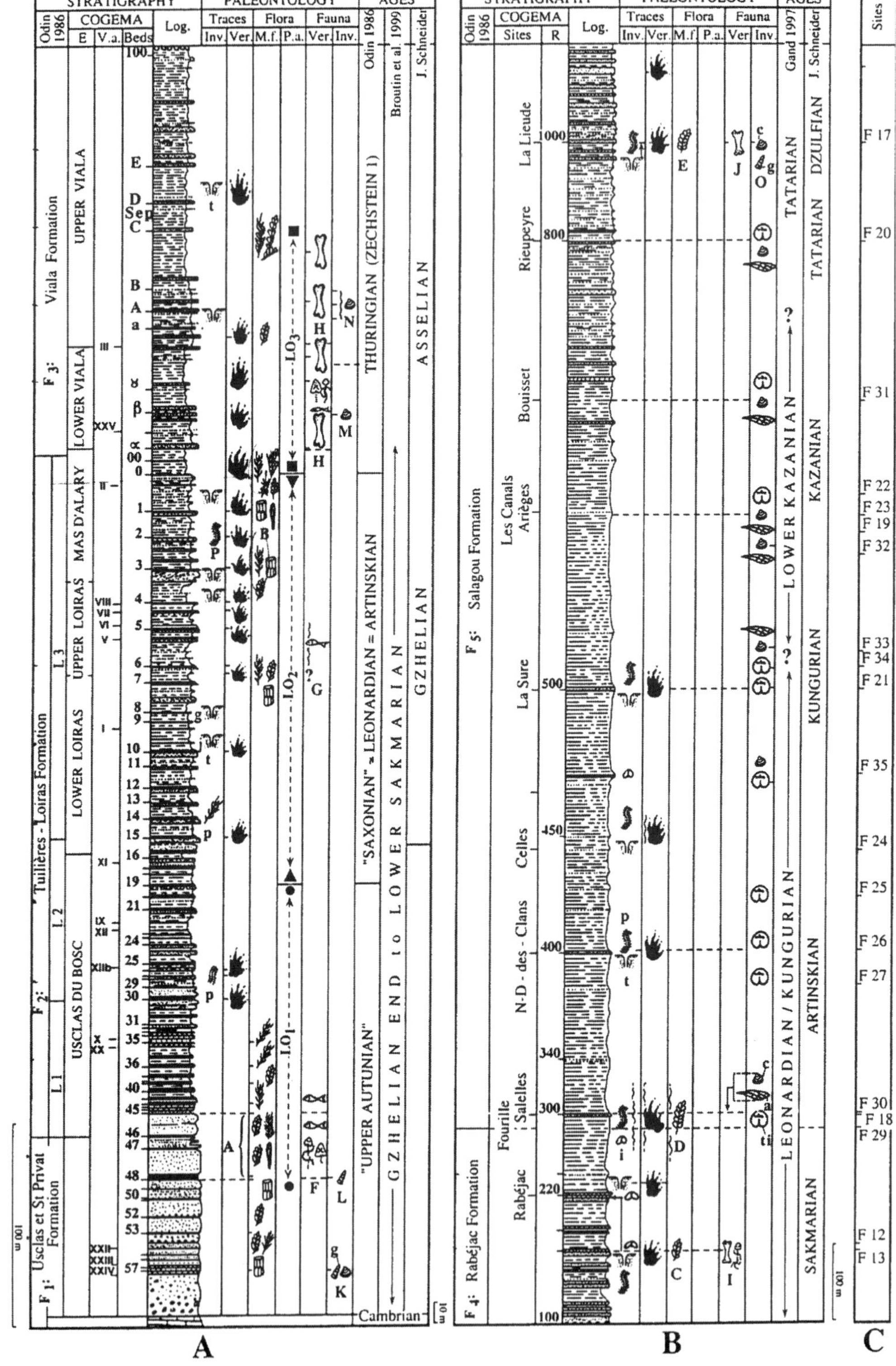
STRATIGRAPHY
PALEONTOLOGY
AGES
Odin 1986
COGEMA
E
V.a.
Beds
Log.
Traces
Inv.
Ver.
Flora
M.f.
P.a.
Fauna
Ver.
Inv.
Odin 1986
Broutin et al. 1999
J. Schneider
F 3: Viala Formation
UPPER VIALA
LOWER VIALA
F 2: Tuilières - Loiras Formation
MAS D'ALARY
UPPER LOIRAS
LOWER LOIRAS
USCLAS DU BOSC
L 3
L 2
L 1
F 1: Usclas et St Privat Formation
THURINGIAN (ZECHSTEIN 1)
"SAXONIAN" = LEONARDIAN = ARTINSKIAN
"UPPER AUTUNIAN"
GZHELIAN END to LOWER SAKMARIAN
ASSELIAN
GZHELIAN
Cambrian
10 m
100 m
A
Sites
R
Gand 1997
F 5: Salagou Formation
F 4: Rabéjac Formation
La Lieude
Rieupeyre
Bouisset
Les Canals Ariéges
La Sure
Celles
N-D - des - Clans
Fourille Salelles
Rabéjac
TATARIAN
DZULFIAN
LOWER KAZANIAN
KAZANIAN
KUNGURIAN
LEONARDIAN / KUNGURIAN
ARTINSKIAN
SAKMARIAN
B
C

In the Rabejac Formation, in addition to *Dimetropus leisnerianus*, one notices the occurrence of *Hyloidichnus major, Varanopus curvidactylus* and *Dromopus didactylus*, which point to a 'post-Autunian to pre-Thuringian interval' (equivalent to the so-called 'Saxonian'). Higher, in the Salagou Formation ('La Lieude' fossil site), *Brontopus giganteus, Merifontichnus thalerius, Lunaepes ollierorum* and *Planipes antecursor* indicate a younger age than that suggested for the Rabejac Formation, related to the Late Permian *sensu lato* (Gand 1987; Gand *et al.* 2000).

Therefore, the so-called 'Saxonian' or 'undated' Salagou Formation has been considered, except for the 'La Lieude' locality, as virtually azoic due to the arid climate that prevailed during deposition. Recently, when Dr Lapeyrie's collection (assembled over more than 20 years from unprospected fossiliferous siltstone channels, incising the red pelites) was taken into account, relevant biostratigraphical data came to light (Gand *et al.* 1997, 2001). The conchostracans found immediately below and above the basal fanglomerates of the La Lieude member indicate that the time span from the upper Lower Permian (Artinskian) to the Upper Permian (according to a two-fold subdivision of this System) could probably be covered from the base of the Rabejac Formation to the top of the Salagou Formation (former 'Saxonian Group' *sensu* Gand *et al.* 1997). More precisely, the occurrence of characteristic ornamented conchostracans suggests a 'Tatarian' age for the uppermost part of the Salagou Formation, corresponding to the 'La Lieude mb.' (Schneider, in Gand *et al.* 1997; Körner *et al.* 2003). In addition, the presence of many orders of insects could generally indicate a Late Cisuralian to Early Lopingian temporal distribution for large parts of this formation (Gand *et al.* 1997; Nel *et al.* 1999*a*, *b*; Béthoux *et al.* 2002; Körner *et al.* 2003).

Furthermore, from the greenish grey, fine-grained lenses located just above the unconformity between the red Permian pelites, at the very base of the Buntsandstein conglomerates, a very rich palynological assemblage has been discovered that clearly points to an Anisian age. In particular, the co-occurrence of *Hexasaccites muelleri* and *Illinites chitonoides, I. kosankei, Concentricisporites nevesi, Microcachryidites sittleri, Protodiploxypinus gracilis* and *Punctatisporites triassicus* allow assignment of an early Pelsonian age to the lowermost Buntsandstein deposits. It is worth mentioning that such an age is totally different from the 'Thuringian' age of the basal 'Buntsandstein' facies observed in the Spanish Permo-Triassic basins, such as near as in the Catalonian Pyrenees (Díez Ferrer 2000; Durand 2006).

Provence

In southern Provence (Fig. 8 & Table 5), the Permian System is widespread, and may reach more than 2000 m in thickness (Toutin-Morin & Vinchon 1989; Toutin-Morin, in Cassinis *et al.* 1992, 1995; Durand 2001, 2006; Deroin *et al.* 2001; Durand *et al.* 2002). In the Early Permian, some general W–E grabens (Toulon-Cuers, Luc, Bas-Argens, Estérel), which are related to extensional movements, began to open on the margin of horsts trending roughly N–S/W–E along the border of the Variscan Maures massif. The oldest, green and red fluviolacustrine clastic deposits (L'Avellan Formation, Estérel) are tilted in a very small ESE–WNW intermontane graben and are assumed to be 'Autunian' in age (*sensu* Broutin *et al.* 1999).

Upward, above an almost general unconformity located on the top of the Avellan Formation, there follow sequences of alluvial fan conglomerates and breccias, as well as fluvial sandstones and mostly reddish and green mudstones, assigned to the 'Saxonian' and 'Thuringian'. These thick continental siliciclastic fills, organized in fining upward sequences, evolved from fanglomeratic to meandering stream deposits or playa mudstones in which

Fig. 9. Permian composite section of the Lodève Basin (Languedoc), drawn up by a recent French-German collaboration (simplified after Gand *et al.* 2001), showing: A, Lower 'Autunian Group' (with F1–F3 formations); B, Upper 'Saxonian Group' (with F4, F5 formations), according to Odin (1986), which are both separated by an unconformity (not indicated by the authors in the composite stratigraphic column); C, F12–F35 'Saxonian' fossiliferous sites. Stratigraphy (due to COGEMA Co.): E, Members/Formations; V.a., volcanic ashes, indicated by roman numerals; α, β, γ, bone beds. Palaeontology: TRACES: Inv., invertebrates – with p, tubes and burrows; P, exogene resting, furrowing, walking/crawling tracks; i, *Isopodichnus*; Ver., footprints. FLORA: M.f., macroflora with A–E fossil sites reported from COGEMA (unpublished) and various authors; P.a., palynology, including the LO1, LO2, LO3 Doubinger *et al.* (1987) associations. FAUNA: Ver., vertebrates with F–J fossil sites from various authors; Inv., invertebrates – with ostracans; g, gasteropodes; c, conchostracans; ti, triopsids; a, insects. For further details see the original figure-caption compiled by Gand *et al.* (2001). Ages according to Odin (1986), Broutin *et al.* (1999), Gand (1997) and J. Schneider. However, dating and correlation with the ICS global scale of all the represented lithostratigraphical units are in progress.

Table 5. *Timing of sedimentary, tectonic and magmatic Late Palaeozoic–Early Mesozoic events in some areas of Southern France (Languedoc and Provence basins), from Late Carboniferous ('Stephanian') to Middle Triassic (Anisian)*

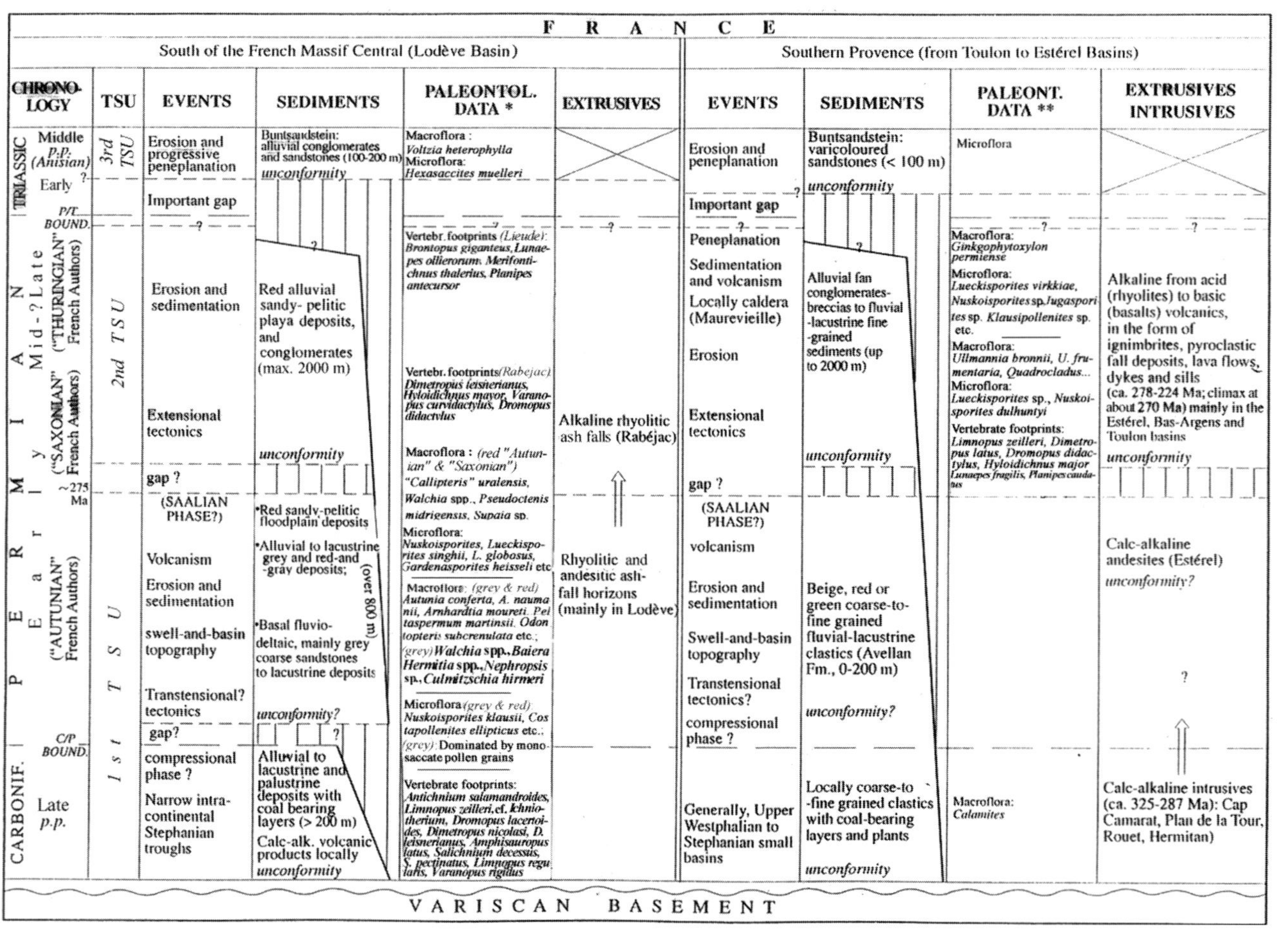

		FRANCE							
		South of the French Massif Central (Lodève Basin)				Southern Provence (from Toulon to Estérel Basins)			
CHRONOLOGY	**TSU**	**EVENTS**	**SEDIMENTS**	**PALEONTOL. DATA ***	**EXTRUSIVES**	**EVENTS**	**SEDIMENTS**	**PALEONT. DATA ****	**EXTRUSIVES INTRUSIVES**
TRIASSIC: Middle *p.p.* (*Anisian*); Early ?	*3rd TSU*	Erosion and progressive peneplanation; Important gap; ?	Buntsandstein: alluvial conglomerates and sandstones (100-200 m); *unconformity*	Macroflora: *Voltzia heterophylla*; Microflora: *Hexasaccites muelleri*		Erosion and peneplanation; Important gap; ?	Buntsandstein: varicoloured sandstones (< 100 m); *unconformity* ?	Microflora	
P/T BOUND.		?	?	?	?	?	?	?	?
PERMIAN: Mid-?Late ("SAXONIAN" "THURINGIAN" French Authors)	*2nd TSU*	Erosion and sedimentation; Extensional tectonics	Red alluvial sandy-pelitic playa deposits, and conglomerates (max. 2000 m); *unconformity*	Vertebr. footprints *(Lieude)*: *Brontopus giganteus, Lunaepes ollierorum, Merifontichnus thalerius, Planipes antecursor*; Vertebr. footprints *(Rabejac)*: *Dimetropus leisnerianus, Hyloidichnus mayor, Varanopus curvidactylus, Dromopus didactylus*; Macroflora: *(red "Autunian" & "Saxonian")* *"Callipteris" uralensis, Walchia* spp., *Pseudoctenis midrigensis, Supaia* sp.	Alkaline rhyolitic ash falls (Rabéjac)	Peneplanation; Sedimentation and volcanism; Locally caldera (Maurevieille); Erosion; Extensional tectonics	Alluvial fan conglomerates-breccias to fluvial-lacustrine fine-grained sediments (up to 2000 m); *unconformity*	Macroflora: *Ginkgophytoxylon permiense*; Microflora: *Lueckisporites virkkiae, Nuskoisporites* sp., *Jugasporites* sp. *Klausipollenites* sp. etc.; Macroflora: *Ullmannia bronnii, U. frumentaria, Quadrocladus*...; Microflora: *Lueckisporites* sp., *Nuskoisporites dulhuntyi*; Vertebrate footprints: *Limnopus zeilleri, Dimetropus latus, Dromopus didactylus, Hyloidichnus major Lunaepes fragilis, Planipes caudatus*	Alkaline from acid (rhyolites) to basic (basalts) volcanics, in the form of ignimbrites, pyroclastic fall deposits, lava flows, dykes and sills (ca. 278-224 Ma; climax at about 270 Ma) mainly in the Estérel, Bas-Argens and Toulon basins; *unconformity*
~275 Ma		gap ? (SAALIAN PHASE?)			⇧	gap ? (SAALIAN PHASE?)			
PERMIAN: Early ("AUTUNIAN" French Authors)	*1st TSU*	Volcanism; Erosion and sedimentation; swell-and-basin topography; Transtensional? tectonics; gap?	•Red sandy-pelitic floodplain deposits; •Alluvial to lacustrine grey and red-and-gray deposits; •Basal fluvio-deltaic, mainly grey coarse sandstones to lacustrine deposits (over 800 m); *unconformity?* ?	Microflora: *Nuskoisporites, Lueckisporites singhii, L. globosus, Gardenasporites heisseli* etc; Macroflora: *(grey & red)* *Autunia conferta, A. naumannii, Arnhardtia moureti, Peltaspermum martinsii, Odontopteris subcrenulata* etc.; *(grey) Walchia* spp., *Baiera Hermitia* spp., *Nephropsis* sp., *Culmitzschia hirmeri*; Microflora *(grey & red)*: *Nuskoisporites klausii, Costapollenites ellipticus* etc.; *(grey)*: Dominated by monosaccate pollen grains	Rhyolitic and andesitic ash-fall horizons (mainly in Lodève)	volcanism; Erosion and sedimentation; Swell-and-basin topography; Transtensional tectonics?; compressional phase ?	Beige, red or green coarse-to-fine grained fluvial-lacustrine clastics (Avellan Fm., 0-200 m); *unconformity?*		Calc-alkaline andesites (Estérel); *unconformity?*; ? ⇧
C/P BOUND.									
CARBONIF.: Late *p.p.*		compressional phase ?; Narrow intra-continental Stephanian troughs	Alluvial to lacustrine and palustrine deposits with coal bearing layers (> 200 m); Calc-alk. volcanic products locally; *unconformity*	Vertebrate footprints: *Antichnium salamandroides, Limnopus zeilleri*, cf. *Ichniotherium, Dromopus lacertoides, Dimetropus nicolasi, D. leisnerianus, Amphisauropus latus, Salichnium decessus, S. pectinatus, Limnopus regularis, Varanopus rigidus*		Generally, Upper Westphalian to Stephanian small basins	Locally coarse-to-fine grained clastics with coal-bearing layers and plants; *unconformity*	Macroflora: *Calamites*	Calc-alkaline intrusives (ca. 325-287 Ma): Cap Camarat, Plan de la Tour, Rouet, Hermitan)
VARISCAN BASEMENT									

TSU, Tectono-Stratigraphic Unit.
Data from Broutin *et al.* (1992*a*) (South of the French Massif Central), and Toutin-Morin in Cassinis *et al.* (1992, 1995) (southern Provence).

paludal to lacustrine carbonates can occur (Deroin *et al.* 2001). Alkaline volcanics, from acidic (rhyolites, tuffs) to basic (basalts) in composition, are locally intercalated in the form of flows, ignimbrites, dikes and pyroclastics. There is little evidence of Permian plutonism.

A 'Thuringian' age is assigned to the Muy Formation (Bas-Argens Basin) by its *Ulmannia*-rich macrofloral assemblage, and by a *Lueckisporites virkkiae*- and *Nuskoisporites dulhuntyi*-dominated microflora, quite similar to the Zechstein associations and thus representing the youngest forms known within the entire French Permian deposits. Numerous tetrapod footprints have also been reported from the Pradinaux Formation (St-Raphaël), such as *Lunaepes fragilis, Planipes caudatus*, as well as from the overlying units, such as *Hyloidichnus major, Varanopus curvidactylus/Microsauripus acutipes, Dromopus didactylus*, etc. All these footprints can be correlated with some specimens already recorded from the TSU 2 of the Lodève Basin and thus probably attributed to a 'Kazanian–Dzhulfian' interval.

As in the Lodève and Saint-Affrique basins, the Triassic System begins with the Buntsandstein, unconformably overlying Permian or older rocks; the age of this unit is generally Anisian (Durand *et al.* 1988). However, according to a general review by Durand (2006), the Buntsandstein sedimentation started in France, as well as in other parts of Europe, as early as Late Permian time.

Spain

During the Permian, the Iberian microplate was affected by tectonic movements which were related to the post-collisional stages of the Variscan orogeny. Intracontinental basins, filled by clastic sediments and volcanic rocks, were formed (Virgili *et al.* 1973, 1976). There are also some beds of lacustrine limestones, and carbonaceous beds are frequent, but rarely with coal layers that can be exploited (Virgili *et al.* 1980; Virgili 1983). There are no formations of evaporites on the surface, nor marine sediments, although occasionally marine sediments have erroneously been described. Some radiometric dating of volcanic rocks (Hernando *et al.* 1980; Lago *et al.* 2004*a*, *b*) and Autunian and Thuringian macro- and microfloral assemblages (Sopeña *et al.* 1995; Virgili in Cassinis *et al.* 1995) are the most valid chronostratigraphical references, but it should not be forgotten that these two terms correspond to floral groups and not to temporal units. Thuringian microflora has also been found in the Zechstein and in the Changshingian and therefore its attribution to the Upper Permian is of no doubt. Although the Autunian flora is equivalent to what is contained in the Autunian type succession, it presents more problems in relation to the boundary between the Permian and the Carboniferous Periods (Becq-Giraudon 1993; Broutin *et al.* 1990, 1999).

As can be seen in the attached diagram (Fig. 10), Carboniferous to Triassic continental successions appear in various parts of the Iberian Peninsula with differing characteristics (Virgili, in Cassinis *et al.* 1995; López-Gómez *et al.* 2002). They are made up of several lithostratigraphical units, separated by unconformities and disconformities but, as in a large part of the above-examined regions of central and western Mediterranean, it is not easy to recognize three well-defined TSUs.

The lower TSU (or TSU 1) includes layers yielding Stephanian and Autunian floral assemblages, and rests unconformably on older Palaeozoic rocks. At its top there is an important regional unconformity and an evident sedimentary interruption (Virgili 1989). It consists of fluvial and lacustrine red, grey or black, fine- to coarse-grained sediments locally rich in andesitic and rhyolitic volcanic products of calc-alkaline type. Autunian macro- and microfloras found in some parts support an Early Permian age for this lower TSU; however, it is very difficult to place the boundary with the Upper Carboniferous because, in some places, the late Stephanian and Autunian floras appear mixed or alternating in the transitional layers (Broutin 1977; Broutin *et al.* 1990, 1999). The composition, number of tectono-stratigraphical bodies and total thickness differ greatly between the various outcrops. The carbonaceous levels and the pebbles of the breccias and conglomerates are polygenic, and local lithic elements are important. They were deposited in small basins formed between mountains, generally controlled by strike-slip faults and isolated from each other (Arche & López-Gómez 1996; López-Gómez *et al.* 2002; Sopeña 2004).

On the other hand, the Late Permian stratigraphic succession is tied to an extensional regime, which in a way heralds the Mesozoic evolution of the western Mediterranean, namely, the start of the 'Alpine Cycle' (Virgili *et al.* 1983). Fluvial clastic red beds and red pelites are spread across wide areas that extend well beyond the basins of the first TSU and cover a large part of the relief. Their composition and thickness (50–600 m) are also much more uniform, and their internal unconformities are scarce (Virgili

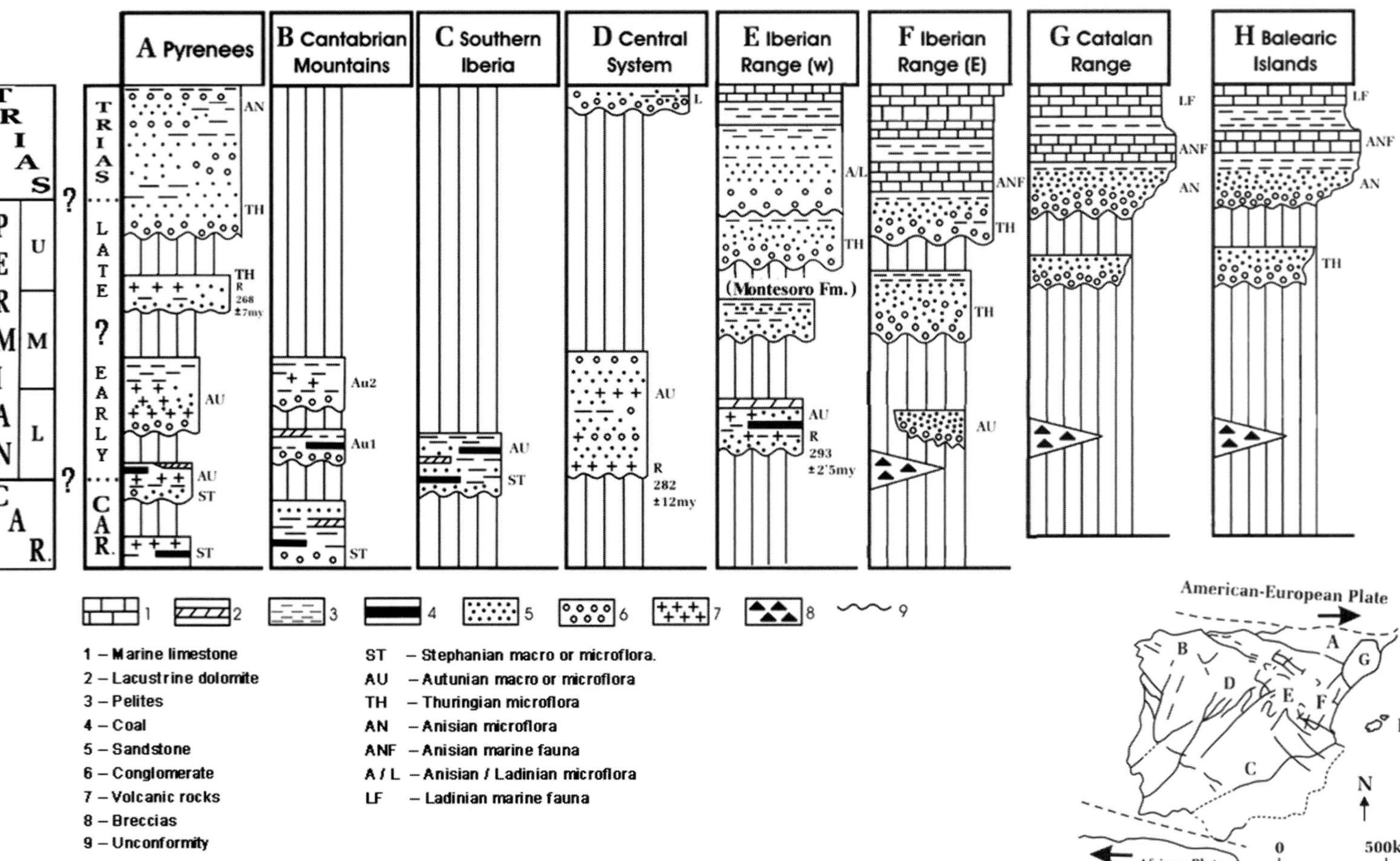

Fig. 10. Upper Carboniferous, Permian and Triassic of selected areas of Spain. Floral assemblages from different areas are indicated in Table 6. Schematic stratigraphic columns are not to scale.

Table 6. Macro- and microfloras (the former underlined) in different areas of Spain. Symbols as in Figure 10.

(A) **PYRENEES: AN** – *Triadispora staplini, T. falcata, Illinites kosankei, Stellapollenites thiergartii,...* **TH** – *Falcisporites zapfei, Limitisporites parvus, Gardenasporites* sp., *Jugasporites* sp., *Lueckisporites* sp., *Endosporites* sp., *Nuskoisporites dulhuntyi, Lueckisporites parvus, L. virkkiae,...* **AU** – *Odontopteris* sp., *Mixoneura* sp., *Cordaites* sp., *Potonieisporites* sp., *Cordaitina* sp., *Gardenasporites* sp., *Costapollenites ellipticus,...* **AU/ST** – *Taeniopteris* sp., *Neuropteris neuropteroides, Walchia* spp., *Ernestiodendron filiciforme,...*; (B) **CANTABRIAN MTS.: AU 2** – *Culmitzschia parvifolia, Taeniosporites* cf. *fallax, Autunia conferta, Neuropteris* sp. **AU 1** – *Autunia conferta, Sphenopteris minutisecta, Pecopteris hemitelioides, Annularia stellata,...*; (C) **SOUTHERN IBERIA: AU/ST** *Potonieisporites* sp., *Nuskoisporites* sp., *Gardenasporites* sp., *Jugasporites* sp., *Hamiapollenites* sp., *Verrucosisporites* sp., *Florinites eremus, Lueckisporites singhii, Thymospora* sp., *Lycospora* sp.,...; (D) **CENTRAL SYSTEM: AU** – *Autunia conferta, Walchia* spp., *Pecopteris* spp., *Odontopteris* sp., *Sphenophyllum* sp., *Annularia stellata, Potonieisporites* spp., *Vittatina costabilis,...*; (E) **IBERIAN RANGES (W): A/L** - *Verrucosisporites remyanus, Triadispora staplini, T. falcata, Hexasaccites muelleri, Alisporites* cf. *grauvogeli,...* **TH** - *Lueckisporites virkkiae, Nuskoisporites dulhuntyi, Falcisporites schaubergeri, Paravesicaspora splendens, Jugasporites delasaucei,...* **AU** – *Cyclogranisporites pergranulus, Potonieisporites novicus, Gardenasporites delicatus, Vittatina costabilis, Pityosporites* sp.; (F) **IBERIAN RANGES (E): AN** – *Alisporites* sp., *Triadispora staplini, T. crassa, Voltziaceaesporites heteromorpha*; **TH** - *Lueckisporites virkkiae, Lunatisporites albertae, L. noviaulensis, Limitisporites* sp., *Vesicaspora ovata*; **AU** – *Autunia conferta, Cathaysiopteris whitei, Equisetites elongates, Lebachia piniformis, Umbellaphyllites annularoides, Gigantonoclea lagrelli*; (G) **CATALAN RANGE: AN** – *Cycadopites* sp., *Lundbladispora* sp., *Triadispora crassa, Stellatopollenites thiergarti, Voltziaceaesporites heteromorpha*; (H) **BALEARIC ISLANDS: AN** – *Porcellispora longdonensis, Sulcosaccispora minuta, Triadispora staplini, Alisporites grauvogeli*; **TH** – *Lueckisporites virkkiae, L. singhii, Lunatisporites* cf. *noviaulensis, Nuskoisporites dulhuntyi, Paravesicaspora splendens, Klausipollenites schaubergeri, Falcisporites stabilis, F. zapfei, Illinites unicus...*

1987). Generally, the pebbles of the conglomerates are well rounded and made of quartz or quartzite. The volcanic products appear only in the lower levels and are scarce, although in the Pyrenees they have a certain importance. Their geochemical characteristics are different from the underlying TSU. This unit (Pyrenees, Iberian Range and Balearic Islands) contains greyish fine pelitic intercalations rich in Thuringian microflora, with *Lueckisporites virkkiae, Nuskoisporites dulhuntyi, Falcisporites schaubergeri* and other forms (e.g. Ramos & Doubinger 1979; Broutin *et al.* 1988).

Consequently, as also reported in Figure 10 and Table 6, palynological discoveries demonstrate that the sedimentation of the Spanish Buntsandstein clearly started during the Permian and continued throughout the Triassic. It stretches across a large part of the centre of the Peninsula and extends to the Middle Triassic, but in the easternmost part of the Iberian Range and in Catalonia it is interrupted by the Anisian marine transgression. It is, therefore, a strongly diachronous unit comprising formations with different characteristics (Arche *et al.* 2004), but with much more unity than the lower TSU 1 succession.

Generally, this 'Buntsandstein' starts with a thick series of conglomerates with a high quartz content. In some cases, lying below an unconformity, there is an important sequence of fluvial red pelites, breccias and sandstones, which in the western Iberian Range is known as 'Formación Montesoro' and was also named later 'Saxonian' (Sopeña *et al.* 1988; López-Gómez *et al.* 2002), considering it equivalent in part to the Middle Permian. However, at least for the Iberian Peninsula, this appellation should be rejected, especially because in other localities where the term 'Saxoniense' has been cited, it contains Thuringian microflora. Therefore, it must be recognized that the unconformity that appears at the base of the conglomerates of the 'Buntsandstein' could generally be considered as intra-Upper Permian (according to the bipartite subdivision of this System), and is probably connected with Middle Permian tectonics (Virgili 2001).

The most characteristic lithostratigraphical successions and their age assessment are represented in Figure 10. The thickest, most complex and fossiliferous section occurs in the Pyrenees (Broutin *et al.* 1988; Virgili 1989; Lago *et al.* 2004*a*). The lower layers show a rich Upper Stephanian flora, on top of which rests another unit, also black and rich in volcanic products, containing Stephanian and Autunian floral assemblages. The overlying deposits include only an Autunian flora; on the whole this aforementioned succession pertains to the lower or first TSU. Above lies a red-coloured series with volcanic intercalations and Thuringian flora, and then, above an unconformity, the 'Buntsandstein'-type red beds that yield Thuringian palynomorphs in the lower part and Anisian flora in the uppermost layers.

In the Central System all of the Permian is represented by a thick detrital unit that is very rich in volcanic products. The area of Atienza (Guadalajara) includes two thick rhyolitic units and reaches a total thickness of more than 500 m. Radiometric measurements give an age of the basal part of 282 ± 12 Ma (Hernando *et al.* 1980; Lago *et al.* 2004*a*, *b*). The rhyolitic units are unconformably overlain by a Buntsandstein containing a Ladinian microflora, which seals a very important gap. The TSU 1 megasequence is also represented in the south of the Iberian block and in Guadalcanal, near Sevilla, where the coexistence of Autunian and Stephanian floras was discovered for the first time (Broutin 1977).

In the Iberian Range, there are more extensive outcrops which are more characteristic and have been better studied (Doubinger *et al.* 1990; López-Gómez & Arche 1993; Díez *et al.* 1996). In its western part, Molina de Aragón (Guadalajara), there is a blackish Lower Permian detrital sequence (TSU 1) with volcanic products and rich in Autunian flora. It is covered by red detrital layers (Montesoro Formation, TSU 2?) lying on an unconformity and, in turn, unconformably overlain by the Buntsandstein conglomerate, which extends over the whole TSU. In the first interpretation it was considered that the Triassic started with the Buntsandstein conglomerates and the intermediate formations of red detritic layers corresponded to the 'Saxoniense', supposed to be Middle Permian in age. The discovery of Thuringian microflora, first in the lower levels of the 'Buntsandstein' (Ramos & Doubinger 1979) and later in the intermediate red detrital formation (Pérez-Arlucea & Sopeña 1985) demonstrated that both units corresponded in part to the Upper Permian. For this reason we think that, in Spain, it is not easy to recognize three well-defined TSUs. For the moment the Thuringian microflora is considered to be of Late Permian, and in Spain any palaeontological evidence of the Middle Permian is not yet clearly documented. Indeed, it is possible that the Middle Permian corresponds totally or partially to a sedimentary gap in connection with the 'Middle Permian Event' (MPE: Deroin & Bonin 2003).

Further to the east, in Catalonia, the Permian still has not been well characterized. It is probably represented by formations of red breccias that lie unconformably below a Buntsandstein-type succession (Arche *et al.* 2004). The starting time of this sedimentation is not known as no flora has been found, and the first ages correspond to the marine sediments cropping out above, which yield an Anisian fauna. The Scythian has not been identified. In the Balearic Islands (especially in Minorca) the Permian is much more developed and, as in the majority of the Iberian Peninsula, there is a Thuringian microflora at the base of the 'Buntsandstein' (Ramos & Doubinger 1989; Broutin *et al.* 1992*b*; Arche *et al.* 2002).

Conclusions

Stratigraphy

Three TSUs, separated by gaps of as yet uncertain duration, occur generally throughout the Carboniferous to Triassic geological evolution of southwestern Europe. The lower unit, or TSU 1, (calc-alkaline intermediate-to-acidic volcanic and alluvial-to-lacustrine deposits) possibly ranges from 'Westphalian'/'Stephanian' to 'Autunian', that is from Late Carboniferous to Early Permian, and perhaps locally (as in Tregiovo, southern Alps) also to slightly younger ages. The middle unit, or TSU 2 (alluvial red clastics, sometimes associated with igneous bimodal products, which are reminiscent of the 'Saxonian'/'Thuringian' succession of western Europe), is 'post-Autunian', that is, from late Early Permian to Mid- and presumed Late Permian time (such as in southern France). The upper unit, or TSU 3, ('Buntsandstein') was traditionally dated as Triassic, but began in places during the Late Permian ('Thuringian'), according to some authors (e.g. Virgili in Cassinis *et al.* 1992, 1995; Broutin *et al.* 1994; Durand 2006).

Following the aforementioned tectono-stratigraphical subdivision, the onset of the Permian 'Buntsandstein' seems to correspond significantly to that of the Verrucano Lombardo-Val Gardena Sandstone in the Southern Alps, which lack volcanic products but have been grouped so far into a marked 'second cycle', persisting from Late Permian to early Middle Triassic time. The shifting of these TSUs could be interpreted by considering the latter Alpine red beds as the basal part of the French TSU 3, for example in Provence, where neither the presence of the Upper Permian (i.e. Lopingian) nor of the Lower Triassic (i.e. Scythian) have been demonstrated thus far (Durand 2006). It was probably caused by another palaeogeographical evolution of the Southern Alps, perhaps linked to the Permo-Triassic marine ingressions of Palaeotethys. As a consequence, in the Southern Alps the Middle Permian would seem to have been affected by a very important stratigraphical gap, probably as long as 20 Ma or more (Cassinis *et al.* 1999*a*, 2002; Cassinis & Ronchi 2001), which can be conveniently regarded as being responsible for the lack of a pre-Upper Permian TSU 2. In the South-Alpine context, however, if the discovery of the Illawarra Reversal by Mauritsch & Becke in the Paularo section (Carnic Alps) and by Dachroth in the Balest section near Ortisei (western Dolomites) can be confirmed from new investigations, the Val Gardena Sandstone evidently started, at least in some areas, during Middle Permian times, and therefore, in this case, the suggested South-Alpine tectono-stratigraphical evolution could be rearranged again in conformity with the traditional subdivision of the Permian into two major intervals (i.e. TSU 1 and TSU 2, in place of the Cycles 1 and 2 generally adopted by Italian authors). In addition, as a result of current research data, the MPE could be also regarded as being responsible for some geological evidence (such as the unconformable superposition of 'Thuringian' on 'Autunian' layers, separated by more or less important gaps) in different areas of Spain, between Early and Late Permian time.

Magmatism

Permian calc-alkaline to alkaline volcanics, acidic and basic in composition, crop out widely in the regions examined, in particular throughout the Corsica–Sardinia block, southern France and the Pyrenees. In northern Italy the latter products are generally absent or very subordinate. Intrusive bodies also occur. According to various authors (e.g. Broutin *et al.* 1994; Toutin-Morin, in Cassinis *et al.* 1992 etc.), the calc-alkaline activity took place in a transtensional and extensional regime and may have been caused by earlier subduction processes during the Carboniferous. In contrast, the alkaline magmatism is classically considered as anorogenic, that is, 'unrelated to orogenic disturbance' (Bonin 1988; Broutin *et al.* 1994). In southern Europe, several alkaline complexes were emplaced during the Middle Permian, at about 270 Ma.

Tectonics

Vast areas were affected by wrench tectonics from the Late Carboniferous to Early Permian (e.g. Ziegler & Stampfli 2001), and locally also to slightly younger times. Such a geological scenario appears compatible with the tectonic setting of late Variscan Europe, generally characterized by extended dextral, lateral and extensional movements with an approximately E–W orientation (Arthaud & Matte 1977; Blès *et al.* 1989). Pull-apart and strike-slip basins were formed. The transtensional regime was associated with the onset of regional uplift, with unroofing, collapse and stretching of the Variscan orogen, upwelling of the asthenosphere and intrusion of granitic melts into the crust. In essence, this period could be interpreted as the final act of the 'Hercynian Cycle'.

The Late Permian marked the development of a large-scale extensional tectonism, probably related to the opening of new oceans (such as Meliata-Maliak and Neotethys). According to many authors (e.g. Hoffmann *et al.* 1989; Cassinis *et al.* 1995; Ziegler & Stampfli 2001; Deroin & Bonin 2003), it is interpreted as the beginning of the 'Alpine Cycle'.

Pangaea configuration

The kinematics of the Pangaea time interval is presently a topic of lively discussion. Irving (1977), Morel & Irving (1981) and Torcq *et al.* (1997) suggested that Gondwana and Laurasia were arranged in a Pangaea 'B' configuration during late Palaeozoic times, and the transformation into the almost universally accepted Wegenerian model of Pangaea, also known as Pangaea 'A', occurred essentially in the Triassic. However, Muttoni *et al.* (1996, 2003) pointed out that this change mainly took place during the Permian and represents the final stages of the Variscan orogeny. Evidence for an Early Permian Pangaea 'B' is supported by Late Permian palaeomagnetic data that allow a Pangaea 'A' configuration, and from a review of the geological literature, which is in support of an intra-Pangaea dextral megashear system of Laurasia versus Gondwana.

This scenario is notably in keeping with the geological data highlighted from a large number of papers by Permian researchers, widely published on the occasion of national and international meetings during the last 50 years. The Middle Permian Episode (MPE) was a determinant in western Europe for the transformation from Pangaea 'B' to Pangaea 'A'. Such a striking event can be generally placed slightly below and during the Middle Permian. It was characterized by specific tectonic, magmatic, thermal and basinal features. According to a remarkable synthetic paper by Deroin & Bonin (2003), the large-scale strike-slip regime evolved from dextral shear in the Late Carboniferous and Early Permian to sinistral in the Mid- and Late Permian. Magmatism changed abruptly at about 275 Ma from high-K calc-alkaline geochemistry to markedly alkaline compositions in southern and western Europe. Metamorphism and high heat flow took place, notably along the shear zones. In the southern Alps, Provence, Languedoc and a number of Spanish regions, one or more unconformities mark the so-called 'Saalian impulses', which affected the late-Variscan interval. This tectonic activity occurred in a large number of regions of Laurasia and northern Gondwana as well, such as the German basins (Schneider *et al.* 1995), the southern Alps (where the unconformity at the base of the Val Gardena Sandstone was identified with the 'Palatine phase' by Kozur (1980) and can be connected with the Altmark movements, interpreted by Hoffmann *et al.* (1989) in the NE German Depression as a post-Saalian tectonic 'phase' that ends the Variscan cycle and gives rise to the Alpine one), the French Massif Central, Provence, Corsica, Lorraine, the Moroccan Meseta, and so on.

This work was partly supported by a grant of the MIUR-COFIN Project 2003 (code 2003048211) on the general research: *Stratigraphy and palaeogeography in the late- and post-Hercynian basins of the Southern Alps, Tuscany and Sardinia, Italy. Comparisons with other areas of the western Mediterranean and geodynamic hypotheses* (responsible G. Cassinis). Earlier version of this paper benefited from the constructive comments of J. Schneider (Freiberg/Saxony), who reviewed the manuscript, and M. Durand (Nancy). The authors wish to thank G. Santi for his technical help.

References

ARCHE, A. & LÓPEZ-GÓMEZ, J. 1996. Origin of the Permian-Triassic Iberian Basin, central-eastern Spain. *Tectonophysics*, **266**, 443–464.

ARCHE, A., LÓPEZ-GÓMEZ, J. & VARGAS, H. 2002. Propuesta de correlación entre los sedimentos Pérmicos y Triasicos de la Cordillera Ibérica y de las Islas Baleares. *Geogaceta*, **32**, 275–278.

ARCHE, A., LÓPEZ-GÓMEZ, J., MARZO M. & VARGAS H. 2004. The siliciclastic Permian–Triassic deposits in Central and Northeastern Iberian Peninsula (Iberian, Ebro and Catalan Basins): a proposal for correlation. *Geologica Acta*, **2**, 305–320.

ARTHAUD, F. & MATTE, PH. 1977. Late Paleozoic strike-slip faulting in southern Europe and northern Africa: results of a right-lateral shear zone between the Apalachians and Urals. *Geological Society of America Bulletin*, **88**, 1305–1320.

ASSERETO, R., BOSELLINI, A., FANTINI SESTINI, N. & SWEET, W.C. 1973. The Permian–Triassic boundary in the Southern Alps (Italy). *In*: LOGAN, A. & HILLS, L. V. (eds) *The Permian and Triassic Systems and Their Mutual Boundary*. Canadian Society of Petroleum Geologists, Memoir, **2**, 176–199.

BAGNOLI, G., GRANELLI, G., PUXEDDU, M., RAU, A., SQUARCI, P & TONGIORGI, M. 1979. A tentative stratigraphic reconstruction of the Tuscan Paleozoic basement. *Memorie della Società Geologica Italiana*, **20**, 99–116.

BARCA, S. & COSTAMAGNA, L. G. 2003. The Upper Carboniferous S. Giorgio succession (Iglesiente, SW Sardinia): stratigraphy, depositional setting and evolution of a late to post-Variscan molassic basin. *In*: DECANDIA, F. A., CASSINIS, G. & SPINA, A. (eds) *Special Proceedings of the Scientific Meeting Late Palaeozoic to Early Mesozoic Events of Mediterranean Europe, and Additional Regional Reports, Siena, 30 April–7 May 2001*. Bollettino della Società Geologica Italiana, Volume speciale, **2**, 89–98.

BARCA, S., CARMIGNANI, L., ELTRUDIS, A. & FRANCESCHELLI, M. 1995. Origin and evolution of Permo-Carboniferous basin of Mulargia Lake (South-Central Sardinia, Italy) related to the Late-Hercynian extensional tectonics. *Comptes Rendus de l'Académie des Sciences, Paris, Série II*, **321**, 171–178.

BARGOSSI, G. M., KLÖTZLI, U. S., MAIR, V., MAROCCHI, M. & MORELLI, C. 2004. The Lower Permian Athesian volcanic group (AVG) in the Adige Valley between Merano and Bolzano: a stratigraphic, petrographic and geochronological outline. *32nd International Geological Congress, Florence, 2004*. Scientific Sessions. Abstracts, **1**, 187.

BECQ-GIRAUDON, J. F. 1993. Problèmes de la bio-estratigraphie dans le Paléozoïque supérieur continental (Stéphanien-Autunien) du Massif Central. *Geodinamica Acta*, **6**, 219–224.

BÉTHOUX, O., NEL, A., GAND, G., LAPEYRIE, J & GALTIER, J. 2002. Discovery of the genus *Iasvia* Zalessky, 1934 in the Upper Permian of France (Lodève Basin) (Orthoptera, Ensifera, Oedischiidae). *Geobios*, **35**, 293–302.

BLÈS, J. L., BONIJOLY, D., CASTAING, C. & GROS, Y. 1989. Successive post-Variscan stress fields in the French Massif Central and its borders (Western European plate): comparison with geodynamic data. *Tectonophysics*, **169**, 79–111.

BODECHTEL, J. 1964. Stratigraphie und Tektonik der Schuppenzone Elbas. *Geologische Rundschau*, **53**, 25–41.

BONIN, B. 1988. From orogenic to anorogenic environments: evidence from associated magmatic episodes. *Schweizerische Mineralogische und Petrographische Mitteilungen*, **68**, 301–311.

BOSELLINI, A. 1964. Stratigrafia, petrografia e sedimentologia delle facies carbonatiche al limite Permiano-Trias nelle Dolomiti Occidentali. *Memorie del Museo di Storia Naturale della Venezia Tridentina*, **15**, 59–110.

BROGLIO LORIGA, C., NERI, C., PASINI, M. & POSENATO, R. 1988. Marine fossil assemblages from Upper Permian to lowermost Triassic in the western Dolomites. *In*: CASSINIS, G. (ed.) *Proceedings of the Field Conference on 'Permian and Permian-Triassic Boundary in the South-Alpine Segment of the Western Tethys, and Additional Regional Reports': SGI. and IGCP. Project 203, Brescia, 4–12 July 1986*. Memorie della Società Geologica Italiana, **34** (1986), 5–44.

BROUTIN, J. 1977. Nouvelles données sur la flore des bassins autuno-stéphaniens des environs de Guadalcanal (Province de Seville, Espagne). *Cuadernos de Geología Ibérica*, **4**, 91–98.

BROUTIN, J. & DURAND, M. 1995. First paleobotanical and palynological data on the 'Les Salettes Formation' uppermost member (Permian Toulon Basin, Southeastern France). *XIIIth International Congress on the Carboniferous-Permian, Kraków, 1995, Abstracts*, 15–16.

BROUTIN, J., DOUBINGER, J., GISBERT, J. & SATTA-PASINI, S. 1988. Premières datations palynologiques dans le faciès Buntsandstein des Pyrenees catalanes espagnoles. *Comptes Rendus de l'Académie des Sciences, Paris, Série II*, **306**, 159–163.

BROUTIN, J., DOUBINGER, J. *et al.* (1990). Le renouvellement des flores au passage Carbonifère-Permien: approches stratigraphique, biologique, sédimentologique. *Comptes Rendus de l'Académie des Sciences, Paris, Série II*, **311**, 1563–1569.

BROUTIN, J., CHÂTEAUNEUF, J. J. & MATHIS, V. 1992*a*. Permian basins in the French Massif Central, Chapter I: the Lodève basin. *Cahiers de Micropaléontologie, Nouvelle Série*, **7**, 107–122.

BROUTIN, J., FERRER, J., GISBERT, J. & NMILA, A. 1992*b*. Première découverte d'une microflore thuringienne dans le faciès saxonien de l'île de Minorque (Baléares, Espagne). *Comptes Rendus de l'Académie des Sciences, Paris, Série II*, **315**, 117–122.

BROUTIN, J., CABANIS, B., CHÂTEAUNEUF, J. J. & DEROIN, J. P. 1994. Evolution biostratigraphique, magmatique et tectonique du domain paléotéthysien occidental (SW de l'Europe): implications paléogéographiques au Permien inférieur. *Bulletin de la Société Géologique de France*, **165**, 63–179.

BROUTIN, J., CASSINIS, G., CORTESOGNO, L., GAGGERO, L., RONCHI, A. & SARRIA, E. 1996. Research in progress on the Permian deposits of Sardinia (Italy). *Permophiles*, **28**, 45–48.

BROUTIN, J., CHÂTEAUNEUF, J. J., GALTIER, J. & RONCHI, A. 1999. L'Autunien d'Autun reste-t-il une référence pour les dépôts continentaux du Permien inférieur d'Europe? Apport des données paléobotaniques. *Géologie de la France*, **2**, 17–31.

CASSINIS, G. 1996. Upper Carboniferous to Permian stratigraphic framework of southwestern Europe and its implications. An overview. *In*: MOULLADE, N. & NAIRN A. E. M. (eds) *'The Phanerozoic Geology of the World. I: The Palaeozoic, B'*. Elsevier, Amsterdam, 109–180.

CASSINIS, G. & DOUBINGER, J. 1991. On the geological time of the typical Collio and Tregioro continental beds in the southalpine Permian (Italy), and some additional observations. *Atti Ticinensi di Scienze della Terra (Pavia)*, **34**, 1–20.

CASSINIS, G. & RONCHI, A. 2001. Permian chronostratigraphy of the Southern Alps (Italy): an update. *In*: WEISS, R. H. (ed.) *Contribution to Geology and Palaeontology of Gondwana in Honour of Helmut Wopfner*. Geological Institute, University of Cologne, 73–88.

CASSINIS, G. & RONCHI, A. 2002. The (late-) post-Variscan continental succession of Sardinia. *Rendiconti della Società Paleontologica Italiana*, **1**, 77–92.

CASSINIS, G., TOUTIN-MORIN, N. & VIRGILI, C. 1992. Permian and Triassic events in the continental domains of Mediterranean Europe. *In*: SWEET, W. C., YANG, Z. Y., DICKINS, J. M. & YIN, H. F. (eds) *Permo-Triassic Events in the Eastern Tethys*. Cambridge University Press, Cambridge, 60–77.

CASSINIS, G., TOUTIN-MORIN, N. & VIRGILI, C. 1995. A general outline of the Permian continental basins in southwestern Europe. *In*: SCHOLLE, P. A., PERYT, T. M. & ULMER-SCHOLLE, D. S. (eds) *The Permian of Northern Pangea. Vol. 2*. Springer Verlag, Berlin, 137–157.

CASSINIS, G., CORTESOGNO, L., GAGGERO, L., RONCHI, A. & VALLONI, R. 1996. Stratigraphic and petrographic investigations into the Permo-Triassic continental sequences of Nurra (NW Sardinia). *Cuadernos de Geología Ibérica, Special Issue*, **21**, 149–169.

CASSINIS, G., CORTESOGNO, L., GAGGERO, L. & RONCHI, A. 1998. Osservazioni preliminari su alcune successioni continentali permiane della Sardegna. *Rendiconti dell' Istituto Lombardo, Accademia di Scienze & Lettere, B*, **130** (1996), 177–205.

CASSINIS, G., CORTESOGNO, L., GAGGERO, L., MASSARI, F., NERI, C. NICOSIA, U. & PITTAU, P. (eds) 1999*a*. *Stratigraphy and Facies of the Permian deposits between Eastern Lombardy and the Western Dolomites. Field Trip Guidebook, 23–25 September 1999*. International Field Conference on 'The continental Permian of the Southern Alps and Sardinia (Italy). Regional reports and general correlations', Brescia, 15–25 September 1999. Earth Science Department, Pavia University.

CASSINIS, G., CORTESOGNO, L., GAGGERO, L., PITTAU, P., RONCHI, A. & SARRIA, E. (eds) 1999*b*. *Late Palaeozoic Continental Basins of Sardinia. Field Trip Guidebook, 15–18 September 1999*. International Field Conference on 'The Continental Permian of the Southern Alps and Sardinia (Italy). Regional reports and general correlations'. Brescia, 15–25 September 1999. Earth Science Department, Pavia University.

CASSINIS, G., DI STEFANO, P., MASSARI, F., NERI, C. & VENTURINI, C. 2000*a*. Permian of south Europe and its interregional correlation. *In*: YIN, H. F., DICKINS, J. M., SHI, G. R. & TONG, J. (eds) *Permian-Triassic Evolution of Tethys and Western Circum-Pacific*. Elsevier Science, Amsterdam, 37–70.

CASSINIS, G., NICOSIA, U., RONCHI, A. & SANTI, G. 2000*b*. Impronte di tetrapodi nel Permiano orobico e loro implicazioni stratigrafiche. Nota preliminare. *Rendiconti dell'Istituto Lombardo, Accademia di Scienze & Lettere, B*, **132** (1998), 197–217.

CASSINIS, G., NICOSIA, U., PITTAU, P. & RONCHI, A. 2002. Palaeontological and radiometric data from the Permian continental deposits of the central-eastern Southern Alps (Italy), and their stratigraphic implications. *Mémoire de l'Association Géologues du Permien*, **2**, 53–74.

CASSINIS, G., CORTESOGNO, L., GAGGERO, L., RONCHI, A., SARRIA, E., SERRI, R. & CALZIA, P. 2003*a*. Reconstruction of igneous, tectonic and sedimentary events in the latest Carboniferous–Early Permian Seui Basin (Sardinia, Italy), and evolutionary model. *In*: DECANDIA, F. A., CASSINIS, G. & SPINA, A. (eds) *Special Proceedings of the Scientific Meeting 'Late Palaeozoic to Early Mesozoic Events* *of Mediterranean Europe, and Additional Regional Reports', Siena, 30 April–7 May 2001*. Bollettino della Società Geologica Italiana, Volume speciale, **2**, 99–117.

CASSINIS, G., DURAND, M. & RONCHI, A. 2003*b*. Permian-Triassic continental sequences of Northwest Sardinia and South Provence: stratigraphic correlations and palaeogeographical implications. *In*: DECANDIA, F. A., CASSINIS, G. & SPINA, A. (eds) *Special Proceedings of the Scientific Meeting 'Late Palaeozoic to Early Mesozoic events of Mediterranean Europe, and Additional Regional Reports', Siena, 30 April–7 May 2001*. Bollettino della Società Geologica Italiana, Volume speciale, **2**, 119–129.

CEOLONI, P., CONTI, M. A., MARIOTTI, N. & NICOSIA, U. 1988. New Late Permian tetrapod footprints from Southern Alps. *Memorie della Società Geologica Italiana*, **34** (1986), 45–65.

CEOLONI, P., CONTI, M. A., MARIOTTI, N., MIETTO, P. & NICOSIA, U. 1987. Tetrapod footprints from Collio Formation (Lombardy, Northern Italy). *Memorie di Scienze Geologiche*, **39**, 213–233.

CHÂTEAUNEUF, J. J. & FARJANEL, G 1989. Synthèse Géologique des Bassins Permiens Français. *Mémoires du Bureau de Recherches Géologiques et Minières*, **128**.

CIRILLI, S., DECANDIA, F. A., LAZZAROTTO, A., PANDELI, E., RETTORI, R., SANDRELLI, F. & SPINA, A. 2002. Stratigraphy and depositional environment of the Mt. Argentario sandstone Formation (southern Tuscany, Italy). *In*: BARCHI, M. R., CIRILLI, S. & MINELLI, G. (eds) *Proceedings of the Scientific Meeting 'Geological and Geodynamic evolution of the Apennines', Foligno (Pg), 16–18 February 2000, in Memory of G. Pialli*. Bollettino della Società Geologica Italiana, Volume speciale, **1**, Parte II, 489–498.

COCOZZA, T. 1967. Il Permo-Carbonifero del bacino di San Giorgio (Iglesiente, Sardegna sud occidentale). *Memorie della Società Geologica Italiana*, **6**, 607–642.

COCOZZA, T., LAZZAROTTO, A. & PASINI M. 1975. Segnalazione di una fauna triassica nel conglomerato di Monte Quoio (Verrucano del Torrente Farma – Toscana meridionale). *Rivista Italiana di Paleontologia e Stratigrafia*, **81**, 425–436

COCOZZA, T., DECANDIA, F. A., LAZZAROTTO, A., PASINI, M. & VAI, G. B. 1987. The marine Carboniferous sequence in southern Tuscany: its bearing for Hercynian paleogeography and tectofacies. *In*: FLUEGEL, H. W., SASSI, F.P. & GRECULA, P. (eds) *Pre-Variscan and Variscan events in the Alpine Mediterranean mountain belts*. Mineralia Slovaca Monography, 135–144.

CONTI, M. A., LEONARDI, G., MARIOTTI, N. & NICOSIA, U. 1977. Tetrapod footprints of the 'Val Gardena Sandstone' (north Italy): their palaeontological, stratigraphic and palaeoenvironmental meaning. *Paleontographia Italica, Nuova Serie*, 40, **70**, 1–91.

CONTI, M. A., FONTANA D. *et al.* 1986. The Bletterbach-Butterloch section (Val Gardena Sandstone and Bellerophon Formation). *In*: Italian IGCP-203 Group (ed.) *Permian and Permian-Triassic Boundary in the South-Alpine Segment of the Western Tethys, Field Guidebook*. Field Conference, Brescia, July 1986. Tipolitografia Commerciale, Pavia, 99–119.

CONTI, M. A., MARIOTTI, N., MIETTO, P. & NICOSIA, U. 1991. Nuove ricerche sugli icnofossili della Formazione di Collio in Val Trompia (Brescia). *Natura Bresciana*, Annali del Museo Civico di Storia Natural Brescia, **26**, 109–119.

CONTI, M. A., MARIOTTI, N., NICOSIA, U & PITTAU, P. 1997. Succession of selected bioevents in the continental Permian of the Southern Alps (Italy): improvements of intrabasinal and interregional correlations. *In*: DICKINS, J. M., YANG, Z. Y., YIN, H. F., LUCAS, S. G. & ACHARYYA, S. K. (eds) *late Palaeozoic and Early Mesozoic Circum-Pacific Events and Their Global Correlation'*, World and Regional Geology Series, Cambridge University Press, **10**, 51–65.

CONTI, M. A., MARIOTTI, N., MANNI, R. & NICOSIA, U. 1999. Tetrapod footprints in the Southern Alps: an overview. *In*: CASSINIS, G., COSTESOGNO, L., GAGGERO, L., MASSARI, F., NERI, C., NICOSIA, U. & PITTAU, P. (eds) *Stratigraphy and Facies of the Permian deposits between Eastern Lombardy and the Western Dolomites. Field Trip Guidebook, 23–25 September 1999*. International Field Conference on 'The Continental Permian of the Southern Alps and Sardinia (Italy). Regional reports and general correlations', Brescia, 15–25 September 1999. Earth Science Department, Pavia University, Appendix, 137–138.

CORTESOGNO, L., CASSINIS, G. *et al.* 1998. The post-Variscan volcanism in Late Carboniferous-Permian sequences of Ligurian Alps, Southern Alps and Sardinia. *Lithos*, **45**, 305–328.

COSTAMAGNA, L. G., BARCA, S., DEL RIO, M. & PITTAU, P. 2000. Stratigrafia, paleogeografia ed analisi di facies deposizionale del Trias del Sarcidano-Gerrei (Sardegna SE). *Bollettino della Società Geologica Italiana*, **119**, 473–496.

COSTANTINI, A., DECANDIA, F. A., LAZZAROTTO, A. & SANDRELLI, F. 1988. L'unità di Monticiano-Roccastrada fra la Montagnola Senese e il Monte Leoni (Toscana meridionale). *Atti Ticinensi di Scienze della Terra*, **31** (1987–88), 382–420.

COSTANTINI, A., ELTER, F. M., PANDELI, E. & SANDRELLI, F. 1991. Geological framework of the Iano metamorphic sequence (Southern Tuscany, Italy). *In: Geologia del Basamento Italiano*, Accademia Fisiocritici, Siena, Abstract volume, 143–144.

COSTANTINI, A., ELTER, F. M., PANDELI, E. & SANDRELLI, F 1998. Geologia dell'area di Iano (Toscana meridionale, Italia). *Bollettino della Società Geologica Italiana*, **117**, 187–218.

DACHROTH, W. 1988. Gesteinsmagnetischer Vergleich permischer Schichtenfolgen in Mitteleuropa. *Zeitschrift für Geologische Wissenschaften*, **16**, 959–968.

DEL RIO, M. 1973. Palinologia di un livello Permo-Carbonifero del bacino di San Giorgio (Iglesiente, Sardegna sud-occidentale). *Bollettino della Società Geologica Italiana*, **93**, 113–124.

DEL RIO, M. & PITTAU, P. 1999. The Upper Carboniferous of the San Giorgio Basin. Excursion 1: The Upper Carboniferous to Permian deposits in southwestern Sardinia (Iglesiente and Sulcis).

In: CASSINIS, G., CORTESOGNO, L., GAGGERO, L., PITTAU, P., RONCHI, A. & SARRIA, E. (eds) *Late Palaeozoic Continental Basins of Sardinia. Field Trip Guidebook, 15–18 September. 1999*. International Field Conference on 'The Continental Permian of the Southern Alps and Sardinia (Italy): Regional reports and general correlations', Brescia, 15–25 September 1999. Earth Science Department, Pavia University, 37–43.

DEROIN, J. P. & BONIN, B. 2003. Late Variscan tectonomagmatic activity in western Europe and surrounding areas: the Mid-Permian Episode. *In*: DECANDIA, F. A., CASSINIS, G. & SPINA, A. (eds) *Special Proceedings of the Scientific Meeting 'Late Palaeozoic to Early Mesozoic events of Mediterranean Europe, and Additional Regional Reports', Siena, 30 April–7 May 2001*. Bollettino della Società Geologica Italiana, Volume speciale, **2**, 169–184.

DEROIN J. P., BONIN, B. *et al.* 2001. The Permian of Southern France: an overview. *In*: CASSINIS, G. (ed.) *Permian Continental Deposits of Europe and Other Areas. Regional Reports and Correlations.* Natura Bresciana, Monografia, **25**, 189–202.

DÍEZ FERRER, J. B. 2000. *Geología y Paleobotánica de la facies Buntsandstein en la Rama Aragonesa de la Cordillera Ibérica. Implicaciones bioestratigráficas en el Peritheris Occidental.* PhD thesis, University Paris/ Zaragoza University.

DÍEZ, J. B., GRAUVOGEL-STAMM, L., BROUTIN, J., FERRER, J., GISBERT, J. & LIÑÁN, E. 1996. Première découverte d'une paléoflore anisienne dans le faciès Buntsandstein de la Branche Aragonaise de la Cordillère Ibérique (Espagne). *Comptes Rendus de l'Académie des Sciences, Paris*, Série II, **323**, 341–347.

DOUBINGER, J., ODIN, B. & CONRAD, G. 1987. Les associations sporopolliniques du Permien continental du bassin de Lodève (Hérault, france): caractérisation de l'Autunien supérieur, du 'Saxonien' et du Thuringien. *Annales de la Société Géologique du Nord*, **106**, 103–109.

DOUBINGER, J., LÓPEZ-GÓMEZ J. & ARCHE, A. 1990. Pollen and spores from the Permian and Triassic sediments of the southeastern Iberian Ranges, Cueva de Hierro (Cuenca) to Chelva-Manzanera (Valencia-Teruel) in Spain. *Review of Palaeobotany and Palynology*, **66**, 25–45.

DURAND, M. (with a contribution by GAND, G.) 2001. *The Continental Permian-Triassic Series of Provence (Southeast France). Field Trip Guidebook, 5–6 May, 2001*. International Meeting on 'The Stratigraphic and Structural Evolution on the Late Carboniferous to Triassic Continental and Marine Successions in Tuscany (Italy). Regional reports and general correlation', Siena, 30 April–7 May 2001.

DURAND, M. 2006. The problem of the transition from the Permian to the Triassic Series in southeastern France: companison with other Peritethyan negions. *In*: LUCAS, S. G., CASSINIS, G. & SCHNEIDER, J. W. (eds) *Non-Marine Permian Biostratigraphy and Biochronology.* Geological Society, London, Special Publications, **265**, 281–296.

DURAND, M., AVRIL, G. & MEYER, R. 1988. Paléogéographie des premiers dépôts triasiques dans les Alpes externes méridionales: Importance de la Dorsale delphino-durancienne. *Comptes Rendus de l'Académie des Sciences, Paris, Série II*, **306**, 557–560.

DURAND, M., GAND, G. & CHÂTEAUNEUF, J. J. 2002. *Les bassins permiens de Provence*. Livret-guide de la 16e excursion annuelle de l'Association Géologues du Permien 9–11 Mai.

ELTER, F. & PANDELI, E. 1990. Alpine and Hercynian orogenic phases in the basement rocks of the Northern Apennines (Larderello geothermal field, southern Tuscany, Italy). *Eclogae Geologicae Helvetiae*, **83**, 241–264.

ENGELBRECHT, H. 1984. *Bericht über die Kartierung und tektonische sowie sediment-petrographische Untersuchung eines Gebiets südlich Monticiano (Toscana)*. Institut Allgemeine und Angewandte Geologie Universität, München.

ENGELBRECHT, H., KLEMM, D. D. & PASINI, M. 1989. Preliminary notes on the tectonics and lithotypes of the 'Verrucano *s.l.*' in the Monticiano area (Southern Tuscany, Italy) and the findings of fusulinids within the M.te Quoio Formation (Verrucano Group). *Rivista Italiana di Paleontologia e Stratigrafia*, **94**, 361–382.

FARABEGOLI, E. & PERRI, M. C. 1998. Permian/Triassic boundary and Early Triassic of the Bulla section (Southern Alps, Italy): lithostratigraphy, facies and conodont biostratigraphy. *Giornale di Geologia, Bologna, Serie 3, Special Issue*, **60**, 292–311.

FONTANA, D., GELMINI, R. & LOMBARDI, G. 1982. Le successioni sedimentarie e vulcaniche carbonifere e permo-triassiche della Sardegna. *In*: CARMIGNANI, L., COCOZZA, T., GHEZZO, C., PERTUSATI, P. C. & RICCI, C. A. (eds) *Guida alla Geologia del Paleozoico sardo*. Guide Geologiche Regionali, Società Geologica Italiana, 183–292.

FONTANA, D., NERI, C., RONCHI, A. & STEFANI, C. 2001. Stratigraphic architecture and composition of the Permian and Triassic siliciclastic succession of Nurra (Northwestern Sardinia). *In*: CASSINIS, G. (ed.) *Permian Continental Deposits of Europe and Other Areas. Regional reports and Correlations.* Natura Bresciana, Monografia, **25**, 149–161.

FRANCAVILLA, F., CASSINIS, G. *et al.* 1977. Macroflora e datazione di alcuni affioramenti (tardo) postercinici presso il Lago di Mulargia (Sardegna sud-orientale). *Consiglio Nazionale della Ricerche, Bollettino del Gruppo di Lavoro sul Paleozoico*, **2**, 31–33.

GAND, G. 1987. *Les traces de vertébrés tétrapodes du Permien français. Paléontologie, stratigraphie, paléoenvironnements.* PhD thesis, University of Bourgogne.

GAND, G., LAPEYRIE, J., GARRIC, J., NEL, A., SCHNEIDER, J. & WALTER, H. 1997. Découverte d'Arthropodes et de Bivalves inédits dans le Permien continental (Lodévois, France). *Comptes Rendus de l'Académie des Sciences, Paris, Série I*, **325**, 891–898.

GAND, G., GARRIC, J., DEMATHIEU, G. & ELLENBERGER, P. 2000. La palichnofaune de vertébrés tétrapodes du Permien Supérieur du bassin de Lodève (Languedoc, France). *Palaeovertebrata*, **29**, 1–82.

GAND, G., GALTIER, J., GARRIC, J., SCHNEIDER, J., KÖRNER, F. & BÉTHOUX, O. 2001. *The Graissessac Carboniferous and Lodève Permian Basins (Languedoc-France). Field Trip Guidebook, 7 May, 2001.* International Meeting on 'The Stratigraphic and Structural Evolution on the Late Carboniferous to Triassic Continental and Marine Successions in Tuscany (Italy). Regional Reports and General Correlation', Siena, 30 April–7 May 2001.

GASPERI, G. & GELMINI, R. 1980. Ricerche sul Verrucano. 4. Il Verrucano della Nurra (Sardegna nord-occidentale). *Memorie della Società Geologica Italiana*, **20** (1979), 215–231.

HAUBOLD, H. & KATZUNG, G. 1975. Die Position der Autun/Saxon-Grenze (Unteres Perm) in Europa und Nordamerika. *Schriftenreihe Geologische Wissenshaften, Berlin*, **3**, 87–138.

HERNANDO, S., SCHOTT, J. J., THUIGAT, R. & MONTIGNY, R. 1980. Âge des andesites et des sediments interstratifiés de la région d'Atienza (Espagne): Ètude stratigraphique et paléomagnetique. *Bulletin de Sciences Géologiques*, **33**, 119–128.

HOFFMANN, N., KAMPS, H.-J. & SCHNEIDER, J. 1989. Neuerkenntnisse zur Biostratigraphie und Paläodynamik des Perms in der Nordostdeutschen Senke – ein Diskussionbeitrag. *Zeitschrift angewandte Geologie*, **35**, 198–207.

IRVING, E. 1977. Drift of the major continental blocks since the Devonian. *Nature*, **270**, 304–309.

ITALIAN IGCP-203 GROUP (ed.) 1986. *Permian and Permian-Triassic Boundary in the South-Alpine Segment of the western Tethys. Field Guidebook.* Field Conference, Brescia, July 1986. Tipolitografia Commerciale, Pavia.

KERP, H., PENATI, F., BRAMBILLA, G., CLEMENT-WESTERHOF, J. A. & VAN BERGEN, P. F. 1996. Aspects of Permian palaeobotany and palynology. XVI. Three-dimensionally preserved stromatolite-incrusted conifers from the Permian of the western Orobic Alps (northern Italy). *Review of Palaeobotany and Palynology*, **91**, 63–84.

KÖRNER, F., SCHNEIDER, J.W., HOERNES, S., GAND, G. & KLEEBERG, R. 2003. Climate and continental sedimentation in the Permian of Lodève Basin (Southern France). *In*: DECANDIA, F.A., CASSINIS, G. & SPINA, A. (eds) *Special Proceedings of the Scientific Meeting 'Late Palaeozoic to Early Mesozoic Events of Mediterranean Europe, and Additional Regional Reports', Siena, 30 April–7 May 2001.* Bollettino della Società Geologica Italiana, **2**, 185–191.

KOZUR, H. 1980. The significance and stratigraphic position of the 'Saalic' and Palatine phases. *In*: VOZÁR, J. & VOZÁROVÁ, A. (eds) *Permian of the West Carpathians*. Geologický ústav Dionýza Štúra, Bratislava, 65–72.

LAGO, M., ARRANZ, E., POCOVÍ, A., GALÉ, C. & GIL, A. 2004*a*. Permian magmatism and basin dynamics in the southern Pyrenees: a record of the transition from late Variscan transtension to early Alpine extension. *In*: WILSON, W., WEUMANN, E-R., DAVIES, G. R., TIMMEVMAN, J., HEEREMAUS, M. & LARSEN, B. T. (eds) *Permo-Carboniferous Magmatism and Rifting in Europe*. Geological Society, London, Special Publications, **223**, 430–464.

LAGO, M., ARRANZ, E., POCOVÍ, A., GALÉ, C. & GIL, A. 2004*b*. Lower Permian magmatism of the Iberian Chain, Central Spain, and its relationships to extensional tectonics. *In*: WILSON, W., WEUMANN, E-R., DAVIES, G. R., TIMMEVMAN, J., HEEREMAUS, M. & LARSEN, B. T. (eds) *'Permo-Carboniferous Magmatism and Rifting in Europe'*. Geological Society, London, Special Publications, **223**, 465–490.

LAURO, G., NEGRETTI, G. C. & SBARRACANI, M. 1963. Contributo alla conoscenza delle formazioni permiane di Seui (Barbagia, Sardegna). *Giornale di Geologia*, **31**, 281–309.

LAZZAROTTO, A., ALDINUCCI, M. *et al.* 2003. Stratigraphic correlation of the Upper Palaeozoic-Triassic successions in southern Tuscany, Italy. *In*: DECANDIA, F. A., CASSINIS, G. & SPINA, A. (eds) *Special Proceedings of the Scientific Meeting 'Late Palaeozoic to Early Mesozoic events of Mediterranean Europe, and Additional Regional Reports', Siena, 30 April–7 May 2001.* Bollettino Società Geologica Italiana, Volume speciale, **2**, 25–35.

LOMBARDI, G., COZZUPOLI, D. & NICOLETTI, M. 1974. Notizie geopetrografiche e dati sulla cronologia K-Ar del vulcanesimo tardo-paleozoico sardo. *Periodico di Mineralogia*, **43**, 221–312.

LÓPEZ-GÓMEZ, J. & ARCHE, A. 1993. Sequence stratigraphy, analysis and paleogeographic interpretation of the Buntsandstein and Muschelkalk facies (Permo-Triassic) in the SE Iberian Ranges, eastern Spain. *Palaeogeography Palaeoclimatology, Palaeoecology*, **103**, 347–361

LÓPEZ-GÓMEZ, J., ARCHE, A. & PÉREZ-LÓPEZ, A. 2002. Permian and Triassic. *In*: GIBBONS, W. & MORENO, T. (eds) *Geology of Spain*, Geological Society, London, 185–212.

MASSARI, F. & NERI, C. 1997. The infill of a supradetachment (?) basin: the continental to shallow-marine Upper Permian succession in the Dolomites and Carnia (Italy). *Sedimentary Geology*, **110**, 181–221.

MASSARI, F., CONTI, M. A. *et al.* 1988. The Val Gardena Sandstone and Bellerophon Formation in the Bletterbach gorge (Alto Adige, Italy): biostratigraphy and sedimentology. *Memorie di Scienze Geologiche*, **40**, 229–273.

MASSARI, F., NERI, C., PITTAU, P., FONTANA, D. & STEFANI, C. 1994. Sedimentology, palynostratigraphy and sequence stratigraphy of a continental to shallow-marine rift-related succession: Upper Permian of the eastern Southern Alps (Italy). *Memorie di Scienze Geologiche*, **46**, 119–243

MAURITSCH, H. J. & BECKE, M. 1983. A magnetostratigraphic profile in the Permian (Gröden beds, Val Gardena Formation) of the Southern Alps near Paularo (Carnic Alps, Friuli, Italy). *International Geological Correlation Programme, Project 5, Newsletter*, **5**, 80–86.

MENNING, M. 1995. A numerical time scale for the Permian and Triassic Periods. An integrated time analysis. *In*: SCHOLLE, P., PERYT, T. M. & ULMER-SCHOLLE, S. (eds) *Permian of the Northern Continents. Vol. 1*. Springer Verlag, Berlin, 77–97.

MENNING, M. 2001. A Permian time scale 2000 and correlation of marine and continental sequences using the Illawarra Reversal (265 Ma). *In*: CASSINIS, G. (ed.) *Permian Continental Deposits of Europe and Other Areas. Regional Reports and Correlations.* Natura Bresciana, Monografia, **25**, 355–362.

MÉNOT, R. P. & ORSINI, J. B. 1990. Evolution du socle antéstéphanien de Corse: événements magmatiques et métamorphiques. *Schweizerische Mineralogische und Petrographische Mitteilungen*, **70**, 35–53.

MOREL, P. & IRVING, E. 1981. Paleomagnetism and the evolution of Pangaea. *Journal of Geophysical Research*, **86**, 1858–1987.

MUTTONI, G., KENT, D. V. & CHANNEL, J. E. T. 1996. The evolution of Pangaea: paleomagnetic constraints from the Southern Alps, Italy. *Earth and Planetary Science Letters*, **140**, 97–112.

MUTTONI, G., KENT, D. V., GARZANTI, E., BRACK, P., ABRAHAMSEN, N. & GAETANI, M. 2003. Early Permian Pangaea 'B' to Late Permian Pangaea 'A'. *Earth and Planetary Science Letters*, **215**, 379–394.

NEL, A., GAND, G., FLECK, G., BÉTHOUX, O., LAPEYRIE, J. & GARRIC, J. 1999*a*. *Saxonagrion minutus* nov. gen. nov. sp., the oldest damselfly from the Upper Permian of France (Odonatoptera, Panodonata, Saxonagrionidae fam. nov. I.) *Géobios*, **32**, 883–888.

NEL, A., GAND, G., GARRIC, J. & LAPEYRIE, J. 1999*b*. The first recorded Protozygopteran insects from the Upper Permian of France. *Palaeontology*, **42**, 83–97.

NICORA, A. & PERRI, M. C. 1999. The P/T boundary in the Tesero section, western Dolomites (Trento). 3.3. Bio- and chronostratigraphy. Conodonts. *In*: CASSINIS, G., CORTESOGNO, L. GAGGERO, L., MASSARI, F., NERI, C., NICOSIA, U. & PITTAU, P. (eds) *Stratigraphy and Facies of the Permian Deposits between Eastern Lombardy and the Western Dolomites, Field Trip Guidebook, 23–25 September 1999.* International Field Conference on 'The Continental Permian of the Southern Alps and Sardinia. Regional reports and general correlations', Brescia, 15–25 September 1999. Earth Science Department, Pavia University, 97–100.

ODIN, B. 1986. Les formations permiennes, Autunien supérieur à Thuringien, du 'bassin' de Lodève (Hérault, France): stratigraphie, minéralogie, paléoenvironnement, corrélations. PhD thesis, University of Marseille.

OLIVET J. L. 1996. La cinématique de la microplaque Ibérique. Bulletin des Centres de Recherches Exploration-Production Elf-Aquitaine, **20**, 131–195.

ORCHARD, M. 2001. Executives Notes. From the Chair. *Albertiana*, **25**, 3.

PASINI, M. 1991. Residual evidences of Permian carbonatic platform within the Apennine sequences (Italy). *Bollettino della Società Geologica Italiana*, **110**, 843–848.

PECORINI, G. 1974. Nuove osservazioni sul Permo-Trias di Escalaplano (Sardegna sud-orientale). *Bollettino della Società Geologica Italiana*, **93**, 991–999.

PÉREZ-ARLUCEA, M. & SOPEÑA, A. 1985. Estratigrafía del Pérmico y Triásico en el sector central de la Rama Castellana de la Cordillera Ibérica (provincias de Guadalajara y Teruel). *Estudios Geológicos*, **41**, 207–222.

PERRI, M. C. & FARABEGOLI, E. 2003. Conodonts across the Permian-Triassic boundary in the Southern Alps. *In*: MAWSON, R. & TALENT, J. A (eds) *Contributions to the Second Australian Conodont Symposium (Auscos II)*, Courer Forschung Institut Senckenberg, **245**, 281–313.

PITTAU, P. 1999*a*. The Tregiovo area and related volcanics in the Tregiovo section. 1.3. Palynology. *In*: CASSINIS, G., CORTESOGNO, L., GAGGERO, L., MASSARI, F., NERI, C., NICOSIA, U. & PITTAU, P. (eds) *Stratigraphy and Facies of the Permian Deposits between Eastern Lombardy and the Western Dolomites. Field Trip Guidebook, 23–25 September 1999.* International Field Conference on 'The Continental Permian of the Southern Alps and Sardinia (Italy). Regional reports and general correlations', Brescia, 15–25 September 1999, Brescia. Earth Science Department, Pavia University, 83–89.

PITTAU, P. 1999*b*. The Bletterbach section (Val Gardena Sandstone and Bellerophon Formation). 3. Palynology. *In*: CASSINIS, G., CORTESOGNO, L., GAGGERO, L., MASSARI, F., NERI, C., NICOSIA, U. & PITTAU, P. (eds) *Stratigraphy and Facies of the Permian Deposits between Eastern Lombardy and the Western Dolomites, Field Trip Guidebook, 23 25 September 1999.* International Field Conference on 'The continental Permian of the Southern Alps and Sardinia (Italy). Regional reports and general correlations', Brescia, 15–25 September 1999, Earth Science Department, Pavia University, 112–125.

PITTAU, P. 1999*c*. The Triassic succession of Northwest Sardinia: data from the subsurface. *In*: CASSINIS, G., CORTESOGNO, L., GAGGERO, L., PITTAU, P., RONCHI, A. & SARRIA, E. (eds) *Late Palaeozoic Continental Basins of Sardinia, Field Trip Guidebook, 15–18 September. 1999.* International Field Conference on 'The Continental Permian of the Southern Alps and Sardinia (Italy). Regional reports and general correlations'. Brescia, 15–25 September. 1999. Earth Science Department, Pavia University, 106–107.

PITTAU, P. 2001. Correlation of the Upper Permian sporomorph complexes of the Southern Italian Alps with the Tatarian complexes of the Stratotype region. *In*: CASSINIS, G. (ed.) *Permian Continental Deposits of Europe and Other Areas. Regional Reports and Correlations.* Natura Bresciana, Monografia, **25**, 109–116.

PITTAU, P., DEL RIO, M., PUTZU, M. T. & BARCA, S. 1999*a*. The Permian of the Guardia Pisano Basin (Sulcis). Excursion 1: The Upper Carboniferous to Permian deposits in Southwestern Sardinia (Iglesiente and Sulcis). *In*: CASSINIS, G., CORTESOGNO, L., GAGGERO, L., PITTAU, P., RONCHI, A. & SARRIA, E. (eds) *Late Palaeozoic Continental Basins of Sardinia, Field Trip Guidebook, 15–18 September 1999.* International Field Conference on 'The Continental Permian of the Southern Alps and Sardinia (Italy): regional reports and general correlations', Brescia, 15–25 September

1999, Earth Science Department, Pavia University, 44–57.

PITTAU, P., DEL RIO, M. & RONCHI, A. 1999*b*. The Middle Triassic of Escalaplano (central Sardinia). Excursion 2: Early Permian basins in Southeastern Sardinia (Gerrei and Ogliastra). *In*: CASSINIS, G., CORTESOGNO, L., GAGGERO, L., PITTAU, P., RONCHI, A. & SARRIA, E. (eds) *Late Palaeozoic Continental Basins of Sardinia, Field Trip Guidebook, 15–18 September 1999.* International Field Conference on 'The Continental Permian of the Southern Alps and Sardinia (Italy). Regional reports and general correlations', Brescia, 15–25 September 1999. Earth Science Department, Pavia University, 65–72.

PITTAU, P., BARCA, S., COCHERIE, A., DEL RIO, M., FANNING, M. & ROSSI, P. 2002. Le bassin permien de Guardia Pisano (Sud-Ouest de la Sardaigne, Italie): palynostratigraphie, paléophytogéographie, corrélations et âge radiométrique des produits volcaniques associés. *Géobios*, **35**, 561–580.

POSENATO, R. 2001. The athyridoids of the transitional beds between Bellerophon and Werfen Formations (uppermost Permian, Southern Alps, Italy). *Rivista Italiana di Paleontologia e Stratigrafia*, **107**, 197–226.

POSENATO, R. & PRINOTH, F. 1999. Discovery of *Paratirolites* from the Bellerophon Formation (Upper Permian, Dolomites, Italy). *Rivista Italiana di Paleontologia e Stratigrafia*, **105**, 129–134.

RAMOS, A. & DOUBINGER, J. 1979. Découverte d'une microflore thuringienne dans le Buntsandstein de la Cordillère Ibèrique Espagne). *Comptes Rendus de l'Académie des Sciences, Paris*, Série B, **289**, 525–529.

RAMOS, A. & DOUBINGER, J. 1989. Permian datations dans le faciès Buntsandstein de Majorque. *Comptes Rendus de l'Académie des Sciences, Paris*, Série II, **309**, 1089–1094.

RAU, A. & TONGIORGI, M. 1974. Geologia dei Monti Pisani a Sud-Est della Valle del Guappero. *Memorie della Società Geologica Italiana*, **13**, 227–408.

RAU, A., TONGIORGI, M. & MARTINI, I. P. 1988. La successione di Punta Bianca: un esempio di rift "abortivo" nel Trias medio del dominio Toscano. *Memorie della Società Geologica Italiana*, **30** (1985), 115–125.

RONCHI, A. 1997. *I prodotti sedimentari e vulcanici dei bacini permiani di Escalaplano e Perdasdefogu nella Sardegna sudorientale: stratigrafia e loro inquadramento nell'evoluzione tardo-paleozoica del settore sudeuropeo*. PhD thesis, University of Parma.

RONCHI, A., BROUTIN, J., DIEZ, J. B., FREYTET, P., GALTIER, J. & LETHIERS, F. 1998. New paleontological discoveries in some Early Permian sequences of Sardinia. Biostratigraphic and paleogeographic implications. *Comptes Rendus de l'Académie des Sciences, Paris*, Série IIA, **327**, 713–719.

SANTI, G. 2001. Icniti di tetrapodi permiani nelle Alpi Bergamasche. Nota preliminare. *Rendiconti dell'Istituto Lombardo, Accademia di Scienze & Lettere, B*, **133** (1999), 41–50.

SARRIA, E. 1987. Ricerche nel bacino antracitifero di Perdasdefogu – Relazione geologica. *In: Progetto di ricerca Antracite, Sardegna centrale*, Progemisa SpA, relazione interna, Cagliari.

SARRIA, E. & SERRI, R. 1986. Tettonica compressiva tardo paleozoica nel bacino antracitifero di Seui (Sardegna Centrale). *Rendiconti della Società Geologica Italiana*, **9**, 7–10.

SCHNEIDER, J. W., RÖSSLER, R. & GAITZSCH, B. 1995. Time lines of the Late Variscan volcanism – a holostratigraphic synthesis. *Zentral blatt für Geologie und Paläontologie*, **1** (1994), 477–490.

SOPEÑA, A. (ed.) 2004. Cordillera Ibérica i Costero Catalana. *In*: VERA, J. A. (ed.) *Geología de España* Sociedad Géologica de Espana, Instituto Geologico Minero España, Madrid, 587–634.

SOPEÑA, A., LÓPEZ-GÓMEZ, J., ARCHE, A., PÉREZ-ARLUCEA, M., RAMOS, A., VIRGILI, C. & HERNANDO, S. 1988. Permian and Triassic rift basins of the Iberian Peninsula. *In*: MANSPEIZER, W. (ed.) *Triassic-Jurassic Rifting. Continental break-up and the origin of the Atlantic Ocean and passive margins, part B.* (Development in Geotectonics, 22B) Elsevier, Amsterdam, 757–786.

SOPEÑA, A., DOUBINGER, J., RAMOS, A. & PÉREZ-ARLUCEA, M. 1995. Palynologie du Permien et du Trias dans le centre de la Péninsule Ibérique. *Bulletin de Sciences Géologiques*, **48**, 119–157.

SPINA, A. 2005. *Stratigrafia delle successioni pre-Verrucano appartenenti al dominio Toscano e correlazioni con il Sudalpino*. PhD thesis, University of Siena.

TORCQ, F., BESSE, J., VASLET, D., MARCOUX, J., RICOU, L. E., HALAWANI, M. & BASAHEL, M. 1997. Paleomagnetic results from Saudi Arabia and the Permo-Triassic Pangaea configuration. *Earth and Planetary Sciences Letters*, **148**, 553–567.

TOUTIN-MORIN, N. & VINCHON, C. 1989. Le bassins permiens du Sud-Est. *In*: CHÂTEAUNEUF, J. J. & FARJANEL, G. (eds) *Synthèse géologique des bassins permiens français*. Mémoires du Bureau de Recherches Géologiques et Minières, **128**, 114–121.

VARDABASSO, S. 1950. Il Permico in Sardegna. *Atti della Accademia Nazionale dei Lincei, Rendiconti, Classe di Scienze Fisiche, Matematiche & Naturali, serie 8*, **20**, 540–545.

VARDABASSO, S. 1966. Il Verrucano sardo. *Atti del Symposium sul Verrucano, Pisa, Settembre 1965. Società Toscana di Scienze Naturali*, 293–310.

VELLUTINI P. J. 1977. *Le magmatism permien de la Corse du Nord-Ouest*. PhD thesis, University of Marseille.

VENTURINI, C. 1990. Geologia delle Alpi Carniche centro-orientali. *Museo Friulano di Storia Naturale, Pubblicazione*, **36**, 1–220.

VIRGILI, C., 1983. El Pérmico en España: Introducción X Congreso Internacional de Estratigrafía y Geología del Carbonífero. *In*: VIRGILI, C. & RAMOS, A. (eds) *El Pérmico, Instituto Geológico y Minero de España*, 385–388.

VIRGILI, C., 1987. Problemática del Trías y Pérmico superior del bloque ibérico. *Cuadernos de Geologia Ibérica*, **11**, 39–52.

VIRGILI, C., 1989. Permian subdivision in the Iberian microplate (western Tethys). *XI Congrès International de Stratigraphie et de Géologie du Carbonifere, Beijing 1987, Compte Rendu*, **2**, 277–285.

VIRGILI, C., 2001. La chronostratigraphie du Permien et les series continentales de l'Europe Occidentale. *In*: *Crise Permienne, 16ª Journée Thématique de l'Association des Géologues du Permien, Paris, 15 Juin 2001*, Résumés, 5–6.

VIRGILI, C., HERNÁNDO, S., RAMOS, A. & SOPEÑA, A. 1973. La sédimentación permienne au centre de l'Espagne, *Comptes Rendus Sommaire de la Société Géologique de France*, **15**, 109–111.

VIRGILI, C., HERNÁNDO, S., RAMOS, A. & SOPEÑA, A. 1976. Le Permien en Espagne. *In*: FALKE, H. (ed.) *The Continental Permian in Central, West, and South Europe.* Proceedings of the Nato Advanced Study Institute held at the Johannes Gutenberg University, Mainz, F.R.G. 23 September–4 October, Series C., **22**, 91–109, D. Reidel Publishing Company, Dordrecht-Holland.

VIRGILI, C., SOPEÑA, A., RAMOS, A., HERNÁNDO, S. & ARCHE, A. 1980. El Pérmico en España. *Revista Española de Micropaleontologia*, **12**, 255–262.

VIRGILI, C., SOPEÑA, A., RAMOS, A., ARCHE, A. & HERNÁNDO, S. 1983. El relleno posthercínico y el comienzo de la sedimentación mesozoica. *Libro Jubilar J.M. Rios: Geologia de España, Instituto Geológico y Minero de España*, **2**, 25–36.

VISSCHER, H., KERP, K., CLEMENT-WESTERHOF, J. A. & LOOY, C. V. 1999. Permian floras of the Southern Alps. *In*: CASSINIS, G., CORTESOGONO, L., GAGGERNO, L., MASSARI, F., NERI, C., NICOSIA, U. & PITTAU, P. (eds) *Stratigraphy and Facies of the Permian deposits between Eastern Lombardy and the Western Dolomites, Field Trip Guidebook, 23–25 September 1999*. International Field Conference on 'The Continental Permian of the Southern Alps and Sardinia (Italy). Regional reports and general correlations', Breacia, 15–25 September 1999. Earth Science Department, Pavia University, Appendix, 139–146.

WESTPHAL, M., BARDON, C., BOSSERT, A. & HAMZEH, R. 1973. A computer fit of Corsica and Sardinia against southern France. *Earth and Planetary Sciences Letters*, **18**, 137–140.

YIN, H. F. 2000. Result of the vote on GSSP of Permian-Triassic boundary. *Albertiana*, **24**, 8–9.

ZIEGLER, P. A. & STAMPFLI, G. M. 2001. Late Palaeozoic-Early Mesozoic plate boundary reorganization: collapse of the Variscan orogen and opening of Neotethys. *In*: CASSINIS, G. (ed.) *Permian Continental Deposits of Europe and Other Areas. Regional Reports and Correlations*. Natura Bresciana, Monografia, **25**, 17–34.

Late Permian to Early Triassic transition in central and NE Spain: biotic and sedimentary characteristics

ALFREDO ARCHE & JOSE LÓPEZ-GÓMEZ

Instituto de Geología Económica, C.S.I.C.-U.C.M., Facultad de Geología, Universidad Complutense, Jose Antonio Novais s/n, 28040 Madrid, Spain (e-mail: aarche@geo.ucm.es)

Abstract: The Late Permian to Early Triassic (P/T) transition represents one of the most important Phanerozoic mass extinction episodes. Data from this transitional period are very scarce in continental basins, and reliable correlation with marine series is still a matter of debate. In this paper, information on the P/T transition in the continental series of central and NE Spain and the Balearic Islands is presented and compared with some coeval western European basins. The Iberian Ranges sections contain detailed information on the P/T transition, with sediments interpreted as alluvial fans, sandy and gravelly braided rivers and high sinuosity rivers with extensive floodplains, dated by means of pollen and spore assemblages. However, the fossil record contains two barren intervals throughout the study area, one in the Late Permian (Thuringian) and another during the latest Permian to Early Triassic. The possible causes of these gaps include the very likely relationship with the emplacement of the basaltic large igneous provinces (LIP) of SE China and western Siberia during this period of time.

The Permian–Triassic sediments of central and NE Spain and the Balearic Islands (Fig. 1) can be broadly described by the classic Germanic trilogy: Buntsandstein, Muschelkalk and Keuper, capped by a Late Triassic to Early Jurassic carbonate-evaporite complex. These sediments and the associated Early Permian and Late Triassic volcanic rocks were deposited during an extensional regime after the Hercynian orogeny. The rift basins went through several synrift and post-rift cycles, and there is evidence of frequent marine transgressive-regressive cycles from the Middle-Triassic (Early Anisian) time until the Triassic–Jurassic transition. These cycles are especially well documented in the Iberian Ranges (Figs 2 & 3).

A more detailed examination of the sedimentary record reveals a much more complex situation due to lateral changes of facies, basin compartmentalization and the absence of some units in vast domains; the main characteristics of the Permian–Triassic sediments of Spain are now well known after a wealth of data have been published since the 1970s (Virgili *et al.* 1976, 1979, 1983; Hernando 1977, 1980; Sopeña 1979, 1980; Ramos 1979, 1980; Arribas 1985; Ortí 1987; Sopeña *et al.* 1988; Jurado 1990; López-Gómez & Arche 1993; Calvet & Marzo 1994; Ortí & Pérez-López 1994; López-Gómez *et al.* 1998, 2002; Sopeña & Sánchez-Moya 2004; Sánchez-Moya & Sopeña 2004; Arche *et al.* 2004).

The long and complex extensional tectonic regime following the Hercynian orogeny spans the Early Permian to the Late Cretaceous. During the Late Permian to Early Triassic, a series of continental rift basins evolved in central and NE Spain; their main basin boundary faults (Fig. 3) trended NW–SE and were Hercynian or older lineaments reactivated as arcuate, listric normal faults with NE–SW trend and associated N–S strike-slip fault systems acting as transfer fault systems accommodating different extension rates along segments of the main, normal basin boundary faults (Arche & López-Gómez 1996). The origin and evolution of these basins have been studied by Salas & Casas (1993), Doblas *et al.* (1993), Arche & López-Gómez (1996), Van Wees *et al.* (1998), López-Gómez *et al.* (2002) and Sánchez-Moya & Sopeña (2004).

During the Alpine orogeny, many of the normal faults were reactivated as thrust faults in a general compressional tectonic regime leading to widespread inversion processes. It is very important to separate the earlier extensional basins (Iberian Basin, Ebro Basin and Catalan Basin) from the more recent folded and thrusted alpine structures (Iberian Ranges and Catalan Ranges), which are also not coincident geographically. The Iberian Basin was bounded by the Serranía de Cuenca and the Ateca Palaeozoic highs (Fig. 3), the Ebro Basin by the Ateca and Lérida Palaeozoic highs and an ill-defined Palaeozoic high in the Pyrenean domain (Fig. 3), and the Catalan Basin by the Lérida and Gerona Palaeozoic highs (Fig. 3). These Palaeozoic highs or basin shoulders were created in the footwall

From: LUCAS, S. G., CASSINIS, G. & SCHNEIDER, J. W. (eds) 2006. *Non-Marine Permian Biostratigraphy and Biochronology*. Geological Society, London, Special Publications, **265**, 261–280.

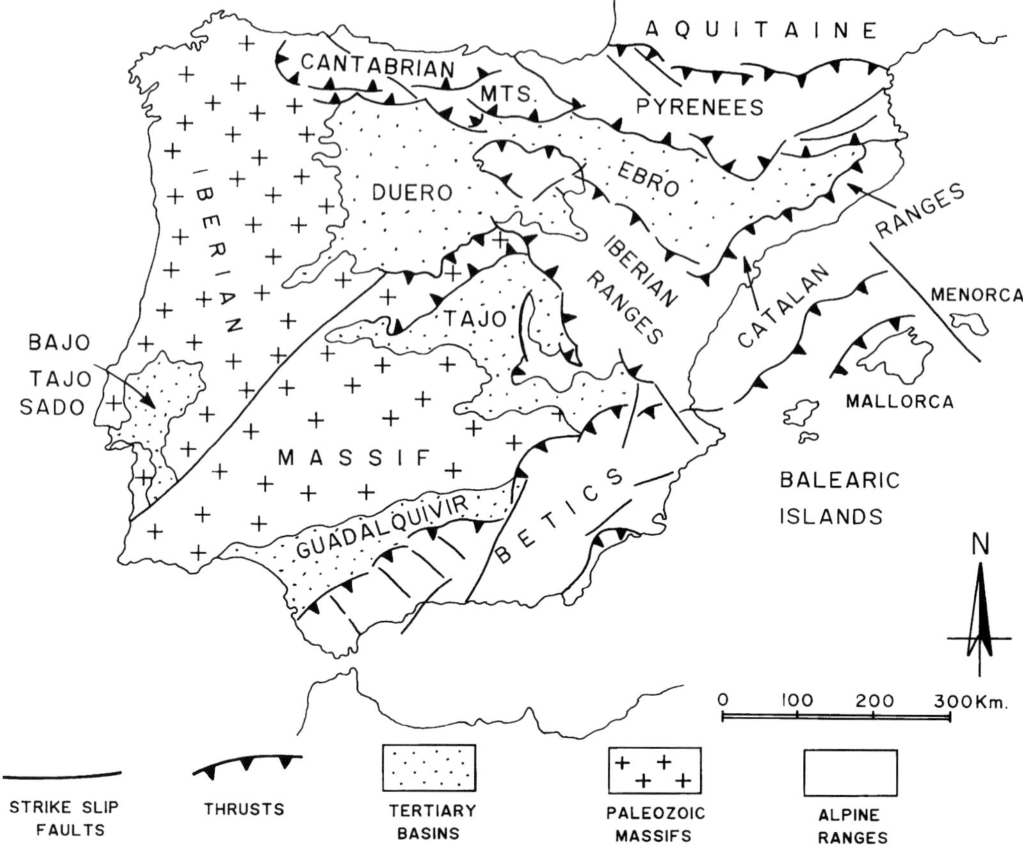

Fig. 1. The Iberian Peninsula and its major geological units.

and hanging-wall blocks as the extensional basin boundary faults, and their configuration changed with time. They were partially drowned by the first marine transgression of the Tethys during the Middle Triassic (Anisian) and ceased to exist as basin divides at the end of the Middle Triassic (late Ladinian).

During the Late Permian to Early Triassic, the eastern margin of the Iberian microplate was located 400–800 km west of the shore of the Neotethys sea. Topographic elevations in this portion of the collapsing Hercynian range are estimated at 2 000–3 000 m (Fluteau *et al.* 2001), and Iberia was at an estimated latitude of 10–15°N (Ziegler & Stampfli 2001; Eren *et al.* 2004), which is in an area dominated by a monsoon regime and important topographically generated rains (Barron & Fawcett 1995). In this paper we will deal mostly with the sedimentary record of the Iberian Basin, where most of the available biostratigraphical data have been obtained and draw correlations with the coeval basins in Spain, based on biostratigraphical data, presence of regional angular unconformities and/or hiatuses and sedimentological and petrological data for each basin.

The sedimentary record

The limitations of the classic lithostratigraphical terminology of the Permian and Triassic were obvious since the first modern studies (Virgili 1979; Sopeña *et al.* 1988) for two main reasons: the presence of major angular unconformities and hiatuses inside the sedimentary record and the lateral changes of facies between shallow marine carbonate and siliciclastic deposits and continental siliciclastic deposits. The sedimentary record of the Permian–Triassic in the study area has been subdivided into several formations, based upon biostratigraphical, lithological and sedimentological data (Sopeña *et al.* 1988; Calvet

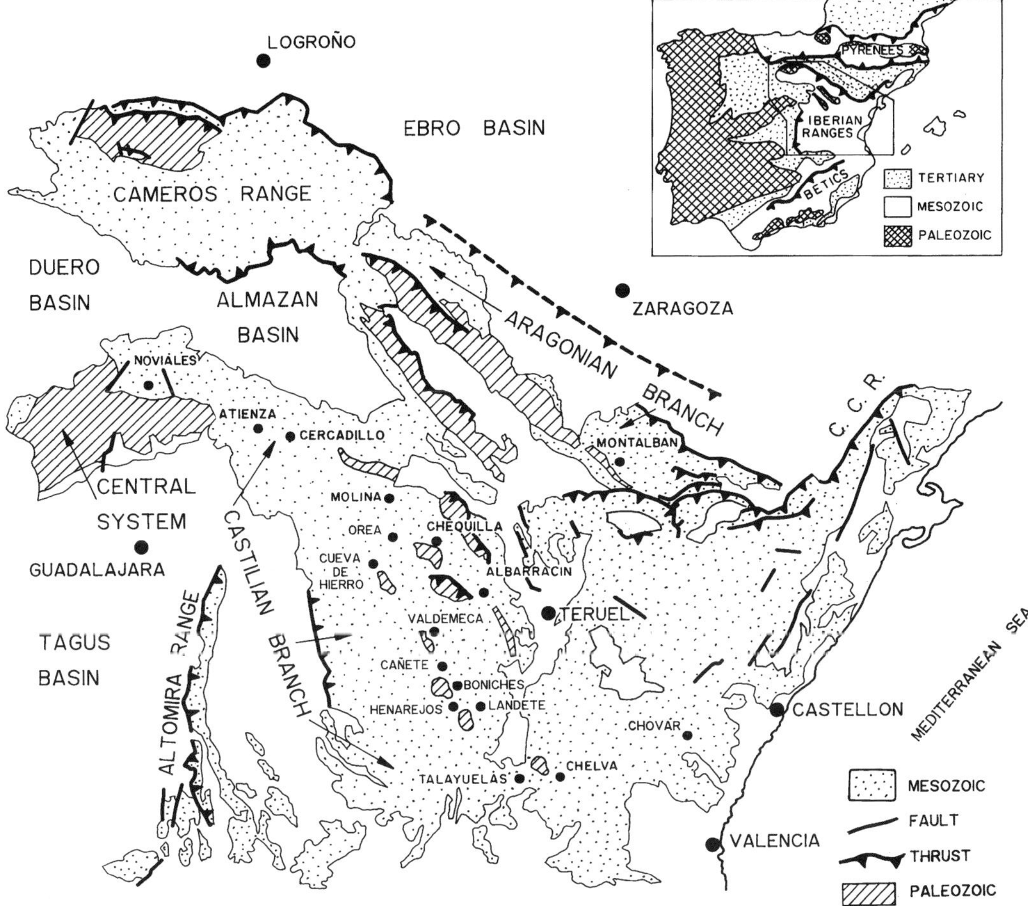

Fig. 2. Present-day configuration of the Iberian Ranges and main localities cited in the text.

& Marzo 1994; López-Gómez *et al.* 1998, 2002; Arche *et al.* 2004).

Iberian Ranges

The present-day Iberian Ranges (Fig. 2) can be subdivided into two main structural units: the Castilian Branch to the SW and the Aragonese Branch to the NE, with the Tertiary Calatayud–Teruel Basin in between. During the Permian and most of the Triassic, the Iberian Basin was bounded by the Serranía de Cuenca and Ateca Palaeozoic highs (Fig. 3), the latter now partially exposed along the Aragonese branch; they were part of the footwall and hanging-wall block of the syn-sedimentary basin boundary fault systems; the structural framework of the extensional basin also included oblique transfer fault systems trending NNE–SSW to N–S, which were responsible for some prominent transverse highs such as the Cercadillo, Cueva de Hierro and Teruel highs (Fig. 3). The Permian–Triassic sedimentary record has been subdivided into seven major sedimentary cycles (Fig. 4) (López-Gómez *et al.* 2002; Arche *et al.* 2004), but only the second and third ones are studied in this paper, with marginal mentions of the first one.

First sedimentary cycle

The oldest sediments, of Early Permian (Autunian) age, in some locations associated with andesitic volcanic rocks, were deposited in small, isolated half-graben basins along the trace of the future Iberian Basin. The continental deposits show a variety of facies, from lacustrine to coarse alluvial fans and thickness ranging from 15 m to more than 700 m, because local tectonic control determined the creation of accommodation space for the sediments and the amount of clastic supply to the basins (Fig. 4). In

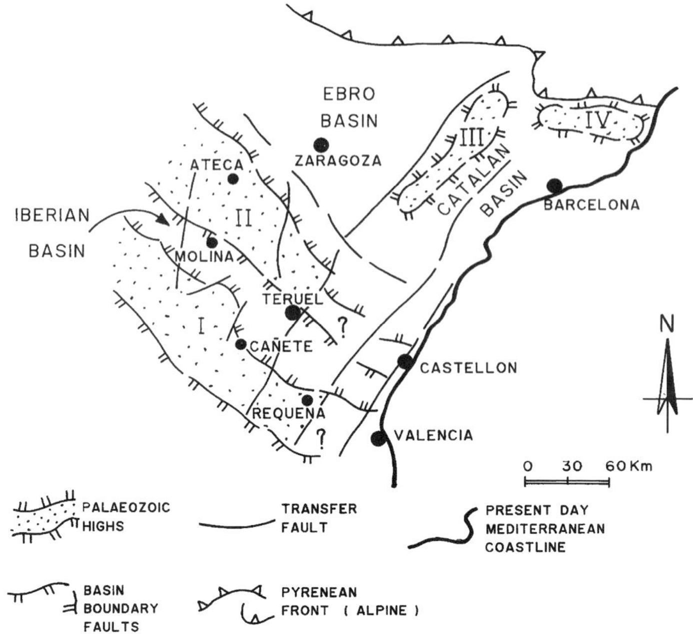

Fig. 3. Palinspastic reconstruction of the continental rift basins in central and NE Spain during the Late Permian and associated Palaeozoic highs. I, Serranía de Cuenca high; II, Ateca high; III, Lérida high; IV, Gerona high. Note the protruding Orea-Cueva del Hierro transversal high between Molina and Cañete.

some areas, the lower part of the succession contains andesitic rocks, as in Atienza, Molina de Aragón, Orea and Montalbán (Fig. 2), followed by red breccias, red sandstones and/or black shales (Hernando 1977, 1980; Sopeña 1979, 1980; Ramos 1979, 1980; Pérez-Arlucea & Sopeña 1985). In other areas, such as Boniches, only a few metres of red breccias are found (López-Gómez 1985; López-Gómez & Arche 1994). Most of the outcrops have been dated by means of pollen and spore assemblages dominated by the presence of *Vittatina* and *Potonieisporites* (Sopeña *et al.* 1995). The Minas de Henarejos outcrop is an exception in the Iberian Ranges because it contains coal measures exploited commercially, with a rich macroflora of Stephanian C (?)-Autunian(?) age (Wagner *et al.* 1983; Melendez *et al.* 1983).

The Early Permian (Autunian) sedimentary cycle always lies unconformably on the Hercynian basement and is unconformably covered by the Late Permian to Early Triassic second or third sedimentary cycles. Calc-alkaline volcanism is represented in both the Castilian and Aragonian branches of the Iberian Ranges (Muñoz *et al.* 1985; Lago *et al.* 2002, 2004) with volcaniclastic deposits, sills, dikes and lava flows. They were emplaced in two phases, represented by amphibolic andesites and daci-andesites (first phase) and gabbros, pyroxenic andesites, basalts and rhyolites (second phase). Absolute ages for these volcanic rocks range from 293 ± 2.5 Ma to 283 ± 2.5 Ma (Hernando *et al.* 1980; Conte *et al.* 1987; Lago *et al.* 1991), which fall within the Sakmarian stage of the Cisuralian (Lower Permian), according to the scales of Gradstein *et al.* (2004) and Menning (2001).

Second sedimentary cycle

The second sedimentary cycle is bounded by two angular unconformities (Figs 4 & 5) and was deposited in a single symmetric graben basin of complex longitudinal geometry (Arche & López-Gómez 1996). It consists of quartzite conglomerates at the base, which are of limited lateral extent (Boniches Formation) and is conformably overlain by red mudstones, sandstones and rare conglomerates of the Alcotas Formation (Ramos 1980; Pérez-Arlucea & Sopeña 1985; López-Gómez & Arche 1993). Its age is Late Permian (Thuringian, *sensu* Visscher 1971), as will be discussed in the biostratigraphy section of this paper. This cycle has been compared to the 'Saxonian' facies in central Europe, but this term is not precise and should be abandoned in Iberia, because the age of the central European

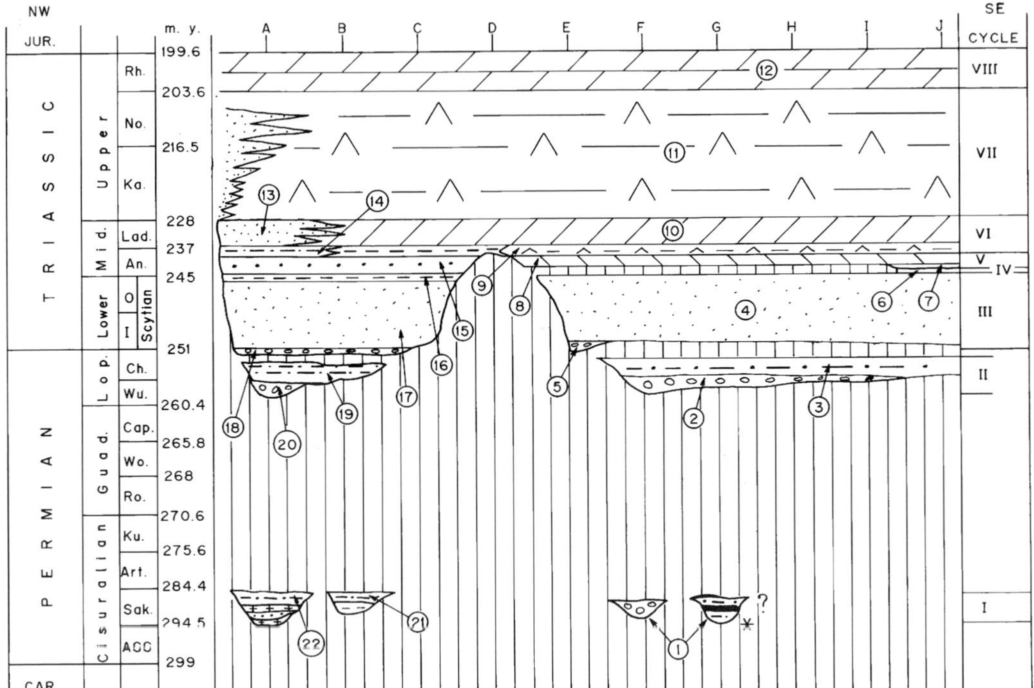

Fig. 4. Sketch of the lateral distribution of the Permian–Triassic sedimentary cycles (I–VIII) in the Iberian Basin. Locations: A, Noviales; B, Molina de Aragón; C, Chequilla; D, Cueva del Hierro; E, Valdemeca; F, Cañete-Boniches; G, Henarejos; I, Chelva; J, Chóvar. Chronostratigraphical units and absolute ages according to Gradstein *et al.* (2004). Formations: 1, Unnamed units in Henarejos and Tabarreña Breccias; 2, Boniches Conglomerates; 3, Alcotas Mudstones and Sandstones; 4, Cañizar Sandstones; 5, Valdemeca Conglomerates; 6, Eslida Sandstones; 7, Marines Mudstones; 8, Landete Dolomites (=Albarracín Dolomites); 9, El Mas Mudstones; 10, Cañete Dolomites (=Tramacastilla and Royuela Dolomites); 11, Keuper Facies; 12, Imón Dolomites; 13, Cuevas de Ayllón Sandstones; 14, Carrascosa Mudstones; 15, Cercadillo Sandstones; 16, Prados Sandstones and Mudstones; 17, Rillo de Gallo Sandstones; 18, Hoz del Gallo Conglomerates; 19, Montesoro Mudstones (=Cañamares Mudstones); 20, Cañamares Conglomerates; 21, Ermita Mudstones; 22, Cañamares Conglomerates and Sandstones. Based on Arche *et al.* (2004). Star indicates coal measures.

sediments are ill-determined or range from Lower to Upper Permian.

The Boniches Formation lies with angular unconformity on the Hercynian basement or on Lower Permian sediments. Sedimentation of this formation was controlled by the activity of the Serranía de Cuenca fault system (Fig. 3) (López-Gómez & Arche 1993), which created steep relief in the footwall block and short, steep transverse drainage networks. Laterally, thickness changes from 130 m in the central part of the basin (Henarejos) to less than 30 m in the NW and SE extremes (Valdemeca and Chelva). Its age is Late Permian (Thuringian) according to pollen and spore associations found in the upper part of the formation in Talayuelas (Doubinger *et al.* 1990).

The Boniches Formation has been subdivided into four members. The lowermost, composed of fining-upwards sequences of massive and cross-stratified conglomerates with palaeocurrents transverse to the basin axis, is interpreted as proximal alluvial fan facies dominated by channel longitudinal bars formed by diffuse gravel sheets migrating over the original bar core; they evolve in time to bar-and-channel complexes with superimposed high-and low-stage deposits (Miall 1981; Crowley 1983, Ashmore 1991; López-Gómez & Arche 1997).

The two middle members consist of fining-upward sequences with gravel cross-stratified basal bodies and increased sandy bodies at the top. They are interpreted as transverse and composite mid-channel bars and associated low-stage sand bars in medial and distal parts of a braided alluvial fan system. The uppermost member consists of thin, fining-upward sequences of

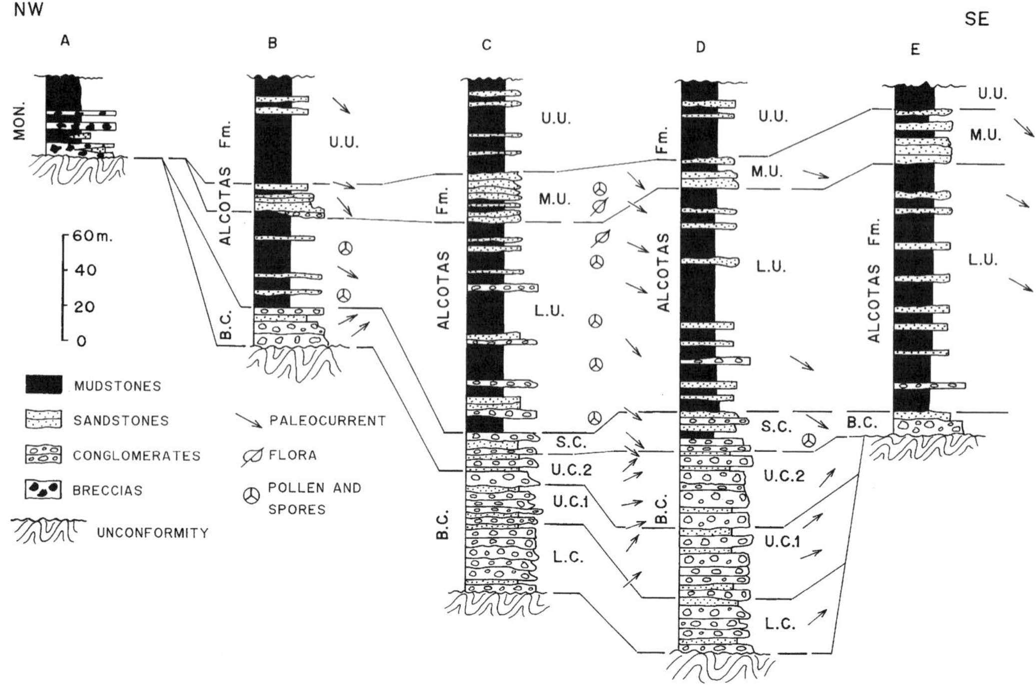

Fig. 5. Second sedimentary cycle correlation along the central and SE Iberian Basin and main pollen and spores localities. A, Molina de Aragón (Barranco de la Hoz section); B, Albarracín (Fuente de la Señora section); C, Cañete (Cañizar section); D, Talayuelas (Arroyo de la Vid section); E, Chelva (Barranco de Alcotas section). Data from Ramos (1979), Temiño (1984), Pérez-Arlucea (1985) and López-Gomez (1985). B.C., Boniches Conglomerates; L.C., Lower Conglomerates; U.C.1, Upper Conglomerates 1; U.C.2, Upper Conglomerates 2; S.C., Sandy Conglomerates; L.U., Lower Unit; M.U., Middle Unit; U.U., Upper Unit; MON, Montesoro Formation.

gravels and sands, interpreted as distal reaches of a braided, mixed load fluvial system. In this member, the transverse drainage pattern changes into a longitudinal drainage system.

The Alcotas Formation lies conformably above the Boniches Formation and consists of red mudstones and sandstones with some conglomerate lenses (Fig. 4).Thickness ranges from 82 m in Valdemeca to 168 m in Chelva, and the formation crops out along the Iberian Basin except in the Cueva de Hierro Palaeozoic high. Red mudstones are the dominant lithology (about 70%) consisting of illite, kaolinite and quartz (Alonso-Azcárate *et al.* 1997) and traces of feldspar and hematite. The illite/kaolinite ratio increases to the SE. Red to pink sandstones (about 27%) consist of subrounded quartz and feldspar grains, mica flakes and slate fragments, with clay matrix and quartz cements. Bulk lithology changes from arkose in the Landete area to protoquartzites and greywackes in the Boniches area and protoquartzites in the Chelva area (López-Gómez 1985).

The age of the Alcotas Formation is well constrained by several pollen and spore assemblages found in the central and SE Iberian Ranges in the lower and middle part of the unit; the upper part is always barren (Boulouard & Viallard 1971; Ramos & Doubinger 1979; Doubinger *et al.* 1990; Sopeña *et al.* 1995). The formation can be subdivided into three fining-upwards units or members separated by sharp erosive surfaces of regional lateral extent. The lower member is 20–40 m thick and consists of lenses of conglomerates, 4–6 m thick, which are embedded in red mudstones; the conglomerate bodies present cross-stratification, lateral accretion surfaces and reactivation surfaces. The mudstone facies are usually massive, with rippled intervals and some dolomitic horizons; calcic palaeosols with laminar and nodular structures are found in this member. It is interpreted to represent gravelly braided river deposits with high avulsion rates and broad, fine-grained floodplains. The instability of the channels can be interpreted as the response to active synsedimentary tectonics leading to constant changes in local slope (Smith 1970; Miall 1987), with sediment supply keeping pace or exceeding the rate of subsidence (Alexander & Leeder 1987).

The middle member shows a sudden change in lithofacies and fluvial style: sandstones are the dominant lithology (up to 70% in some sections). These single-storied sandstone bodies show tabular geometry, internal structures dominated by trough cross stratification at the base and current ripples at the top and sometimes lateral accretion surfaces littered with comminuted plant remains (López-Gómez & Arche 1994) and some large silicified tree trunks. This member can be interpreted as a transition from distal sandy braided rivers to high sinuosity, meandering rivers; reactivation surfaces point to marked seasonality in flow (Puigdefábregas 1973; Allen 1983; López-Gómez 1985; Ashmore 1991). The banks were probably vegetated by tree-sized plants, sometimes uprooted by the lateral migration of the channels; the fine-grained deposits are interpreted as floodplain deposits.

The upper member, 20–35 m thick, consists of red mudstones (about 85%) containing thin, lenticular sandstone bodies up to 1.5 m thick; no palaeosols or macro-or microflora have been found in this interval. This unit is interpreted as a distal, very low-energy, sandy braided river system with high avulsion rates and marked seasonality, flowing in an extensive floodplain subject to frequent crevassing and flooding (Collinson 1970; Cant & Walker 1978; Walker & Cant 1979; O'Brian & Wells 1986).

Third sedimentary cycle

The third cycle consists of two formations (Figs 4 & 6): the Hoz del Gallo Formation and the Cañizar (= Rillo de Gallo) Formation, and is limited by an angular unconformity at the base and a hiatus or a conformable contact at the top. The P/T transition lies inside this cycle, as will be discussed later in this paper. This cycle can be compared to the Buntsandstein facies *sensu stricto* of central Europe.

The Hoz del Gallo Formation lies unconformably on the Hercynian basement or on the Alcotas Formation. Its age is Late Permian (Thuringian) at the base, well constrained by several pollen and spore assemblages found in its

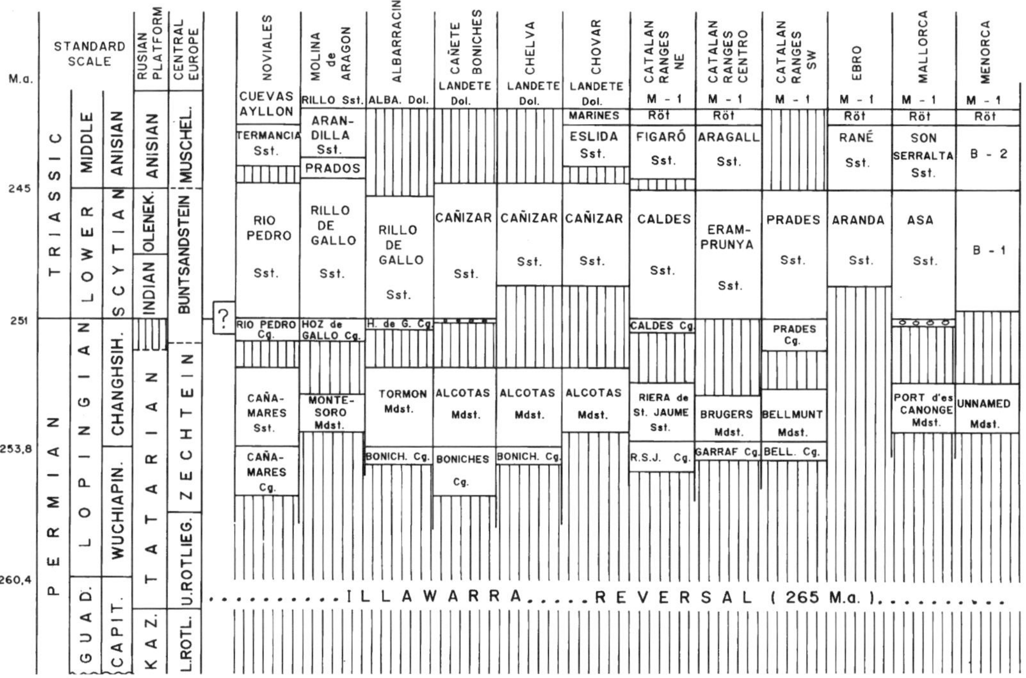

Fig. 6. Third sedimentary cycle correlation along the central and SE Iberian Basin and pollen and spores localities. A, Molina de Aragón; B, Albarracín; C, Valdemeca; D, Boniches; E, Chelva. Data from Ramos (1979), Temiño (1984), Pérez-Arlucea (1985), Horra (2004) and López-Gómez (1985). L.C., Lower Conglomerates; U.C., Upper Conglomerates.

lower part (Ramos 1979; Ramos & Doubinger 1979; Ramos & Sopeña 1983; Pérez-Arlucea 1985; Ramos *et al.* 1986). Thickness of the Hoz del Gallo Formation can change from a maximum of 150 m in the Molina de Aragón area to 60–80 m in the Albarracín area and only 5–10 m in the Orea and Valdemeca-Boniches areas; it pinches out in the Talayuelas area.

A major erosive bounding surface divides the formation into two members: the lower and upper conglomerates: The upper conglomeratic member is more laterally extensive than the lower and is found from Valdemeca to Boniches, lying on older sediments or the Hercynian basement. The lower member is composed of truncated, fining-upward conglomeratic, cross-stratified sequences with erosive, concave bases, internal trough and planar cross-stratification and some sandstone lenses at the top. Fines of overbank origin are seldom preserved, but they contain the pollen and spore assemblages found in the formation. This member has been interpreted as bar-and-channel associations in a proximal to distal alluvial fan system with frequent lateral switches of the active channels (Ramos *et al.* 1986).

The upper member is dominated by tabular sets of massive and trough cross-stratified conglomerates bounded by extensive planar erosion surfaces. The sudden change in fluvial style is associated with the presence of white, igneous derived vein quartz pebbles, ventifacts (=dreikanters) and a switch of palaeocurrents from the NE to the SE (Ramos & Sopeña 1983; Ramos *et al.* 1986; M. Durand, pers. comm.). This member is interpreted as a more stable, coarse-grained, braided alluvial system with clear lateral aggradation, deeper channels and a more distal source area containing igneous and/or metamorphic rocks, in contrast with the monomictic quartzite composition of the lower member, derived from local sources.

The Cañizar Formation (Figs 4 & 6) lies in sharp, conformable contact with the Hoz del Gallo Formation or unconformably on the Alcotas Formation. The age of its upper part is early Middle Triassic (early Anisian), but the age of the base is unknown up to now; the age of the lower part of the underlying, conformable Hoz del Gallo Formation is Late Permian; the P/T transition lies somewhere between the upper part of the Hoz del Gallo Formation and the lower part of the Cañizar Formation (Arche *et al.* 2004).

The Cañizar Formation was deposited all over the Iberian Basin, and its thickness ranges from 80 to 170 m; its petrographical composition changes from arkoses in the NW to protoquartzites in the SE, the change taking place in the Cañete–Landete area, probably due to feldspar abrasion during downstream transport (López-Gómez & Arche 1994). As the palaeocurrents point to the S and SE (Fig. 6) with dispersions that rarely exceed 50 degrees, the drainage was parallel to the axis of the basin, and the source area was far away to the NW. The basin configuration was a symmetric graben limited by the basin boundary fault systems, the Serranía de Cuenca to the SW and the Ateca to the NE (Fig. 3).

Within these sandy braided river deposits, almost devoid of fine materials, six multilateral, multi-storey sandstone sheets have been identified, separated by planar erosive surfaces of regional extent (López-Gómez & Arche 1993). The lower two units are made up of truncated, fining upward-sandstone sequences up to 1.2 m thick with thin pebble lags at the base, organized in amalgamated vertical sequences. They are interpreted as sandy, braided channel fill with a very high width/depth ratio (over 25), comparable to the present-day North Platte River (Miall 1978; Crowley 1983) or the South Saskatchewan River (Cant & Walker 1978).

The three middle units mark a progressive change towards more complex, sandy amalgamated units with cross-stratification, reactivation surfaces and downward accretion structures showing overpassing; they are interpreted as sand flats of composite bars infilling the active channels of broad, braided river systems, comparable to some Old Red Sandstone examples (Campbell 1976). The two uppermost units reflect a sudden energy increase in the braided river depositional system, with complex sand flat and composite bar infillings of the broad active channels. Frequent convex reactivation surfaces point to repeated flood and dry period cycles with marked discharge fluctuations (Jones 1977). The laterally equivalent, coeval Rillo de Gallo Formation in the NW Iberian Basin has been interpreted broadly in the same terms by Ramos *et al.* (1986) and Pérez–Arlucea & Sopeña (1985).

Catalan Ranges

The three sedimentary cycles described in the Iberian Ranges can be identified in the Catalan Ranges (Fig. 1). During the Permian to Lower Triassic interval, the Catalan Basin was subdivided into three tectono-sedimentary domains: Montseny-Llobregat (NW), Garraf (central) and Miramar-Prades-Priorat (SW), partially bounded by the Lérida and Gerona Palaeozoic highs (Figs 3 & 6) (Marzo 1980).

The siliciclastic deposits of the Catalan Basin considered here remain undated by biostratigraphical methods, so the correlations with other areas are based only on lithostratigraphical considerations and must be considered provisional (Arche *et al.* 2004). Basal breccias are found in small outcrops in the NE and SW domains, lying unconformably on the Hercynian basement and unconformably overlain by the second or third sedimentary cycles. The lithological characteristics and tectono-stratigraphical position of these unnamed breccias is identical to the Tabarreña Breccias in the Iberian Basin (Fig. 4), so a Lower Permian age is attributed to these sediments.

A second sedimentary cycle, bounded by angular unconformities, is found in the three domains, but its lateral extent is limited (Fig. 6). It consists of the Riera de Sant Jaume, Garraf and Bellmunt quartzitic conglomerates, conformably overlain by the Riera de Sant Jaume, Brugers and Bellmunt red mudstones and sandstones. Again, even in the absence of biostratigraphical data, a tentative correlation can be proposed using lithological and tectono-stratigraphical analogies: the basal conglomeratic units are the time equivalents of the Boniches Formation, and the overlying red mudstones and sandstones, of the Alcotas Formation (Arche *et al.* 2004).

A new sedimentary cycle, found all along the Catalan Ranges (Fig. 7), lies unconformably on the Hercynian basement, on the first or the second sedimentary cycles. The third sedimentary cycle consists of quartzitic conglomerate in the NE and SW domains (Caldes and Prades units) conformably overlain by arkosic pink sandstones of multi-storied, amalgamated internal geometry (Caldes, Eramprunyá and Prades units). The top of the cycle is marked in the SW domain by a well-developed hiatus surface enriched in Fe oxides.

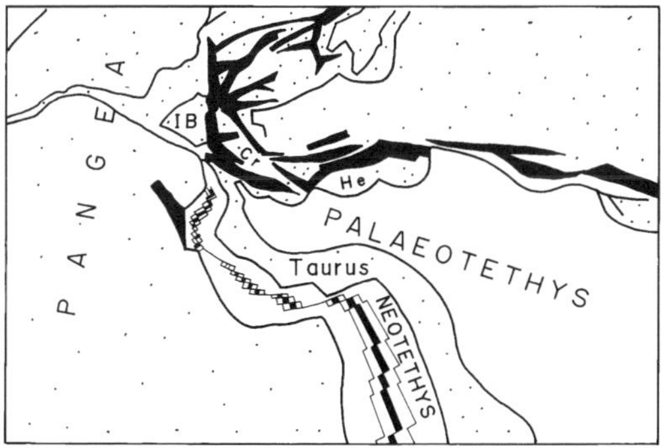

Fig. 7. Reconstruction of the central western Tethys at the end of the Permian; in black, the main continental rifts. From Eren *et al.* (2004), slightly modified. IB, Iberia; Cr, Carnian Alps; He, Hellenides.

The correlation with the Iberian Basin is reasonably drawn as follows: the conglomeratic units can be the time equivalents of the Hoz del Gallo Formation and the sandstone units, of the Cañizar (=Rillo de Gallo) Formation. Synsedimentary extensional tectonics is evident in the Catalan Basin (Marzo 1980), and differential subsidence along the axis of the basin created the three different domains and the lateral changes of facies and thickness.

Ebro Basin

Duing the Permian, the Ebro Basin was separated from the Iberian Basin by the Ateca Palaeozoic high (Fig. 3) and was connected with the Catalan Basin at its SE corner. The SW margin of the Ebro Basin is now exposed along the Aragonese branch of the Iberian Ranges (Fig. 2), but most of it is now covered by Tertiary sediments, so essential information can be obtained only from commercial oil well logs and cores (Arribas 1984, 1985; Jurado 1988, 1990) (Fig. 6).

Isolated outcrops of volcaniclastic rocks associated with grey mudstones and lying unconformably on the Hercynian basement are found from Reznos to Montalban (Fig. 2). They have been termed the Arroyo Riduero Formation by Rey & Ramos (1991) and contain rich pollen and spore assemblages of Lower Permian (Autunian) age (De la Peña *et al.* 1977). The straightforward correlation with the Ermita Formation and equivalents in the Iberian Basin, that is, with the first sedimentary cycle, is based on similar palynofloras, identical volcanic rocks and identical tectono-stratigraphical position at the base of the Permian sedimentary record. A second sedimentary cycle lies unconformably on the Hercynian basement or on the first sedimentary cycle (Arribas 1984, 1985). It consists of the Moncayo Conglomerates Formation and the Tabuenca Mudstones Formation. Although it is not dated, its correlation with the second sedimentary cycle of the Iberian Basin, that is, the Boniches and Alcotas formations, is most plausible.

Lying unconformably on the second cycle, a third cycle, composed of one sedimentary unit, the Aranda Formation, is found throughout the basin (Arribas 1985), capped by a hiatus surface or conformably overlain by the Carcalejos Formation and the Rané Mudstones and Sandstones Formation. This cycle is not dated, but its petrographical, sedimentological and tectono-stratigraphical characteristics are identical to the Cañizar Formation of the Iberian Basin.

Balearic Islands

The Balearic Islands show a good sedimentary record of the Late Permian–Triassic time interval (Rodriguez-Perea *et al.* 1987; Ramos & Doubinger 1989; Ramos 1995), but the correlation with mainland Iberia has been seldom attempted (Arche *et al.* 2002), probably due to its present-day position in the Betic alpine orogenic belt (Fig. 1). The siliciclastic deposits of Mallorca (Fig. 6), lying unconformably on the Hercynian basement on the northern coast of the island, have been subdivided into three formations by Ramos (1995). From base to top they are: Port d'es Canonge Formation, Asá Formation and Son Serralta Formation. There is a sharp contact between the first and second formation and a transitional one between the second and the third ones.

A similar succession has been described in Menorca (Bourrouilh 1973; Rosell *et al.* 1988; Broutin *et al.* 1992), but no formal stratigraphical units have been described on this island. Above a few metres of polymictic breccias, a red mudstone unit equivalent to the Port d'es Canonge Formation in lithology and tectono-stratigraphical position is covered unconformably by massive red sandstones (B1 unit) and alternating mudstones and sandstones (B2 unit), equivalent to the Asá and Son Serralta formations.

Several palynological assemblages have been found in both islands and are described in detail in a later section of this paper. The correlation with the Iberian Basin (Arche *et al.* 2002) is clear according to lithological, sedimentological and biostratigraphical criteria: the Port d'es Canonge Formation and the basal red unit of Menorca are the time-equivalents of the Alcotas (=Tormón and Montesoro) Formation, the Asá Formation and the B1 unit correlate with the Cañizar (=Rillo de Gallo) Formation and the Son Serralta Formation and the B2 Unit with the Eslida (=Prados and Arandilla) Formation.

Biostratigraphical data

The Permian and Lower Triassic continental sediments of the Iberian Basin and the Balearic Islands have yielded numerous pollen and spore assemblages in many localities since the 1970s. Detailed descriptions and references can be found in Bourrouilh (1973), Ramos & Doubinger (1989), Doubinger *et al.* (1990), Broutin *et al.* (1992), Sopeña *et al.* (1995) and Díez (1997). The scarce palynological data for the Catalan Basin (Middle and Upper Triassic only) are found in Solé de Porta *et al.* (1985, 1987). The Ebro Basin sediments remained undated, as the scarce attempts to find microfossils in borehole samples have been unsuccessful up to now.

Iberian Basin

The sediments of the first sedimentary sequence (Fig. 7), not studied in this paper, have yielded rich palynological and macrofloral assemblages of typical Lower Permian (Autunian) age (Ramos *et al.* 1976; Doubinger *et al.* 1978; Ramos & Doubinger 1979; Sopeña 1979; Melendez *et al.* 1983; Wagner *et al.* 1983), in good agreement with the absolute ages obtained for the interstratified volcanic rocks: 283 ± 2.5 to 293 ± 2 Ma, Sakmarian stage (Hernando *et al.* 1980; Lago *et al.* 2004).

The sediments of the second and third sedimentary sequences (Figs 5 & 6) have yielded rich palynological assemblages of Late Permian (Thuringian) age in the Boniches, Alcotas (=Tormón and Montesoro) and Hoz de Gallo formations. Boulouard and Viallard (1971) found a pollen and spores assemblage in the Alcotas Formation of the Landete section (Fig. 2) composed of: *Lueckisporites virkkiae, Taeniasporites albertae, T. noviaulensis, Limitisporites sp., Pilasporites calculus, Nuskoisporites dulhuntyi, Jugasporites delasaucei, Vesicaspora ovata* and *Platysaccus umbratus.* These are of Late Permian (Thuringian) age.

In the region of Molina de Aragón (Fig. 2), Ramos & Doubinger (1979) found a palynological assemblage near the base of the Hoz de Gallo Formation composed of: *Punctatisporites* sp., *Endosporites* sp., *Trizonaesporites grandis, Nuskoisporites dulhuntyi, Cordaitina* sp., *Lueckisporites virkkiae, Paravesicaspora splendens, Jugasporites delasaucei, Protohaploxipinus microcorpus, Striatopodocarpites* sp., *Gardenasporites heisseli, Falcisporites schaubergerii* and *Cycadopites* sp. These are also of Late Permian (Thuringian) age, but this formation lies unconformably on the Alcotas Formation in this section. This apparent anomaly remained unexplained for a decade.

In the SE domain of the Iberian Basin, Temiño (1982) described a palynological assemblage in the middle part of the Alcotas Formation near Albarracín (Fig. 2), composed of *Lueckisporites virkkiae, Nuskoisporites dulhuntyi, Falcisporites schaubergerii* and *Bisaccates* sp., of

Late Permian (Thuringian) age. In the same region, Pérez-Arlucea (1985), Pérez-Arlucea & Sopeña (1985) and Sopeña *et al.* (1995) found palynological assemblages in several localities, both in the Hoz del Gallo and Tormón (=Montesoro and Alcotas) formations, composed of: *Angustiporites anguinus, Densoisporites* sp., *Converrucosisporites deasauey, C. eggeri, Uvaesporites* sp., *Acanthotriletes* sp., *Lycospora* sp., *Nuskoisporites dulhuntii, N. klausii, Trizonaesporites grandis, Wilsonites* sp., *Klausipollenites schaubergeri, Paravesicaspora splendens, Jugasporites delasaucei, Jugasporites* sp., *Alisporites* sp., *Sulcatisporites ovatus, Gardenasporites* sp., *Pytiosporites* sp., *Falcisporites zapfei, Lueckisporites virkkiae, Crucisaccites variosulcatus, Protohaploxipinus sewardi, Striatopodocarpites rarus, Lunatisporites* sp. and *Crustaesporites* sp. All of them are of Late Permian (Thuringian) age.

In the Cañete-Boniches area (Fig. 2), López-Gómez *et al.* (1984) and Doubinger *et al.* (1990) found a palynological assemblage in the upper part of the Boniches Formation composed of: *Klausipollenites schaubergeri, Lueckisporites virkkiae, Cedripites* sp., *Lycospora* sp., *Nuskoisporites dulhuntyi, Paravesicaspora splendens, Potonieisporites* sp., *Protohaploxipinus microcorpus, Verrucosisporites* sp. and *Vittatina* sp. Typical Thuringian forms, such as *Lueckisporites virkkiae* and *Nuskoisporites dulhuntyi* are associated with typical Autunian forms such as *Vittatina* sp. and *Potonieisporites* sp. This indicates an older age than the assemblages of the Alcotas Formation, but still in the Thuringian.

The same authors have found in the middle part of the Alcotas Formation of the Talayuelas section (Fig. 2) an assemblage composed of: *Lueckisporites virkkiae, L. microgranulatus, Klausipollenites schaubergerii, Nuskoisporites dulhuntyi, Protohaploxipinus microcorpus, Trizonaesporites grandis, Falcisporites zapfei* and *Plaifordiaspora crenulata.* They are of Late Permian (Thuringian) age. In the lower part of the Alcotas Formation, in the Minas de Henarejos section (Fig. 6), the same authors have found an assemblage composed of: *Lueckisporites virkkiae, L. microcorpus, Nuskoisporites dulhuntyi, Paravesicaspora splendens, Lundbladispora* sp., *Klausipollenites schaubergeri, Falcisporites zapfei, Protohaploxipinus sevardi, Platisaccus papillionis, Jugasporites perspicus, Gardenasporites oberrauchi* and *Converrucosisporites eggeri.* They are of Late Permian (Thuringian) age. In the classic section of Landete (Fig. 2), they found in the Alcotas Formation an assemblage composed of: *Lueckisporites virkkiae, Nuskoisporites dulhuntiy, Klausipollenites schaubergeri, Falcisporites nuttialensis, Platysaccus papillionis, Playfordiaspora crenulata, Jugasporites perspicuous* and *Corisaccites* sp. They are of Late Permian (Thuringian) age.

In the upper part of the Cañizar Formation of the Cañete section (Fig. 2), Doubinger *et al.* (1990) found a palynological assemblage composed of: *Allisporites toralis, Falcisporites cf. stabilis, Leiotriletes* sp. and *Lycospora* sp. These are of early Middle Triassic (early Anisian) age.

Catalan Ranges

The palynological data from this region are scarce (Visscher 1967; Solé de Porta *et al.* 1985, 1987). The Figaró Formation has yielded a poor assemblage, containing *Lundbladispora* sp. and *Cycadopytes* sp., of Early Triassic (?) age. Some units of the Muschelkalk and Keuper facies have yielded palynological assemblages of Mid-Late Triassic age.

Balearic Islands

The Late Permian to Early Triassic sediments of Mallorca and Menorca have yielded some rich palynological assemblages. Ramos (1979) and Ramos & Doubinger (1989) found an assemblage near the base of the Asá Formation composed of: *Nuskoisporitres dulhuntyi, Lueckisporites virkkiae, Klausipollenites schaubergeri, Falcisporites zapfei, Protohaploxipinus microcorpus, Paravesicaspora splendens, Lunasporites delasaucei, Crucisaccites variosulcatus, Endosporites velatus* and *Crustaesporites* sp. These are of Late Permian (Thuringian) age. Also, near the top of the Son Serralta Formation, there is an assemblage dominated by *Porcellispora longdonensis* and *Sulcosaccispora minuta*, which is of early Middle Triassic (Anisian) age.

The same formations can be identified in Menorca, where Bourrouilh (1973) found in the unnamed red mudstones at the base of the sequence an assemblage very similar to the one in the Port d'es Canonge Formation, which includes: *Lueckisporites virkkiae, Falcisporites schaubergeri, Nuskoisporites dulhuntyi* and *Taeniasporites angulistriatus.* They are of Late Permian (Thuringian) age. In the same outcrops on the northern coast of the island, Broutin *et al.* (1992) found in the same red mudstones an assemblage composed of: *Lueckisporites virkkiae, L. singhii, Lunatisporites* cf. *novicus, Klausipollenites schaubergeri, Falcisporites stabilis, Illinites unicus* and *Striatoabietites richteri.* They are of Late Permian (Thuringian) age.

Biostratigraphical significance of the Late Permian and Middle Triassic palynofloras

The palynofloras found in the Alcotas and Hoz del Gallo Formations contain the typical elements of the Thuringian Stage *sensu* Visscher (1971) (=Zechstein), such as *Lueckisporites virkkiae, Nuskoisporites dulhuntyi* and *Paravesicaspora splendens*, that he correlates with the Tatarian Stage of the Upper Permian of the Russian Platform. This assemblage is also comparable to Zone 29 of Gorsky *et al.* (2003), who also studied the palynofloras of this classic (Russian) area. There are numerous common elements, such as *Lycospora* sp., *Protohaploxipinus sevardi, Platysaccus papilionis, Lueckisporites virkkiae* and *Paravesicaspora* sp. The age of this assemblage is late Thuringian (=late Tatarian). As stated before, it is important to point out that the upper part of the Alcotas Formation is barren of organic remains

More problematic is the attribution of an age to the assemblage found in the Boniches Formation. The coexistence of typical Late Permian forms, such as *Lueckisporites virkkiae* and *Nuskoisporites dulhuntyi*, with typical Early Permian (Autunian) forms, such as *Vittatina* sp. and *Potonieisporites* sp., is found in Zones 30 and 31 of the Permian of the Russian Platform (Gorsky *et al.* 2003), which is from the lower Tatarian (Upper Permian) to the upper Kazanian (Middle Permian).

The sedimentary continuity between the top of the Boniches Formation and the base of the Alcotas Formation is evident in any field section. As the former is only 80–100 m thick, it is reasonable to assume that it could accumulate in only a few thousand years if we take into account that present-day deposits of comparable environment, thickness and extension can accumulate in about 30 000 years (Harvey 1990) or in 10 000–170 000 years in well-calibrated ancient examples, such as in Montserrat and Sant Llorens de Munt Fans, Eocene, Catalunya, Spain (López-Blanco *et al.* 2000). If this assumption is correct, only the top of Zone 30 of the Upper Permian of Gorsky *et al.* (2003) is recorded in these sediments.

As the duration of the Tatarian is 15 Ma (265–251 Ma: Menning 1995, 2002), only its uppermost part is represented in the Hoz del Gallo, Alcotas and Boniches formations, which is in the Thuringian of western Europe, as defined by Visscher (1971) and Uttig & Piasecki (1995). There is also a major hiatus between the Lower Permian (=Autunian) and the Upper Permian (Thuringian) deposits. This hiatus comprises part of the Sakmarian, the Artinskian, Kungurian and Kazanian stages, in sharp contrast with the more or less complete record of the Russian Platform. According to these data, Spain was part of the western European realm (Visscher 1971; Gorsky *et al.* 2003).

From the data, it is evident that the Permian–Triassic transition must lie somewhere between the upper part of the Hoz del Gallo Formation and the lower part of the Cañizar Formation, as will be discussed at length in the next section. The Illawarra magnetic anomaly, dated as 265 Ma (Menning 2002), which is, in the Guadalupian (Middle Permian), is much older than the base of the second sedimentary cycle and therefore is not recorded in Spain (Fig. 6).

The location of the Permian–Triassic transition in the Iberian Basin

As stated earlier in this paper, the Iberian Ranges hold one of the best-dated records of the Late Permian to Middle Triassic in the Iberian microplate, so the location of the Permian–Triassic transition will be discussed here, and, then, correlations with other areas will be drawn. The second sedimentary cycle is of Late Permian (Thuringian) age, but it is important to point out that the upper third of the Alcotas Formation is devoid of all organic remains, soil profiles and thick sandstone bodies.

A die-off of the flora and the absence of soil profiles could be related to local tectonic or fluvial sedimentology factors, but as this is a general feature over more than 400 km in the Iberian Ranges and a general feature in the Ebro and Catalan basins, it can only be explained by an extra-basinal, general process, not by local tectonic, climatic or taphonomic causes, as no geochemical, mineralogical, tectonic or sedimentological changes within the middle and upper parts of the formation have been identified. The most plausible cause is the coeval emplacement of the China Emeishan basalts, a large igneous province of moderate size (Thompson *et al.* 2001). If this hypothesis can be confirmed, a good chronostratigraphical reference horizon will be established in the search for the position of the Permian–Triassic transition in the Iberian basin and the rest of central and NE Iberia.

The absolute age of the Emeishan basalts is now well established at 259 ± 3 Ma for the main phase of volcanism (Zhou *et al.* 2002), which is clearly in the Tatarian stage. The widely quoted analytical results of Lo *et al.* (2002), giving an absolute age of 252.8 ± 1.3 Ma for the same levels, have been questioned and rejected by Courtillot & Renne (2003) because of flawed and incorrect analytical procedures.

The Emeishan basalts are older than the Permian–Triassic transition in the standard section of Meishan, China (Mundil *et al.* 2001*a*,*b*, 2004; Yin *et al.* 2001) and their emplacement should be related to the so-called 'end-Guadalupian' biotic crisis (Ali *et al.* 2002; Thompson *et al.* 2001; Courtillot & Renne 2003), in spite of the small discrepancy between the accepted age for the Guadalupian–Lopingian boundary and the youngest part of the Emeishan basalts (Courtillot *et al.* 1999; Wignall 2001). The upper part of the Alcotas Formation could be coeval with the emplacement of the Emeishan basalts because pollen and spore assemblages of Zone 29 of the upper Tatarian of the Russian Platform (Gorsky *et al.* 2003) are found immediately beneath it, and the proposed absolute ages of this zone are similar to the above-mentioned ones for the basaltic flows of SE China. The causal relationship between the volcanism and the absence of any organic remains after a severe biotic crisis caused by atmospheric pollution by aerosol and gas poisoning of the atmosphere during the deposition of the upper part of the Alcotas Formation is at least reasonable, if not proven beyond any doubt.

The third sedimentary cycle, lying unconformably on the second one after a phase of uplifting and erosion more marked in the NW part of the Iberian Basin, started with renewed extensional tectonics along the SW basin boundary fault (Serranía de Cuenca fault), a situation comparable to the beginning of the second sedimentary cycle.

The presence of Late Permian (Thuringian) pollen and spore assemblages in the lower part of the Hoz del Gallo Formation, at the base of the cycle (Fig. 7), mark a biotic recovery after the barren period at the top of the underlying Alcotas Formation and different preservation conditions (Ramos & Doubinger 1979; Ramos & Sopeña 1983; Sopeña *et al.* 1995). Renewed tectonic activity created high topographical areas and energetic relief that could provide a refuge for vegetation, since they created topographical rains at a time when the Late Permian icecaps melted away completely (Beauchamps & Baud 2002), increasing the temperatures at the equatorial belt and enhancing the cyclonic storms in the Western Tethys (Stampfli & Borel 2002).

The age of the palynological assemblages found in the Hoz del Gallo Formation is identical to the ones found in the Alcotas Formation, so the unconformity at the base of the third sedimentary cycle is intra-Late Permian, and the Permian–Triassic must lie above it and the lower part of the Hoz del Gallo Formation. The upper part of the Hoz del Gallo Formation is barren of organic remains and marks a sudden change in palaeocurrents, sedimentological and petrological characteristics (Ramos & Sopeña 1983; Ramos *et al.* 1986), caused by another tectonic pulse on the Serranía de Cuenca fault system leading to the enlargement of the rift basin and the tapping of new source areas to the NW containing igneous and metamorphic rocks.

The youngest formation considered in this paper, the Cañizar Formation, in well-defined, conformable contact with the underlying Hoz del Gallo Formation or unconformably on the Alcotas Formation, is interpreted as deposits of sandy, braided rivers of extreme channel instability and frequent shifting of the active depositional zone (Arche & López-Gómez 1999). These fluvial systems found no obstacles for the lateral migration of active channels across a wide, flat alluvial plain, in spite of small to moderate stream power, as indicated by the metric scale of the internal structures. No organic remains have been found in this formation except a lower Anisian pollen and spore assemblage at the top (Doubinger *et al.* 1990).

The vertical transition from Late Permian high-sinuosity rivers, with thick floodplain deposits, rich in plant and pollen and spore remains to Late Permian – Early Triassic sandy, braided river systems devoid of fines and organic remains is a remarkable feature of very distant, coeval sedimentary basins, such as the Karoo Basin, South Africa (Smith 1995; Ward *et al.* 2000; Smith & Ward 2001; Retallack *et al.* 2003), the Sydney Basin, Australia (Miall & Jones 2003), the Sanga do Cabral-Santa Maria Basin, Brazil (Zerfass *et al.* 2003), the Collio and Orobic basins, Italian Alps (Ronchi & Santi 2003) and the South Devon Basin, England (Audley-Charles 1970; Smith *et al.* 1974; Laming 1980), among many others. According to the biostratigraphical data presented previously, the Permian–Triassic transition can be placed between the upper part of the Hoz del Gallo Formation and an undetermined level of the lower part of the Cañizar Formation, but it is possible to constrain the possible location to a narrower bracket if we consider global palaeogeographic and geodynamic aspects of this period. The presence of alluvial plains without significant vegetation cover, increased bed-load, a substantial decrease in the fine supply from the source area and the high instability of the active channels over thousands of kilometres cannot be explained by just local or even regional tectonic activity or regional climatic change because of the large area where these features are present; a global cause may be the explanation.

The hypothesis of a rapid episode of plant die-off at the Permian–Triassic transition has been proposed by Smith (1995) and Ward *et al.* (2000) for the Karoo Basin, combined with substantial decrease of oxygen and parallel increase of CO_2 in the atmosphere (Retallack *et al.* 2003). These phenomena could explain the general vertical transition from the Alcotas Formation to the Hoz del Gallo and Cañizar formations. In any case, the ultimate cause of the die-off needs an explanation.

Once again it is very tempting to invoke a major volcanic event, such as the emplacement of the West Siberia Basaltic LIP. These basaltic rocks, up to 6500 m thick, covered more than 4 million km^2 in about 600 000 years, from 251.7 ± 0.4 Ma to 251.1 ± 0.3 Ma (Renne *et al.* 1995; Reichow *et al.* 2002; Büchl & Gier 2003; Kamo *et al.* 2003). If the age of the Permian–Triassic Transition at Meishan, SE China has been established at 251.4+−0.2 Ma (Yin *et al.* 2001; Menning 2001; Wardlaw & Schiappa 2002), the gigantic outburst of basaltic magma was coeval with the Permian–Triassic transition. It is now very important to establish some precision in the absolute age of this transition at the standard section of Meishan, because the value of 251 ± 0.2 Ma (Menning 2001), in total coincidence with the ages proposed for the Western Siberia basalts (Renne *et al.* 1995; Reichow *et al.* 2002; Büchl & Gier 2003; Kamo *et al.* 2003) has been revised by Mundil *et al.* (2004) and corrected to 252.1 ± 1.6 Ma. Even if this new age for the transition is accepted, the biotic crisis at the base of the Triassic and the emplacement of the Western Siberia Basalts coincide, if we consider the error margins of the analytical methods and the minor inconsistencies in the comparison of absolute age datasets obtained by different geochemical procedures (Bowring *et al.* 1998; Metcalfe & Mundil 2001; Metcalfe *et al.* 2001; Mundil *et al.* 2001*a,b*). The enormous amounts of SO_2, CO_2, HF, Cl and other gases released in a brief period of time into the atmosphere would trigger a rapid climatic change with increased temperatures, poisoning by acid rain and disruption of the food chains, leading to major biotic extinctions both on land and in the sea (Renne & Basu 1991; Erwin 1994; Renne *et al.* 1995; Twitchett *et al.* 2001). The die-off of most terrestrial vegetation, from peat to conifers, would cause a switch to a braided configuration of the alluvial channels due to the loss of bank stability normally provided by the riparian vegetation. Bedload would increase because of the rapid denudation of unprotected regoliths in source areas, geochemical cycles leading to the formation of clays would virtually cease in absence of humic acids and the impact on terrestrial vertebrates would be enormous (Michaelsen 2002). For a period of time around and immediately after the Permian–Triassic transition, extensive areas of semi-desert or desert climate were created in equatorial and tropical Pangaea. In the continental rift basins of the Western Tethys realm (Fig. 7), sheet-like, sandy braided river deposits were accumulating not only in the small Iberian and Catalan basins, but in many other interconnected or isolated basins.

As the emplacement of the Western Siberia Basalts is coeval with the Permian–Triassic boundary, and one of the consequences of the die-off of most of the terrestrial vegetation would be the establishment of sandy braided river systems, then the Permian–Triassic transition can be placed in a narrow vertical bracket (Fig. 6) where the fluvial style changes: the top of the Hoz del Gallo Conglomerates Formation and the first few metres of the Cañizar Sandstones Formation. If this causal connection is correct, it would also explain the absence of any organic remain in most of the third sedimentary cycle. At the beginning of the Middle Triassic (early Anisian), the recovery of the environment allowed a radiation of new plants, and pollen and spores are found again at the top of the Cañizar Formation (Doubinger *et al.* 1990), and a new change in fluvial style led to the deposition of a new sedimentary cycle (the fourth) (Fig. 4), composed of more than 60% red mudstones and isolated simple sand ribbons or laterally restricted multistorey channel fills (Eslida Formation, not examined in this paper, also dated as Anisian, marking the recovery of the terrestrial ecosystems: López-Gómez *et al.* 2002).

If a constant rate of accumulation is assumed for the Cañizar Sandstones Formation, and the age of its top is placed at the early Anisian (about 244 Ma), and the age of the lower part of the Hoz del Gallo Conglomerates Formation is estimated as latest Permian (about 252 Ma), this rate can be estimated at 15 m per million years. In this case the probable position of the Permian–Triassic boundary can be placed in the lower 10 m of the Cañizar Formation, but uncertainties about the exact age of the topmost beds of the Hoz del Gallo Formation cannot rule out the possibility of this boundary being there, although, if this is the case, extremely conservative rates of accumulation should be assumed for the latter unit.

If the proposed position of the Permian–Triassic Basin is correct, then the P/T transition in other basins of the Iberian microplate (Fig. 8) can be located with some precision: the top of the Rio Pedro Conglomerates (T1.1) and the base of the Rio Pedro Sandstones Formations (T1.2)

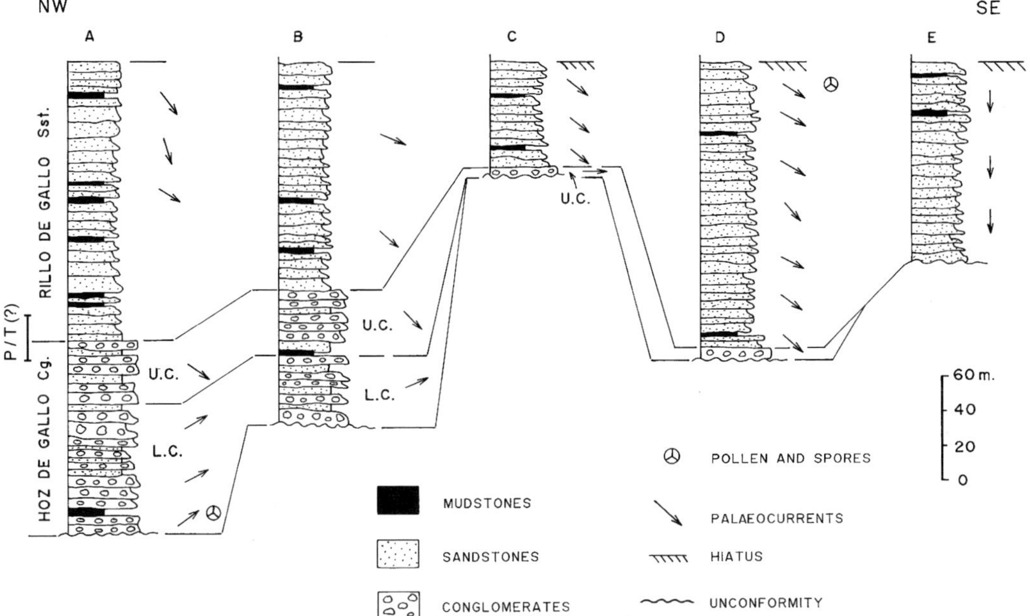

Fig. 8. Proposal of correlation of the Late Permian–Mid-Triassic sediments in central and NE mainland Spain and the Balearic Islands, and the most probable position of the Permian–Triassic transition. Note the uncertainty bracket on the left. Modified from Arche *et al.* (2004). A, Molina de Aragón; B, Albarracin; C, Valdemeca; D, Cañete; E, Chelva; U.C., Upper Conglomerates; L.C., Lower Conglomerates.

(Hernando 1977, 1980) in the NW Iberian Basin (Fig. 4); the top of the Prades and Caldes Conglomerates and the base of the Prades, Eramprunyá and Caldes formations in the Catalan Basin (Fig. 6); the base of the Aranda Formation in the Ebro Basin (Arribas 1984, 1985) (Fig. 6) or its subsurface equivalents; the B-1 conglomerates and the base of the B-2 sandstones (Jurado 1988) and the base of the Asa Formation in Mallorca (Ramos 1995) and the B-1 unit of Menorca (IGME 1989). This is according to the general correlation for the Permian and Triassic deposits of central and Spain proposed by Arche *et al.* (2002). It is evident that further palynological, vertebrate and palaeomagnetic research is needed to confirm this correlation and to locate more precisely the P/T transition, combined with geochemical and ichnofossil data. However, the available data allow for a great advance in comparison with previous studies.

Finally, there is a fact that needs further consideration: the apparent heterochrony between the sedimentological change from high-sinuosity, clay-rich fluvial environments of the Late Permian (middle and upper units of the Alcotas Formation) to the low-sinuosity, sandy braided river deposits of the latest Permian to Early Triassic (lower part of the Cañizar Formation), a sedimentological evolution found in several basins around the world. The Permian–Triassic transition is located in the upper part of the clay-rich unit in the Karoo Basin, South Africa (Retallack *et al.* 2003), before the sedimentological change, but it is located in the overlying sandy unit or 'Verrucano Sardo' in Sardinia, Italy (Fontana *et al.* 2001), in well-dated sections such as Lu Caparoni and Cala Viola, the Iberian Basin of Spain, of the Serbian Karpathos (Maslarevic & Krstic 2001). These problems can be solved if we consider the difficulties in establishing detailed correlations based in different fossil groups of distant faunal and floristic provinces. Another clue to this apparent anomaly is the often neglected fact that there are two successive extinction events very close in time: the 'end-Guadalupian' one, related to the SE China basalt flows, and the 'end-Permian' one, related to the Western Siberia basalt flows.

Conclusions

1. The most complete sections and most accurate biostratigraphical data to investigate the Permian–Triassic transition in central

and NE Spain are found in the Iberian Ranges.

2. An angular unconformity of Late Permian age separates the second and third sedimentary cycles in the Iberian Ranges and other basins of Spain, so the P/T transition lies above it.
3. Biostratigraphical data show two barren intervals in the sedimentary record: one at the top of the Alcotas Formation (second sedimentary cycle), during the Late Permian, and another at the top of the Hoz del Gallo Formation and most of the Cañizar Formation (third sedimentary cycle), during the latest Permian to Early Triassic.
4. The two barren intervals are possibly the consequence of the emplacement of the SE China (Emeishan) and Western Siberia Large Igneous Basaltic Provinces.
5. The P/T transition in the Iberian Ranges can be placed in a narrow sedimentary interval between the upper part of the Hoz del Gallo Formation and the lowermost part of the Cañizar Formation, according to biostratigraphical data and calculations of absolute rates of sedimentation in this period.
6. The barren interval in the Cañizar Formation, spanning about 4 Ma, agrees well with the slow recovery of the terrestrial environments after the end-Permian extinction event.
7. The probable position of the P/T transition in other parts of central and NE mainland Spain and the Balearic Islands is proposed based on a general correlation among these basins.

We thank C. Sanchez and M. Escudero for their help with the illustrations and the S. Lucas for his encouragement and patience during the preparation of this paper and his careful revision of an early version of this paper. The authors thank L. Tanner and K. E. Ziegler for their careful revision and comments that improved the final version of this paper. This is a contribution to Project CGL2005-01520 BTE of the Spanish Ministry of Science and Technology.

References

ALEXANDER, J. & LEEDER, M. R. 1987. Active tectonic control on alluvial architecture. *In*: ETHRIDGE, F. G., FLORES, R. M. & HARVEY, M. D. (eds) *Fluvial Sedimentology*. Society of Economic Petrology and Mineralogy, Special Publications, **39**, 243–252.

ALI, J. R., THOMPSON, G. M., SONG, X. & WANG, Y. 2002. Emeishan basalts (SW China) and the 'end-Guadalupian' crisis: magnetobiostratigraphic constraints. *Journal of the Geological Society, London*, **159**, 21–29.

ALLEN, J. R. L. 1983. Studies in fluviatile sedimentation: bars, bar complexes and sandstone sheets. *Sedimentary Geology*, **33**, 237–293.

ALONSO-AZCÁRATE, J., ARCHE, A., BARRENECHEA, J. F., LÓPEZ-GÓMEZ, J., LUQUE, F. J. & RODAS, M. 1997. Paleogeographical significance of the clay minerals in the Permian and Triassic sediments of the SE Iberian Ranges, Spain. *Palaeogeography, Palaeoclimatology, Palaeoecology*, **136**, 309–330.

ARCHE, A. LÓPEZ-GÓMEZ, J. 1996. Origin of the Permian–Triassic Basin, central Spain. *Tectonophysics*, **266**, 433–464.

ARCHE, A. & LÓPEZ-GÓMEZ, J. 1999. Subsidence rates and fluvial architecture of rift-related Permian and Triassic sediments of the SE Iberian Range, eastern Spain. *In*: SMITH, N. D. & ROGERS, N. (eds) *Fluvial Sedimentology Vol 6*. International Association of Sedimentology, Special Publications, **28**, 283–304.

ARCHE, A., LÓPEZ-GÓMEZ, J. & VARGAS, H. 2002. Propuesta de correlación entre los sedimentos pérmicos y triásicos de la Cordillera Ibérica Este y de las Islas Baleares. *Geogaceta*, **32**, 275–278.

ARCHE, A., LÓPEZ-GÓMEZ, J., MARZO, M. & VARGAS, H. 2004. The siliciclastic Permian–Triassic deposits in central and northeastern Iberian Peninsula (Iberian, Ebro and Catalan Basins): a proposal for correlation. *Geologica Acta*, **2**, 305–320.

ARRIBAS, J. 1984. *Sedimentología y diagénesis del Buntsandstein y Muschelkalk de la Rama Aragonesa de la Cordillera Ibérica (Provincias de Soria y Zaragoza)*. PhD thesis, Universidad Complutense, Madrid.

ARRIBAS, J. 1985. Base litostratigráfica de las facies Buntsandstein y Muschelkalk de la Rama Aragonesa de la Cordillera Ibérica (Zona Norte). *Estudios Geológicos*, **41**, 47–57.

ASHMORE, P. E. 1991. How do gravel-bed rivers braid? *Canadian Journal of Earth Sciences*, **28**, 326–341.

AUDLEY-CHARLES, M. D. 1970. Stratigraphical correlation of the Triassic rocks of the British Isles. *Quarterly Journal of the Geological Society, London*, **126**, 19–47.

BARRON, E. J. & FAWCETT, P. J. 1995. The climate of Pangea: a review of climate model simulations of the Permian. *In*: SCHOLLE, P. A., PERYT, T. M. & ULLMER-SCHOLLE, D. S. (eds) *The Permian of Northern Pangea. Vol 1*. Springer Verlag, Berlin, 37–52.

BEAUCHAMP, B. & BAUD, A. 2002. Growth and demise of Permian biogenic chert along northwest Pangea: evidence for end-Permian collapse of thermohaline circulation. *Palaeogeography, Palaeoclimatology, Palaeoecology*, **184**, 37–63.

BOULOUARD, CH. & VIALLARD, P. 1971. Identification du Permien dans la Chaîne Ibérique. *Comptes Rendues de l'Académie des Sciences, Paris, Serie B*, **273**, 2441–2444.

BOURROUILH, R. 1973. *Stratigraphie, sédimentologie et tectonique de l'île de Minorque et du NE de Majorque (Baleares, Espagne)*, PhD thesis, Université Pierre et Marie Curie, Paris.

BOWRING, S. A., ERWIN, D. H., JIN, Y. G., MARTIN, M. W., DAVIDEK, K. & WANG, W. 1998. U/Pb zircon geochronology and tempo of the end-Permian mass extinction. *Science*, **280**, 1039–1045.

BROUTIN, J., FERRER, J., GISBERT, J. & NMILA, A. 1992. Prémière découverte d'une microflore thuringienne dans le faciès saxonien de l'île de Minorque (Baleares, Espagne). *Compte Rendues de l'Académie des Sciences, Paris, Serie B*, **315**, 117–122.

BÜCHL, A. & GIER, S. 2003. Petrogenesis and alteration of tuffs associated with continental flood basalts from Putorana, northern Siberia. *Geological Magazine*, **140**, 649–659.

CALVET, F. & MARZO, M. 1994. *El Triásico de las Cordilleras Costero Catalanas*. Field Guide. III Coloquio Estratigrafía y Sedimentología del Triásico y Pérmico de España, Cuenca.

CAMPBELL, C. V. 1976. Reservoir geometry of a fluvial sheet sandstone. *American Association of Petroleum Geologists Bulletin*, **60**, 1009–1020.

CANT, D. J. & WALKER, R. G. 1978. Fluvial processes and facies sequences in the sandy braided South Saskatchewan River, Canada. *Sedimentology*, **25**, 625–648.

COLLINSON, J. D. 1970. Bedforms of the Tana River, Norway. *Geographyska Annaler*, **52A**, 31–55.

CONTE, J. C., GASCÓN, F., LAGO, M. & CARLS, P. 1987. Materiales stephano-pérmicos en la Fosa de Fombuena (provincia de Zaragoza). *Boletín del Instituto Geológico y Minero de Espana*, **98**, 460–470.

COURTILLOT, V. & RENNE, P. 2003. On the age of flood basalt events. *Comptes Rendus de Geosciences*, **335**, 113–140.

COURTILLOT, V., JAUPART, I. MANIGHETTI, P., TAPPONIER, J. & BESSE, J. 1999. On causal links between flood basalts and continental breakup. *Earth and Planetary Science Letters*, **166**, 177–195.

CROWLEY, K. D. 1983. Large-scale bed configurations (macroforms), Platte River Basin, Colorado and Nebraska: primary structures and formative processes. *Geological Society of America Bulletin*, **94**, 117–133.

DE LA PEÑA, J. A., FONOLLÁ, P., RAMOS, J. L. & MARFIL, R. 1977. Identificación del Autuniense en la Rama Aragonesa de la Cordillera Ibérica (Provincia de Soria). *Cuadernos de Geología Ibérica*, **4**, 123–134.

DÍEZ, J. B. 1997. *Geología y Paleobotánica de las facies Buntsandstein en la Rama Aragonesa de la Cordillera Ibérica: implicaciones biostratigráficas en el Peritethys occidental.* PhD thesis, Universidad de Zaragoza–Université Pierre et Marie Curie, Paris.

DOBLAS, M., OYARZUN, R., SOPEÑA, A. & SÁNCHEZ-MOYA, Y. 1993. Variscan–Late Variscan–Early Alpine progressive extensional collapse of central Spain. *Geodinamica Acta*, **7**, 1–14.

DOUBINGER, J., ADLOFF, M. C., RAMOS, A., SOPEÑA, A. & HERNANDO, S. 1978. Primeros estudios palinológicos en el Pérmico y Triásico del Sistema Ibérico y bordes del Sistema Central. *Revista de Palinología*, **1**, 27–33.

DOUBINGER, J., LÓPEZ-GÓMEZ, J. & ARCHE, A. 1990. Pollen and spores from the Permian and Triassic sediments of the SE Iberian Ranges, Cueva de Hierro (Cuenca) to Chelva-Manzanera (Valencia) region, Spain. *Review of Paleobotany and Palynology*, **66**, 25–45.

EREN, Y., KURT, H., ROSSELET, F. & STAMPFLI, G. M. 2004. Paleozoic deformation and magmatism in the Northern area of the Anatolide block (Konya): witness of the Paleotethys active margin. *Eglogae Geologica Helveticae*, **97**, 293–306.

ERWIN, D. H. 1994. The Permo-Triassic extinction. *Nature*, **367**, 231–236.

FONTANA, D., NERI, C., RONCHI, A. & STEFANI, C. 2001. Stratigraphic architecture and composition of the Permian and Triassic succession of Nurra (NW Sardinia). *In*: CASSINIS, G. (ed.) *Permian Continental Deposits of Europe. Reports and Correlation*, Natura Bresciana, Monografia, **25**, 149–161.

FLUTEAU, F., BESSE, J., BROUTIN, J. & RAMSTEIN, G. 2001. The Late Permian climate. What can be inferred from climate modelling concerning Pangea scenarios and Hercynian range altitude? *Palaeogeography, Palaeoclimatology, Palaeoecology*, **167**, 39–71.

GORSKY, V. P., GUSSEVA, E. A., CRASQUIN-SOLEAU, S. & BROUTIN, J. 2003. Stratigraphic data of the Middle–Late Permian of the Russian Platform. *Géobios*, **36**, 533–556.

GRADSTEIN, T., OGG, J. & SMITH, A. 2004. *A Geologic Time Scale 2004.* Cambridge University Press, Cambridge.

HARVEY, A. M. 1990. Factors influencing Quaternary alluvial fan development in SE Spain. *In*: RACHOCKY, A. H. & CHURCH, M. (eds) *Alluvial Fans: A Field Approach,* John Wiley, Chichester, 247–261.

HERNANDO, S. 1977. *Pérmico y Triásico de la región de Ayllón-Atienza (provincias de Segovia, Soria y Guadalajara).* Seminarios de Estratigrafía, Monografias, **2**.

HERNANDO, S. 1980. Mapa geológico del Pérmico y Triásico de la región de Ayllón-Atienza. *Cuadernos de Geología Ibérica,* **6**, 21–54.

HERNANDO, S., SCHOTT, J. J., THUIZAT, R. & MONTIGNY, R. 1980. Age des andésites et des sediments interstratifies de la region d'Atienza (Espagne): etude stratigraphique, géochronologique et paléomagnetique. *Sciences Geologiques Bulletin,* **33**, 117–128.

HORRA, R., DE LA, LÓPEZ-GÓMEZ, J. & ARCHE, A. 2005. Caracterización de la unidad Conglomerados de Valdemeca en la transición Pérmico-Triásico de la Cordillera Iberica centro-oriental. *Geo-Temas,* **8**, 141–145.

IGME 1989. Maó, Hoja 647, Mapa Geologica de España 1:50 000, Madrid.

JONES, C. M. 1977. Effects of varying discharge regimes on bedform sedimentary structures in modern rivers. *Geology,* **5**, 567–570.

JURADO, M. J. 1988. *El Triásico del subsuelo de la Cuenca del Ebro.* PhD thesis, Universidad de Barcelona.

JURADO, M. J. 1990. El Triásico y el Jurásico basal evaporíticos del subsuelo de la Cuenca del Ebro. *In*: ORTÍ, F. & SALVANY, J. (eds) *Formaciones Evaporíticas de La Cuenca del Ebro y Cadenas*

Periféricas y de La Zona de Levante. Enresa, Madrid, 21–28.

KAMO, S. L., CZAMANSKE, G. K., AMELIN, Y., FEDORENKO, V. A., DAVIS, D. W. & TROFIMOV, V. R. 2003. Rapid eruption of Siberian flood-volcanic rocks and evidence for coincidence with the Permian–Triassic boundary and mass extinction at 251 M.a. *Earth and Planetary Science Letters*, **214**, 75–91.

LAGO, M., POCOVI, A., ZACHMANN, D., ARRANZ, E., CARLS, P., TORRES, J. A. & VAQUER, R. 1991. Comparación preliminar de las manifestaciones magmáticas calcoalcalinas stephanienses-pérmicas de la Cadena Ibérica. *Cuadernos Geológicos de Laxe*, **16**, 95–107.

LAGO, M., ARRANZ, E. & GALÉ, C. 2002. Stephanian-Permian volcanism of the Iberian Ranges and Atienza. *In*: GIBBONS, W. & MORENO, T. (eds) *The Geology of Spain.* Geological Society, London, 126–128.

LAGO, M., ARRANZ, E., POCOVI, A., GALÉ, C. & GIL, A. 2004. Lower Permian magmatism of the Iberian Chain, Central Spain, and its relationship to extensional tectonics. *In*: WILSON, M., NEUMANN, E. R., DAVIES, G., TIMMERMANN, M., HEEREMANS, M. & LARSEN, B. T. (eds) *Permo-Carboniferous Magmatism and Rifting*, Geological Society, London, Special Publications, **223**, 465–491.

LAMING, D. J. C. 1980. The new Red sandstone. *In*: DURRANCE, E. M. & LAMING, D. J. C. (eds) *The Geology of Devon.* Exeter University Press, Exeter, 148–178.

LO, C., CHING, S., LEE, T. & WU, G. 2002. Age of the Emeishan flood magmatism and relations to Permian–Triassic boundary events. *Earth and Planetary Science Letters*, **198**, 449–458.

LÓPEZ-BLANCO, M., MARZO, M. & PIÑA, J. 2000. Transgressive-regressive sequence hierarchy of foreland fan-delta clastic wedges (Montserrat and Snat Llorenc del Munt, Middle Eocene, Ebro Basin, NE Spain). *Sedimentary Geology*, **138**, 41–69.

LÓPEZ-GÓMEZ, J. 1985. *Sedimentología y Estratigrafía de los materiales pérmicos y triásicos del sector SE de la Cordillera Ibérica entre Cueva de Hierro y Chelva (provincias de Cuenca, Guadalajara y Valencia).* Seminarios de Estratigrafía, Monografías, **11**.

LÓPEZ-GÓMEZ, J. & ARCHE, A. 1993. Sequence stratigraphy analysis and paleogeographic interpretation of the Buntsandstein and Muschelkalk facies (Permian–Triassic) in the Iberian Ranges. *Palaeogeography, Palaeoecology, Palaeoclimatology*, **103**, 179–201.

LÓPEZ-GÓMEZ, J. & ARCHE, A. 1994. Pérmico y Triásico del SE de la Cordillera Ibérica. Field guide *IV Coloquio de Estratigrafía y Sedimentología del Pérmico y Triásico de España, Cuenca.*

LÓPEZ-GÓMEZ, J. & ARCHE, A. 1997. The Late Permian Boniches Conglomerates Formation: evolution from alluvial fan to fluvial system environments and accompanying tectonic and climatic controls in the SE Iberian Ranges, central Spain. *Sedimentary Geology*, **114**, 267–298.

LÓPEZ-GÓMEZ, J., ARCHE, A. & DOUBINGER, J. 1984. El Triásico del Anticlinorio de Cueva de Hierro (Serranía de Cuenca), España. *Revista Española de Micropaleontología*, **16**, 19–42.

LÓPEZ-GÓMEZ, J., ARCHE, A., CALVET, F. & GOY, A. 1998. Epicontinental marine carbonate sediments of the Middle and Upper Triassic in the westernmost part of the Tethys Sea. *Zentralblatt für Geologie und Paläontologie*, **1**(9–10), 1033–1084.

LÓPEZ-GÓMEZ, J., ARCHE, A., PÉREZ-LÓPEZ, 2002. Permian and Triassic. *In*: GIBBONS, W. & MORENO, T. (eds) *The Geology of Spain.* Geological Society, London, 185–212.

MARZO, M. 1980. *El Buntsandstein de los Catalánides: estratigrafía y procesos de sedimentación.* PhD thesis, Universidad de Barcelona.

MASLAREVIC, L. & KRSTIC, B. 2001. Continental Permian and Triassic red beds of the Serbian Karpato-Balkanides. *In*: CASSINIS, G. (ed.) *Permian Continental Deposits of Europe. Regional Reports and Correlations.* Natura Bresciana, Monografia, **25**, 245–252.

MELÉNDEZ, B., TALENS, J., FONOLLÁ, J. F. & ALVAREZ-RAMIS, C. 1983. Las cuencas carboníferas del sector central de la Cordillera Ibérica (Henarejos y Montalbán). *In*: MARTINEZ, C. (ed.) *Carbonífero y Pérmico de España.* Ministerio de Industria, Madrid, 207–220.

MENNING, M. 1995. A numerical time scale for the Permian and Triassic periods: an integrated time analysis. *In*: SCHOLLE, P. A., PERYT, T. M. & ULMER-SCHOLLE, D. S. (eds) *The Permian of Northern Pangea.* Springer, Berlin, 77–97.

MENNING, M. 2001. A Permian time scale 2001 and a correlation of marine and continental sequences using the Illawarra reversal (265 M.a.). *In*: CASSINIS, G. (ed.) *Permian Continental Deposits. Regional Reports and Correlations.* Natura Bresciana, Monografia, **25**, 355–362.

METCALFE, I. & MUNDIL, R. 2001. Age of the Permian–Triassic boundary and mass extinction. *Permophiles*, **39**, 11–12.

METCALFE, I., NICOLL, R. S. *et al.* 2001. The Permian–Triassic boundary and mass extinction in China. *Episodes*, **24**, 239–244.

MIALL, A. D. 1978. *Fluvial Sedimentology: A Historical Review.* Canadian Journal of Petroleum Geology, Memoir, **5**, 1–47.

MIALL, A. D. 1981. *Alluvial Sedimentary Basins: Tectonic Setting and Basin Architecture.* Geological Association of Canada, Special Paper, **23**, 1–33.

MIALL, A. D. 1987. Recent developments in the study of fluvial facies models. *In*: ETHRIDGE, F. G. & FLORES, R. M. (eds) *Recent Developments in Fluvial Sedimentology*, Society of Economic Petrology and Mineralogy, Special Publications, **39**, 1–9.

MIALL, A. D. 1992. Alluvial deposits. *In*: WALKER, R. G. & JAMES, N. P. (eds) *Facies Models.* Geological Association of Canada, 119–142.

MIALL, A. D. & JONES, B. G. 2003. Fluvial architecture of the Hawkesbury Sandstone (Triassic), near Sydney, Australia. *Journal of Sedimentary Research*, **73**, 531–545.

MICHAELSEN, P. 2002. Mass extinction of peat-forming plants and the effect on fluvial styles across the Permian–Triassic boundary, northern Bowen Basin, Australia. *Palaeogeography, Palaeoclimatology, Palaeoecology*, **179**, 173–188.

MUNDIL, R., METCALFE, I., LUDWIG, K. R., RENNE, P. R., OBERLI, F. & NICOLL, R. S. 2001*a*. Timing of the Permian–Triassic biotic crisis: implications from new zircon U/Pb data (and their limitations). *Earth and Planetary Science Letters*, **187**, 131–145.

MUNDIL, R., METCALFE, I., LUDWIG, K.R., RENNE, P.R., OBERLI, F. & NICOLL, R.S. 2001*b*. Timing of the Permian–Triassic biotic crisis: implications from new U/Pb dating of zircon data and their limitations. *Earth and Planetary Science Letters*, **187**, 131–145.

MUNDIL, R., LUDWIG, K. R., METCALFE, I. & RENNE, P. R. 2004. Age and timing of the Permian mass extinctions: U/Pb dating of closed-system zircons. *Science*, **305**, 1760–1763.

MUÑOZ, M., ANCOCHEA, E., SAGREDO, J., DE LA PEÑA, J. A., HERNAN, F., BRANDLE, J. L. & MARFIL, R. 1985. Vulcanismo permo-triásico de la Cordillera Ibérica. *VIII Congrès International du Carbonifère, Comptes Rendues*, Ministerio de Industria, Madrid, **3**, 27–52.

O'BRIEN, P. E. & WELLS, A. T. 1986. A small crevasse splay. *Journal of Sedimentary Petrology*, **56**, 876–879.

ORTÍ, F. 1987. Aspectos sedimentológicos de las evaporitas del Triásico y el Liásico inferior en el Este de la Península Ibérica. *Cuadernos de Geología Ibérica*, **11**, 837–858.

ORTÍ, F. & PÉREZ-LÓPEZ, A. 1994. El Triásico Superior de Levante. Field Guide. *III Congreso de Estratigrafía y Sedimentología del Pérmico y Triásico de España, Cuenca*.

PÉREZ-ARLUCEA, M. 1985. *Estratigrafía y sedimentología del Pérmico y Triásico en el sector Molina de Aragón-Albarracín (Provincias de Guadalajara y Teruel)*. PhD thesis, Universidad Complutense, Madrid.

PÉREZ-ARLUCEA, M. & SOPEÑA, A. 1985. Estratigrafía del Pérmico y Triásico en el sector central de la Rama Castellana de la Cordillera Ibérica (Provincias de Guadalajara y Teruel). *Estudios Geológicos*, **41**, 207–222.

PUIGDEFÁBREGAS, C. 1973. Miocene point bar deposits in the Ebro Basin, northern Spain. *Sedimentology*, **20**, 133–144.

RAMOS, A. 1979. *Estratigrafía y paleogeografía del Pérmico y Triásico al oeste de Molina de Aragón (Provincia de Guadalajara)*. Seminarios de Estratigrafía, Monografías, **6**.

RAMOS, A. 1980. Mapa geológico del Pérmico y Triásico de la región de Alcolea del Pinar-Molina de Aragón. *Cuadernos de Geología Ibérica*, **6**, 55–72.

RAMOS, A. 1995. Transition from alluvial to coastal deposits (Permian–Triassic) on the island of Mallorca, western Mediterranean. *Geological Magazine*, **132**, 435–447.

RAMOS, A. & DOUBINGER, J. 1979. Découverte d'une microflore thuringienne dans le Buntsandstein de la Cordillère Ibérique. *Comptes Rendues de l'Academie des Sciences, Paris, Serie B*, **289**, 525–528.

RAMOS, A. & DOUBINGER, J. 1989. Prémières datations palinologiques dans les faciès Buntsandstein de l'île de Majorque (Baleares, Espagne). *Comptes Rendues de l'Academie des Sciences, Paris, Serie B*, **309**, 1089–1094.

RAMOS, A. & SOPEÑA, A. 1983. Gravel bars in low sinuosity streams, Permian and Triassic, central Spain. *In*: COLLINSON, J. D. & LEWIN, J. (eds) *Modern and Ancient Fluvial Systems*. International Association of Sedimentology, Special Publications, **6**, 301–312.

RAMOS, A., DOUBINGER, J. & VIRGILI, C. 1976. El Pérmico Inferior de Rillo de Gallo (Guadalajara). *Acta Geologica Hispanica*, **11**, 65–70.

RAMOS, A., SOPEÑA, A. & PÉREZ-ARLUCEA, M. 1986. Evolution of Buntsandstein fluvial sedimentation in NW Iberian Ranges (central Spain). *Journal of Sedimentary Geology*, **56**, 862–875.

RETALLACK, G. J., SMITH, R. M. H. & WARD, P. D. 2003. Vertebrate extinction across the Permian-Triassic boundary in the Karoo Basin, South Africa. *Geological Society of America Bulletin*, **115**, 1133–1152.

REICHOW, M. K., SAUNDERS, A. W., WHITE, R. V., PRINGLE, M. S., AL'MUKHAMEDER, A. I., MEDVEDEV, A. J. & KIDD, N. P. 2002. Ar^{40}/Ar^{39} dates from Western Siberia Basin: Siberian flood basalt province doubled. *Science*, **296**, 1846–1849.

RENNE, P. R. & BASU, A. R. 1991. Rapid eruption of the Siberian Trapps flood basalts at the Permian–Triassic boundary. *Science*, **253**, 176–179.

RENNE, P. R., ZANG, Z., RICHARDS, M. A., BLACK, M. T. & BASU, A. R. 1995. Synchrony and causal relations between Permian–Triassic boundary crisis and Siberian flood volcanism. *Science*, **269**, 1413–1416.

REY, D. & RAMOS, A. 1991. Estratigrafía y sedimentología del Pérmico y Triásico del sector Deza – Castejón (Soria). *Revista Sociedad Geologica España*, **4**, 105–126.

RODRIGUEZ-PEREA, A., RAMOS-GUERRERO, E., POMAR, L., PANIELLO, X., OBRADOR, A. & MARTÍ, J. 1987. El Triásico de la Baleares. *Cuadernos de Geología Ibérica*, **11**, 295–321.

RONCHI, A. & SANTI, G. 2003. Non-marine biota from the Lower Permian of central Southern Alps (Orobic and Collio basins, N. Italy). *Géobios*, **36**, 749–760.

ROSELL, J., ARRIBAS, J., ELÍZAGA, E. & GÓMEZ-GRAS, D. 1988. Carcterización sedimentológica y petrográfica de la serie roja permo-triásica de la isla de Menorca. *Boletín del Instituto Geológico y Minero de España*, **99**, 71–82.

SALAS, R. & CASAS, A. 1993. Mesozoic extensional tectonics, stratigraphy and crustal evolution during the Alpine cycle of the Eastern Iberian Basin. *Tectonophysics*, **228**, 35–55.

SÁNCHEZ-MOYA, Y. & SOPEÑA, A. 2004. Máxima expansión del rift, sucesiones continentales, marinas y evaporíticas. *In*: VERA, J. A. (ed.) *Geología de España*. Instituto Geológico y Minero de España-Sociedad Geológica de España, 489–492.

SMITH, D. B., BRUNSTROM, R. G., MANNING, P. I., SIMPSON, S. & SHOTTON, F. W. 1974. *A Correlation of the Permian Rocks in the British Isles*. Geological Society, London, Special Report, **5**.

SMITH, N. D. 1970. The braided steam deposition environment: comparison of the Platte River with some Silurian clastic rocks, north central Appalachians. *Geological Society of America Bulletin*, **81**, 2993–3014.

SMITH, R. M. H. 1995. Changing fluvial environments across the Permian–Triassic boundary in the Karoo Basin, South Africa, and possible causes of tetrapod extinctions. *Paaleogeography, Palaeoclimatology, Palaeoecology,* **117**, 81–104.

SMITH, R. M. H. & WARD, P. D. 2001. Pattern of vertebrate extinctions across an event bed at the Permian–Triassic boundary in the Karoo Basin of South Africa. *Geology*, **29**, 1147–1150.

SOLÉ DE PORTA, N., CALVET, F. & TORRENTÓ, L. 1987. Análisis palinológico del Triásico de los Catalánides (NE de España). *Cuadernos de Geología Ibérica,* **11**, 237–254.

SOLÉ DE PORTA, N., TORRENTÓ, L. & CALVET, F. 1985. Sucesiones microflorísticas en el Triásico de los Catalánides (NE de España). *Revista de Investigacions Geológiques, Barcelona*, **41**, 17–22.

SOPEÑA, A. 1979. *Estratigrafía del Pérmico y Triásico del NW de la Provincia de Guadalajara.* Seminarios de Estratigrafía, Monografías, **5**.

SOPEÑA, A. 1980. Mapas geológicos del borde SE del Sistema Central. *Cuadernos de Geología Ibérica,* **6**, 73–95.

SOPEÑA, A. & SÁNCHEZ-MOYA, Y. 2004. Las cuencas continentales del fin de la orogenia varisca. *In*: VERA, J. A. (ed.) *Geología de España.* Instituto Geológico y Minero de España, Sociedad Geológica de España, 479–481.

SOPEÑA, A., LÓPEZ-GÓMEZ, J., ARCHE, A., PÉREZ-ARLUCEA, M., RAMOS, A. VIRGILI, C. & HERNANDO, S. 1988. Permian and Triassic rift basins of the Iberian Peninsula. *In*: MANSPEIZER, W. (ed.) *Triassic–Jurassic Rifting.* Elsevier, Amsterdam, 757–786.

SOPEÑA, A., DOUBINGER, J., RAMOS, A. & PÉREZ-ARLUCEA, M. 1995. Palynologie du Permien et du Triassique dans le centre de la Péninsule Iberique. *Sciences Geologiques Bulletin,* **48**, 119–157.

STAMPFLI, G. M. & BOREL, G. D. 2002. A plate tectonic model for Paleozoic and Mesozoic constrained by dynamic plate boundaries and restored synthetic oceanic isochrones. *Earth and Planetary Science Letters,* **196**, 17–33.

TEMIÑO, J. 1982. *Estudio estratigráfico del Pérmico y Triásico de Sierra Carbonera, Albarracín, Teruel.* BSc thesis, Universidad Complutense, Madrid.

THOMPSON, G. M., ALI, J. R., SONJ, X. & JOLLEY, D. W. 2001. Emeishan Basalts, SW China: reappraisal of the formations type area stratigraphy and a discussion of its significance as a large igneous province. *Journal of the Geological Society, London,* **158**, 593–599.

TWITCHETT, R. J., LOOY, C., MORANTE, R., VISSCHER, H. & WIGNALL, P. B. 2001. Rapid and synchronous collapse of marine and terrestrial ecosystems during the end-Permian biotic crisis. *Geology*, **29**, 351–354.

UTTIG, J. & PIASECKI, S. 1995. Palynology of the Permian of the Northern Continent: a review. *In*: SCHOLLE, P. A., PEYRT, T. M. & ULLMER-SCHOLLE, D. S. (eds) *The Permian of Northern Pangea. Vol. 1.* Springer, Bellin, 236–261.

VAN WEES, J. D., ARCHE, A., BEIJDORFF, C., LÓPEZ-GÓMEZ, J. & CLOETINGH, S. 1998. Temporal and spatial variations in tectonic subsidence in the Iberian Basin (eastern Spain): inferences from automated forward modelling of high-resolution stratigraphy. *Tectonophysics*, **300**, 285–310.

VIRGILI, C. 1979. Le Trias du Nord de l'Espagne. *Bulletin du Bureau de Recherches Géologiques et Minières,* **4**,12–30.

VIRGILI, C., HERNANDO, S., RAMOS, A. & SOPEÑA, A. 1976. Le Permien en Espagne. *In*: FALKE, H. (ed.) *Continental Permian in Central, West and South Europe.* Reidel, Dordrecht, 91–109.

VIRGILI, C., SOPEÑA, RAMOS, A., HERNANDO, S. & ARCHE, A. 1979. El Pérmico en España. *Revista Española de Micropaleontología,* **12**, 255–262.

VIRGILI, C., SOPEÑA, A., RAMOS, A., ARCHE, A. & HERNANDO, S. 1983. El relleno post-hercínico y el comienzo de la sedimentación mesozoica. *In*: COMBA, J. (ed.) *Libro Jubilar de José María Ríos*, **2**, 25–36. Ministerio de Industria, Madrid.

VISSCHER, H. 1971. *The Permian and Triassic of the Kingscourt outlier, Ireland: A Palynological Investigation Related to Regional Stratigraphic Problems in the Permian and Triassic of Western Europe.* Geological Survey of Ireland, Special Papers, **1**.

VISSCHER, H. 1967. Permian and Triassic palynology and the concept of Triassic twist. *Palaeogeography, Palaeoclimatology, Palaeoecology*, **3**, 157–166.

WAGNER, R., TALENS, J. & MELÉNDEZ, B. 1983. Upper Stephanian stratigraphy and megaflora of Henarejos (Province of Cuenca), Cordillera Ibérica, central Spain. *Annales Faculdade de Ciencias de Porto*, **64** (Supplement), 445–480.

WALKER, R. G. & CANT, D. J. 1979. Sandy fluvial systems. *In*: WALKER, R. G. (ed.) *Facies Models.* Geosciences Canada Reprint Series, **1**, 23–31.

WARD, P. D., MONTGOMERY, D. R. & SMITH, R. 2000. Altered river morphology in South Africa related to the Permian–Triassic extinction. *Science*, **289**, 1740–1743.

WARDLAW, B. R. & SCHIAPPA, T. A. 2001. *In*: CASSINIS, G. (ed.) *Permian Continental Deposits of Europe. Regional Reports and Correlations.* Towards a refined Permian chronostratigraphy. Natura Bresciana, Monografia, **25**, 363–365.

WIGNALL, P. B. 2003. Large igneous provinces and mass extinctions. *Earth-Science Reviews*, **53**, 1–33.

YIN, H., ZHANG, K., TONG, J., YANG, Z. & WU, S. 2001. The global stratotype section and point (GSSP) of the Permian–Triassic boundary. *Episodes*, **24**, 102–114.

ZIEGLER, P. A. & STAMPFLI, G. M. 2001. *In*: CASSINIS, G. (ed.) *Permian Continental Deposits. Regional Reports and Correlations.* Late Paleozoic–Early Mesozoic plate boundary reorganization: collapse of the Variscan orogen and opening of the Neotethys. Natura Bresciana, Monografia, **25**, 12–30.

ZERFASS, H., LAVINIA, E. L., SCHULTY, C. L., GARCÍA, A. J. V., FACCINI, V. F. & CHEMALE, F. 2003. Sequence stratigraphy of continental Triassic strata of southernmost Brazil. *Sedimentary Geology*, **161**, 85–105.

ZHOU, M. F., MALPAS, J. *et al.* 2002. A temporal link between Emeishan large igneous province (SW China) and the end-Guadalupian mass extinction. *Earth and Planetary Science Letters*, **196**, 113–122.

The problem of the transition from the Permian to the Triassic Series in southeastern France: comparison with other Peritethyan regions

MARC DURAND

47 rue de Lavaux, F-54520 Laxou, France (e-mail: mada.durand@wanadoo.fr)

Abstract: In the French sedimentary basins, widespread alluvial deposits sealing narrow Permian troughs are referred to as 'Buntsandstein'. An Early Triassic age is generally put forward despite a lack of any Scythian biochronological elements. In Provence (Southeast Basin) some doubts remain about the age of the latest Permian deposits, and the oldest Triassic fossils (Anisian palynomorphs) appear in the upper 'Buntsandstein'. Three types of contact occur: disconformity overlain by a quartz-conglomerate, apparent transition, and angular unconformity, according to an increasing basal incompleteness of the 'Buntsandstein'. Whereas the conglomerate was deposited under arid conditions, the overlying fluvial deposits indicate a marked climate change. A transect from France up to the Germanic Basin centre shows that the 'French Buntsandstein' cycle might begin considerably before the end of the Permian; the Early Triassic arid 'event' is Dienerian/Smithian in age; and the Provençal 'basal' conglomerate corresponds to the uppermost part of a coeval subordinate cycle, and thus the underlying sub-Triassic unconformity represents a hiatus estimated at 10–15 Ma. Works in progress confirm that sedimentary climate indicators constitute powerful tools for correlations within non-marine formations devoid of biostratigraphical marker that straddle the Permian–Triassic boundary, at least on the scale of the Western European Plate.

In the three main French sedimentary basins (the Paris, Southeast and Aquitaine basins) the Mesozoic cycle begins, at least in some parts, with widespread alluvial deposits burying relatively narrow Permian troughs, and frequently referred to as the 'Buntsandstein' Group. The basal beds of this 'Buntsandstein' were dated in one place only: above the small Lodève Permian basin, where they yielded a Mid- Anisian palynoflora (Broutin *et al.* 1992). In all other areas where the Middle Triassic was recognized – in the upper part of the 'Buntsandstein' or higher – an Early Triassic age is generally accepted for these red sandy and gravelly units despite the fact that, up to now, no Scythian biochronological indicators have been found. Thus, in actual fact the position of the Permian–Triassic Boundary (PTB) with respect to the sub-'Buntsandstein' unconformity is not straightforward.

For regions stretching further southwest ('North Iberian domain', including the Balearic Islands: Broutin *et al.* 1992) a Late Permian age is adopted for that unconformity, the lower 'Buntsandstein' still yielding 'Thuringian' (*sensu* Visscher 1971) palynofloras. A similar context prevails in the very NE corner of France (Dachroth 1985; Durand *et al.* 1994). Where the non-marine sedimentation seems to have been continuous from the Permian to the Triassic, the PTB is readily believed to coincide with an abrupt change in fluvial style, sometimes related to the global-scale end-Permian extinction (Ward *et al.* 2000). In other regions, it is generally admitted that the sub-'Buntsandstein' unconformity actually fits with the boundary between the Permian and Triassic series. Nevertheless, in both cases the lack of typically Early Triassic fossil remains can be explained either by conditions unfavourable to life and preservation (i.e. 'desert' environments) or by a true stratigraphical gap.

The purpose of this paper is for the first time to test these alternatives on case studies in Provence, where the largest outcrop area of Permian–Triassic deposits in the French Southeast Basin is present (Fig. 1), by using sedimentological climate indicators and step-by-step geometric correlations towards sections that have increasing completeness. These results will be used to tentatively propose a more or less new correlation scheme for other European regions.

The PTB problem in Provence

On the regional scale there is no difficulty distinguishing the 'Buntsandstein' Group (Fig. 1). Constituting the base of the Mesozoic section, it crops out along a narrow, practically continuous belt from Sanary (SW) to Cannes (NE). It reflects a clear change from local drainage systems in several distinct basins, a characteristic of the Permian palaeogeography, to a single widespread system that is typical of the Triassic,

From: LUCAS, S. G., CASSINIS, G. & SCHNEIDER, J. W. (eds) 2006. *Non-Marine Permian Biostratigraphy and Biochronology*. Geological Society, London, Special Publications, **265**, 281–296.
0305-8719/06/$15.00

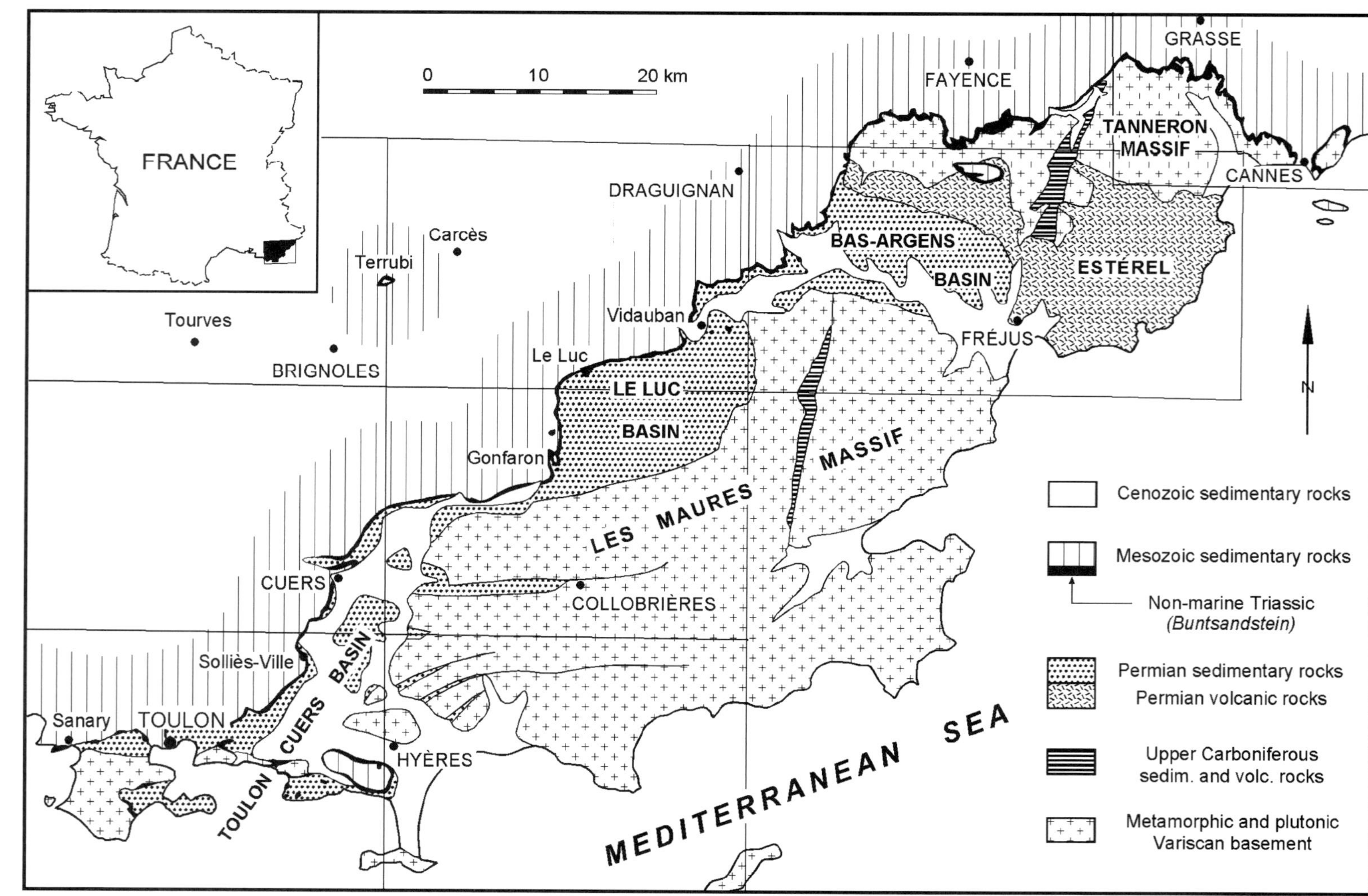

FRANCE
0
10
20 km
GRASSE
FAYENCE
TANNERON MASSIF
CANNES
DRAGUIGNAN
Carcès
Terrubi
BAS-ARGENS BASIN
ESTEREL
Tourves
BRIGNOLES
Vidauban
Le Luc
FRÉJUS
N
LE LUC BASIN
Gonfaron
MASSIF
LES MAURES
CUERS
COLLOBRIÈRES
Solliès-Ville
BASIN
CUERS
Sanary
TOULON
HYÈRES
TOULON
MEDITERRANEAN SEA
Cenozoic sedimentary rocks
Mesozoic sedimentary rocks
Non-marine Triassic
(Buntsandstein)
Permian sedimentary rocks
Permian volcanic rocks
Upper Carboniferous
sedim. and volc. rocks
Metamorphic and plutonic
Variscan basement

and thus constitutes the major criterion used to define the 'Buntsandstein' in France. It is noteworthy that the main Triassic palaeocurrents flowed to the SW, along the Maures Massif, that is – at least for the Toulon–Cuers area – in a direction opposite to the Permian palaeoflow (Durand *et al.* 1989; Durand 1993).

Dating elements

The period during which the sub-'Buntsandstein' pediplanation surface, assumed to be the regional PTB, developed is very loosely appraised. On one hand some discrepancies persist between different dating elements from the Permian series, and on the other hand only Mid- Triassic palaeontological data were obtained from the 'Buntsandstein' so far, in its upper part.

Nevertheless, the very local occurrence of the Les Arcs Formation (Fig. 2) – the Permian or Triassic age of which is debatable – provides evidence that the sub-'Buntsandstein' unconformity probably corresponds to a major hiatus. In this connection, it is perhaps useful to report that two independent methods (geometric extrapolation and organic matter study) show that, above the rather nearby Lodève Basin (about 100 km WNW of Toulon), nearly 1500 m of Permian sediments could have been eroded before the Middle Anisian (Lopez *et al.* 2005).

Upper part of the Permian series

The main stratigraphical marker through the Permian basins of Provence is an acidic lava-flow succession called the 'A7 Rhyolite' (Fig. 2). The last and more reliable isotopic dating gives it an age of 272.5 ± 0.3 Ma (Zheng *et al.* 1992), which is clearly Kungurian according to the latest geological time scale (Gradstein *et al.* 2004). Its thickness ranges from 150 to 300 m in the Estérel and Bas-Argens basins, but it does not encroach very much on the NE part of the Le Luc Basin.

In the Toulon–Cuers Basin, the sole dated unit in the Permian Series is the lacustrine 'Calcaires du Bau Rouge', which is the uppermost member of the Les Salettes Formation, about two-thirds above its base. It yields well-preserved macroscopic plant remains: *Ullmannia frumentaria* (dominant), *U. bronnii, Pseudovoltzia libeana, Lesleya* (alias *Taeniopteris*) *eckardtii, 'Sphenopteris' dichotoma* and *Odontopteris osmundaeformis*. The palynoflora is composed of scarce microspores, such as *Calamospora* spp. and *Converrucosporites eggeri*, and numerous pollen taxa: *Potonieisporites 'novicus-bhardwaji', Nuskoisporites dulhuntyi, Plicatipollenites* spp., *Vesicaspora–Scheuringipollenites* complex, *Gardenasporites leonardii, Vitreisporites pallidus, Protohaploxipinus microcorpus, Striatoabietites richteri, Lunatisporites* sp., *Lueckisporites virkkiae, Vittatina costabilis, Costapollenites ellipticus*, among others. This palaeofloristic assemblage led to assignment of a 'post-Kungurian / pre-Tatarian' age for the Bau Rouge Member (Broutin & Durand 1995). More precisely, subsequent comparative data (e.g. Poort *et al.* 1997) suggest a Wordian age.

The 'A7 rhyolite' is overlain by the Pradineaux Formation (Fig. 2), up to 200 m thick, which includes detrital to pyroclastic sedimentary deposits, as well as mafic and felsic volcanic-flow intercalations. This formation is the richest in biostratigraphically significant fossils, distributed on three principal levels, but it is developed mainly in the Estérel Basin, where the uppermost part of the Permian and the whole Triassic Series have been removed by Quaternary erosion. On the other hand, primary pinching-out and secondary faults hamper the reconstruction of the actual succession.

The lower fossiliferous beds ('11f unit' of Boucarut 1971) crop out near Agay. These lacustrine shales yielded mainly macrofloral remains such as cf. *Sphenopteris kukukiana*, cf. *Pseudoctenis middridgensis, Ullmannia bronnii, U. frumentaria, Quadrocladus orobiformis* and cf. *Q. solmsii* and cf. *Culmitzschia florinii*; conversely, very few palynomorphs are preserved (only *Lueckisporites virkkiae* and *Nuskoisporites dulhuntyi* were recognized). This assemblage led Visscher (1968) to propose a 'Thuringian age' for that part of the series. In agreement with Menning (1995), it must be emphasized that the term 'Thuringian' is misleading in several respects. For many French geologists it keeps its original sense (Renevier 1874) – equivalent of the Zechstein – whereas in Visscher's opinion it corresponds to the vertical range of *L. virkkiae*, that is, it begins already with the Kazanian (Utting *et al.* 1997, Gorsky *et al.* 2003). In fact, the Agay fossil-bearing beds (Fig. 2) could be only a little younger than those from the Bau Rouge Member (J. Broutin, pers. comm.).

The second fossiliferous beds (Pra Baucous locality) are thin lacustrine limestone layers interstratified with siltstones ('13a unit' of Boucarut 1971). They yielded an ostracode assemblage with *Iniella kutznetskiensis* (Spizharsky 1937), *Paleodarwinula acervalis* (Mandelstam 1956),

Fig. 1. Distribution of the non-marine Permian–Triassic outcrop areas in Provence. Fine straight lines are boundaries between sheets of the 'Carte géologique de la France à 1 / 50 000'. Each sheet is referred to by the town name in upper case letters.

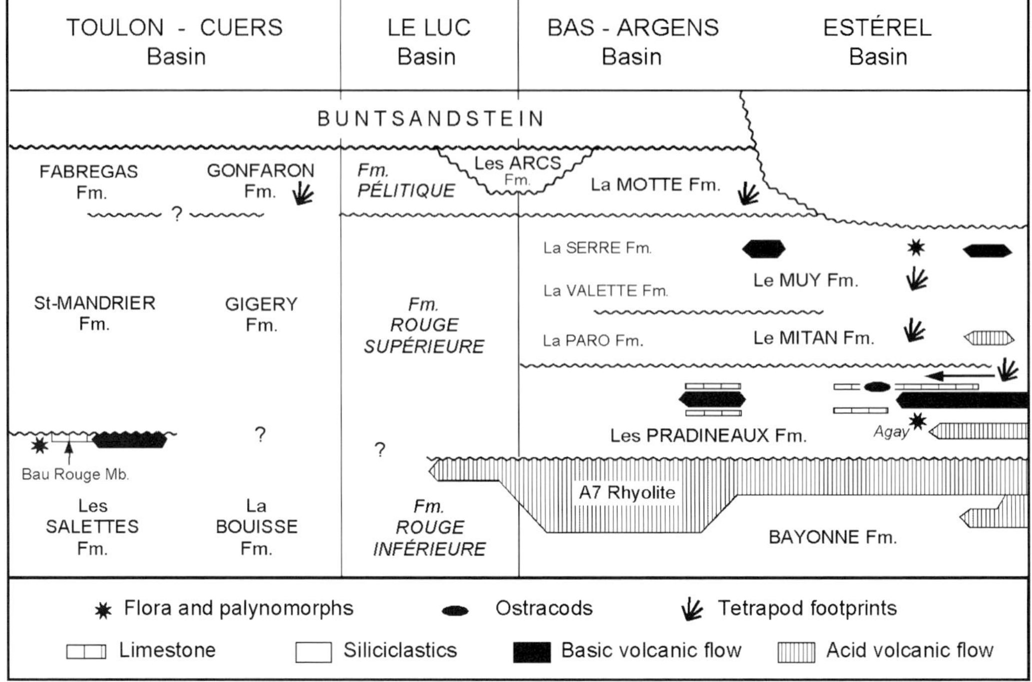

Fig. 2. Lithostratigraphical scheme for the upper part of the Permo-Triassic succession in Provence (modified after Toutin-Morin & Vinchon 1989). Without scale.

and *Paleodarwinula alexandrinae* (Belousova 1956). According to the known stratigraphical ranges of these taxa in Russia, this association was assigned to the earliest Tatarian (Lethiers *et al.* 1993); subsequent data show that the last taxon actually seems to be a good marker of the *P. fainaie* Zone, that is, Late Kazanian of the type area (Crasquin-Soleau 2003).

The third fossiliferous beds of the Pradineaux Formation in the Estérel, at a slightly higher level ('13c unit' of Boucarut 1971), are rhyolitic tuffs with fossil ground-surfaces, which, in the town of Saint-Raphaël (St-Sébastien disused quarry), display many tetrapod tracks studied by Gand *et al.* (1995). The ichnotaxon assemblage expresses a more advanced stage of evolution than the Cisuralian ones, but is also distinct from those typical of the Lopingian (Gand & Durand 2006); through the occurrence of the therapsid tracks *Lunaepes* and *Planipes* it can be referred to the 'tapinocephalid stage', which corresponds to the North American Roadian and Wordian (Cassinis *et al.* 2002; Lozovsky 2003).

From the revision of the previous age determinations just carried out, it can be concluded that the Pradineaux Formation as a whole is of Wordian age. The erosional unconformity at the top of the 'A7 Rhyolite' could have been formed during the Roadian.

It is higher in the series that problems persist. The Le Muy Formation provided, in a quarry since filled up, a great deal of coalified plant remains (Germain 1968; Visscher 1968) identified as: *Calamites* sp., *Annularia* sp., *Cordaites* sp., *Ullmannia* cf. *lycopodioides, U. bronnii*, silicified woods such as *Ginkgophytoxylon permiense* (Vozenin-Serra *et al.* 1991), and palynomorphs with a quantitative predominance of *Lueckisporites virkkiae*, *Nuskoisporites dulhuntyi* and *Falcisporites zapfei*, as well as *Jugasporites delasaucei, Strotersporites richteri* and *Paravesicaspora splendens*. For Visscher (1968) that assemblage is comparable with those from the German Zechstein *sensu stricto* (i.e. Upper Permian), and it is thus regarded as the most recent known so far in the French Permian (Broutin, in Toutin-Morin *et al.* 1994). Conversely, the tetrapod footprint assemblages found in the Le Mitan, Le Muy and even La Motte formations do not differ notably from those of the Pradineaux Formation (Gand & Durand 2006).

Furthermore, the first palaeomagnetic study carried out on the uppermost part of the Permian sedimentary series in Provence (the pelitic La Motte Formation and its equivalents in the Le Luc and Cuers basins) detected a reversed component in about 20% of the samples, and thus referred these formations to the Illawarra Mixed Polarity Superchron (Merabet & Daly 1986). But, it should be emphasized that similar results were also obtained at that time on older sediments and the A7 Rhyolite, whereas more recent works conclude on the contrary that all the Estérel volcanics (including the basalts occurring in Le Mitan and Le Muy formations: Fig. 2) are representative of the Kiaman Reversed Polarity Superchron (Rochette *et al.* 1997) and therefore should be older than 265 Ma (Menning 2001). In fact, several clues suggest that the Estérel rocks might be affected by remagnetization (Vlag *et al.* 1997).

Triassic 'Buntsandstein'

The deposits referred to as 'Buntsandstein' are very much less developed in thickness (maximum 80 m). The main part, beginning in many places with pebble beds, is everywhere devoid of any fossils, with the exception of invertebrate traces (*Scoyenia*, *Beaconites*, *Phycodes*, *Arenicoloides*, etc.) deserving further study but without biochronological significance. The only palaeontological elements that enable dating of that unit are palynomorphs from the uppermost part. These are: *Triadispora staplini, T. falcata, Alisporites grauvogeli, Microcacrhyidites fastidoides, M. sittleri, Pityosporites* sp., *Angustisulcites klausii, Voltziacaesporites heteromorpha, Illinites kosankei* and *Hexasaccites muelleri* (syn. *Stellapollenites thiergartii*). This assemblage, very close to those found in the Grès à Voltzia Formation of NE France, allows assignment of an early Anisian age (Adloff in Durand *et al.* 1989). Very comparable associations also represent the oldest Triassic palynofloras in the upper Buntsandstein of the southern Catalonian Pyrenees (Broutin *et al.* 1988), and NW Sardinia (Pittau 2002). Occasional occurrences of tetrapod footprints were noted, moreover, although in general rather badly preserved with compared with the Permian ones; they are mostly chirotheroid traces (*Chirotherium*, *Brachychirotherium*) with some *Rhynchosauroides* and *Capitosauroides.* They are much less discriminating than palynomorphs, but their stratigraphical ranges also encompass the early Anisian (Demathieu & Durand 1991).

At the end of this review, one can conclude that neither the presence of the Upper Permian (i.e. Lopingian) nor of the Lower Triassic (i.e. Scythian) are proven in Provence up to now. Other criteria must be used to try to restrict the zone of uncertainty.

Outcrop configurations

Three different main types of contact between the Permian sedimentary deposits and the Triassic 'Buntsandstein' can be observed in the field over a short distance: disconformity overlain by a 'basal' conglomerate, apparent transition, and angular unconformity (Fig. 3). They will herein be called 'Sanary', 'Gonfaron' and 'Vidauban' type respectively.

The 'Sanary type'

In most cases, the very even surface regarded as the regional PTB is blanketed with an oligomictic orthoconglomerate composed of exclusively siliceous pebbles (mainly vein quartz, scarce lydites and quartzites) with a variable content of quartz-sand matrix (Fig. 3a) The thickness of this 'Poudingue de Port-Issol'(Glinzbœckel & Durand 1984) is commonly about 1 m and reaches a maximum of 8 m at the type section near Sanary. Sedimentary structures evoke longitudinal bars with lateral sand-wedges in a braided-channel river, and indicate palaeocurrents flowing along the NW border of the present Maures Massif.

The great majority of the pebbles were well rounded by long (polycyclic?) fluvial transport, but many display secondary ridges fashioned by wind-blown sand shortly before their last reworking (Fig. 4). Such clasts are usually known by the German term 'dreikanter', which refers to a specific shape that is never dominant among them, and furthermore can form in a different environment (Jones 1953); this is why the term 'ventifact' (Evans 1911), which can apply to all wind-worn pebbles whatever their shape, should be used in preference. The systematic occurrence of ventifacts in the Poudingue de Port-Issol joined to the lack of transverse supply (as shown by an unexpected decrease of pebble size towards the borders) testify that the depositional area underwent clearly arid conditions; the catchment area of the 'Port-Issol wadi' could be located in mountainous areas farther north (Durand *et al.* 1988, 1989).

Another very important distinctive feature of the Port-Issol Formation is its sharp upper boundary, showing truncation of previous sedimentary structures (see Fig. 3a), and marking an abrupt change in depositional style. The appearance of many indices of biotic activity conveys a climatic evolution into less extreme conditions, of semi-arid type. They are especially caliche nodules (*in situ* and reworked) and, in terminal fan facies of the downstream part (Toulon area), vegetation-induced primary sedimentary structures (Rygel *et al.* 2004). New sediment

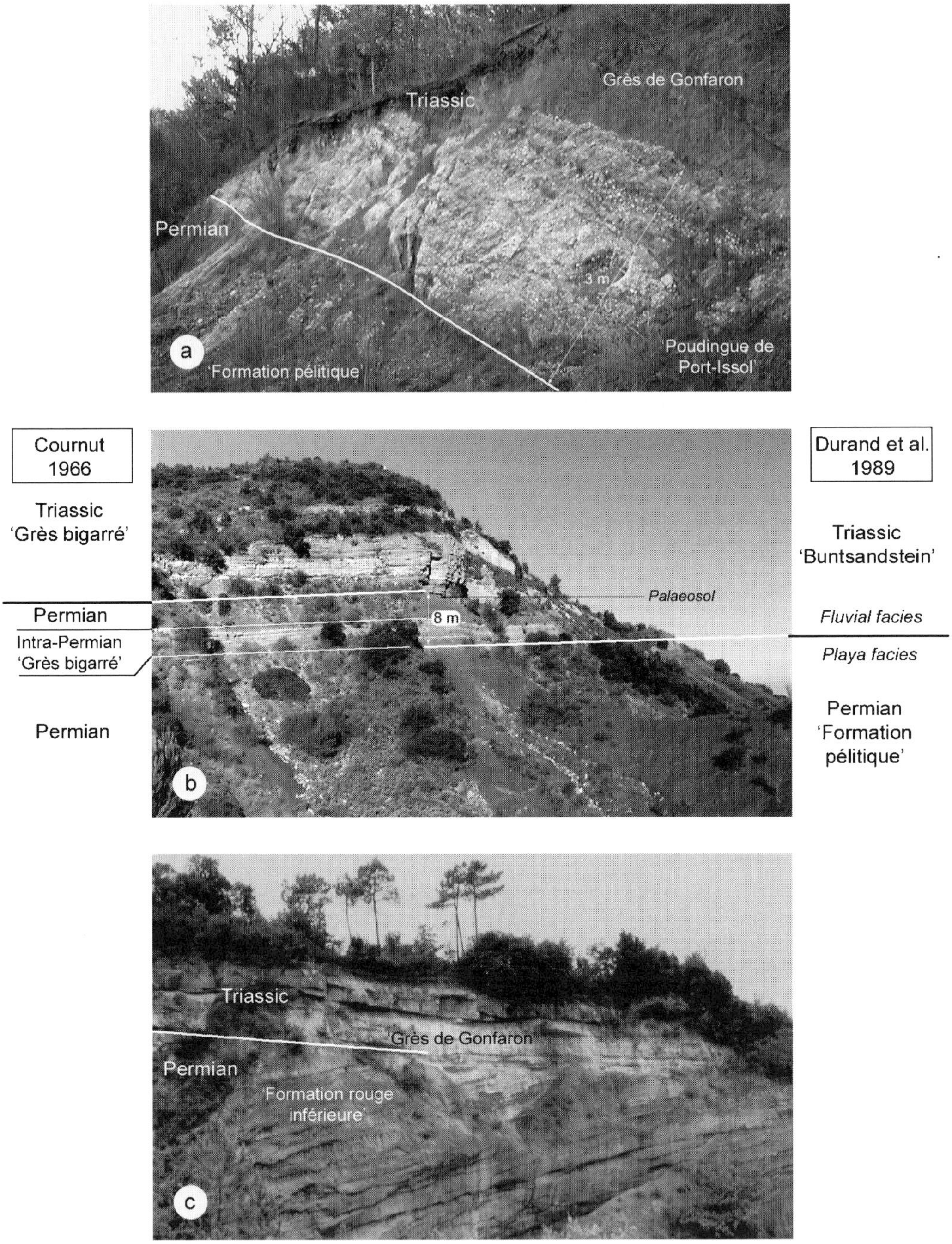

Cournut 1966
Triassic
'Grès bigarré'
Permian
Intra-Permian
'Grès bigarré'
Permian
Palaeosol
8 m
b
Durand et al. 1989
Triassic
'Buntsandstein'
Fluvial facies
Playa facies
Permian
'Formation pélitique'

Triassic
'Grès de Gonfaron'
Permian
'Formation rouge inférieure'
c

Fig. 3. The three main types of Permian–Triassic boundary on the outcrops of Provence. (a) Disconformity overlain by a quartz-conglomerate (Cuers); (b) Apparent transition (Gonfaron); (c) Angular unconformity (La Garduère near Vidauban).

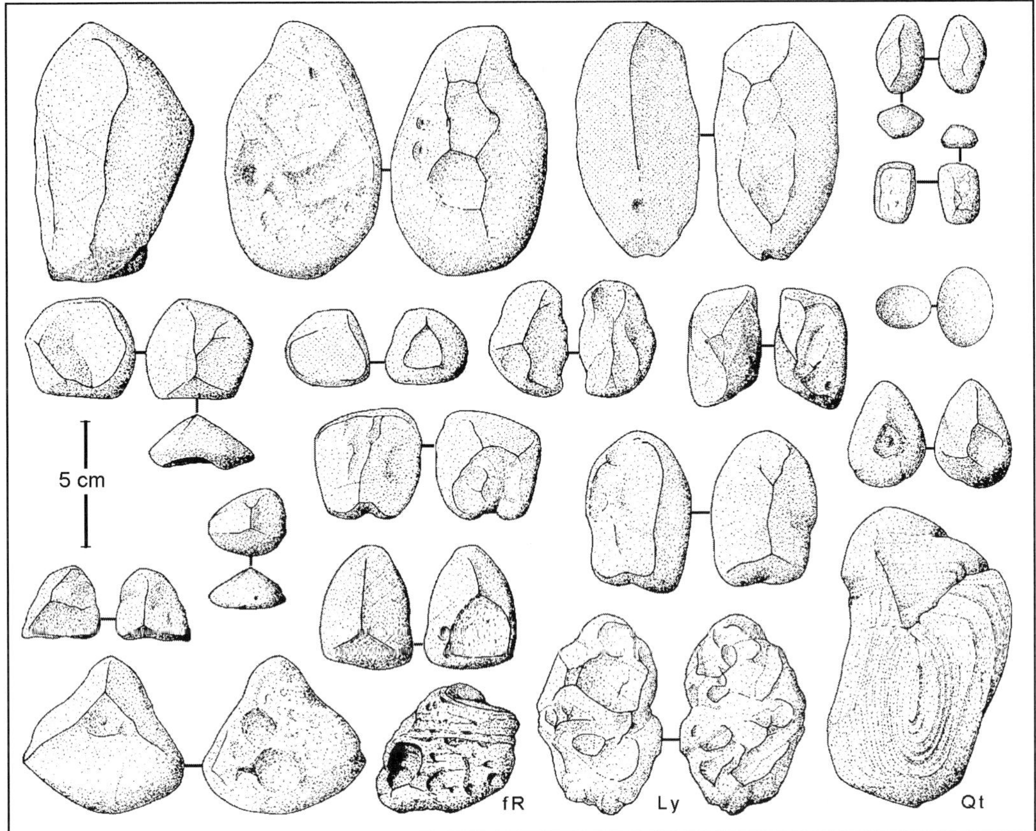

Fig. 4. Main types of ventifact from the basal beds of the 'Buntsandstein' of Provence (modified after Durand *et al.* 1989). All these specimens are quartz, except: fR, rhyolite with fluidal structure; Ly, lydite; Qt, quartzite.

supply reached the basin by the Bas-Argens zone, perhaps coming from NE Corsica (Durand *et al.* 1989). The apparently rapid character of the recorded climate change is probably due to only a more or less significant depositional hiatus, as will be seen later. Moreover, several sections in the neighbourhood of Toulon (Fabregas, Solliès-Ville) show the development of a palaeosol at the expense of the uppermost conglomerate layers; in other places (La Garonne beach) the Buntsandstein begins directly with a discontinous dolocrete overlain by a thin conglomerate produced mainly by the reworking of the 'Poudingue de Port-Issol'.

The 'Gonfaron type'

Less frequent than the former, this type corresponds to an apparent transition from Permian to Triassic sedimentary deposits because of a lack of any conglomerate. On the Gonfaron outcrops (Fig. 3b), a 'precursory' sandstone facies typical of the 'Buntsandstein' seems to be inserted in Upper Permian red silt-clay deposits ('Intra-Permian *Grès bigarré*' of Cournut 1966; see Fig. 3). For a time, the regional PTB was believed to be located at the top of the uppermost thick silty unit because of the development at this level of a spectacular palaeosol with long, drab-haloed root traces and yellowish subvertical caliche nodules (Cournut 1966; Toutin-Morin 1986). Subsequently, a careful examination of the fine-grained facies revealed significant differences between the lower playa sediments and the upper floodplain deposits. Conversely, the sandstone units are clearly related; they contain particularly subspherical (wind-worn) coarse quartz grains concentrated mostly at the very base of the lower unit. This is why it is currently believed that this level is actually the base of the Triassic Series (Durand *et al.* 1989).

It is clear that an erosional unconformity, which may correspond to a hiatus of several million years, is likely to pass unnoticed much more easily than a variegated palaeosol, even though it is rather moderately developed, such as that described above. Two superimposed lithological units of very different ages can present on the outcrop a 'mimetic' aspect, and it may be that detailed petrographic studies are necessary to differentiate them.

The 'Vidauban type'

The least frequent type of appearance of the PTB in Provence is that of an angular unconformity. At the best outcrop (Fig. 3c), SE of Vidauban, the angle between strata of the Permian and Triassic sandstones approximates 20°, and the truncation reached a rather deep level of the Permian Series ('Formation rouge inférieure': Fig. 2). Such an unconformity is not the expression of some latest Permian or earliest Triassic tectonic movements; it results from intra-Permian tilting (Baudemont 1988). There, on the Vidauban swell between the Bas-Argens and Le Luc basins, transtensional deformations generated progressive unconformities, some of which, at a very short distance, are sealed by the uppermost Permian unit in this region ('Formation pélitique').

One can conclude that the three types of contact between the Permian and Triassic series noticed in Provence depend on their location being more and more distant from the axis of the Triassic depositional basin (Fig. 5). Attention must be drawn to the fact that the apparent transition does not correspond to the shorter basal gap.

The 'Buntsandstein' hyper-arid period

Whereas the other non-marine Triassic siliciclastics of Provence are relatively rich in traces of animal and plant activity, especially caliche nodules (*in situ* and/or reworked) typical of semi-arid soils, the Poudingue de Port-Issol lacks such features. Conversely, it is characterized by the occurrence of numerous, very well-preserved windkanters. The shaping of such gravels could be carried out only on the surface of a ground free from any vegetation during many thousands of years (Wright *et al.* 1991). Thus, it can be inferred that this conglomerate formed during one particular period characterized by hyper-arid climatic conditions. Dating that period would make it possible, on the one hand, to reduce the margin of uncertainty left by biochronological data for the estimate of the age of the regional PTB in Provence and, on the other hand, to attempt correlation with distant regions. That could be carried out starting from a comparison with the series of NE France.

Tentative dating

The northeastern France context

To observe, in other localities of France, relationships between Permian and Triassic series very similar to those described in Sanary, it is necessary to move over 500 km northwards to the Lure area (Franche-Comté province), on the southernmost slope of the Vosges Massif (see location on Fig. 7). From there it is possible to follow an evolution in continuity SW–NW (at first towards the axis of the Lorraine Basin, then downstream) as far as the much more complete series of SW Germany (Fig. 6).

The thin Triassic conglomerate with ventifacts that lies directly above the Permian red siltites in the Lure area is nothing other than the 'Conglomérat principal' of the Vosges (Durand

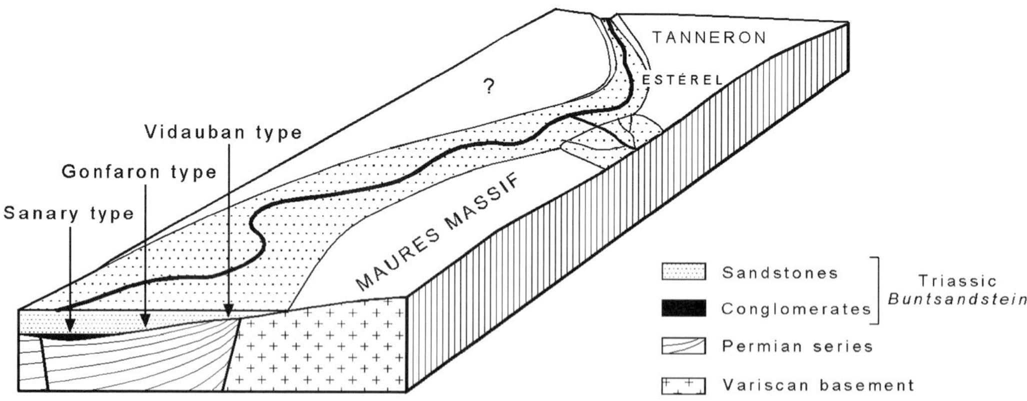

Fig. 5. Schematic diagram showing the locations, within a Lower Triassic palaeogeographical context, of the three forms presented by the PTB in Provence. Not to scale.

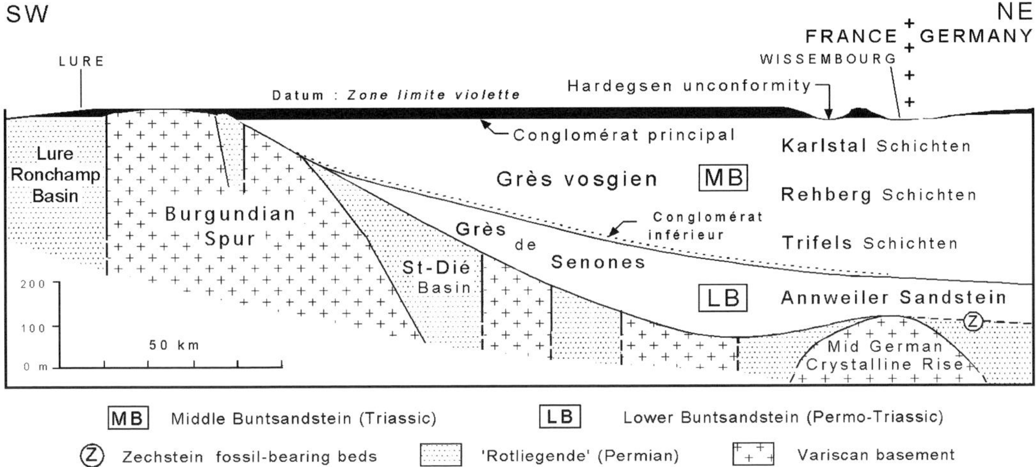

Fig. 6. Simplified lithostratigraphical cross-section across the Lower and Middle Buntsandstein of NE France, from Franche-Comté (SW) to Palatinate (NE, Germany) through the Vosges Massif in Lorraine and Alsace.

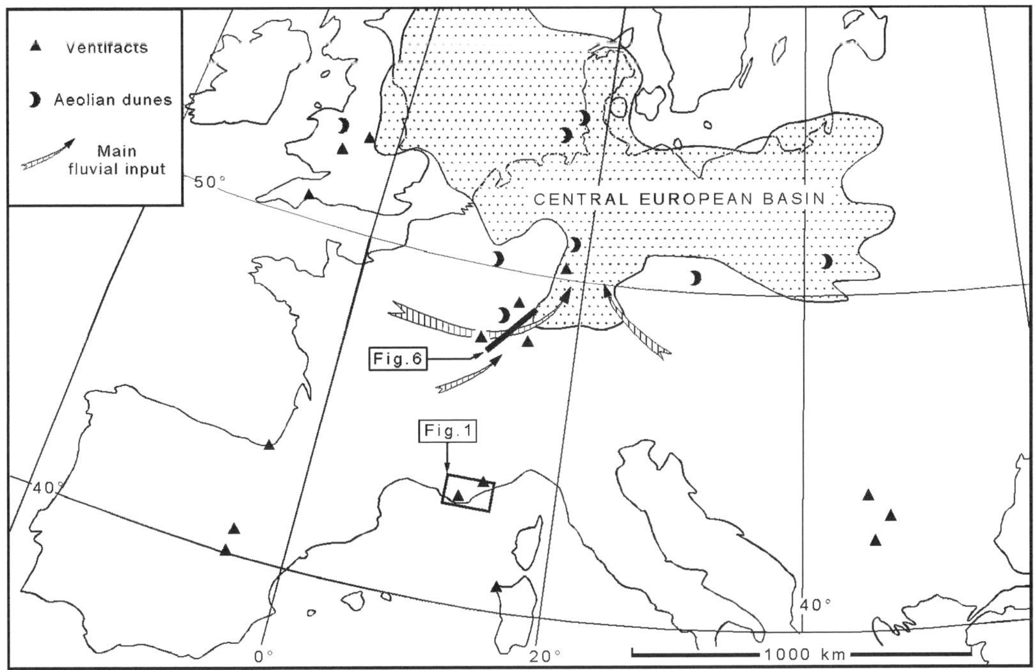

Fig. 7. Geographical distribution of the sedimentary features ascribed to the Dienerian–Smithian arid period in SW Europe. Since the maximum extension of coeval deposits is yet known with inadequate precision, only that of the uppermost Permian 'Zechstein', indicating roughly the central part of the Lower Triassic basin, is represented here (dotted area).

1978). On most of the Vosges Massif this unit, generally about 20 m thick, tops a sandy formation ('Grès vosgien') constituting the major part of the Buntsandstein, where remains of large aeolian dunes were seldom preserved by the wandering of fluvial braided channels, and where the pebbly basal member ('Conglomérat inférieur') is already rich in ventifacts (Durand *et al.* 1994). As the same types of pebbles and the same sand fraction characterize the 'Conglomérat principal' and the whole underlying 'Grès vosgien' as well, both formations are considered as part of the same sedimentary cycle, up to 350 m thick, that was entirely deposited during the Buntsandstein arid period; it is referred to as the 'Middle Buntsandstein' by French geologists (Courel *et al.* 1980).

There is, nevertheless, a difference between the setting of the conglomerates in Provence and the Lure area. In the first case the conglomerate remained in the axial part of the basin; in the latter it resulted from an overfilling of the Lorraine Basin, and real divergence on its margin towards the Bresse–Jura Triassic Basin (Durand 1978). But it can be inferred from well-log data that, beneath the Mesozoic cover of the present Paris Basin, the 'Conglomérat principal' does occur upstream from the 'Grès vosgien' depositional area, in a configuration similar to that observed in Provence, about 150 km west of the transect in Figure 6. However, it may be that both conglomerates mark the latest part of the Buntsandstein arid period.

Unfortunately, there is practically no element for direct dating of that period in the French Buntsandstein. The whole Middle Buntsandstein has so far yielded a single body fossil, in the upper 'Grès vosgien': a conchostracan identified as *Cornia* sp. by Kozur (1993), who assumed a Dienerian age. It should, however, be pointed out that this genus is known also from the Smithian (Sludkian Horizon) of the Moscow syneclise (Lozovsky 1993).

The French Middle Buntsandstein is well delimited between two units that present several petrographical and sedimentological characters in common: much lower mineralogical and textural maturity, and presence of palaeosols. Below, under a very sharp boundary, the 'Grès de Senones' (Fig. 6) allows reconstruction of a palaeocurrent system already very comparable with that of the 'Grès vosgien', and rests on various kinds of typically Permian strata ('Rotliegend' of the German geologists) or on the older basement; this is why it is referred to the Lower Buntsandstein. It is devoid of fossils (apart from scarce *Scoyenia*), but near the French–German border the base of its time equivalent 'Annweiler Sandstein' contains a malacofauna that is indicative of the first cycle of the German 'Zechstein' (Forche 1935); thus, the entire Lower Buntsandstein of the Vosges could be Late Permian.

The top of the Middle Buntsandstein is generally marked by a palaeosol complex, with dolocretes and silcretes – the 'Zone limite violette' – which developed in the course of a very long period without noticeable detrital supply following the 'Conglomérat principal' deposition. In most places where this key bed does not occur, one can show that the gap results from a subsequent erosion, having reached the 'Grès vosgien' in places (Durand *et al.* 1994), which very probably corresponds to the Hardegsen disconformity known in many parts of the Triassic Central European Basin (CEB), that is, the Germanic Basin, and formed during the Middle Spathian (Kozur 1999).

Finally, observations limited to NE France do not allow a more precise dating than Early Triassic (Scythian) for the arid period during which formed the conglomerate overlying the sub-Buntsandstein unconformity in Provence.

Comparison with the central CEB

The continuity of exposure towards the central part of the Germanic Basin is unfortunately interrupted by the Rhine Graben. On its eastern side the Triassic series is rather more complete, but correlations with NE France and northern Germany as well are still debated. This is why recognition of a time equivalent of the 'French Middle Buntsandstein' was directly sought among the five formations located under the Hardegsen unconformity in the centre of the basin. The required stratigraphical unit was one where no palaeosol was known, but from where conversely the greatest number of aeolian features (ventifacts and sand dunes) were reported, and where non-marine biotas were rare and the least diversified. The best candidate is, without question, the Volpriehausen Formation, especially as it displays the same sequential evolution (Aigner & Bachmann 1992) as the Palatine coeval deposits of the 'Grès vosgien'.

The Volpriehausen Formation yielded several fossil taxa of unequal biochronological interest. The conchostracan fauna is typically Smithian, with the first Spathian element, *Liograpta (Magniestheria) deverta,* appearing in its uppermost part (Kozur 1999, 2003). Miospores are primarily represented by *Densoisporites nejburgii* with subordinate *Endosporites papillatus*, that is: the 'PI' subzone of the *D. nejburgii* Zone (Orłowska-Zwolińska 1985) – that is the 'GTr3' Palynozone (Heunisch 1999) – ascribed to the Smithian stage by its author, the Dienerian by Reitz (1988), and the upper Griesbachian to lower Smithian by Fijałkowska-Mader (1999). Megaspores are usually lacking: 'Barren interzone Ib1' of Fugliewicz (1980). Vertebrates are extremely rare but invaluable; the labyrinthodont *Parotosuchus helgolandicus* was also found in the Torrey Member of the Moenkopi

Formation (SE Utah), which is inserted between ammonoid-bearing units, allowing it to be referred to the Smithian (Lucas & Schoch 2002).

In addition, the Volpriehausen Formation is bracketed between formations richer in palynomorphs. Below, the German 'Lower Buntsandstein' (i.e. Calvörde and Bernburg formations, which are not represented west of the Rhine) is characterized by the *Otynisporites eotriassicus* megaspore Zone beginning in the Late Permian (Fuglewicz 1980). But other palynomorphs were used to define the typically Triassic *Lundbladispora obsoleta – Protohaploxypinus pantii* Zone (Orłowska-Zwolińska 1985), namely the *Endosporites papillatus – Densoisporites playfordi* Zone (Reitz 1988) ≈ 'GTr2' palynozone (Heunisch 1999), belonging mainly to the Griesbachian stage. Kozur (1999) characterizes the whole Lower Buntsandstein by the *Lundbladispora willmotti – Lunatisporites hexagonalis* miospore Zone ranging the entire Induan. Above the Volpriehausen Formation, but still in the Middle Buntsandstein, the *Trileites polonicus – Pusulosporites populosus* megaspore Zone (Fuglewicz 1980) includes a score of taxa. Straddling the Detfurth and Hardegsen formations, the *Densoisporites nejburgii* PII subzone (Orłowska-Zwolińska 1985) = *D. nejburgiii* Zone (Reitz 1988) also has a rich vegetation that already characterizes the lower Spathian.

Recent magnetostratigraphic studies (Szurlies 2004) show that the Volpriehausen Formation is dominated by a reversed polarity record (upper part of his sr2 magnetozone and the entire sr3), with a short normal-polarity interval (sn4) around its lower third, and another (sn5) including the uppermost minor cycle 6 and partly the first of the following Detfurth Formation. It could be that the systematically normal polarities found in the uppermost part of the Middle Buntsandstein on the Beckenhof section (Palatinate) by Burek (1970) belong to the magnetozone sn5. Szurlies (2004) concludes that, in spite of some problems that remain, the Volpriehausen Formation may correlate to the late Dienerian to early Smithian interval; and this is what will be retained herein, in the current state of knowledge, for dating the Early Triassic arid period.

Correlation potential

It would be beyond the scope of this paper to make a review of all sedimentary features ascribable to the Early Triassic arid period, and Figure 7 does not claim to be exhaustive; it only aims at giving an idea of the wide dispersion of the observation points. Aeolian dune deposits are used, as for Poland and the Czech Republic (Gradziński *et al.* 1979; Uličný 2004), while bearing in mind that they are worse climatic indicators than the ventifacts. At first I will briefly discuss a few cases where recent publications introduce a certain confusion; I will then follow with a short presentation of the most recent results in southern Europe.

Hounslow & McIntosh (2003) suggest, after a palaeomagnetic study, that the ventifact-bearing Budleigh Salterton Pebble Beds (south Devon, United Kingdom) could be of Late Spathian to Early Aegean age (i.e. the same age as the Solling Formation and lower Röt Formation in Germany: Kozur 1999) and extend their conclusion to the 'Conglomérat principal' of the Vosges because both these conglomerates are responses to the same geodynamic events (Smith & Edwards 1991). The last assertion is obvious, but the probable Smithian age of the French conglomerate proposed here is more strongly constrained, which is why both would be better ascribed to the mid-Scythian arid period.

Ptaszyński & Niedźwiedzki (2004) assign the aeolian Tumlin Sandstone (Holy Cross Mountains, Poland) to the Late Permian on the basis of vertebrate track studies. Although aeolian dune deposits are frequent below the Zechstein of the CEB, the combined regional results of palynology (Fijałkowska-Mader 1999) and magnetostratigraphy (Nawrocki *et al.* 2003) show clearly that a correlation with the Volpriehausen Formation, proposed initially by Fuglewicz (1980), is most probable. The reason why ichnology seems to indicate a Permian rather than a Triassic age could be linked with the fact that the European lower and middle Scythian vertebrate footprints are very poorly documented (Demathieu & Haubold 1972; Avanzini *et al.* 2001).

Italy

In Nurra (NW Sardinia), on a relatively limited surface, a thick Permian–Triassic siliciclastic series crops out that shows remarkable similarities to that of the Toulon–Cuers Basin in Provence, allowing us to regard both areas as parts of the same basin, that initially closely faced each other (Cassinis *et al.* 2003). The presence of ventifacts scarcely reworked in the local 'Conglomerato del Porticciolo' is one of the arguments used to correlate this one with the Provençal 'Poudingue de Port-Issol'.

Otherwise, as in many other areas of limited exposure, the Nurra provides an excellent example of different 'Buntsandstein concepts', more or less explained, and likely to create serious problems in palaeogeographical syntheses. Sciunnach (2001) isolates under the name of 'Buntsandstein' the higher part of the 'Verrucano Sardo' (Gasperi & Gelmini 1979), which encompassed all the post-Autunian siliciclastic units. His 'Lower Buntsandstein' (i.e. Cala del Vino Formation: Cassinis *et al.* 2003) can be equated with the St-Mandrier Formation (more than 700 m thick) of the Toulon area (Fig. 2). Costamagna & Barca (2002) call 'Buntsandstein' the whole 'Verrucano Sardo'. Conversely, Cassinis *et al.* (2003), using the same concept as in Provence, restrict the 'Buntsandstein' to the uppermost part, starting from the 'Conglomerato del Porticciolo'.

Spain

It is generally admitted that, in the Castilian branch of the Iberian Ranges, the PTB lies somewhere in the lower part of the Cañizar Sandstones Formation, or its time equivalent, the Hoz de Gallo Conglomerates and Rillo de Gallo Sandstones formations, constituting the lower (and major) part of the 'Buntsandstein *sensu stricto*' (López-Gómez *et al.* 2002; Arche *et al.* 2004). In fact, the Hoz de Gallo Conglomerates include two distinct lithostratigraphical units separated by a discontinuity (Ramos 1979) whose importance has been underestimated. The lower unit, which yields Permian palynomorphs, ends locally with sandstones and then a silcrete, while a few hundred metres further these beds were eroded, so that the two conglomerates are in direct contact. Much further SE the upper conglomerate, yielding ventifacts, is the only unit present and constitutes the basal unit of the Cañizar Formation. This is why it is supposed here that the unconformity in question is an equivalent of that below the 'Grès vosgien' and represents the regional PTB.

The Balearic Islands are also believed to belong to the domain where the age of the Buntsandstein ranges from 'Thuringian' to Anisian (Broutin *et al.* 1992; Arche *et al.* 2002). Nevertheless, in Minorca, above the formation yielding Permian palynomorphs, and a clear unconformity, the lower Buntsandstein unit (B1) begins with a quartz conglomerate (Gómez-Gras & Alonso-Zarza 2003) that deserves further study but brings to mind the basal conglomerate of Provence and Sardinia. Since no conglomerate occurs in Majorca, the contact between dated Permian and Triassic formations appears transitional, but as the series is thinner it could be that the local PTB setting correponds to the 'Gonfaron type' and to a hiatus longer than in Minorca.

Bulgaria

In NW Bulgaria the Peri-Tethyan Lower Triassic appears under the Buntsandstein facies (Petrohan terrigenous Group) and usually begins with a basal oligomictic siliceous conglomerate; in its uppermost part occur marine intercalations of Spathian age (Belivanova 2000). For Zagorchev & Budurov (1997), the basal gap could correspond to a major part of the Lower Triassic, whereas for Yanev *et al.* (2001) it would be more related to the Upper Permian. The basal conglomerates recently provided many ventifacts, tending to show a general lack of the lowermost Triassic, but they appear in various settings. On the Noevtsi section (Kraishte Unit), the thin basal conglomerate is abruptly overlain by the palaeosol-rich sandstones of the Murvodol Formation; such a relationship calls to mind that of the 'Poudingue de Port-Issol' in Provence. In contrast, the conglomerate of the Smolyanovtsi section (Prebalkan Unit) passing upwards to the Belogradchik Formation, devoid of any pedogenic features, evokes very much the basal conglomerates of the 'Grès vosgien' or the Cañizar Formation, and thus could be a little older.

Discussion and conclusions

The 'French Buntsandstein' sedimentary cycle may begin considerably before the end of the Permian. Its basal unconformitiy ('Pfälzer Diskordanz', i.e. Palatine unconformity *sensu* Dachroth 1985) can be followed beneath the whole CEB. It is the 'Altmark unconformity' below the Upper Rotliegend II (Hoffmann *et al.* 1989; Schneider & Gebhart 1993) of northern Germany – where it can come down to the Illawarra Reversal – and below the Silverpit Claystone Formation of the North Sea and the Piła Claystone Formation in Poland (Karnkowski 1994); it also corresponds to the basal unconformity of the Val Gardena Sandstone and the Verrucano Lombardo of the Southern Alps. Intra-Zechstein tectonic movements were limited ('Tubantian' faultings of Geluk 1999). As pointed out by Fuglewicz (1980), the true 'Pfälzic' = 'Palatine' unconformity (i.e. between the Permian and Triassic in Palatinate) is nothing other than the 'Volpriehausen unconformity' below the time equivalents of the 'Grès vosgien' (i.e. Dachroth's 'Lauter' unconformity).

Nevertheless, during the Late Permian most areas in France, and elsewhere in SW Europe, experienced a more or less deep erosion forced by exceptional drops of global sea level (Ross & Ross 1987; Hallam & Wignall 1999; Seidler 2000; Heydari *et al.* 2001), and became only bypass zones for sediments during the earliest Triassic. Therefore, the time gap corresponding to the sub-Triassic unconformity actually encompasses the PTB, and this gap is increasingly long as one approaches the edges of the basin. For the Provence realm, it cannot be excluded that Les Arcs Formation results from the earliest Triassic infilling (transgressive system tract) of a palaeovalley incised during the Late Permian.

The aforementioned scenario seems characteristic of basins open towards the world ocean. Conversely, it is only in closed basins that one can expect a continuity of sedimentation from the Permian to the Triassic. Indeed, the only places on the West European Plate where such a continuity could be proven correspond to the deeper part of the endorheic CEB (Nawrocki 2004).

It must be emphasized that, everywhere in the Northern Hemisphere where such a continuity was demonstrated, the climate seems to have evolved towards rather more humid conditions (e.g. Fuglewicz 1980; Kozur 2003); macroflora is still of a prevailing Permian character and palynoflora shows a transitional nature (Shu & Norris 1999; Lozovsky *et al.* 2001). Nothing indicates a sudden collapse of the terrestrial ecosystem. This is why, when outcrops display an abrupt change in depositional style or palaeoclimatic conditions at the PTB boundary, a more or less significant gap should be suspected.

The first and main Early Triassic arid climatic phase in Europe is of Dienerian–Smithian age. The 'basal' conglomerate of the 'Buntsandstein' in Provence ('Poudingue de Port-Issol') is nothing but the uppermost part of a subordinate cycle corresponding to that phase and is coeval with the middle Buntsandstein of the Vosges. Therefore, the sub-Triassic unconformity represents there a hiatus probably as long as 10–15 Ma, encompassing at least the entire Lopingian and the majority of the Induan.

Much confusion currently exists about the concept of 'Buntsandstein' in certain regions of discontinuous exposure, where a major unconformity may be overlooked. This problem must be borne in mind every time different palaeogeographical domains seem likely to be distinguished. Works in progress confirm that the careful recognition of unconformities and the use of sedimentary indicators of a clearly arid climate constitute powerful tools for correlation within the non-marine rock units, devoid of any biostratigraphical markers, which are straddling the PTB, at least on the scale of the West European Plate.

The author is grateful to G. Cassinis for his invitation to contribute this paper. He would like also to thank J. W. Schneider, S. G. Lucas, A. Heckert and an anonymous referee for their valuable suggestions for improving the manuscript.

References

AIGNER, T. & BACHMANN, G. H. 1992. Sequence-stratigraphic framework of the German Triassic. *Sedimentary Geology*, **80**, 115–135.

ARCHE, A., LÓPEZ-GÓMEZ, J. & VARGAS, H. 2002. Propuesta de correlación entre los sedimentos Pérmicos y Triásicos de la Cordillera Ibérica Este y de las Islas Baleares. *Geogaceta*, **32**, 275–278.

ARCHE, A., LÓPEZ-GÓMEZ, J., MARZO, M. & VARGAS, H. 2004. The siliciclastic Permian–Triassic deposits in central and northeastern Iberian peninsula (Iberian, Ebro and Catalan basins): a proposal for correlation. *Geologica Acta*, **2**, 305–320.

AVANZINI, M., CEOLONI, P. *et al.* 2001. Permian and Triassic tetrapod ichnofaunal units of Northern Italy: their potential contribution to continental biochronology. *In*: CASSINIS, G. (ed.) *Permian Continental Deposits of Europe and Other Areas. Regional Reports and Correlations.* Natura Bresciana, Monografia, **25**, 89–107.

BAUDEMONT, D. 1988. Discordances angulaires multiples dans le Permien de Provence (France). Tectonique extensive antémésozoïque avec effondrements diachrones. *Comptes rendus de l'Académie des Sciences, Paris, Série II*, **306**, 149–152.

BELIVANOVA, V. 2000. Triassic in the Golo Bardo Mountain – one example for the Balkanide facial type of Triassic in Bulgaria. *In*: BACHMANN, G. H. & LERCHE, I. (eds) *Epicontinental Triassic.* Vol. 2. *Zentralblatt für Geologie und Paläontologie, Teil I*, **1998**(9–10), 1105–1121.

BOUCARUT, M. 1971. Etude volcanologique et géologique de l'Estérel (Var, France). PhD. thesis, Nice University.

BROUTIN, J. & DURAND, M. 1995. First paleobotanical and palynological data on the 'Les Salettes Formation' uppermost member (Permian Toulon Basin, southeastern France). XIIIth International Congress on Carboniferous-Permian, Kraków, *Abstracts*, 15–16.

BROUTIN, J., DOUBINGER, J., GISBERT, J. & SATTA-PASINI, S. 1988. Permières datations palynologiques dans le faciès Buntsandstein des Pyrénées catalanes espagnoles. *Comptes rendus de l'Académie des Sciences, Paris,* Série *II*, **306**, 159–163.

BROUTIN, J., FERRER, J., GISBERT, J. & NMILA, A. 1992. Première découverte d'une microflore thuringienne dans le faciès saxonien de l'île de Minorque (Baléares, Espagne). *Comptes rendus de l'Académie des Sciences, Paris, Série II*, **315**, 117–122.

BUREK, P. J. 1970. Magnetic reversals: their application to stratigraphic problems. *American Association of Petroleum Geologists Bulletin*, **54**, 1120–1139.

Cassinis, G., Durand, M. & Ronchi, A. 2003. Permian-Triassic continental sequences of northwest Sardinia and south Provence: stratigraphic correlations and palaeogeographical implications. *Bolletino della Società Geologica Italiana, Volume speciale*, **2**, 119–129.

Cassinis, G., Nicosia, U., Lozovsky, V. R. & Gubin, Y. M. 2002. A view of the Permian continental stratigraphy of the Southern Alps, Italy, and general correlation with the Permian of Russia. *Permophiles*, **40**, 4–16.

Costamagna, L. G. & Barca, S. 2002. The 'Germanic' Triassic of Sardinia (Italy): a stratigraphic, depositional and palaeogeographic review. *Rivista Italiana di Paleontologia e Stratigrafia*, **108**, 67–100.

Courel, L., Durand, M., Maget, P., Maiaux, C., Ménillet, F. & Pareyn, C. 1980. Trias. *In*: Mégnien, C. & Mégnien, F. (eds) *Synthèse Géologique du Bassin de Paris*. Vol. 1: *Stratigraphie et Paléogéographie*. Mémoires du Bureau de Recherches Géologiques et Minières, **101**, 37–74.

Cournut, A. 1966. Contribution à l'étude sédimentologique et métallogénique du Grès bigarré de la région du Luc-en-Provence (Var). PhD thesis, Nancy University.

Crasquin-Soleau, S. 2003. Middle and Late Permian correlations on the Russian Platform: significance of ostracodes. *Revue de Micropaléontologie*, **46**, 23–33.

Dachroth, W. 1985. Fluvial sedimentary styles and associated depositional environments in the Buntsandstein west of River Rhine, in Saar area and Pfalz (F.R. Germany) and Vosges (France). *In*: Mader, D. (ed.) *Aspects of fluvial Sedimentation in the Lower Triassic Buntsandstein of Europe*. Lecture Notes in Earth Sciences, **4**, 197–248.

Demathieu, G. & Durand, M. 1991. Les traces de pas de Tétrapodes dans le Trias détritique du Var et des Alpes maritimes (France). *Bulletin du Museum National d'Histoire Naturelle, Paris, C*, **13**, 115–133.

Demathieu, G. & Haubold, H. 1972. Stratigraphische Aussagen der Tetrapodenfährten aus der terrestrischen Trias Europas. *Geologie*, **21**, 802–836.

Durand, M. 1978. Paléocourants et reconstitution paléogéographique: l'exemple du Buntsandstein des Vosges méridionales (Trias inférieur et moyen continental). *Sciences de la Terre*, **22**, 301–390.

Durand, M. 1993. Un exemple de sédimentation continentale permienne dominée par l'activité de chenaux méandriformes: la Formation de Saint-Mandrier (Bassin de Toulon, Var). *Géologie de la France*, **2**, 43–55.

Durand, M., Avril, G. & Meyer, R. 1988. Paléogéographie des premiers dépôts triasiques dans les Alpes externes méridionales: importance de la Dorsale delphino-durancienne. *Comptes rendus de l'Académie des Sciences, Paris, Série II*, **306**, 557–560.

Durand, M., Meyer, R. & Avril, G. 1989. *Le Trias détritique de Provence, du dôme de Barrot et du Mercantour*. Publications de l'Association des Sédimentologistes Français, **6**.

Durand, M., Chrétien, J.-C. & Poinsignon, J.-M. 1994. Des cônes de déjection permiens au grand fleuve triasique: evolution de la sédimentation continentale dans les Vosges du Nord autour de – 250 Ma. National congress of the APBG, field-trip guide. Pierron, Sarreguemines.

Evans, J. W. 1911. Dreikanter. *Geological Magazine*, **8**, 334–335.

Forche, F. 1935. Stratigraphie und Paläogeographie des Buntsandstein im Umkreis der Vogesen. *Mitteilungen aus dem Geologischen Staatinstitut in Hamburg*, **15**, 15–55.

Fijałkowska-Mader, A. 1999. Palynostratigraphy, palaeoecology and palaeoclimatology of the Triassic in South-Eastern Poland. *In*: Bachmann, G. H. & Lerche, I. (eds) *Epicontinental Triassic*. Vol. 1. *Zentralblatt für Geologie und Paläontologie, Teil I*, **1998**(7–8), 601–627.

Fuglewicz, R. 1980. Stratigraphy and palaeogeography of the Lower Triassic in Poland on the basis of megaspores. *Acta Geologica Polonica*, **30**, 417–470.

Gand, G. & Durand, M. 2006. Tetrapod footprint ichnoassociations from French Permian basins: comparisons with other Euramerican ichnofaunas. *In*: Lucas, S. G., Cassinis, G. & Schneider, J. W. (eds) *Non-Marine Permian Biostratigraphy and Biochronology*. Geological Society, London, Special Papers, **265**, 157–177.

Gand, G., Demathieu, G. & Ballestra, F. 1995. La palichnofaune de vertébrés tétrapodes du Permien supérieur de l'Estérel (Provence, France). *Palaeontographica, Abteilung A*, **235**, 97–139.

Gasperi, G. & Gelmini, R. 1979. Ricerche sul Verrucano -4- Il Verrucano della Nurra (Sardegna NW). *Memorie della Società Geologica Italiana*, **20**, 215–231.

Geluk, M. 1999. Late Permian (Zechstein) rifting in the Netherlands: models and implications for petroleum geology. *Petroleum Geoscience*, **5**, 189–199.

Germain, D. 1968. Au sujet des bois silicifiés du Permien supérieur du Muy (massif de l'Estérel, Var, France). *Bulletin de la Société Belge de Géologie, Paléontologie et Hydrologie*, **57**, 203–215.

Glinzbœckel, C. & Durand, M. 1984. Trias: Provence et chaînes subalpines méridionales. *In*: Debrand-Passard, S., Courbouleix, S. & Lienhardt, M.-J. (eds) *Synthèse géologique du Sud-Est de la France. Vol. 1: Stratigraphie et paléogéographie*. Mémoire du Bureau de Recherches Géologiques et Minières, **125**, 99–100.

Gómez-Gras, D. & Alonso-Zarza, A. M. 2003. Reworked calcretes: their significance in the reconstruction of alluvial sequences (Triassic, Minorca, Balearic Islands, Spain). *Sedimentary Geology*, **158**, 299–319.

Gorsky, V. P., Gusseva, E. A., Crasquin-Soleau, S. & Broutin, J. 2003. Stratigraphic data of the Middle – Late Permian on Russian Platform. *Geobios*, **36**, 533–558.

Gradstein, F. M., Ogg, J., Smith, A. G., Bleeker, W. & Lourens, L. J. 2004. A new geologic time scale with special reference to Precambrian and Neogene. *Episodes*, **27**, 83–100.

Gradziński, R., Gągol, J. & Ślączka, A. 1979. The Tumlin Sandtone (Holy Cross Mts, Central

Poland): Lower Triassic deposits of aeolian dunes and interdune areas. *Acta Geologica Polonica*, **29**, 151–175.

Hallam, A. & Wignall, P. B. 1999. Mass extinctions and sea-level changes. *Earth-Science Reviews*, **48**, 217–250.

Heydari, E., Wade, W.J. & Hassanzadeh, J. 2001. Diagenetic origin of carbon and oxygen isotope compositions of Permian–Triassic boundary strata. *Sedimentary Geology*, **143**, 191–197.

Heunisch, C. 1999. Die Bedeutung der Palynologie für Biostratigraphie und Fazies in der Germanischen Trias. *In*: Hauschke, N. & Wilde, V. (eds) *Trias: Eine ganz andere Welt*. Friedrich Pfeil, München, 207–220.

Hoffmann, N., Kamps, H-J. & Schneider, J. 1989. Neuerkentnisse zur Biostratigraphie und Paläodynamik des Perms in der Nordostdeutschen Senke: ein Diskussionbeitrag. *Zeitschrift für Angewandte Geologie*, **35**, 198–207.

Hounslow, M. W. & McIntosh, G. 2003. Magnetostratigraphy of the Sherwood Sandstone Group (Lower and Middle Triassic), south Devon, UK: detailed correlation of the marine and non-marine Anisian. *Palaeogeography, Palaeoclimatology, Palaeoecology*, **193**, 325–348.

Jones, D. J. 1953. Tetrahedroid pebbles. *Journal of Sedimentary Petrology*, **23**, 196–201.

Karnkowski, P. H. 1994. Rotliegend lithostratigraphy in the central part of the Polish Permian Basin. *Geological Quarterly*, **38**, 27–42.

Kozur, H. 1993. Annotated correlation tables of the Germanic Buntsandstein and Keuper. *In*: Lucas, S. G. & Morales, M. (eds) *The Nonmarine Triassic*. New Mexico Museum of Natural History and Science Bulletin, **3**, 243–248.

Kozur, H. 1999. The correlation of the Germanic Buntsandstein and Muschelkalk with the Tethyan scale. *In*: Bachmann, G. H. & Lerche, I. (eds) *Epicontinental Triassic. Vol. 1. Zentralblatt für Geologie und Paläontologie, Teil I*, **1998**(7–8), 701–725.

Kozur, H. 2003. Integrated ammonoid, conodont and radiolarian zonation of the Triassic and some remarks to stage/substage subdivision and the numeric age of the Triassic stages. *Albertiana*, **28**, 57–74.

Lethiers, F., Damotte, R. & Sanfourche, J. 1993. Premières données sur les ostracodes du Permien continental de l'Estérel (SE de la France): systématique, biostratigraphie, paléoécologie. *Géologie Méditerranéenne*, **20**, 109–125.

López-Gómez, J., Arche, A. & Pérez-López, A. 2002. Permian and Triassic. *In*: Gibbons, W. & Moreno, T. (eds) *The Geology of Spain*. Geological Society, London, 185–212.

Lopez, M., Gand, G., Garric, J. & Galtier, J. 2005. *Playa environments in the Lodève Permian Basin and the Triassic Cover (Languedoc – France)*. Publications de l'Association des Sédimentologistes Français, **49**.

Lozovsky, V. R. 1993. The most complete and fossiliferous Lower Triassic section of the Moscow syneclise: The best candidate for a nonmarine global time scale. *In*: Lucas, S. G. & Morales, M. (eds) *The Nonmarine Triassic*. New Mexico Museum of Natural History and Science Bulletin, **3**, 293–295.

Lozovsky, V. R. 2003. Correlation of the continental Permian of northern Pangea: a review. *Bolletino della Società Geologica Italiana, Volume speciale*, **2**, 239–244.

Lozovsky, V. R., Krassilov, V. A., Afonin, S. A., Burov, B. V. & Yaroshenko, O. P. 2001. Transitional Permian–Triassic deposits in European Russia, and non-marine correlations. *In*: Cassinis, G. (ed.) *Permian Continental Deposits of Europe and Other Areas. Regional Reports and Correlations*. Natura Bresciana, Monografia, **25**, 301–310.

Lucas, S. G. & Schoch, R. R. 2002. Triassic temnospondyl biostratigraphy, biochronology and correlation of the German Buntsandstein and North American Moenkopi Formation. *Lethaia*, **35**, 97–106.

Menning, M. 1995. A numerical time scale for the Permian and Triassic periods: an integrated time analysis. *In*: Scholle, P. A., Peryt, T. M. & Ulmer-Scholle, D. S. (eds) *The Permian of Northern Pangea*. Vol. 1. Springer Verlag, Berlin, 77–97.

Menning, M. 2001. A Permian time scale 2000 and correlation of marine and continental sequences using the Illawarra Reversal (265 Ma). *In*: Cassinis, G. (ed.) *Permian Continental Deposits of Europe and Other Areas. Regional Reports and Correlations*. Natura Bresciana, Monografia, **25**, 355–362.

Merabet, N. & Daly, L. 1986. Détermination d'un pôle paléomagnétique et mise en évidence d'aimantations à polarité normale sur les formations du Permien supérieur du Massif des Maures (France). *Earth and Planetary Science Letters*, **80**, 156–166.

Nawrocki, J. 2004. The Permian–Triassic boundary in the Central European Basin: magnetostratigraphic constraints. *Terra Nova*, **16**, 139–145.

Nawrocki, J., Kuleta, M. & Zbroja, S. 2003. Buntsandstein magnetostratigraphy from the northern part of the Holy Cross Mountains. *Geological Quarterly*, **47**, 253–260.

Orłowska-Zwolińska, T. 1985. Palynological zones of the Polish epicontinental Triassic. *Bulletin of the Polish Academy of Sciences*, **33**, 107–117.

Pittau, P. 2002. Palynofloral biostratigraphy of the Permian and Triassic sequences of Sardinia. *Rendicotti della Società Paleontologica Italiana*, **1**, 93–109.

Poort, R. J., Clement-Westerhof, J. A., Looy, C. V. & Visscher, H. 1997. Aspects of Permian palaeobotany and palynology. XVII. Conifer extinction in Europe at the Permian–Triassic junction: morphology, ultrastructure and geographic/stratigraphic distribution of *Nuskoisporites dulhuntyi* (prepollen of *Ortiseia*, Walchiaceae). *Review of Palaeobotany and Palynology*, **97**, 9–39.

Ptaszyński, T. & Niedźwiedzki, G. 2004. Late Permian vertebrate tracks from the Tumlin Sandstone, Holy Cross Mountains, Poland. *Acta Palaeontologica Polonica*, **49**, 289–320.

Ramos, A. 1979. *Estratigrafía y paleogeografía del Pérmico y Triásico al oeste de Molina de Aragón (Provincia de Guadalajara)*. Seminarios de Estratigrafía, Monografías, **6**.

REITZ, E. 1988. Palynostratigraphie des Buntsandsteins in Mitteleuropa. *Geologisches Jahrbuch Hessen*, **116**, 105–112

RENEVIER, E. 1874. *Tableau des Terrains Sédimentaires Formés Pendant les Époques de la Phase Orogénique du Globe Terrestre*. Rouge & Dubois, Lausanne.

ROCHETTE, P., BEN ATIG, F., COLLOMBAT, H., VANDAMME, D. & VLAG, P. 1997. Low paleosecular variation at the equator: a paleomagnetic pilgrimage from Galapagos to Esterel with Allan Cox and Hans Zijderveld. *Geologie en Mijnbouw*, **76**, 9–19.

ROSS, C. A. & ROSS, J. R. P. 1987. Late Paleozoic sea levels and depositional sequences. *In*: ROSS, C. A. & HAMAN, D. (eds) *Timing and Depositional History of Eustatic Sequences: Constraints on Seismic Stratigraphy*. Cushman Foundation for Foraminiferal Research, Special Publications, **24**, 137–149.

RYGEL, M. C., GIBLING, M. R. & CALDER, J. H. 2004. Vegetation-induced sedimentary structures from fossil forests in the Pennsylvanian Joggins Formation, Nova Scotia. *Sedimentology*, **51**, 531–552.

SCHNEIDER, J. & GEBHART, U. 1993. Litho- und Biofaziesmuster in intra- und extramontanen Senken des Rotliegend (Perm, Nord- und Ostdeutschland). *Geologische Jahrbuch, (A)*, **131**, 57–98.

SCIUNNACH, D. 2001. Heavy mineral provinces as a tool for palaeogeographic reconstruction: a case study from the Buntsandstein of Nurra (NW Sardinia, Italy). *Eclogae Geologicae Helvetiae*, **94**, 197–211.

SEIDLER, L. 2000. Incised submarine canyons governing new evidence of Early Triassic rifting in East Greenland. *Palaeogeography, Palaeoclimatology, Palaeoecology*, **161**, 267–293.

SHU, O. & NORRIS, G. 1999. Earliest Triassic (Induan) spores and pollen from the Junggar Basin, Xinjiang, northwestern China. *Review of Palaeobotany and Palynology*, **106**, 1–56.

SMITH, S. A. & EDWARDS, R. A. 1991. Regional sedimentological variations in Lower Triassic fluvial conglomerates (Budleigh Salterton Pebble Beds), southwest England: some implications for palaeogeography and basin evolution. *Geological Journal*, **26**, 65–83.

SZURLIES, M. 2004. Magnetostratigraphy: the key to a global correlation of the classic Germanic Trias – Case study Volpriehausen Formation (Middle Buntsandstein), central Germany. *Earth and Planetary Science Letters*, **227**, 394–410.

TOUTIN-MORIN, N. 1986. Quelques exemples de discontinuités sédimentaires dans le domaine continental provençal (France). *Archives des Sciences, Genève*, **39**, 67–78.

TOUTIN-MORIN, N. & VINCHON, C. 1989. Les bassins permiens du Sud-Est. *In*: CHÂTEAUNEUF, J.-J. & FARJANEL, G. (eds) *Synthèse géologique des bassins permiens français*. Mémoire du Bureau de Recherches Géologiques et Minières, **128**, 114–121.

TOUTIN-MORIN, N., BONIJOLY, D. *et al.* 1994. *Notice explicative: Carte géologique de la France à 1150 000, feuille Fréjus-Cannes n° 1024*. BRGM, Orléans.

ULIČNÝ, D. 2004. A drying-upward aeolian system of the Bohdašín Formation (Early Triassic), Sudetes of NE Czech Republic: record of seasonality and long-term palaeoclimate change. *Sedimentary Geology*, **167**, 17–39.

UTTING, J., ESAULOVA, N. K., SILANTIEV, V. V. & MAKAROVA, O. V. 1997. Late Permian palynomorph assemblages from Ufimian and Kazanian type sequences in Russia, and comparison with Roadian and Wordian assemblages from the Canadian Arctic. *Canadian Journal of Earth Sciences*, **34**, 1–16.

VISSCHER, H. 1968. On the Thuringian age of the Upper Palaeozoic sedimentary and volcanic deposits of the Estérel, southern France. *Review of Palaeobotany and Palynology*, **6**, 71–83.

VISSCHER, H. 1971. *The Permian and Triassic of the Kingscourt Outlier, Ireland: A Palynological Investigation Related to Regional Stratigraphic Problems in the Permian and the Triassic of western Europe*. Geological Survey of Ireland, Special Papers, **1**.

VLAG, P., VANDAMME, D., ROCHETTE, P. & SPINELLI, C. 1997. Paleomagnetism of the Esterel rocks: a revisit after the thesis of Hans Zijderveld. *Geologie en Mijnbouw*, **76**, 21–33.

VOZENIN-SERRA, C., BROUTIN, J. & TOUTIN-MORIN, N. 1991. Bois permiens du Sud-Ouest de l'Espagne et du Sud-Est de la France. Implications pour la taxonomie des Gymnospermes paléozoïques et la phylogénie des Ginkgophytes. *Palaeontographica, Abteilung B*, **221**, 1–26.

WARD, P. D., MONTGOMERY, D. R. & SMITH, R. 2000. Altered river morphology in South Africa related to the Permian–Triassic extinction. *Science*, **289**, 1740–1743.

WRIGHT, V. P., MARRIOTT, S. B. & VANSTONE, S. D. 1991. A reg palaeosol from the Lower Triassic of south Devon: stratigraphic and palaeoclimatic implications. *Geological Magazine*, **128**, 517–523.

YANEV, S., POPA, M., SEGHEDI, A. & OAIE, G. 2001. Overview of the continental Permian deposits of Bulgaria and Romania. *In*: CASSINIS, G. (ed.) *Permian Continental Deposits of Europe and Other Areas. Regional Reports and Correlations*. Natura Bresciana, Monografia, **25**, 269–279.

ZAGORCHEV, I. & BUDUROV, K. 1997. Outline of the Triassic palaeogeography of Bulgaria. *Albertiana*, **19**, 12–24.

ZHENG, J. S., MERMET, J.-F., TOUTIN-MORIN, N., HANES, J., GONDOLO, A., MORIN, R. & FERAUD, G. 1992. Datation $^{40}Ar–^{39}Ar$ du magmatisme et de filons minéralisés permiens en Provence orientale (France). *Geodinamica Acta*, **5**, 203–215.

New continental Carboniferous and Permian faunas of Morocco: implications for biostratigraphy, palaeobiogeography and palaeoclimate

D. HMICH[1], J. W. SCHNEIDER[2], H. SABER[3], S. VOIGT[2] & M. EL WARTITI[1]

[1]*Département de Géologie, Université Mohammed V, BP 1014, Rabat, Morocco*

[2]*Institut für Geologie, TU Bergakademie Freiberg, B.v.Cotta-Strasse 2, 09596 Freiberg, Germany (schneidj@geo.tu-freiberg.de)*

[3]*Département de Géologie, Université Chouaib Doukkali, BP 20, 24000 El Jadida, Morocco*

Abstract: Late Palaeozoic sediments in central Morocco and the High Atlas Mountains document the development of this area during the formation of the Mauretanide part of the Hercynian orogeny. Continental basins formed during the Stephanian and Permian. Although scattered in time, they provide valuable biogeographical and climatic information for the Mauretanides as a link between the Variscides in the east, the Appalachians in the west and the Karoo in the south. New blattid insects in the Souss Basin enable correlation to Early Stephanian B. Furthermore, we document the oldest African tetrapod tracks (*Batrachichnus, Dromopus*). Litho- and biofacies indicate seasonally wet and dry phases. Wet red beds of the Khenifra Basin have produced tetrapod bones and the tracks *Limnopus, Batrachichnus* and *Dromopus*. Macrofloras give a transitional Autunian/Saxonian age. This fits well into the Artinskian wet phase. Similar facies pattern in the Tiddas Basin are correlated by tetrapod tracks as transitional Artinskian to Kungurian. Advanced tetrapod tracks of *Synaptichnium* and *Rhynchosauroides* were discovered in the Ikakern Formation of the Argana Basin, dated by pareiasaur remains as Wuchiapingian. Red beds of similar type are known in Europe, for example, from the Late Permian of the Lodève Basin. They originated during the Wuchiapingian wet phase.

Late Palaeozoic sediments, magmatites and volcanites crop out in the central Moroccan Meseta and High Atlas Mountains. They document the development of this area during the formation of the Mauretanide part of the Hercynian orogen (Fig. 1). The Hercynian orogen consists of the Variscides in Europe and the Mauretanides in North Africa, as well as the Alleghanian and Ouachita orogens in North America (Fig. 2). In the central Moroccan Western Meseta, late Visean to early Westphalian marine turbidite sequences mark the early stages of foreland basin development, which is interpreted as the southern prolongation of the European Variscan belt (Ben Abbou *et al.* 2001). Pure continental intramontane basins developed during Stephanian and Permian times (Saber *et al.* 1995). They are preserved mostly as fault-bounded basin remnants of restricted extent (Fig. 1). The largest one, the Chougrane Basin, recently mapped for the first time (Hmich 2004), has an extent of only 24×7.5 km. The sedimentary and volcanic infill of these basins is of special interest for understanding the development of late Palaeozoic climate, environments and biota, because the Mauretanides form the link between the Variscides in the east and the Alleghanians in the west, as well as to the Gondwana Karoo system in the South.

Traditionally, the biostratigraphy of the most important continental Carboniferous–Permian basins in Morocco has been based exclusively on macrofloras and microfloras (e.g. Jongmans 1950; Doubinger & Roy-Dias 1985; Broutin *et al.* 1989). Animal fossils have been discovered rarely and have not been used for biostratigraphy, apart from rare exceptions (e.g. Jalil & Dutuit 1996). Fieldwork since 2001 has produced new fossil sites. First reports of the newly discovered fossil insects are given by Hmich *et al.* (2002, 2003). In the following, we provide a synthesis of these new and the older discoveries, together with the palaeobotanical data, to produce a greater understanding of biostratigraphy, palaeobiogeography and palaeoclimatology. We use the modified time scale of Menning & German Stratigraphic Commission (2002) and the holostratigraphic correlation charts of Lützner *et al.* (2003) and Roscher & Schneider (2005).

From: LUCAS, S. G., CASSINIS, G. & SCHNEIDER, J. W. (eds) 2006. *Non-Marine Permian Biostratigraphy and Biochronology*. Geological Society, London, Special Publications, **265**, 297–324.
0305-8719/06/$15.00

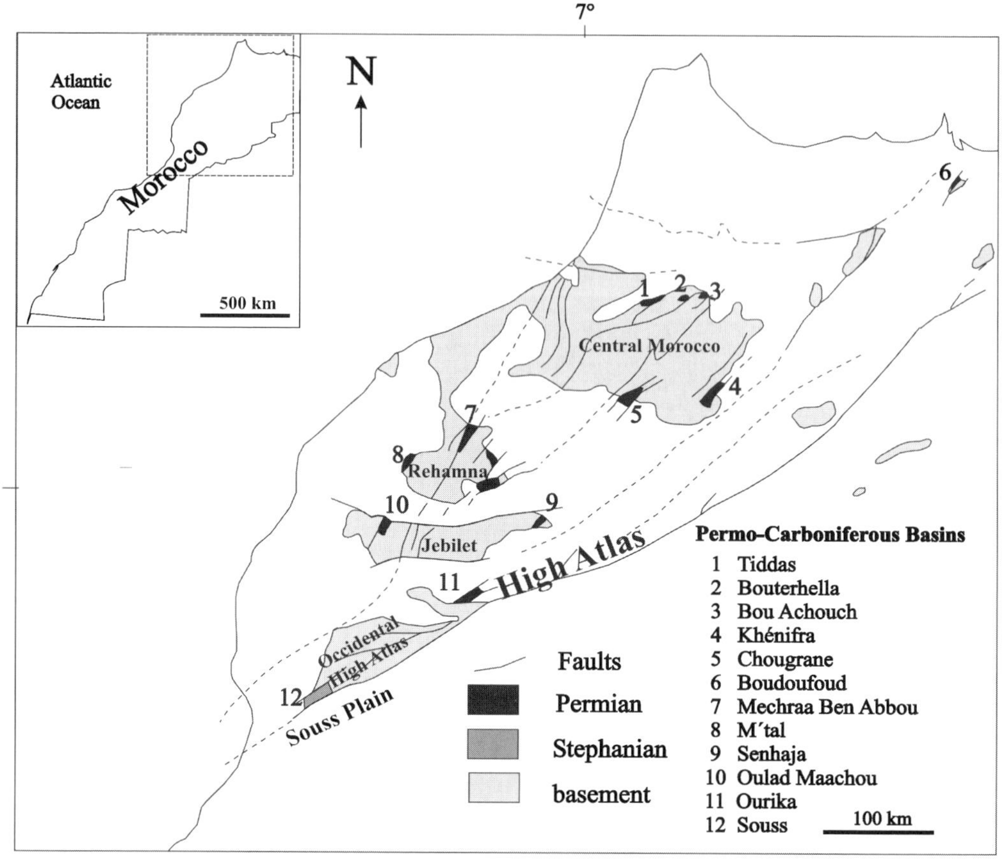

Fig. 1. Generalized map of northern Morocco with the most important Late Carboniferous and Permian basins.

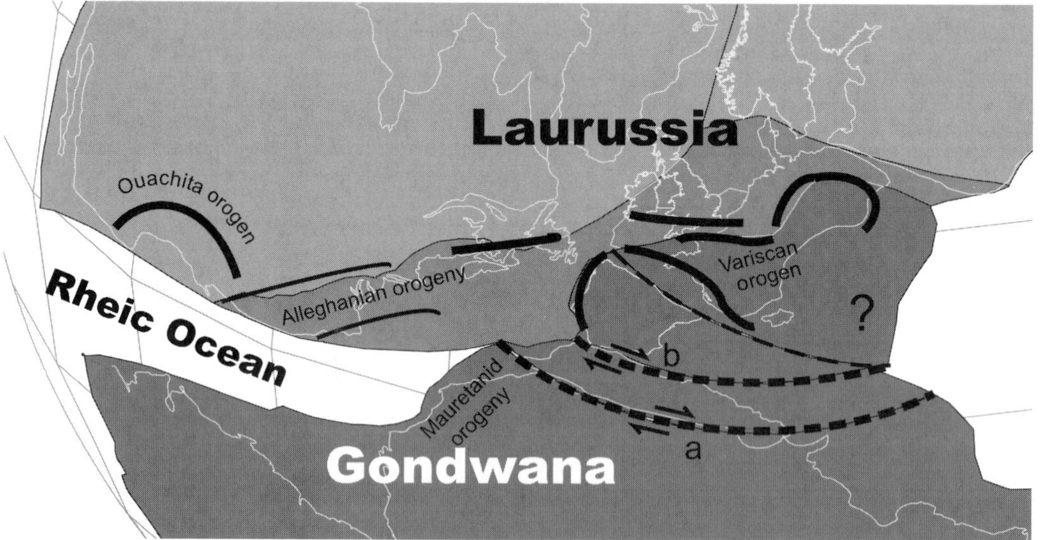

Fig. 2. The Hercynian orogenic system by about 300 Ma (Pennsylvanian) at the collision front of Gondwana and Laurussia, consisting of (east to west) the Variscan, the Mauretanid, the Alleghanian and Ouachita orogens.

Souss Basin

Coal exploration in the Souss Basin resulted in the first comprehensive descriptions (De Koning 1957; Feys & Greber 1963), and, with regard to the flora, by Jongmans (1950). New data have been added by Broutin *et al.* (1989), Saber (1994) and Saber *et al.* (1995, 2001).

Geological setting

The Souss Basin is the southwesternmost basin in the western High Atlas Mountains. It consists of the two sub-basins of Ida Ou Zal and Ida Ou Ziki (Fig. 3). They are fault-bounded remnants of a former larger basin that was filled with Late Pennsylvanian grey and Permian red sediments up to 2600 m thick (Saber 1994, 1998; Saber *et al.* 2001). These sediments cover an occasionally distinct palaeorelief of the deformed and metamorphosed basement. In both of the sub-basins, sedimentation started with a basal fining-upward mega-sequence of conglomerates and sandstones about 400–600 m thick (Figs 4 & 6). They are overlain by up to 1200 m of grey, alluvial plain sediments with fluvial sandstones, lacustrine black shales and local decimetre-thick coal seams. Previously, only plant fossils were known from this second unit, apart from ostracods and some conchostracans (see below). Subsequently, a variety of animal fossils were discovered in both sub-basins that, together with the palaeobotanical data, provide new insights into biotic environments and biostratigraphy.

Ida Ou Ziki Sub-Basin

The basal coarse clastics of the Ida Ou Ziki Sub-Basin (Fig. 4), the Tajgaline Formation, are about 600 m thick. They consist of a monotonous sequence of 2–15-m-thick, fining-upward cycles, with metre-thick fluvial channel conglomerates at the base, overlain by decimetre-thick fluvial sandstones and metre-thick siltstones at the top. These distal fan deposits are conformably overlain by the 500-m-thick, fossiliferous

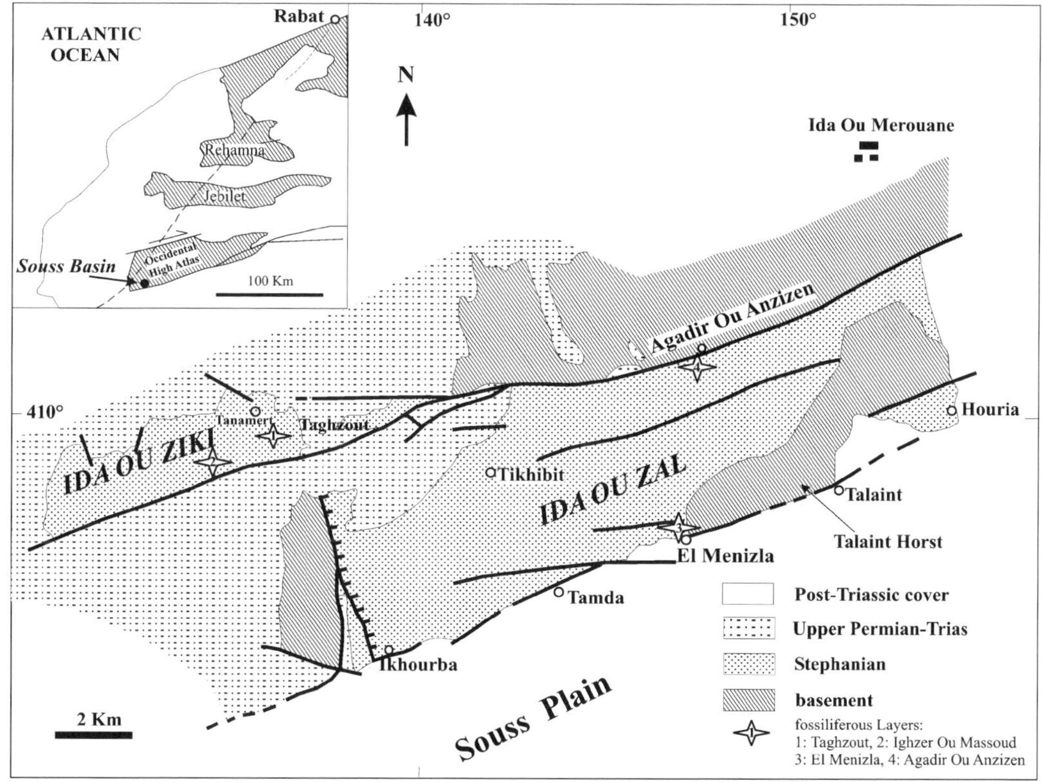

Fig. 3. Sketch map of the Stephanian (Late Pennsylvanian) Souss Basin with the sub-basins of Ida Ou Ziki and Ida Ou Zal, western High Atlas Mountains. Stars indicate recently discovered insect sites. Fossiliferous sites: 1, Taghzout; 2, Ighzer Ou Massoud; 3, El Menizla; 4, Agadir Ou Anzizen.

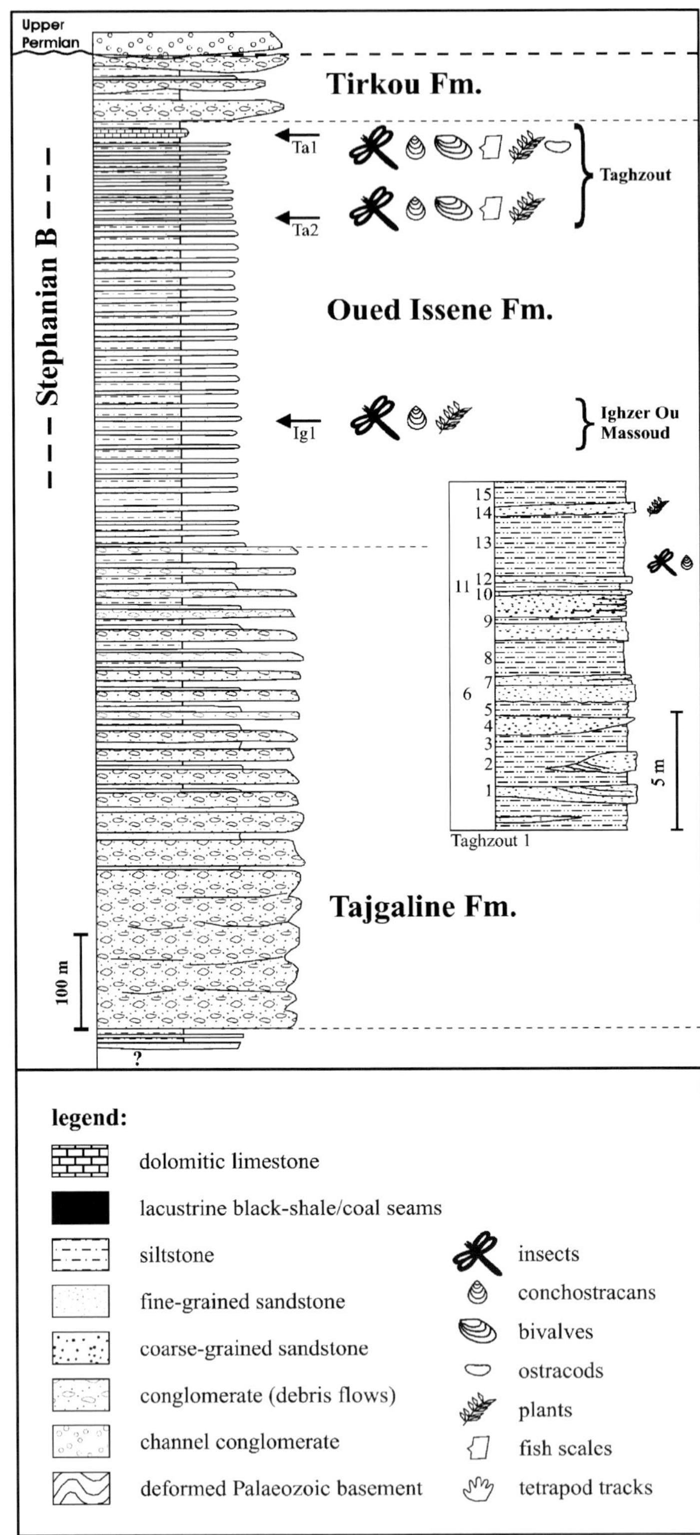
Upper Permian
Tirkou Fm.
Ta1
Ta2
Taghzout
Oued Issene Fm.
Stephanian B
Ig1
Ighzer Ou Massoud
15
14
13
12
11
10
9
8
7
6
5
4
3
2
1
5 m
Taghzout 1
Tajgaline Fm.
100 m
?
legend:
dolomitic limestone
lacustrine black-shale/coal seams
siltstone
fine-grained sandstone
coarse-grained sandstone
conglomerate (debris flows)
channel conglomerate
deformed Palaeozoic basement
insects
conchostracans
bivalves
ostracods
plants
fish scales
tetrapod tracks

alluvial plain/flood plain/flood basin sediments of the Oued Issène Formation (Saber *et al.* 1995; the Amlal Formation of Errami *et al.* 2002 is a junior synonym). Again, this formation appears relatively monotonous, consisting of fining up-cycles of metre to decimetre scale. Each cycle starts with fluvial, small-scale trough cross-bedded to planar horizontally bedded sandstones at the base, followed by interbedding of sandstones with siltstones and claystones in decimetre to centimetre scale (Fig. 5a). The sandstones contain plant detritus and plant trunks, the claystones plant leaf fragments and sometimes very common insect wings (see below). Lacustrine limestones, 1–2-m-thick with ostracods and occasional microbial mat structures, appear close to the top of the Oued Issène Formation. The Tirkou Formation, which is up to 80-m-thick, lies above an erosional unconformity. The formation consist mainly of metre to decimetre-thick channel conglomerates, sometimes stacked, with locally interbedded decimetre-thick sandstones, containing silicified wood and metre to decimetre-thick overbank siltstones containing plant detritus.

In the Ida Ou Ziki Sub-Basin, Middle to Late Permian red beds of the Ikakern Formation (the T1 unit or Ait Driss Member) and the overlying Middle Triassic deposits (T4 unit or Aglegal Member) rest with angular disconformity on the eroded Tirkou Formation (Fig. 5c). The Middle to Late Permian age of the Ait Driss Member (T1) is based on the discovery of Permian tetrapods at the level of T2 unit or the Tourbihine Member, respectively, of the Argana Basin (see below).

Fossil content of the Ida Ou Ziki Sub-Basin

The fossiliferous Oued Issène Formation is well exposed in the Oued Issène Canyon (Fig. 5a–d). Plant remains from these outcrops and the siltstone intercalations of the Tirkou Formation have been described by Bertrand (in Clariond 1932), Jongmans (1950), Feys & Greber (1963) and Broutin *et al.* (1989). Based on such taxa as *Annularia stellata, A. sphenophylloides, Pecopteris candolleana, P. hemitelioides, P. cyathea, P. monyi, P. pinnatifida, Sphenopteris matheti, Odontopteris obtusa, Mixoneura neuropteropteroides, Stigmaria* roots and other plant remains (Broutin *et al.* 1989, p. 1511), this flora has been regarded as middle to late Stephanian in age since Bertrand (in Clariond 1932) and Jongmans (1950) (Ait Moussa in Jongmans 1950 is now the Ida Ou Ziki Sub-Basin). Broutin *et al.* (1989) designated this flora as a Stephanian ecotype. Only one fossiliferous horizon is known from the uppermost part of the profile: the Tirkou Formation. Broutin *et al.* (1989) report an interesting flora there, consisting of *Sphenophyllum oblongifolium, Alethopteris subelegans, Odontopteris* cf. *subcrenulata*, cf. *Taeniopteris* gr. *jejunata, Neuropteris cordata, N. neuropteroides, Poacordaites, Walchia piniformis, Otovicia hypnoides*, and some *Culmitzschia* species, as well as *Sphenobaiera* sp. Because of the large number of conifers and the associated occurrence of *Sphenobaiera* sp., an Early Permian or Early Autunian age was adopted by Broutin *et al.* (1989). Aassoumi *et al.* (2003) reported a microflora from the middle Oued Issène Formation dominated by trilete and monolete spores, such as *Laevigatosporites, Calamospora, Lycospora, Reticulatisporites* and others. Monosaccate and bisaccate pollen are very rare. Further samples from the top of the Oued Issène Formation just below the conglomerates of the Tirkou Formation contain *Reticulatisporites, Lycospora, Thymospora, Leiotriletes* and others. Monosaccate pollen of a *Potonieisporites* type are less abundant here but more common than in older samples.

Some metres below, in an interval of plant-bearing siltstone and fine sandstone horizons in the Oued Issène Formation at Taghzout close to Tanamert village (Figs 3, fossil site 1; Fig. 5d), a 20-cm-thick lacustrine black claystone (Fig. 4, inset profile, base of horizon 13) yielded in 2001 the first late Palaeozoic insects found in North Africa (Hmich *et al.* 2002, 2003, 2005). Later, additional insect sites were discovered in metre-thick lacustrine horizons some tens of metres downwards in the profile between Tanamert and Ighzer Ou Massoud villages (Fig. 3, 5b). The entomofauna of these sites consist of blattids ('cockroaches') only and is dominated by poroblattinids and spiloblattinids (Fig. 8c, d). Mylacrids are remarkably rare in contrast to the Ida Ou Zal Sub-Basin (Hmich *et al.* 2003, 2005). Aquatic fauna, such as ostracods and sporadic conchostracans, are very common, pelecypods are not rare, and minute fish scales are scarce. The flora of the lake horizons is clearly dominated by walchians and *Odontopteris*; callipterids are rare, and fragments of long needles could belong to *Dicranophyllum*.

Fig. 4. Stephanian profile of the Ida Ou Ziki Sub-Basin of the Souss Basin (cf. Fig. 3). Levels of fossil insects are indicated. Inset profile: insect locality Taghzout 1; the basal part of the lacustrine black claystones of the horizon no. 13 is very rich in insect wings; fluvial fine-grained sandstones of horizon no. 14 contain plant remains, such as leaf fragments and small trunks.

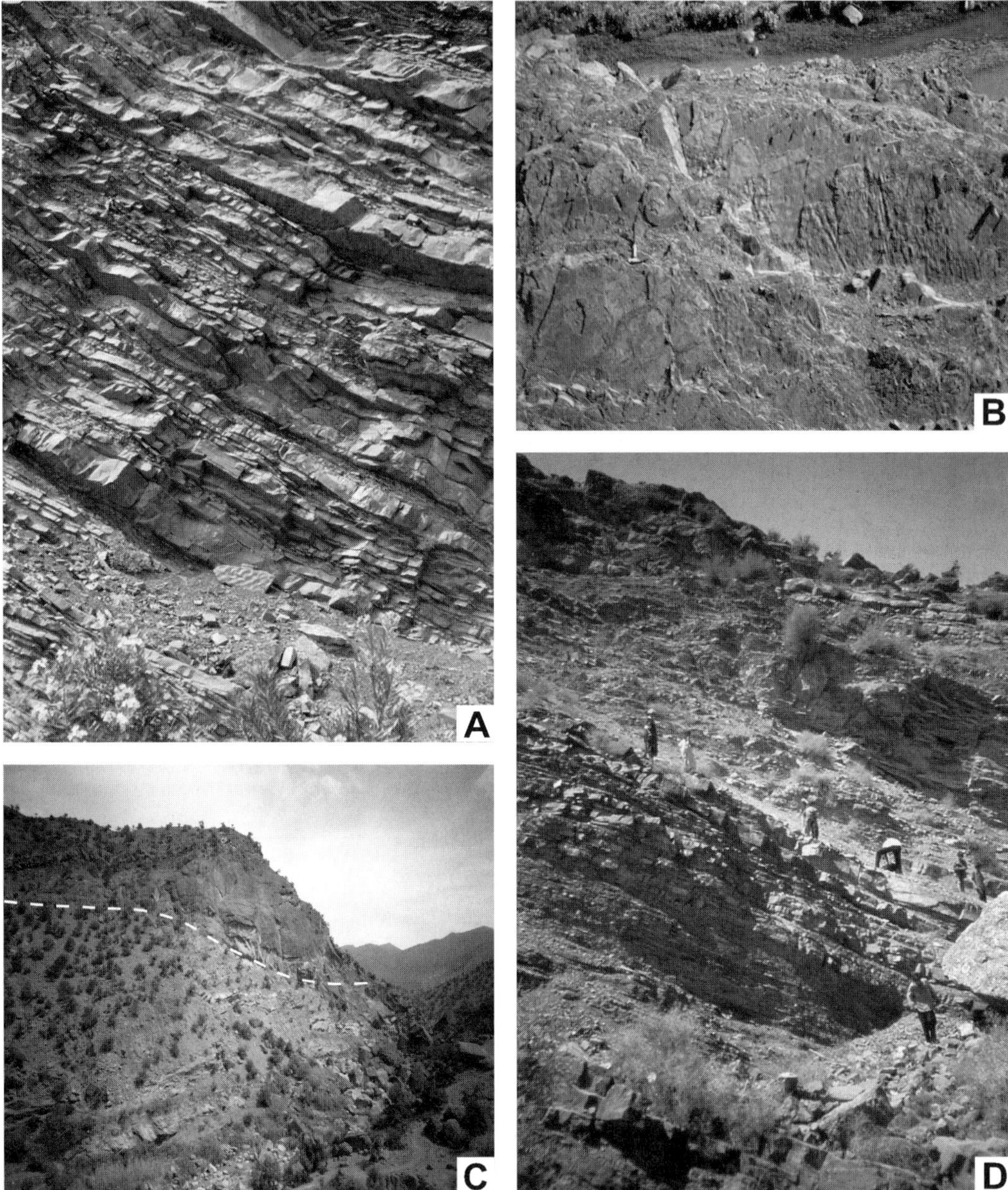

Fig. 5. Typical outcrops in the type area of the Oued Issène Formation in the Ida Ou Ziki Sub-Basin of the Souss Basin. (**A**) Monotonous interbedding of shallow channels of fluvial sandstones with overbank siltstones and thin lacustrine black shales; Oued Issène canyon between Taghzout and Ighzer Ou Massoud (cf. Fig. 3). (**B**) Shallow lacustrine to fluvial fine sandstones with desiccation cracks (below hammer), flood casts, ripple and sole marks (right); intercalated are fossiliferous lacustrine grey siltstones and black claystones; Oued Issene canyon, insect site Taghzout 2. (**C**) Angular unconformity (white broken line) between the Middle to Late Permian Ait Driss Member, Ikakern Formation, and the Late Pennsylvanian (Stephanian) Oued Issène Formation; the red fluvial conglomerates and pebbly sandstones of the Ait Driss Member are deeply cut into the grey sediments of the Oued Issène Formation; Qued Issène Canyon near Taghzout. (**D**) Exposure of fluvial channel sandstones, floodplain siltstones and lacustrine black shales; insect site Taghzout 1 in the Oued Issène canyon close to Tanamert village; people in the middle of the picture stand in the level of the fossiliferous horizon no. 13, (cf. inset profile in Fig. 4).

Ida Ou Zal Sub-Basin

In the Ida Ou Zal Sub-Basin, the basal conglomerates (= Ikhourba Formation) have a thickness of 400–600 m (Fig. 6). They consist of a monotonous sequence of 1–15-m-thick fining-upward cycles, with up to 5-m-thick fluvial channel conglomerates at the base, overlain by up to 1-m-thick fluvial sandstones and centimetre-thick siltstones at the top. Debris flow conglomerates and carbonaceous horizons up to a maximum of 0.5-m-thick are intercalated. These distal fan deposits are overlain by 1200 m of fossiliferous alluvial plain/flood plain/flood basin sediments

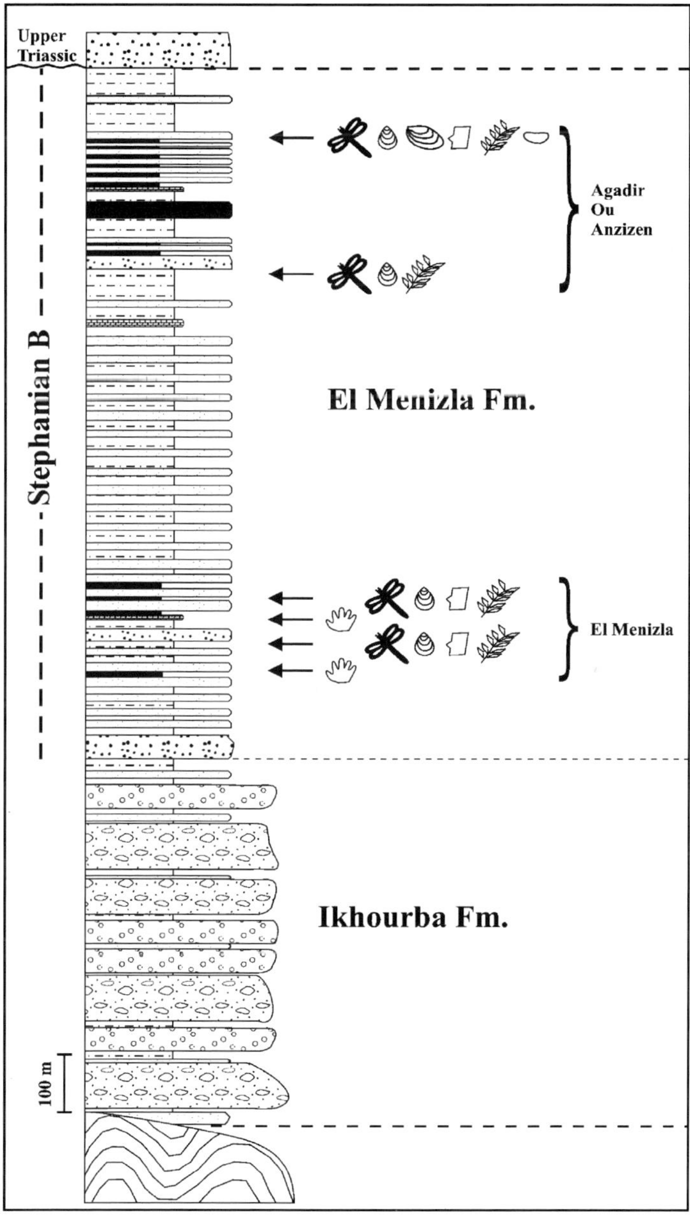

Fig. 6. Stephanian profile of the Ida Ou Zal Sub-Basin of the Souss Basin (cf. Fig. 3). Levels of fossil insects are indicated. For legend see Figure 4.

A
B
C

of the El Menizla Formation. They form a monotonous sequence of fining-upward cycles 0.5–15 m thick (Fig. 7a). Each cycle starts with massive to trough cross-bedded channel sandstones of 0.1–2 m (rarely 10 m) thickness, overlain by decimentre to metre-thick siltstones. Commonly, the cycle-top consists of fossiliferous laminated black lacustrine claystones of decimetre to metre thickness (Fig. 7b, c). Plant trunks and leaf fragments occur most commonly in the sandstones. The claystones contain, in addition to leaf fragments, insects and aquatic animals. With an upward decrease in grain size there are rare intercalations of centimetre to decimetre-thick, laminated bituminous dolomitic limestones, containing ostracods. Carbonaceous siltstones and impure coal seams of 10 cm thickness occur particularly in the lower and upper part of the formation (Saber 1998).

Red sediments were observed at only one location. They crop out with decametre thickness at the mountain path to Agadir Ou Anzizen at 30°38.239′N, 009°005′435′W (Système Géodesique Merchich). The facies architecture is quite the same as in the grey facies, except that channel sandstones about 1 m thick occur with greater frequency. The sandstones are pale red-brown, and the intercalated clayey-silty layers are dark red-brown. Above an erosional unconformity, the El Menizla Formation is overlain by Late Triassic sediments of the Timesgadiouine Formation (T5 or Irohalene Member).

Fossil content of the Ida Ou Zal Sub-Basin

The type section of the El Menizla Formation crops out along the El Menizla River at El Menizla village near the southern border of the Ida Ou Zal Sub-Basin (Fig. 7b). From here the following flora is reported by Saber *et al.* (1995) and Saber (1998): *Calamites, Asterophyllites equisetiformis, Annularia, Macrostachya, Sphenophyllum oblongifolium, Lepidodendron, Pecopteris arborescens, P. hemitelioides, P. unita, P. paleacea, Asterotheca* and *Odontopteris. Walchia* was reported as rare. A more or less identical hydromorphic floral association together with leaiaid conchostracans has been reported from the 'Assise productive' (now El Menizla Formation) of this region by Feys & Greber (1963).

Fossiliferous, laminated lacustrine black shales of the lower El Menizla Formation are exposed at the water basin of El Menizla village, reaching a thickness of up to 10 m. They are intercalated in a grey-facies sequence of alluvial sandstones and siltstones (Fig. 7b). The aquatic fauna consists of pelecypods, pseudesteriid and leaiaid conchostracans, ostracods and very rare, millimetre-sized fish scales. Coprolites of about 3–5 cm in length indicate the presence of larger aquatic animals. Insect remains are unusually common, but up to now only blattids have been found. The entomofauna (Fig. 8) is absolutely dominated by the mylacrid species *Opsiomylacris thevenini*; spiloblattinids, poroblattinids and phylloblattids are rare (Hmich *et al.* 2003, 2005). Fine sandstones with raindrop and ripple marks contain tetrapod tracks tentatively assigned to *Batrachichnus* Woodworth 1900 and *Dromopus* Marsh 1894. These two ichnogenera, referred to small temnospondyls and araeoscelids, respectively, are typical elements of a latest Pennsylvanian (Stephanian) to Lower Permian tetrapod footprint association. These are widespread in fluvial to marginal lacustrine deposits of Europe and North America (Haubold *et al.* 1995; Voigt 2005). This is the first report of Carboniferous vertebrate tracks from the African continent.

Along the mountain path to Agadir Ou Anzizen village, close to the northern border of the sub-basin, decimetre-thick lacustrine black clayey siltstones are intercalated in different levels of alluvial plain channel sandstones and overbank siltstones of the higher El Menizla Formation (Fig. 7a, b). Generally, the facies architecture is very monotonous, consisting of decimetre-thick, rarely metre-thick channel and crevasse splay sandstones interbedded with decimetre to metre-thick siltstones and centimetre to decimetre-thick, occasionally metre-thick, lacustrine varved black claystones. These lacustrine fine clastics contain plant and insect remains as well as pelecypods, conchostracans, ostracods and rare, minute fish scales. Very common are wings of the mylacrid species *Opsiomylacris thevenini*, but spiloblattinids and poroblattinids are rather rare. Possibly the

Fig. 7. Typical outcrops in the type area of the El Menizla Formation between El Menizla and Agadir Ou Anzizen in the Ida Ou Zal Sub-Basin of the Souss Basin. (**A**) Alluvial plain to floodplain sediments of the upper El Menizla Formation as monotonous sequences of fining upward cycles, each 0.5–15 m thick, with channel sandstones at the base and fluvial to lacustrine siltstones and lacustrine black shales on the top; mountain path to Agadir Ou Anzizen. (**B**) Section of grey to black lacustrine siltstones bearing plant remains and insect wings; two fine sandstone beds intercalated; close-up view of the horizon just right to the person in A; scale 1 m. (**C**) About 10-m-thick sequence of laminated lacustrine black shales (bottom right up to the right corner of the water basin) overlain by greenish to grey alluvial plain siltstones with intercalated channel sandstones; insect site at the water basin of El Menizla village, lower El Menizla Formation; folding rule bottom right 2 m.

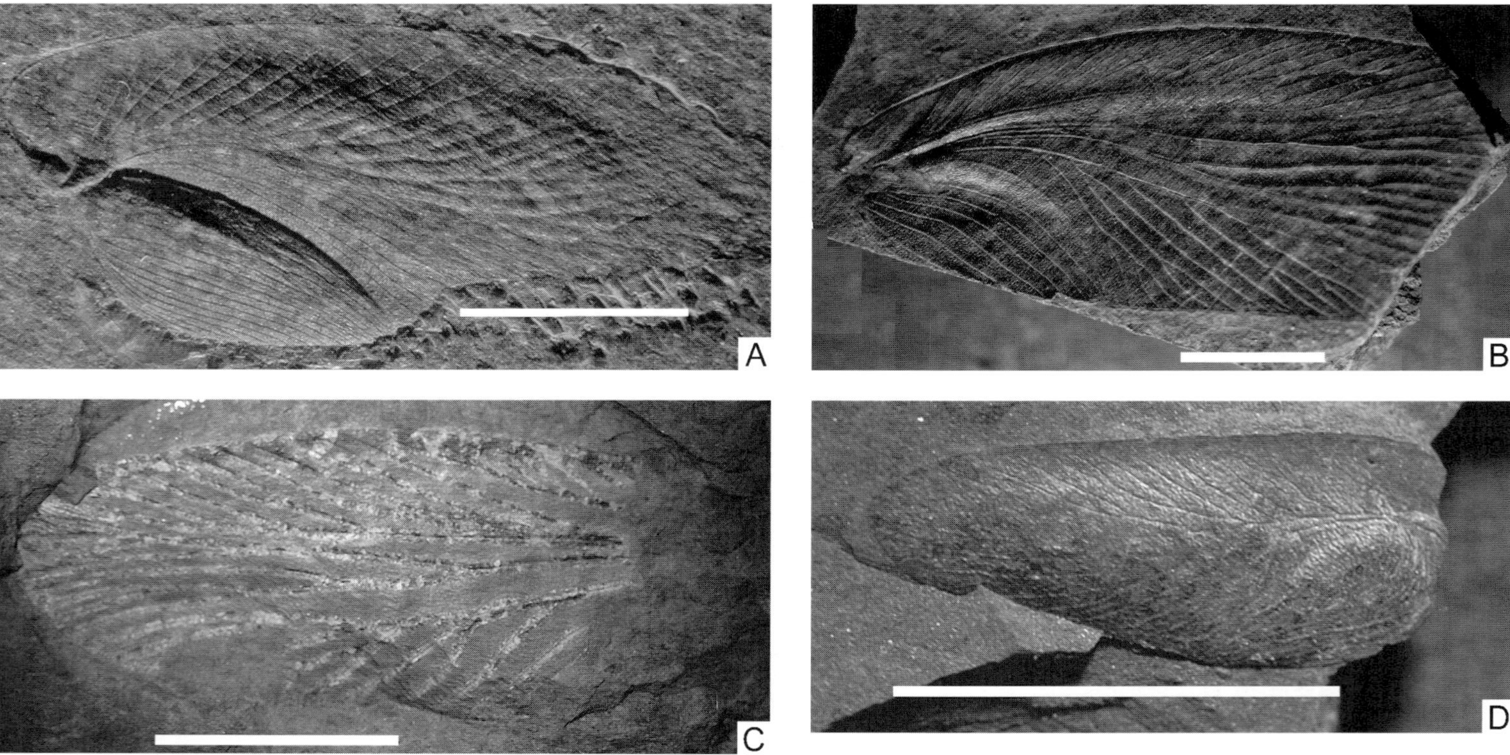

Fig. 8. Typical blattid insects of the El Menizla and Oued Issène formations of the Souss Basin, Stephanian B. Scale bars 5 mm. (**A**) The mylacrid *Opsiomylacris thevenini* (Meunier); Ida Ou Zal Sub-Basin, El Menizla water reservoir; M 1/20; 1.6×0.7 cm. (**B**) The phyloblattid *Anthracoblattina* sp.; Ida Ou Zal Sub-Basin, El Menizla water reservoir; M4–56; 2.5×1.0 cm. (**C**) The spiloblattinid *Sysciophlebia* cf. *grata*; Ida Ou Ziki Sub-Basin, Ighzer Ou Massoud; Ig 1–17; 1.3×0.65 cm. (**D**) *Poroblattina* sp.; Ida Ou Ziki Sub-Basin, Taghzout; Ta 2–11; 0.65×0.25 cm.

esteriid conchostrans mentioned in Figure 2 and described by Feys & Greber (1963, p. 34) came from there. Tasch (1987, p. 49) described *Leaia bertrandi* from Hamalou-Agadir as a new species; the real taxonomic position remains open.

The macroflora of the lacustrine laminated to varved black claystones is clearly dominated by conifers, such as *Otovicia hypnoides* and *Ernestiodendron filiciforme*. Similarly common is cf. *Lepidostrobophyllum*; rarer are *Odontopteris subcrenulata* and *Autunia* cf. *A. conferta*. This suggests xeric conditions around the lakes.

Biostratigraphy, palaeoclimatology and palaeobiogeography of the Moroccan non-marine Late Carboniferous

Plant remains from the El Menizla and Oued Issène formations provide no indisputable age. Based on the macroflora, an age ranging from Stephanian to Autunian was formerly adopted (Jongmans 1950; Saber *et al.* 1995). The samples described by Feys & Greber (1963) and Saber *et al.* (1995) were collected close to the coal seams. They represent the wet environments of local swamps. The new samples from lacustrine horizons (determined by H. Kerp) contain, apart from typical Stephanian elements, such as cf. *Lepidostrobophyllum* and *Odontopteris subcrenulata*, the callipterid *Autunia* cf. *A. conferta*. The First Occurrence Datum (FOD) of *Autunia conferta* is Stephanian B, the Last Occurrence Datum (LOD) is the Upper Rotliegend I or Saxonian I, respectively (Kerp 1996). The transitional Stephanian/Permian character of the flora is related to the climatic situation of the Moroccan basins (see below).

Hmich *et al.* (2003) have proposed a middle Stephanian age based on the common occurrence of *Opsiomylacris thevenini* in the Oued Issène and the El Menizla formations. The type horizon of *O. thevenini* is the lacustrine black shale of the Grande Couche in the Commentry Basin of the French Massif Central. Based on the macroflora and spiloblattinid zonation of Schneider (1982), this level belongs to the *Sysciophlebia praepilata*-insect zone of Stephanian B/C age. Meanwhile, the determination of the spiloblattinid zone species *Spiloblattina pygmaea* and *Sysciophlebia* cf. *S. grata* at five insect sites of the Souss-Basin enables the exact biostratigraphical correlation with the early Stephanian B of Europe (Hmich *et al.* 2005). The type horizon of *Spiloblattina pygmaea* is the lowermost part of the Heusweiler Formation of the Saar–Nahe Basin, Germany (Table 1); this formation is early Stephanian B (Gzhelian) in age from plant remains. The type horizon of *Sysciophlebia grata* is the Hredle Member of the Slaný Formation of the Kladno Basin in Bohemia, Czech Republic, which is dated by macro- and microfloras as Stephanian B (Pešek 2004). In comparison to biostratigraphical data and isotopic ages of the profiles of the Thuringian Forest Basin and the Saar–Nahe Basin (Lützner *et al.* 2003), which represent the reference sections for the European continental Late Carboniferous and the Permian (Schneider *et al.* 1995), an estimated age interval of about 303–300 Ma is calculated for the *Spiloblattina pygmae*/*Sysciophlebia grata* zones (Hmich *et al.* 2005). Compared to the global marine scale, this indicates a latest Kasimovian to middle Gzhelian age for the El Menizla and Oued Issène formations (Table 1).

Lithofacially, the Oued Issene and El Menizla formations are dominated by uniform, monotonous interbedding of fluvial fine- to medium-grained channel sandstones and grey overbank siltstones with intercalated lacustrine black shales. Notable red and violet sandstones were known only from the higher part of the profile in the SE of the Ida Ou Zal basin (Feys & Greber 1963). Tentatively, we interpreted the environment as a wide alluvial plain/floodplain with very shallow relief gradients and braided to anastomosing river courses. The groundwater level was relatively high, based on the complete absence of *Scoyenia* and *Planolites montanus* burrows. In parts of the profile, as at El Menizla village and at Tanamert insect-site 1 (Figs 5d, 7b), the facies architectures are more varied. Up to metre-thick large-scale, trough-cross-bedded channel sandstones, sometimes with extra-basinal (quartz, metamorphites) and intraclasts, are intercalated in metre to decametre-thick greenish grey floodplain siltstones. Calcisols are nearly absent, whereas hydromorphic root horizons are common (simple roots, of 0.1–1 cm diameter); stigmarian roots occur sporadically. Brown-red fine sandstones and siltstones of decimetre thickness are known from El Menizla only. The depositional environment is interpreted as proximal alluvial plain with braided to meandering rivers as well as local lakes and swamps.

The fossil-bearing lake horizons of the Souss Basin at El Menizla, Agadir Ou Anzizen and Tanamert consist of decimetre to metre-thick, finely laminated lacustrine black shale intercalated within fluvio-lacustrine plant-bearing sequences (e.g. Fig. 7c). Varvite-like fine lamination of lacustrine black shales in the European Permo-Carboniferous basins is commonly interpreted as a result of a seasonal climate with annual dry and rainy periods (Schneider *et al.* 1982; Schäfer *et al.* 1990; Broutin *et al.* 1990; Clausing & Boy 2000).

Table 1. *Stratigraphic correlation chart of Europe, North and South Africa*

Geographic region	Column	Units (top to bottom, c. 250–310 Ma)
Stratigraphy	System / Series / Stage	Triassic; Permian: Lopingian (Changhsingian, Wuchiapingian), Guadalupian (Capitanian, Wordian, Roadian), Cisuralian (Kungurian, Artinskian, Sakmarian, Asselian); Carboniferous: Pennsylvanian (Gzhelian, Kasimovian, Moscovian)
Stratigraphy	Regional	Zechstein; Upper Rotliegend II; Upper Rotliegend I; Lower Rotliegend; Stephanian C, B; Barruelian (A); Westphalian D, C, B
France	Lodève	Triassic; La Lieude; Salagou: Merifons Mbr., Octon Mbr.; Rabejac; Viala; Tuilières-Loiras; Usclas-St. Privat; Graiss.
Germany	Saar Nahe Basin	Buntsandstein; ?; Nahe Subgroup: Standenbühl, Kreuzn., Wadern, Donnersberg; Glan Subgroup: Thallich., Oberkirchen, Disibodenb., Meisenheim, Lauterecken, Quirnbach, Wahnwegen, Altenglan, Remigiusberg; Ottweiler Subgroup: Breitenbach, Heusweiler, Dilsburg, Göttelborn; Saarbrücken Subgroup
Germany	Thuringian Forest Basin	Buntsandstein; Zechstein (Z7, Z6, Z5, Z4, Z3, Z2, Z1); Förtha; ?; Eisenach; Tambach; Rotterode; Oberhof; Goldlauter; Manebach; Ilmenau; Gehren Sbgr.
Germany	Saale Basin	Buntsandstein; Zechstein (Z7, Z6, Z5, Z4, Z3, Z2, Z1); Eisleben; ?; Brachwitz; U.Hornburg; L.Hornburg; Halle; Mansfeld Subgroup: Siebigerode, Rothenburg, Gorenzen; Roitzsch
	International scale	Triassic; Changhsingian; Wuchiapingian; Capitanian; Wordian; Roadian; Kungurian; Artinskian; Sakmarian; Asselian; Gzhelian; Kasimovian; Moscovian
Czech Republic	Central and Western Bohemia	Líně; Slaný; Týnec; Kladno Fm.: Nýřany Mbr., Radnice Mbr.
Morocco	Argana Basin Sous Basin	Tanamert; ?; Ikakern Fm.: Tourbihine Member (T2), Ait Driss Member (T1); ?; ?; Tirkou; Oued Issène; Tajgaline; ?
southern Africa	Karoo Basin	Burgersdorp Katberg; Beaufort Group: Balfour, Middelton, Teekloof, Abrahamskraal, Koonap; Ecca Group: Waterford, Fort Brown, Ripon, Collingham, Whitehill, Volksrust, Vryheid, Pietermaritzburg; Dwyka Group: Deglaciation cycle IV, Hardap Sh. Mbr., III, Ganigobis Sh. Mbr., II, I
	International scale	Triassic; Changhsingian; Wuchiapingian; Capitanian; Wordian; Roadian; Kungurian; Artinskian; Sakmarian; Asselian; Gzhelian; Kasimovian; Moscovian
biostratigraphy	Insect zonation	*Sysciobiatta n.sp. Obora*; *Sysciophlebia n.sp. B*; *Sysciophlebia alligans*; *Sysciophlebia balteata Form-G*; *Sysciophlebia balteata Form-H*; *Sysciophlebia ilfeldensis*; *Sysciophlebia euglyptica*; *Sysciophlebia rubida*; *Sysciophlebia grata*; *Spiloblattina pygmaea*; *Syciophlebia n.sp. A*
	International scale	Triassic; Changhsingian; Wuchiapingian; Capitanian; Wordian; Roadian; Kungurian; Artinskian; Sakmarian; Asselian; Gzhelian; Kasimovian; Moscovian

The flora of both sub-basins is clearly dominated by conifers, such as *Otovicia hypnoides* and *Ernestiodendron filiciforme*. This indicates xeric conditions around the lakes, which is well supported by the dominance of *Opsiomylacris* in the entomofauna (cf. Schneider 1989). Next in abundance to the conifers is *Autunia*, a meso- to xerophilous floral element of well-drained areas (e.g. fluvial sand bars) and micro- to macroclimatically drier environments. The *Autunia*-containing and conifer-dominated samples of 'Autunian' aspect represent the flora around the lakes, and of the hinterland, from where they were washed in. Contrary to the autochthonous hygro- to hydrophilous typical Stephanian flora of the local swamps, the allochthonous floral remains of river sediments and lake horizons are representative of larger areas and consequently of the meso- to macroclimate as well. The meso- to macro-climate was obviously seasonally dry, as indicated by the plants, the mylacrid insects and the varved lake sediments.

A similar relatively dry climate is known from the Middle and early Late Stephanian of the Saar Nahe Basin (Heusweiler Formation) and the Saale Basin (Rothenburg and Siebigerode formations, Mansfeld Subgroup) of Stephanian B and early C age in Germany (Schneider *et al.* 1984; Gaitzsch *et al.* 1999; Schneider *et al.* 2005). There, red alluvial fan and wet alluvial plain to floodplain red beds of the *Scoyenia* ichnofacies are the dominant basin fill; lacustrine black shales and minor coal seams are restricted to local depocentres. In the sequence of the Souss Basin, extensive red beds are missing in the El Menizla and Oued Issène formations (see above). The facies pattern of these formations with dominant grey facies is similar to the Stephanian A of the Saar–Nahe Basin and the Stephanian A of the French Massif Central, but with more pronounced changes of dry and wet seasons, as indicated by the varvites, the entomofauna and the macroflora. In this respect, the El Menizla and Oued Issène formations are transitional between the Early Stephanian wet phase and the Middle Stephanian dry phase. As a first approximation, we infer that the Moroccan Souss Basin during this time could have been situated in the southern subtropical summer-wet belt (biome 2 of Ziegler 1990), whereas the Massif Central was situated at this time in the tropical ever-wet belt (biome 1), and the German basins were in the northern subtropical summer-wet belt during the Middle and Late Stephanian.

Saber *et al.* (1995), Broutin *et al.* (1998) and El Wartiti *et al.* (1990) have documented the close relationship of the Carboniferous and Permian macro- and microflora of Morocco and Europe. Broutin *et al.* (1995) and Berthelin *et al.* (2003) have discussed the occurrence of mixed Gondwanan/Euramerican and Gondwanan/Euramerican/Cathaysian floras in the Permian of southern Spain, northwestern Gondwana (Morocco, Niger, Gabon) and eastern Gondwana (Oman). It was shown by these authors that the Early Carboniferous (Visean–Namurian) microfloras of Tarat in Niger, West Africa, are typical Gondwanan, whereas in the Kungurian to Kazanian upper Tarat sequence a mixed Gondwanan/Euramerican microflora, together with an earliest Permian (Autunian) European macroflora, appears. Together with the Kungurian mixed Gondwanan/Euramerican/Cathaysian floras of Morocco and southern Spain, this leads to the conclusion (Broutin *et al.* 1998) that:

(1) During the Early Permian, Euramerican gymnosperms and pteridosperms extended progressively southwards into the northern Gondwana domain;
(2) During late Early Permian time Gondwanan elements migrated into the Euramerican floral province through North Africa up to southwestern Spain.

Our data on the insect fauna and macroflora from Morocco indicate that the appearance of Euramerican forms in northern Gondwana had started, at the latest, in the Early Stephanian (Late Kasimovian to Early Gzhelian).

Some questions still remain open. Where were the evolutionary centres situated from which the migration of 'Euramerican' meso- to xerophilous plants and insects took place? Where are those extra-basinal areas, in which their evolution is assumed (Kerp 1996, 2000)? Were they actually situated in the classical (well-investigated) palaeotropical to subtropical Euramerican regions, or were they located in the drier areas bordering the Euramerican realm, such as the unknown large uplands of the Fennoscandian Peneplain, and the extensive (since the end of the Westphalian) increasingly drier lowlands of the northern and (?)southern foreland of the Hercynides? Or was it the dry extra-basinal areas within the Euramerican region, such as the (mostly not preserved) montane and submontane biotopes of the Hercynian mountain chain, or the still poorly investigated transitional areas of the inland desert (biome 3) between tropical to subtropical Europe and the cool temperate (biome 6) region of Gondwana (Ziegler 1990)? However, such migrations are climate-governed and occur at several times when the climate changes. It is known from well-investigated and well-dated central European basins, such as the Thuringian

Forest Basin, the Saale Basin and Saar–Nahe Basin, that the general trend to increasingly drier environments during the Lower Rotliegend and the Upper Rotliegend I (Early to Middle Permian, or Cisuralian to Guadalupian) was interrupted by some wet periods (e.g. Schneider 1989, 2001; Schneider *et al.* 2006; Roscher & Schneider 2006).

Khenifra Basin

From the Khenifra Basin, mapped and described by El Wartiti (1990), only invertebrate burrows were known until 2000, when the first vertebrate bone was discovered by J.W.S. during an international excursion, guided by Moroccan and French geologists.

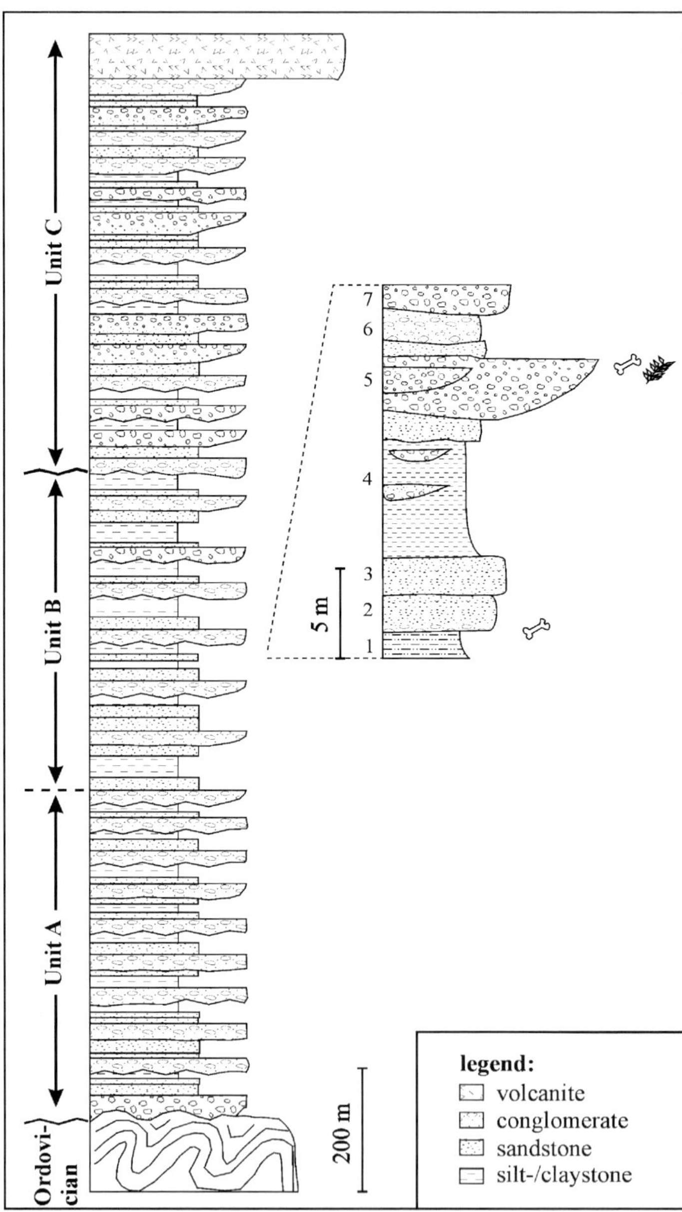

Fig. 9. Permian profile of the Khenifra Basin. Inset: profile of the abandoned sandstone quarry west of Ait Khouya in the El Messalla area (cf. Fig. 9, locality 4) with the vertebrate bone-containing horizons.

Geological setting

The tectonically bounded remnants of the Khenifra Basin, cropping out over an area of about 100 km^2, are situated in eastern central Morocco (Fig. 1) (El Wartiti 1990; Saber *et al.* 1995; Broutin *et al.* 1998). It is filled by about 1800 m of clastic sediments, resting unconformably on folded Ordovician to Visean marine sediments. In the northwestern part of the basin, acidic and basic calc-alkaline volcanics, which are more than 160 m thick (Youbi & Cabanis 1995), have no exposed direct contact with the sediments. Inside the basin, rhyolithic volcanites overlie the sediments. Three sedimentary sequences are distinguished (Fig. 9; El Wartiti 1990, Broutin *et al.* 1998). The lower sequence (unit A in Fig. 9) of about 600 m thickness is represented by red alluvial fan conglomerates as well as sandstones and mudstones of minor thickness. Intercalated are debris flow fanglomerates of distal fans. The second sequence (unit B in Fig. 9) consists mainly of red alluvial plain to floodplain siltstones and claystones; channel sandstones and fanglomeratic conglomerates are rarely intercalated. Locally, transitions from red to grey floodplain fine sandstones and siltstones rich in plant remains are observed (El Wartiti 1990; Broutin *et al.* 1998), for example, at Bouzouggagh 1 km NE of Khenifra town in the SW of El Messalla area (Fig. 10, locality 2). The upper sequence (unit C in Fig. 9) again consists of coarse clastics, genetically comparable to the lower sequence. Conglomerates are more common, but the size of the rounded pebbles is smaller. Intercalated siltstones locally have a pedogenic overprint, indicated by colour mottling, immature calcic soils and calcareous rhizoconcretions as in the whole red-bed sequence of the basin.

Fossil content of the Khenifra Basin

As in most of the Carboniferous and Permian basins of Morocco, most attention has been paid

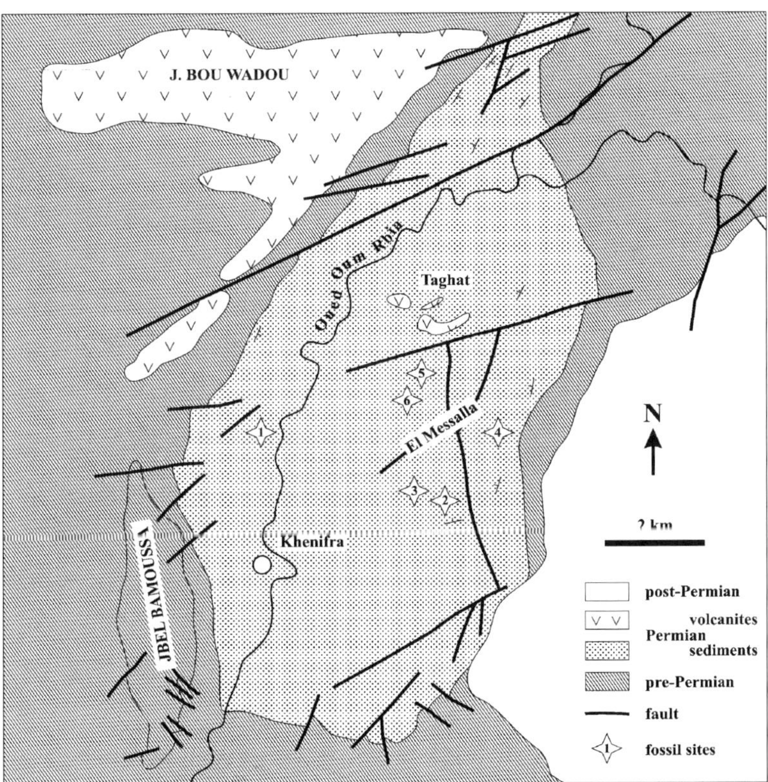

Fig. 10. Sketch map of the Permian Khenifra Basin. Stars indicate the fossil sites discussed in the text: 1, Nkhilat; 2, Bouzouggagh; 3, Behar Ikkourn; 4, Ait Khouya; 5, Ifri Ou Châou; 6, Dahra hill.

to fossil floras of the Khenifra Basin (Broutin *et al.* 1998).

Flora
Sheetflood-like deposits and stacked, large-scale trough cross-bedded channel conglomerate form metre to some decametre-thick red-bed sequences of unit A resting on folded Visean sediments at the northeasternmost end of the basin. Red silty fine sandstones and fine sandy siltstones are intercalated and contain sparse conifer twigs ('*Walchia*'). Conglomeratic horizons of reworked calcisoil nodules, decimetre thick are typical, as throughout the Khenifra Basin.

A rich, conifer-dominated flora is known from different sites in unit B, for example, at the river banks of Oued Oum Rbia north of Khenifra, Nkhilat and at Bouzouggagh (Fig. 10; for details and the microflora see Broutin *et al.* 1998, table 1). Typical are different *Culmitzschia* species, *Otovicia, Ernestiodendron, Walchia* and *Feysia*, as well as the wet-adapted *Calamites*, the hygro- to mesophilous *Calamites gigas* and *Cordaites*, the mesophilous pteridosperms *Sphenopteris* and *Neuropteris,* the mesophilous callipterid *Lodevia nicklesii* and *Ginkgophyllum*. Of special interest is *Sphenopteris pseudogermanica*, a Cathaysian element. The microflora exhibits the same spectrum of wet (*Endosporites*) and dry (*Potonieisporites*) elements. The (often red oxidized) plant remains are concentrated in greyish to whitish fluvial siltstones. Northeast of Behar Ikkourn (Fig. 10, locality 3), decametre-thick fluvial grey sediments intercalated in red beds could be traced over a lateral extent of 250 m. The former consist of pale greyish to yellowish fluvial channel sandstones. The well-organized trough cross-bedded single channels have an average thickness of about 0.5 m and a width of 10 m. Climbing ripple sandstones and sandy overbank siltstones with plant remains are intercalated. In contrast to the red beds, *Scoyenia* burrows are missing. Red beds in the neighbourhood of the grey facies contain up to 1-m-thick, yellowish mottled red-brown, pedogenic dolocretes with root traces.

Fauna: Ait Khouya quarry
Vertebrate remains were first discovered in the abandoned sandstone quarry west of Ait Khouya in the El Messalla area (Fig. 10, locality 4). The quarry is situated in the lower half of unit B (Fig. 9). Red-brown floodplain sandy siltstones of the *Scoyenia* ichnofacies are exposed. Typical are immature nodular dolocretes. Intercalated are up to 3.2-m-thick, horizontally stratified matrix-supported fanglomeratic conglomerate channels, up to 25 m wide (mean pebble size 1–2 cm), as well as sandstone channels about 2–3 m thick and tens of metres wide. The sandstone channels display large- to small-scale trough cross bedding. Some metre-thick, fanglomeratic sheet-flood deposits (e.g. horizon 7 in inset of Fig. 9) interfinger vertically and laterally with pebbly sandstones. They can be traced as weathering-resistant NE–SW-trending topographical ridges that are kilometres long (Fig. 11a). One 7.7-cm-long bone (possibly a rib fragment) was found at the top of a conglomerate channel (horizon 5: inset of Fig. 9). One 2-cm-long vertebra was discovered in a 35-cm-thick horizon of reworked pedogenic dolomite nodules, 0.1–2 cm in size, and siltstone rip-up clasts up to 4 cm long (base of horizon 2: inset of Fig. 9).

Fauna: Ifri Ou Châou hill and clay pit
East of Ifri Ou Châou village, northeast of Khenifra town, a 44-m-thick profile of floodplain fine clastics and channel sandstones is exposed at the slope of a prominent hill and at a clay pit in this hill (Fig. 10, locality 5). This series belongs to unit B (Fig. 9). The detailed profile description will be given in a forthcoming paper. At the foot of the hill, 8–9 m of clayey to fine sandy siltstones are exposed. The exposed section is structureless because of strong rooting. The branched, pale green to whitish root traces are millimetres in diameter. Colour mottling of red-brown to violet and yellowish, as well as green speckling, is very common. Occasionally, layers of 1–3-cm-diameter, large dolomitic concretions are observed, representing possible groundwater dolocretes. The top of the clay pit is formed by a 3-m-thick, grey-violet, internally large-scale, cross-bedded channel sandstone. Bedding planes with mud drapes and raindrop imprints point to multistorey channels. Inside the channel, 1-cm-wide and 10-cm-long plant axes occur. At the channel bottom, trunks of trees up to 2 m long and 20 cm thick have been found (Fig. 11c). On a dump of large blocks of channel sandstones at the border of the clay pit, a number of tetrapod tracks have been discovered (Fig. 11b). Most probably, they come from a second channel horizon 1.4 m above the former one. This channel consists of fine- to medium-grained sandstones

Fig. 11. Exposures and fossils of the Permian Khenifra Basin. (**A**) El Messalla area, parallel to the eastern basin border NNE–SSW-trending ridges of outcropping distal fan conglomerates, indicating a source area of the bajada-like fans in the east; red siltstones are intercalated. (**B**) 14-cm-long tetrapod track (?cf. *Limnopus*), a very common track morph at the base of channel sandstones in the Ifri Ou Châou clay pit (Fig. 9, locality 5); scale 10 cm. (**C**) Tree trunks, 2 m long and about 20 cm wide, at the bottom of a channel sandstone, Ifri Ou Châou clay pit; scale 1 m.

A
B
C

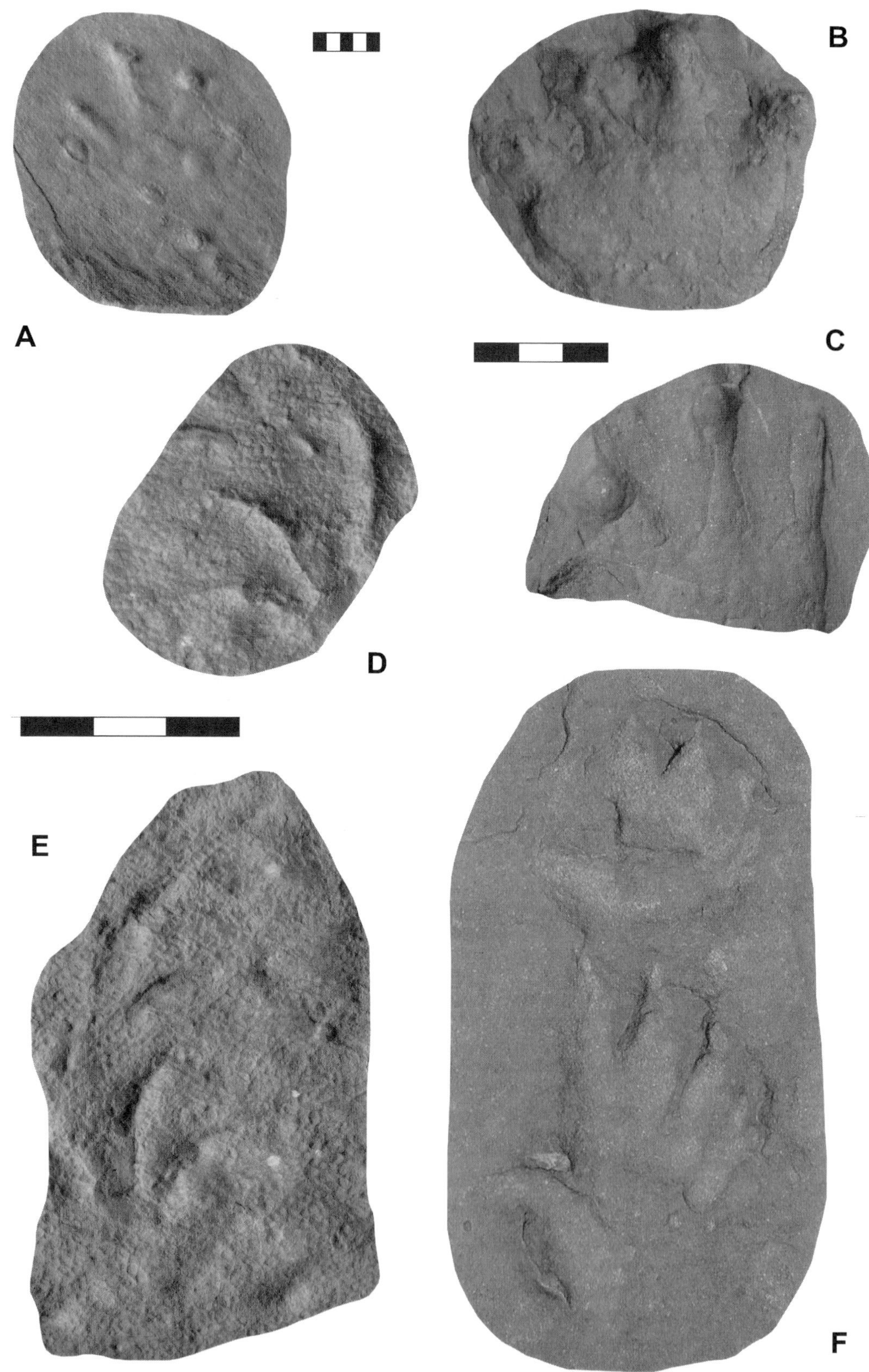
A
B
C
D
E
F

horizontally laminated in 30–40-cm sets. The track assemblage is monospecific, representing large imprints of *Limnopus* Marsh, 1894 only (Fig. 12b, c). This ichnogenus is referred to larger-sized temnospondyls, such as eryopoids, which grew in lakes or streams, but are thought to have spent their adult life outside of the water on wet land (Baird 1965; Haubold *et al.* 1995).

A second tetrapod track horizon, with layers of conifer twigs, horizontal root systems and raindrop marks, crops out at the slope of the hill 4 m higher. It can be traced along the slope for some 100 m with a nearly constant thickness of 40 cm. These fine sandstones are planar horizontal to flaser-bedded, with occasional flood casts. The tracks are identified as cf. *Batrachichnus* and *Dromopus* (Fig. 12a). Sporadically intercalated small channels filled with dolomitic nodules of reworked soils and siltstone intraclasts contain centimetre-scale angular bone splinters. *Scoyenia* is common in silty fine sandstones on the hill slope above the clay pit. One indeterminable conchostracan has been found in pond claystones. Unfortunately, these claystones are strongly tectonically overprinted.

Fig. 12. Tetrapod footprints from Permian red beds of Morocco: A–E Khenifra Basin; F Argana Basin. (**A**) cf. *Batrachichnus* Woodworth 1900 from the slope of Ifri Ou Châou hill. Left manus and pes couple, convex hyporelief. The tetradactyle manus track shows a short digit IV, which is a little bit longer than digit I, both framing the longer digits II and III. Digit tips are bluntly rounded. A striking pad is seen proximal to digit I. Outline of the palm is missing. The pes track is closely arranged behind the manus, however, incompletely preserved by showing bluntly rounded tips of digits I–IV, only. Scale in millimetres. (**B**) and (**C**) cf. *Limnopus* Marsh 1894 from Ifri Ou Châou clay pit. Manus tracks of the right side, convex hyporelief. *Limnopus* shows tetradactyle manus tracks, which differ from *Batrachichnus* Woodworth 1900 by a more extended digit IV about as long as digit II. It is well explained by the specimen figured in C. Scale in centimetres. (**D**) and (**E**) cf. *Dromopus* Marsh 1894 from Dahra hill near El Messalla village. Undifferentiated manus and pes tracks, convex hyporelief. The tracks are characterised by slender and curved digits diverging distally remarkable. Both specimens show tracks on uneven surface of wrinkled texture (‘elephant skin’) indicating that microbial mats were components of the original sediment. Scale in centimetres. (**F**) cf. *Synaptichnium* Nopcsa 1923 from the side-valley of the Ait Messaoud River close to Aguenza village, Argana Basin (cf. Fig. 12**a**). Left manus and pes couple, convex hyporelief. Pentadactyle manus and pes, sole two to three times longer than palm. The well-preserved pes track shows a continuous increase of digit length from I to IV and digit V is more laterally arranged behind the front digits than proximally. Digits of the manus track are distinctly shorter than those ones of the pes imprint. For scale see Figures B and C.

Fauna: Dahra hill near El Messalla village

In the area of Dahra hill, well-developed fluvial megacycles are exposed (Fig. 10, locality 6). Each cycle starts with decametre-thick, fanglomeratic conglomerates. They occur in beds 1–2 m thick that have indistinct horizontal stratification and are matrix- to clast-supported. Pebbles are subangular to subrounded. The conglomerates interfinger and grade upward into approximately 1.5-m-thick, low-angle, trough cross-bedded to planar cross-bedded sandstones, which grade into 10-cm-thick floodplain fine clastics. At the top of the silt- and claystones, just below the overlying conglomeratic horizon, metres of vertisols with minor pale red mottling and nodular calcretes up to 1 m thick occur. At the foot of the hill, a 20-m-thick series of silt- and claystones crops out. Unfortunately, they have undergone a slight pedogenic overprint as indicated by yellowish, millimetre-thick root traces and brecciation. Bedding planes of clay- and siltstones frequently show raindrop marks as well as the typical ‘elephant skin’ structure created by biomats (Fig. 12d, e). Obviously, these fine clastics were deposited in ephemeral shallow pools or lakes on the floodplain. Imprints of walchian twigs and cones, as well as tetrapod footprints assigned to cf. *Batrachichnus* and *Dromopus* (Fig. 12d, e), are very common. Unfortunately, complete tracks could not be extracted because the claystones break easily into small shards.

Conclusions on biostratigraphy, environment and palaeoclimatology

Modern elements of the flora, such as *Culmitzschia*, *Lodevia* and *Feysia*, give an Autunian, Early Permian, age for the deposits of the Khenifra Basin. The microflora contains some elements, such as *Gardenasporites* and *Striatoabieites*, which point to a somewhat younger, possibly Saxonian or Late Rotliegend age. A Kungurian age was adopted by Broutin *et al.* (1998). The dominance of drought-adapted mesophilous to xerophilous conifers, as well as the *Scoyenia* ichnofacies, the vertisols and the common occurrence of (immature) dolomitic calcisols, point to a semihumid climate with periodic droughts for the origin of these red beds. Seasonal precipitation with occasionally heavy rainfalls is indicated by the dominance of matrix-rich fluvial conglomerates, matrix-supported coarse clastic debris flow and sheetflood beds.

The angular bone splinters resulted from physical weathering of bones exposed to prolonged subaerial exposure during drought phases.

The well-organized grey fluvial sediments with features of a meandering river system argue for a more or less constant water flow, contrary to the foregoing climatic interpretation. A closer look at the facies architecture might explain this paradox. NNE–SSW-trending ridges of distal fan conglomerates cropping out parallel to the eastern basin border (Fig. 11a) indicate a source area of the bajada-like fans in the east. The grey fluvial facies sediments are intercalated in red floodplain sediments in front of these distal fans. Clearly, the grey sediments were deposited in a drainage system perpendicular to the distal fans. Coarse clastic fans can store water long after rainy periods and deliver it to a drainage system during the drier season. Similar wet red beds of the *Scoyenia* ichnofacies associated with common calcisols are well known from some German basins. They occur, for example, in the Heusweiler Formation of the Saar–Nahe Basin, Stephanian B, the Rothenburg and Siebigerode formations of the Saale Basin, Stephanian B and C, and in the Leukersdorf Formation of the Erzgebirge Basin, Lower/Upper Rotliegend (Lower Permian; cf. Table 1).

In all these basins, grey fluvial facies with local swamps and lacustrine black shale facies formed in back swamps of river courses and in scattered local depocentres that were situated at or below the groundwater level. From surface outcrops and drill cores in the Saale Basin, the lateral transition from workable coal seams into red beds is well documented (Schneider *et al.* 1984, 2005). The Saar–Nahe Basin has an extent of 300×100 km, the Saale Basin of 150×90 km and the Erzgebirge Basin of 70×30 km. The tectonically bounded remnants of the Khenifra Basin have an extent of only 18×7 km. The steep facies gradients from the red coarse clastics of the distal fans to the grey fluvial facies in the basin centre indicate a significantly smaller basin compared to the German ones. In the best case, this basin was twice as large as it is as present. To generate the meandering river patterns of the grey fluvial facies, a larger catchment area outside the Khenifra Basin would have been necessary. Possibly, a river flowed through a basin whose catchment area was situated far outside the basin in mountain regions that acted as a rain trap, as do the High Atlas Mountains for the scarce perennial rivers of modern Morocco.

Tiddas Basin

This basin will be discussed briefly with respect to tetrapod track biostratigraphy only because, as yet, we have no new fossil finds.

Geological setting

The sediments of the Tiddas Basin (Larhrib 1996; Broutin *et al.* 1998) are preserved in a NE–SW half-graben structure, 20×2.5 km in extent, below Triassic to Miocene cover (Fig. 1). The basement consists of folded Visean flysch deposits. The basin fill, estimated at 300–500 m, is informally subdivided into three units (Larhrib 1996; Table 1). The 'Lower Formation' (F1) consists of about 100–180 m of interbedded matrix- or component-supported red conglomerates (sub-rounded pebbles) and red to violet silty sandstones. The 'Middle Formation' (F2) is about 120 m thick and comprises mainly red clayish siltstones with intercalations of channel conglomerates and sandstones as well as some tuff beds. Red sandy to clayey siltstones and fine sandstones, horizons of grey siltstones, locally carbonaceous, as well as decimetre-thick lacustrine limestones and nodular calcisols are typical of the 80-m-thick 'Upper Formation' (F3).

Fossil content and biostratigraphy of the Tiddas Basin

A conifer-dominated flora was reported by El Wartiti *et al.* (1986), Larhrib (1996) and Broutin *et al.* (1998) (for details see these) from different levels of carbonaceous grey siltstones of the 'Upper Formation'. Most important are the conifers *Culmitzschia*, *Ernestiodendron*, *Walchia*, *Otovicia*, *Feysia* and *Darneyella*, the callipterid *Rhachiphyllum*, hygro- to mesophilous calamite species, some Cathaysian elements, such as *Annularia* cf. *hunanensis*, *Pecopteris* cf. *chihliensis*, *Protoblechnum* cf. *wongii* etc. and Euramerican *Pecopteris* and *Odontopteris* species. Based on the macroflora, a Kungurian age was proposed by Broutin *et al.* (1998). Based on the flora, silicified wood, pedogenic features, etc., 'a warm humid climate with somewhat irregularly distributed drought periods' is assumed by the same authors (Broutin *et al.* 1998, p. 272).

Invertebrate ichnia of the *Scoyenia* ichnofacies are widespread in the red beds of the Tiddas Basin. Tetrapod tracks are known only from the 'Upper Formation'. Vertebrate footprints of the Tiddas Basin were reported initially by El Wartiti *et al.* (1986). The tracks were assigned to *Amphisauroides discessus* Haubold 1970, *Gilmoreichnus brachydactylus* (Pabst 1900) and *Hyloidichnus* Gilmore 1927 (Broutin *et al.* 1987; Larhrib 1996). The only published footprint slab shows five footprints of two distinct trackways preserved on a rippled surface, which is densely covered by raindrop marks (Broutin *et al.* 1987, pl. 1, fig. 1). Judged from the photography, the specimen represents two manus-pes couples of *Hyloidichnus* Gilmore 1927 and an isolated track of uncertain ichnotaxonomic position. The type specimens of *Amphisauroides discessus* Haubold 1970 and *Gilmoreichnus brachydactylus* (Pabst 1900) from the Lower Permian of the Thuringian Forest Basin of Germany are now considered as synonyms of *Amphisauropus kablikae* (Geinitz & Deichmüller 1882) (Voigt 2005). However, the occurrence of *Amphisauropus* cannot be confirmed by the published specimen of the Tiddas Basin. *Hyloidichnus* is well known from the Hermit Formation of the Grand Canyon, Arizona (Gilmore 1926, 1927), and from the Rabejac Formation of the Lodève Basin, southern France (Gand 1988). Based on vertebrate tracks, a correlation of the 'Upper Formation' of the Tiddas Basin with the late Early Permian (Cisuralian, Artinskian–Kungurian) of Europe and North America is suggested (Table 1).

Argana Basin

The Argana Basin is located at the western edge of the High Atlas Mountain between Marrakech and Agadir (in Fig. 1 west of the Souss Basin). The Permian to Early Jurassic continental basin fill is well exposed over an area of about 70×15 km. The first lithostratigraphical subdivision goes back to Tixeront (1974). His T1 to T8 units were later formally named by Brown (1980). The T1 unit or the Ait Driss Member and the T2 unit or the Tourbihine Member together form the Permian Ikakern Formation (Table 1). Whereas the Permian deposits are not studied in detail, extensive sedimentological investigations have been carried out in the Triassic to Early Jurassic red beds (Hofmann *et al.* 2000).

Geological setting

The Ikakern Formation is restricted to the Argana graben in the central part of the Argana Valley (Brown 1980). Deposition of the Ikakern Formation took place syntectonically in ENE–WSW-trending half-grabens (Medina 1991, 1995). Maximum thickness of 2000 m is reached in the SE. The basal conglomerates of the Ait Driss Member rest unconformably on Cambrian to Devonian folded metasediments, and locally with an angular unconformity on Late Carboniferous sediments (see above, Ida Ou Ziki Sub-Basin). This 25-m-thick member consists of alluvial fan conglomerates, which grade vertically and laterally into fluvial channel sandstones and floodplain siltstones of the usually up to 200-m-thick Tourbihine Member. Thin intercalations of lacustrine micritic limestone appear close to the top of the member. A major tectonic phase with normal faulting, uplift and erosion of Permian sediments occurred prior to the sedimentation of the overlying (?Middle) Triassic braided river conglomerates of the Tanamert Member (T3) of the Timesgadiouine Formation (Medina 1995; Hofman *et al.* 2000).

Fossil content of the Permian of the Argana Basin

Spectacular tetrapod skeletal remains were discovered by Dutuit (1988) in the Tourbihine Member. They were identified as diplocaulid nectrideans (*Diplocaulus minimus*), the captorhinid *Acrodonta* and a moradisaurine (Jalil & Dutuit 1996). Besides these skeletal remains, only rhynchosauroid tracks (Jones 1975), the conifer *Voltzia heterophylla* from Jebel Tafilalt (De Koning 1957) and supposed pelecypods were mentioned from the Tourbihine Member.

The bone-bearing red beds of the Tourbihine Member around the vertebrate site near Tikida, 200 m south of Irerhi village, belong to the *Scoyenia* ichnofacies type of floodplain sediments. Typical are centimetre-scale horizontally bedded siltstones, sometimes brecciated, with intercalations of up 0.5-m-thick sandstone channels and flat channels of fine sandy siltstones. Calcic soils are rare. Only one 1-m-thick sandstone horizon with masses of claystone rip-up clasts was observed, possibly generated by a sheetflood event. The siltstones contain tetrapod footprints, but up to now only indeterminate specimens have been found.

In the area north of Irehri village, well-preserved tetrapod tracks were discovered on a loose block of 2×1.6 m in a side-valley of the Ait Messaoud River close to Aguenza village (Fig. 13). The convex lower surface of this sandstone block exhibits the fill of a channel bottom (Fig. 13a). The track-makers moved along this channel, partly in opposite directions. Two tetrapod ichnotaxa are distinguished: *Synaptichnium* Nopcsa 1923, represented by well-preserved imprints (Fig. 12f), and *Rhynchosauroides* Maidwell 1911, represented by some vague imprints of 50–60 mm length. Possible track-makers are thecodontian archosaurs and prolacertiform reptiles. This block comes from a 1.5-m-thick channel horizon some tens of metres higher up in the valley (Fig. 13b). The lower 20–30 cm of this horizon consist of low-angle, trough cross-bedded sandstone; the remaining

B

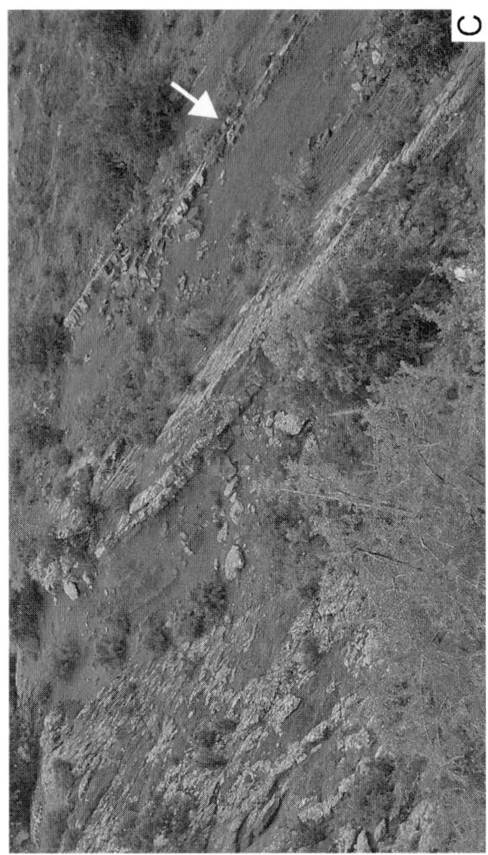
C

A

part consists of changing component/matrix-supported channel conglomerates.

Above follow silty and sandy floodplain deposits. A further tetrapod track horizon was discovered by one of us (DH) at the southern slope of the side-valley, opposite to the former one. At this slope 3–4-m-thick floodplain fine clastics are interbedded with 0.5–1.5-m-thick fluvial conglomerates (Fig. 13c). A mass tetrapod track site occurs at the bottom of a 0.5-m-thick, coarse clastic horizon. The horizon consists of alternating clast- to matrix-supported conglomerates with well-rounded pebbles, mostly of quartz. The tetrapod track-bearing surfaces represent monospecific assemblages of about 2–3-cm-long footprints. However, ichnotaxonomic assignation of the given specimens has been impossible up to now due to the coarse-grained sediment, the extensively trampled surfaces, the small size and the incomplete preservation of the imprints.

Conclusions on biostratigraphy, environment and palaeoclimatology

Based on the vertebrate bones, the Tourbihine Member was dated as Kazanian (Middle Permian, Guadalupian) after Dutuit (1976, 1988) and Jalil & Dutuit (1996). Ongoing re-investigations of the tetrapod bones have revealed pareiasaur remains, which are closely related to those from the Late Permian of Elgin, Scotland, and the Zechstein of Richelsdorf, Germany (after Jalil, pers. comm. 2002), which is Middle Wuchiapingian.

The lithostratigraphical position of both of the new tetrapod track sites remains somewhat unclear up to now. After the map of Tixeront (1974), they should be situated in the Permian Tourbihine Member, possibly close to the contact with the (?Middle) Triassic Tanamert Member of the Timesgadiouine Formation. The facies pattern of the first tetrapod track site (the loose block) at the northern flank of the side-valley, with overwhelming floodplain fine clastics, points to a position in the Tourbihine Member (T2). The second track site at the opposite valley slope is separated from the first one by a fault. The facies patterns generally fit together. The pebble content of the conglomerates on both sides of the fault is dominated by quartz, quartzites and cherts (lydites).

Synaptichnium and *Rhynchosauroides* are typical and well-known elements of Early and Middle Triassic tetrapod footprint assemblages in North America and Europe (e.g. Peabody 1948; Demathieu & Haubold 1982; Avanzini 2000). Both ichnogenera have also been reported from the Late Permian Val Gardena Sandstone of the Gröden Formation, Southern Alps, northern Italy (Conti *et al.* 1977). The Gröden Formation is considered to be of (?)late Guadalupian to early Lopingian age (Conti *et al.* 1999). Certainly, determination of *Synaptichnium* and *Rhynchosauroides* in the Val Gardena Sandstone is questionable (cf. Haubold 1984; Avanzini *et al.* 2001). However, the recently recovered Permian track sites of the Moroccan Argana Basin give occasion to scrutinize the presence of late Palaeozoic precursors of typical Triassic tetrapod track morphs. Thus, the tetrapod ichnotaxa from the Argana Basin mentioned above are most common in Triassic deposits, but currently it cannot be excluded that they first appeared during the Late Permian.

The lithofacies characters of the red beds in the Tiddas, Khenifra and Argana basins are nearly the same. In all these basins the *Scoyenia* ichnofacies and nodular calcisoils of wet red beds are typical. Only grey facies and larger rhizoconcretions are completely missing in the Late Permian of the Argana Basin. Wet red beds similar to those in the Argana Basin are known in Europe from the La Lieude Formation, Late Permian of the Lodève Basin, southern France (see Roscher & Schneider, 2006), as well as from the Gröden Formation of the Southern Alps above the 'Cephalopod horizon' in the Bletterbach Valley. In the Lodève Basin, as in the Bletterbach profile of the Gröden Formation, these wet red beds overlie sediments deposited under drier conditions: in the Lodève Basin over the playa deposits of the Salagou Formation (Table 1) and in the Bletterbach Valley over

Fig. 13. Exposures and fossils of the Late Permian Argana Basin in the Western High Atlas Mountains. (**A**) Close-up of the convex lower surface of a sandstone channel bottom with tetrapod tracks of *Synaptichnium* and some vague *Rhynchosauroides* imprints (not well visible), loose block of 2×1.6 m extent in a side-valley of the Ait Messaoud River close to Aguenza village; scale bar 10 cm. (**B**) Exposure of 1.5-m-thick channel horizon intercalated in silty and sandy floodplain deposits, the 20–30-cm-thick lower part of the horizon consists of low-angle, trough cross-bedded sandstone overlain by component- to matrix-supported channel conglomerates; most probably, the loose block with the tetrapod tracks (Fig. 13**a**) comes from the lower part of this horizon. (**C**) Southern slope of the side-valley of the Ait Messaoud River; interbedding of 3–4 m floodplain fine clastics and 0.5–1.5-m-thick fluvial conglomerates; at the bottom of the conglomerate bed in the right half of the picture (arrow), masses of about 2–3-cm-long tetrapod tracks occur.

evaporitic sabkha deposits. Based on the macro- and microflora and tetrapod tracks, the above discussed sediments of the Khenifra and the Tiddas basins were deposited under semihumid conditions of the late Artinskian wet phase (Schneider *et al.* 2006; Roscher & Schneider 2006). Similar climatic conditions recur after the Kungurian/Roadian drought maximum in the Wuchiapingian wet phase, which started in the Capitanian, late Middle Permian. Based on biostratigraphical data and facies patterns, the Late Permian Ikakern Formation with the tetrapod-bearing Tourbihine Member belongs in the latter wet phase.

General conclusions

The climatic development of the Moroccan basins discussed here, which are scattered through time, is comparable to the processes observed in the continuous Late Pennsylvanian to Middle Cisuralian profiles of the central European basins and the continuous middle Cisuralian to Late Permian Lopingian sequence of the southern French Lodève Basin. The general aridization trend during the Pennsylvanian and Permian is interrupted by wet phases (Schneider 1996, Schneider *et al.* 2006; Roscher & Schneider 2006). During these wet phases, similar facies pattern recur, but are stepwise modified with decreasing age. For example, extensive coal seams, typical of the Early Asselian wet phase, are commonly missing in the Asselian/Sakmarian wet phase (with the exception of Buxieres-les-Mines, French Massif Central; Roscher & Schneider 2006) and lacustrine black shale facies, typical of the latter wet phase, are very rare in the following late Artinskian wet phase. These climatic changes had strong influences on the evolution of organisms and the composition of their associations, or, in other words, on the composition of ecosystems. With the disappearance of swamp as well as permanent river and perennial lake environments in the palaeo-equatorial belt, organisms adapted to these biotopes disappeared. Aquatic invertebrates and vertebrates, as well as hydro- to hygromorphic adapted plants, were strongly affected up to extinction long before the Permo-Triassic event. Refuges for these organisms inside the equatorial belt or the Euramerican region became increasingly rare; most possibly they disappeared completely during the Kungurian/Roadian maximum aridity. Refuges outside the Euramerican region have been described for the Mid- to Late Permian mixed floras of the Arabian Peninsula, such as the Oman Gharif mixed flora (Berthelin *et al.* 2003). Euramerican elements in this flora are (significantly) the mesophilous to xerophilous *Otovicia hypnoides* and *Calamites gigas* (for the latter cf. Rössler 1999) as well as *Calamostachys dumasii* and *Baieroxylon implexum*. One exceptional relict fauna of temnospondyl amphibians at the outermost border of the Euramerican area has been recently reported from Niger (Sidor *et al.* 2005).

In contrast, organisms adapted to seasonal or longer lasting drought periods underwent a strong radiation, such as the pteridosperms and especially the conifers (Kerp 1996, 2000) among the plants, and terrestrially adapted amphibians and the reptiles among the animals (Berman *et al.* 1997). This is well supported by the appearance of mylacrid insects together with conifers in the Early Stephanian B of Morocco, the relatively common occurrence of ginkgophytes in the late Early Permian, Artinskian, of the Tiddas, Khenifra and Bou Achouch basins (Broutin *et al.* 1998), and the *Synaptichnium* and *Rhynchosauroides* tracks of advanced thecodontian archosaurs and prolacertiform reptiles in the Middle to Late Permian Argana Basin. A similar, very modern tetrapod track association of Mesozoic appearance is known from the Late Permian Val Gardena sandstone of the Southern Alps (Avanzini *et al.* 2001). Last, but not least, are:

(1) the discovery of dicynodonts in the Early Lopingian Hopeman Sandstone (Clark 1999) in the Moray Firth, NE Scotland, 10 km north of Elgin;
(2) the occurrence of dicynodonts and a pareiasaur in the possibly contemporaneous Cutties Hillock Sandstone at Elgin;
(3) the occurrence of South Africa related cynodonts and dicynodonts in the Middle Lopingian Zechstein of Germany (Sues & Munk 1996);
(4) the appearance of 'typical' Euramerican archegosaurids in the Late Permian of Kashmir (Werneburg & Schneider 1996);
(5) the appearance of tracks and skeletal remains of large, herbivorous pelycosaurs (cf. *Cotylorhynchus*) in the Late Permian (Wuchiapingian) La Lieude Formation of the Lodève Basin (Schneider *et al.* 2006)

These indicate, together with the foregoing, a Late Cisuralian to Early Lopingian radiation and migration event of tetrapods, heralding the uniform tetrapod faunas of the Early Mesozoic. Migration pathways were possibly coastal regions of the Tethys and of the transgressing Late Permian epicontinental seas, such as the

Zechstein and the Bellerophon seas, with a more equable climate caused by a maritime imprint on the otherwise strongly continental climate (Schneider 1989; Legler *et al.* 2005). A similar, very rapid radiation and migration, not hampered by climatically governed floral provincialism, is reported for peltasperm plants from the pre-Urals (there together with Angara floral elements), Germany (Saar Basin), China (Turpan–Hami Basin) and Morocco (Bou Achouch Basin) by Kerp *et al.* (2001). Morocco seems to be one of the key areas for understanding the evolution of ecosystems in relation to changing environmental conditions during the late Palaeozoic/early Mesozoic icehouse/greenhouse transition and the faunal and floral migrations and exchanges across Pangaea.

The support of the German Academic Exchange Survey (DAAD) in respect of PhD-project A/01/00754 is gratefully acknowledged by D. Hmich as well as the study visits grant A/05/06579 for joint research work at the TU Bergakademie Freiberg by H. Saber. We thank F. Körner (Freiberg) the discoverer of the Ihreri tracks, for joint fieldwork in Morocco, M. Roscher (Freiberg) for technical support, J. Broutin (Paris) for palaeobiogeographical discussion, as well as R. Rössler (Chemnitz) and C. Hartkopf-Fröder (Krefeld) for palaeoecological and biostratigraphical discussions. We also thank H. Kerp (Münster) for the determination of plant remains and biostratigraphical discussions. R. Werneburg provided information on vertebrates. The senior author, J.W.S., wishes to thank all his friends of the French Association of Permian and Triassic Geologists (AAGPT) for the introduction to the French and Moroccan Permian basins, and M. Eaton and S. G. Lucas for improving the English of this manuscript. The reviewers are acknowledged for their critical evaluation and improvement of the manuscript. This publication results partly from the following projects: Menning and Schneider Me 1134/5, 'Devonian – Carboniferous – Permian Correlation Chart', Schn 408/7 'Reference Profile Lodève', in respect of the main project 'SPP 1054 Evolution of the System Earth' and Schn 408/10 'Permian Playa to Sabkha' in respect of the main project SPP 1135 'Central European Basin System' of the Deutsche Forschungsgemeinschaft (DFG). It is also a contribution to the tasks of the working group 'Marine – Non-marine Correlations' of the Subcommission on Permian Stratigraphy of the IUGS.

References

AASSOUMI, H., SABER, H., BROUTIN, J. & EL WARTITI, M. 2003. First spore, pollen and acritarch associations in the Ida Ou Ziki basin (southern slope of the Western High Atlas, Morocco). *XVth International Congress on Carboniferous and Permian Stratigraphy, Utrecht, 2003. Abstracts*, **195**, 1–2.

AVANZINI, M. 2000. *Synaptichnium* tracks with skin impressions from the Anisian (Middle Triassic) of the Southern Alps (Val di Non–Italy). *Ichnos*, **7**, 243–252.

AVANZINI, M., CEOLONI, P. *et al.* 2001. Permian and Triassic tetrapod ichnofaunal units of northern Italy: their potential contribution to continental biochronology. *In*: CASSINIS, G. (ed.) *Permian Continental Deposits of Europe. Regional Reports and Correlations.* Natura Bresciana, Monografia, **25**, 89–107.

BAIRD, D. 1965. Early Permian Vertebrates from the Cutler Formation of the Placerville area Colorado. U.S. Geological Survey Professional Paper, **503**, 47–50.

BEN ABBOU, M., SOULA, J.-C. *et al.* 2001. Contrôle tectonique de la sédimentation das le système de basins d'avant-pays de la Meseta marocaine. *Comptes Rendus de l'Académie des Sciences, Paris, Serie II*, **332**, 703–709.

BERMAN, D. S., SUMIDA, S. S. & LOMBARD, R. E. 1997. Biogeography of primitive amniotes. *In*: SUMIDA, S. S. & MARTIN, K. L. M. (eds) *Amniote Origins. Completing the Transition to Land.* Academic Press, San Diego, 85–139.

BERTHELIN, M., BROUTIN, J., KERP, H., CRASQUIN-SOLEAU, S., PLATEL, J.-P. & ROGER, J. 2003. The Oman Gharif mixed paleoflora: a useful tool for testing Permian Pangea reconstructions. *Palaeogeography, Palaeoclimatology, Palaeoecology*, **196**, 85–98.

BROUTIN, J., EL WARTITI, M., FREYTET, P., HEYLER, D., LARHRIB, M. & MOREL, J.-L. 1987. Nouvelles découvertes paléontologiques dans le basin détritique carbonaté permien de Tiddas (Maroc central). *Comptes Rendus de l'Académie des Sciences, Paris, Serie II*, **305**, 143–148.

BROUTIN, J., FERRANDINI, J. & SABER, H. 1989. Implications stratigraphiques et paléogéographiques de la découverte d'une flore permienne euraméricaine dans le Haut-Atlas occidental (Maroc). *Comptes Rendus de l'Académie des Sciences, Paris, Serie II*, **308**, 1509–1515.

BROUTIN, J., DOUBINGER, J. *et al.* 1990. Le renouvellement des flores au passage Carbonifère-Permien: approches stratigraphique, biologique, sédimentologique. *Comptes Rendus de l'Académie des Sciences*, **311**, 1563–1569.

BROUTIN, J., ROGER, J. *et al.* 1995. The Permian Pangea. Phytogeographic implications of new paleontological discoveries in Oman (Arabian Peninsula). *Comptes Rendus de l'Académie des Sciences, Paris, Serie II*, **321**, 1069–1086.

BROUTIN, J., AASSOUMI, H., EL WARTITI, M., FREYTET, P., KERP, H., QUESADA, C. & TOUTIN-MORIN, N. 1998. The Permian Basins of Tiddas, Bou Achouch and Khenifra (central Morocco). Biostratigraphic und palaeophytogeographic implications. *In*: CRASQUIN-SOLEAU, S. & BARRIER, É. (eds) *Peri-Tethys Memoir 4: Epicratonic Basins of Peri-Tethyan platforms. Mémoires du Muséum National d'Histoire Naturelle, Paris*, **179**, 257–278.

BROWN, R. H. 1980. Triassic rocks of Argana Valley, southern Morocco and their regional structural implications. *American Association of Petroleum Geologists Bulletin*, **64**, 988–1003.

CLARIOND, L. 1932. Sur le Stéphanien des Ida ou Zal (Maroc occidental). *Comptes Rendus de l'Académie des Sciences, Paris, Serie II*, **195**, 62–64.

CLARK, N. D. L. 1999. The Elgin marvel. *OUGS Journal*, Symposium Edition, **20**(2), 16–18.

CLAUSING, A. & BOY, J. A. 2000. Lamination and primary production in fossil lakes: Relationship to plaeaoclimate in the Carboniferous-Permian transition. *In*: HART, M. B. (ed.) *Climates: Past and Present*. Geological Society, London, Special Publications, **181**, 5–16.

CONTI, M. A., LEONARDI, G., MARIOTTI, N. & NICOSIA, U. 1977. Tetrapod footprints of the Val Gardena Sandstone (North Italy). Their paleontological, stratigraphic and paleoenvironmental meaning. *Palaeontographia Italica*, **70**, 1–79.

CONTI M. A., MARIOTTI N., MANNI R. & NICOSIA U. 1999. Tetrapod footprints in the Southern Alps: an overview. *In*: CASSINIS G., CORTESOGNO, L., GAGGERO, L., MASSARI, F., NERI, C., NICOSIA, U. & PITTAU, P. (eds) *Stratigraphy and Facies of the Permian Deposits Between Eastern Lombardy and the Western Dolomites. Field Trip Guidebook 15–18 September 1999*. International Field Conference. The Continental Permian of the Southern Alps and Sardinia (Italy), Brescia, 15–25 September 1999. Earth Science Department, Pavia University, Appendix, 137–138.

DE KONING, G. 1957. *Géologie des Ida ou Zal (Maroc). Stratigraphie, pétrographie et tectonique de la partie Sud-Ouest du bloc occidental du Massif ancien du Haut-Atlas (Maroc)*. Leidse Geologische Mededelingen, **23**.

DEMATHIEU, G. & HAUBOLD, H. 1982. Reptilfährten aus dem Mittleren Buntsandstein von Hessen (BRD). *Hallesches Jahrbuch für Geowissenschaften*, **7**, 97–110.

DOUBINGER, J. & ROY-DIAS, C. 1985. La paléoflore du Stéphanien de l'Oued Zat (Haut-Atlas de Marrakech – versant Nord-Maroc). *Geobios*, **18**, 573–586.

DUTUIT, J.-M. 1976. Découverte d'Amphibiens Lépospondyles dans la série inférieure de la Formation rouge d'Argana (Atlas occidental marocain). *Comptes Rendus de l'Académie des Sciences, Paris, Serie D*, **283**, 1733–1734.

DUTUIT, J.-M. 1988. *Diplocaulus minimus* n. sp. (Amphibia: Nectridea), Lépospondyle de la Formation d'Argana, dans l'Atlas marocain. *Comptes Rendus de l'Académie des Sciences, Paris, Serie II*, **307**, 851–854.

EL WARTITI, M. 1990. *Le Permien du Maroc mésétien: Étude géologique et implications paléogéographiques*. Thèse d'État, Université Mohamed V, Rabat.

EL WARTITI, M., BROUTIN J. & FREYTET, P. 1986. Premières découvertes paléontologiques dans les séries rouges carbonatées permiennes du bassin de Tiddas (Maroc Central). *Comptes Rendus de l'Académie des Sciences, Paris, Série II*, **303**, 263–268.

EL WARTITI, M., BROUTIN, J., FREYTET, P., LAHRHIB, M. & TOUTIN-MORIN, N. 1990. Continental deposits in Permian basins of the Mesetian Morocco: geodynamic history. *Journal of African Earth Sciences*, **10**, 361–368.

ERRAMI, A., MEDINA, F., HOEPFNER, C. AHMANOU, M. F. & BESAHAL, A. 2002. Etude structurale des formations stéphano-autuniennes d'El Menizla et nouvelle chronologie des phases tardi-hercyniennes dans le Haut Atlas occidental (Maroc). *Africa Geoscience Review*, **9**, 157–170.

FEYS, R. & GREBER, CH. 1963. Le Stéphanian et l'Autunien du Souss dans les Ida ou Zal (Haut-Atlas occidental – Maroc). *Notes et Mémoires du Service Géologique, Maroc*, **22**(170), 19–43.

GAITZSCH, B., RÖSSLER, R., SCHNEIDER, J. W. & SCHRETZENMAYR, ST. 1999. Neue Ergebnisse zur Verbreitung potentieller Muttergesteine im Karbon von Nord- und Mitteldeutschland. *Geologisches Jahrbuch, Reine A*, **149**, 25–58.

GAND, G. 1988. *Les traces de vertébrés tétrapodes du Permien francais*. PhD thesis, University of Bourgogne.

GILMORE, CH. W. 1926. Fossil footprints from the Grand Canyon. *Smithsonian Miscellaneous Collections*, **77**(9), 1–41.

GILMORE, CH. W. 1927. Fossil footprints from the Grand Canyon. II. *Smithsonian Miscellaneous Collections*, **80**(3), 1–78.

HAUBOLD, H. 1984. *Saurierfährten*. Ziemsen-Verlag, Wittenberg, Die Neue Brehm-Bücherei, **479**.

HAUBOLD, H., HUNT, A. P., LUCAS, S. G. & LOCKLEY, M. G. 1995. Wolfcampian (Early Permian) vertebrate tracks from Arizona and New Mexico. *In*: LUCAS, S. G. & HECKERT, A. B. (eds) *Early Permian Footprints and Facies*. New Mexico Museum of Natural History and Science Bulletin, **6**, 135–165.

HMICH, D. 2004. *Le Stéphanien du Haut Atlas et le Permien du Maroc Central : étude biostratigraphique et implications paléogéographiques (Maroc)*. PhD thesis, Université Mohammed V.

HMICH, D., SCHNEIDER, J., SABER, H., EL WARTITI, M. & BROUTIN, J. 2002. First record on late Palaeozoic insect faunas from Morocco: biostratigraphic and biogeographic implications. *19th Colloquium of African Geology, El Jadida, Abstracts*, 101–102.

HMICH, D., SCHNEIDER, J. W., SABER, H. & EL WARTITI, M. 2003. First Permocarboniferous insects (blattids) from North Africa (Morocco): implications on palaeobiogeography and palaeoclimatology. *Freiberger Forschungshefte, Hefte C*, **499**, 117–134.

HMICH, D., SCHNEIDER, J. W., SABER, H. & EL WARTITI, M. 2005. Spiloblattinidae (Insecta, Blattida) from the Carboniferous of Morocco, North Africa: implications on biostratigraphy. *In*: LUCAS, S. G. & ZEIGLER, K. E. (eds) *The Nonmarine Permian. New Mexico Museum of Natural History and Science Bulletin*, **30**, 111–114.

HOFMANN, A., TOURANI, A. & GAUPP, R. 2000. Cyclicity of Triassic to Lower Jurassic continental red beds of the Argana Valley, Morocco: implications for palaeoclimate and basin evolution. *Palaeogeography, Palaeoclimatology, Palaeoecology*, **161**, 229–266.

JALIL, N. E., & DUTUIT, J. M. 1996. Permian Captorhinid reptiles from the Argana Formation, Morocco. *Palaeontology*, **39**, 907–918.

JONES, D. F. 1975. *Stratigraphy, Environments of Deposition, Petrology, Age, and Provenance, Basal Red Beds of the Argana Valley, Western High Atlas Mountains, Morocco*. M.S. thesis, New Mexico Institute of Mining and Technology, Socorro.

JONGMANS, W. J. 1950. Notes sur la flore du Carbonifère du versant Sud du Haut-Atlas. *Notes et Mémoires du Service Géologique, Maroc*, **76**, 155–172.

KERP, H. 1996. Post-Variscan late Palaeozoic Northern Hemisphere gymnosperms: the onset to the Mesozoic. *Review of Palaeobotany and Palynology*, **90**, 263–285.

KERP, H. 2000. The modernization of landscapes during the late Paleozoic–early Mesozoic. *In*: GASTALDO, R. A. & DIMICHELE, W. A. (eds) *Phanerozoic Terrestrial Ecosystems*. Paleontological Society, Lawrence, Paleontological Society Papers, **6**, 79–113.

KERP, H., BROUTIN, J., LAUSBERG, S. & AASSOUMI, H. 2001. Discovery of Latest Carboniferous–Early Permian radially symmetrical peltaspermaceous megasporophylls from Europe and North Africa. *Comptes Rendus de l'Académie des Sciences, Paris, Serie II*, **332**, 513–519.

LARHRIB, M. 1996. Flore fossile et séquences des formations rouges fluviatiles du basin autunien de Tiddas-Sebt Aït Ikkou (NW du Maroc central). *In*: MEDINA, F. (ed.) *Le Permien et le Trias du Maroc: état des connaissances*. Editions Pumag, Marrakech, 19–29.

LEGLER, B., GEBHARDT, U. & SCHNEIDER, J. W. 2005. Late Permian Non-Marine – Marine Transitional Profiles in the Central Southern Permian Basin, Northern Germany. 9:*International Journal of Earth Sciences*, DOI: 10.1007/s00531-005-0002-5.

LÜTZNER, H., MÄDLER, J., ROMER, R. L. & SCHNEIDER, J. W. 2003. Improved stratigraphic and radiometric age data for the continental Permocarboniferous reference-section Thüringer-Wald, Germany. *XVth International Congress on Carboniferous and Permian Stratigraphy, Utrecht, 2003, Abstracts*, 338–341.

MEDINA, F. 1991. Superimposed extensional tectonics in the Argana Triassic formations (Morocco), related to the early rifting of the central Atlantic. *Geological Magazine*, **128**, 525–536.

MEDINA, F. 1995. Syn- and postrift evolution of the El Jadida–Agadir Basin (Morocco): constraints for the rifting models of the central Atlantic. *Canadian Journal of Earth Sciences*, **32**, 1273–1291.

MENNING, M. & German Stratigraphic Commission 2002. A geologic time scale 2002. *In*: German Stratigraphic Commission (ed.) *Stratigraphic Table of Germany 2002*. Potsclam.

PEABODY, F. E. 1948. Reptile and amphibian trackways from the Lower Triassic Moenkopi Formation of Arizona and Utah. *Department of Geological Sciences, University of California, Bulletin*, **27**(57), 295–467.

PEŠEK, J. 2004. Late Palaeozoic limnic basins and coal deposits of the Czech Republic. *Folia Musei Rerum Naturalium Bohemiae Occidentalis: Geologica, Editio Specialis*, **1**.

ROSCHER, M. & SCHNEIDER, J. W. 2005. An annotated correlation chart for continental Late Pennsylvanian and Permian basins and the marine scale. *In*: LUCAS, S. G. (ed.) *The Continental Permian. New Mexico Museum of Natural History and Science Bulletin*, **30**, 282–291.

ROSCHER, M. & SCHNEIDER, J. W. 2006. Permo-Carboniferous climate: Early Pennsylvanian to Late Permian climate development of central Europe in a regional and global context. *In*: LUCAS, S. G., CASSINIS, G. & SCHNEIDER, J. W. (eds) *Non-Marine Permian Biostratigraphy and Biochronology*. Geological Society, London, Special Publications, **265**, 95–136.

RÖSSLER, R. 1999. Neue Organzusammenhänge eines Calamiten: taphonomische Beobachtungen im Oberkarbon des Saar-Nahe-Beckens. *Freiberger Forschungsheft, Hefte C*, **481**, 49–61.

SABER, H. 1994. Données sédimentologiques et évidence d'une tectonique tardi-hercynienne dans le bassin stéphano-permien des Ida Ou Ziki. Sud Ouest du Massif Ancien du Haut Atlas (région d'Argana, Maroc). *Journal of African Earth Sciences*, **19**, 99–108.

SABER, H. 1998. *Le Stéphano-Permien du Haut Atlas occidental: étude géologique et évolution géodynamique (Maroc)*. Thèse d'Etat ès-Sciences, Université Chouaib Doukkali.

SABER, H., EL WARTITI, M. & BROUTIN, J. 2001. Dynamique sédimentaire comparative dans les bassins Stéphano-Permiens des Ida Ou Zal et Ida Ou Ziki, Haut Atlas Occidental, Maroc. *Journal of African Earth Sciences*, **32**, 573–594.

SABER, H., EL WARTITI, M., BROUTIN, J. & TOUTIN MORIN, N. 1995. L'intervalle Stephano-Permien (fin du cycle varisque au Maroc). *Gaia*, **11**, 57–71.

SCHÄFER, A., RAST, U. & STAMM, R. 1990. Lacustrine paper shales in the Permocarboniferous Saar–Nahe Basin (west Germany): depositional environment and chemical characterization. *In*: HELING, D. (ed.) *Sediments and Environmental Geochemistry*. Springer Verlag, Berlin, Heidelberg, New York, London, 220–238.

SCHNEIDER, J. 1982. Entwurf einer biostratigraphischen Zonengliederung mittels der Spiloblattinidae (Blattodea, Insecta) für das kontinentale euramerische Permokarbon. *Freiberger Forschungshefte, Hefte C*, **375**, 27–34.

SCHNEIDER, J. 1989. Basic problems of biogeography and biostratigraphy of the Upper Carboniferous and Rotliegendes. *Acta Musei Reginaehradecensis, Serie A, Scientiae Naturales*, **XXII**, 31–44.

SCHNEIDER, J. 1996. Biostratigraphie des kontinentalen Oberkarbon und Perm im Thüringer Wald, SW-Saale-Senke – Stand und Probleme. *Beiträge zur Geologie von Thüringen, Neue Folge*, **3**, 121–151.

SCHNEIDER, J. W. 2001. Rotliegendstratigraphie: Prinzipien und Probleme. *Beiträge zu Geologie von Thüringen, Neue Folge*, **8**, 7–42.

SCHNEIDER, J., WALTER, H. & WUNDERLICH, J. 1982. Zur Biostratinomie, Biofazies und Stratigraphie des Unterrotliegenden der Breitenbacher Mulde (Thüringer Wald). *Freiberger Forschungshefte, Hefte C*, **366**, 65–84.

Schneider, J., Siegesmund, S. & Gebhardt, U. 1984. Paläontologie und Genese limnischer Schill und Algenkarbonate in der Randfazies der kohleführenden Wettiner Schichten (Oberkarbon, Stefan C) des NE Saaletroges. *Hallesches Jahrbuch für Geowissenschaften*, **9**, 35–51.

Schneider, J. W., Rössler, R. & Gaitzsch, B. 1995. Proposal for a combined reference section of the central European continental Carboniferous and Permian for correlations with marine standard sections. *Permophiles*, **26**, 26–31.

Schneider, J. W., Rössler, R., Gaitzsch, B., Gebhardt, U. & Kampe, A. 2005. Saale-Senke. *In*: Deutsche Stratigraphische Kommission (ed.) *Stratigraphie von Deutschland*. V. Oberkarbon. Courier Forschungsinstitut Senkenberg, **254**, 419–440.

Schneider, J. W., Körner, F., Roscher, M. & Kroner, U. 2006. Permian climate development in the northern peri-Tethys area: the Lodève Basin, French Massif Central, compared in a European and global context. *Palaeogeography, Palaeoclimatology, Palaeoecology*, (in press).

Sidor, C. A., O'Keefe, Damiani, R., Steyer, J. S., Smith, R. M. H., Larsson, H. C. E., Sereno, P. C., Ide, O. & Maga, A. 2005. Permian tetrapods from the Sahara show climate-controlled endemism in Pangea. *Nature*, **434**, 886–889.

Sues, H.-D. & Munk, W. 1996. A remarkable assemblage of terrestrial tetrapods from the Zechstein (Upper Permian: Tatarian) near Korbach (northwest Hesse). *Paläontologische Zeitschrift*, **1/2**, 213–223.

Tasch, P. 1987. *Fossil Conchostraca of the Southern Hemisphere and Continental Drift. Paleontology, Biostratigraphy and Dispersal*. Geological Society of America, Memoire, **165**.

Tixeront, M. 1974. Carte géologique et minéralogique du Couloir d'Argana, 1/100 000. *Edition du Sérvice Géologique du Maroc, Notes et Mémoires*, **205**.

Voigt, S. 2005. *Die Tetrapodenichnofauna des kontinentalen Oberkarbon und Perm im Thüringer Wald: Ichnotaxonomie, Paläoökologie und Biostratigraphie*. Cuvillier Verlag, Göttingen.

Werneburg, R. & Schneider, J. 1996. The Permian temnospondyl amphibians of India. *Special Papers in Palaeontology*, **52**, 105–128.

Youbi, N. & Cabanis, B. 1995. Histoire volcano-tectonique du massif permien de Khenifra (Sud-Est du Maroc central). *Geodinamica Acta*, **8**, 158–172.

Ziegler, A. M. 1990. Phytogeographic patterns and continental configurations during the Permian Period. *In*: McKerrow, W. S. & Scotese, C. R. (eds) *Palaeozoic Palaeogeography and Biogeography*. Geological Society, London, Memoir, **12**, 363–379.

Insect biostratigraphy of the Euramerican continental Late Pennsylvanian and Early Permian

JOERG W. SCHNEIDER[1] & RALF WERNEBURG[2]

[1]*TU Bergakademie Freiberg, B.v. Cotta-Strasse 2, D-09596 Freiberg, Germany (schneidj@geo.tu-freiberg.de)*

[2]*Naturhistorisches Museum Schloss Bertholdsburg, Burgstrasse 6, D-98553 Schleusingen, Germany (museum.schleusingen@gmx.de)*

Abstract: An insect zonation with a time resolution of 1.5–2 Ma for Late Pennsylvanian to Early Permian (Kasimovian to Artinskian) non-marine deposits is presented. The zonation is based on the directed morphogenetic evolution of colour pattern in the forewings of the blattid (cockroach) family Spiloblattinidae. This evolution is observed in lineages of succeeding species of three genera. All three genera are widely distributed in the palaeo-equatorial zone from Europe to North America, that is, in the Euramerican biota province. Increasing reports of spiloblattinid zone species in conodont-bearing, interfingered marine/continental strata of North American Appalachian, Mid-Continent and West Texas basins could be the key to direct biostratigraphical correlations of pure continental profiles, as are present in the most parts of the Hercynides, to the global marine scale.

Palaeozoological biostratigraphy of the continental late Palaeozoic

Traditionally, the biostratigraphy of continental Carboniferous and Permian deposits has been based on macro- and micro-floras. However, decreasing marine transgressions during the Pennsylvanian and the aridization during the Late Pennsylvanian and the Permian (cf. Roscher & Schneider 2006) generated a change from inter-regional, balanced wet macro- and meso-climates (with a maritime imprint of some degree) to increasingly drier continental climates with stronger seasonality and stronger accentuation of meso- and microclimatic effects. Consequently, edaphic differentiation of the floral associations occurred. The persistence of conservative Carboniferous hydro- to hygrophilous floral elements into Permian (local) wet biotopes and the local appearance of modern typical Permian meso- to xerophilous floral elements in the Carboniferous thus underlies the well-known problems of biostratigraphy with macro- and microfloras (e.g. Broutin *et al.* 1990; DiMichele *et al.* 1996; Kerp 1996).

During the last two decades a number of palaeozoological biostratigraphies were developed and tested. Tetrapod tracks, common in alluvial grey and red sediments, display a wide range in time and space (Haubold 1980, Gand & Haubold 1988, Haubold & Lucas 2003). Unfortunately, their time resolution is very low (Lucas 1998; Voigt 2005). Conchostracans (Spinicaudata) were successfully tested for surface outcrops and drill cores as well (Schneider *et al.* 2005). They have a very high distribution potential because of their minute, drought-resistant and wind-transportable eggs, and they often form mass occurrences in lacustrine environments of grey and red facies. Hence, conchostracans belong to the most common animal fossils of the continental late Palaeozoic. Regrettably, the time range of single species is not well known at present (Martens 1983; Schneider *et al.* 2005).

Xenacanth shark teeth were applied to regional correlations between some neighbouring European basins. Their wider use is limited because the migration of fishes is restricted to joint river systems connecting the basins (e.g. Schneider 1996; Schneider *et al.* 2000). The fish zonation of Zajic (2000) is thus more a local ecostratigraphy than a biostratigraphy of some Bohemian basins.

Biostratigraphic zonations using osteologic species of aquatic or semi-aquatic amphibians were presented by Boy (1987) and Werneburg (1989*a*, *b*, 1996). The inter-regional amphibian zones of Werneburg (1996) are based on species-chronoclines with a time resolution of about 1.5–3.0 Ma (for details see Werneburg & Schneider 2006). For large-scale inter-regional subdivisions and correlations, Lucas (1998, 2002, 2006) has presented land-vertebrate faunachrons based on amphibian and reptile skeletons.

The first serious attempts to use insect wings for biostratigraphy were made by Scudder as early as 1879. He recognized the common

From: LUCAS, S. G., CASSINIS, G. & SCHNEIDER, J. W. (eds) 2006. *Non-Marine Permian Biostratigraphy and Biochronology*. Geological Society, London, Special Publications, **265**, 325–336.

occurrence of genera and species of blattid insects (cockroaches) in North America and Europe and their potential for 'delicate discriminations of the age of rock deposits' (Scudder 1895). Later, Durden (1969, 1984) proposed blattid zonations for the Pennsylvanian and Permian, but his correlations remain very doubtful because of inadequate classifications. A revised classification of Pennsylvanian and Permian blattids was presented by Schneider (1983*a* ff.) based on comparative investigations of individual, sexual, intraspecific and interspecific variation of the wing venation of modern *Periplaneta americana* and *P. australasiae*, as well as fossil blattids (Schneider 1977, 1978). From these results came the first proposals of spiloblattinid zones (Schneider 1982, Schneider & Werneburg 1993) and later of archimylacrid/spiloblattinid/conchostracan zones (Schneider & Rössler, unpubl. internal report for gas exploration companies), which were tested and improved for the Early Pennsylvanian (Westphalian A, Late Bashkirian) through the late Early Permian (Cisuralian, Artinskian). The archimylacrid/conchostracan zonation for Westphalian time is published in Schneider *et al.* (2005). Here, new results and a completed spiloblattinid-zonation for the time interval from the Late Pennsylvanian Kasimovian up to the Middle Cisuralian Artinskian are presented. For species authors, synonymies and further taxonomic information see Schneider (1982, 1983*a*) and Schneider & Werneburg (1993).

Occurrence, taphonomy and palaeoecology of spiloblattinids

Insect remains occur in all fine clastics, from claystones and siltstones to silty fine sandstones and volcanic ashes of different continental to nearshore marine depositional environments, ranging from swamps and lakes of the grey facies to alluvial plain, playa and sabkha deposits of the red-bed facies. By far the most common insects of the late Palaeozoic are blattids. Best preserved are the tegmina-like, strengthened forewings because of the generally high preservation potential of chitinous substances under subaerial and subaquatic conditions as well as their relatively high resistance to physical forces during water or wind transport.

Generally, insects have a high distribution potential because of active flight and passive distribution by air currents. Among blattid insects, the family Spiloblattinidae forms an exceptional group, characterized by their extensively coloured wings (Fig. 1; Schneider 1983*a*, 1984*a*). Based on venation and colour patterns of the 1.5–2.5-cm-long forewings, three genera of spiloblattinids are distinguished (Schneider & Werneburg 1993): *Sysciophlebia* Handlirsch, 1906, *Spiloblattina* Scudder, 1885, and *Syscioblatta* Handlirsch, 1906. All three genera are widely distributed in the palaeo-equatorial zone from Europe to North America, that is, in the Euramerican biotic province. In some places they occur together in one and the same horizon. Until now, a distinct biotope preference is not evident. Spiloblattinids are associated with hydro- to hygrophilous floras of the roof shales of coal seams, as in the Breitenbach Formation of the Saar–Nahe Basin and the Wettin Subformation of the Saale Basin, both in Germany, or in the roof shale of the Waynesburg coal, Dunkard Group, of West Virginia. Associated with meso- to xerophilous floral elements, such as callipterids and walchians, they are not rare in black-shale lake deposits, as in the Homigtal Lake of the Thuringian Forest Basin and the Svitavka-Zbonek Lake, Letovice Formation of the Boskovice Graben, Czech Republic, as well as in the lake deposits of the El Menizla and Oued Issene formations of the Moroccan Souss Basin (Hmich *et al.* 2005, 2006). *Syscioblatta* is unusually common in brackish–marine deposits of the Wild Cow Formation of New Mexico, and some spiloblattinids have been found, together with walchians and cordait leaves, in the nearshore lagoonal Red Tanks Member, Bursum Formation, of the Lucero Basin, New Mexico (Schneider *et al.* 2004 and new observations).

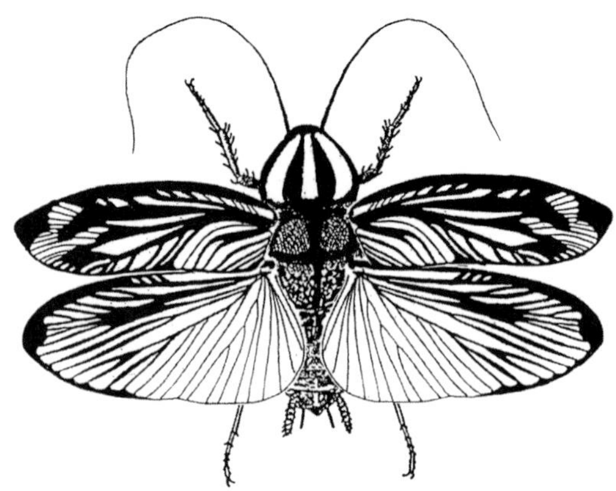

Fig. 1. Life reconstruction of *Syscioblatta dohrni* (Scudder 1879), male, based on the holotype and additional specimens (Schneider & Werneburg 1993, figs 4a, b). The reconstruction shows the colour pattern of fore- and hindwings as well as of the head shield (pronotum), which is typical of blattids ('cockroaches') of the family Spiloblattinidae (Handlirsch 1906*a*). Length of forewings 2.5 cm.

At present, spiloblattinids are absent in playa red beds, as in the still blattid-rich and very diverse insect faunas of the Salagou Formation of the Lodève Basin, southern France, and the playa lake deposits of the Wellington Formation of Kansas and Oklahoma. Obviously, spiloblattinids occur in a wide range of biotopes from humid to semi-arid conditions. They seem to disappear in Euramerica with the maximum of arid climate during the Kungurian (see Roscher & Schneider 2006). Typical of semi-arid to arid environments of this time in Euramerica are xeromorphic mylacrid blattids, as some opsiomylacrids and the genus *Moravamylacris* (Schneider 1980, Hmich *et al.* 2003). Because of the co-occurrence of *Moravamylacris* species with spiloblattinids, they are used to complete the spiloblattinid zones for the time of the general transition from interfingering grey/red to exclusively red beds in the Middle Cisuralian.

Basics of spiloblattinid zonation

The early evolution of spiloblattinids during the Westphalian is not really well known (Schneider 1984*a*). In the Late Westphalian some blattids appear with sparse venation and widened fields between the main veins. These features are very similar to the venation ground plan of spiloblattinids, but the characteristic light spots or patches on the dark wing surface are still missing. In this regard, '*Syscioblatta*' *corsini* Laurentiaux 1950 from the latest Westphalian or early Cantabrian (Loire Basin, France, Carrière de l'Èparre, Eighth seam) is transitional between '*Kinklidoblatta*' *morini* (Pruvost 1912) of Westphalian D age (northern France, top Faisceau de Dusoich to Faisceau d'Edouard) and the typical Early Stephanian spiloblattinids. The first unquestionable members of this family, detected by their typical wing colouration, appear in the Stephanian A (Barruelian, Kasimovian) (Fig. 2: *Sysciophlebia* n. sp. A and *Syscioblatta intermedia*). The colour pattern originates from the appearance of sharply bounded light patches on the wing. In the area of these patches the normally dark black, coalified wing surface lacks the coaly substance, and therefore the normal sediment colour and the sediment particles are visible. If the organic substance is oxidized or missing in any way, the patches appear somewhat rough compared to the wing surface outside the spots. It is assumed that the wing colour depends on the deposition of coloured organic substances (pigments), as in extant cockroaches.

The number, size and arrangement of light spots (maculae) and bands (fasculae) changed through time in continuous, directed sequences. First, they increased in number and extent, and later they decreased. The arrangement of the colour pattern during increase and decrease of light patches is quite different (Fig. 2: compare the increase in the sequence *Sysciophlebia* n. sp. A to *Sysciophlebia ilfeldensis* and the following decrease from *S. ilfeldensis* to *Sysciophlebia* n. sp. B).

Interestingly, in the genus *Sysciophlebia*, some differences in the wing colouration of both sexes are observed for *S. ilfeldensis* and *S. balteata* Form H. The stouter (supposed) female forewings (Fig. 2, nos 6 & 7, right) exhibit a stronger decrease of light patches than the contemporaneous forewings of (supposed) males (Fig. 2, nos 6 & 7, left). Such differences must be taken into account, in order to prevent incorrect stratigraphic conclusions. As far as is known, in the genera *Spiloblattina* and *Syscioblatta* only minor differences occur, which can be ignored (see Schneider & Werneburg 1993, figs 7 & 8).

Spiloblattinid specimens, sampled in a dense vertical sequence of lake horizons in the Goldlauter Formation (Figs 2 & 3, nos 7 & 8), indicate that the change of the colour patterns is a very continuous process. The distinct differences, which can be seen between the forms in Fig. 2, result simply from a discontinuous fossil record in continental deposits. Consequently, each of the sequences of 'species' of all the three genera in Figure 2 is in reality a continuous lineage of only one chronospecies. To handle the single forms as 'species' is merely a pragmatism due to their practical use in biostratigraphy. Regarded as formal species or biospecies, each of them has a First and Last Appearance Datum (FAD and LAD). Despite the theoretical questions of classification and systematics, these phylomorphogenetic lineages provide very precise biostratigraphic data for correlations.

Definition of the spiloblattinid insect zones

The zonation is based mainly on the *Sysciophlebia* lineage, the most complete lineage known thus far. If one of the *Sysciophlebia* zone species occurs together with species of the other two genera in the same lithostratigraphic horizon, this co-occurrence is used for the definition of the respective zone. If the co-occurrence is inferred only, the respective species will be mentioned as 'inferred accompanying' form. The respective base and top of any zone is defined by the FAD of the zone species. Because real species do not exist in lineages of continuously evolving features (see above), it is nearly insignificant if the rate of feature changes is different between the zone

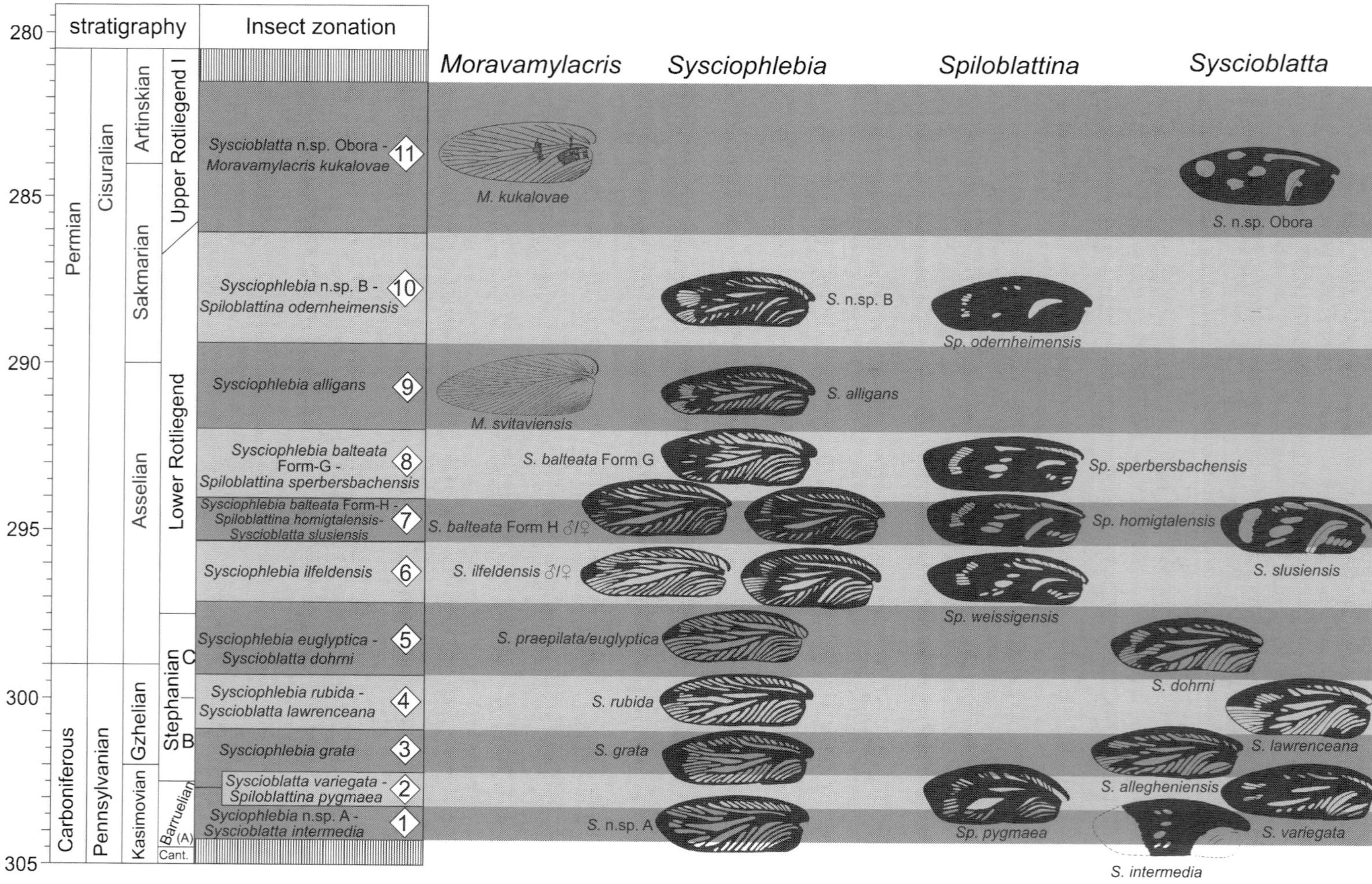
stratigraphy
Insect zonation
Moravamylacris
Sysciophlebia
Spiloblattina
Syscioblatta
280
285
290
295
300
305
Permian
Carboniferous
Cisuralian
Pennsylvanian
Artinskian
Sakmarian
Asselian
Gzhelian
Kasimovian
Upper Rotliegend I
Lower Rotliegend
Stephanian
C
B
Barruelian
(A)
Cant.
Syscioblatta n.sp. Obora - Moravamylacris kukalovae
11
Sysciophlebia n.sp. B - Spiloblattina odernheimensis
10
Sysciophlebia alligans
9
Sysciophlebia balteata Form-G - Spiloblattina sperbersbachensis
8
Sysciophlebia balteata Form-H - Spiloblattina homigtalensis- Syscioblatta slusiensis
7
Sysciophlebia ilfeldensis
6
Sysciophlebia euglyptica - Syscioblatta dohrni
5
Sysciophlebia rubida - Syscioblatta lawrenceana
4
Sysciophlebia grata
3
Syscioblatta variegata - Spiloblattina pygmaea
2
Syciophlebia n.sp. A - Syscioblatta intermedia
1
M. kukalovae
S. n.sp. Obora
S. n.sp. B
Sp. odernheimensis
M. svitaviensis
S. alligans
S. balteata Form G
Sp. sperbersbachensis
S. balteata Form H ♂/♀
Sp. homigtalensis
S. ilfeldensis ♂/♀
S. slusiensis
Sp. weissigensis
S. praepilata/euglyptica
S. dohrni
S. rubida
S. lawrenceana
S. grata
S. allegheniensis
S. n.sp. A
Sp. pygmaea
S. variegata
S. intermedia

species of different lineages (which will normally be the case). The numbers in the following descriptions refer to the numbers in the synthetic spiloblattinid zonation (Fig. 2), and in the correlation chart for most of the basins mentioned in the text as well as the stratigraphic levels of spiloblattinid occurrences (Fig. 3).

1 *Sysciophlebia* n. sp. A – *Syscioblatta intermedia* zone

Definition: From FAD of *Sysciophlebia* n. sp. A and/or *Syscioblatta intermedia* to FAD of *Sysciophlebia grata*.

Occurrences: Stephanian A (Barruelian, Kasimovian), Saar Basin, Germany, lower Ottweiler Subgroup, Göttelborn Formation.

Remarks: The typical colour design of spiloblattinids appears suddenly in the earliest Stephanian (Barruelian, Kasimovian) without transitional forms. However, some late Westphalian/earliest Cantabrian (Moscovian/Kasimovian) blattids, such as '*Syscioblatta*' *corsini*, *Kinklidoblatta morini* and *K. zavjalovensis*, could be precursors of true spiloblattinids (see above).

2 *Syscioblatta variegata* zone

Definition: From FAD of *Syscioblatta variegata* to FAD of *Syscioblatta allegheniensis*.

Occurrences: Virgilian, Appalachian Basin, Ohio, roof of the Ames Limestone; conodonts of the Ames Limestone (Ritter 1995), such as *Streptognathodus simulator* and *Streptognathodus elegantulus*, indicate an Early Gzhelian age.

Inferred accompanying species: *Spiloblattina pygmaea*, Stephanian A/B (Kasimovian), Saar–Nahe Basin, Germany, middle Ottweiler Subgroup, lowermost Heusweiler Formation; early Stephanian B (Kasimovian), Morocco, Souss Basin, Ida Ou Zal Sub-Basin, upper El Menizla Formation; Ida Ou Ziki Sub-Basin, upper Oued Issene Formation.

Remarks: *S. variegata* is the successor of *S. intermedia* and the precursor of *S. allegheniensis*. However, up to now, we have no proof of the co-occurrence of *S. variegata* and *Sp. pygmaea*, and no co-occurring *Sysciophlebia* species are known so far. The range of *Sp. pygmaea* could overlap the range of *Sysciophlebia* n. sp. A and/or *S. grata*. Therefore, this zone is sandwiched with some overlap between the *Sysciophlebia* n. sp. A – *Syscioblatta intermedia* zone and the *Sysciophlebia grata* zone.

3 *Sysciophlebia grata* zone

Definition: From FAD of *Sysciophlebia grata* to FAD of *Sysciophlebia rubida* and/or FAD of *Syscioblatta lawrenceana*.

Occurrences: Stephanian B (Kasimovian/Gzhelian), Kladno Basin, Czech Republic, Slaný Formation, Hředle Member, early Stephanian B (Kasimovian/Gzhelian), Morocco, Souss Basin, Ida Ou Ziki Sub-Basin, upper Oued Issene Formation (together with *Spiloblattina pygmaea*: see above).

Inferred accompanying species: *Syscioblatta allegheniensis*, Virgilian, Appalachian Basin, Pennsylvania, Conemaugh Formation, Duquesne Shale.

Discussion: Based on colour pattern, *S. allegheniensis* is intermediate between the preceding *S. variegata* and the following *S. lawrenceana*. The latter co-occurs with the succeeding species of *S. grata*, namely *S. rubida*. Therefore, it could be assumed that *S. allegheniensis* covers a time interval similar to that of *S. grata*.

4 *Sysciophlebia rubida* – *Syscioblatta lawrenceana* zone

Definition: From FAD of *Sysciophlebia rubida* and/or *Syscioblatta lawrenceana* to FAD of *Sysciophlebia euglyptica* and/or *Syscioblatta dohrni*.

Occurrences: Both species occur together at the type locality of *S. rubida*, in the Late Stephanian B (Gzhelian) Plouznice Horizon of the Semily Formation in the Krkonoše-Piedmont Basin, Czech Republic; *S. lawrenceana*, Virgilian, Mid-Continent Basin, Kansas, Lower Douglas Group, Lawrence Shale, *Streptognathodus firmus*-conodont zone after Ritter (1995), Early to Middle Gzhelian; *S. lawrenceana* together with

Fig. 2. Zone species of the spiloblattinid insect zonation of Late Carboniferous (Pennsylvanian) and Early Permian (Cisuralian). Figured are the three species lineages of the spiloblattinid genera *Sysciophlebia*, *Spiloblattina* and *Syscioblatta* as known from the fossil record. Biostratigraphically significant is the evolution of colour pattern of the approximately 1.5–2.5-cm-long forewings. In the *Sysciophlebia* lineage, a well-expressed sexual dimorphism is observed in the colour pattern as shown for *S. ilfeldensis* and *S. balteata* Form H (male forewing left, female forewing right). Numbers in diamonds refer to the zone descriptions in the text. Numerical ages are based on Ogg (2004), Menning *et al.* (2003) and Lützner *et al.* (2003).

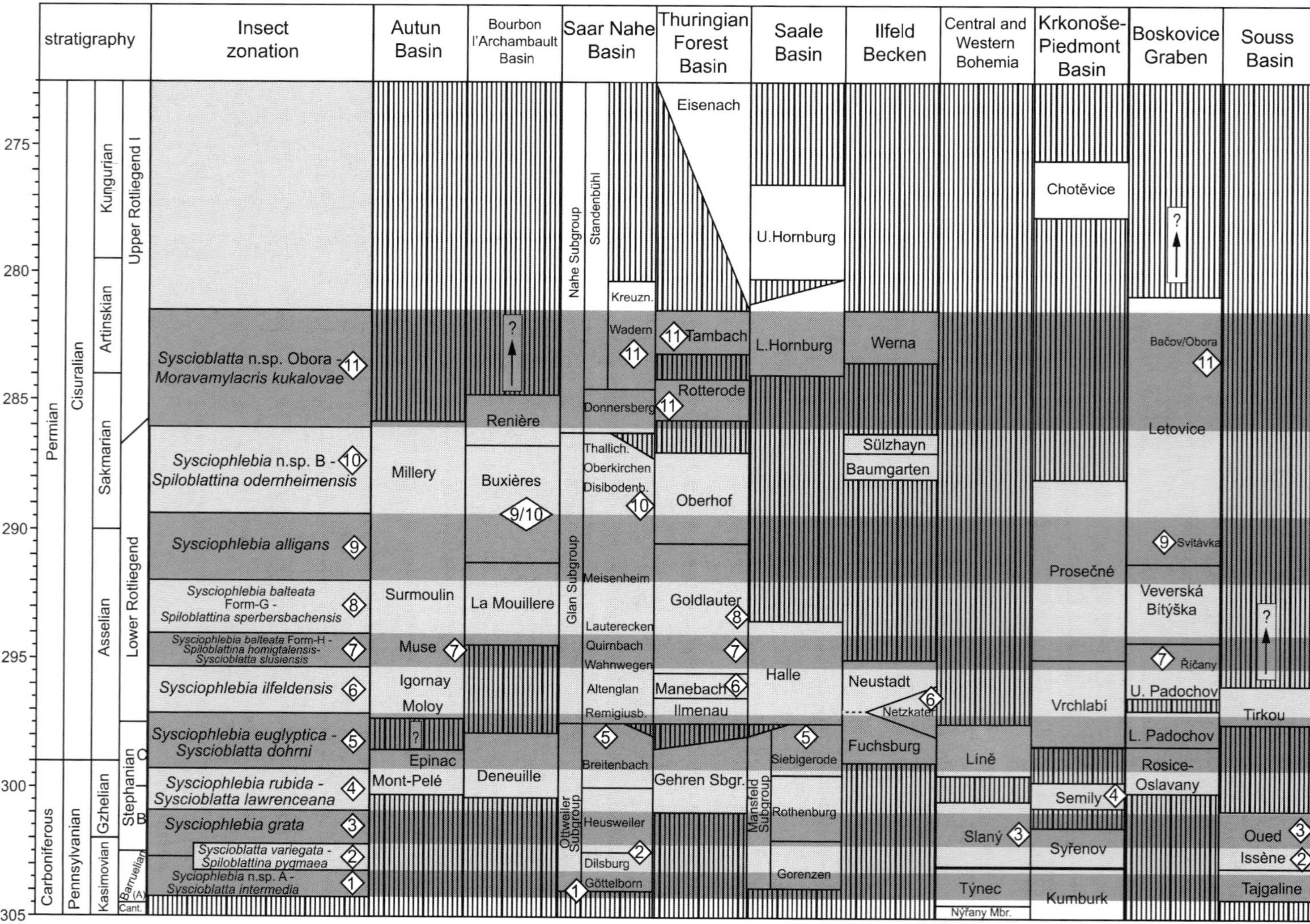
stratigraphy
Insect zonation
Autun Basin
Bourbon l'Archambault Basin
Saar Nahe Basin
Thuringian Forest Basin
Saale Basin
Ilfeld Becken
Central and Western Bohemia
Krkonoše-Piedmont Basin
Boskovice Graben
Souss Basin
Permian
Cisuralian
Kungurian
Artinskian
Sakmarian
Asselian
Upper Rotliegend I
Lower Rotliegend
Carboniferous
Pennsylvanian
Kasimovian
Gzhelian
Stephanian
Barruelian
Cant.
Syscioblatta n.sp. Obora - Moravamylacris kukalovae
Sysciophlebia n.sp. B - Spiloblattina odernheimensis
Sysciophlebia alligans
Sysciophlebia balteata Form-G - Spiloblattina sperbersbachensis
Sysciophlebia balteata Form-H - Spiloblattina homigtalensis - Syscioblatta slusiensis
Sysciophlebia ilfeldensis
Sysciophlebia euglyptica - Syscioblatta dohrni
Sysciophlebia rubida - Syscioblatta lawrenceana
Sysciophlebia grata
Syscioblatta variegata - Spiloblattina pygmaea
Syciophlebia n.sp. A - Syscioblatta intermedia
Millery
Surmoulin
Muse
Igornay
Moloy
Epinac
Mont-Pelé
Renière
Buxières
La Mouillere
Deneuille
Nahe Subgroup
Standenbühl
Kreuzn.
Wadern
Donnersberg
Thallich.
Oberkirchen
Disibodenb.
Glan Subgroup
Meisenheim
Lauterecken
Quirnbach
Wahnwegen
Altenglan
Remigiusb.
Breitenbach
Ottweiler Subgroup
Heusweiler
Dilsburg
Göttelborn
Eisenach
Tambach
Rotterode
Oberhof
Goldlauter
Manebach
Ilmenau
Gehren Sbgr.
U.Hornburg
L.Hornburg
Halle
Siebigerode
Mansfeld Subgroup
Rothenburg
Gorenzen
Werna
Sülzhayn
Baumgarten
Neustadt
Netzkater
Fuchsburg
Líně
Slaný
Týnec
Nýřany Mbr.
Chotěvice
Prosečné
Vrchlabí
Semily
Syřenov
Kumburk
Bačov/Obora
Letovice
Svitávka
Veverská Bítýška
Říčany
U. Padochov
L. Padochov
Rosice-Oslavany
Tirkou
Oued
Issène
Tajgaline
275
280
285
290
295
300
305

a form intermediate between *S. grata* and *S. rubida*, Stephanian B/C transition (Gzhelian), Blanzy-Montceau Basin, French Massif Central, Montceau Formation.

5 *Sysciophlebia euglyptica – Syscioblatta dohrni* zone

Definition: From FAD of *Sysciophlebia euglyptica* and/or *Syscioblatta dohrni* to FAD of *Sysciophlebia ilfeldensis*.

Occurrences: Both the zone species are known from the Stephanian C, Gzhelian/Asselian transition, Saale Basin, Germany, Siebigerode Formation, Wettin Subformation, and the Saar Basin, Germany, Ottweiler Subgroup, Breitenbach Formation; *S. dohrni* is common in the marine-lagoonal Wild Cow Formation, Pine Shadow Member, Virgilian, Estancia Basin, New Mexico; marine limestones of this member contain the conodonts *Adetognathus lautus* and *Idiognathodus delicatus* (Krukowski 1992); spiloblattinid fragments from the Bursum Formation, Red Tanks Member, of the Lucero Basin, New Mexico, most probably belong to *S. dohrni* and *S. euglyptica*; the first conodont investigations of this member give a Virgilian age (Orchard *et al.* 2004).

Discussion: *Sysciophlebia praepilata* from the Grande Couche, Commentry Basin (Schneider 1983*b*), French Massif Central, the famous insect locality with *Meganeura*, is very close to *S. euglyptica*. Both species can hardly be distinguished by small differences in the colour pattern (see Basics of spiloblattinid zonation). No new material from Commentry is known for further study, so a separate *praepilata* zone will not be used. The occurrence of *S. praepilata* covers the latest Stephanian B to possibly earliest Stephanian C.

Accompanying species: A common mylacrid species in this zone is *Dictyomylacris densistriata*, the successor species of *D. insignis* from the late Stephanian B of Commentry (Schneider 1983*b*); a very typical accompanying species of *Spiloblattina praepilata* at Commentry is *Opsiomylacris thevenini*, which occurs in Morocco together with *Spiloblattina pygmaea* in the Oued Issene and El Menizla formations of the Souss Basin (Hmich *et al.* 2003, 2005).

6 *Sysciophlebia ilfeldensis* zone

Definition: From FAD of *Sysciophlebia ilfeldensis* to FAD of *Sysciophlebia balteata* and/or *Spiloblattina homigtalensis*.

Occurrences: Lower Rotliegend, Asselian, Ilfeld Basin, Germany, Netzkater Formation; Lower Rotliegend, Asselian, Lower Austria, Permian of Zöbing (southern continuation of the Czech Boskovice Graben); fragments of *S. ilfeldensis* are known from the Manebach Formation, Lower Rotliegend, Asselian, of the Thuringian Forest Basin, Germany.

Accompanying species: In the Netzkater Formation the zone species occur together with *Opsiomylacris phyloblattoidea*.

Inferred accompanying species: *Spiloblattina weissigensis*, Lower Rotliegend, Asselian, Weissig Basin near Dresden, Germany; *Sp. weissigensis*, Wolfcampian, Asselian, Midcontinent Basin, Colorado, Fairplay, Maroon Formation, Pony Spring Member.

Discussion: Based on the colour pattern, *Sp. weissigensis* is the immediate precursor form of *Sp. homigtalensis*, which co-occurs with *S. balteata*, the succeeding form of *S. ilfeldensis*. Because of these relationships, it is concluded that *Sp. weissigensis* covers a time interval similar to that of *S. ilfeldensis*. *S. ilfeldensis* marks the turning point in the development of colour pattern in the *Sysciophlebia* lineage. The size of the light areas increases from *Sysciophlebia* n. sp. A up to *S. ilfeldensis*; the next form, *S. balteata*, indicates the subsequent decrease of these light areas.

7 *Sysciophlebia balteata* Form H – *Spiloblattina homigtalensis – Syscioblatta slusiensis* zone

Definition: From FAD of *Sysciophlebia balteata* Form H and/or *Spiloblattina homigtalensis* and/or *Syscioblatta slusiensis* to FAD of *Sysciophlebia balteata* Form G and/or *Spiloblattina sperbersbachensis*.

Occurrences: All three zone species occur together in the Homigtal Lake Horizon of the Lower Goldlauter Formation, Lower Rotliegend, Asselian, Thuringian Forest Basin, Germany; *S. balteata*, Autunian, Asselian,

Fig. 3. Correlation chart of the most important European basins and the Moroccan Souss Basin, where spiloblattinid insect zone species were collected. Correlations are based on the insect biostratigraphy presented here and the amphibian biostratigraphy of Werneburg & Schneider (2006). Numbers in diamonds show the lithostratigraphical levels of respective basin profiles in which zone species occur and refer to the zone descriptions in the text. Numerical ages are after Ogg (2004), Menning *et al.* (2003) and Lützner *et al.* (2003).

Autun Basin, French Massif Central, Muse Formation; *S. balteata*, Wolfcampian, Asselian, Appalachian Basin, West Virginia, Dunkard Group, Cassville Shale, roof of the Waynesburg Coal; *Sp. homigtalensis*, Lower Rotliegend, Asselian, Boskovice Graben, Czech Republic, Padochov Formation, Říčany Horizon.

Accompanying species: *Opsiomylacris procerus* occurs with the zone species in the Homigtal Lake Horizon of the Thuringian Forest Basin; in the Říčany Horizon, Boskovice Graben, *Sp. homigtalensis* is accompanied by *Moravamylacris ricanyensis*.

Discussion: The holotype specimen of *S. balteata* (Scudder 1879) is not well preserved (see Schneider & Werneburg 1993, pl. 1, (Fig. 2a, b,) text-Fig. 2), therefore the certain assignment of the specimens from the lower Goldlauter Formation to this species is not absolutely clear. Nevertheless, this species designation is used to prevent too many new species names. For the same reason, immediately succeeding forms of *S. balteata* in the *Sysciophlebia* lineage are designated as 'Forms'.

8 *Sysciophlebia balteata* Form G – *Spiloblattina sperbersbachensis* zone

Definition: From FAD of *Sysciophlebia balteata* Form G and/or *Spiloblattina sperbersbachenis* to FAD of *Sysciophlebia alligans*.

Occurrences: Both zone species occur together in the Acanthodes Lake Horizons, lower Goldlauter Formation, Lower Rotliegend, Asselian, Thuringian Forest Basin, Germany.

Accompanying species: *Opsiomylacris procerus* occurs together with the zone species in the Acanthodes Lake Horizons.

9 *Sysciophlebia alligans* zone

Definition: From FAD of *Sysciophlebia alligans* to FAD of *Sysciophlebia* n. sp. B and/or *Spiloblattina odernheimensis*.

Occurrences: Lower Rotliegend, Asselian/Sakmarian, Boskovice Graben, Czech Republic, Letovice Formation, Svitavka–Zboněk Horizon; fragments of forms transitional between *S. alligans* and *Sysciophlebia* n. sp. B were found in redeposited fossiliferous pyroclastics above the Lien Vert volcanic ash horizon of the Buxieres Formation, Upper Autunian, Bourbon l'Archambault Basin, French Massif Central.

Accompanying species: *Moravamylacris svitaviensis* occurs with the zone species in the Svitavka–Zboněk Horizon.

10 *Sysciophlebia* n. sp. B – *Spiloblattina odernheimensis* zone

Definition: From FAD of *Sysciophlebia* n. sp. B and/or *Spiloblattina odernheimensis* to FAD of *Syscioblatta* n. sp. Obora.

Occurrences: Both the zone species occur together in the Humberg Lake Horizon, upper Meisenheim Formation, and Körborn Lake Horizon, lower Disibodenberg Formation, upper Lower Rotliegend, Sakmarian, Saar–Nahe Basin, Germany.

Discussion: Because of some small differences in the colour pattern, Schneider & Werneburg (1993) distinguished *Sp. odernheimensis* Form H from the Humberg Lake Horizon and *Sp. odernheimensis* Form K (Fig. 2, no. 10, right) from the Körborn Lake Horizon. In the forewings of the latter, the light spot between the medialis and radius veins is missing, which is still present in Form H. At this time, it was not clear whether this is simple variability or of stratigraphical significance. A newly discovered wing from the uppermost Humberg Lake level show the same more reduced pattern as Form K, therefore both forms are biostratigraphically significant (Poschmann & Schindler 2004).

11 *Syscioblatta* n. sp. Obora – *Moravamylacris kukalovae* zone

Definition: From the FAD of *Syscioblatta* n. sp. Obora and/or *Moravamylacris kukalovae* to LAD of *Syscioblatta* n. sp. Obora and/or *Moravamylacris kukalovae*.

Occurrences: *Syscioblatta* n. sp. Obora occurs together with *M. kukalovae* in the Upper Rotliegend I, Sakmarian/Artinskian, Boskovice Graben, Czech Republic, Letovice Formation, Obora Horizon; *M. kukalovae* alone in the Upper Rotliegend I, Sakmarian/Artinskian, Boskovice Graben, Czech Republic, Letovice Formation, Bačov Horizon; *M. kukalovae* is widespread in red beds of the Upper Rotliegend I, Late Sakmarian to Artinskian, as the Rotterode and Tambach Formations, Thuringian Forest Basin, Germany, and the Sobernheim Horizon, Wadern Formation, Nahe Subgroup, Saar–Nahe Basin, Germany.

Discussion: *Syscioblatta* n. sp. Obora was primary regarded as a Permian species of the Triassic genus *Subioblatta* Lin 1978, and a new family Subioblattidae was introduced by Schneider (1983*a*). Now it seems more likely that the late Early Triassic to Jurassic *Subioblatta* is related to Permian spiloblattinids of the genus *Syscioblatta*.

If the similarity in the venation ground plan and the arrangement of colour pattern (compare Schneider 1984*a*, pl. 6, figs 1 & 2; Papier *et al.* 1994, fig. 23a–d) is not the result of homeomorphism, the spiloblattinid zonation could possibly be extended through the Mid- and Late Permian into the Mesozoic. In this regard, the late Cisuralian to Lopingian entomofauna of the Lodève Basin (Gand *et al.* 1997) will provide promising new data.

State of the art and perspectives on insect biostratigraphy – conclusions

As shown above, the insect zonation as well as the amphibian zonation of Werneburg (1996) and Werneburg & Schneider (2006) are primarily based on the sequence of species in absolutely reliable vertical successions of collecting horizons of the Thuringian Forest Basin and, for insects, of the Saar–Nahe Basin, especially (Figs 2–3). They become increasingly completed with the addition of forms from other basins. This is based on the assumption that, if parts of an evolutionary or morphogenetic lineage are well known from one vertical sequence, than this lineage can be completed with forms that fit into the morphogenetic trend from elsewhere. Sometimes, insects and amphibians were found in the same lake horizon or in different succeeding lake horizons of the same basin profile. Such cases are used for cross-checks of the biostratigraphical age delivered by each method. If available, isotopic ages and xenacanth shark teeth were included in these cross-checks. For examples see Werneburg (2003) and Werneburg & Schneider (2006).

From the synthesis of isotopic ages and biostratigraphical data, it is learned that isotopic ages *per se* give only very rough ideas of the real age. This has been demonstrated by the repeated publications of Menning (e.g. 1995) and Menning *et al.* (1997, 2000, 2005) on attempts to calibrate time scales. Menning *et al.* (2005, p. 189) show that isotopic ages of Carboniferous and Permian rocks give in the best case a 2 σ – error of $\pm$2.4 to $\pm$7.4 Ma. These are graphical error bars of 4.8–14.8 Ma long! The Cisuralian has a duration of roughly 22 Ma, the Guadalupian only 12 Ma, the Asselian 9 Ma and the Artinskian about 5 Ma. These uncertainties result partly from repeated volcano-tectonic reactivations in the European Variscides and multiple Mesozoic thermal events, which have upset the geochronological systems throughout large areas.

On a very tight sequence of isotopic ages in the profile of the Thuringian Forest Basin, Lützner *et al.* (2003, 2006) have demonstrated that only the careful proof of consistency of isotopic ages and the cross-check against independent data, for example, from biostratigraphy, can deliver reliable numerical ages. The time resolution of the spiloblattinid zonation is on average 1.5–2 Ma and of the amphibian zonation (Werneburg & Schneider 2006) on average 1.5–3 Ma, which is very well suited for such cross-checks and the calibration of numerical time scales.

In Figures 2 and 3, the insect zones and the basin profiles are correlated with the series and stages of the global marine scale. Until now, those non-marine/marine correlations are based nearly exclusive on isotopic ages from volcanites of the continental basins, not on biostratigraphy. On the other hand, most of the numerical ages used for the stage boundaries are estimated ages only, not really measured in the marine sections (see Ogg 2004; Menning 1995; Menning *et al.* 2000). Possibly, the insect biostratigraphy can deliver a solution to this problem. An increasing number of marine/non-marine interfingerings in brackish marine-lagoonal and estuarine settings with specimens of the spiloblattinid-zone species and conodonts have been discovered, such as the Ames limestone, Virgilian of Ohio, the Lawrence Shale, Virgilian of Kansas as well as the Wild Cow Formation and Bursum Formation of New Mexico. The conodonts and spiloblattinids of both the New Mexico occurrences could be the long-sought tools for reliable correlations of the marine Carboniferous/Permian boundary into the purely continental sections of the Euramerian Hercynides (Schneider *et al.* 2003, 2004). Very promising prospects for marine/non-marine correlations of the Permian stages are also the mixed marine/continental deposits of the North American Mid-Continent Basin and of the Volga-Kama region of Tatarstan in eastern Europe. Future research should focus on those sections.

Both zonations, the amphibian-based one and the spiloblattinid zonation, cover the Late Pennsylvanian Kasimovian and Gzhelian up to the late Early Permian (Cisuralian) Artinskian. After the Early Artinskian wet phase (Roscher & Schneider 2006), the biotopes of spiloblattinids and the aquatic to semi-aquatic amphibians, used for the zonation, disappear by increasing aridization in the central Euramerican region. Therefore, xerophilous insects, such as some mylacrid blattids, could be useful to extend the insect zonation into the Guadalupian. First in the Moroccan Lower Stephanian (Kasimovian/Gzhelian transition), and later in the European Lower Rotliegend (Asselian), mylacrids of the genus *Opsiomylacris* became increasingly common besides phylloblattids and the spiloblattinid guide forms (Hmich *et al.* 2003, 2006). A further genus, *Moravamylacris*, displays strong, sclerotized, beetle-like elytra (Fig. 2, nos 9 & 11, left). These, together with their common occurrence in red beds, demonstrate that *Moravamylacris* was a xeromorphic blattid. Species of both genera are mentioned as accompanying species or zone species in the description of the spiloblattinid zones above.

Additionally, advanced phylloblattids of the genus *Phyloblatta* appear in the Artinskian, as in the Elmo insect bed of the Wellington shales of Kansas and Oklahoma, the Obora insect beds of the Boskovice Graben (Schneider 1984*b*) and the Tambach vertebrate site (Werneburg & Schneider 2006) in the Thuringian Forest Basin. Together with other insects of a diverse entomofauna, they are common throughout the Late Cisuralian and Guadalupian succession of the Lodève Basin (Gand *et al.* 1997, Bethoux *et al.* 2002). Species of the phyloblattid *Aisoblatta*, which are common in the Kungurian and Tatarian of eastern Europe, China and the German marine Zechstein, could be useful for subdivision and correlation of Late Guadalupian and Lopingian time.

Generally, in late Palaeozoic continental settings each method – tetrapod track biostratigraphy, reptilian zonations, macro- and micropalaeobotany, insect or amphibian zonations etc. – has its own advantages and disadvantages. We know lake sediments with thousands of amphibian skeletons but not one single insect wing, red beds with hundreds of tetrapod tracks but no bones. So, any biostratigraphical information is of importance for local to inter-regional correlations, especially in the red beds of the Mid- and Late Permian. Stepwise improved annotated correlation charts based on multidisciplinary information, as presented by Roscher & Schneider (2005), will give a realistic picture of progress, precision and further demands for stratigraphical research.

We thank M. Roscher (Freiberg), for technical support and, together with H. Kerp (Münster), and S. Voigt (Halle) for biostratigraphical discussions. We also thank colleagues of the New Mexico Museum of Natural History, especially Lucas S. G., for current extensive cooperative work in collections and the field. These investigations are supported by the DFG grant SCHN 408/12–1 (Schneider) and WE 2833/3–1 (Werneburg). This publication is a contribution to the project Menning and Schneider Me 1134/5, 'Devonian – Carboniferous – Permian Correlation Chart' and the tasks of the working group 'Marine – non-marine correlations' of the Subcommission on Permian Stratigraphy of the IUGS. The DFG grant SCHN 408/8 enables the revision of Euramerican Carboniferous and Permian conchostracans by J. Goretzki and the investigation of co-occurring insects by J. Schneider. The reviewers, S.G. Lucas and one anonymous, are thanked for critical comments and the improvement of the English.

References

Bethoux, O., Nel, A., Gand, G., Lapeyrie, J. & Galtier, J. 2002. Discovery of the genus *Iasvia* Zalessky, 1934 in the Upper Permian of France (Lodéve Basin) (Orthoptera, Ensifera, Oedischiidae). *Géobios*, **35**, 293–302.

Boy, J. A. 1987. Studien über die Branchiosauridae (Amphibia: Temnospondyli; Ober-Karbon-Unter-Perm). 2. Systematische Übersicht. *Neues Jahrbuch für Geologie und Paläontologie, Abhandlungen*, **174**, 75–104.

Broutin, J., Doubinger, J. *et al.* 1990. Le renouvellement des flores au passage carbonifère-permien: approches stratigraphique, biologique, sédimentologique. *Comptes Rendus – Academie des Sciences, Paris, Série II*, **311**, 1563–1569.

DiMichele, W. A., Pfefferkorn, H. W. & Phillips, T. L. 1996. Persistence of Late Carboniferous tropical vegetation during glacially driven climatic and sea-level fluctuations. *Palaeogeography, Palaeoclimatology, Palaeoecology*, **125**, 105–128.

Durden, C. J. 1969. Pennsylvanian correlation using blattoid insects. *Canadian Journal of Earth Sciences*, **6**, 1159–1177.

Durden, C. J. 1984. North American provincial insect ages for the continental last half of the Carboniferous and first half of the Permian. *Neuvième Congrès International de Stratigraphie et de Géologie du Carbonifère, Nanking, 1979, Comptes Rendus*, **2**, 606–612.

Gand, G. & Haubold, H. 1988. Permian tetrapod footprints in Central Europe: stratigraphical and palaeontological aspects. *Zeitschrift für Geologische Wissenschaften*, **16**, 885–894.

Gand, G., Lapeyrie, J., Garric, J., Nel, A., Schneider, J. & Walter, H. 1997. Découverte d'Arthropodes et de bivalves inédits dans le Permien continental (Lodévois, France). *Comptes Rendus – Academie des Sciences, Paris, Série II*, **325**, 891–898.

Handlirsch, A. 1906*a*. Revision of American paleozoic insects. *Proceedings U.S. National Museum*, **29**, 661–820.

Handlirsch, A. 1906*b*. *Die fossilen Insekten und die Phylogenie der rezenten Formen*. Verlag Engelmann, Leipzig (1906–1908).

Haubold, H. (1980). Die biostratigraphische Gliederung des Rotliegenden (Permosiles) im mittleren Thüringer Wald. *Schriftenreihe für Geologische Wissenschaften*, **16**, 331–356.

Haubold, H. & Lucas, S. G. 2003. Tetrapod footprints of the Lower Permian Choza Formation. *Paläontologische Zeitschrift*, **77**, 247–261.

Hmich, D., Schneider, J. W., Saber, H. & El Wartiti, M. 2003. First Permocarboniferous insects (blattids) from North Africa (Morocco): implications on palaeobiogeography and palaeoclimatology. *Freiberger Forschungshefte, Hefte C*, **499**, 117–134.

HMICH, D., SCHNEIDER, J. W., SABER, H. & EL WARTITI, M. 2005. Spiloblattinidae (Insecta, Blattida) from the Carboniferous of Morocco, North Africa: implications on biostratigraphy. *In*: LUCAS, S. G. & ZEIGLER, K. E. (eds) *The Nonmarine Permian*. New Mexico Museum of Natural History and Science Bulletin, **30**, 111–114.

HMICH, D., SCHNEIDER, J. W., SABER, H., VOIGT, S. & EL WARTITI, M. 2006. New continental Carboniferous and Permian faunas of Morocco: implications for biostratigraphy, palaeobiogeography and palaeoclimate. *In*: LUCAS, S. G., CASSINIS, G. & SCHNEIDER, J. W. (eds) *Non-Marine Permian Biostratigraphy and Biochronology*. Geological Society, London, Special Publications, **265**, 297–324.

KERP, H. 1996. Post-Variscan late Palaeozoic Northern Hemisphere gymnosperms: the onset to the Mesozoic. *Review of Palaeobotany and Palynology*, **90**, 263–285.

KRUKOWSKI, S. T. 1992. Conodont platform elements from the Madera Formation (Pennsylvanian) at the Kinney Brick Company Quarry, Manzanita Mountains, New Mexico. *New Mexico Bureau of Mines & Mineral Resources Bulletin*, **138**, 143–144.

LAURENTIAUX, D. 1950. Les insectes des bassins houillers du Gard et de la Loire. *Annales de Paléontologie*, **36**, 63–84.

LIN, QI-BIN 1978. On the fossil Blattoidea of China. *Acta Entomologica Sinica*, **21**, 335–344.

LUCAS, S. G. 1998. Toward a tetrapod biochronology of the Permian. *In*: LUCAS, S. G., ESTEP, J. W. & HOFFER, J. M. (eds) *Permian Stratigraphy and Palaeontology of the Robledo Mountains, New Mexico. New Mexico Museum of Natural History and Sciences Bulletin*, **12**, 71–91.

LUCAS, S. G. 2002. Tetrapods and the subdivision of Permian Time. *In*: HILLS, L. V., HENDERSON, C. M. & BAMBER, E. W. (eds) *Carboniferous and Permian of the World*. Canadian Society of Petroleum Geologists, Memoirs, **19**, 479–491.

LUCAS, S. G. 2006. Global Permian tetrapod biostratigraphy and biochronology. *In*: LUCAS, S. G., CASSINIS, G. & SCHNEIDER, J. W. (eds) *Non-Marine Permian Biostratigraphy and Biochronology*. Geological Society, London, Special Publications, **265**, 65–93.

LÜTZNER, H., LITTMANN, S., MÄDLER, J., ROMER, R. L. & SCHNEIDER, J. W. 2006. Radiometric and biostratigraphic data of the Permocarboniferous reference section Thüringer Wald. *15th International Congress on Carboniferous and Permian Stratigraphy, Utrecht, 2003, Proceedings*, **265**, (in press).

LÜTZNER, H., MÄDLER, J., ROMER, R. L. & SCHNEIDER, J. W. 2003. Improved stratigraphic and radiometric age data for the continental Permocarboniferous reference-section Thüringer-Wald, Germany. *15th International Congress on Carboniferous and Permian Stratigraphy, Utrecht, 2003, Abstracts*, 338–341.

MARTENS, T. 1983. Zur Taxonomie und Biostratigraphie der Conchostraca (Phyllopoda, Crustacea) des Jungpaläozoikums der DDR. Teil I. *Freiberger Forschungshefte, Hefte C*, **382**, 7–105.

MENNING, M. 1995. A numerical time scale for the Permian and Triassic periods: an integrated time analysis. *In*: SCHOLLE, P., PERYT, T. M. & SCHOLLE-ULMER, N. (eds) *Permian of the Northern Pangea*. Springer Verlag, Heidelberg, 77–97.

MENNING, M., SCHNEIDER, J. W. *et al.* 2003. A Devonian–Carboniferous–Permian correlation chart 2003 (DCP 2003). *15th International Congress on Carboniferous and Permian Stratigraphy, Utrecht, 2003, Abstracts*, 350–351.

MENNING, M., WEYER, D., DROZDZEWSKI, G. & VAN AMEROM, H. W. J. 1997. Carboniferous time scale revised 1997. Time scale A (min. ages) and time scale B (max. ages). Use of geological time indicators. *Newsletters on Carboniferous Stratigraphy*, **15**, 26–28.

MENNING, M., WEYER, D., DROZDZEWSKI, G., VAN AMERON, H. W. J. & WENDT, F. 2000. A Carboniferous time scale 2000: discussion and use of geological parameters as time indicators from central and western Europe. *Geologisches Jahrbuch, A*, **156**, 3–44.

MENNING, M., WEYER, D., WENDT, I. & DROZDZEWSKI, G. 2005. 3.3 Eine numerische Zeitskala für das Pennsylvanium in Mitteleuropa. *In*: DEUTSCHE STRATIGRAPHISCHE KOMMISSION & WREDE, V. (eds) *Stratigraphie von Deutschland. V: Das Oberkarbon (Pennsylvanium) in Deutschland*. Courier Forschungsinstitut Senckenberg, **254**, 181–198.

OGG, J. G. 2004. Status of divisions of the International Geological Time Scale. *Lethaia*, **37**, 183–199.

ORCHARD, M. J., LUCAS, S. G. & KRAINER, K. 2004. Conodonts and the age of the Red Tanks Member of the Bursum Formation at Carizzo Arroyo, central New Mexico. *In*: LUCAS, S. G. & ZEIGLER, K. E. (eds) *Carboniferous–Permian Transition at Carrizo Arroyo, Central New Mexico*. New Mexico Museum of Natural History and Science Bulletin, **25**, 123–126.

PAPIER, F., GRAUVOGEL-STAMM, L. & NEL, A. 1994. *Subioblatta undulata* n. sp., une nouvelle blatte (Subioblattidae SCHNEIDER) du Buntsandstein supérieur (Anisien) des Vosges (France). Morphologie, systématique et affinités. *Neues Jahrbuch für Geologie und Paläontologie, Monatshefte*, **5**, 277–290.

POSCHMANN, M. & SCHINDLER, T. 2004. Sitters and Grügelborn, two important Fossil-Lagerstaetten in the Rotliegend (?Late Carboniferous – Early Permian) of the Saar–Nahe Basin (SW Germany), with the description of a new palaeoniscoid (Osteichthyes, Actinopterygii). *Neues Jahrbuch für Geologie und Paläontologie, Abhandlungen*, **232**, 283–314.

PRUVOST, P. 1912. Les insectes houillers du Nord de la France. *Annales de la Société Géologique du Nord*, **41**, 323–380.

RITTER, S. M. 1995. Upper Missourian–Lower Wolfcampian (Upper Kasimovian–Lower Asselian) conodont biostratigraphy of the Midcontinent, U.S.A. *Journal of Paleontology*, **69**, 1139–1154.

ROSCHER, M. & SCHNEIDER, J. W. 2005. An annotated correlation chart for continental Late Pennsylvanian and Permian basins and the marine scale. *In*: LUCAS, S. G. & ZEIGLER, K. E. (eds) *The Nonmarine*

Permian. New Mexico Museum of Natural History and Science, Bulletin, **30**, 282–291.

Roscher, M. & Schneider, J. W. 2006. Permo-Carboniferous climate: Early Pennsylvanian to Late Permian climate development of Central Europe in a regional and global context. *In*: Lucas, S. G., Cassinis, G. & Schneider, J. W. (eds) *Non-Marine Permian Biostratigraphy and Biochronology*. Geological Society, London, Special Publications, **265**, 95–136.

Schneider, J. 1977. Zur Variabilität der Flügel paläozoischer Blattodea (Insecta). Teil 1. *Freiberger Forschungshefte, Hefte C*, **326**, 87–105.

Schneider, J. 1978. Zur Variabilität der Flügel paläozoischer Blattodea (Insecta). Teil 2. *Freiberger Forschungshefte, Hefte C*, **334**, 21–39.

Schneider, J. 1980. Zur Entomofauna des Jungpaläozoikums der Boskovicer Furche (CSSR), Teil 1: Mylacridae (Insecta, Blattodea). *Freiberger Forschungshefte, Hefte C*, **357**, 43–55.

Schneider, J. 1982. Entwurf einer Zonengliederung für das euramerische Permokarbon mittels der Spiloblattinidae (Blattodea, Insecta). *Freiberger Forschungshefte, Hefte C*, **375**, 27–47.

Schneider, J. 1983*a*. Die Blattodea (Insecta) des Paläozoikum. Teil 1: Systematik, Ökologie und Biostratigraphie. *Freiberger Forschungshefte, Hefte C*, **382**, 106–145.

Schneider, J. 1983*b*. Taxonomie, Biostratigraphie und Palökologie der Blattodea-Fauna aus dem Stefan von Commentry (Frankreich): Versuch einer Revision. *Freiberger Forschungshefte, Hefte C*, **384**, 77–100.

Schneider, J. 1984*a*. Die Blattodea (Insecta) des Paläozoikum. Teil 2: Morphogenese des Flügelstrukturen und Phylogenie. *Freiberger Forschungshefte, Hefte C*, **391**, 5–34.

Schneider, J. 1984*b*. Zur Entomofauna des Jungpaläozoikums der Boskovicer Furche (CSSR). Teil 2: Phyloblattidae (Insecta, Blattodea). *Freiberger Forschungshefte, Hefte C*, **395**, 19 – 37.

Schneider, J. W. 1996. Xenacanth teeth: a key for taxonomy and biostratigraphy. *Modern Geology*, **20**, 321–340.

Schneider, J. & Werneburg, R. 1993. Neue Spiloblattinidae (Insecta, Blattodea) aus dem Oberkarbon und Unterperm von Mitteleuropa sowie die Biostratigraphie des Rotliegend. *Naturhistorisches Museum Schloss Bertholdsberg, Schleusingen, Veröffentlichungen*, **7/8**, 31–52.

Schneider, J. W., Hampe, O. & Soler-Gijón, R. 2000. The Late Carboniferous and Permian aquatic vertebrate zonation in southern Spain and German basins. *In*: Blieck, A. & Turner, S. (eds) *Palaeozoic Vertebrate Chronology and Global Marine/Nonmarine Correlation*. Courier Forschungsinstitut Senckenberg, **223**, 543–561.

Schneider, J. W., Werneburg, R., Lucas, S. G. & Bethoux, O. 2003. Insect biochronozones: a powerful tool in the biostratigraphy of the Upper Carboniferous and the Permian. *Permophiles*, **42**, 11–13.

Schneider, J. W., Lucas, S. G. & Rowland, J. M. 2004. The Blattida (Insecta) fauna of Carrizo Arroyo, New Mexico: biostratigraphic link between marine and non-marine Pennsylvanian/Permian boundary profiles. *In*: Lucas, S. G. & Zeigler, K. E. (eds) *Carboniferous–Permian Transition at Carrizo Arroyo, Central New Mexico*. New Mexico Museum of Natural History and Science, Bulletin, **25**, 247–261.

Schneider, J. W., Goretzki, J. & Rößler, R. 2005. Biostratigraphisch relevante nicht-marine Tiergruppen im Karbon der variscischen Vorsenke und der Innensenken. *In*: Deutsche Stratigraphische Kommission & Wrede, V. (eds) *Stratigraphie von Deutschland. V: Oberkarbon (Pennsylvanium) in Deutschland*. Courier Forschungsinstitut Senkenberg, **254**, 103–118.

Scudder, S. H. 1879. Paleozoic cockroaches: a complete revision of the species of both worldt, with an essay toward their classification. *Memoirs of the Boston Society of Natural History*, **3**, 23–134.

Scudder, S. H. 1885. New genera and species of fossil cockroaches, from the older American rocks. *Proceedings of the Academy of Natural Sciences, Philadelphia*, **1885**, 34–39.

Scudder, S. H. 1895. *Revision of the American Fossil Cockroaches with Description of New Forms*. Bulletin of the United States Geological Survey, **124**.

Voigt, S. 2005. *Die Tetrapodenichnofauna des kontinentalen Oberkarbon und Perm im Thüringer Wald: Ichnotaxonomie, Paläoökologie und Biostratigraphie*. Cuvillier Verlag Göttingen.

Werneburg, R. 1989*a*. Labyrinthodontier (Amphibia) aus dem Oberkarbon und Unterperm Mitteleuropas: Systematik, Phylogenie und Biostratigraphie. *Freiberger Forschungshefte, Hefte C*, **436**, 7–57.

Werneburg, R. 1989*b*. Some notes to systematic, phylogeny and biostratigraphy of labyrinthodont amphibians from the Upper Carboniferous and Lower Permian in Central Europe. *Acta Musei Reginaehradecensis, Serie A, Scientiac Naturates*, **XXII** (1989), 117–129.

Werneburg, R. 1996. Temnospondyle Amphibien aus dem Karbon Mitteldeutschlands. *Naturhistorisches Museum Schloss Bertholdsburg, Schleusingen, Veröffentlichungen*, **11**, 23–64.

Werneburg, R. 2003. The branchiosaurid amphibians from the Lower Permian of Buxières- les-Mines, Bourbon I'Archambault Basin (Allier, France) and its biostratigraphic significance. *Bulletin de la Société Géologique de France*, **174** (4), 1–7.

Werneburg, R. & Schneider, J. W. 2006. Amphibian biostratigraphy of the European Permo-Carboniferous. *In*: Lucas, S. G., Cassinis, G. & Schneider, J. W. (eds) *Non-Marine Permian Biostratigraphy and Biochronology*. Geological Society, London, Special Publications, **265**, 201–215.

Zajic, J. 2000. Vertebrate zonation of the non-marine Upper Carboniferous–Lower Permian basins of the Czech Republic. *In*: Blieck, A. & Turner, S. (eds) *Palaeozoic Vertebrate Chronology and Global Marine/Non-Marine Correlation*. Courier Forschungsinstitut Senckenberg, **223**, 563–575.

Index

Page numbers in *italic* denote figures. Page numbers in **bold** denote tables.